TM 9-2320-272-24-4
5 Ton M939 Series Truck
Direct and General Support
Maintenance Manual
Vol 4 of 4
June 1998

This manual contains maintenance information for the 5 ton M939 US Military Trucks. This is volume 4 of 4 in the Direct / General Support Manual Series. M939 series trucks are a 5 ton heavy duty 6x6 truck. Cargo versions were designed to transport 10,000 pounds of cargo in all terrain and all weather conditions. Originally designed in the 1970's to replace the M39 and M809 series of vehicles. 44,590 units were produced. This manual is printed to help private owners in the maintenance of their vehicles.

Should you have suggestions or feedback on ways to improve this book please send email to Books@OcotilloPress.com

Edited 2021 Ocotillo Press
ISBN 978-1-954285-68-2

Ocotillo Press
Houston, TX 77017
Books@OcotilloPress.com

Disclaimer: The user of this book is responsible for following safe and lawful practices at all times. The publisher assumes no responsibility for the use of the content of this book. The publisher has made an effort to ensure that the text is complete and properly typeset, however omissions, errors, and other issues may exist that the publisher is unaware of.

ARMY TM 9-2320-272-24-4
AIR FORCE TO 36A12- 1 C- 1155-2-4

This publication supersedes TM 9-2320-272-20- 1, October 1985, and changes 1 through 4; TM 9-2320-272-20-2, October 1985, and changes through 3; 9-2320-272-34-1, 1986, changes through TM 9-2320-272-34-2, June 1986, and changes 1 and 2; and TM 9-2320-358-24&P, October 1992.

TECHNICAL MANUAL
VOLUME 4 OF 4

UNIT, DIRECT SUPPORT, AND GENERAL SUPPORT MAINTENANCE MANUAL FOR

TRUCK, 5-TON, 6X6, M939, M939A1, M939A2 SERIES TRUCKS (DIESEL)

GENERAL SUPPORT (GS) MAINTENANCE	5-1
SHIPMENT AND LIMITED STORAGE INSTRUCTIONS	6-1
REFERENCES	A-1
MAINTENANCE ALLOCATION CHART (MAC)	B-1
EXPENDABLE/DURABLE SUPPLIES AND MATERIALS LIST	C-1
MANDATORY REPLACEMENT PARTS	D-1
TOOL IDENTIFICATION LIST	E-1
ILLUSTRATED LIST OF MANUFACTURED ITEMS	F-1
TORQUE LIMITS	G-1
SCHEMATIC AND WIRING DIAGRAMS	H-1

TRUCK, CARGO: 5-TON, 6X6, DROPSIDE,
M923 (2320-01-050-2084) (EIC: BRY); M923A1 (2320-01-206-4087) (EIC: BSS); M923A2 (2320-01-230-0307) (EIC: B57);
M925 (2320-01-047-8769) (EIC: BRT); M925Al (2320-01-206-4088) (EIC: BST); M925A2 (2320-01-230-0308)(EIC: BS8);

TRUCK, CARGO: 5-TON, 6X6 XLWB,
M927 (2320-01-047-8771) (EIC: BRV); M927Al (2320-01-206-4089) (EIC: BSW); M927A2 (2320-01-230-0309) (EIC: BS9);
M928 (2320-01-047-8770) (UC: BRU); M928Al (2320-01-206-4090) (EIC: BSX); M928A2 (2320-01-230-0310) (EIC: BTM);

TRUCK, DUMP: 5-TON, 6X6,
M929 (2320-01-047-8756) (EIC: BTH); M929Al (2320-01-206-4079) (EIC: BSY); M929A2 (2320-01-230-0305) (EIC: BTN);
M930 (2320-01-047-8755) (EIC: BTG); M930Al (2320-01-206-4080) (EIC: BSZ); M930A2 (2320-0l-230-0306) (EIC: BTO);

TRUCK, TRACTOR: 5-TON, 6X6,
M931 (2320-01-047-8753) (EIC: BTE); M931A1 (2320-01-206-4077) (EK: BS2); M931A2 (2320-01-230-0302) (EIC: BTP);
M932 (2320-01-047-8752) (EIC: BTD); M932Al (2320-01-205-2684) (EIC: 855); M932A2 (2320-01-230-0303) (EIC: BTQ);

TRUCK, VAN, EXPANSIBLE: 5-TON, 6X6,
M934 (2320-01-047-8750) (EIC: BTB); M934A1 (2320-01-205-2682) (EIC: BS4); M934A2 (2320-0l-230-0300) (EIC: BTR);

TRUCK, MEDIUM WRECKER: 5-TON, 6X6,
M936 (2320-01-047-8754) (EIC: BTF); M936Al (2320-0l-206-4078) (EIC: BS6); M936A2 (2320-01-230-0304) (EIC: BTT).

DEPARTMENTS OF THE ARMY AND THE AIR FORCE

JUNE 1998

WARNING

EXHAUST GASES CAN KILL

1. DO NOT operate vehicle engine in enclosed area.
2. DO NOT idle vehicle engine with windows closed.
3. DO NOT drive vehicle with inspection plates or cover plates removed.
4. BE ALERT at all times for odors.
5. BE ALERT for exhaust poisoning symptoms. They are:
   - Headache
   - Dizziness
   - Sleepiness
   - Loss of muscular control
6. IF YOU SEE another person with exhaust poisoning symptoms:
   - Remove person from area
   - Expose to open air
   - Keep person warm
   - Do not permit person to move
   - Administer artificial respiration or CPR, if necessary*
   * For artificial respiration, refer to FM 21-11.
7. BE AWARE: The field protective mask for Nuclear, Biological, or Chemical (NBC) protection will not protect you from carbon monoxide poisoning.
   THE BEST DEFENSE AGAINST EXHAUST POISONING IS ADEQUATE VENTILATION.

WARNING SUMMARY

- Hearing protection is required for the driver and passenger. Hearing protection is also required for all personnel working in and around this vehicle while the engine is running (AR-40-5 and TB MED 501).

- If required to remain inside vehicle during extreme heat, occupants should follow the water intake, work/rest cycle, and other stress preventive measures (FM 21-10, Field Hygiene and Sanitation).

- If NBC exposure is suspected, all air filter media should be handled by personnel wearing protective equipment. Consult with your unit NBC officer or NBC NCO for appropriate handling or disposal instructions.

- This vehicle has been designed to operate safely and efficiently within the limits specified in this TM. Operation beyond these limits is prohibited by IAW AR 70-1 without written approval from the commander, U.S. Army Tank-automotive and Armaments Command, ATTN: AMCPEO-CM-S, Warren, MI 48397-5000.

- Never work under dump body unless safety braces are properly positioned. Failure to do this will result in injury to personnel.

- During winching operation, never stand between vehicles. Assistant must remain in secondary vehicle to engage service brake if cable snaps or automatic brake fails while towing vehicle. Failure to do this may result in injury to personnel.

- Accidental or intentional introduction of liquid contaminants into the environment is in violation of state, federal, and military regulations. Refer to Army POL (para. 1-7) for information concerning storage, use, and disposal of these liquids. Failure to do so may result in injury or death.

- Cleaning solvents are flammable and toxic. Do not. use near open flame and always have a fire extinguisher nearby when solvents are used. Use only in well-ventilated places, wear protective clothing, and dispose of cleaning rags in approved container. Failure to do this will result in injury to personnel and/or damage to equipment.

- Eyeshields must be worn when cleaning with compressed air. Compressed air source will not exceed 30 psi (207 kPa). Failure to do so may result in injury to personnel.

- Extreme care should be taken when removing surge tank filler cap if temperature gauge reads above 175°F (79°C). Steam or hot coolant under pressure will cause injury.

- Alcohol used in the alcohol evaporator is flammable, poisonous, and explosive. Do not smoke when removing alcohol evaporator or adding fluid, and do not drink fluid. Failure to do this will result in injury or death.

- Do not perform electrical circuit testing fuel tank with fill cap or sending unit removed. Fuel may ignite, causing injury to personnel.

- When performing battery maintenance, ensure batteries are seated and clamped down, all rubber boots are installed, clamps are well down on battery posts, and all battery cables lie flat against the top of the batteries. Failure to do this may result in injury to personnel and/or damage to equipment.

- Ensure companion seatbelts are not caught inside battery box. This will cause belts to rot which may lead to injury of personnel.

- On M936/Al/A2 model vehicles, remove spare tire prior to changing tire and install tire in spare tire carrier after tire change is complete. Operation of crane and/or vehicle engine while vehicle is on jacks may result in injury to personnel or damage to equipment.

- Never assemble or disassemble tire and rim assembly while inflated, use inflation to seat lockring on split rim or tire on two-piece rim, or inflate a tire without a tire inflation cage. Injury to personnel may result.

- Do not disconnect air lines or hoses, remove safety valves or CTIS components, or perform brake chamber repairs before draining air reservoirs. Small parts under pressure may shoot out with high velocity, causing injury to personnel.

WARNING SUMMARY (Contd)

- Remove all jewelry when working on electrical circuits. Jewelry coming in contact with electrical circuits may produce a short circuit, causing extreme heat, explosions, and fling particles of metal. Failure to do so will result in injury or death and damage to equipment.

- Use eyeshields and follow instructions carefully when performing assembling, disassembling, or maintenance on this device. Components of this device are under spring tension and may shoot out at a high velocity. Failure to do so will result in injury to personnel.

- Do not remove hoses with engine running or start engine with hoses removed. High-pressure fluids may cause hoses to whip violently and spray randomly. Failure to do so may result in injury to personnel.

- Keep hands out from between metal surfaces when removing heavy components. Failure to do so may result in injury to personnel.

- Keep personnel out from under equipment and components of equipment when supported by only a lifting device. Sudden loss of lifting power or shift in load may result in injury or death.

- Do not drain engine, transmission, or radiator fluids, or remove lines containing these fluids, when hot. Doing so may result in injury to personnel.

- Vehicle will become charged with electricity if it contacts or breaks high-voltage wires. Do not leave vehicle while high-voltage lines are in contact with vehicle. Failure to do so may result in injury to personnel.

- Wear hand protection when handling lifting and winching cables, hot exhaust components, and parts with sharp edges. Failure to do so may result in injury to personnel.

- Do not perform fuel system procedures while smoking or within 50 ft (15.2 m) of sparks or open flame. Diesel fuel is highly flammable and can explode easily, causing injury or death to personnel and/or damage to equipment.

- Ensure drainvalve on aftercooler is open when filling cooling system. Failure to do so may result in injury to personnel.

- Turbocharger intake fins are extremely sharp and turn at very high rpm. Keep hands and loose items away from intake openings. Failure to do so may result in injury to personnel.

- Do not place hands between frame and radiator when removing screws from trunnion or lifting radiator. Sudden changes in support may cause the radiator to shift, causing injury to personnel.

- Air pressure may create airborne debris. Use eye protection or injury to personnel may result.

- Air system components are subject to high pressure. Always relieve pressure before loosening or removing air system components.

- Wear safety goggles when using a hammer.

- Ether is extremely flammable. Do not perform ether start system procedures near fire. Injury to personnel may result.

ARMY TM 9-2320-272-24-4
AIR FORCE TO 36A12-1C-1155-2-4

TECHNICAL MANUAL
NO.9-2320-272-24-4

TECHNICAL ORDER
NO. 36A12-1C-1155-2-4

HEADQUARTERS
DEPARTMENTS OF THE ARMY AND THE AIR FORCE
Washington, D.C., 30 JUNE 1998

TECHNICAL MANUAL
VOLUME 4 OF 4
UNIT, DIRECT SUPPORT, AND
GENERAL SUPPORT MAINTENANCE MANUAL
FOR
TRUCK, 5-TON, 6X6, M939, M939A1, M939A2 SERIES TRUCKS (DIESEL)

TRUCK	MODEL	EIC	NSN WITHOUT WINCH	NSN WITH WINCH
Cargo, Dropside	M923	BRY	2320-01-050-2084	
Cargo, Dropside	M923Al	BSS	2320-01-206-4087	
Cargo, Dropside	M923A2	BS7	2320-01-230-0307	
Cargo, Dropside	M925	BRT		2320-01-047-8769
Cargo, Dropside	M925Al	BST		2320-01-206-4088
Cargo, Dropside	M925A2	BS8		2320-01-230-0308
Cargo	M927	BRV	2320-01-047-8771	
Cargo	M927Al	BSW	2320-01-206-4089	
Cargo	M927A2	BS9	2320-01-230-0309	
Cargo	M928	BRU		2320-01-047-8770
Cargo	M928Al	BSX		2320-01-206-4090
Cargo	M92842	BTM		2320-01-230-0310
Dump	M929	BTH	2320-01-047-8756	
Dump	M929Al	BSY	2320-01-206-4079	
Dump	M929A2	BTN	2320-01-230-0305	
Dump	M930	BTG		2320-01-047-8755
Dump	M930Al	BSZ		2320-01-206-4080
Dump	M930A2	BTO		2320-01-230-0306
Tractor	M931	BTE	2320-01-047-8753	
Tractor	M931Al	BS2	2320-01-206-4077	
Tractor	M931A2	BTP	2320-01-230-0302	
Tractor	M932	BTD		2320-01-047-8752
Tractor	M932Al	BS5		2320-01-205-2684
Tractor	M932A2	BTQ		2320-01-230-0303
Van, Expansible	M934	BTB	2320-01-047-8750	
Van, Expansible	M934Al	BS4	2320-01-205-2682	
Van, Expansible	M934A2	BTR	2320-01-230-0300	
Medium Wrecker	M936	BTF		2320-01-047-8754
Medium Wrecker	M936Al	BS6		2320-01-206-4078
Medium Wrecker	M936A2	BTT		2320-01-230-0304

REPORTING OF ERRORS AND RECOMMENDING IMPROVEMENTS

You can help improve this manual. If you find any mistakes or if you know of a way to improve the procedures, please let us know. Mail your letter or DA Form 2028 (Recommended Changes to Publications and Blank Forms), or DA Form 2028-2 located in back of this manual, directly to: Director, Armament and Chemical Acquisition and Logistics Activity, ATTN: AMSTA-AC-NML, Rock Island, IL 61299-7630. A reply will be furnished to you. You may also provide DA Form 2028-2 information via datafax or e-mail:
- E-mail: amsta-ac-nml.@ria-emh2.army.mil
- Fax: DSN 783-0726 or commercial (309) 782-0726

---

*This publication supersedes TM 9-2320-272-20-1,24 October 1985, and changes 1 through 4; TM 9-2320-272-20-2,25 October 1985, and changes 1 through 3; TM 9-2320-272-34-1, 10 June 1986, and changes 1 through 2; TM 9-2320-272-34-2, 10 June 1986, and changes 1 and 2; and TM 9-2320-358-24&P, 21 October 1992.

This publication is published in four volumes. TM 9-2320-272-24-1 contains chapters 1,2, and 3 (through section IX). TM 9-2320-272-24-2 contains chapters 3 (sections X through XVI) and 4 (sections I through III). TM 9-2320-272-24-3 contains chapter 4 (sections IV through XVI). TM 9-2320-272-24-4 contains chapters 5 and 6 and appendices A through H. Volume 1 contains a table of contents for the entire manual. Volumes 1,2, and 3 contain an alphabetical index covering tasks found in their respective volume. Volume 4 contains an alphabetical index covering all tasks found in the entire manual.

TABLE OF CONTENTS

VOLUME 4 OF 4

Section I. ENGINE (M939/A1) MAINTENANCE

5-1. GENERAL

5-2. ENGINE (M939/A1) MAINTENANCE INDEX

5-3. CYLINDER HEAD REPAIR

THIS TASK COVERS:

a. Disassembly   c. Inspection
b. Cleaning   d. Assembly

INITIAL SETUP:

APPLICABLE MODELS
M939/A1

SPECIAL TOOLS
Head holding future (Appendix E, Item 7)
Cleaning brush (Appendix E, Item 28)
Valve guide arbor (Appendix E, Item 153)
Gauge block (Appendix E, Item 52)
Crosshead guide puller (Appendix E, Item 106)
Crosshead guide spacer (Appendix E, Item 33)

TOOLS
General mechanic's tool kit (Appendix E, Item 1)
Outside micrometer (Appendix E, Item 80)
Depth micrometer (Appendix E, Item 81)
Torque wrench (Appendix D, Item 145)
Soft-faced hammer

MATERIALS/PARTS
Six freeze plugs (Appendix D, Item 133)
Sixteen half-keepers (Appendix D, Item 253)
Two O-rings (Appendix D, Item 438)
Two screw-assembled lockwashers (Appendix D, Item 578)
Gasket (Appendix D, Item 94)
Freeze plug (Appendix D, Item 132)
Two freeze plugs (Appendix D, Item 134)
Lubricating oil (Appendix C, Item 50)
Prussian blue (Appendix C, Item 54)
Sealing compound (Appendix C, Item 61)
Antiseize tape (Appendix C, Item 72)
Crocus cloth (Appendix C, Item 20)

PERSONNEL REQUIRED
TWO

REFERENCES (TM)
TM 9-2320-272-24P
TM 9-247

EQUIPMENT CONDITION
Cylinder head removed (para. 4-12).

GENERAL SAFETY INSTRUCTIONS
When cleaning with compressed air, wear eyeshields and ensure source pressure does not exceed 30 psi (207 kPa).

a. Disassembly

1. Remove sixteen half-keepers (1) from valve springs (3) and cylinder head (7). Discard sixteen half-keepers (1).

NOTE
Tag springs for installation

2. Remove eight upper spring guides (2), valve springs (3), and lower spring guides (4) from valve guide (5) and cylinder head (7).

3. Tap eight valve (6) stems down lightly to loosen and remove from cylinder head (7). Place on numbered valve board and hold for inspection.

5-3. CYLINDER HEAD REPAIR (Contd)

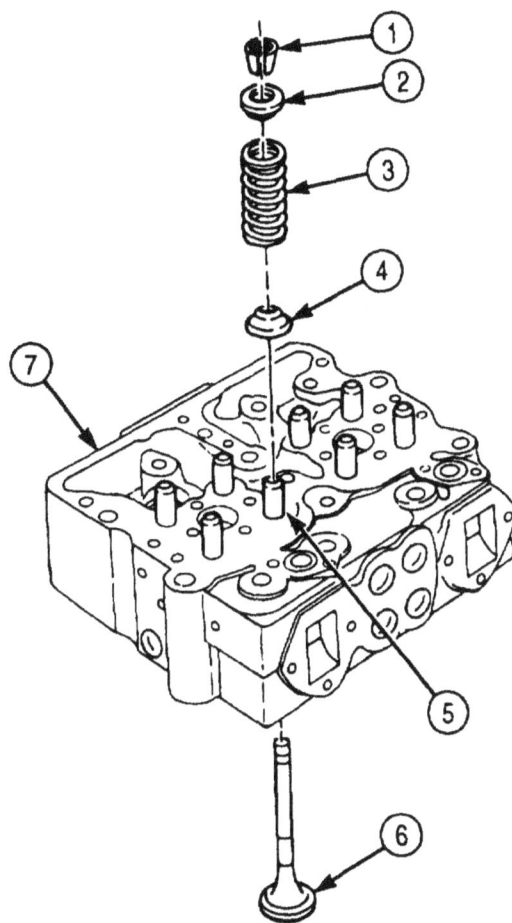

5-3. CYLINDER HEAD REPAIR (Contd)

4. Remove two screw-assembled lockwashers (2), plate (3), and two O-rings (4) from cylinder head (11) and fuel crossover connection (5). Discard O-rings (4) and screw-assembled lockwashers (2).

5. Remove pipe plugs (10) from front and rear face of cylinder head (11). Hold pipe plugs (10) for installation.

6. Remove four freeze plugs (8) from exhaust ports (17) on cylinder head (11). Discard freeze plugs (8).

7. Remove two freeze plugs (6) from front and rear face of cylinder head (11). Discard freeze plugs (6).

8. Remove two freeze plugs (1) from cylinder head (11). Discard freeze plugs (1).

9. Remove freeze plug (9) from front of cylinder head (11). Discard freeze plug (9).

5-3. CYLINDER HEAD REPAIR (Contd)

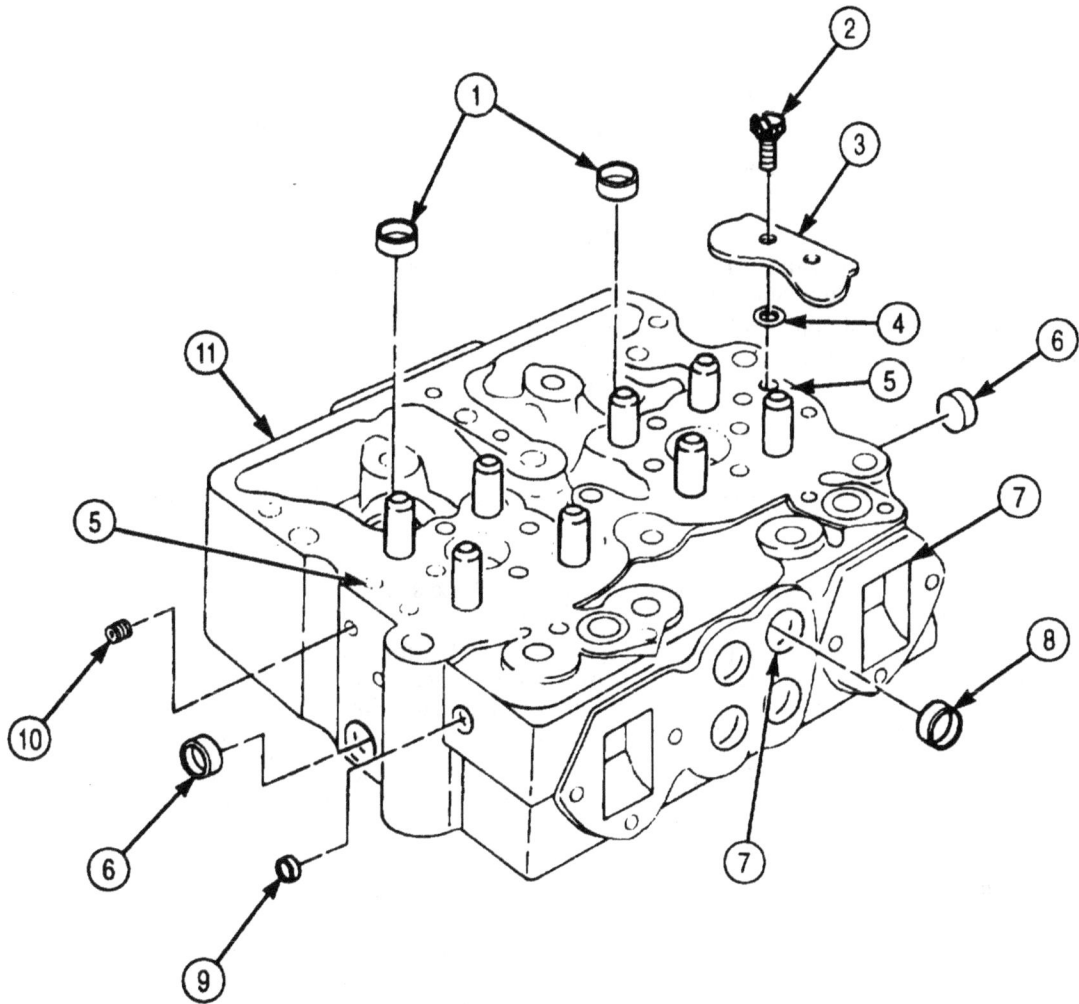

5-3. CYLINDER HEAD REPAIR (Contd)

b. Cleaning

Clean all cylinder head (1) components (TM 9-247).

c. Inspection

1. Install four pipe plugs (11) in front and rear face of cylinder head (1).

NOTE

Apply sealing compound to outer diameter of freeze plugs before installation.

2. Install new freeze plug (10) in front face of cylinder head (1).
3. Install two new freeze plugs (2) in cylinder head (1).
4. Install two new freeze plugs (7) in front and rear face of cylinder head (1).
5. Install four new freeze plugs (9) in exhaust ports (8) on cylinder head (1).
6. Install two new O-rings (5), plate (4), and two new screw-assembled lockwashers (3) on fuel crossover connection (6) in cylinder head (1).

CAUTION

Do not use sander to polish cylinder heads. Serious damage to gasket sealing surfaces can result.

7. Clean cylinder head mating surfaces (12) lightly enough to remove all gasket remains and carbon deposits. Inspect in accordance with instructions in para. 5-25.

NOTE

Instructions for use of portable magnetic tester are included with the tester.

8. Inspect valve seats (13) and injector ports (14) on cylinder head (1) for cracks. If cracks are found, replace cylinder head (1).

NOTE

The following examples of cylinder head defects are provided to assist in determining causes of failures.

9. Check cylinder head valve seats (13) and injector ports (14) for hot spots and correct probable causes. If this condition exists, probable causes are overheating, loss of coolant, coolant flow stoppage, over-fueling, tight injector holddowns, incorrect injector sleeve installation, defective casting, hot shutdowns, and incorrect insert fittings. If hot spots are found, replace cylinder head (1).
10. Check cylinder head (1) and water passage holes (15) for pits and scratches. If pits and scratches are less than .003 in. (0.08 mm), remove with crocus cloth. If pits and scratches are more than 0.003 in. (0.08 mm) deep in the area 0.-625-0.156 in. (1.59-3.97 mm) from edge of water passage hole (15), replace cylinder head (1).
11. Check cylinder head surfaces (12) for warped surfaces. If warped surface exceeds 0.002 in. (0.05 mm), replace cylinder head (1).
12. Check cylinder head (1) for required thickness. Cylinder head (1) must measure 4.340 in. (110.24 mm) thick. If less than 4.340 in. (110.24 mm) thick, replace cylinder head (1).

5-3. CYLINDER HEAD REPAIR (Contd)

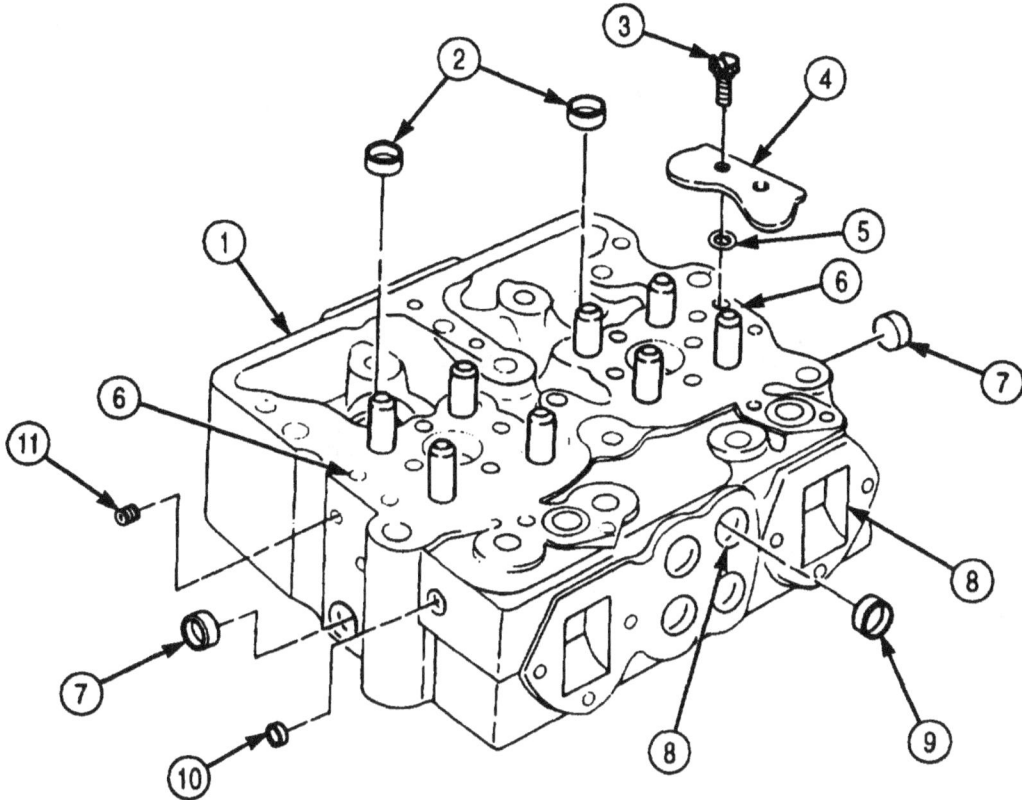

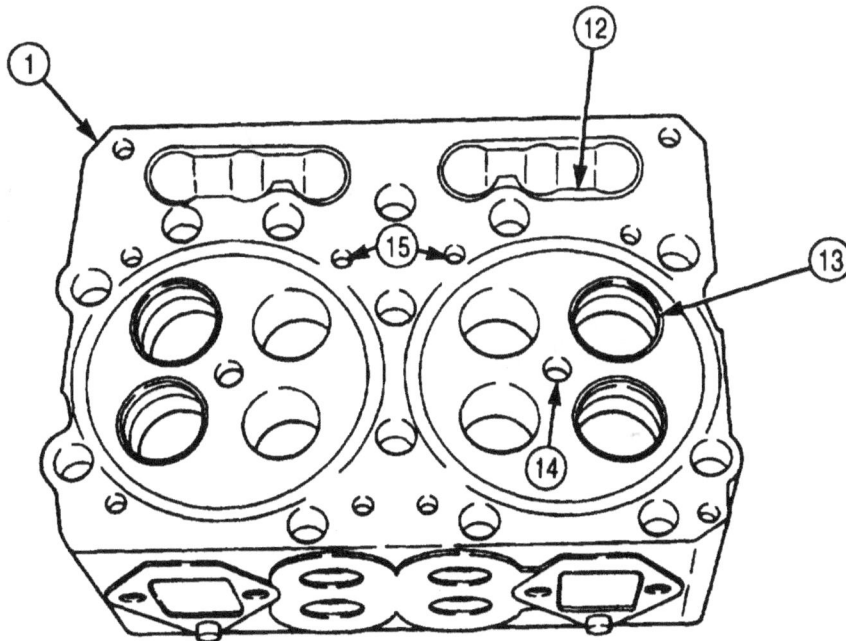

5-3. HEAD REPAIR (Contd)

13. Check valve seat insert (2) and cylinder head (3) for looseness by tapping surface around valve seat insert (2). Replace valve seat inserts (2) that bounce when tapped (para. 5-4).

14. Measure width (6) of valve seat insert (2). If width exceeds 0.125 in. (3.18 mm) at any one point and cannot be narrowed during regrinding, mark valve seat inserts (2) for replacement (para. 5-4).

NOTE
The following examples of valve seat defects are provided to assist in determining causes of failure.

15. Inspect valve seat insert (2) and cylinder (3) for cracks (1) and correct probable causes. If cracks exist, probable causes are improperly machined insert bore, improper fitting of insert in bore, foreign particle under insert, faulty installation, and overheating.

16. Inspect valve seat insert (2) and cylinder head (3) for bums and correct probable causes. If burned, probable causes are carbon or foreign matter that prevents proper seating of valve. If burned, resurface or replace (para. 5-4).

17. Inspect injector sleeves (4) and cylinder head (3) in accordance with instructions in para. 5-6. Check injector sleeves (4) for scratches with bright light. If scratched, mark injector sleeve (4) for replacement (para. 5-6).

18. Check injector cup seating area (5).

    a. Lightly coat injector cup (11) with Prussian blue.

    b. Install injector (9) into sleeve (4) with washer (10), clamp (8), and two screws (7). Tighten alternately in 4 lb-ft (5.4 N.m) steps to 10-12 lb-ft (14-16 N-m).

    c. Remove two screws (7), clamp (8), washer (10), and injector (9).

    d. Check seat pattern in bottom of sleeve (4) and sleeve seating area (5).

    e. Blued band (12) on sleeve (4) in sleeve seating area (5) must be 0.060 in. (1.52 mm) minimum width and be located approximately 0.469 in. (11.91 mm) from bottom of cylinder head (3) surface.

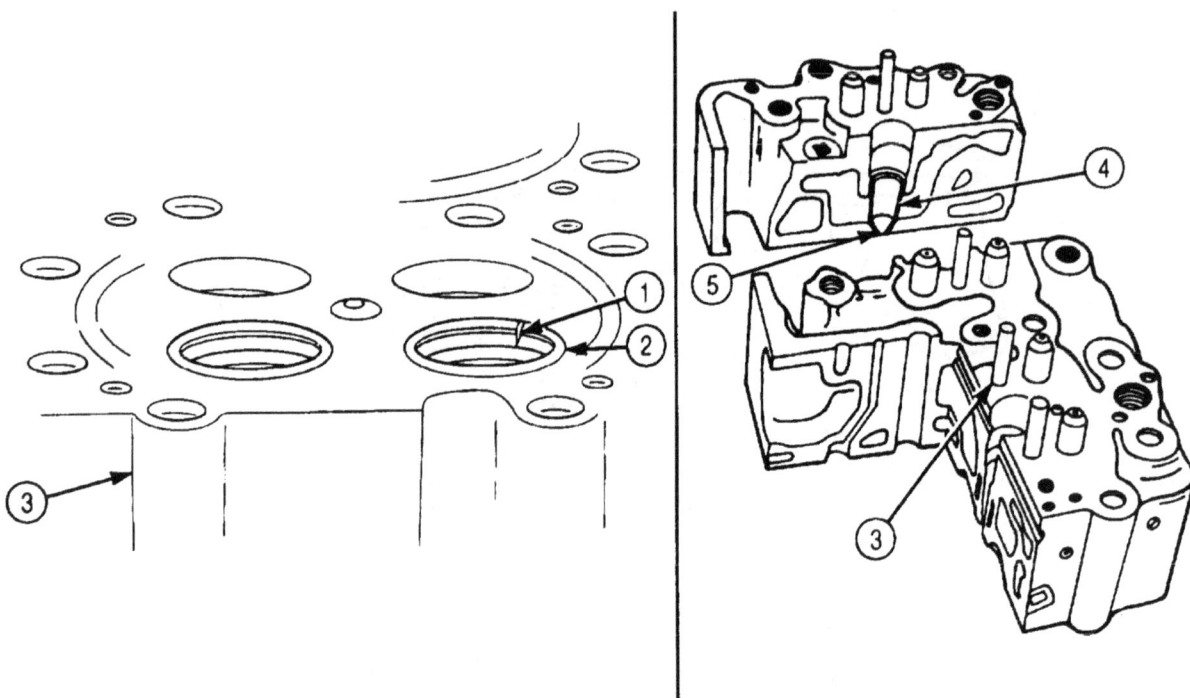

5-3. CYLINDER HEAD REPAIR (Contd)

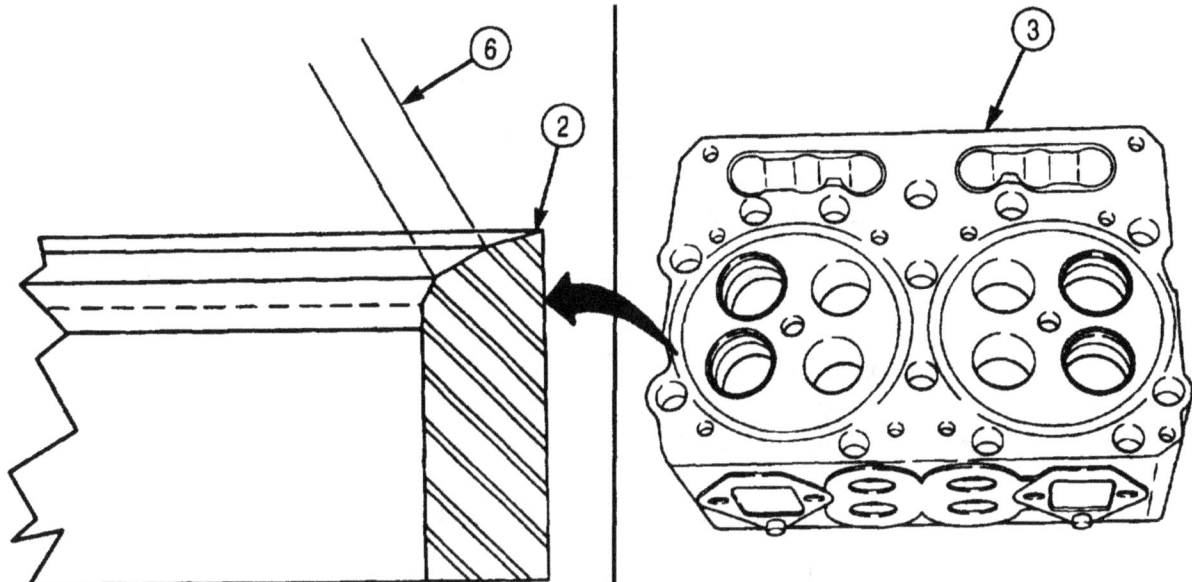

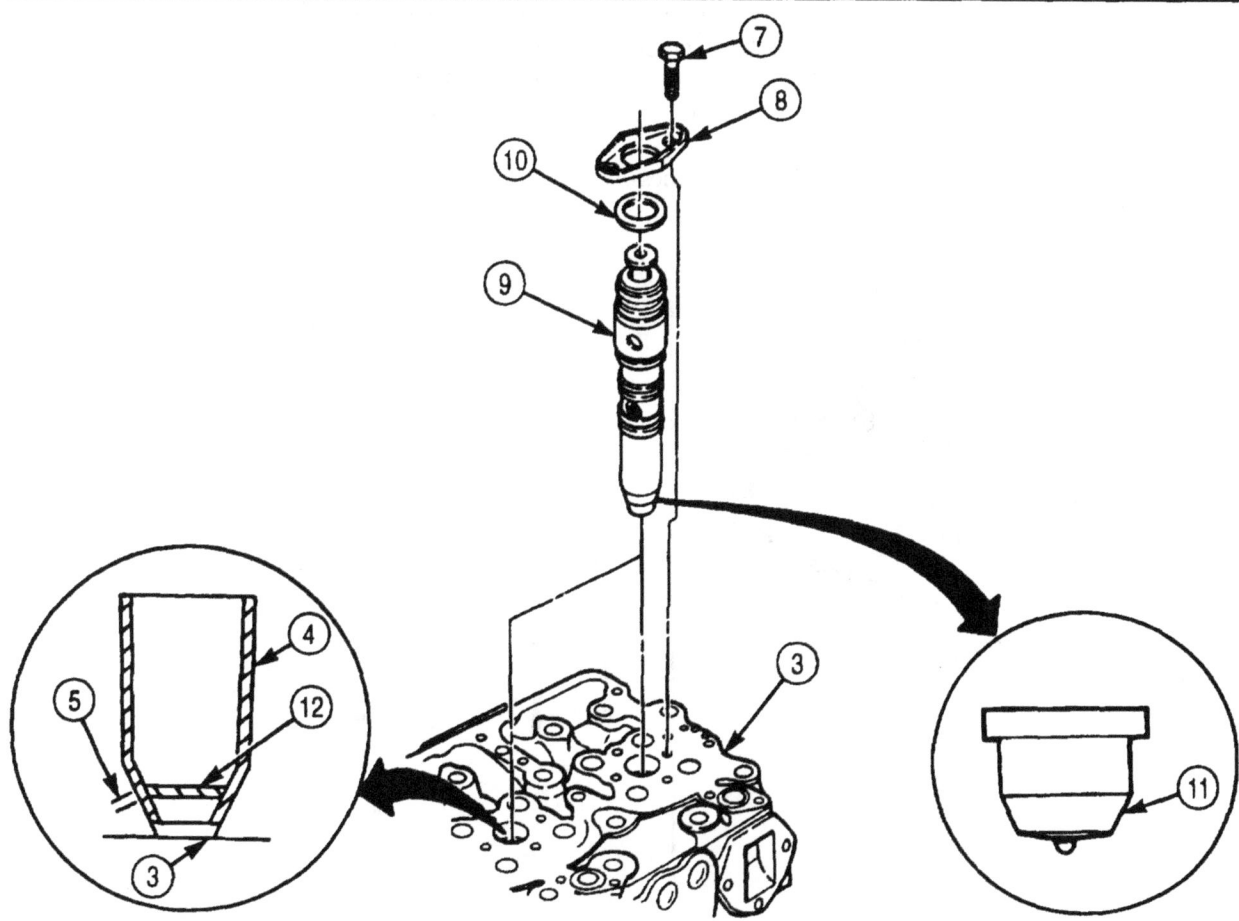

5-3. CYLINDER HEAD REPAIR (Contd)

19. Install injector (18) in cylinder head (3) with washer (19), clamp (10), and two screws (11). Tighten alternately in 4 lb-ft (5.4 N-m) steps to 10-12 lb-ft (14-16 N-m).

20. Measure protrusion of injector tip (8) with gauge block. Protrusion should be 0.060-0.070 in. (1.52-1.78 mm). If not, mark sleeve (9) for replacement (para. 5-6).

21. Air-test fuel inlet passage (4) and fuel outlet passage (5) for leakage and cracks.

    a. Install two O-rings (6), plate (2), and two screw-assembled lockwashers (1).

    b. Install pipe adapter (151, pipe extension (16), and air pressure gauge (17) into fuel outlet passage (5).

    c. Install air hose adapter (121, air pressure control valve (13), and air hose (141 into fuel inlet passage (4).

    d. Open air pressure control valve (13) and apply air pressure until air pressure gauge (17) reads 80-100 psi (550-690 pKa), then close valve (13).

    e. Observe air pressure gauge (17). If pressure drops before fifteen seconds, replace cylinder head (3).

    f. If air pressure holds for fifteen seconds, cylinder head (3) is serviceable.

NOTE
Replace pipe plugs in fuel passages of cylinder head after removal of test adapters.

22. Remove air pressure gauge (17). pipe extension (16), and pipe adapter (15).

23. Remove air hose adapter (12), air pressure control valve (13), and air hose (14) from fuel inlet passage (4)

24. Inspect four crosshead guides (7) in accordance with instructions in para. 2-15.

25. Check outside diameter of four crosshead guides (7) using micrometer or dial gauge. If outside diameter is less than 0.432 in. (10.97 mm), mark guide (7) for replacement.

26. Check four crosshead guides (7) for correct height. If height is not 1.860-1.880 in. (47.24-47.75 mm), mark guide (7) for replacement.

27. Check four crosshead guides (7) for straightness. If guide (7) is not straight, replace guide (7) (subtask d, step 1).

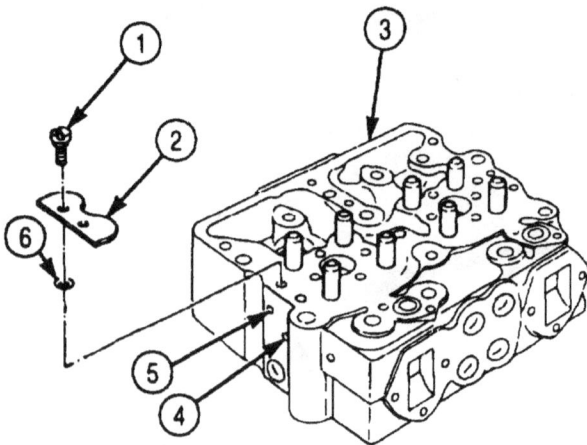

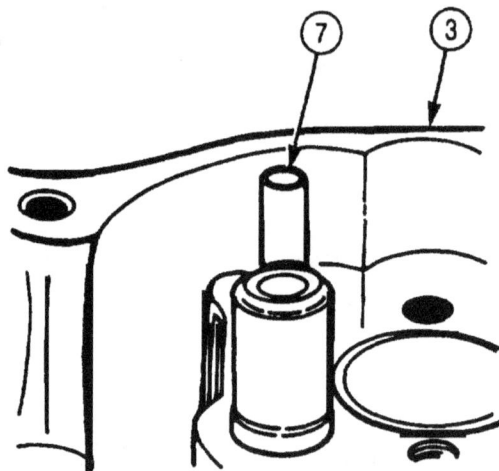

5-3. CYLINDER HEAD REPAIR (Contd)

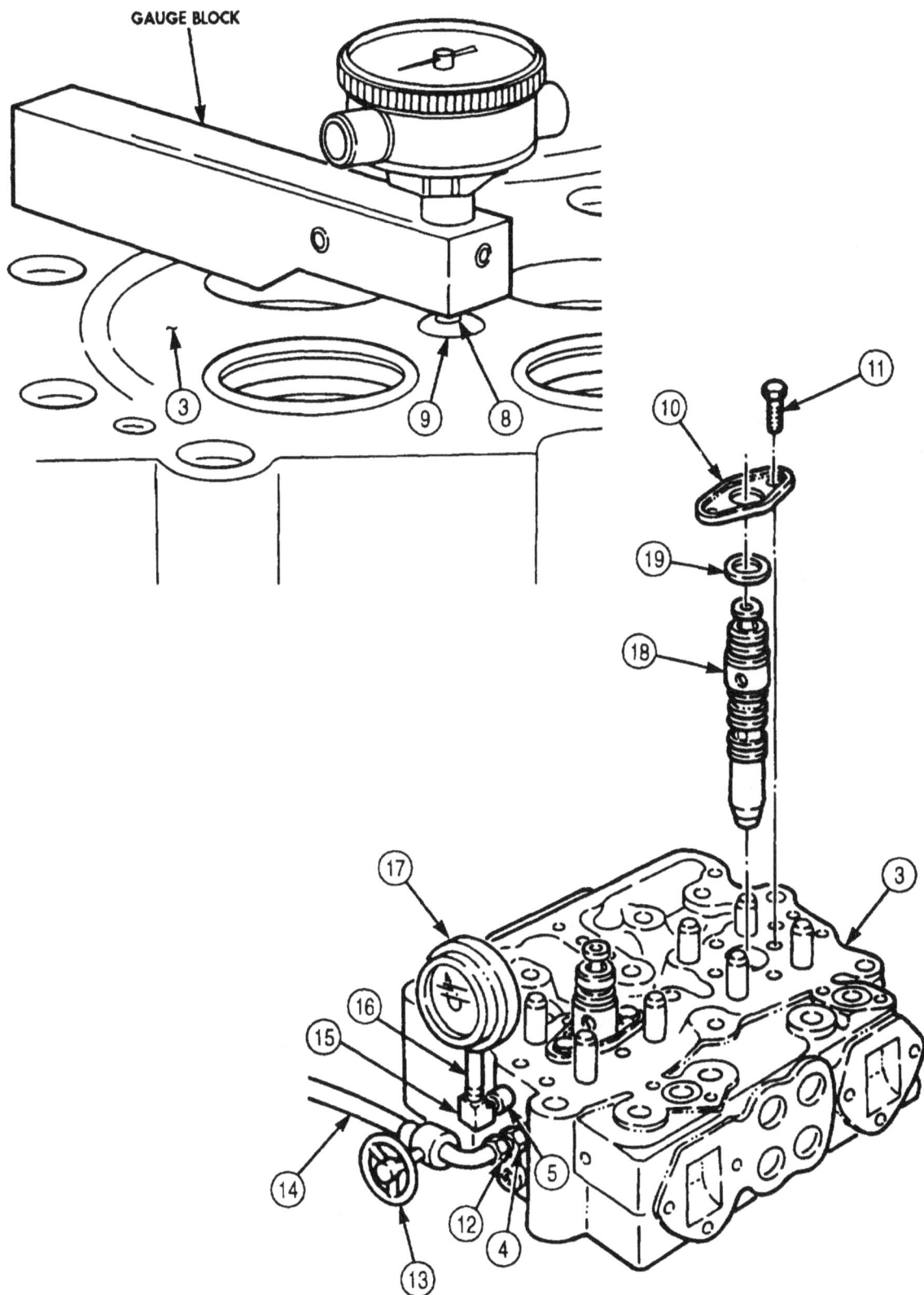

GAUGE BLOCK

5-3. CYLINDER HEAD REPAIR (Contd)

28. Inspect valve crossheads (1) in accordance with instructions in para. 2-15. Discard crossheads (1) if defective.

   a. Check valve crossheads (1) for damaged adjusting screw threads (5) and excessive wear on rocker lever contact area (2).

   b. Using micrometer, set small bore gauge at 0.4402 in. (11.181 mm).

   c. Attempt to insert gauge into bore (4). Discard valve crosshead (1) if bore gauge goes into bore (4).

   d. Check for out-of-round bore (4) by gauging at several points 90° apart. Discard valve crosshead (1) if bore (4) is out of round.

   e. Check valve stem counterbore depth (3). Discard crosshead (1) if depth (3) is not 0.1200-0.1400 in. (3.048-3.556 mm).

29. Inspect eight valve guides (9) in accordance with instructions in para. 2-15. If defective, mark valve guides (9) for replacement.

   a. Check eight valve guides (9) for chips, cracks, burrs, or broken out sections. If chipped, cracked, broken, or burrs are found, mark for replacement.

   b. Check valve guide (9) for protrusion (7). If protrusion (7) is not 1.270-1.280 in. (32.26-32.51 mm) above cylinder head surface (8), mark valve guide (9) for replacement.

   c. Set small bore gauge at 0.4552 in. (11.562 mm) and attempt to insert gauge into guide bore (6). If gauge goes into bore (6), mark guide (9) for replacement.

30. Check valve head (14) and intake and exhaust valves (15) for cracks, warping, pits, burns, or cupping. Discard valve(s) (15) if cracked or warped, pitted, burned, or cupped.

   a. Check rim thickness (13) on intake and exhaust valves (15). Discard valve (15) if rim thickness (13) is less than 0.105 in. (2.67 mm).

   b. Check intake and exhaust valve keeper grooves (11) for wear. Use new keeper to check grooves (11). Discard valve(s) (15) if new keepers fit loosely in grooves (11).

   c. Check valve stem (12) for cracks, scoring, and galling. If cracked, scored, or galled, discard valve (15).

   d. Measure valve stem (10) outside diameter with micrometer. If stem (10) outside diameter is less than 0.449 in. (11.41 mm), discard valve (15).

CAUTION
Use care when selecting replacement valve springs. Intermixing of old and new valve guides in any one cylinder head is permissible only if a specific crosshead has two of the same type or equivalent guides and springs installed under it.

31. Inspect valve springs (16) in accordance with instructions in para. 2-15. Discard valve springs (16) if defective.

   a. Check valve springs (16) for distortions, cracked, or collapsed coils. Discard valve spring (16) if distorted, or if coils are cracked or collapsed.

   b. Check valve spring (16) free length. No. 1 valve spring is 2.29 in. (58 mm) in length. No 2. valve spring is 2.69 in. (68 mm) in length.

   c. Using spring tester, inspect for serviceability by checking load when spring is compressed. Discard spring No. 1 if spring does not give load of at least 150 lb (667 N) when compressed to 1.77 in (45 mm). Discard spring No. 2 if spring does not give load of at least 143 lb (636 N) when compressed to 1.72 in. (44 mm).

5-3. CYLINDER HEAD REPAIR (Contd)

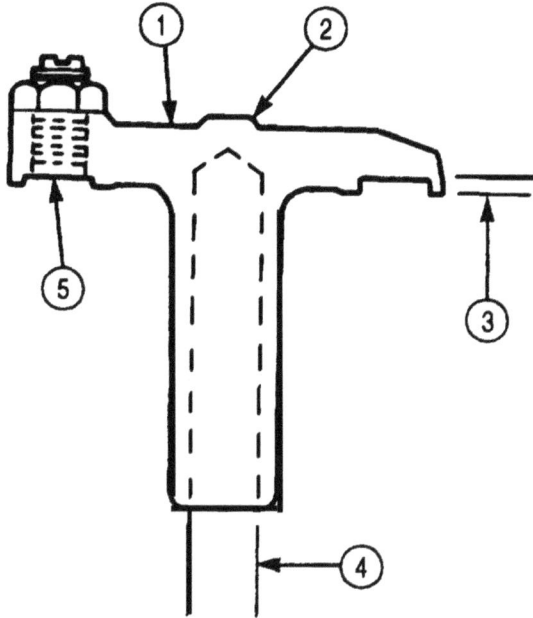

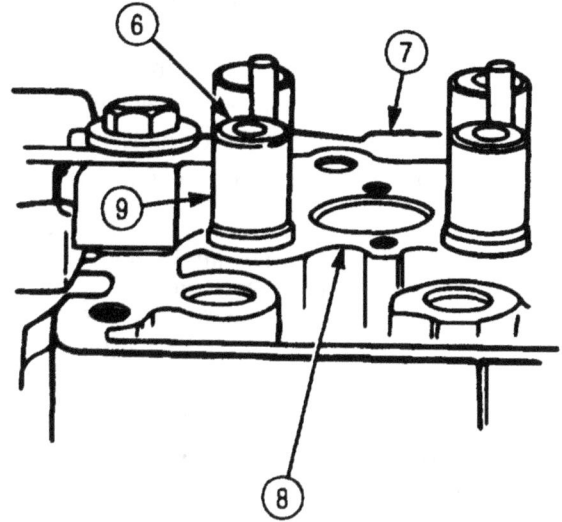

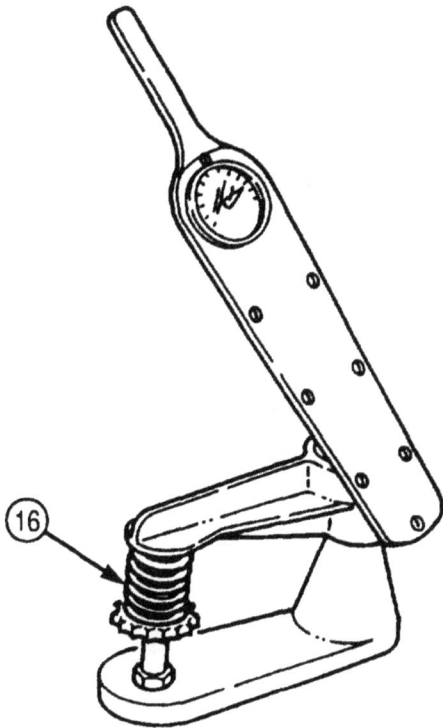

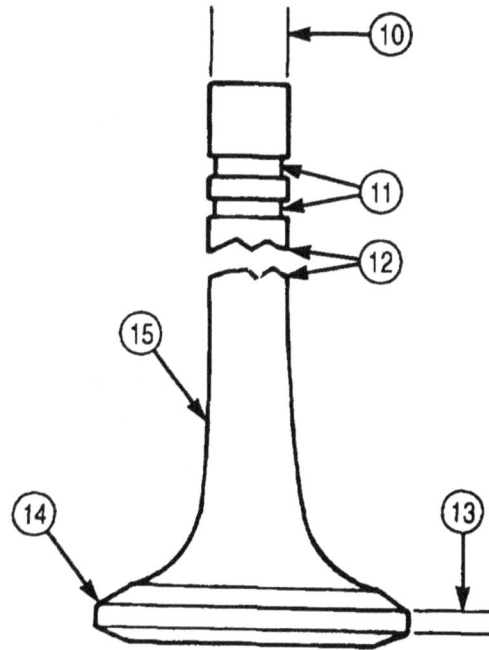

5-3. CYLINDER HEAD REPAIR (Contd)

d. Assembly

NOTE

Use repaired and inspection-approved cylinder heads only.

1. Replace worn valve guides (3). Drive valve guides (3) from underside of cylinder head (2) with hammer and punch.

2. Using arbor press, mandrel, and valve guide arbor, install new valve guides (3).

3. Using crosshead guide puller, remove defective crosshead guides (6) from cylinder head (2).

4. Thoroughly clean crosshead guide (9) holes.

5. Check crosshead guide (6) height. Assembled height must be 1.860-1.880 in. (47.24-47.75 mm).

CAUTION

- To install intake and exhaust valves, position cylinder head on intake port face. Use wooden surface, workbench, or protective surface to prevent damage. Bench must be clean.
- Be sure to install valves in original locations as numbered in step 3.
- Ensure cylinder head is clean

6. Dip valve stem (4) in clean engine oil.

7. Install valve stems (4) through valve guides (3) from face side (5) of cylinder head (2). Ensure valve heads are correctly seated on valve seats (1).

8. Carefully position cylinder head (2) face down on workbench after all valves are installed so valve springs can be installed.

CAUTION

- Two differently-designed valve spring guides have been used in NHC-250 series engines. Part number 128879 spring guide cannot be used with 211999 valve spring. Number 170296 spring guide can be used with either valve spring.
- Reground valve heads seat deeper in cylinder head, causing valve stem to protrude farther above valve guide. This allows valve spring to extend beyond length limits of 2.250 in. (57.150 mm), and causes weak spring action. Use spacers up to 0.0625 in. (15.875 mm) to reduce valve spring to proper height.

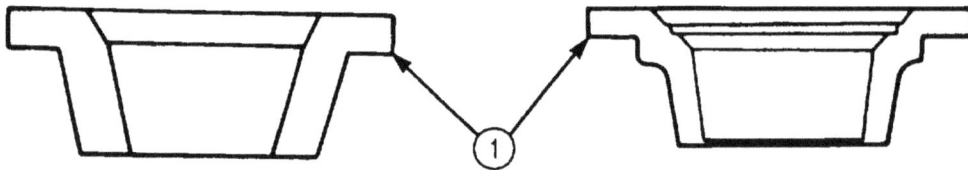

5-3. CYLINDER HEAD REPAIR (Contd)

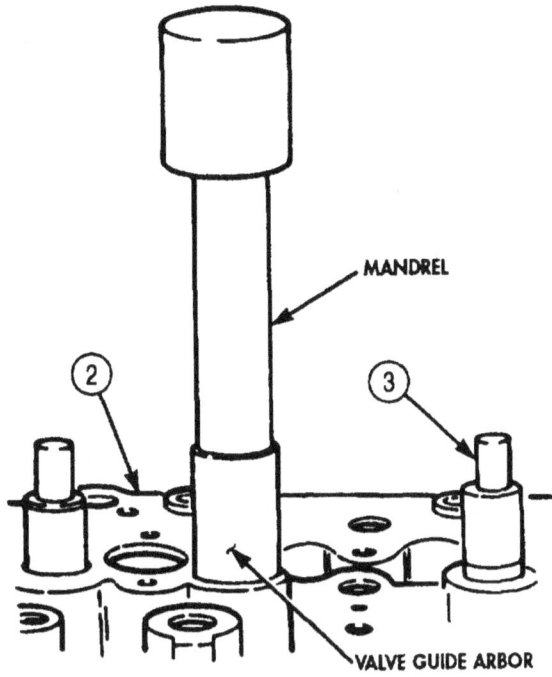

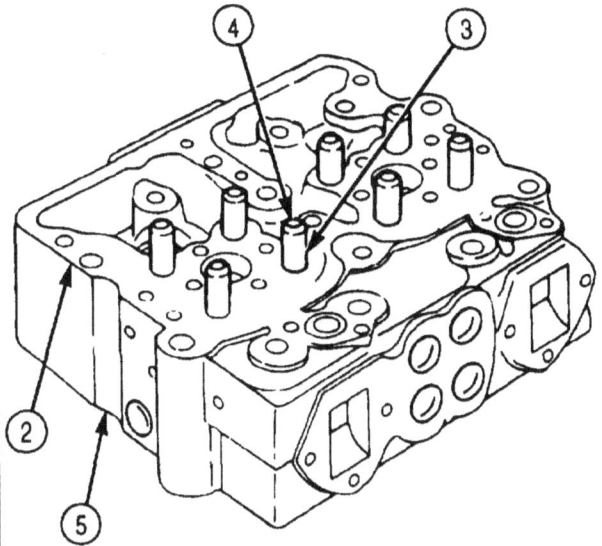

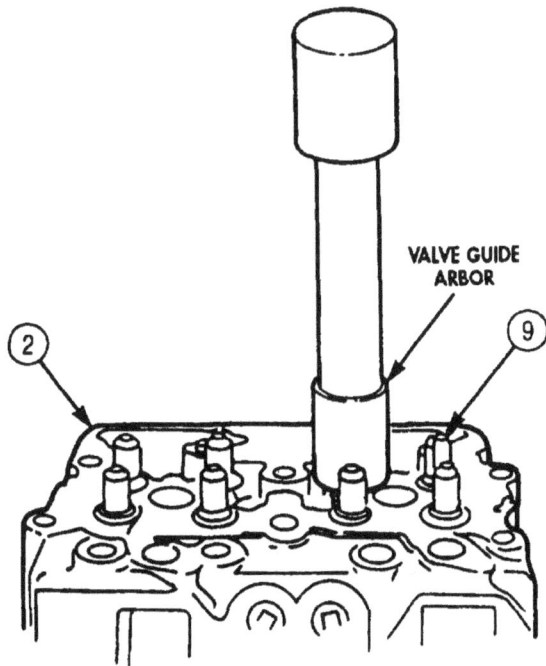

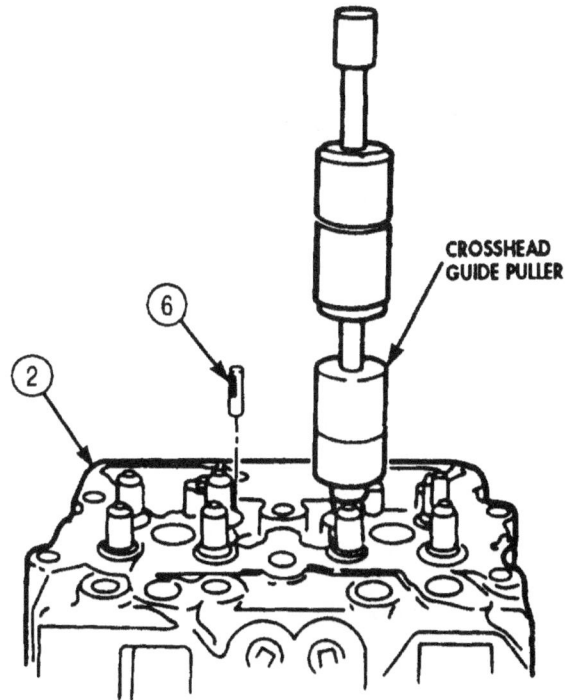

MANDREL

VALVE GUIDE ARBOR

VALVE GUIDE ARBOR

CROSSHEAD GUIDE PULLER

5-3. CYLINDER HEAD REPAIR (Contd)

NOTE

A maximum of two 0.03125 in. (0.794 mm) spacers may be used under lower spring guide when cylinder head has been resurfaced and valve seat insert has been refaced. Do not use spacers to compensate for weak springs.

9. Place eight lower spring guides (18) over valve guides (16) and seat to cylinder head (1).

NOTE

Do not mix intake and exhaust springs. Intake springs are taller.

10. Place eight valve springs (15) on lower spring guides (18).

11. Place eight upper spring guides (14) on top of springs (15) and over stem end of valve (8). Compress spring compressor until keeper grooves (12)on valve (8) are exposed.

12. Install new keepers (13) into valve grooves (12) and slowly release spring compressor. Repeat this step until all valves (8) are locked by keepers (13).

13. Using vacuum tester, test intake and exhaust valves (8) for proper seating.

NOTE

- Valves and valve seats must be dry.
- Grease can be applied to O-ring on vacuum cup for better seal.

a. Select correct size vacuum cup (9) for size valves (8) being tested.

b. Hold vacuum cup (9) over head of valve (8) and seat flat on cylinder head surface (2) surrounding valve (8).

c. Turn tester shutoff valve (6) to open position and hold pushbutton (7) down to operate vacuum pump (4).

d. Operate tester shutoff valve (6) until indicator hand (5) on vacuum gauge (4) stops climbing between 18-25 in. (457-635 mm) of mercury. Close tester shutoff valve (7) and release pushbutton (8).

e. Begin timing as soon as hand (5) reaches 18 in. of mercury on gauge (4).

f. Stop timing as soon as indicator hand (5) reaches 8 in. (20.32 cm) of mercury. If time is less than ten seconds, valve (8) seating is not satisfactory

g. Tap valve stem end with soft-faced hammer and retest by repeating steps a through f.

h. If valve seating is unsatisfactory, proceed to step i.

i. Check for loose connections on tester.

j. Operate vacuum pump (3) with suction cup (9) against a clear glass window.

k. Check indicator hand (5) for movement. If indicator hand (5) moves, there is leakage in the tester.

l. Tighten connections and retest valves (8).

m. Repeat steps b through l.

n. If test fails, regrind valves before retesting (para. 5-4).

5-3. CYLINDER HEAD REPAIR (Contd)

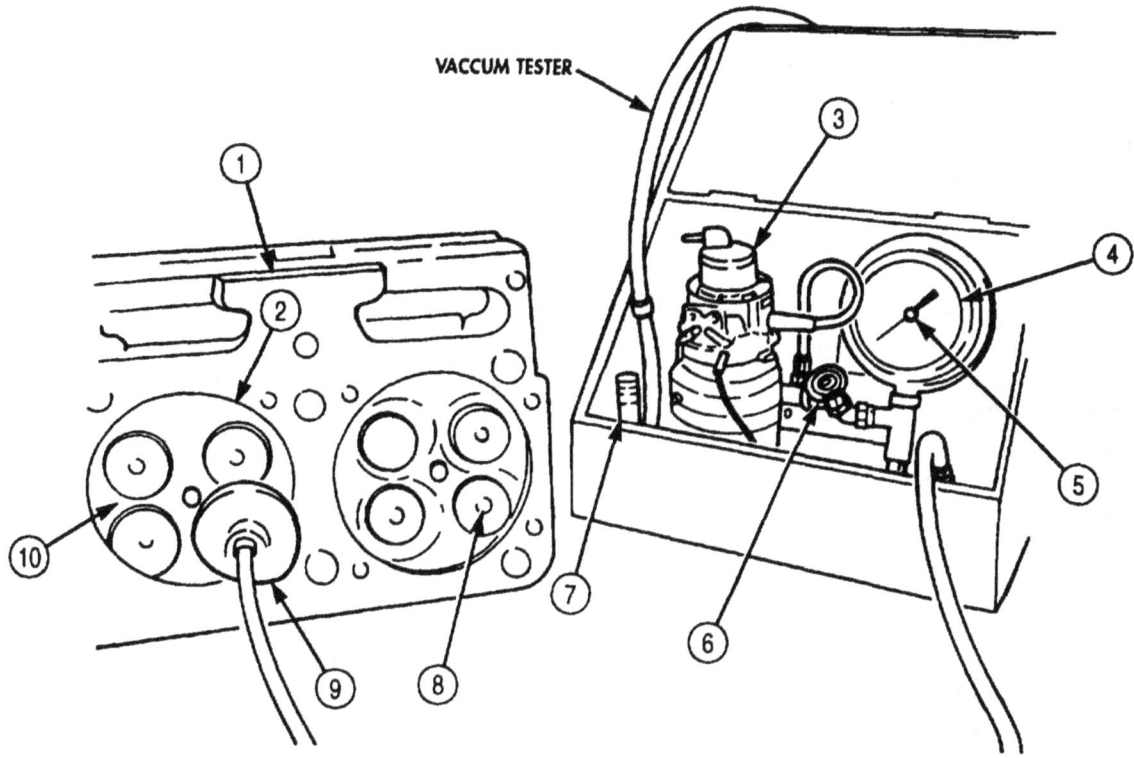

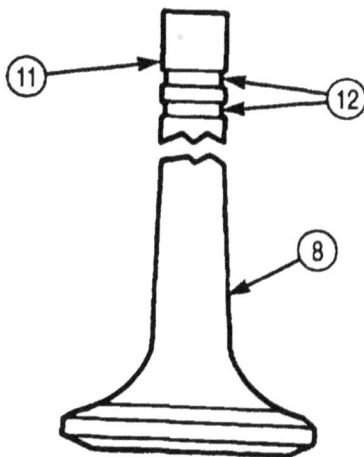

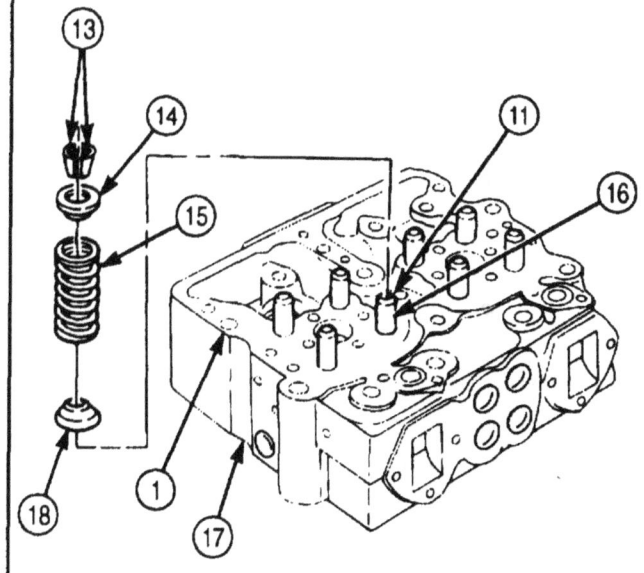

VACCUM TESTER

FOLLOW-ON TASK: Install injectors (para. 4-32)
  • Install cylinder head (para. 4-12).

5-4. INTAKE AND EXHAUST VALVES REFACING

THIS TASK COVERS:

a. Valve Specifications　　　　　　c. Cleaning after Refacing
b. Grinding or Refacing Valves

INITIAL SETUP:

APPLICABLE MODELS
M939/A1

TOOLS
General mechanic's tool kit (Appendix E, Item 1)
Valve refacer

REFERENCES (TM)
TM 9-2320-272-24P
TM 9-4910-484-10

EQUIPMENT CONDITION
Cylinder head disassembled (para. 5-3)

SPECIAL ENVIRONMENTAL CONDITIONS
Well-ventilated work area.

---

a. Valve Specifications

NOTE

- Hard-faced exhaust valves are marked by letters "EX" or "HF" in recessed area of valve head. Intake valves are not marked.
- Use table 5-1 for solid valves
- Sodium-filled valves are not used in Cummins engines.

5-4. INTAKE AND EXHAUST VALVES REFACING(Contd)

Table 5-1. Valve Specifications.

Ref.	Intake and Exhaust Valves	New Minimum	New Maximum	Worn limits
1	Stem	0.4500 in. (11.4300 mm)	0.4510 in. (11.4554 mm)	0.4990 in. (11.4046 mm)
2	Seat angle	30°	30°	
3	Refacing depth	0.0625 in. (1.59 mm)		0.0625 in (1.59 mm)
4	Valve head thickness	0.105 in. (2.67 mm)		0.105 in. (2.67 mm)

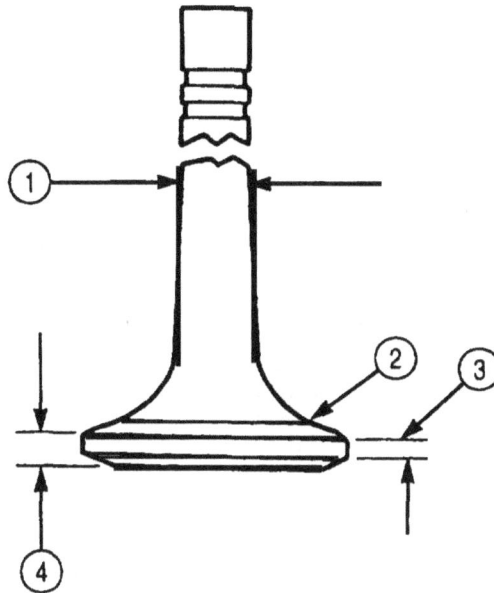

5-4. INTAKE AND EXHAUST VALVES REFACING (Contd)

b. Grinding or Refacing Valves

NOTE

- Mark on ground face of valve. Mark is used to determine head or stem warpage.
- For operating instructions of valve refacer, refer to TM 9-4910-484-10.

1. Install valve (3) stem in chuck (2).

2. Turn switch (5) of valve refacer to ON to start valve grinder electric motor.

NOTE

An out-of-round condition of valve will be marked by a small bright spot on valve seating surface of valve head.

3. Turn grinder wheel handle (6) and very lightly touch valve (3) face with grinder wheel (4).

4. Move grinder wheel (4) away from valve head and switch valve grinder electric motor to OFF with valve refacer switch (5).

5. Indicate location of small bright spot on ground face of valve (3).

6. Rechuck valve (3) 180° from first position and mark new position on valve (3).

7. Turn valve refacer switch (5) to ON to start valve refacer electric motor.

8. Turn grinder wheel (4) handle (6) and very lightly touch valve (3) face with grinder wheel (4).

9. Move grinder wheel (4) away from valve (3) head and switch grinder electric motor to OFF with switch (5).

10. If bright spot is in the same position after both chucking operations, the valve (3) is warped. If valve (3) is warped, replace valve (3).

11. If bright spots occur in different positions, the chuck (2) is out of alignment or the valve (3) is being incorrectly chucked. Run-out should not exceed 0.0001 in. (0.0254 mm).

NOTE

Use valve refacer handle for left and right motion of valve.

12. Wet grind valves (3) to an exact 30° angle from horizontal and check valve (3) head rim thickness with wheel handle (7) controlling depth of grind (table 5-1).

NOTE

Keep valves in order in a numbered valve stick or board.

c. Cleaning after Refacing

NOTE

Do not use cloth to wipe valves clean.

1. Clean refaced valves (3). For general cleaning instructions, refer to para. 2-14.

5-4. INTAKE AND EXHAUST VALVES REFACING (Contd)

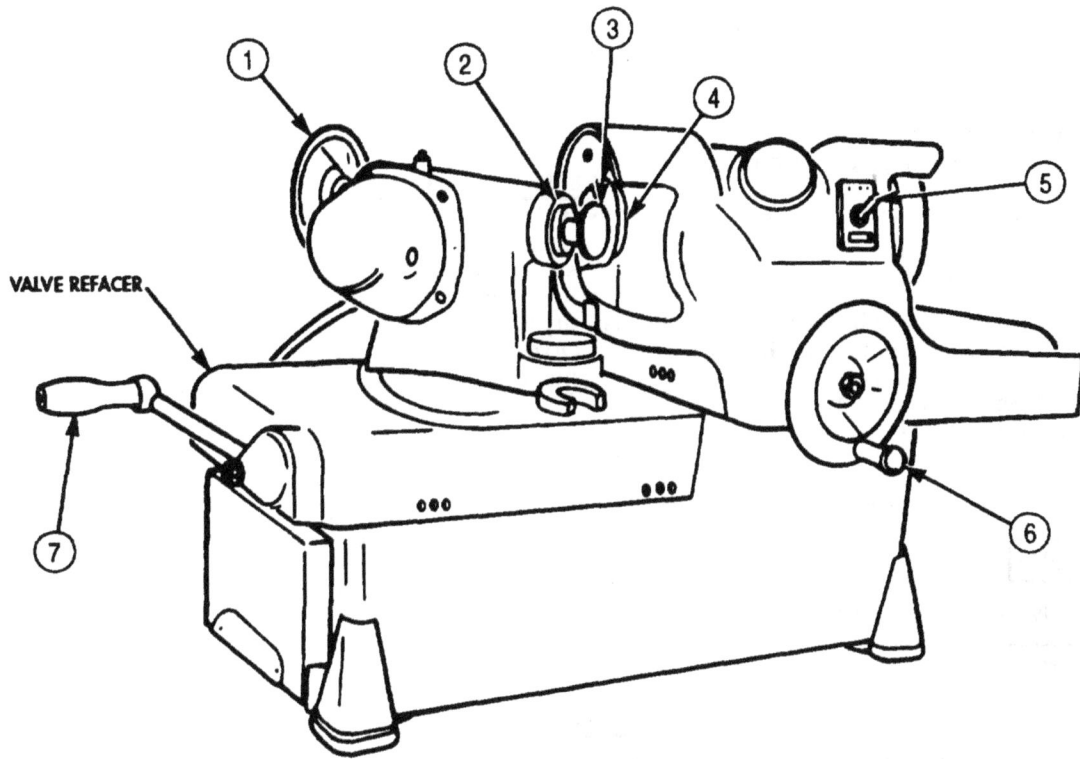

VALVE REFACER

FOLLOW-ON TASK: Assemble cylinder heads (para. 5-3).

5-5. VALVE SEAT INSERTS MAINTENANCE

THIS TASK COVERS:

a. Removal
b. Gauging
c. Counterboring

d. Cleaning
e. Installation

INITIAL SETUP:

APPLIABLE MODELS
M939/Al

SPECIAL TOOLS
Valve seat insert tool (Appendix E, Item 158)
Cutter seat (Appendix E Item 34)
Valve guide arbor (mandrel set) (Appendix E, Item 154)
Tool driver (Appendix E, Item 140)
Valve seat insert staking tool (Appendix E, Item 157)
Valve seat insert extractor (Appendix E, Item 156)

TOOLS
General mechanic's tool kit (Appendix E, Item 1)
Inside micrometer (Appendix E, Item 82)
Outside micrometer (Appendix E, Item 80)
Vernier calipers (Appendix E, Item 159)
Depth gauge (Appendix E, Item 81)
Brush
Electric drill

MATERIALS/PARTS
Valve seat inserts (Appendix D, Item 708)
Drycleaning solvent (Appendix C, Item 71)

REFERENCES (TM)
TM 9-2320-272-24P

EQUIPMENT CONDITION
Cylinder head disassembled (para. 5-3).

GENERAL SAFETY INSTRUCTIONS
• When cleaning with compressed air, wear eyeshields and ensure source pressure does not exceed 30 psi (207 kPa).
• Keep fire extinguisher nearby when using drycleaning solvent.
• Drycleaning solvent is flammable and toxic. Do not use near an open flame.

---

a. Removal

1. Pull valve seat insert (3) from bore (2) on cylinder head (1) with valve seat insert extractor tool.

b. Gauging

CAUTION

Each replacement valve seat insert outside diameter and thickness must be measured and compared to relating valve seat insert counterbore in the cylinder head before counterboring. These measurements will prevent over-boring and damage to the cylinder head.

1. Measure outside diameter of valve seat insert (3) using micrometer. Record reading and compare with specifications in table 5-2.

2. Measure thickness of valve seat insert (3). Record reading and compare with specifications in table 5-2.

3. Measure valve seat counterbore (4) depth (5) with depth micrometer. Record reading and compare to specifications in table 5-2.

4. Measure valve seat insert counterbore (4) inside diameter with inside micrometer. Record reading and compare to specifications in table 5-2.

5-5. VALVE SEAT INSERTS MAINTENANCE(Contd)

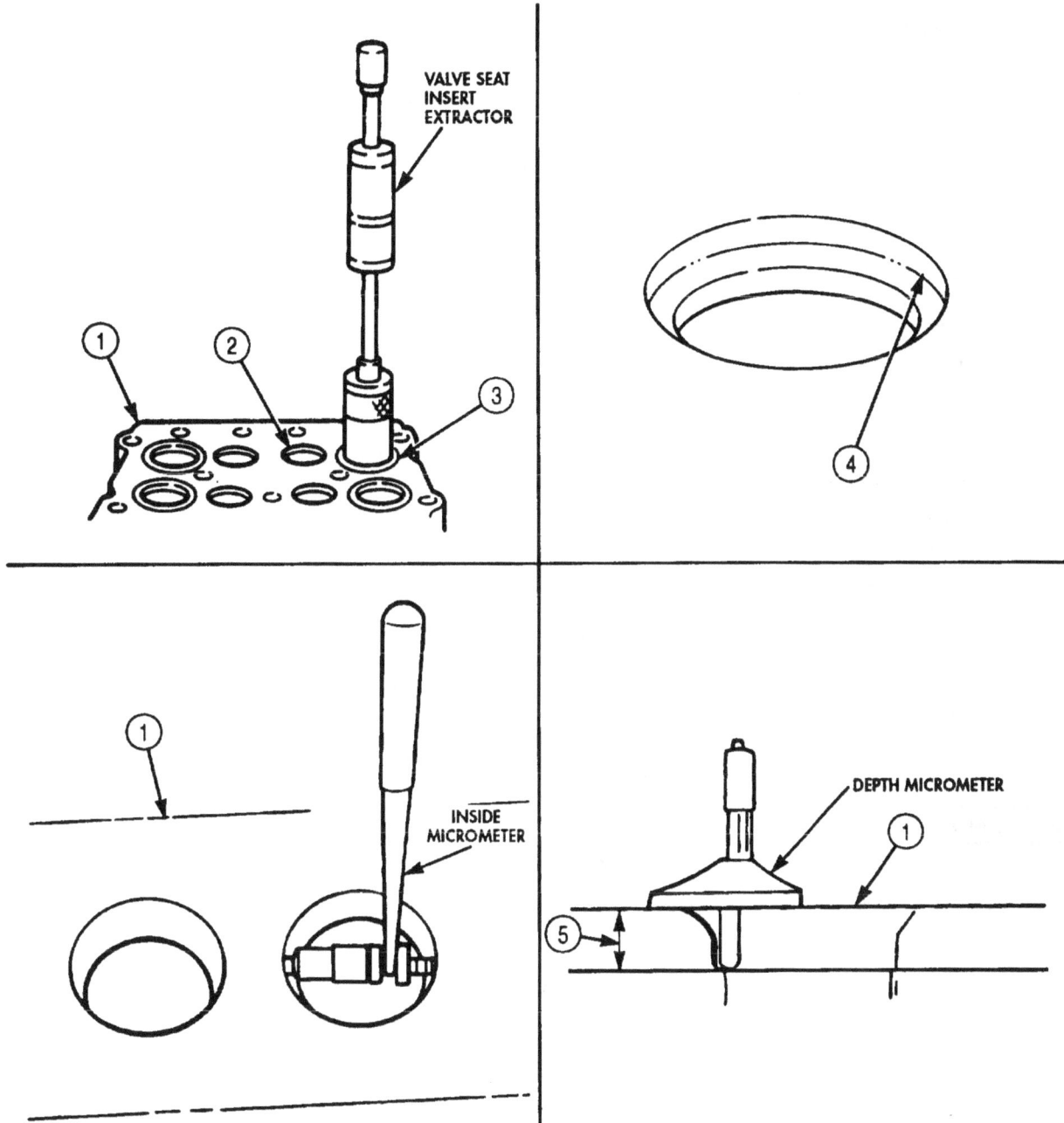

Table 5-2. Valve Seat Specifications.

CYLINDER HEAD VALVE SEAT COUNTERBORE DEPTH	INSERT OUTSIDE DIAMETER	CYLINDER HEAD COUNTERBORE INSIDE DIAMETER	INSERT THICKNESS
Standard	2.0025 - 2.0035 in. (50.864 - 50.889 mm)	1.9995 - 2.0005 in. (50.787 - 50.813 mm)	0.278 - 0.282 in. (7.0612 - 7.1629 mm)

5-5. VALVE SEAT INSERTS MAINTENANCE (Contd)

c. Counterboring

1. Clamp base (2) of counterbore cutter (7) to cylinder head (3) near valve insert seat bore (4). Ensure counterbore cutter (7) is securely clamped before starting electric drill motor (1).

NOTE

Allow cutter to turn several revolutions at exact moment the proper depth in cylinder head is reached to ensure a perfectly flat bottom of bore for valve seats to seat.

2. Center counterbore cutter (7) in valve seat insert bore (4) and valve guide mandrel (6).

3. Cut counterbore (5) 0.006-0.010 in. (0.1524-0.2540 mm) deeper than valve seat insert thickness to allow staking (peening) of cylinder head (3) to secure valve seat insert.

d. Cleaning

WARNING

Eyeshields must be worn when cleaning with compressed air. Compressed air source will not exceed 30 psi (207 kPa). Failure to do so may result in injury to personnel.

1. Blow cylinder head (3) out with compressed air.

WARNING

Drycleaning solvent is flammable and toxic. Do not use near open flame and always have a fire extinguisher nearby when solvents are used. Use only in well-ventilated places, wear protective clothing, and dispose of cleaning rags in approved container. Failure to do this may result in injury or death to personnel and/or damage to equipment.

2. Clean opening and ports of cylinder head (3) with brush and drycleaning solvent.

3. Dry cylinder head (3) with compressed air.

e. Installation

CAUTION

Valve seat inserts may be installed one time only. If valve seat insert is not properly installed the first time, a new valve seat insert must be installed.

NOTE

Keep valve seat inserts in cold storage until ready to install in cylinder head. Install valve seat insert very quickly so room temperature does not have a chance to expand insert and make it difficult to install.

1. Drive valve seat insert (9) into valve seat insert bore (4) with valve seat insert tool until fully seated.

2. Stake (peen) valve seat insert (10) into cylinder head (3) with valve seat insert tool and valve seat insert staking tool.

5-5. VALVE SEAT INSERTS MAINTENANCE (Contd)

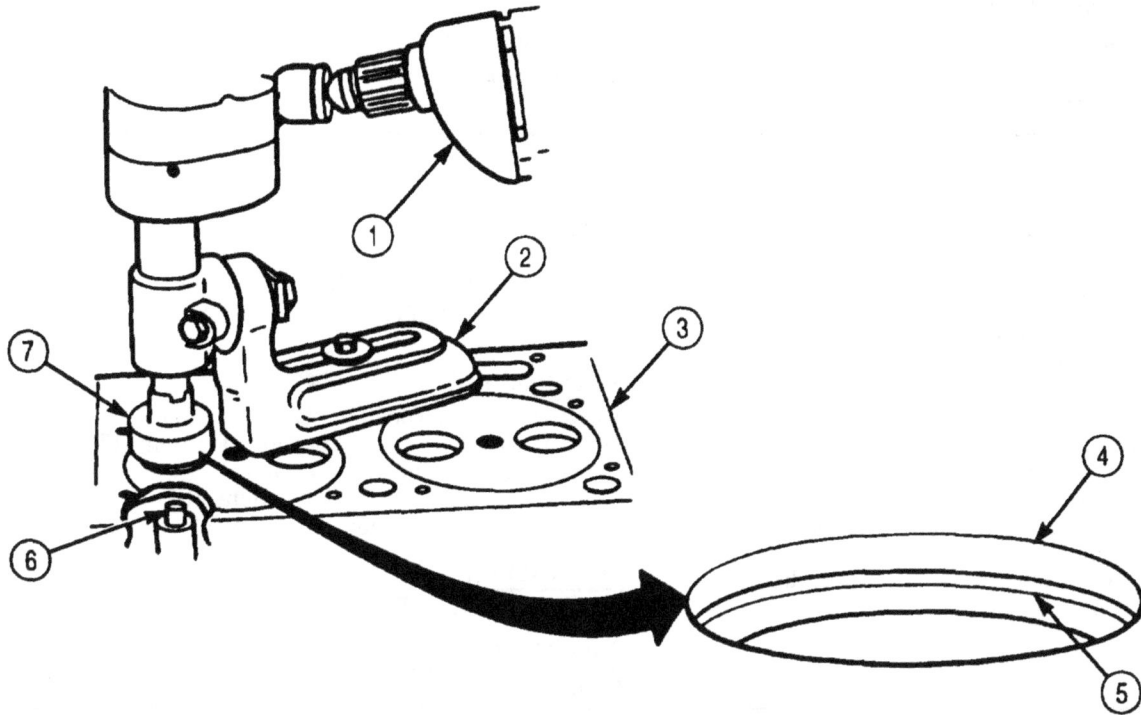

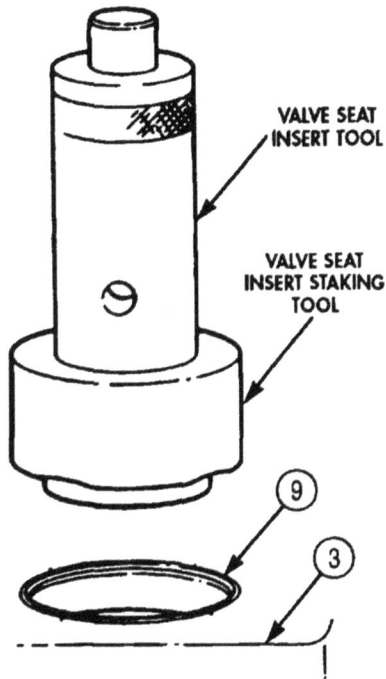

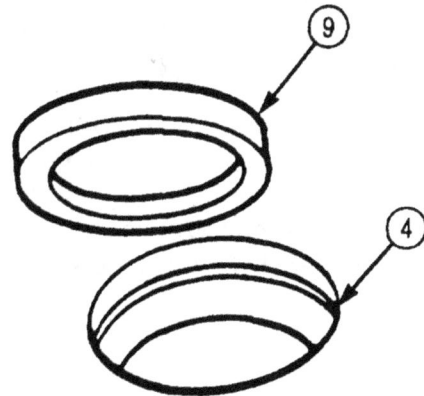

VALVE SEAT
INSERT TOOL

VALVE SEAT
INSERT STAKING
TOOL

FOLLOW-ON TASK: Grind valve seats (para. 5-2).

5-6. FUEL INJECTOR SLEEVE MAINTENANCE

THIS TASK COVERS:

a. Removal
b. Bead Cutting
c. Installation

d. Fitting and Forming
e. Check and Test

INITIAL SETUP:

APPLICABLE MODELS

M9391A1

SPECIAL TOOLS

Bead cutting tool (Appendix E, Item 9)
Injector sleeve expander tool (Appendix E, Item 73)
Injector sleeve cutter (Appendix E, Item 72)
Injector protrusion gauge (Appendix E, Item 53)
Injector sleeve holding tool (Appendix E, Item 74)
Injector sleeve installation mandrel (Appendix E, Item 75)
Injection sleeve extractor (Appendix E, Item 67)

TOOLS

General mechanic's tool kit (Appendix E, Item 1)
Torque wrench (Appendix E, Item 146)
Dial indicator (Appendix E, Item 36)
Drill press

MATERIALS/PARTS

Injector sleeve (Appendix D, Item 257)
Injector sleeve O-ring (Appendix D, Item 478)
Cutting oil (Appendix C, Item 26)
Lubricating oil (Appendix C, Item 50)
Prussian blue (Appendix C, Item 54)

REFERENCES (TM)

TM 9-2320-272-24P

EQUIPMENT CONDITION

Cylinder head disassembled (para. 5-3).

---

a. Removal

1. Place cylinder head (5) on workbench with exhaust manifold side down.
2. Insert tapered end of lower puller collar (7) into injector sleeve (6).
3. Insert upper puller collar (9) on top of collar (7) with threaded end out.
4. Position support bridge (10) over threaded end of collar (9) with one hand holding support bridge (10) legs against cylinder head (5) and install washer (11) and nut (12). Tighten nut (12) finger tight.
5. While holding support bridge (10) insert threaded rod (4) through holes in collars (7) and (9).
6. Install hex nut (1) and extractor tip (3) on threaded rod (4) and tighten. Ensure extractor tip (3) is seated firmly against injector sleeve (6) bottom.
7. Tighten nut (1) to remove injector sleeve (6). Discard injector sleeve (6).
8. Remove O-ring (8) from injector sleeve bore (2). Discard O-ring (8).

5-6. FUEL INJECTOR SLEEVE MAINTENANCE (Contd)

5-6. FUEL INJECTOR SLEEVE MAINTENANCE (Contd)

b. Bead Cutting

1. Insert cylinder head (5) cutter (3) and cutter pilot (2) into injector sleeve bore (6).
2. Position holder (1) to cutter (3).
3. Set cylinder head (5) on table of drill press.

CAUTION

- Tool chatter may occur if drill speed is higher than 75 rpm, causing damage to cylinder head.
- Do not cut more than 0.010 in. (0.254 mm) deep into cylinder head.
- Cutter must be sharp to prevent bead damage.

NOTE

Press may be turned by hand for light cuts or to prevent accidental removal of too much material.

4. Set drill press speed no higher than 75 rpm.
5. Lift holder (1), pilot (2), and cutter (3) and apply cutting oil to lubricate cutter (3).
6. Turn drill press motor on and take a very light cut. When the proper cut depth has been reached, allow cutter (3) to dwell ten seconds to ensure a good seat and to clean the groove.
7. Remove holder (1), pilot (2), and cutter (3) from injector sleeve bore (6).
8. Apply Prussian blue to coat seat end of new injector sleeve (7) and insert new injector sleeve (7) in injector sleeve bore (6) until fully seated.
9. Remove new injector sleeve (7) from injector sleeve bore (6). Ensure blueing shows a complete 360° band on both the injector sleeve (7) and seat (4).

c. Installation

CAUTION

Ensure sleeve seat at bottom of injector bore is free from oil, carbon, or other foreign materials.

1. Apply lubricating oil to coat new O-ring (10).
2. Install new O-ring (10) into groove (11) of injector sleeve bore (6).

NOTE

Ensure mandrel is not struck with a hammer during this operation.

3. Use injector sleeve mandrel to push the new injector sleeve (7) into injector sleeve bore (6) until it bottoms.
4. Remove mandrel from injector sleeve bore (6).
5. Install injector sleeve holding tool (8) in cylinder head (5). Tighten nut (9) 35-40 lb-ft (48-54 N•m)
6. Insert injector sleeve mandrel into injector sleeve bore (6) and drive injector sleeve (7) in with soft-faced hammer until seated.
7. Retighten nut (9) 35-40 lb-ft (48-54 N•m).

5-6. FUEL INJECTOR SLEEVE MAINTENANCE(Contd)

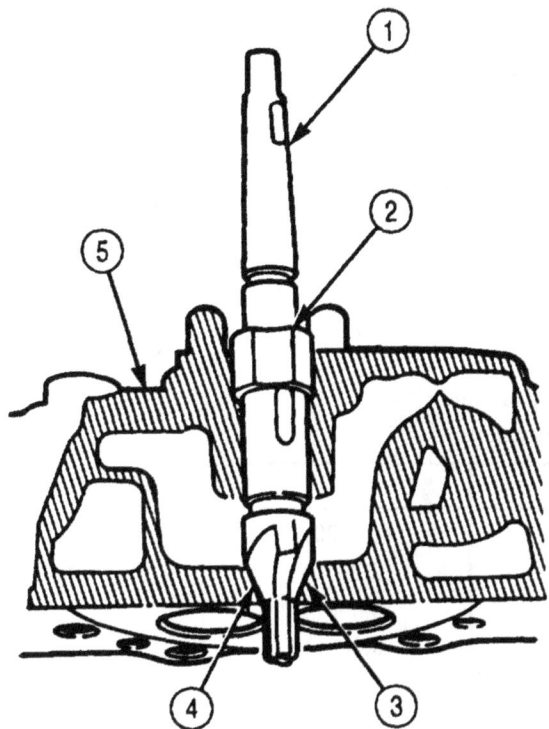

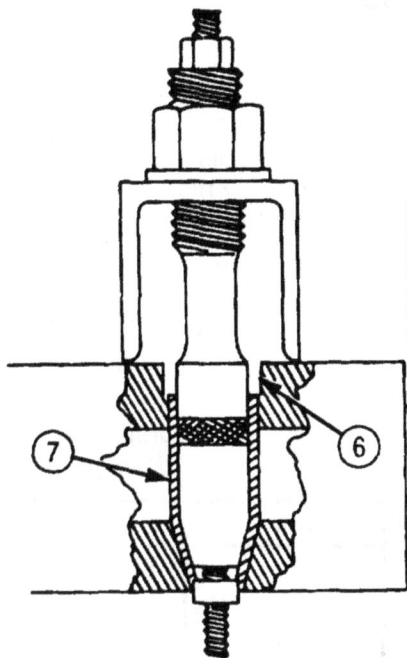

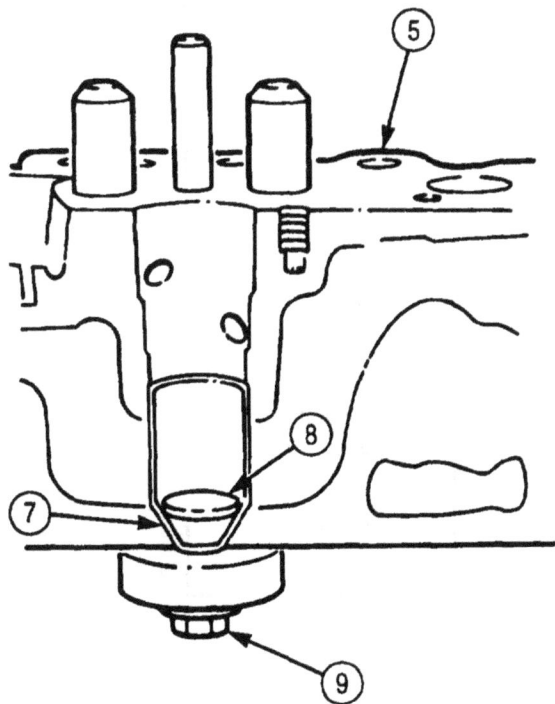

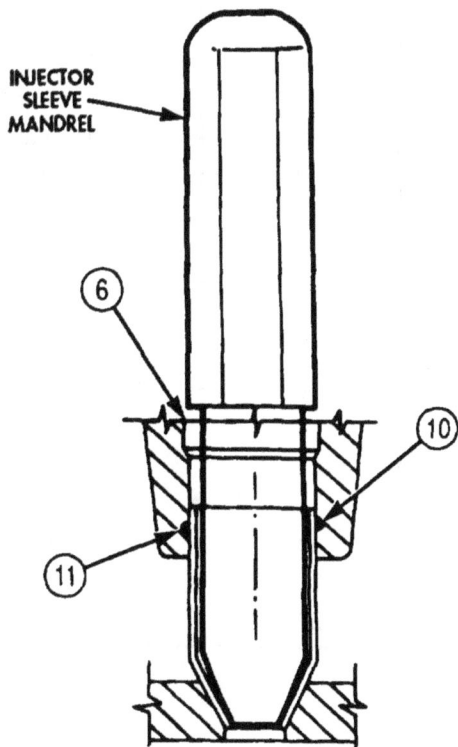

INJECTOR
SLEEVE
MANDREL

5-6. FUEL INJECTOR SLEEVE MAINTENANCE(Contd)

d. Fitting and Forming

CAUTION

- Do not roll lower area of injector sleeve which will cause distortion of total sleeve.
- Over-rolling of injector sleeve will cause deformation of sleeve into O-ring groove.

1. Position injector sleeve expander tool into injector sleeve (2) in cylinder head (1).
2. Turn thrust nut (4) to adjust depth of expander tool so that roller (3) extends 0.5 in. (12.7 mm) into sleeve (2).
3. Lock thrust nut (4) in place using an Allen wrench.
4. Push expander tool down, turn, and adjust until 75 lb-in. (8.5 N-m) maximum is reached.

NOTE

Form cutter per specifications in table 5-3.

Table 5-3. Cutter Specifications.

Item	In.	(MM)
5	0.080	(2.0320)
	0.090	(2.2860)
6	0.0015	(0.0381)
	0.0025	(0.0635)
7	0.0077	(0.1955)
	0.0097	(0.2463)
8	1.0615	(26.9621)
	1.0635	(27.0129)
9	15° angle relief	
10	15° angle relief	
11	30° 9' angle relief	
12	0.375	(9.5250)
13	0.312	(7.9375)
14	0.125	(3.1750)
15	0.375	(9.5250)
16	0.010	(0.2540)
17	0.0937	(2.3812)
18	0.384	(9.7536)
	0.386	(9.8044)
19	0.226	(5.7404)
	0.236	(5.9944)
20	0.3425	(8.6995)
21	1.250	(31.7500)
22	0.375	(9.5250)

5-6. FUEL INJECTOR SLEEVE MAINTENANCE (Contd)

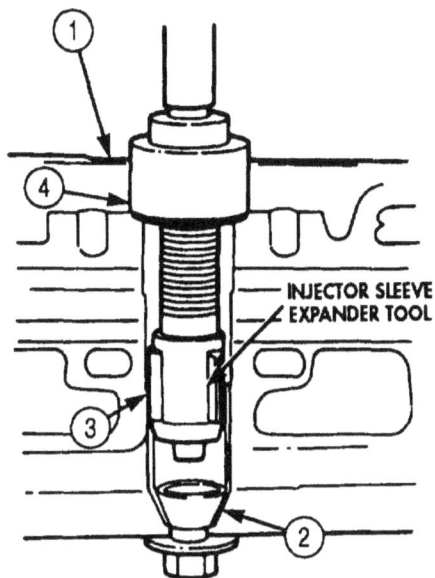

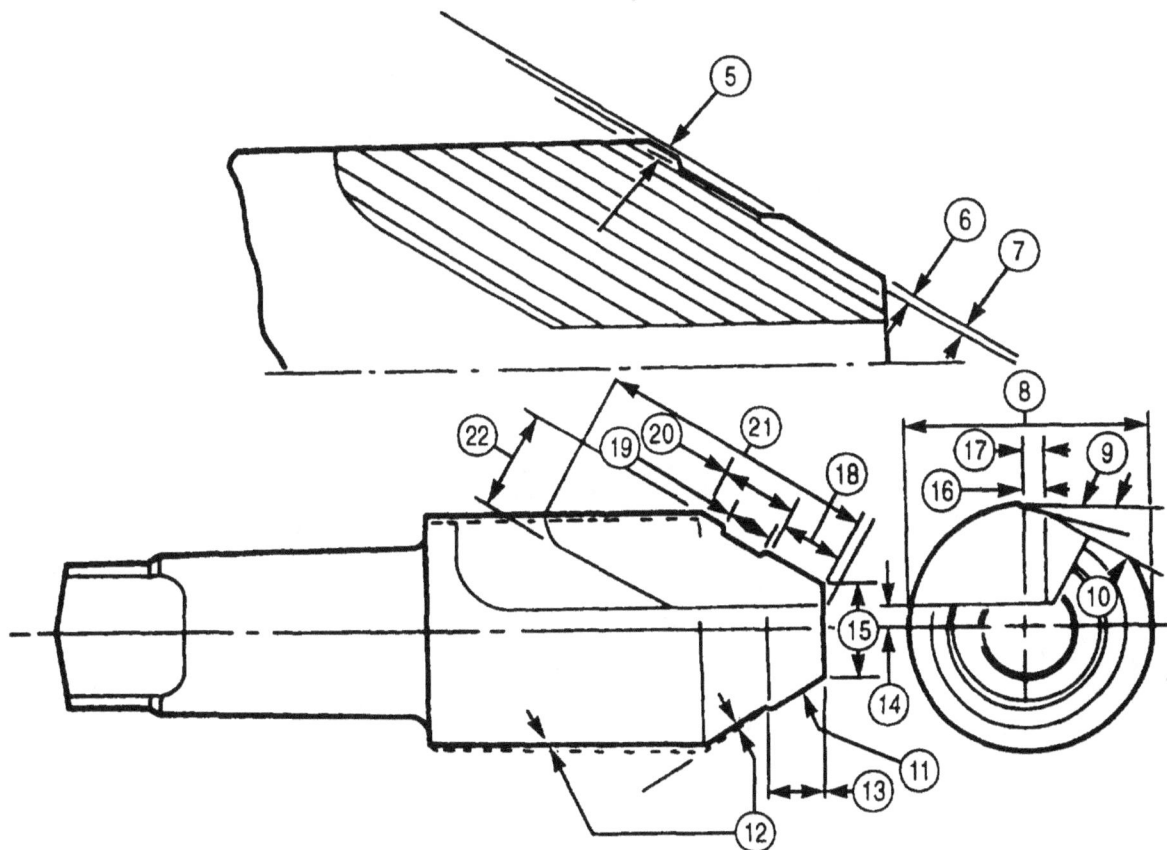

INJECTOR SLEEVE
EXPANDER TOOL

5-6. FUEL INJECTOR SLEEVE MAINTENANCE(Contd)

5. Size, grind, and inspect injector sleeve seat cutter (5) and ensure it is ground to exact contours listed in table 5-3.

NOTE

- Use a solid stream of clean cutting oil to allow injector sleeve seat cutter to cut freely without grabbing.
- Proper seating and protrusion of injector sleeve seat cutter are checked in task e.

6. Install injector sleeve seat cutter (5) on head (6) and use in drill press with pilot (4) to cut injector sleeve (2) just enough to provide for proper seating of injector and to maintain correct injector tip protrusion.

e. Check and Test

NOTE

Steps 1 through 4 check injector cup seating pattern in injector.

1. Apply light coat of Prussian blue to injector cup (11) and install injector assembly (9) into cylinder head (1) with washer (10), clamp (8), and two screws (7). Tighten screws (7) alternately in 4 lb-ft (5 N•m) steps to 10-12 lb-ft (14-16 N•m).

2. Remove two screws (7), clamp (8), washer (10), and injector assembly (9) from cylinder head (1).

3. Check seat pattern in bottom of injector sleeve (2) cup seating area.

4. If blued band (3) on injector sleeve (2) seating area is not at least 0.060 in. (1.52 mm) wide and located approximately 0.469 in. (11.91 mm) from bottom of cylinder head (1) surface, regrind injector sleeve (2) seating area as described in steps 5 and 6 of task f.

5. Install injector assembly (9) in injector sleeve (2) in injector head (1) with clamp (8), washer (10), and two screws (7). Tighten screws (7) alternately in 4 lb-ft (5 N•m) steps to 10-12 lb-ft. (14-16 N•m).

6. Check protrusion of injector tip (13) using dial indicator (12).

   a. If injector tip (13) protrusion exceeds 0.070 in. (1.78 mm), install new injector sleeve (2).

   b. If injector tip (13) protrusion is less than 0.060 in. (1.52 mm), regrind injector sleeve (2).

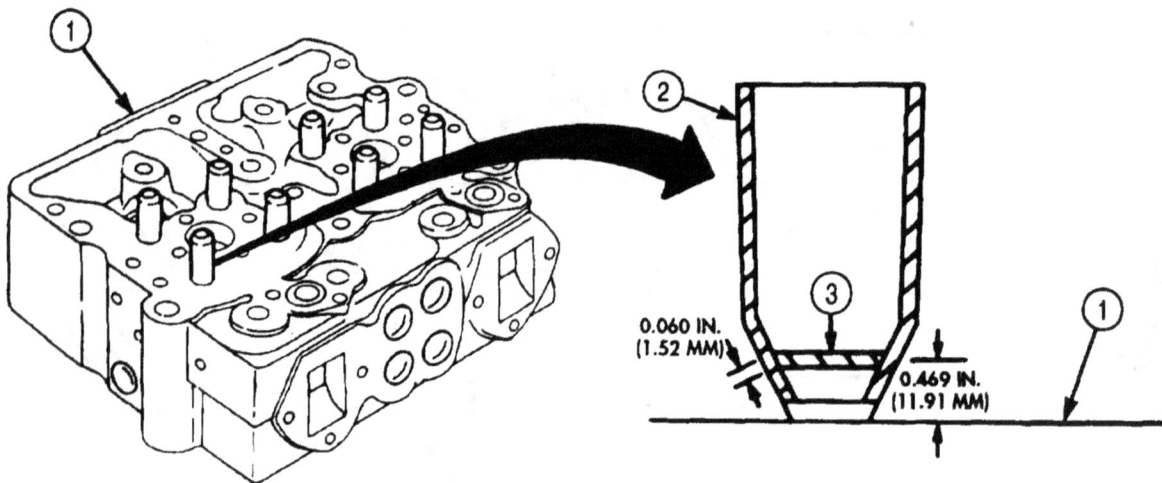

5-6. FUEL INJECTOR SLEEVE MAINTENANCE (Contd)

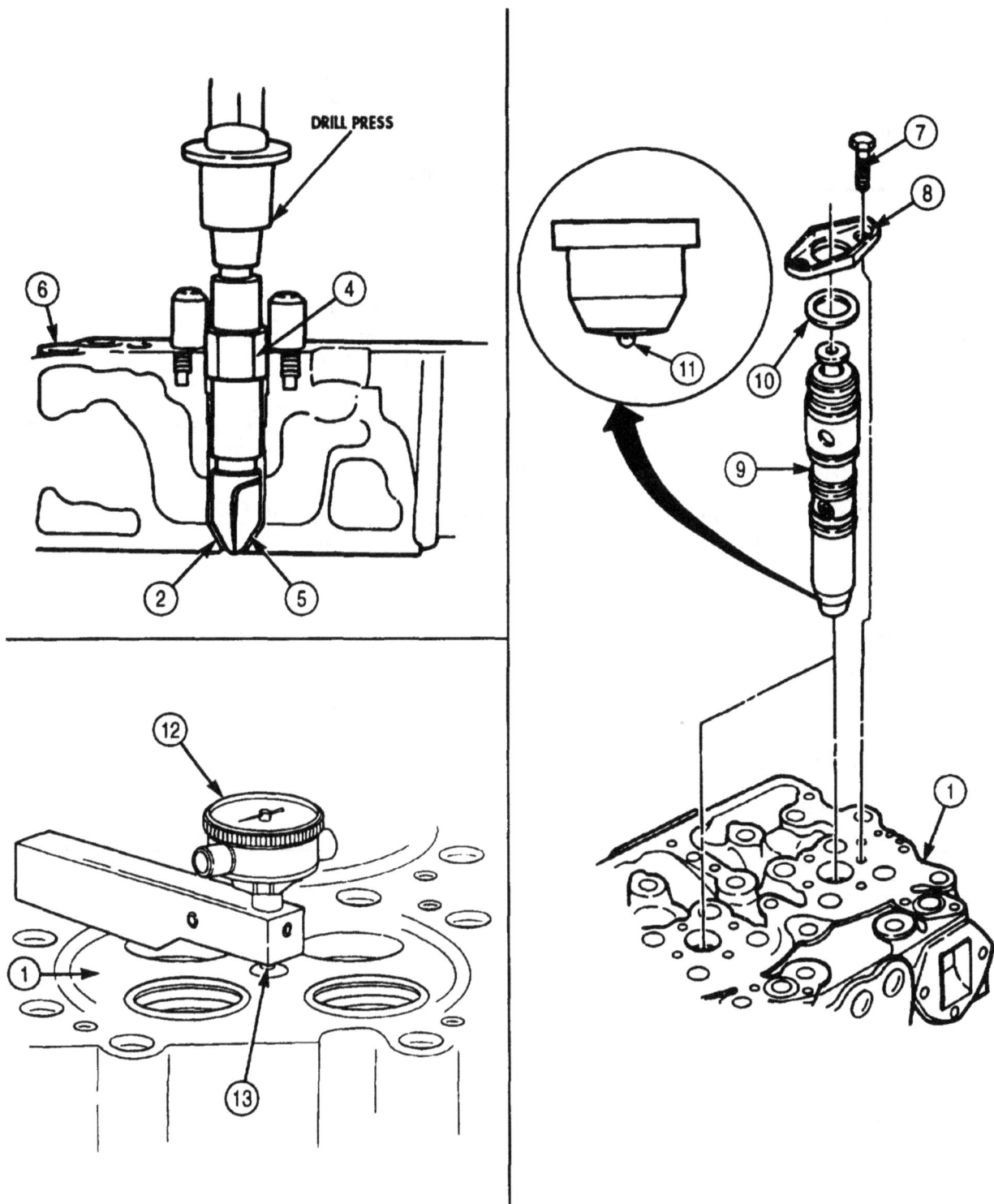

DRILL PRESS

FOLLOW-ON TASK: Assemble cylinder head (para. 5-3).

5-7. CYLINDER LINERS AND BLOCK MAINTENANCE

THIS TASK COVERS:

a. Removal
b. Disassembly

c. Cleaning and Inspection
d. Cylinder Liners Installation

INITIAL SETUP:

APPLICABLE MODELS
M939/A1

SPECIAL TOOLS
Liner clamp set (Appendix E, Item 78)
Liner driver (Appendix E, Item 79)

TOOLS
General mechanic's tool kit (Appendix E, Item 1)
Cylinder liner puller (Appendix E, Item 107)
Outside micrometer (Appendix E, Item 80)
Vernier calipers (Appendix E, Item 159)
Depth gauge (Appendix E, Item 81)
Gauge block (Appendix E, Item 52)
Dial indicator (Appendix E, Item 36)
Telescoping gauge set (Appendix E, Item 136)
Torque wrench (Appendix E, Item 144)
Feeler gauge
Hold-down tool

MATERIAL PARTS
Two lockwashers (Appendix, Item 349)
Suction flange gasket (Appendix D, Item 680)
Two O-rings (Appendix D, Item 460)
Crevice seal (Appendix D, Item 88)
Drycleaning solvent (Appendix C, Item 71)
Antiseize tape (Appendix C, Item 72)
Lint-free cloth (Appendix C, Item 21)
Lubricating oil OE/HDO (Appendix C, Item 50)
Detergent (Appendix C, Item 27)

REFERENCES (TM)
TM 9-2320-272-24P

EQUIPMENT CONDITION
Crankshaft and main bearings removed (para. 5-8).

GENERAL SAFETY INSTRUCTIONS
• Keep fire extinguisher nearby when using drycleaning solvent.
• Drycleaning solvent is flammable and toxic. Do not use near an open flame.
• When cleaning with compressed air, wear eyeshields and ensure source pressure does not exceed 30 psi (207 kPa).

SPECIAL ENVIRONMENTAL CONDITIONS
Work area clean and free from blowing dirt and dust.

---

a. Removal

NOTE
Tag all cylinder liners for installation.

Remove six cylinder liners (1) from cylinder liner bores (2) with cylinder liner puller.

5-7. CYLINDER LINERS AND BLOCK MAINTENANCE (Contd)

CYLINDER
LINER PULLER

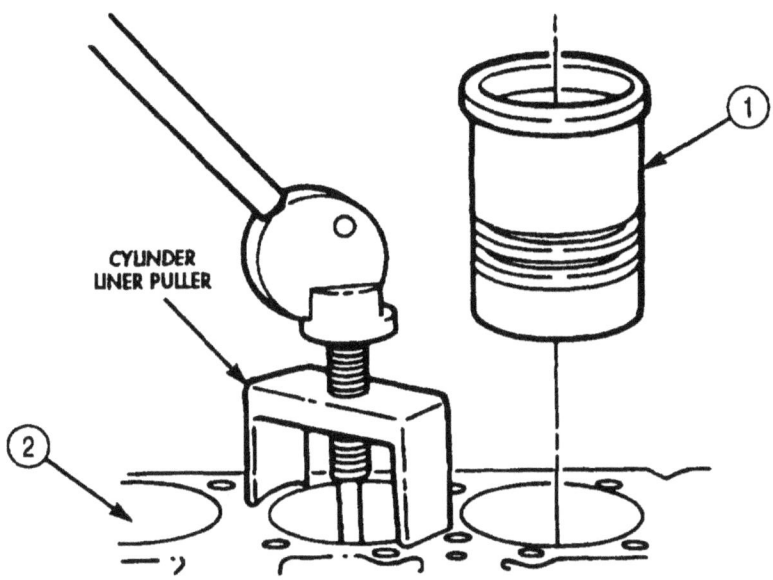

5-7. CYLINDER LINERS AND BLOCK MAINTENANCE (Contd)

b. Disassembly

NOTE

All six cylinder liners are repaired the same way. This procedure covers only one cylinder liner.

1. Remove upper and lower O-rings (4) and (5) and crevice seal (3) from cylinder liner (1). Discard O-rings (4) and (5) and crevice seal (3).

2. Remove shim pack (2) from cylinder liner (1) and measure thickness with micrometer. Record readings so same shim pack (2) thickness can be installed.

3. Tape shim pack (2) together and tag with corresponding cylinder liner (1) number. Hold shim pack (2) for reassembly.

4. Place cylinder liner (1) on numbered rack and hold for cleaning and inspection.

NOTE

Note location of plug removal for installation.

5. Remove four water passageway pipe plugs (6) from engine block (7).

6. Remove eight oil passageway pipe plugs (9) from engine block (7).

7. Remove two water passageway cup plugs (8) from engine block (7).

8. Remove two screws (11), lockwashers (12), suction flange plate (10), and gasket (13) from engine block (7). Discard lockwashers (12) and gasket (13) and clean gasket remains from mating surfaces.

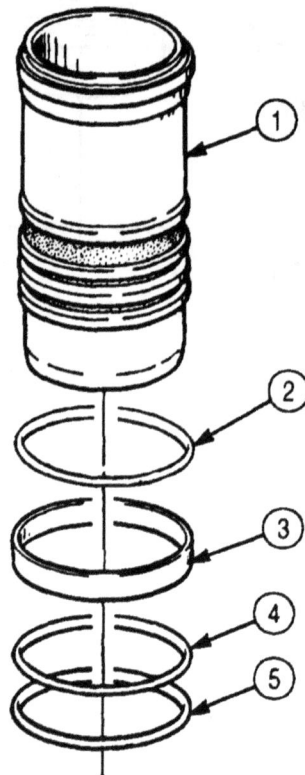

5-7. CYLINDER LINERS AND BLOCK MAINTENANCE (Contd)

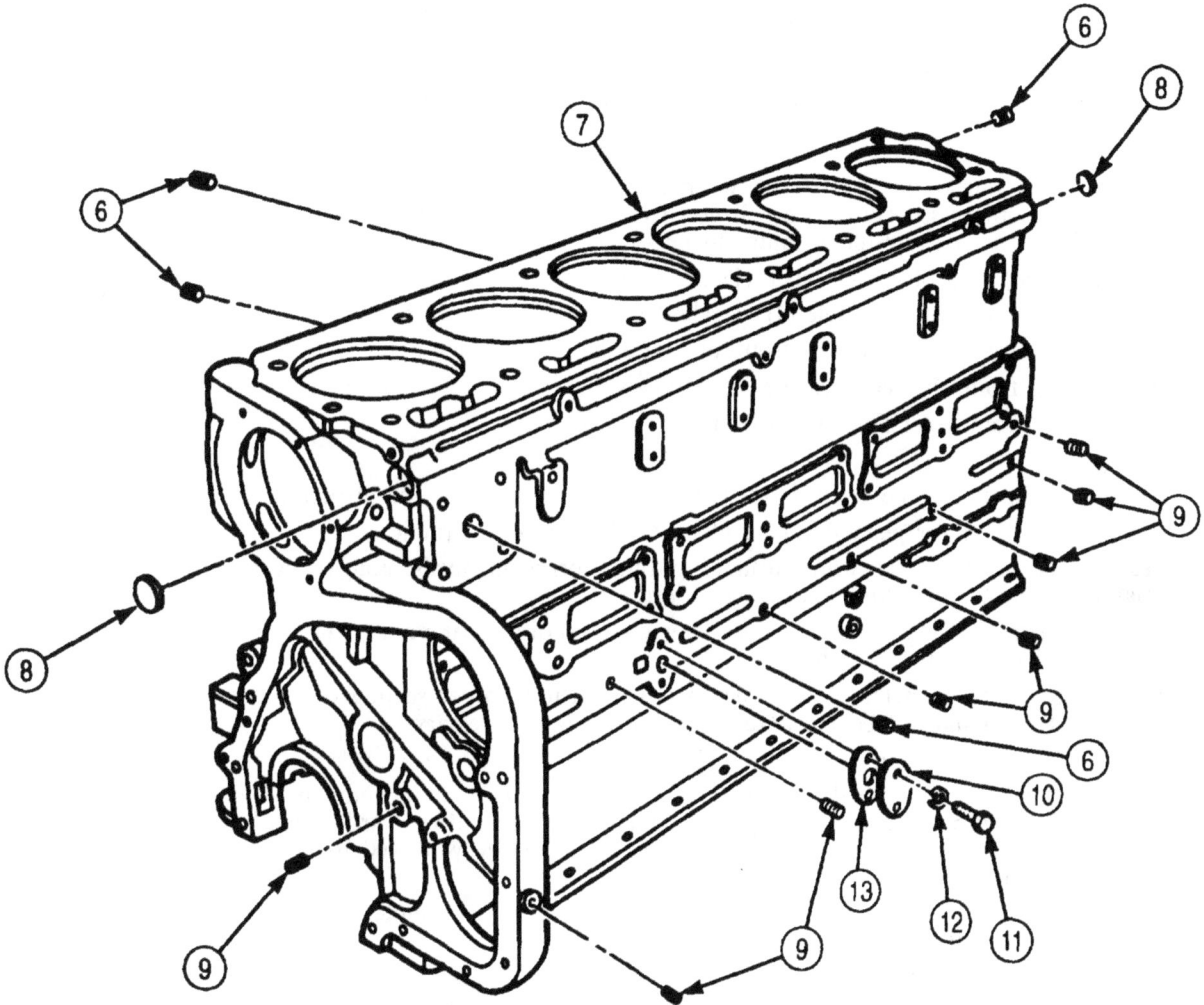

5-7. CYLINDER LINERS AND BLOCK MAINTENANCE (Contd)

c. Cleaning and Inspection

WARNING

Drycleaning solvent is flammable and toxic. Do not use near open flame and always have a fire extinguisher nearby when solvents are used. Use only in well-ventilated places, wear protective clothing, and dispose of cleaning rags in approved container. Failure to do this may result in injury or death to personnel and/or damage to equipment.

1. Clean engine block (1). Refer to para. 2-14 for general cleaning instructions.
2. Run rods (3) with brushes through all oil passages (2) in engine block (1).

WARNING

Eyeshields must be worn when cleaning with compressed air. Compressed air source will not exceed 30 psi (207 kPa). Failure to do so may result in injury to personnel.

3. Blow oil passages (2) of engine block (1) clean with compressed air.
4. Clean water pump air bleed hole (4) in cylinder bore (5) with compressed air.
5. Blow all dirt and cleaning solvent from all engine block (1) screw holes (7).
6. Remove scale from liner counterbore ledge (6).
7. Clean carbon from lower liner bore (8) of engine block (1).

NOTE

Ensure engine block is on flat surface or workbench for inspection.

8. Refer to para. 2-15 for general engine block (1) inspection instructions.
9. Check engine block (1) for worn surfaces, pitting, corrosion, nicks, gouges, burrs, eroded waterholes, damaged threads, distortion, and cracked areas.
10. Inspect counterbore ledge (6) of cylinder liner counterbore (9) at points (A) and (B) to ensure counterbore ledge (6) is 90° to liner bore.
11. Check counterbore diameter (A) and depth (B) to insure proper seating of cylinder liner and to determine if machining of counterbore ledge (6) depth is necessary.

NOTE

• If counterbore depth varies more than 0.001 in. (0.03 mm) at several areas measured or counterbore slants downward toward center of cylinder liner bore, counterbore ledge will need to be refinished.
• Steps 12 and 13 check engine block for possible counter-bore distortion.

12. Position telescoping gauge in counterbore (9) and set dial to zero.
13. Move bore gauge around circumference of counterbore (9) and take readings at three intervals of 120°. The difference in the readings will indicate amount of distortion. If specifications in table 5-4 are exceeded, replace engine block (1).

5-7. CYLINDER LINERS AND BLOCK MAINTENANCE(Contd)

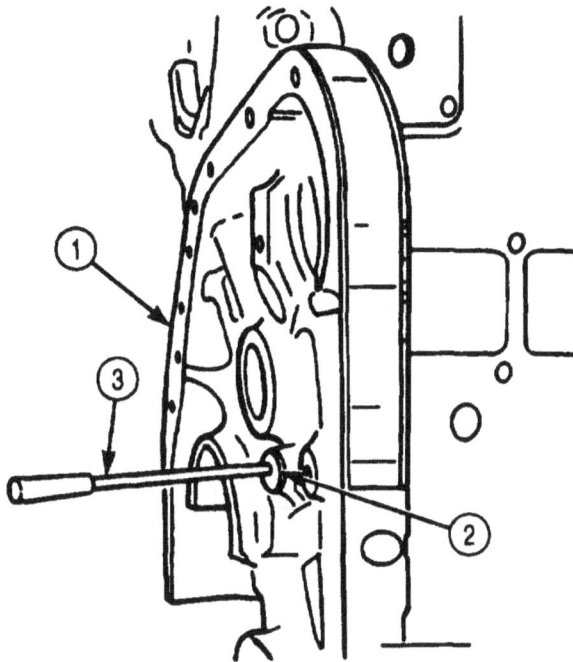

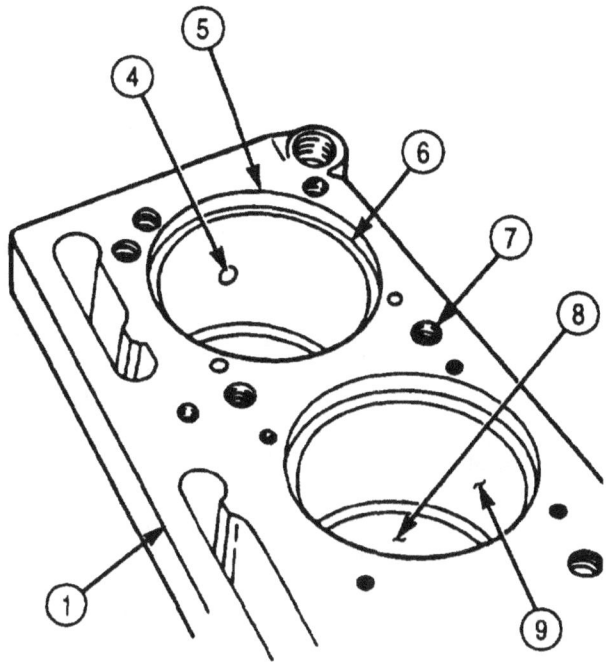

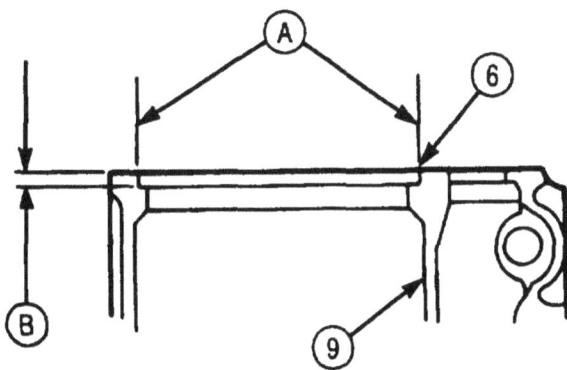

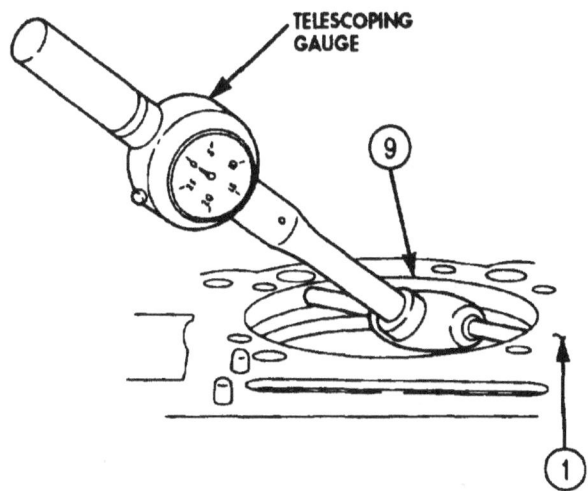

TELESCOPING
GAUGE

5-7. CYLINDER LINERS AND BLOCK MAINTENANCE (Contd)

Table 5-4. Cylinder Liner Counterbore.

	INSIDE DIAMETER	DEPTH
New minimum	6.5615 in. (166.662 mm)	0.350 in. (8.89 mm)
New maximum	6.5635 in. (166.713 mm)	0.352 in. (8.94 mm)
Wear limit		0.412 in. (10.56 mm)

14. Ensure cylinder liners (2) protrude 0.003-0.006 in. (0.08-0.15 mm) above the engine block (1) when they are properly installed.

NOTE

- Shims are used to compensate for counterbore depth wear.
- Steps 15 through 17 check cylinder liner for proper protrusion without installing a cylinder liner.

15. Measure cylinder liner (2) outside flange (4) with micrometer. Do not include bead (3) when taking measurements.

16. Measure counterbore ledge (5) depth with gauge block. Ensure counterbore ledge (5) is smooth and not "cupped" more than 0.0014 in. (.036 mm). Depth must not vary more than 0.001 in. (0.254 mm).

17. Subtract counterbore depth from cylinder liner (2) outside flange (4) depth to determine amount of shims or depth of counterbore cut to be made to obtain 0.0036-0.006 in. (0.08-0.15 mm) liner protrusion.

18. Install cylinder liner (2) and clamps in engine block (1) without O-rings or crevice seal. Ensure clamps are installed so there is equal pressure on liner. Tighten clamps 50 lb-ft (68 N•m).

19. Use feeler gauge to check clearance between lower liner (2) and engine block (1). If clearance is not within limits specified in table 5-5, check lower O-ring groove inside diameter in engine block (1).

20. Remove cylinder liner (2) from engine block (1).

21. Use telescoping gauge to check lower liner bore (6) in engine block (1). Bore (6) should be within limits set in table 5-5.

Table 5-5. Lower Liner Bore Inside Diameter and Block Clearance.

	MINIMUM	MAXIMUM
Lower liner bore inside diameter	.6.124 in. (155.55 mm)	6.126 in. (155.60 mm)
Lower liner to block clearance	0.002 in. (0.05 mm)	0.006 in. (0.15 mm)

5-7. CYLINDER LINERS AND BLOCK MAINTENANCE (Contd)

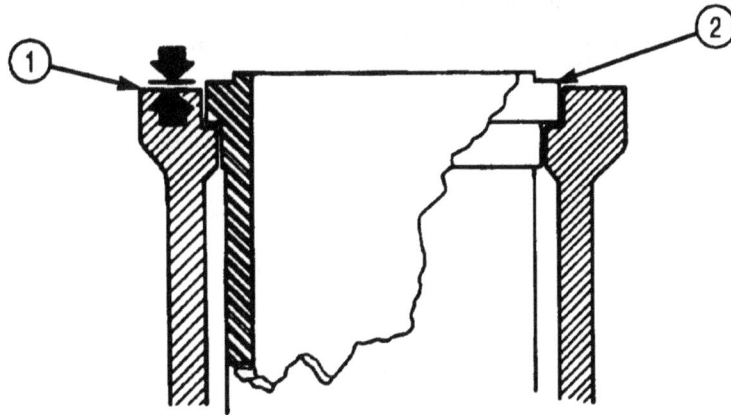

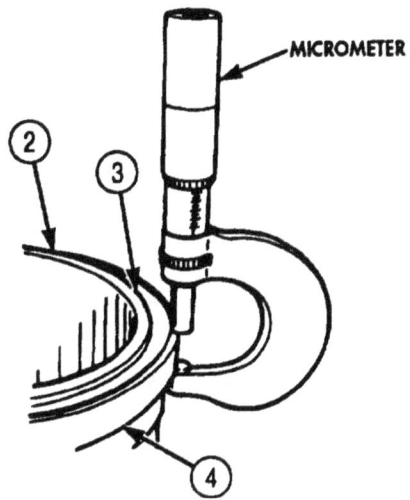

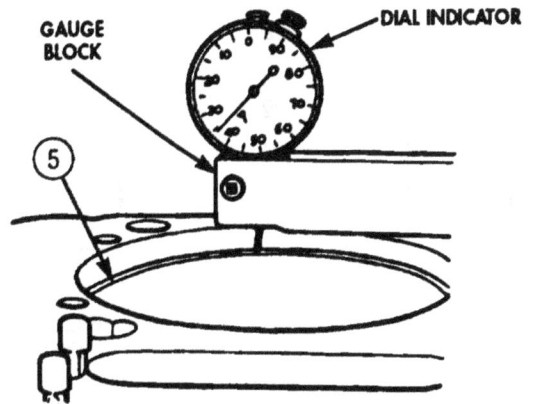

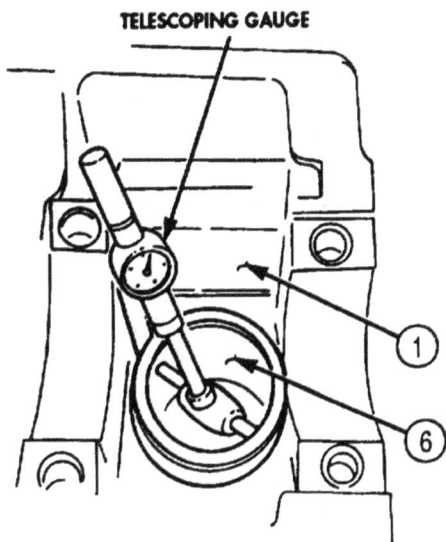

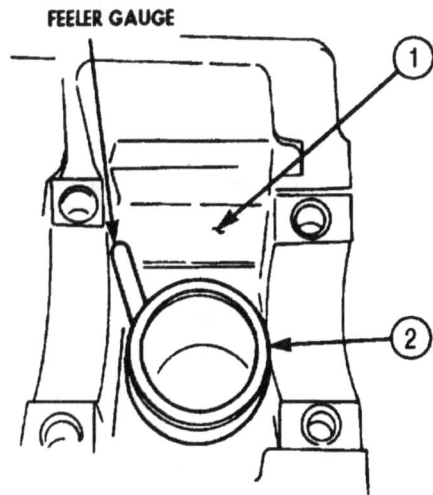

MICROMETER

GAUGE BLOCK

DIAL INDICATOR

TELESCOPING GAUGE

FEELER GAUGE

5-7. CYLINDER LINERS AND BLOCK MAINTENANCE (Contd)

22. Install main bearing caps (2) in engine block (1) without crank or bearing shells.
23. Measure main bearing bore diameter (3) horizontally, vertically, and diagonally with telescoping gauge (4). Ensure bore diameter does not exceed 4.75 in. (120-663 mm).

NOTE

Pipe plugs, cup plugs, and suction flange plate are installed at this time to keep passages clean. Wrap pipe plug threads with sealing tape to prevent leakage.

24. Install two cup plugs (9) in engine block (1).
25. Install four pipe plugs (8) in block (1) and tighten according to table 5-6.
26. Install eight pipe plugs (10) in block (1) and tighten according to table 5-6.
27. Install new gasket (14) and suction flange plate (13) on engine block (1) with two new lockwashers (12) and screws (11). Tighten screws (11) 10-15 lb-ft (14-20 N•m).

Table 5-6. Engine Block Pipe Plug Torque.

PIPE PLUG SIZE	MINIMUM		MAXIMUM	
	LB-FT	(NM	LB-FT	(N•M)
1/8 in.	15	(20)	20	(27)
1/4 in.	30	(41)	35	(47)
3/8 in.	35	(47)	45	(61)
1/2 in.	45	(61)	55	(75)
3/4 in.	60	(81)	70	(95)
1-1/4 in.	75	(102)	85	(115)
1-1/2 in.	90	(122)	100	(136)

NOTE

• Most attempts to hone or deglaze cylinder liners provide worse results than leaving them "as is". Experience and laboratory results indicate that liners do not need to be honed, rotated, or deglazed to provide proper setting.

• It is recommended that cylinder liners be inspected before cleaning so defects can be clearly noted.

• Inspect cylinder liners closely for any of the metal conditions illustrated. If one liner has failed, then other liners in the same engine are likely to have early signs of the same type of failure.

28. Check cylinder liner (5) for scoring or vertical grooving (6) on the inside diameter indicating heavy metal-to-metal contact of piston to cylinder liner (5). If scoring or vertical grooving (6) is present, tag cylinder liner for replacement.
29. Check cylinder liner (5) for cracks (7) indicated by magnetic detection. If cracks (7) are present, tag cylinder liner (5) for replacement.

5-7. CYLINDER LINERS AND BLOCK MAINTENANCE (Contd)

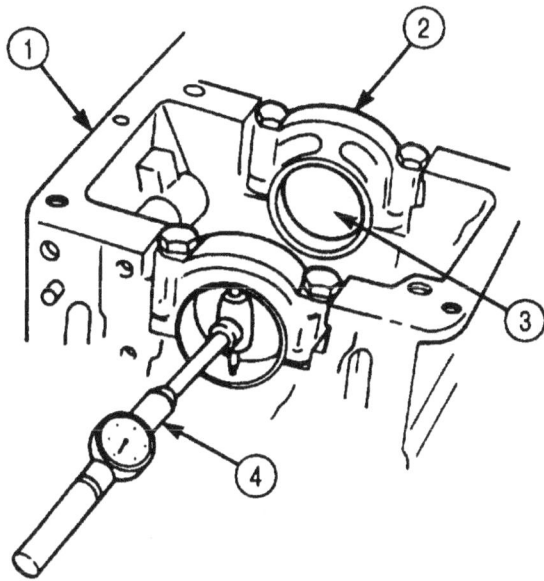

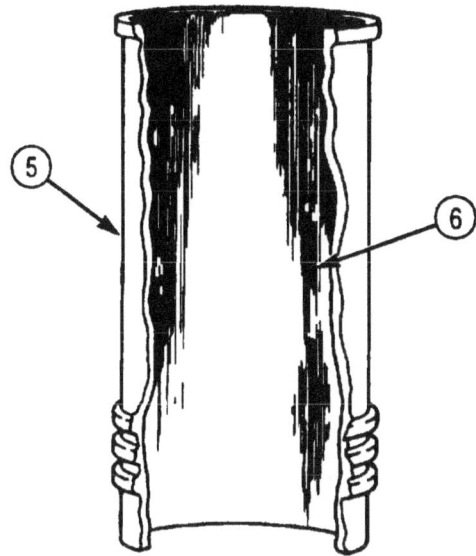

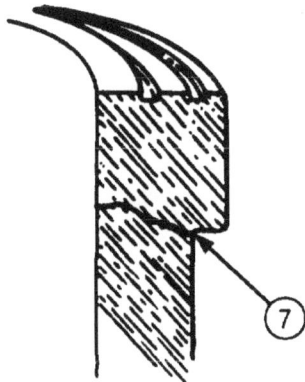

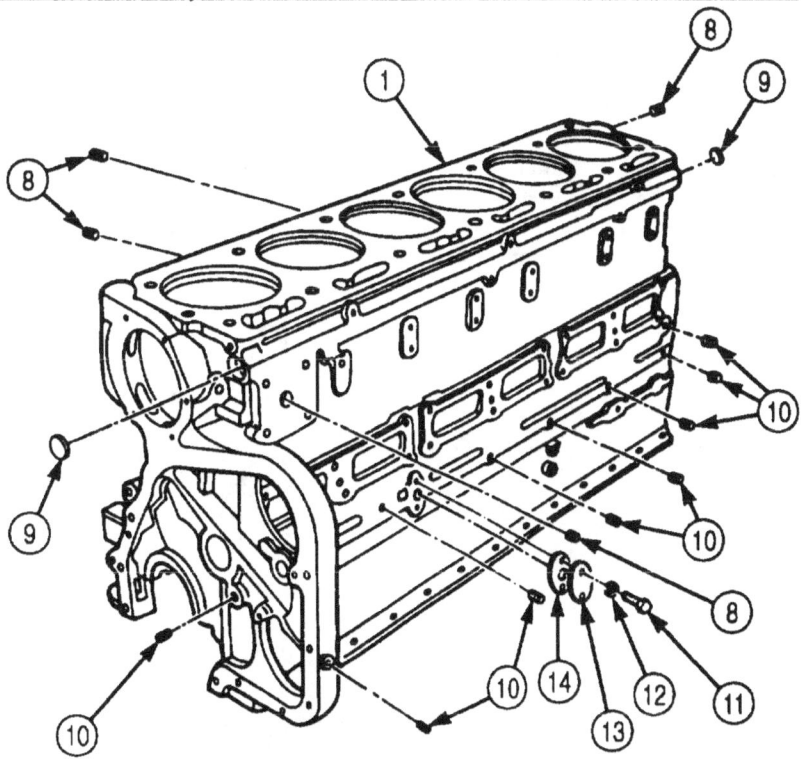

5-7. CYLINDER LINERS AND BLOCK MAINTENANCE (Contd)

30. Check cylinder liner (2) for a series of pit erosions (3) on the thrust or antithrust side of the cylinder liner (2) outside diameter. If pit erosions (3) are present, tag cylinder liner (2) for replacement.

31. Check cylinder liner (2) for visible cracks. As a rule, cylinder liners are highly resistant to vertical cracks or breakage. If cracks are present, tag cylinder liner (2) for replacement.

32. Check cylinder liner (2) for eroded surface (4). Moving coolant contacting the cylinder liner (2) outside diameter erodes the surface away and attacks the crevice seal. If surface is eroded (4), tag cylinder liner (2) for replacement.

33. Check cylinder liner (2) for fretting of surfaces (6), and/or machined area. Top of cylinder liner bead has worn away on illustrated cylinder liner (2). If surfaces (6) have fretted, tag cylinder liner (2) for replacement.

NOTE
Do not use wire brush on cylinder liners.

34. Steam, clean, or wash cylinder liner (2) in hot water and detergent. Scrub with bristle brush.

35. Remove rust, scale, and corrosion from cylinder liner (2). If cylinder liner (2) is excessively rusted, scaled, or corroded, replace.

WARNING
Eyeshields must be worn when cleaning with compressed air.
Compressed air source will not exceed 30 psi (207 kPa). Failure to
do so may result in injury to personnel.

36. Blow cylinder liner (2) dry with compressed air.

37. Apply generous coat of lubricating oil to cylinder liner (2) and let stand for five to ten minutes.

38. Wipe oil from cylinder liner bore (1) with heavy paper towel. Gray and black residue will appear on towels.

39. Repeat steps 37 and 38 until black or gray residue no longer appears on towel.

40. Apply lubricating oil lightly to cylinder liners (2) and wrap cylinder liners (2). Store in clean, dry location until installation.

NOTE
Cylinder liners must be checked at 60°-70°F (16°-21°C). New
cylinder liners with lubrite finish may be 0.0002-0.0006 in.
(0.005-0.015 mm) smaller than indicated due to lubrite coating.

41. Check liner bore (1) with telescoping gauge. If measurement exceeds 5.5050 in. (139.830 mm), tag cylinder liner (2) for replacement.

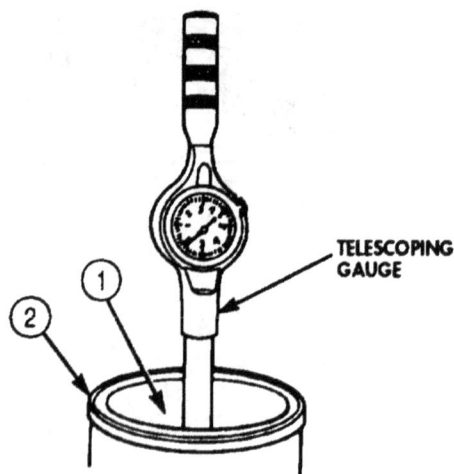

TELESCOPING GAUGE

5-7. CYLINDER LINERS AND BLOCK MAINTENANCE (Contd)

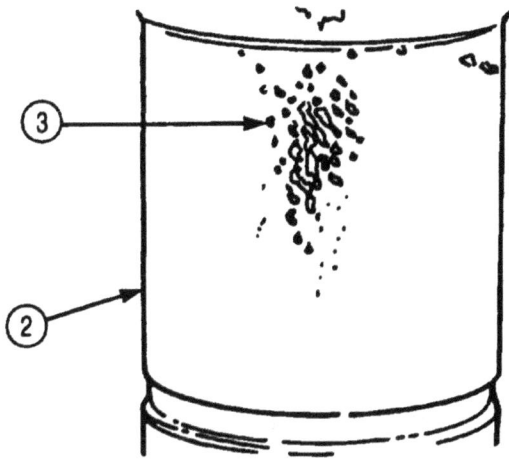

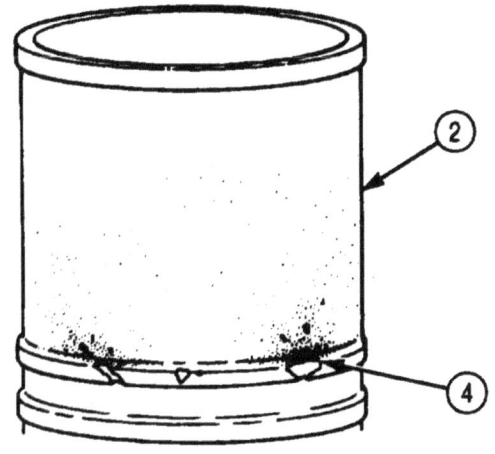

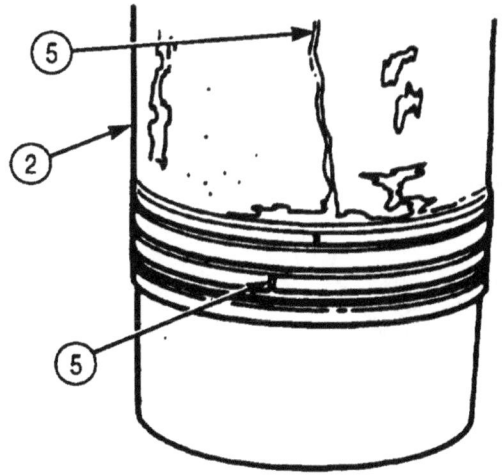

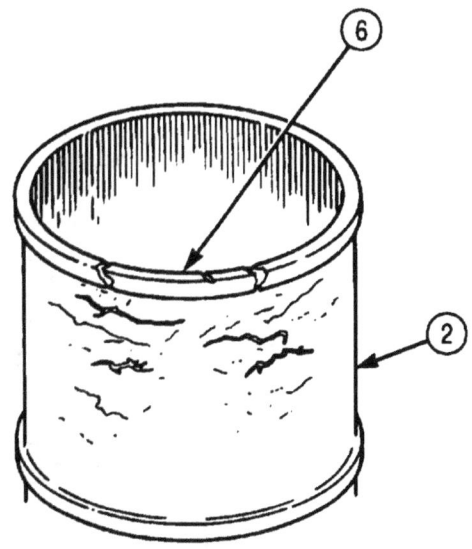

5-7. CYLINDER LINERS AND BLOCK MAINTENANCE (Contd)

d. Cylinder Liners Installation

1. Clean cylinder liner flange (2), crevice seal groove (12), and O-ring grooves (3) and (11) with lint-free cloth.

2. Install new crevice seal (10) in crevice seal groove (12). Ensure crevice seal (10) is straight and not twisted.

3. Install new black O-ring (9) in top O-ring groove (11) using molding mark as a guide. Ensure new black O-ring (9) is straight and not twisted.

4. Install new red O-ring (9) in bottom O-ring groove (3). Ensure new red O-ring (9) is straight and not twisted.

5. Clean cylinder bore (7), counterbore (4), and lower bore (5) and apply lubricating oil.

6. Apply light coat of clean engine oil to crevice seal (11) and O-rings (8) and (9) and position cylinder liner (1) in engine block (6) by hand. Take care not to dislodge O-rings (8) and (9) and crevice seal (10). Press cylinder liner (1) in place using hand pressure.

NOTE
Install cylinder liner without shims until liner protrusion is checked in step 8. It may be necessary to remove liners and add shims.

7. Drive cylinder liner (1) in with liner driver and liner clamp until cylinder liner flange (2) is seated. Hold down with hold-down tool.

8. Position dial indicator and gauge block on cylinder liner (1) and measure cylinder liner (1) protrusion. If protrusion is not 0.003-0.006 in. (0.08-0.15 mm), add or remove shims and repeat task d.

NOTE
When performing step 9, if liner is more than 0.002 in. (0.05 mm) out-of-round in lower bore O-ring area, remove liner and check for cause of distortion. It is permissible to have 0.003 in. (0.08 mm) out-of-round at the top 1 in. (25.4 mm) of liner bore.

9. Position telescoping gauge in cylinder liner (1) and measure at several points within range piston travel for out-of-round condition. If out-of-round, remove cylinder liner (1) and reinstall.

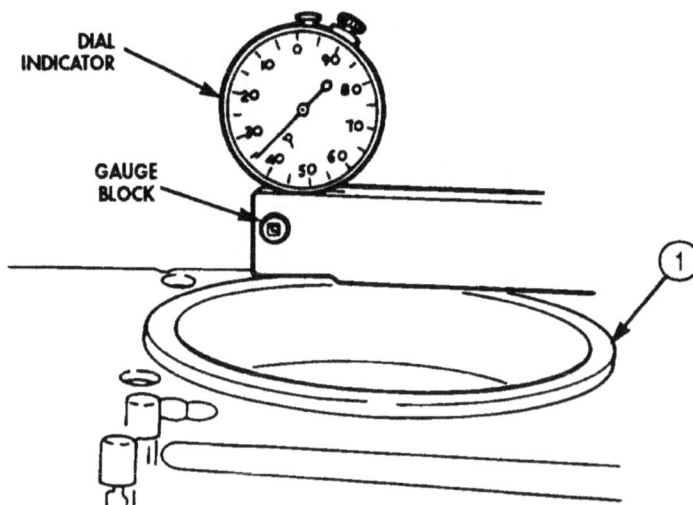

5-7. CYLINDER LINERS AND BLOCK MAINTENANCE (Contd)

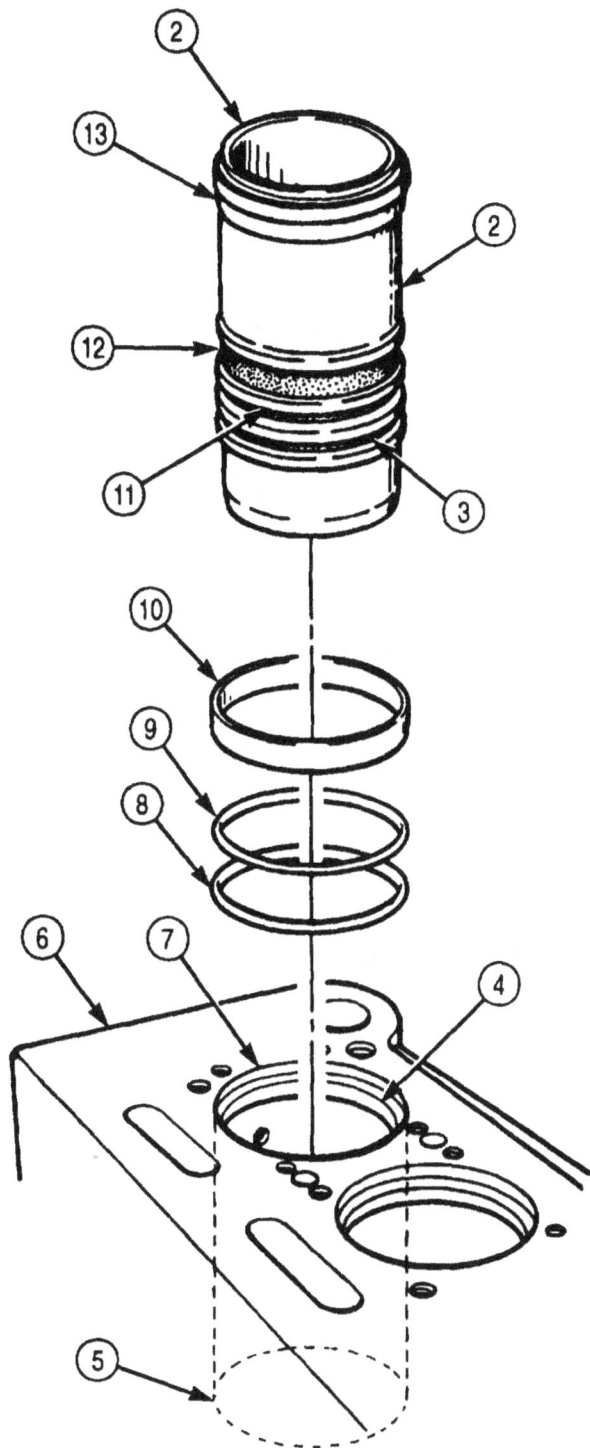

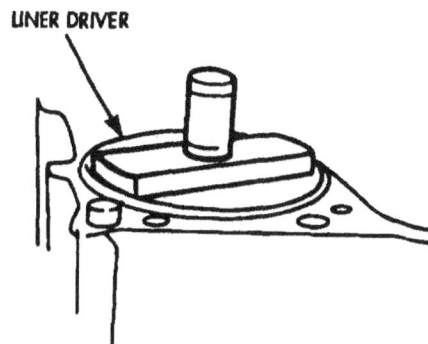

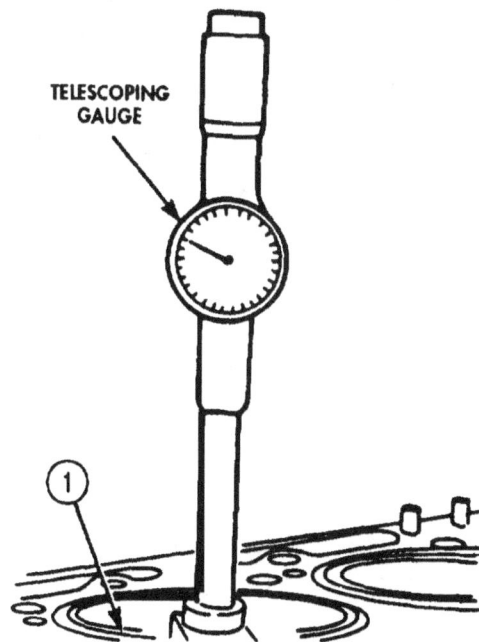

LINER DRIVER

TELESCOPING GAUGE

FOLLOW-ON TASK: Install crankshaft and main bearings (para. 5-8).

5-8. CRANKSHAFT AND MAIN BEARINGS MAINTENANCE

THIS TASK COVERS:

a. Removal
b. Cleaning and Inspection

c. Installation
d. End Play Clearance

INITIAL SETUP:

APPLICABLE MODELS
M939/A1

TOOLS
General mechanic's tool kit (Appendix E, Item 1)
Main bearing cap puller (Appendix E, Item 108)
Outside micrometer (Appendix E, Item 80)
Vernier caliper (Appendix E, Item 159)
Torque wrench (Appendix E, Item 144)
Dial indicator (Appendix E, Item 36)
Hoist
Rubber-protected hooks

MATERIALS/PARTS
Crankshaft main bearing set
  (Appendix D, Item 87)
Thrust ring set (Appendix D, Item 682)
Fourteen lockplates (Appendix D, Item 341)
Woodruff Key (Appendix D, Item 738)
Fourteen dowel rings (Appendix D, Item 100)
Lint-free cloth (Appendix C, Item 21)
GAA grease (Appendix C, Item 28)
Lubricating oil (Appendix C, Item 50)
Drycleaning solvent (Appendix C, Item 71)

REFERENCES (TM)
TM 9-2320-272-24P

EQUIPMENT CONDITION
Connecting rod and pistons removed (para. 5-10).

GENERAL SAFETY INSTRUCTIONS
- Keep fire extinguisher nearby when using drycleaning solvent.
- Drycleaning solvent is flammable and toxic. Do not use near an open flame.
- When cleaning with compressed air, wear eyeshields and ensure source pressure does not exceed 30 psi (207 kPa).

---

a. Removal

1. Turn engine block (1) upside down.

2. Bend down lockplate tab (8) and remove two screws (7) from seven main bearing caps (9). Discard lockplates (8).

NOTE

Tag all bearing caps and bearing shells for installation.

3. Remove main bearing caps (9) from engine block (1) with main bearing cap puller.

4. Remove lower half bearing shell (10) from main bearing caps (9). Discard lower half bearing shell (10).

5. Remove lower half of thrust ring (4) from crankshaft near journal (5). Discard thrust ring (4).

6. Carefully remove crankshaft (6) from engine block (1) using hoist and rubber-protected hooks (11). Place crankshaft (6) on clean, flat surface.

7. Remove upper half bearing shells (2) from engine block (6). Discard upper half bearing shells (2).

8. Remove fourteen dowel rings (3) from engine block (1). Discard dowel rings (3).

9. Remove upper half of thrust ring (4) from engine block (1). Discard thrust ring (4).

5-8. CRANKSHAFT AND MAIN BEARINGS MAINTENANCE (Contd)

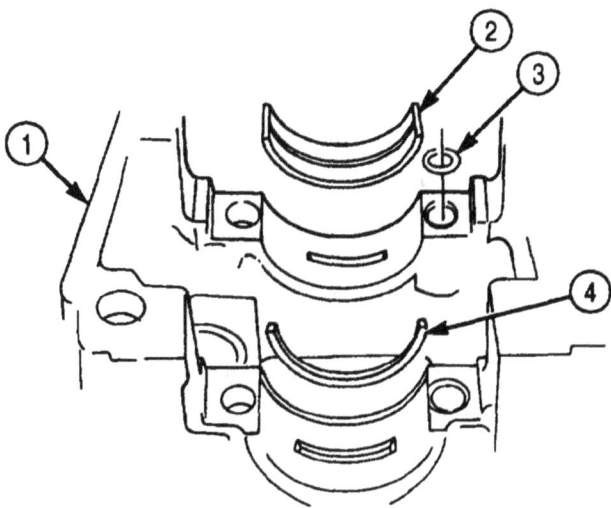

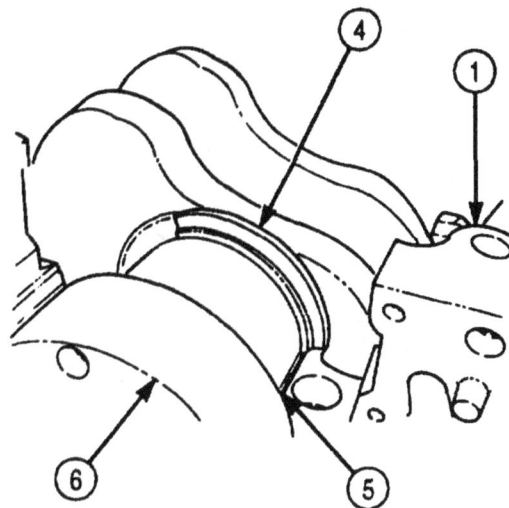

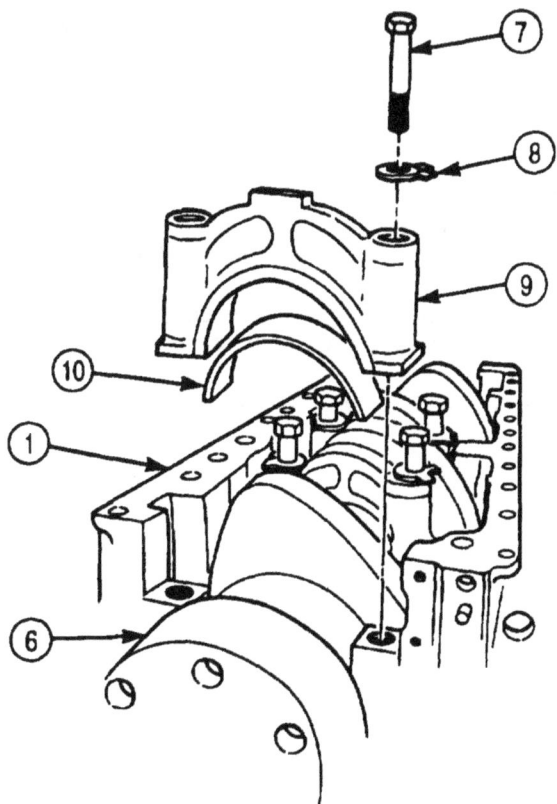

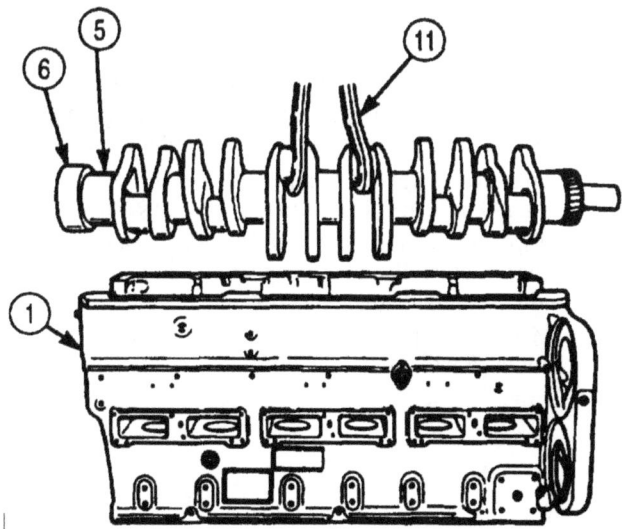

5-8. CRANKSHAFT AND MAIN BEARINGS MAINTENANCE (Contd)

b. Cleaning and Inspection

WARNING

Drycleaning solvent is flammable and toxic. Do not use near open flame and always have a tire extinguisher nearby when solvents are used. Use only in well-ventilated places, wear protective clothing, and dispose of cleaning rags in approved container. Failure to do this may result in injury or death to personnel and/or damage to equipment.

1. Clean crankshaft (2) and crankshaft gear (6) with drycleaning solvent. For cleaning instructions, refer to para. 2-14.

2. Inspect crankshaft gear (6) for breaks, cracks, or chips. If broken, cracked, or chipped, replace crankshaft (2) assembly.

3. Wipe rear No. 7 main bearing journal thrust flange (1) clean and check stamping (4) on shaft web (5). Stamped numbers show standard or oversize thrust rings both front and rear.

4. Inspect rear No. 7 main bearing journal thrust flange (1) for scratches or scoring. If scratched or scored, regrind and stamp shaft web (5) accordingly.

CAUTION

If any one bearing shell half is damaged, all bearing shells must be discarded. Not doing so will vary oil clearance limits when installed in engine and cause lubrication problems.

5. Wipe lower and upper half main bearing shell (7) clean and inspect for pits, chips, and scratches. If pitted, chipped, or scratched, replace bearing set.

6. Measure thickness of lower and upper half main bearing shell (7). If thickness is less than 0.1215 in. (3.086 mm), replace bearing set.

7. Wipe crankshaft main bearing journals (3) clean with lint-free cloth and inspect for out-of-round condition. Measure out-of-round condition with micrometer. If out-of-round more than 0.002 in. (0.05 mm), replace crankshaft (2).

8. Measure outer diameter of crankshaft main bearing journals (3) with micrometer. If outer diameter is less than 4.4975 in. (114.237 mm), repair and stamp on front counterweight (8) accordingly.

9. Wipe lower and upper half rod bearing shell (7) clean and inspect for pits, chips, and scratches. If pitted, chipped, or scratched, replace bearing set.

10. Measure lower and upper half rod bearing shell (7) thickness with micrometer. If thickness is less than 0.071 in. (1.80 mm), replace bearing set.

11. Wipe crankshaft rod journals (9) clean with lint-free cloth and measure outer diameter with micrometer. If outer diameter is less than 3.122 in. (79.30 mm), repair and stamp on front counterweight (8) accordingly.

12. Measure outer diameter of rear counterweight seal flange (10) with micrometer. If outer diameter is under 5.997 in. (152.94 mm), replace crankshaft (2).

5-8. CRANKSHAFT AND MAIN BEARINGS MAINTENANCE (Contd)

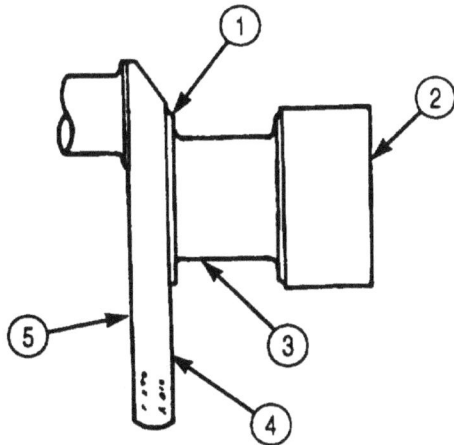

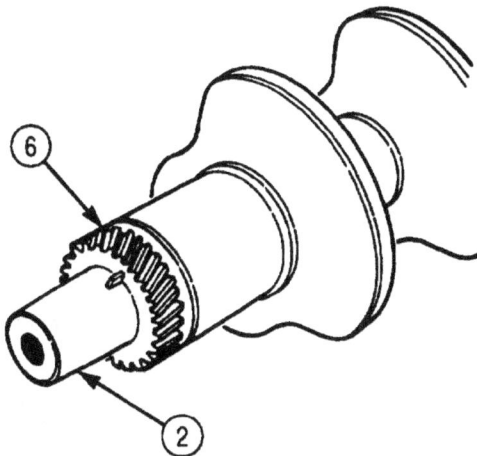

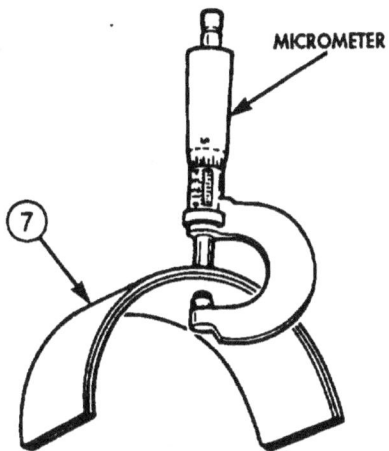

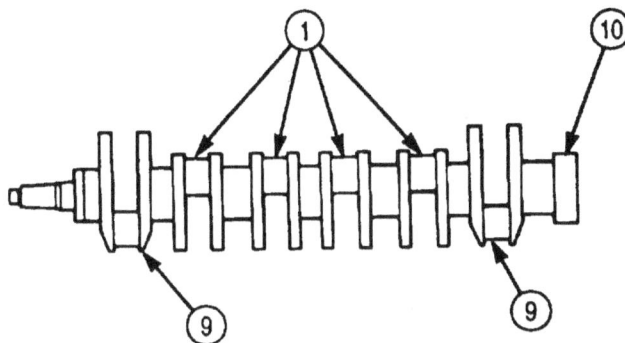

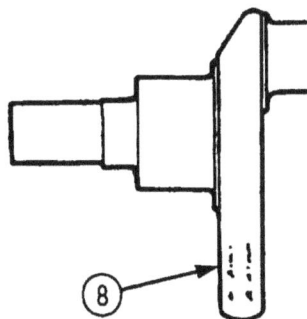

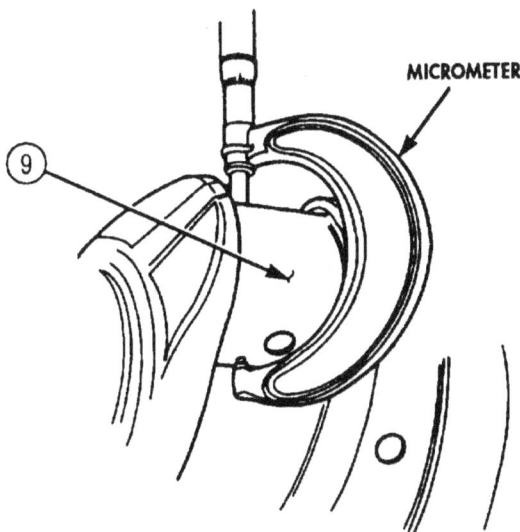

MICROMETER

MICROMETER

5-8. CRANKSHAFT AND MAIN BEARINGS MAINTENANCE (Contd)

c. Installation

WARNING

Eyeshields must be worn when cleaning with compressed air.
Compressed air source will not exceed 30 psi (207 kPa). Failure to
do so may result in injury to personnel.

1. Ensure main bearing bores (3) of engine block (1) are facing up.

2. Clean all engine block screw holes (2) with compressed air.

3. Wipe all engine block bearing bores (3) clean with lint-free cloth.

CAUTION

Touching bearing shell wear surface after shells have been cleaned
will cause shell corrosion, resulting in engine damage.

NOTE

Count new bearing upper shells from front of engine block.

4. Wipe new bearing upper shells (4) Nos. 1, 2, 3, 4, 5, and 6 (4) clean with lint-free cloth.

5. Position new bearing upper shells (4) Nos. 1, 2, 3, 4, 5, and 6 (4) in bearing bores (3) with oil holes
    (5) aligned and press in place. Apply coat of lubricating oil. Bearing shell (4) will project slightly
    above bore (3).

NOTE

Wide portion of upper bearing shell is measured from oil groove to
edge.

6. Install new bearing upper shell No. 7 (7) in bore (10) with wide portion (8) toward flywheel end of
    block (1).

7. Align new bearing upper shell No. 7 oil holes (9) and press in place. Apply coat of lubricating oil.
    Bearing shell (7) will project slightly above bore (10).

8. Install fourteen new dowel rings (6) in engine block (1).

5-8. CRANKSHAFT AND MAIN BEARINGS MAINTENANCE (Contd)

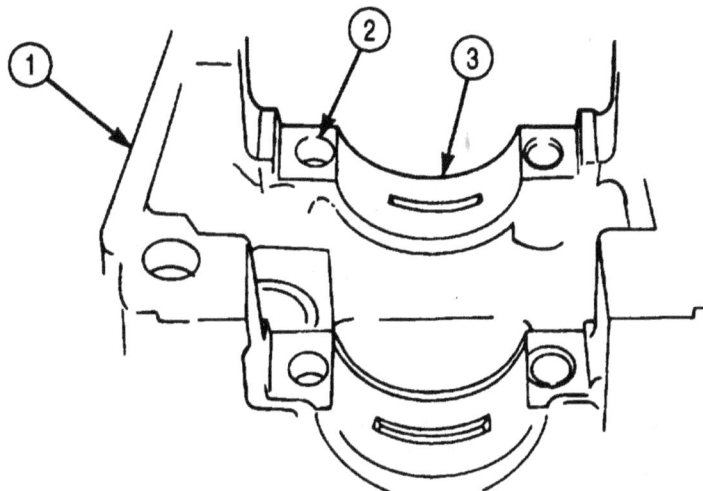

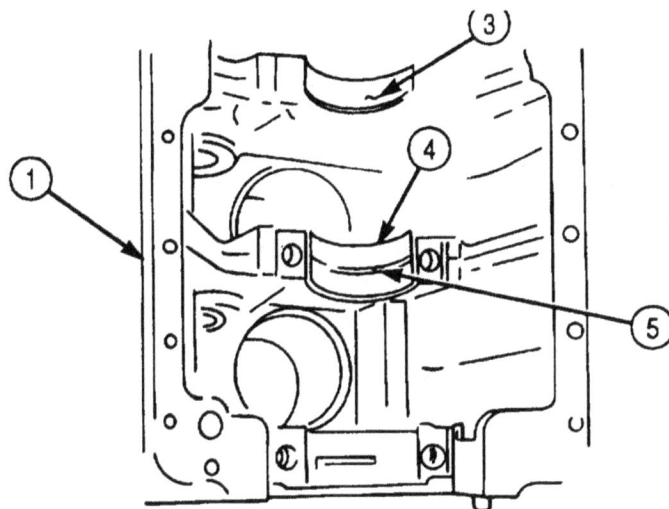

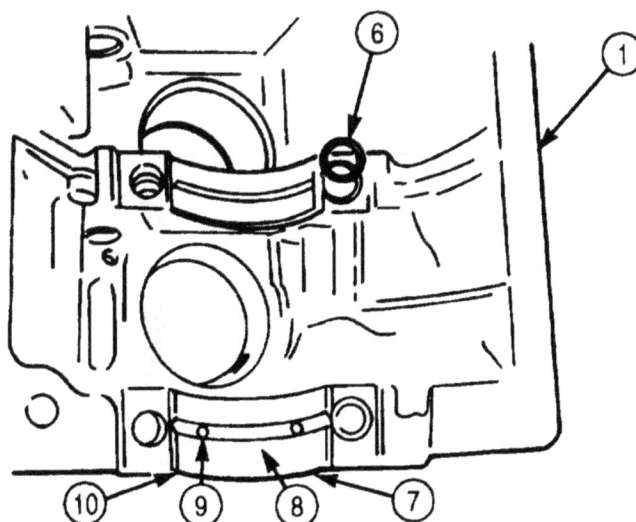

5-8. CRANKSHAFT AND MAIN BEARINGS MAINTENANCE (Contd)

9. Wipe clear surfaces of crankshaft (1) clean with lint-free cloth, and apply lubricating oil.

10. Position crankshaft (1) in engine block (2) with hooks protected with rubber hose or rope sling at two crank throws and rotate crankshaft (1) until rear crankshaft web is visible.

NOTE
Upper thrust rings are not doweled to block. Lower halves are doweled to No. 7 bearing cap.

11. Check markings (12) on rear crankshaft web (10) to determine what size upper thrust rings (7) and (8) are to be placed at front or rear of journal (9).

12. Apply lubricating oil to new upper thrust ring (8) and roll in place. Ensure babbit face or grooved side of new upper thrust ring (8) is next to crankshaft flange (11).

13. Wipe new lower bearing shells (5) clean with lint-free cloth and apply lubricating oil.

14. Insert new lower bearing shells (5) over crankshaft (1).

NOTE
Lower thrust ring and No. 7 bearing cap are installed together.
Lower thrust ring must be located over dowel on bearing cap.

15. Apply lubricating oil to new lower thrust ring (8) and position on No. 7 main bearing cap. Ensure babbit face or grooved side is next to crankshaft flange (11).

16. Wipe wear surfaces of main bearing caps (6) clean with lint-free cloth, and apply coat of lubricating oil.

17. Position main bearing caps (6) in engine block (2) over new lower bearing shells (5). Ensure numbers on caps (6) correspond with numbers on engine block (2) on camshaft side.

CAUTION
- Do not tap main bearing caps to seat. Hammering will jar bearing shells out of position and cause engine damage.
- Main bearing screws must be tightened alternately and slowly to ensure proper seating of bearing caps.

18. Apply lubricating oil to screw (3) threads and new lockplates (4).

19. Install new lockplates (4) and fourteen screws (3) through caps (6) in engine block (2).

20. Tighten each screw (3) in three steps of 100 lb-ft (136 N•m) until 300-310 lb-ft (410-423 N•m) is reached.

21. Loosen all screws (3) three turns and repeat step 20.

NOTE
Do not bend up tabs on lockplates until after crankshaft end clearance is checked.

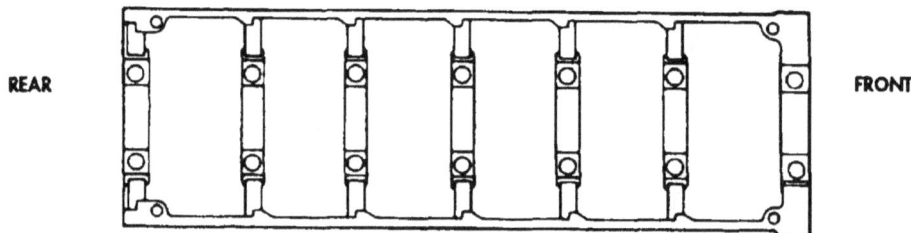

REAR   FRONT

5-8. CRANKSHAFT AND MAIN BEARINGS MAINTENANCE (Contd)

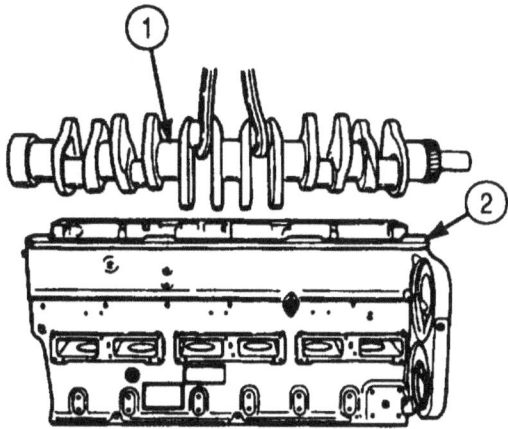

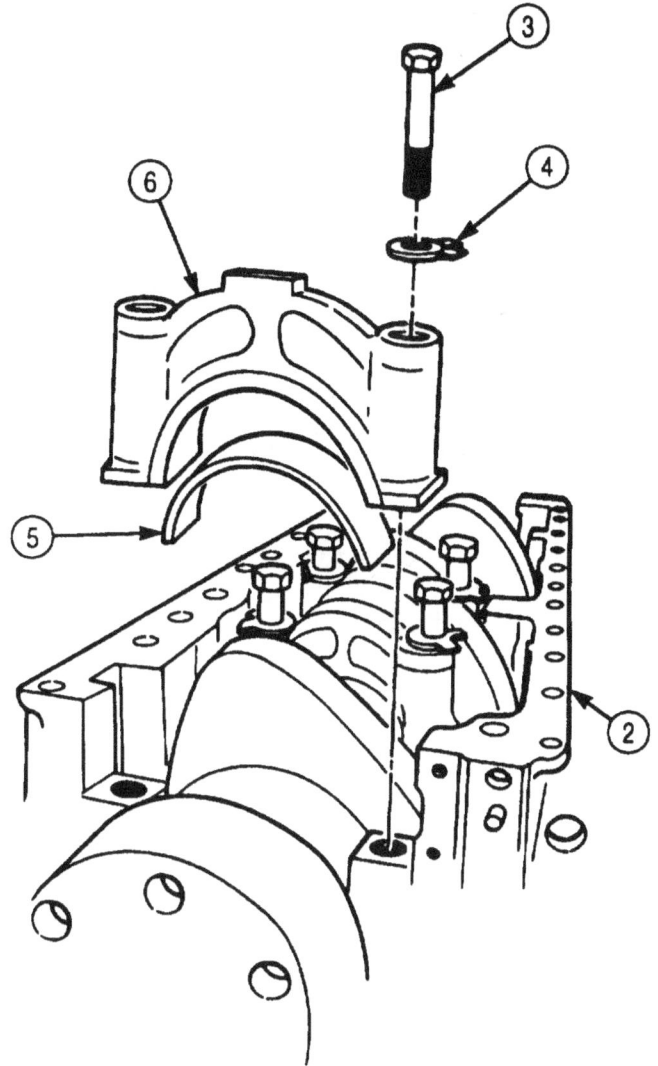

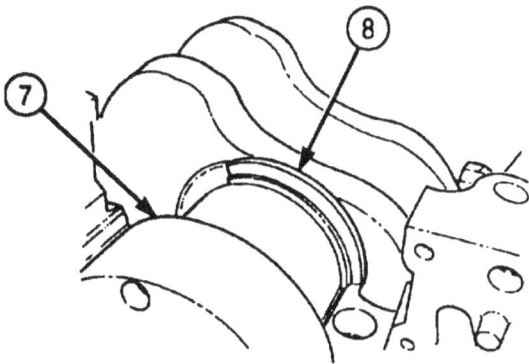

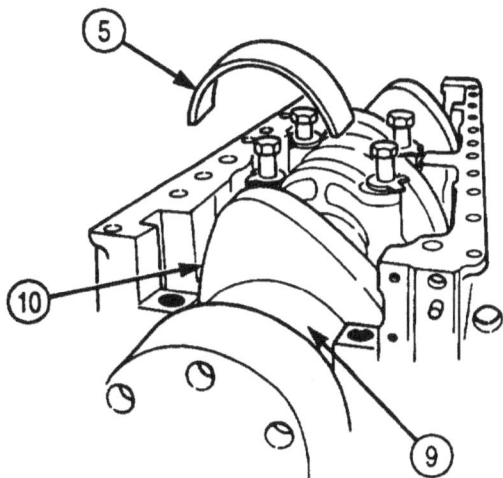

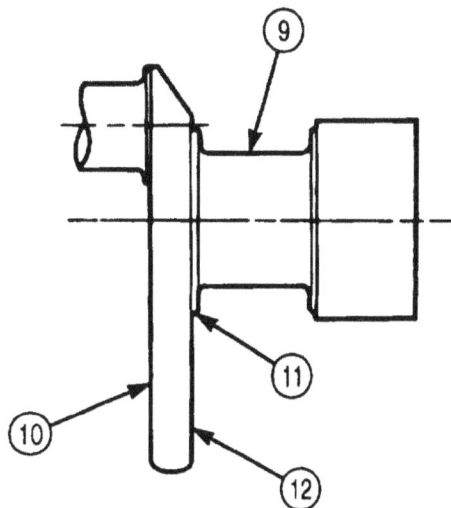

5-8. CRANKSHAFT AND MAIN BEARINGS MAINTENANCE (Contd)

d. End Play Clearance

1. Attach dial indicator to rear of engine block (6) with contact point (4) resting on crankshaft end face (3).
2. Push crankshaft end face (3) toward front of engine block (6) and set dial indicator to zero.
3. Push crankshaft end face (3) back toward rear of engine block (6) and check dial indicator,

NOTE

Complete steps 4 through 6 only if end clearance of crankshaft is less than 0.007 in. (0.18 mm).

4. Loosen main bearing screws (1) one turn and push crankshaft (2) first toward front, then toward rear of engine block (6).
5. Tighten main bearings screws (1) as in task c, steps 18 through 21.
6. Repeat steps 1 through 3.

NOTE

If end clearance is more than 0.22 in. (0.56 mm), crankshaft must be replaced or oversize thrust rings installed. Ensure markings on rear crankshaft counterbalance are correct. (Refer to task a, steps 11 and 12).

7. Bend up tabs on fourteen lockplates (7) to hold screws (1).

5-8. CRANKSHAFT AND MAIN BEARINGS MAINTENANCE (Contd)

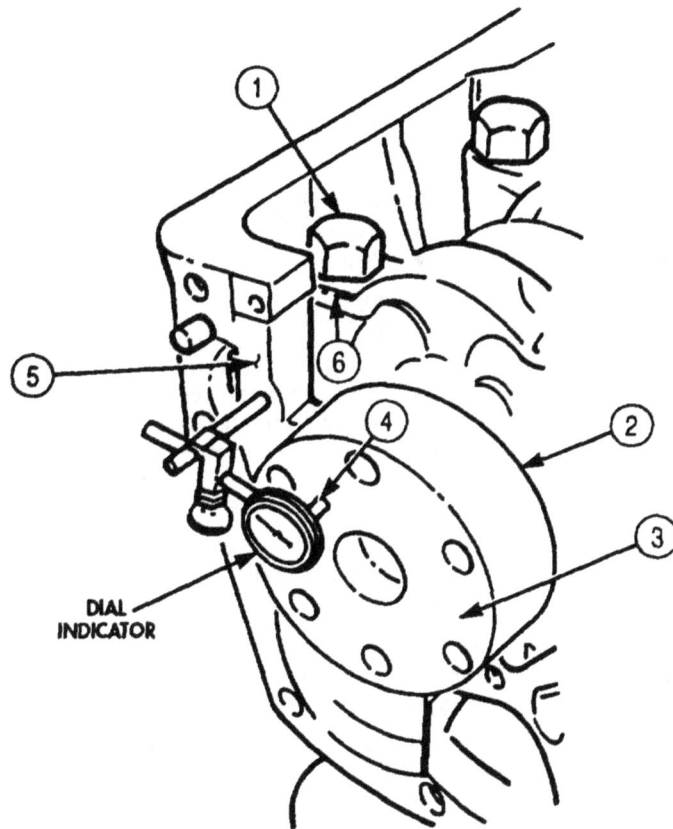

DIAL
INDICATOR

FOLLOW-ON TASK: Install pistons and connecting rods (para. 5-10).

5-9. CAMSHAFT AND GEAR MAINTENANCE

THIS TASK COVERS:

a. Check Backlash
b. Removal
c. Disassembly

d. Cleaning and Inspection
e. Assembly
f. Installation

INITIAL SETUP:

APPLICABLE MODELS
M939/A1

SPECIAL TOOLS
Telescoping gauge (Appendix E, Item 136)
Cam bushing replacement tool (Appendix E, Item 26)

TOOLS
General mechanic's tool kit (Appendix E, Item 1)
Dial indicator (Appendix E, Item 36)
Puller kit (Appendix E, Item 102)
Outside micrometer (Appendix E, Item 80)
Vernier caliper (Appendix E, Item 159)
Brass rod

MATERIALS/PARTS
Seven cam bushings (Appendix D, Item 34)
Camshaft bushing (Appendix D, item 33)
Lubricating oil (Appendix C, Item 50)
Drycleaning solvent (Appendix C, Item 71)

REFERENCES (TM)
TM 9-2320-272-24P

EQUIPMENT CONDITION
● Engine oil pan removed (para. 4-22).
● Piston and connecting rod installed (para. 5-10).

GENERAL SAFETY INSTRUCTIONS
● Keep tire extinguisher nearby when using drycleaning solvent.
● Drycleaning solvent is flammable and toxic. Do not use near an open flame.

---

a. Check Backlash

1. Attach dial indicator to engine block (3) and rotate camshaft gear (2) as far as it will freely move and hold it in place. Ensure crankshaft gear (1) does not move.
2. Position dial indicator arm to camshaft gear (2) tooth and set dial indicator to zero.
3. Rotate camshaft gear (2) in opposite direction and read backlash measurement as rotation stops. If backlash is not 0.002-0.020 in. (0.05-0.51 mm), replace camshaft gear (2).

NOTE
Normal backlash is 0.0045-0.0105 in. (0.114-0.267 mm) on new gear with a minimum of 0.002 in. (0.05 mm).

b. Removal

Rotate camshaft gear (2) and camshaft (4) slightly and remove from engine block (3).

CAUTION
Use care when removing camshaft to avoid damaging bearings.

5-9. CAMSHAFT AND GEAR MAINTENANCE (Contd)

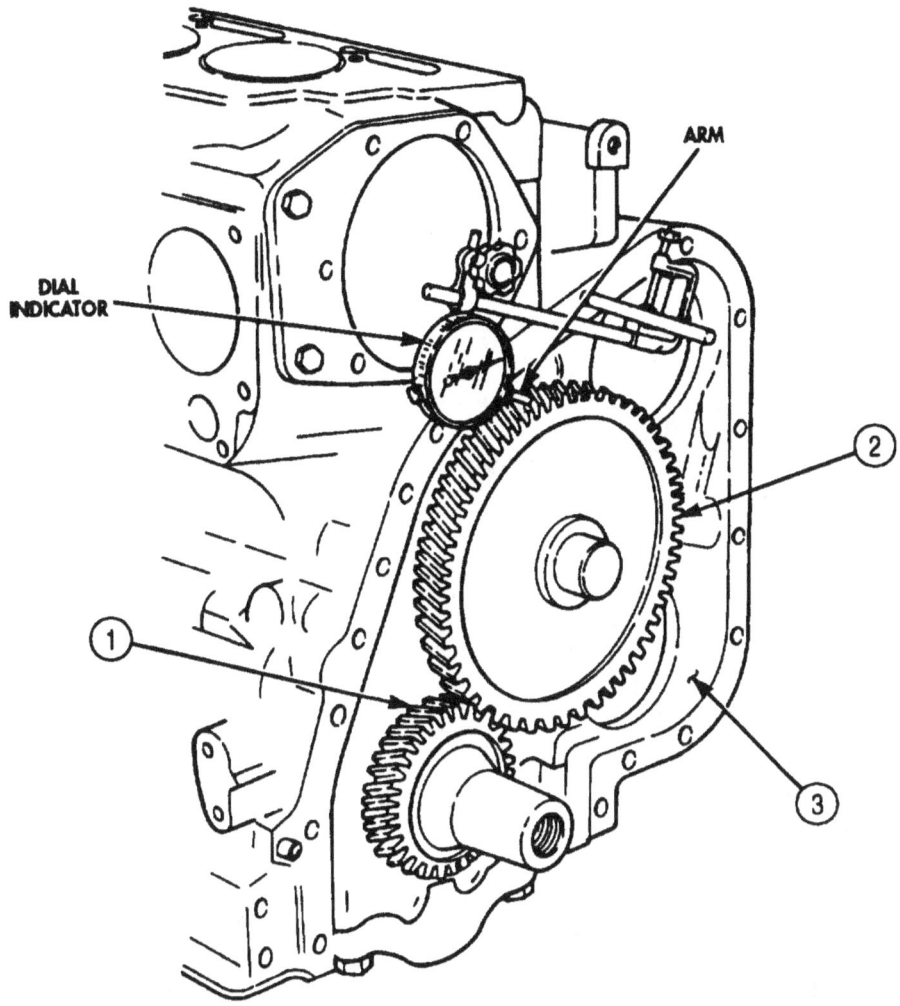

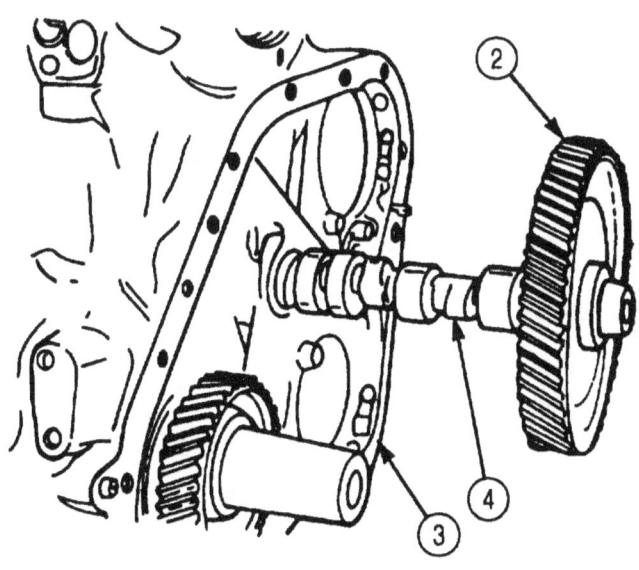

5-9. CAMSHAFT AND GEAR MAINTENANCE (Contd)

c. Disassembly

NOTE

Before disassembling camshaft gear from camshaft, perform task d of this procedure. If, as a result of inspection, camshaft gear must be disassembled from camshaft, perform stops 1 through 4.

1. Place camshaft (2) between V-blocks. V-blocks should support camshaft (2).
2. Remove pipe plug (5) from camshaft (2).
3. Remove camshaft gear (1) from camshaft (2) with puller.
4. Remove woodruff key (4) and thrust ring (3) from camshaft (2).

NOTE

Before removing camshaft bushings from engine block, perform task d of this procedure. If, as a result of inspection, bushings must be replaced, perform step 5.

5. Remove seven camshaft bushings (6) from cylinder block (7) with cam bushing replacer.

d. Cleaning and Inspection

WARNING

Drycleaning solvent is flammable and toxic. Do not use near open flame and always have a fire extinguisher nearby when solvents are used. Use only in well-ventilated places, wear protective clothing, and dispose of cleaning rags in approved container. Failure to do this may result in injury or death to personnel and/or damage to equipment.

1. Clean camshaft (2), camshaft gear (1), and thrust washer (3) with drycleaning solvent and inspect for cracks, breaks, or pits. If cracked, pitted, or broken, discard.
2. Measure outside diameter of camshaft journals (8) using micrometer. If outside diameter is less than 1.996 in. (50.70 mm), replace camshaft (2).
3. Measure thickness of thrust ring (3). If thickness is less than 0.083 in. (2.11 mm), discard thrust ring (3).
4. Clean seven camshaft bushings (6) with drycleaning solvent and inspect for breaks, cracks, or pits. If broken, cracked, or pitted, discard.
5. Measure inner diameter of seven camshaft bushings (6) with telescoping gauge. If inner diameter exceeds 2.001 in. (50.83 mm), discard.

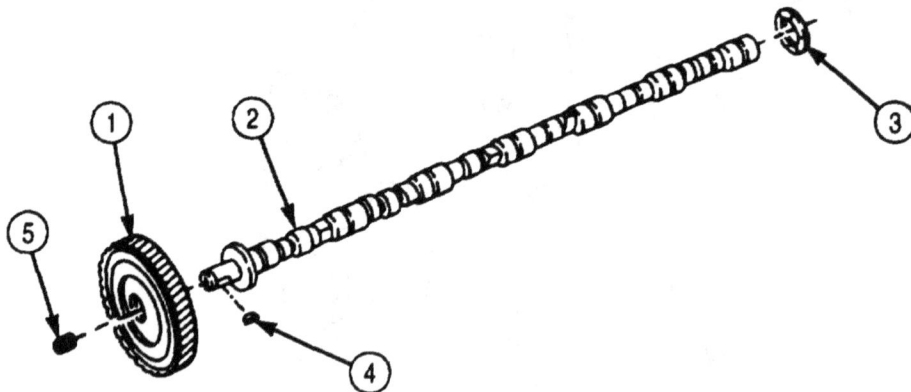

5-9. CAMSHAFT AND GEAR MAINTENANCE (Contd)

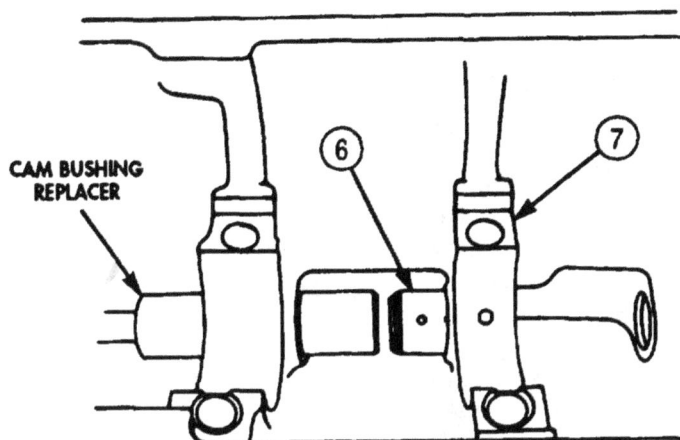

CAM BUSHING
REPLACER

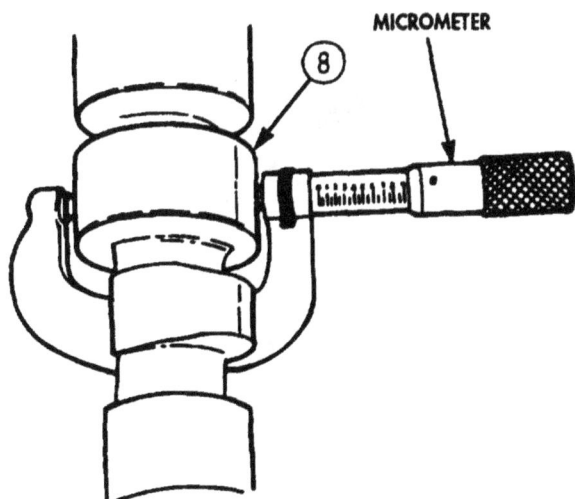

MICROMETER

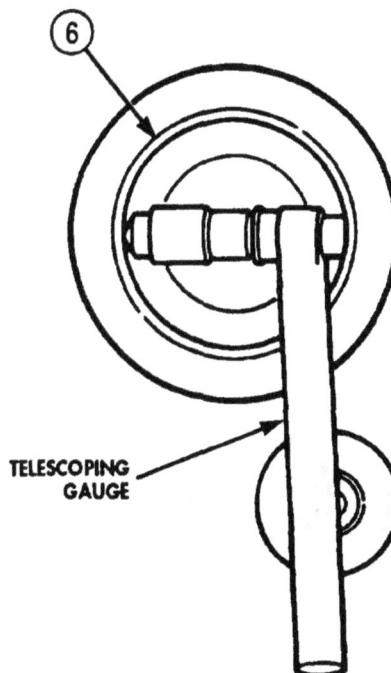

TELESCOPING
GAUGE

5-9. CAMSHAFT AND GEAR MAINTENANCE (Contd)

e. Assembly

Install thrust ring (3), woodruff key (4), camshaft gear (1), and pipe plug (5) in camshaft (2).

f. Installation

CAUTION
Positioning of new camshaft bushing in No. 7 bushing bore (rear of cylinder block) is critical. The new bushing must be pressed in, leaving clearance between bushing and rear face of cylinder block to allow oil to drain from hole at rear of camshaft. Hydraulic lock will occur if oil drain passage is blocked.

NOTE
No. 1 cam bushing (gear end) is wider; all others are the same.

1. Position seven cam bushings (7) on cam bushing replacer (6), aligning oil hole (8) in cam bushing (7) to oil hole in main bearing bore.
2. Press seven cam bushings (7) into position in cam bore and check oil hole (8) alignment with a brass rod through main bearing bore oil hole and cam bushing oil hole (8). If brass rod does not pass through with ease, reposition cam bushings (7).

CAUTION
Use extreme care when installing camshaft to avoid damage to bearings and camshaft lobes.

NOTE
Assistant will help with steps 3 through 5.

3. Apply coat of lubricating oil to camshaft lobes (10) on camshaft (2) and camshaft gear (1).
4. Position camshaft (2) and camshaft gear (1) in engine block (9) by grasping camshaft gear (1) with both hands and gently sliding end into cam bore (11). With aid of assistant, carefully guide lobes (10) through bore (11) as camshaft (2) is installed.
5. Align index mark on camshaft gear (1) with index mark on crankshaft gear (12).
6. Repeat task a. to check backlash.

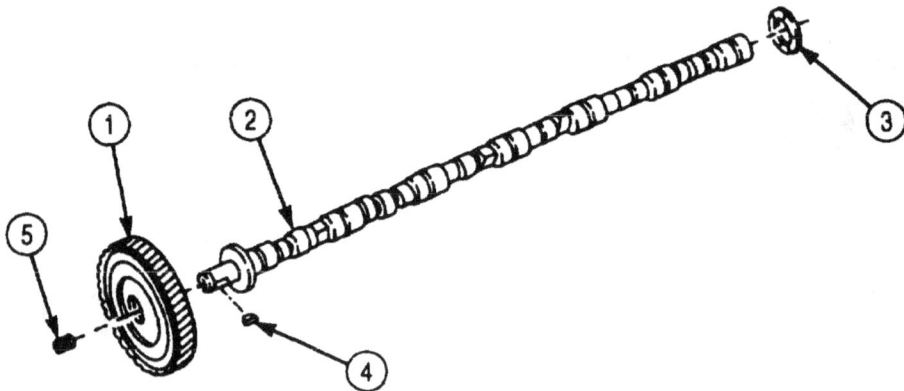

5-9. CAMSHAFT AND GEAR MAINTENANCE (Contd)

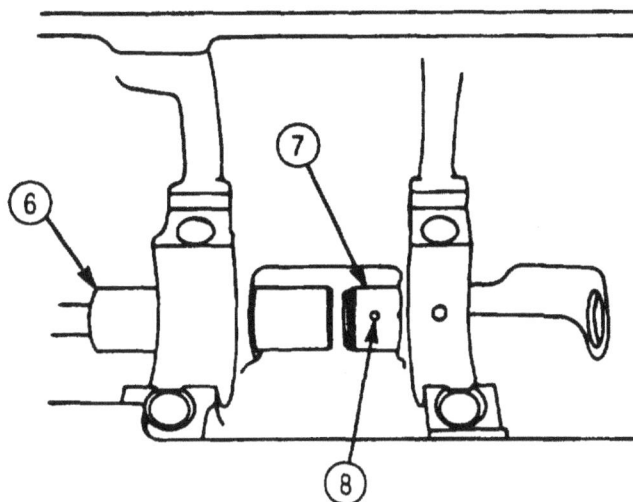

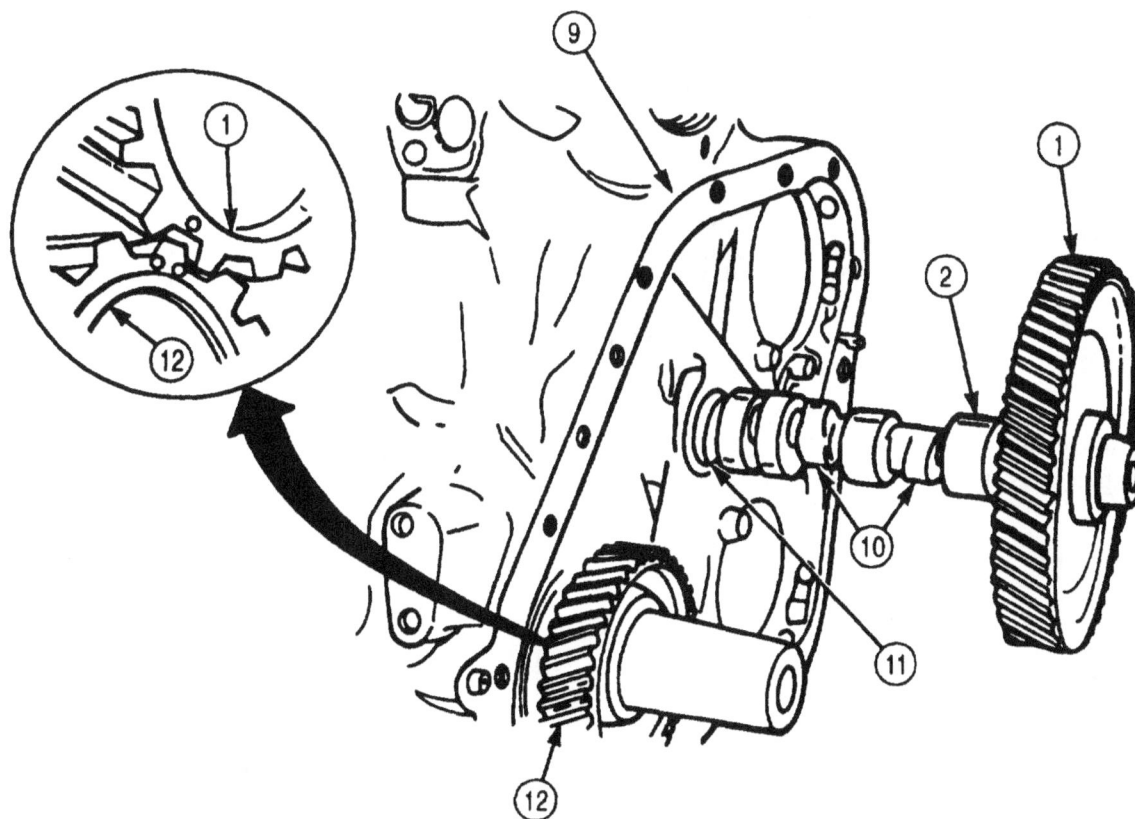

FOLLOW-ON TASKS● Replace engine oil pan (para. 4-22).
● Install cylinder heads (para. 5-10).

5-10. PISTON AND CONNECTING ROD MAINTENANCE

THIS TASK COVERS:

a. **Connecting Rod Side Clearance Check**   d. **Cleaning and Inspection**
b. **Removal**   e. **Assembly**
c. **Disassembly**   f. **Installation**

INITIAL SETUP:

APPLICABLE MODELS
M939/A1

SPECIAL TOOLS
Piston ring groove gauge (Appendix E, Item 97)

TOOLS
General mechanic's tool kit (Appendix E, Item 1)
Torque wrench (Appendix E, Item 144)
Telescoping gauge (Appendix E, Item 136)
Outside micrometer (Appendix E, Item 80)
Inside micrometer (Appendix E, Item 82)
Piston ring expander (Appendix E, Item 96)
Arbor press
Feeler gauge
Soft-faced hammer
Vise
Tape-protected tool

MATERIALS/PARTS
Twelve bearing shells (Appendix D, Item 14)

REFERENCES (TM)
TM 9-2320-272-24P

EQUIPMENT CONDITION
Camshaft and gear removed (para. 5-9).

GENERAL SAFETY INSTRUCTIONS
Gloves must be worn when handling hot pistons.

---

a. Connecting Rod Side Clearance Check

NOTE

Connecting rod should have free movement at crank journal.
Check with hand pressure first. Tap lightly with soft-faced
hammer only if necessary.

1. Move connecting rod (1) up and down on crank journal (3) and measure clearance with feeler gauge. Clearance should be 0.0045-0.13 in. (0.11-0.33 mm). Record clearance of all six connecting rods for reassembly.

NOTE

If clearance is not 0.0045-0.13 in. (0.11-0.33 mm) or rod will not
move, continue with steps 2 and 3.

2. Remove cap (2) and check for improper bearing size, dirt, or burrs.

3. Install cap (2) as outlined in task f.

5-10. PISTON AND CONNECTING ROD MAINTENANCE (Contd)

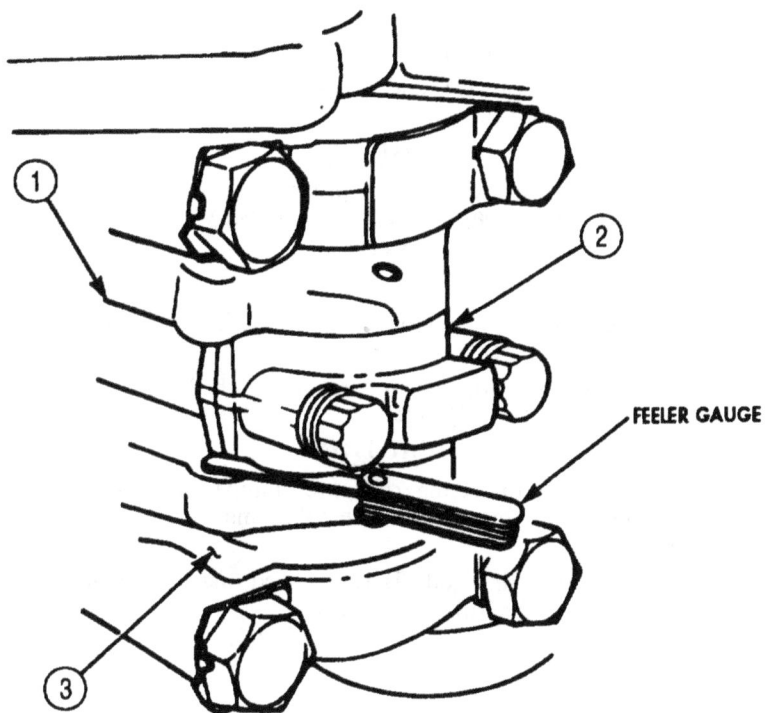

FEELER GAUGE

5-10. PISTON AND CONNECTING ROD MAINTENANCE (Contd)

b. Removal

1. Place cylinder block (1) in vertical position and remove all carbon from upper inside wall of each cylinder liner. Use ridge reamer as necessary.

NOTE

All six connecting rod and bearing caps are removed the same way.
Steps 2 through 5 are for one connecting rod and bearing cap only.

2. Rotate crankshaft (3) until connecting rod bearing caps (5) are above edge of cylinder block flange (2).
3. Loosen two screws (6) on connecting rod bearing cap (5) 3/8 in. (9.5 mm) and tap with soft-faced hammer to free connecting rod bearing cap (5) from connecting rod (7).

NOTE

- Tag and mark bearing cap for installation.
- Do not mix bearing caps.

4. Remove two screws (6), bearing cap (5), and lower half bearing shell (4) from connecting rod (7). Discard lower half bearing shell (4).

CAUTION

Use a tape-protected tool to push piston from block. Failure to do this may result in damage to cylinder liners.

NOTE

- Assistant will help with steps 5 and 6.
- Mark location and position of all pistons before removal from block. Discard upper half bearing shells.

5. Push connecting rods (7), pistons (8), and upper bearing shells (9) out of cylinder block (1) with tape-protected tool. Begin at crankshaft (3) side.

CAUTION

Bearing caps and connecting rods are stamped with same number as the cylinder. Mixing piston assemblies may cause equipment damage.

6. After each piston (8) and connecting rod (7) is removed, assemble connecting rod bearing cap (5) to connecting rod (7) with two screws (6).

5-10. PISTON AND CONNECTING ROD MAINTENANCE (Contd)

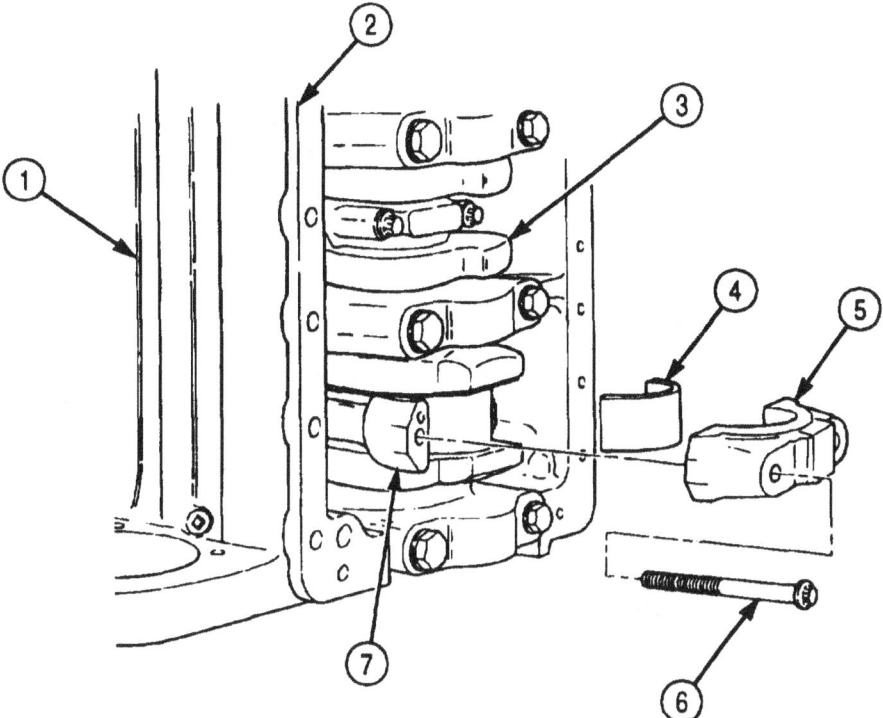

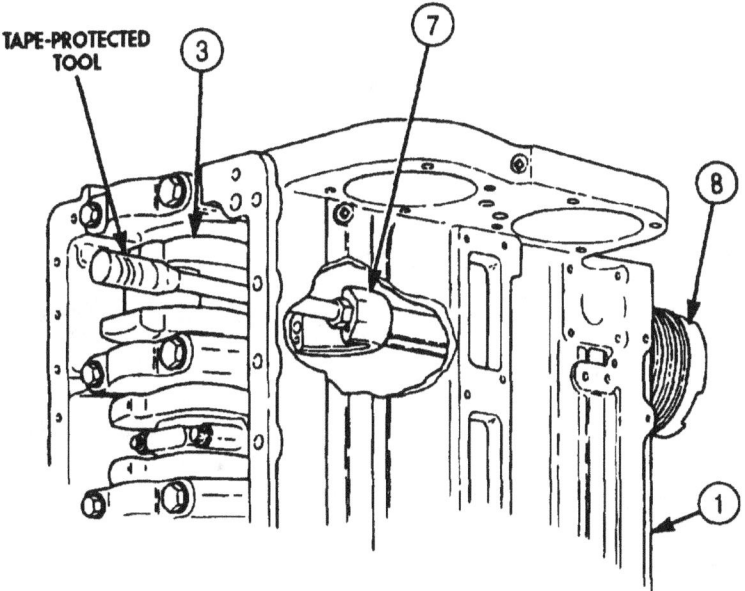

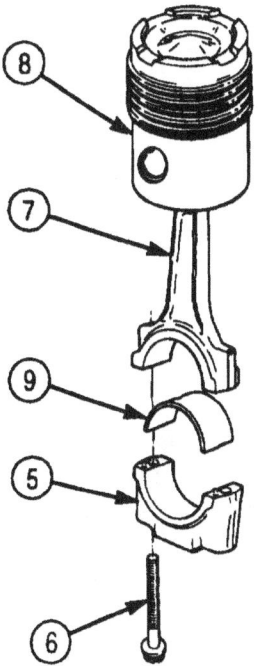

TAPE-PROTECTED
TOOL

5-10. PISTON AND CONNECTING ROD MAINTENANCE (Contd)

c. Disassembly

1. Remove four piston rings (1) from piston (6).
2. Remove two snaprings (2) from piston pin bore (5).

WARNING

Pistons must be heated in hot water to remove piston pins. Do not handle hot pistons with bare hands. Use protective gloves to avoid burning hands.

NOTE

Mark piston and connecting rod so it can be reassembly in same manner.

3. Submerge pistons (6) and connecting rod (4) assemblies in container of hot water for fifteen minutes to allow pistons to expand.
4. Remove pistons (6) and connecting rod (4) assemblies from hot water with tongs or hook and place in vise. Push piston pin (3) out of piston pin bore (5). Do not drive pins (3) out with hammer.
5. Place pistons (6), piston pins (3), and connecting rods (4) on a numbered rack and hold for inspection.

d. Cleaning and Inspection

NOTE

Ensure bearing caps remain assembled to mating connecting rods at all times to avoid mixing.

1. Clean connecting rod assemblies (7) (para. 2-14).
2. Inspect connect rod assemblies (7) (para. 2-15).
3. Visually check I-beam section (8) of connecting rod (4) for nicks, dents, and gouges. If nicks, dents, or gouges are greater than 0.031 in. (0.787 mm), replace connecting rod (4).
4. Inspect connecting rod assemblies (7) for cracks (para. 2-16).

NOTE

To accurately measure all connecting rod bearing cap bores, the screws must be tightened to the operating torque specified in table 5-7.

5. Install bearing caps (9) on connecting rod (4) with screws (10).
6. Alternately tighten screws (10) with torque wrench in accordance with the sequence order in table 5-7.

Table 5-7. Connecting Rod Screw Tightening Sequence.

TIGHTEN SEQUENCE	TORQUE VALUES	
	LB-FT	N•M
Step 1-Tighten to	70-75	(95-102)
Step 2-Tighten to	140-150	(90-203)
Step 3-Loosen to	(0)	(0)
Step 4-Tighten to	25-30	(34-41)
Step 5-Advance to	70-75	(95-102)
Step C--Advance to	140-150	(190-203)

5-10. PISTON AND CONNECTING ROD MAINTENANCE (Contd)

5-10. PISTON AND CONNECTING ROD MAINTENANCE (Contd)

7. Using telescoping gauge, measure inside diameter of crankshaft. journal bore (2) in connecting rod (1) at points A-A and B-B, up to 30° either side of parting line C. Record readings. If crankshaft journal bore (2) diameter is not 3.2725-3.2730 in. (83.122-83.134 mm), replace connecting rod (1).

8. Using telescoping gauge, measure inside diameter of crankshaft journal bore (2) in connecting rod at points D-D and E-E. Record readings. If crankshaft journal bore (2) diameter is not 3.2725-3.2730 in. (83.122-83.134 mm), replace connecting rod (1).

NOTE

Check pin bores with bushings installed.

9. Check inside diameter of piston pin bushing bore (6) in connecting rod (1) with telescoping gauge. If piston pin bushing bore (6) is worn beyond 2.0022 in. (50.856 mm), mark bushings for replacement.

NOTE

- Pistons must be cleaned before inspection.
- Conditions of the pistons and piston rings should be carefully noted, as 'they indicate borderline conditions leading to engine failures.
- Pistons should be inspected at ambient temperatures of 70°- 90°F (21°-32°C).

10. Check piston (5) wear surface for scoring, scuffing, and cracks. If scored, scuffed, or cracked, replace piston (5).

NOTE

Vertical scratching and discoloration of piston ring sealing surface is a major cause of oil consumption and piston and liner scoring.

11. Check piston rings (3) for vertical scratching and discoloration of piston ring sealing surface. If ring sealing surface is scratched or discolored, replace piston rings (3).

12. Check for broken piston rings (3). If broken, replace piston ring (3).

13. Check piston rings (3) for formation of deposits which prevent outward movement of piston rings (3) to seal. If piston ring (3) sticks in piston grooves, replace piston ring (3).

14. Check piston pin bore (4) on piston (5) for fractures. If piston ring bore (4) is fractured, replace piston (5).

NOTE

- Pistons must be inspected at ambient temperatures of 70°-90°F (21°-32°C) to obtain accurate measurement readings.
- First and second piston ring grooves are checked for depth wear the same way. Step 15 checks for ring groove depth wear on the top ring groove.

15. Insert ring groove gauge into top ring groove (7) of pistons (5) and note whether shoulders (6) of gauge touch ring groove (7). If shoulders (6) of gauge touch either ring groove (7), pistons (5) are not serviceable and must be discarded.

16. Repeat step 15 for second ring groove (8).

5-10. PISTON AND CONNECTING ROD MAINTENANCE (Contd)

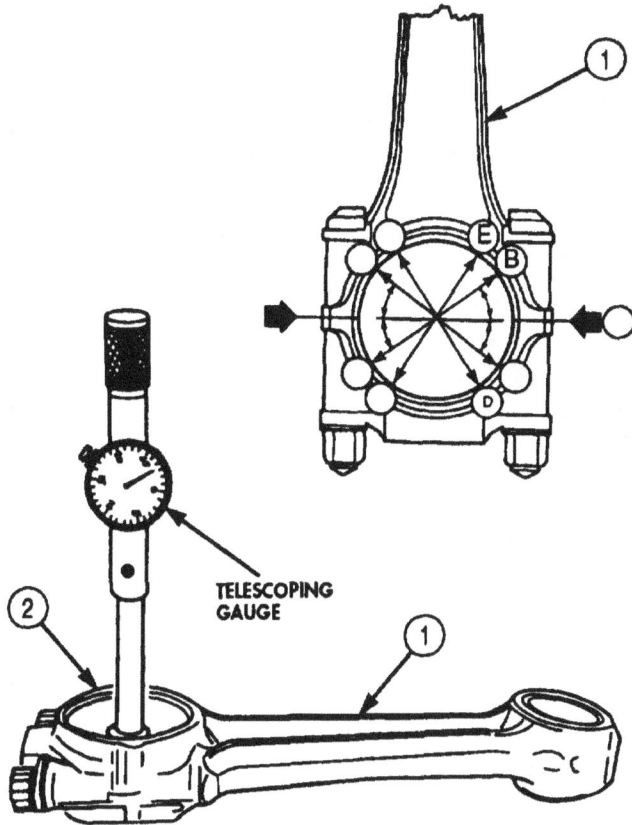

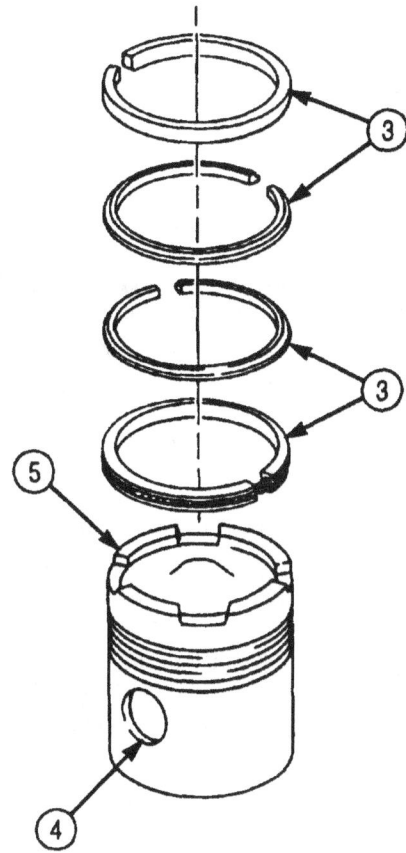

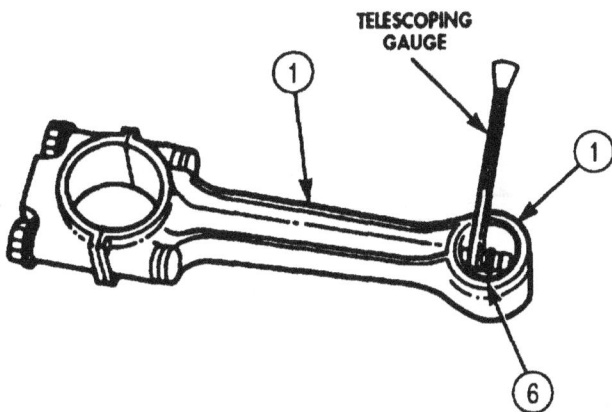

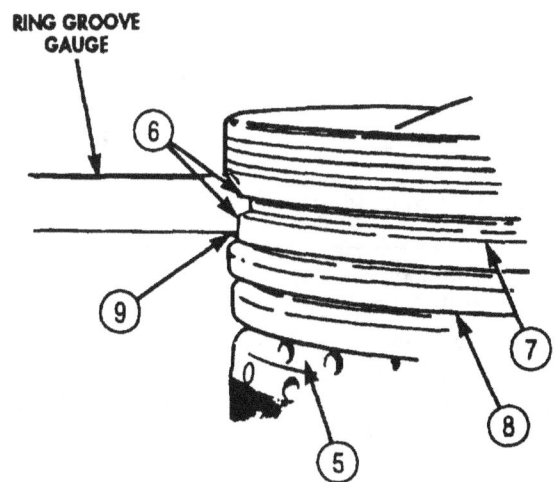

TELESCOPING
GAUGE

TELESCOPING
GAUGE

RING GROOVE
GAUGE

5-10. PISTON AND CONNECTING ROD MAINTENANCE (Contd)

NOTE

First and second piston ring groove widths are checked for wear
the same way. Step 17 checks top ring groove width for wear.

17. Place and hold new piston ring (2) in top ring groove (3) and try to insert 0.006 in. (0.15 mm) feeler gauge into ring groove (3) at top of piston ring (2). If feeler gauge enters either ring groove (3) without force, piston (1) is not serviceable and must be discarded.

18. Repeat step 17 for second ring groove (4).

NOTE

Measurements to check piston skirt outside diameters are taken
at right angle to piston bore.

19. Measure piston (1) upper skirt outside diameter (6) approximately 1 in. (25.4 mm) below bottom ring groove (4) with micrometer.

20. Measure piston (1) bottom skirt outside diameter (7) approximately 1 in. (25.4 mm) above skirt bottom (8). If outside diameters (6) and (7) measure less than 5.483 in. (139.27 mm), discard piston (1).

21. Measure piston pin bore (5) inside diameter. If inside diameter exceeds 1.99 in. (50.775 mm) at 70°F (21°C), discard piston (1). Add .0005 in. (0.13 mm) per each 10°F over 70°F (21°C) up to 90°F (32°C).

22. Check outside diameter of piston pin (9) with micrometer. If outside diameter is worn out-of-round more than 0.001 in. (0.03 mm), discard piston pin (9). If diameter is smaller than 1.99885 in. (50.762 mm), discard piston pin (9).

23. Check all piston (1) parts to be sure they are numbered the same. Number is generally located on the inside of piston (1) skirt.

NOTE

New piston rings must be checked in the cylinder liner in which
they will be used to ensure ring gaps are correct.

24. Insert piston rings (11), (13), (14), and (15) in each mating cylinder liner (10) bore, using head (top) of piston (1) to position piston ring (11) so it seats squarely in piston ring (11) travel area.

25. Measure ring gap (12) with feeler gauge. Ring gap (12) should meet specifications given in table 5-8.

NOTE

Add 0.003 in. (0.08 mm) ring gap to new maximum for each 0.001
in. (0.03 mm) wear in cylinder liner wall. Measurement made in
ring travel area of liner.

Table 5-8. Ring Gap.

No. 1 compression ring (13) gap	0.023-0.033 in. (0.58-0.85 mm)
No. 2 and 3 compression ring (11) and (14) gap	0.019-0.029 in. (0.48-0.74 mm)
No. 4 oil ring (15) gap	0.010-0.025 in. (0.25-0.64 mm)

26. Install piston rings (13), (11), (14), and (15) in piston (1) with piston ring expander. Ensure the word "top" faces toward top of piston (1).

27. Check connecting rod journals (17) on crankshaft assembly (16) for scores and scratches. If scored or scratched, replace crankshaft.

28. Check connecting rod journals (17) for out-of round wear with micrometer. If worn more than 0.002 in. (0.05 mm), replace crankshaft.

5-10. PISTON AND CONNECTING ROD MAINTENANCE (Contd)

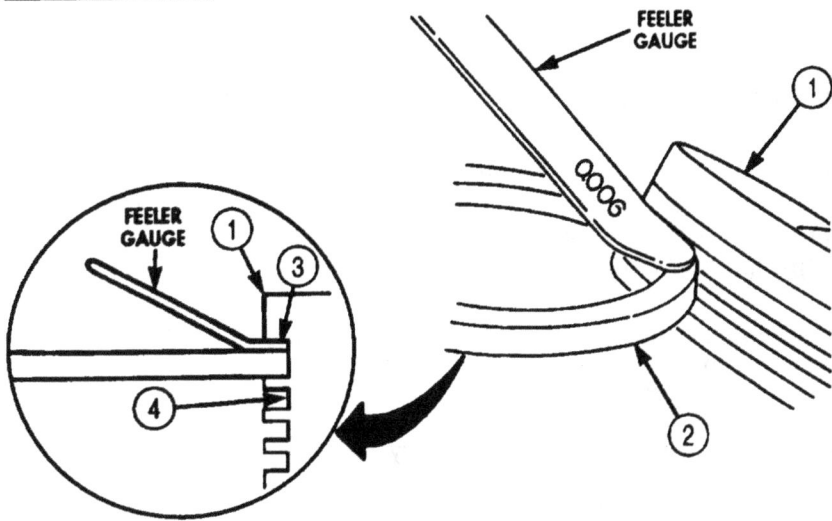

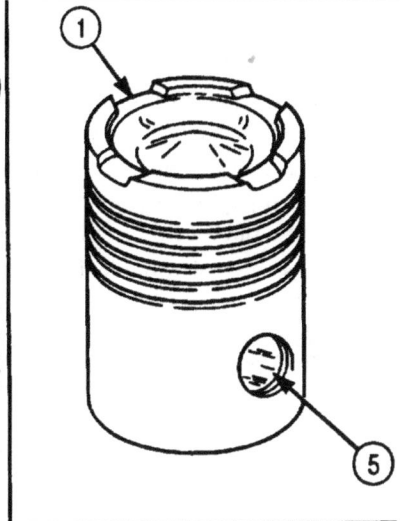

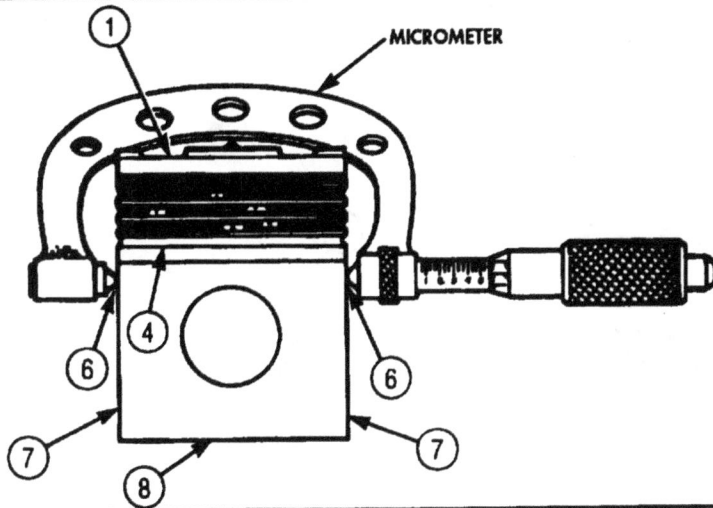

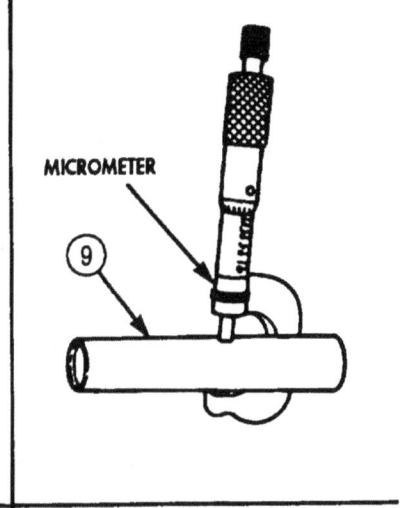

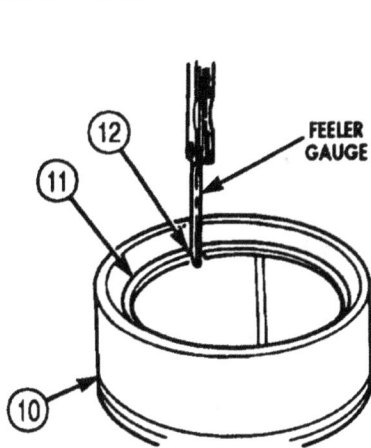

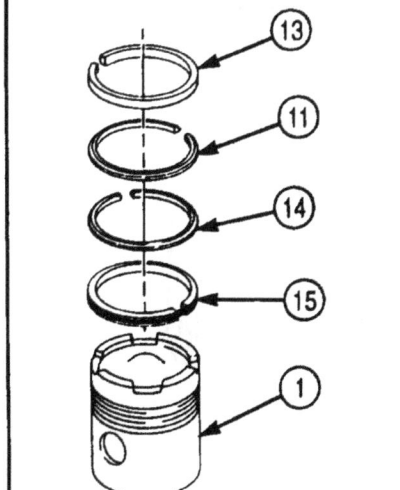

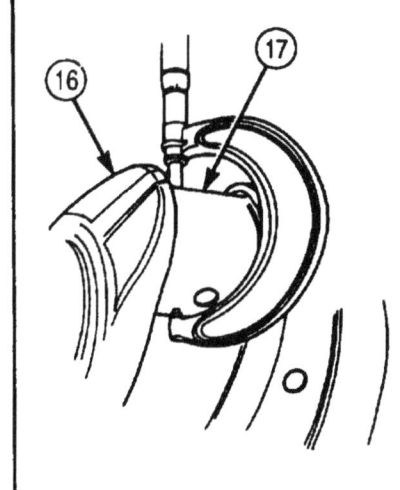

5-10. PISTON AND CONNECTING ROD MAINTENANCE (Contd)

e. Assembly

1. Remove piston pin bushing (10) from connecting rod (5) with mandrel, removal tool, and arbor press.
2. Install sleeve (6), new standard size piston pin bushing (10), and guide sleeve (8) on mandrel.
3. Place connecting rod (5) on block (7) and place in horizontal position.
4. Line up mark on guide sleeve (8) with middle of boss on connecting rod (5). Ensure oil hole in new piston pin bushing (10) and pin bore (9) are lined up.
5. Press new piston pin bushing (10) into pin bore (9) with arbor press until guide sleeve (8) contacts side of rod pin boss.
6. Check inside diameter of new piston rod bushing (10). If inside diameter is not 2.001-2.0015 in. (50.83-50.838 mm), replace piston pin bushing (10).

WARNING

Do not handle hot pistons with bare hands or injury to personnel may result.

CAUTION

Never drive piston pins into pistons. Driving may cause distortion of piston and cause piston seizure in cylinder liner.

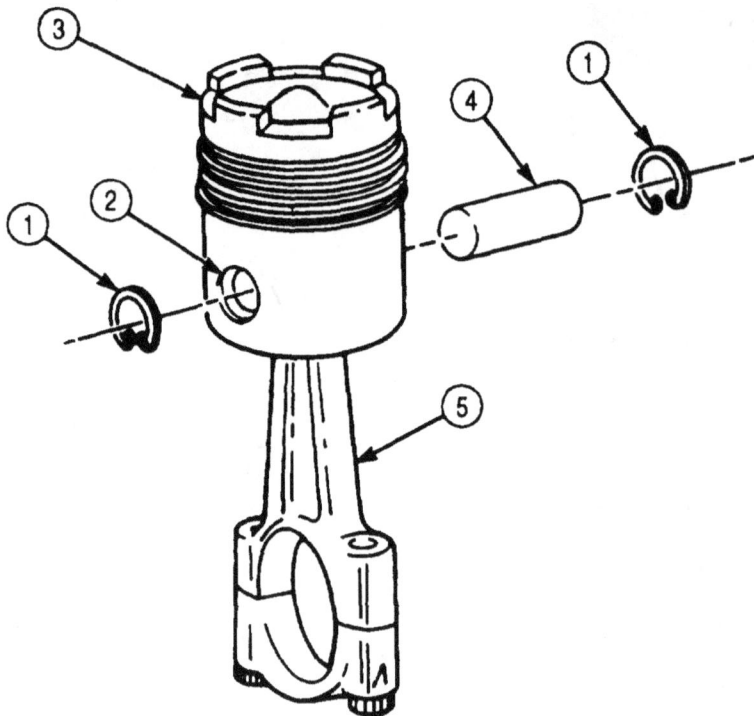

5-10. PISTON AND CONNECTING ROD MAINTENANCE (Contd)

NOTE

- Ensure rod and bearing cap are stamped with the numbers of the cylinders they were removed from. All pistons must have same part number.
- Ensure pistons and connecting rods are assembled with orientation marks matching.

7. Install one piston pin snapring (1) in groove of piston pin bore (2).

8. Heat pistons (3) in hot water.

CAUTION

Do not attempt to install pin after piston has cooled. Pin will not fit.

9. Install piston pin (4) through piston (3) and connecting rod (5) pin bores (9) before piston (3) cools.

10. Install second snapring (1) in piston pin (4) groove at opposite end of pin bore (2).

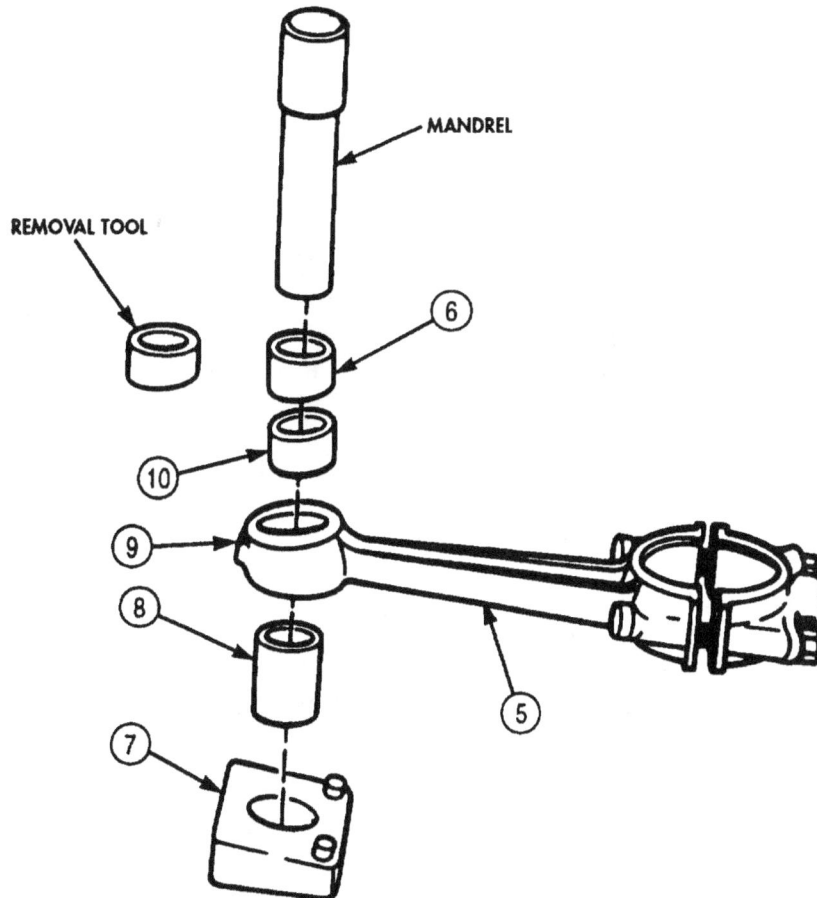

5-10. PISTON AND CONNECTING ROD MAINTENANCE (Contd)

f. Installation

NOTE

- Before installation, ensure all pistons have been properly assembled and lubricated.
- If old pistons are being installed, ensure they are installed in same location from which they were removed.

1. Remove two screws (5) and rod cap (4) from connecting rod (2).
2. Apply coat of lubricating oil to piston (1), connecting rod (2), and upper bearing shell (3).
3. Position tang (6) on new bearing (3) to groove (7) in connecting rod (2) and snap into place.

NOTE

Ensure piston ring gaps are staggered so they are not in line with each other or piston pin.

4. Position ring compressor (10) over piston (1) and tighten with Allen wrench.

CAUTION

Assistant must guide connecting rod through cylinder from oil pan side of block to avoid damaging liner.

NOTE

Ensure numbered side of connecting rod is toward the camshaft side of engine block.

5. Carefully insert connecting rod (2) in cylinder liner (9) and hold ring compressor (10) right and firmly seated against engine block (8). Ensure crankshaft journal (12) is at bottom dead center.

CAUTION

- Ring compressor must be held firmly against engine block to prevent compressor from sipping and causing piston ring breakage when pushing piston into cylinder liner.
- Do not force piston assembly into liner. If piston does not install freely in liner, remove and check for broken rings.

NOTE

Assistant will guide rod onto crankshaft journal to prevent damage to journal and liner.

6. Push piston (1) through ring compressor (10) until all piston rings are well into cylinder liner (9) in engine block (8).
7. Drive piston (1) and rod (2) into cylinder liner (9) with rubber or wooden mallet handle until upper rod bearing shell (3) seats on crankshaft journal (12).
8. Apply lubricating oil to lower connecting rod bearing shell (11).
9. Position new bearing shell (11) in rod cap (4) and install rod cap (4) so numbered side is matched to numbered side of rod (2). Ensure tang (6) on bearing shell (11) is aligned with groove in rod cap (4).
10. Apply lubricating oil to screw (5) threads and install screws (5) through rod cap (4) and tighten (table 5-7).
11. Check connecting rod clearance by following task a.

5-10. PISTON AND CONNECTING ROD MAINTENANCE (Contd)

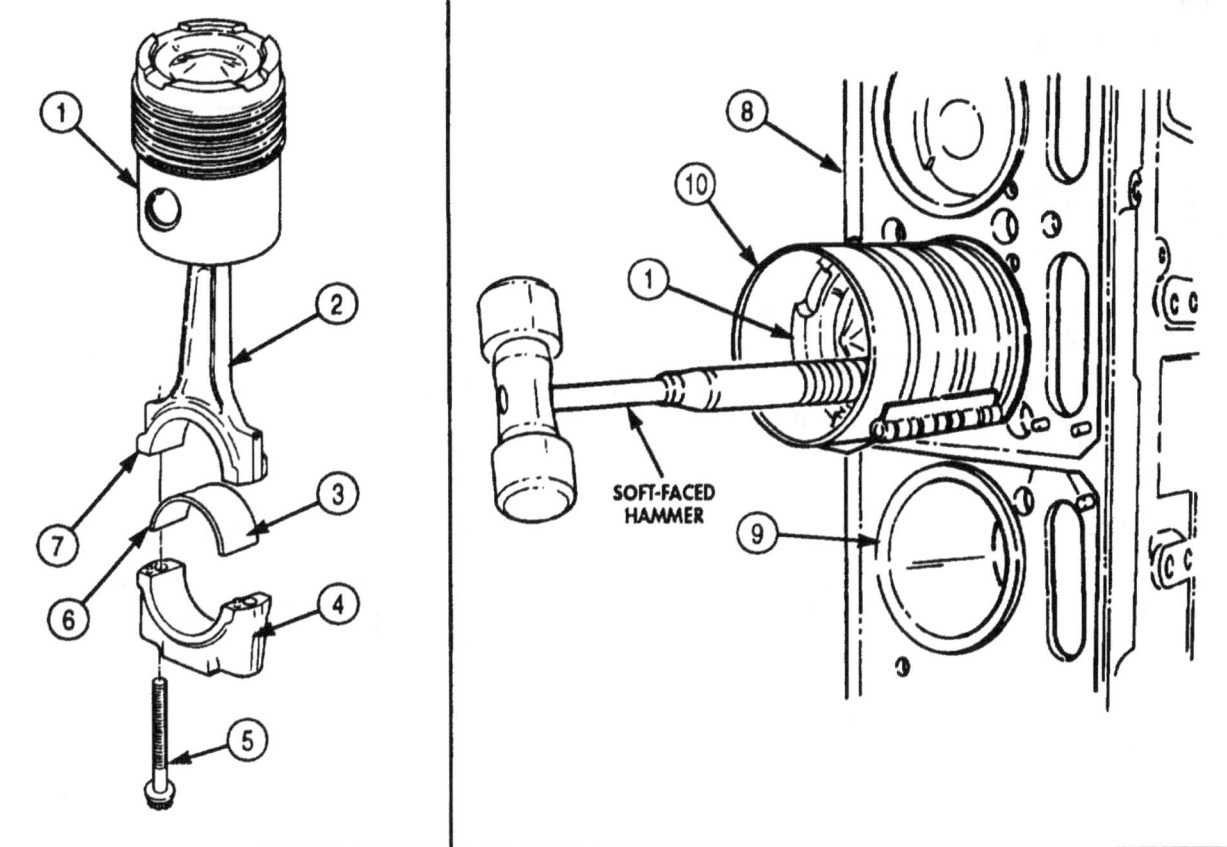

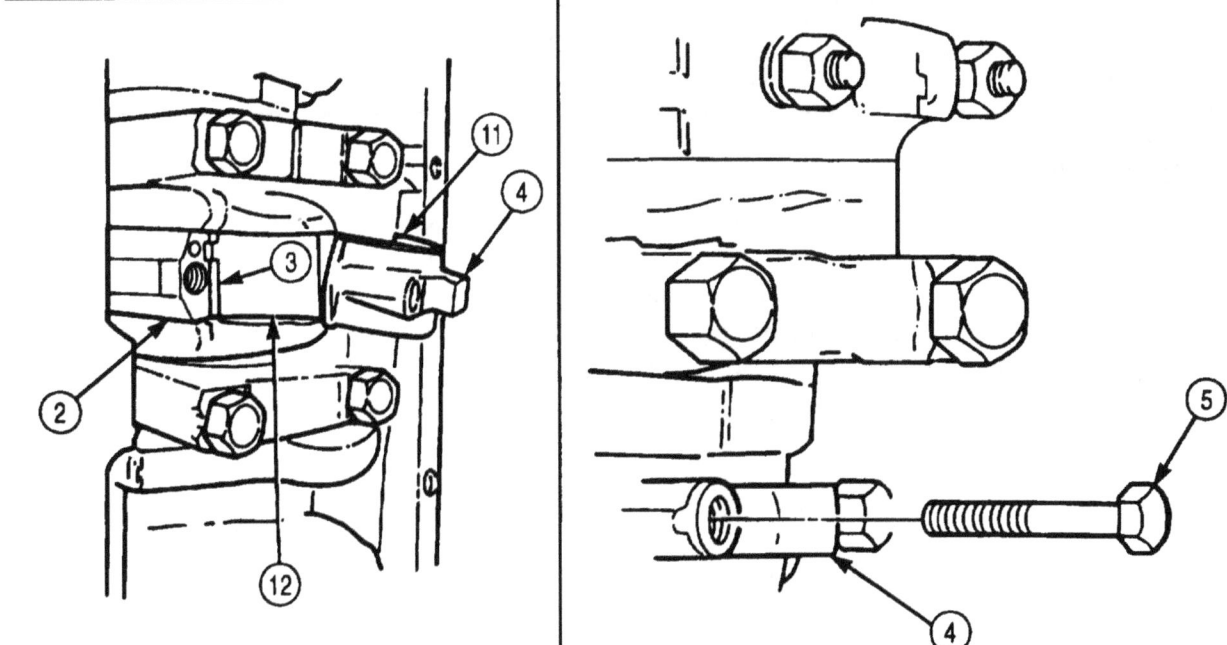

SOFT-FACED HAMMER

FOLLOW-ON TASK: Install camshaft and gear (para. 5-9).

5-11. ENGINE OIL PUMP REPAIR

THIS TASK COVERS:

a. Disassembly

b. Cleaning and Inspection

c. Repair

d. Assembly

INITIAL SETUP:

APPLICABLE MODELS
M939/A1

TOOLS
General mechanic's tool kit (Appendix E, Item 1)
Telescoping gauge (Appendix E, Item 136)
Snap gauge (Appendix E, Item 123)
Boring tool (Appendix E, Item 18)
Torque wrench (Appendix E, Item 144)
Soft-faced hammer
Arbor press

MATERIALS/PARTS
Bypass seat (Appendix D, Item 29)
Bypass spring (Appendix D, Item 30)
Bypass disc (Appendix D, Item 28)
Lockplate (Appendix D, Item 339)
Gasket (Appendix D, Item 176)
Two dowel pins (Appendix D, Item 96)
Two screw-assembled lockwashers
  (Appendix D, Item 590)
Two gaskets (Appendix D, Item 174)
Seven lockwashers (Appendix D, Item 354)
Spring (Appendix D, Item 668)
Three lockwashers (Appendix D, Item 364)
Lubricating oil (Appendix C, Item 50)
Antiseize tape (Appendix C, Item 72)
Drycleaning solvent (Appendix C, Item 71)

REFERENCES (TM)
TM 9-2320-272-24P

EQUIPMENT CONDITION
- Engine oil filter removed (para. 3-5).
- Engine front gearcase cover removed (para. 4-18).
- Engine accessory drive removed (para. 4-26).
- Engine oil pump removed (para. 4-21).

GENERAL SAFETY INSTRUCTIONS
- Keep tire extinguisher nearby when using drycleaning solvent.
- Drycleaning solvent is flammable and toxic. Do not use near an open flame.
- When cleaning with compressed air, wear eyeshields and ensure source pressure does not exceed 30 psi (207 kPa).

a. Disassembly

1. Remove adapter (7), elbow (8), and nipple (6) from oil pump body flange (1).
2. Remove elbow (9) and nipple (10) from oil pump body flange (1).
3. Remove pipe plugs (11) and (12) from oil pump body flange (1).
4. Remove pipe plug (5) from inner body (2).
5. Remove pipe plug (3) from filter head (4).

5-11. ENGINE OIL PUMP REPAIR (Contd)

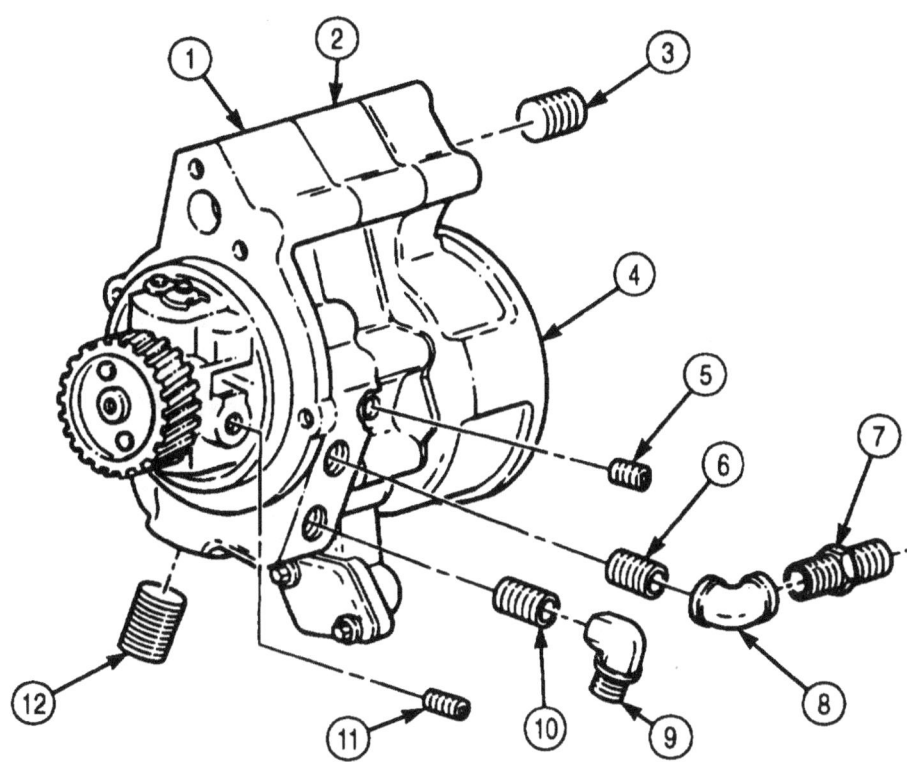

5-11. ENGINE OIL PUMP REPAIR (Contd)

6. Remove two screw-assembled lockwashers (12), flange plate (11), and gasket (10) from inner body (9). Discard gasket (10) and screw-assembled lockwashers (12).

7. Hold retainer (14) down and remove screw (1) from oil pump body (13).

8. Slowly release retainer (14) and remove lockwasher (16), lockplate (15), retainer (14), retainer cap (2), spring (3), and regulator plunger (4) from oil pump body (13). Discard lockwasher (16), lockplate (15), and spring (3).

9. Remove bypass seat (6), disc (7), and bypass spring (8) from filter head (5). Discard bypass seat (6) and bypass spring (8).

5-11. ENGINE OIL PUMP REPAIR (Contd)

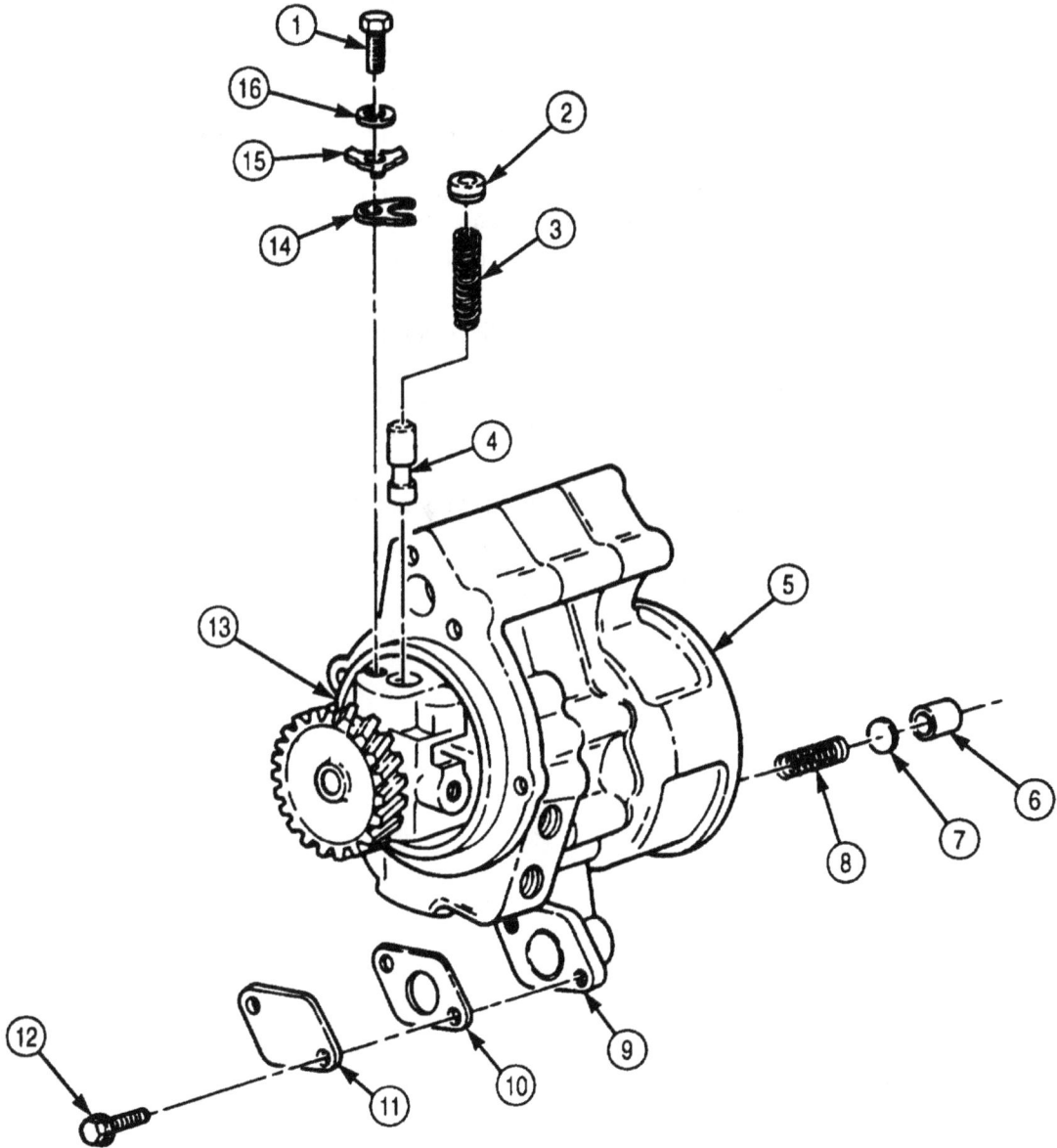

5-11. ENGINE OIL PUMP REPAIR (Contd)

10. Remove two screws (6) and lockwashers (5) from filter head (1). Discard lockwashers (5).

11. Remove six screws (7) and lockwashers (8) from filter head (1). Discard lockwashers (8).

12. Tap filter head (1) with soft-faced hammer to separate from dowel pin (4) and inner body (3) and remove filter head (1) and gasket (2) from inner body (3). Discard gasket (2).

13. Clean gasket remains from mating surfaces of filter head (1) and inner body (3).

14. Remove idler gear (14) from idler shaft (15) and inner body (3).

15. Press driveshaft (10) through driven gear (13) with arbor press and remove driven gear (13) from driveshaft (10).

16. Tap inner body (3) and gasket (12) with soft-faced hammer to separate from dowel pin (11) and remove inner body (3) and gasket (12) from oil pump body (9). Discard gasket (12).

17. Clean gasket remains from mating surfaces of oil pump body (9), inner body (3), and gasket (12).

18. Remove idler gear (17) from idler shaft (15).

19. Press driveshaft (10) through driven gear (15) with arbor press and remove driven gear (15) from driveshaft (10).

20. Remove drive gear (18) from driveshaft (10) with arbor press.

21. Remove idler shaft (15) from oil pump body (9) with arbor press.

NOTE

Perform steps 22 and 23 only if dowel pins are damaged.

22. Remove dowel pin (4) from inner body (3). Discard dowel pin (4).

23. Remove dowel pin (11) from oil pump body (9). Discard dowel pin (11).

5-11. ENGINE OIL PUMP REPAIR (Contd)

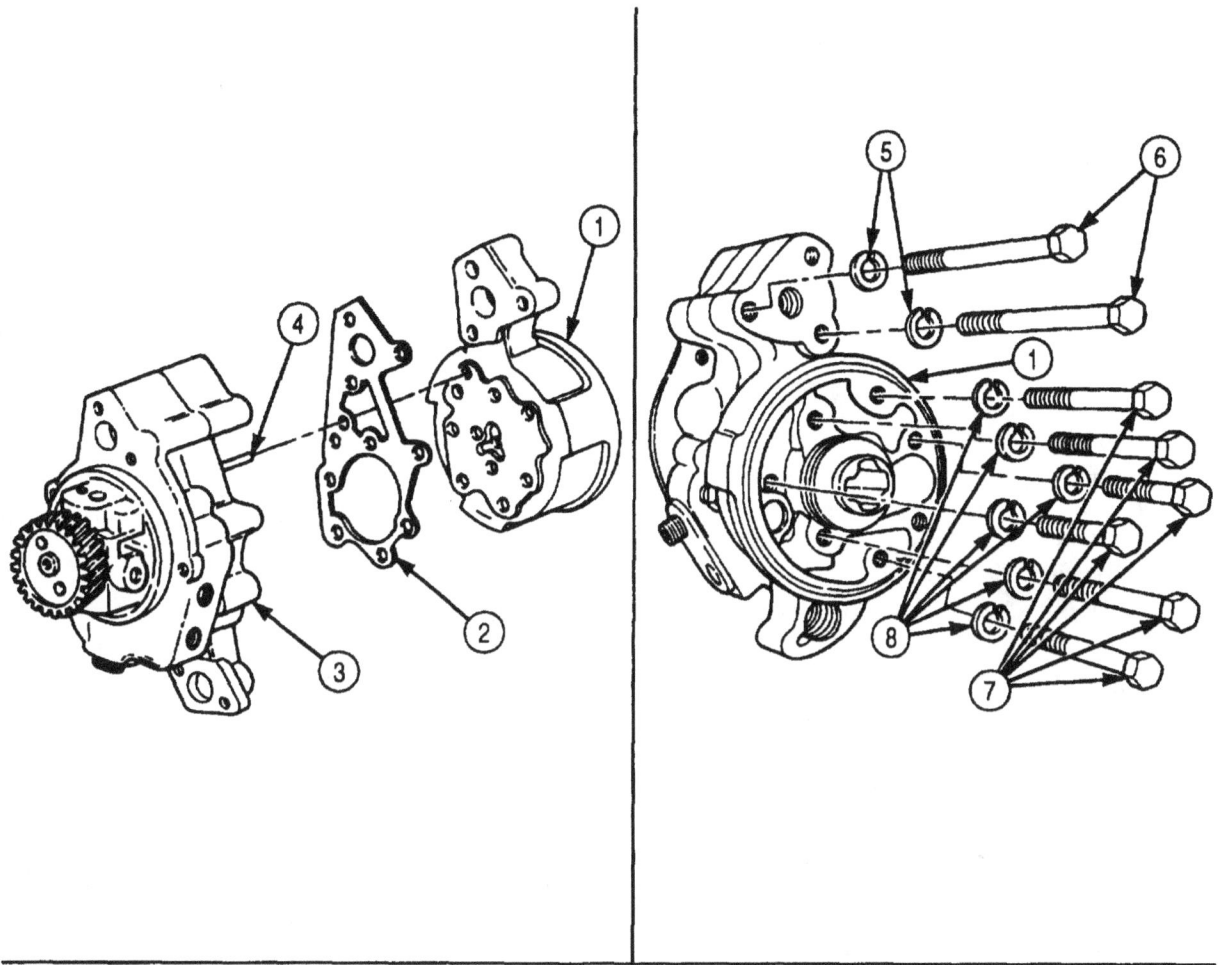

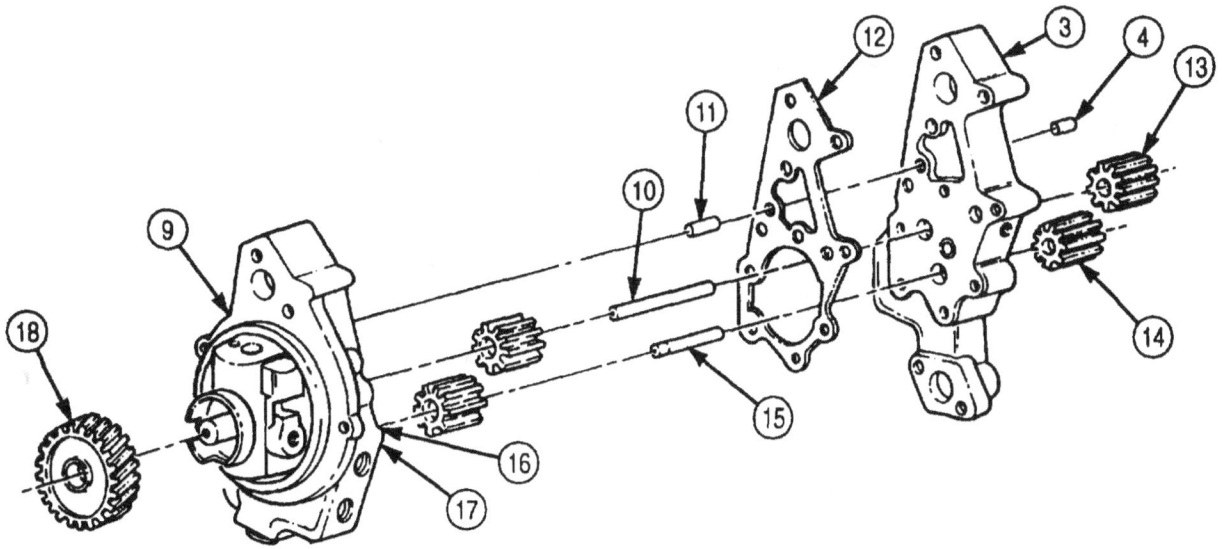

5-11. ENGINE OIL PUMP REPAIR (Contd)

b. Cleaning and Inspection

WARNING

- Drycleaning solvent is flammable and toxic. Do not use near open flame and always have a fire extinguisher nearby when solvents are used. Use only in well-ventilated places, wear protective clothing, and dispose of cleaning rags in approved container. Failure to do this may result in injury or death to personnel and/or damage to equipment.
- Eyeshields must be worn when cleaning with compressed air. Compressed air source will not exceed 30 psi (207 kPa). Failure to do so may result in injury to personnel.

1. Wipe pump body (2) clean with drycleaning solvent.
2. Blow out pump body (2) oil passages with compressed air.
3. Inspect pump body (2) for breaks and cracks. Replace if broken or cracked.
4. Check inside bore diameter of front and rear driveshaft bushings (1). Replace front and rear drive shaft bushings (1) if inside diameter exceeds 0.6185 in. (15.710 mm).
5. Wipe filter head (5) clean with drycleaning solvent.
6. Blow out filter head (5) oil passages with compressed air.
7. Inspect filter head (5) for breaks and cracks. Replace filter head (5) if broken or cracked.
8. Check inside diameter of filter head bushing (4) for wear with dial bore gauge. Replace bushing (4) if inner diameter exceeds 0.6185 in. (15.710 mm).
9. Wipe inner body (3) clean with drycleaning solvent.
10. Blow out inner body (3) oil passages with compressed air.
11. Inspect inner body (3) for breaks and cracks. Replace inner body (3) if broken or cracked.
12. Clean drive gear (11), two driven gears (9), and two idler gears (8) with drycleaning solvent, and inspect for cracked, chipped, or broken teeth. Replace gears if gears (11), (8), and (9) teeth are cracked, chipped, or broken. Replace gears (11), (8), and (9) if gear teeth show pitting over more than 1/4 width of active tooth area.
13. Clean two idler gears (8) and four bushings (10) with drycleaning solvent.
14. Check inside diameter of two idler gears (8) and four bushings (10) with bore gauge. Replace idler gears (8) and bushings (10) if inner diameter exceeds 0.6185 in. (15.71 mm).
15. Inspect pressure regulator plunger (6) to ensure it does not bind in bore (7) of oil pump body (2). Replace pressure regulator plunger (6) if bent or binds.
16. Clean idler shaft (12) and driveshaft (13) with drycleaning solvent.
17. Inspect idler shaft (12) and driveshaft (13) for breaks, cracks, and galling. Replace if broken, cracked, or galled.
18. Check idler shaft (12) and driveshaft (13) outside diameters using dial snap gauge. Replace shafts (12) and (13) if outside diameters are less than 0.6145 in. (15.608 mm).

5-11. ENGINE OIL PUMP REPAIR (Contd)

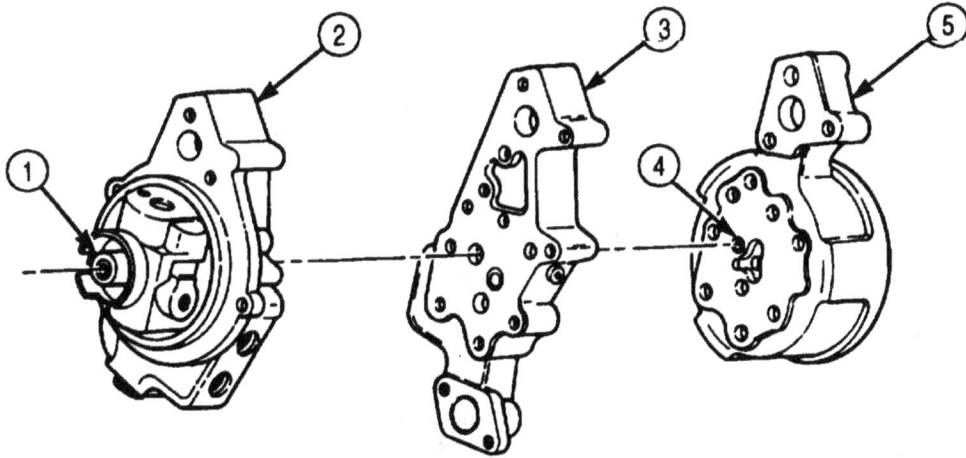

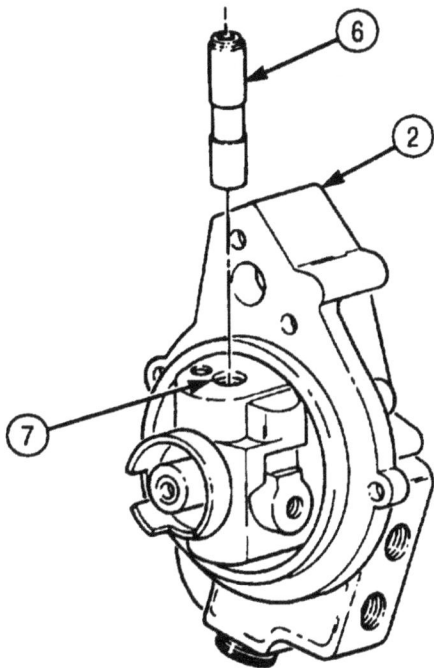

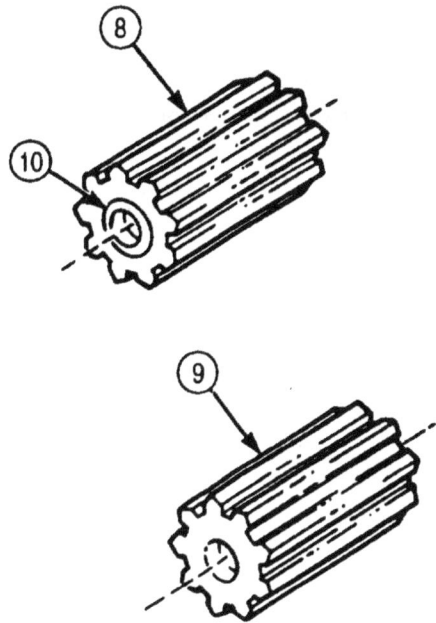

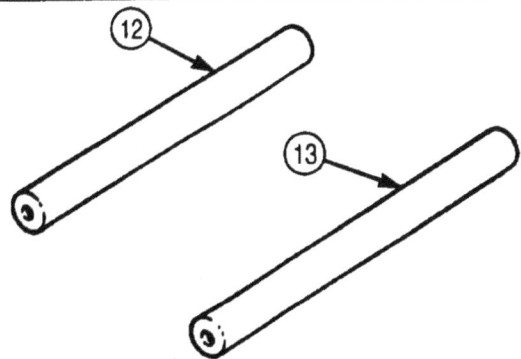

5-11. ENGINE OIL PUMP REPAIR (Contd)

c. Repair

NOTE

- Steps 1 through 9 are performed to replace parts found defective during cleaning and inspection steps 3, 7, 12, and 13.
- Bushings must be replaced as a pair.

1. Using arbor press and mandrel, remove old front and rear driveshaft bushings (2) from oil pump body (1).

2. Install new front and rear driveshaft bushings (2) in oil pump body (1), flush to 0.030 in. (0.78 mm) below surfaces.

3. Bore new front and rear driveshaft bushings (2) inner diameter to 0.6165-0.6175 in. (15.659-15.684 mm).

4. Remove old bushing (3) from filter head (4).

5. Install new bushing (3) in filter head (4) to 0.020 in (0.51 mm) below front surface.

6. Bore new bushing (3) inner diameter to 0.6165-0.6175 in. (15.659-15.684 mm).

7. Using arbor press and mandrel, remove four old bushings (6) from two idler gears (5).

NOTE

Idler gear bushings must be replaced as a pair.

8. Press two new bushings (6) in each idler gear (5) flush to 0.20 in. (0.51 mm) below gear face.

9. Bore inner diameter of four bushings (6) to 0.6165-0.6175 in. (15.659-15.684 mm).

5-11. ENGINE OIL PUMP REPAIR (Contd)

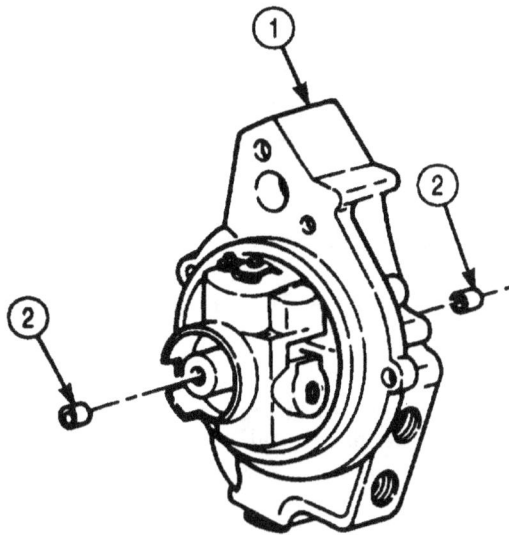

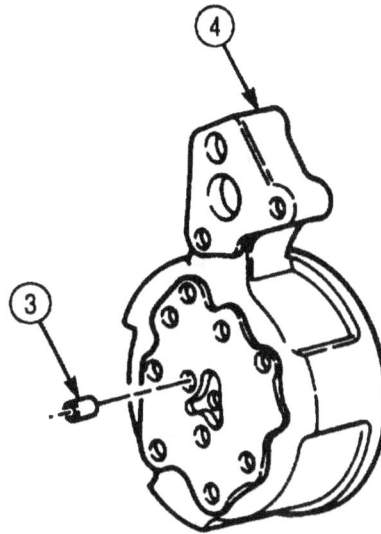

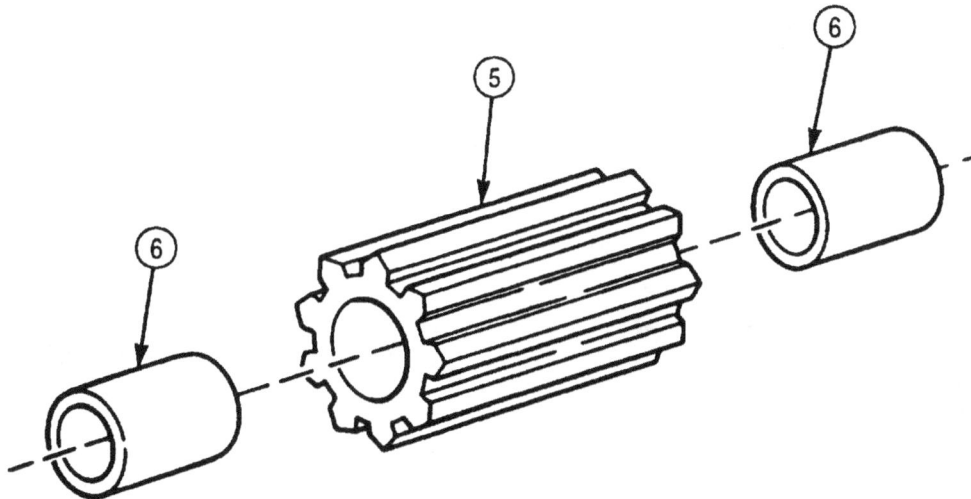

5-11. ENGINE OIL PUMP REPAIR (Contd)

d. Assembly

1. Press drilled end of idler shaft (3) into gear pocket side of oil pump body (1) with arbor press and mandrel until idler shaft (3) protrudes 2.604-2.620 in. (66.04-66.55 mm) above back of body.

2. Press drive gear (5) over drilled end of driveshaft (2) with arbor press and mandrel until driveshaft. (2) end protrudes 0.040-0.060 in. (1.02-1.52 mm) above drive gear (5) face.

3. Apply lubricating oil to driveshaft (2) and insert driveshaft (2) into oil pump body (1).

4. Position 0.012 in. (0.20 mm) shim (4) between back side of drive gear (5) and front of oil pump body (1) and install drive gear (5) against shim (4) until snug. Remove shim (4).

NOTE

Perform step 5 only if dowel pin was removed during disassembly procedure.

5. Install new dowel pin (7) in oil pump body (1) to 0.990-1.010 in. (22.15-25.65 mm) above face.

6. Install idler gear (10) on idler shaft (3).

7. Press driven gear (6) on driveshaft (2) with arbor press and mandrel.

8. Position 0.012 in. (0.30 mm) shim (4) on rear surface of driver gear (6) facing inner body (9) and position inner body (9) over shafts (2) and (3) so it rests on the shim. Install inner body (9) to seat driven gear (6) and remove inner body (9) and shim (4).

9. Apply lubricating oil to gears (6) and (10), shafts (3) and (2), and gear pockets.

10. Install new gasket (8) and inner body (9) on pump body (1), using soft-faced hammer to seat inner body (9) on dowel pin (7). Ensure screw holes are open.

NOTE

Perform step 11 only if dowel pin was removed during disassembly procedure.

11. Install new dowel pin (11) in inner body (9) to 0.990-1.010 in. (25.15 to 25.65 mm) above face.

12. Press driven gear (6) on driveshaft (2) with arbor press and mandrel. Leave 0.002-0.004 in. (0.05-0.10 mm) clearance between bottom of gear pocket and driver gear (6) surface.

13. Install idler gear (10) on idler shaft (3).

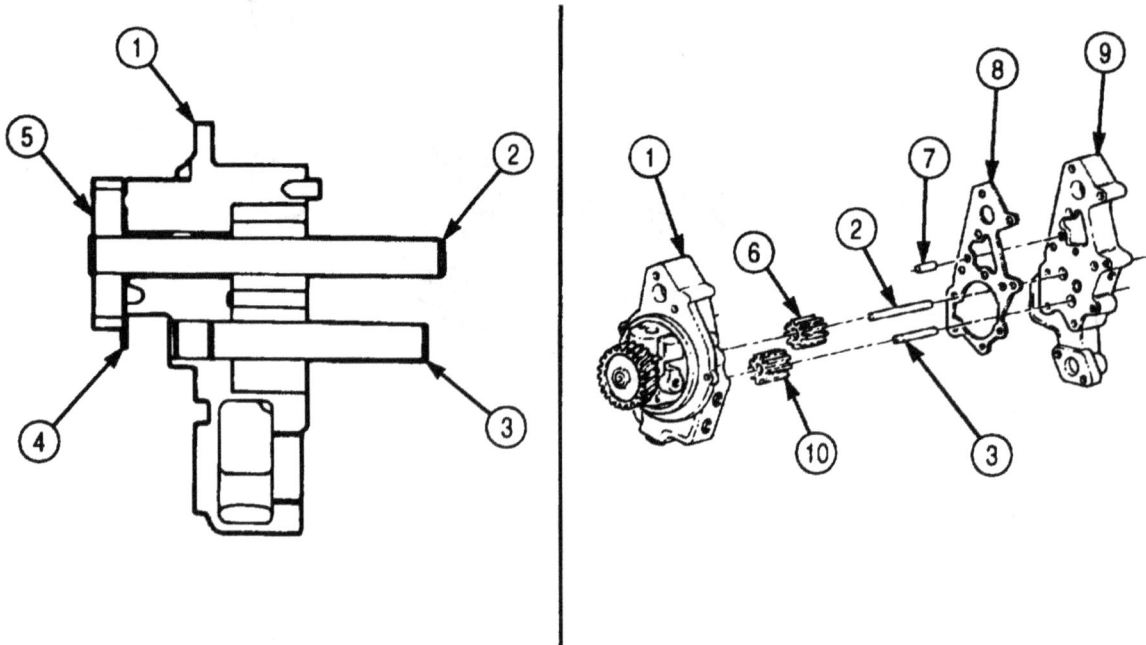

5-11. ENGINE OIL PUMP REPAIR (Contd)

14. Apply lubricating oil to gears (6) and (10), shafts (2) and (3), and gear pockets.

NOTE

Screw is an installation screw and cannot be tightened at this time.

15. Position screw (14) and new lockwasher (15) in inner body (9).

16. Install new gasket (12) and filter head (13) on inner body (9) with two screws (17), new lockwashers (16), six screws (18), and new lockwashers (19). Use soft-faced hammer to seat filter head (13). Tighten screws (17) and (18) 30-35 lb-ft (41-47 N•m).

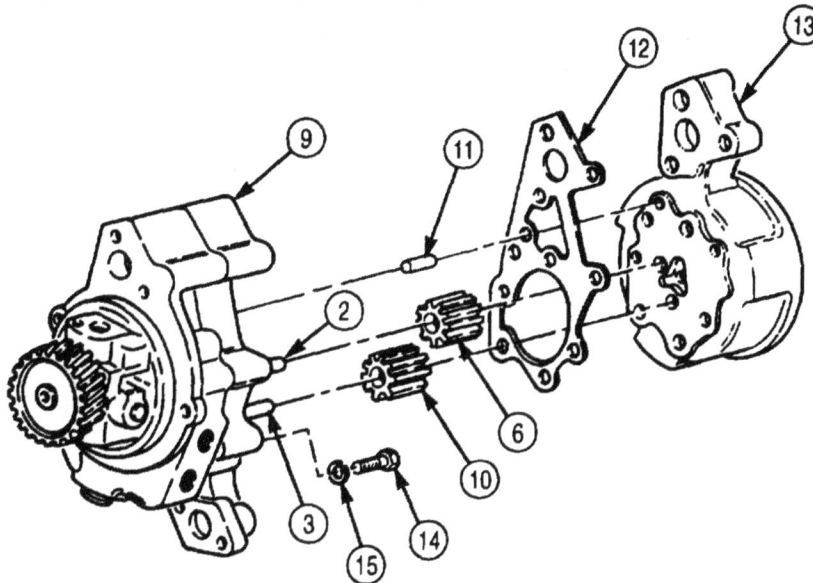

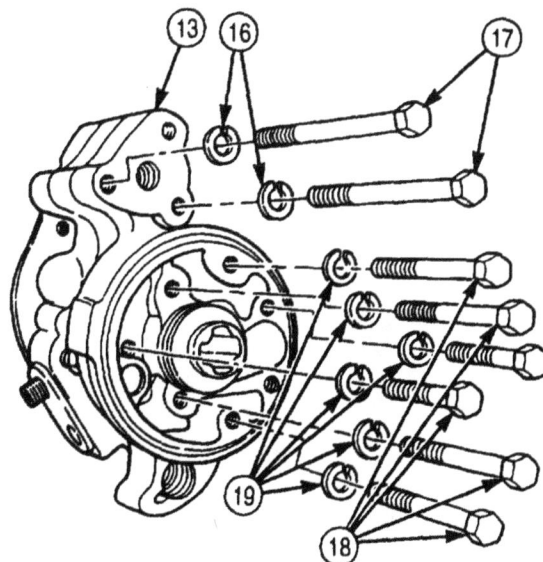

5-11. ENGINE OIL PUMP REPAIR (Contd)

17. Rotate drive gear (17) back and forth to ensure inner gears are free.

18. Check driveshaft (18) end play. End play must be 0.004-0.007 in. (0.10-0.18 mm).

19. Install pressure regulator plunger (15), new spring (5), and retainer cap (4) in pump body (14) with retainer clamp (16), new lockplate (3), new lockwasher (2), and screw (1). Tighten screw (1) 30-35 lb-ft (41-47 N•m).

20. Bend tabs of lockplate (3) with hammer and drift punch.

21. Install new bypass spring (7), disc (9), and new bypass seat (8) in filter head assembly (6).

22. Install new oil pump flange gasket (11) and flange (12) on pump flange (10) with two new lockwashers (14) and screw-assembled washers (13).

5-11. ENGINE OIL PUMP REPAIR (Contd)

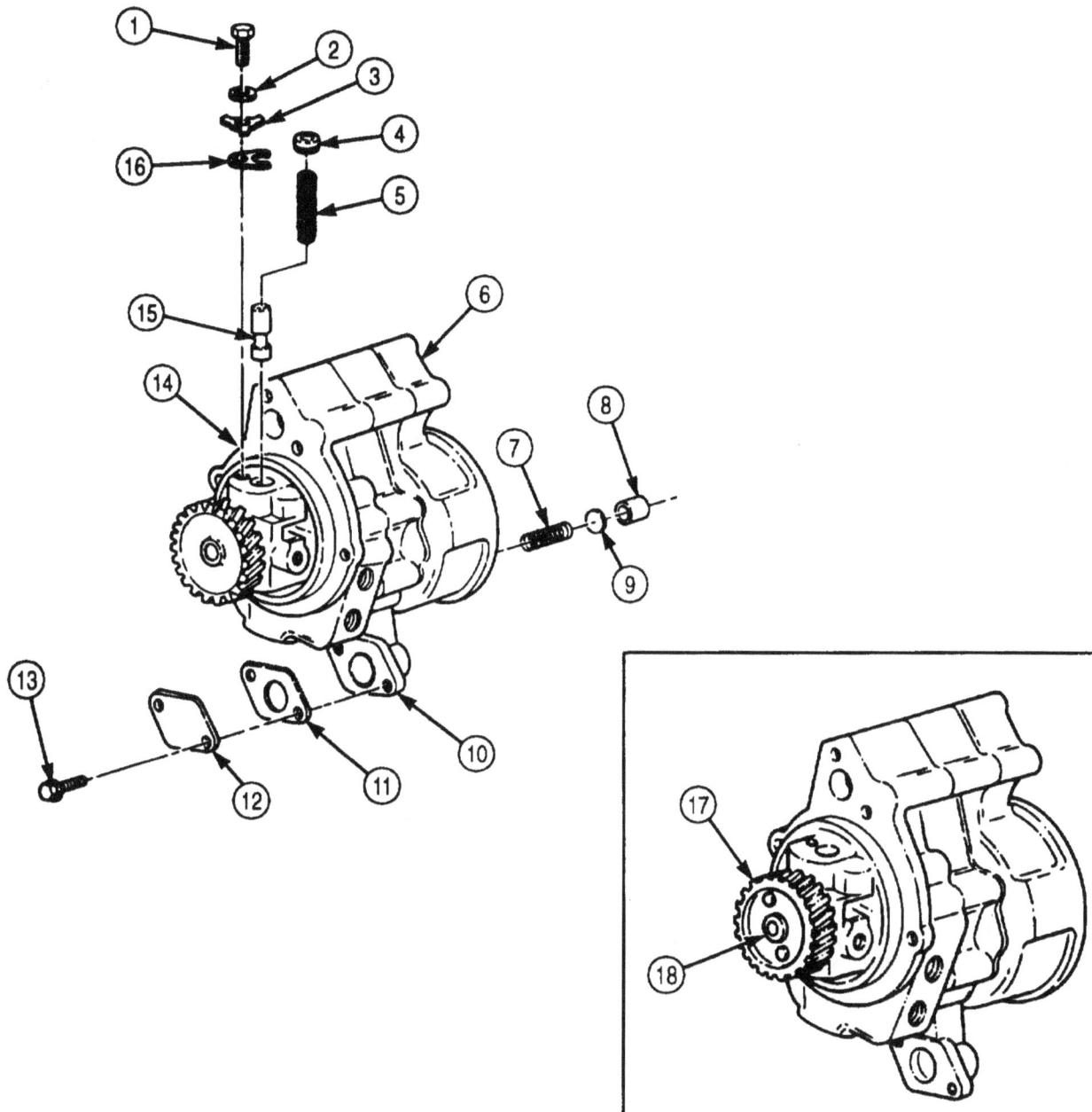

FOLLOW-ON TASKS● Replace engine oil filter (para. 3-5).
● Install engine accessory drive (para. 4-26).
● Install engine front gearcase cover (para. 4-18).
● Install engine oil pump (para. 4-21).

5-12. MOUNTING FUEL PUMP TO HOLDING FIXTURE

THIS TASK COVERS:

a. Removal   b. Installation

INITIAL SETUP:

APPLICABLE MODELS
M939/A1

SPECIAL TOOLS
Mounting plate (Appendix E, Item 84)
Ball joint vise (Appendix E, Item 7)

TOOLS
General mechanic's tool kit (Appendix E. Item 1)

REFERENCES (TM)
TM 9-2320-272-24P

EQUIPMENT CONDITION
• Fuel pump removed from engine (para. 4-35).
• Fuel pump shutoff valve removed (para. 4-36 or 4-37).

a. Removal

1. Remove two nuts (2), screws (5), and fuel pump (3) from mounting plate (1).
2. Remove two screws (4) and mounting plate (1) from ball joint vise.

b. Installation

1. Attach mounting plate (1) to ball joint vise and install two screws (4).
2. Install fuel pump (3) to mounting plate (1) with two screws (5) and nuts (2).

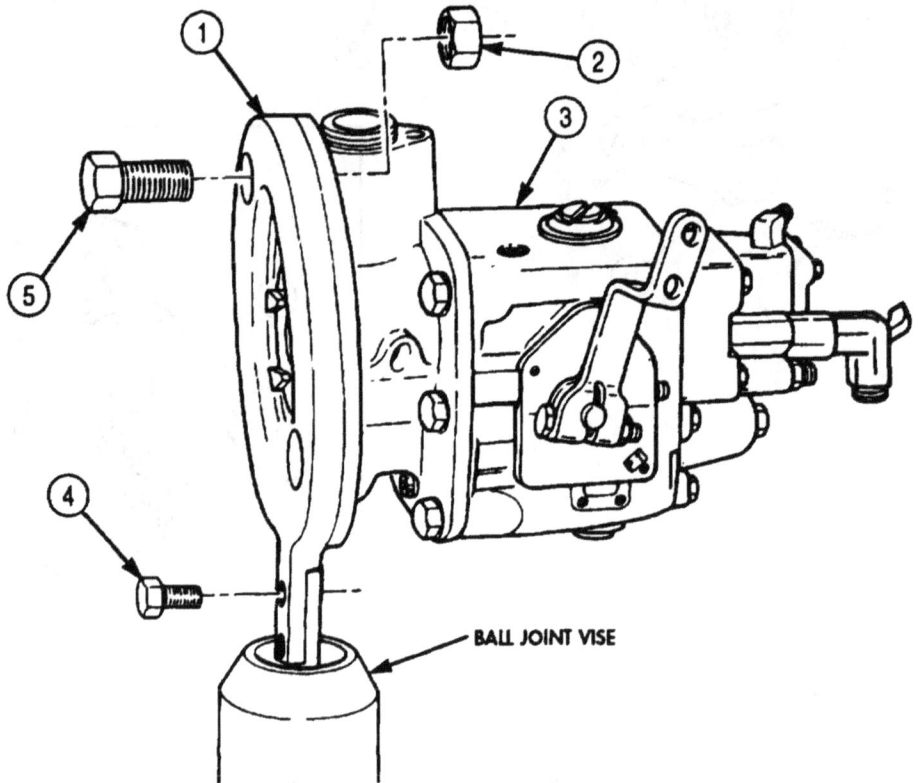

BALL JOINT VISE

FOLLOW-ON TASKS• Install fuel pump on engine (para. 4-35).
        • Install fuel pump shutoff valve (para. 4-36 or 4-37).

5-13. PRESSURE GEAR PUMP REMOVAL

THIS TASK COVERS:
Removal

INITIAL SETUP:

APPLICABLE MODELS
M939/A1

TOOLS
General mechanic's tool kit (Appendix E, Item 1)
Soft-faced hammer

MATERIALS/PARTS
Two lockwashers (Appendix D, Item 407)

REFERENCES (TM)
TM 9-2320-272-24P

Removal

1. Remove two screws (4), lockwashers (3), washers (2), pulsation damper (1), and seal (5) from gear pump (8). Discard lockwashers (3).
2. Remove four screws (10) and lockwashers (9) connecting gear pump (8) to fuel pump housing (6). Discard lockwashers (9).
3. Tap gear pump (8) lightly with soft-faced hammer and remove from fuel pump housing (6).
4. Remove gasket (7) from gear pump (8). Discard gasket (7).
5. Clean gasket remains from gear pump (8) mating surfaces.
6. Install seal (5) and pulsation damper (1) on fuel pump housing (6) with two washers (2), new lockwashers (3), and screws (4).

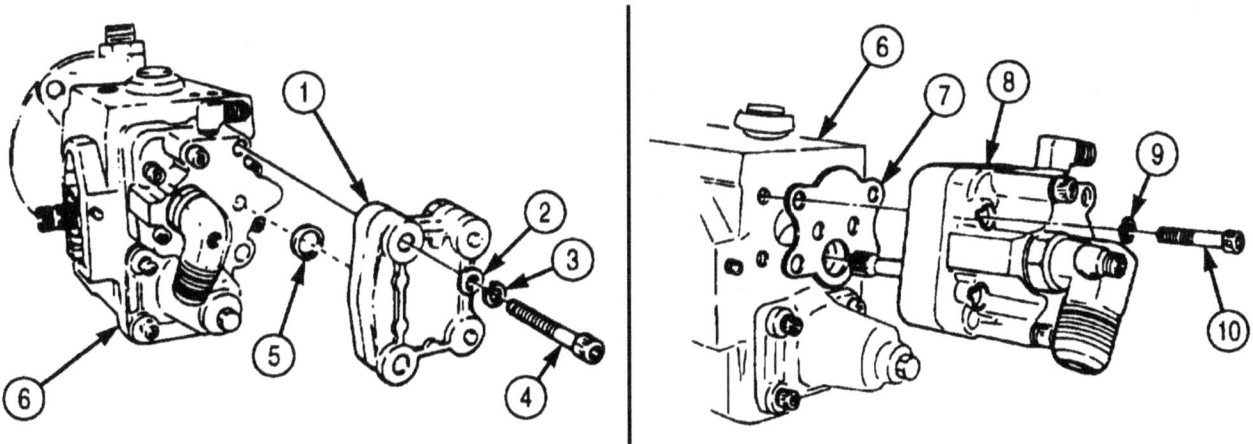

5-14. GOVERNOR SPRING PACK MAINTENANCE

THIS TASK COVERS:

a. Removal   c. Installation
b. Inspection

INITIAL SETUP:

APPLICABLE MODELS   **REFERENCES (TM)**
M939/A1   TM 9-2320-272-24P

TOOLS   **EQUIPMENT CONDITION**
General mechanic's tool kit (Appendix E, Item 1)   Fuel pump mounted on holding fixture (para. 5-12).
Spring tester (Appendix E, Item 131)

MATERIALS/PARTS
Retaining ring (Appendix D, Item 538)
Gasket (Appendix D, Item 223)
Seal (Appendix D, Item 612)

NOTE
Spring packs are maintained basically the same for AFC and
VS fuel pumps. This procedure covers the AFC fuel pump.

a. Removal

1. Remove seal (12) from governor adjusting screw (13) and throttle leakage adjusting screw (18). Discard seal (12).

2. Remove four screws (14), spring pack cover (15), and gasket (16) from fuel pump (1). Discard gasket (16).

3. Remove retaining ring (11), spring retainer (10), shim(s) (9), spring (8), spring guide (5), washer (4), spring (3), and plunger (2) from spring pack barrel (17). Discard retaining ring (11).

4. Remove adjusting screw (7) and spring (6) from spring guide (5).

b. Inspection

1. For general inspection instructions, refer to para. 2-15.

2. Inspect spring pack cover (15) for cracks, breaks, and stripped threads. Replace spring pack cover (15) if cracked, broken, or threads are stripped.

3. Using spring tester, inspect tension of spring (8). Replace spring (8) if tension is not within 0.69-0.85 lb (0.31-0.39 kg) when compressed to 1.0 in. (25.4 mm).

4. Using spring tester, inspect tension of spring (3). Replace spring (3) if tension is not within 16.02-17.78 lb (7.27-8.07 kg) when compressed to 1.025 in. (26.0 mm).

5-14. GOVERNOR SPRING PACK MAINTENANCE (Contd)

| c. Installation |

1. Install spring (6) and adjusting screw (7) on spring guide (5).
2. Install plunger (2), spring (3), washer (4), spring guide (5), spring (8), shim(s) (9), and spring retainer (10) on spring pack barrel (17) with new retaining ring (11).
3. Install new gasket (16) and spring pack cover (15) on fuel pump (1) with four screws (14).

NOTE
Seal is not installed until fuel pump calibration (para. 5-18).

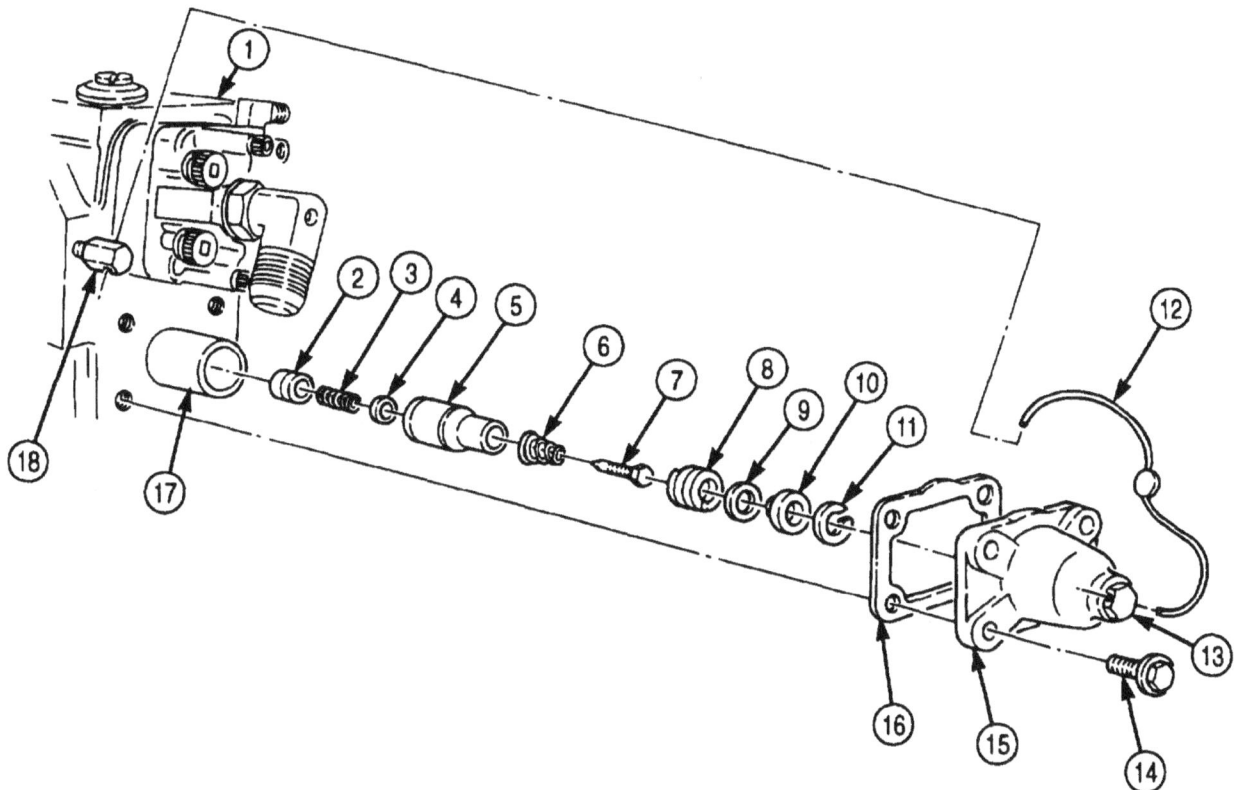

FOLLOW-ON TASKS● Remove fuel pump from holding fixture (para 5-12).
● Calibrate fuel pump (para. 5-18).

5-15. FUEL PUMP HOUSING MAINTENANCE

THIS TASK COVERS:

a. Disassembly   c. Assembly
b. Inspection

INITIAL SETUP:

APPLICABLE MODELS   **REFERENCES (TM)**
M939/A1   TM 9-2320-272-24P

TOOLS   **EQUIPMENT CONDITION**
General mechanic's tool kit (Appendix E, Item 1) • Governor spring pack removed (para. 5-14).
                                   • Throttle shaft removed (para. 5-16).
MATERIALS/PARTS
Seven lockwashers (Appendix D, Item 412)
Gasket (Appendix D, Item 239)
Diesel fuel (Appendix C, Item 41)

---

a. Disassembly

NOTE
Tap edge of hardware lightly with soft-faced hammer to loosen,

1. Remove screw (7), six screws (1), seven lockwashers (2), and washers (3) connecting front drive cover (6) to fuel pump housing (4). Discard lockwashers (2).
2. Pull fuel pump housing (4) straight out from front drive cover (6) to clear dowels.
3. Remove gasket (5) from front drive cover (6). Discard gasket (5).
4. Clean gasket remains from front drive cover (6) mating surfaces.
5. Remove governor plunger (9) and torque spring (8) from fuel pump housing (4).

NOTE
Do not use straight pull on governor plunger torque spring. To remove, twist spring off shoulder.

6. Remove torque spring (8) from governor plunger (9).

b. Inspection

1. Inspect governor plunger (9) for scoring, nicks, and scratches. If scored, nicked, or scratched, replace governor plunger (9).
2. Inspect torque spring (8) for broken coils. If coils are broken, replace torque spring (8).

c. Assembly

NOTE
Parts must submerged in diesel fuel and hands wet with diesel fuel before steps 1 and 2 to prevent damage to close-tolerance parts.

1. Install torque spring (8) on governor plunger (9).
2. Install torque spring (8) and governor plunger (9) in fuel pump housing (4).
3. Install new gasket (5) and fuel pump housing (4) on front drive cover (6) with seven washers (3), new lockwashers (2), six screws (1), and screw (7).

5-15. FUEL PUMP HOUSING MAINTENANCE (Contd)

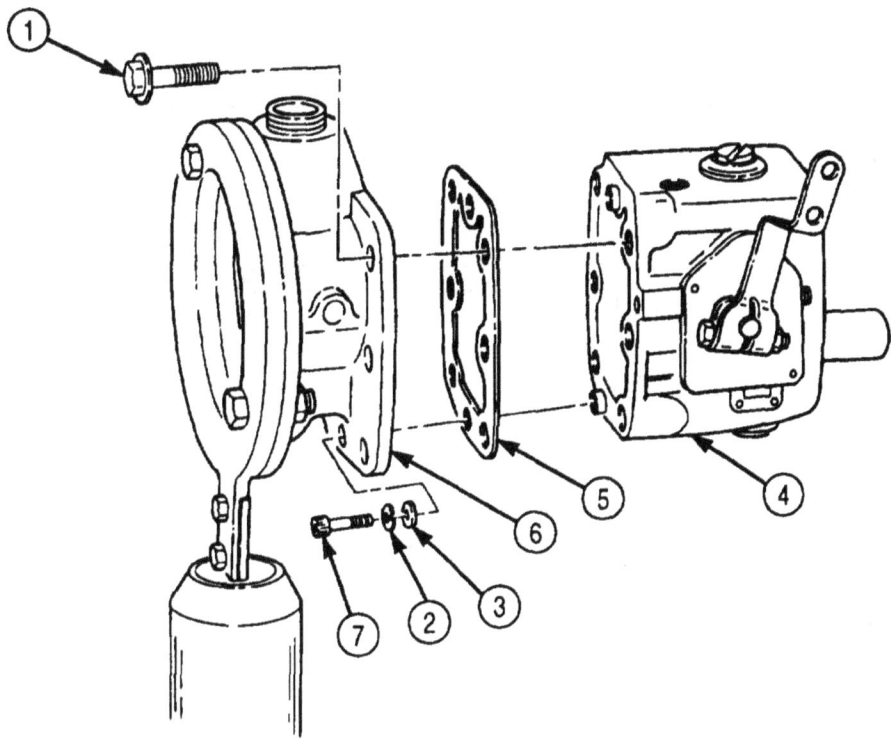

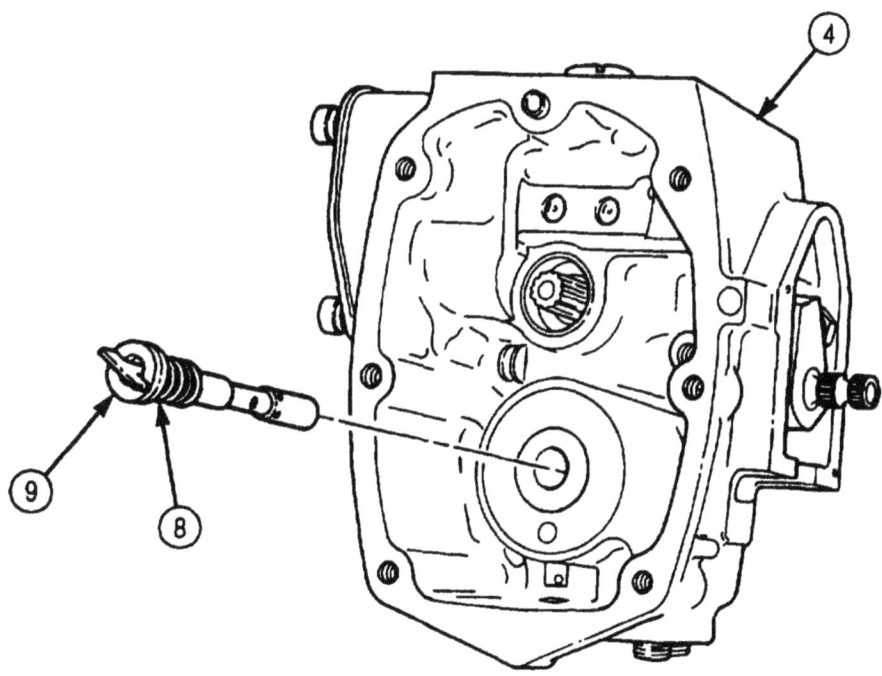

FOLLOW-ON TASKS● Install throttle cover and shaft (para. 5-16).
● Install governor spring pack (para. 5-14).

5-16. THROTTLE COVER AND SHAFT MAINTENANCE

THIS TASK COVERS:

a. Disassembly   c. Assembly
b. Inspection

INITIAL SETUP:

APPLICABLE MODELS   **REFERENCES (TM)**
M939/A1   TM 9-2320-272-24P

TOOLS   **EQUIPMENT CONDITION**
General mechanic's tool kit (Appendix E, Item 1) Fuel pump housing disassembled (para. 5-15).

MATERIALS/PARTS
Two drive pins (Appendix D, Item 101)
O-ring (Appendix D, Item 461)
O-ring (Appendix D, Item 462)
O-ring (Appendix D, Item 463)
Lockwasher (Appendix D, Item 400)
Throttle shaft soft ball bearing (Appendix D,
  Item 666)

---

a. Disassembly

1. Center punch and drill out the two drive pins (8) connecting the fuel pump housing (1) to throttle shaft cover (2). Discard drive pins (8).
2. Remove nut (4), lockwasher (5), washer (6), screw (7), throttle lever (3), and throttle shaft cover (2) from fuel pump housing (1). Discard lockwasher (5).
3. Remove retaining ring (10) and throttle shaft (9) from fuel pump housing (1).
4. Drill and remove soft ball bearing (18) from throttle shaft (9). Discard soft ball bearing (18).
5. Remove two set screws (15) and throttle stop control (14) from throttle shaft (9).
6. Remove O-rings (13) and (16), fuel-adjust screw (17), valve plug (12), and O-ring (11) from throttle shaft (9). Discard O-rings (13), (16), and (11).

b. Inspection

1. Inspect throttle shaft (9) for scoring, nicks, and scratches. If scored, nicked, or scratched, replace throttle shaft (9).

5-16. THROTTLE COVER AND SHAFT MAINTENANCE (Contd)

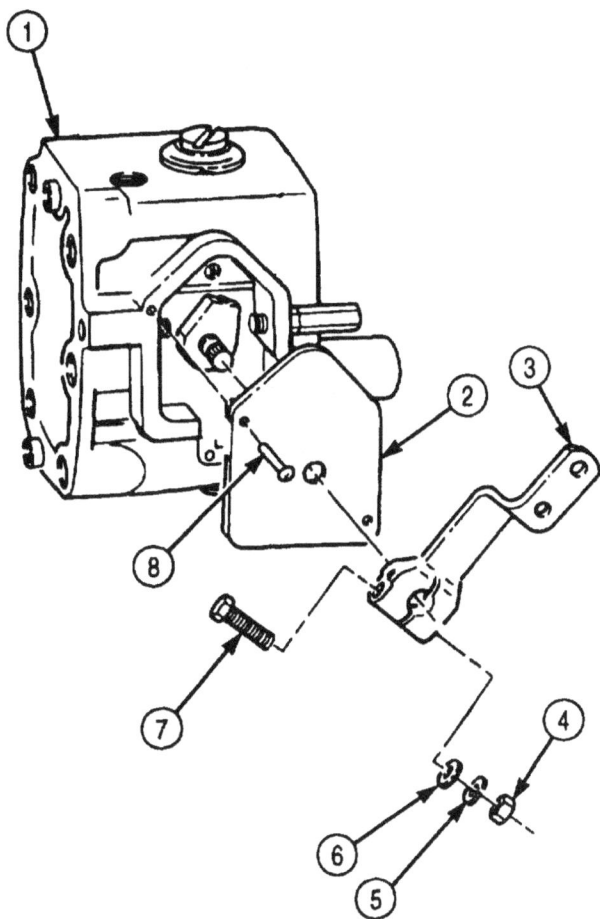

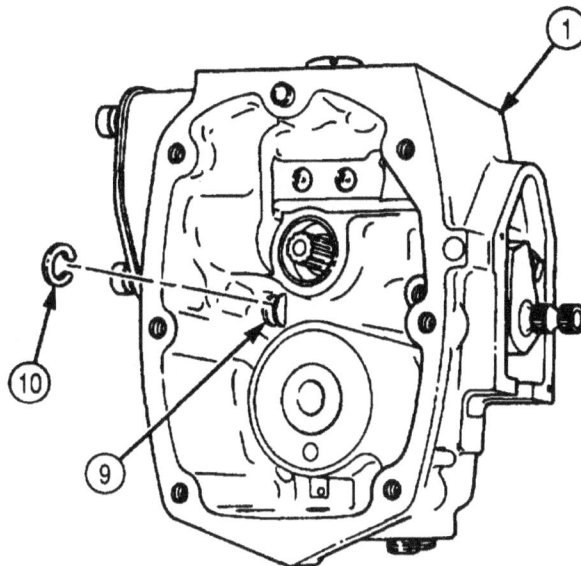

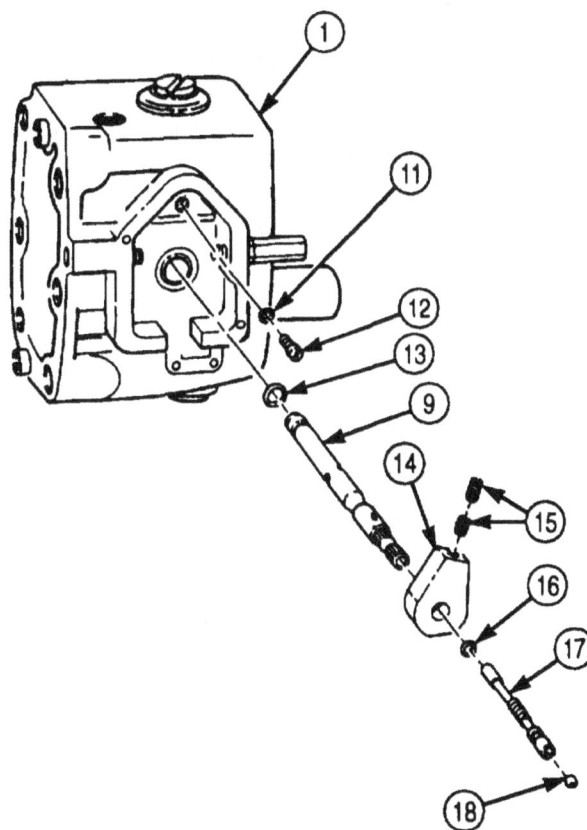

5-16. THROTTLE COVER AND SHAFT MAINTENANCE (Contd)

c. Assembly

NOTE

Governor weight must be installed before assembling throttle shaft cover (para. 5-17).

1. Install new O-ring (11) and valve plug (12) in fuel pump housing (1).

2. Install fuel-adjusting screw (17), new O-rings (13) and (16), throttle stop control (14) and two setscrews (15) on throttle shaft (9).

3. Install new throttle shaft ball (18) in end of throttle shaft (9).

4. Install throttle shaft (9) in fuel pump housing (1) with retaining ring (10).

NOTE

- Do not install throttle shaft cover until fuel pump has been calibrated (para. 5-18).

- Do not perform steps 5 through 7 unless pin holes are damaged beyond use.

5. Position throttle shaft cover (2) on fuel pump housing (1).

CAUTION

Use care when drilling new holes for drive pins. The pump housing is made of cast aluminum and is easily damaged. Do not allow metal particles to enter pump housing.

6. Center punch location of new holes on throttle shaft cover (2). Ensure holes are opposite one another.

7. Carefully drill through throttle shaft cover (2) and into throttle shaft cover flange with 1/15 in. drill bit NO MORE than 1/4 in. (6 mm).

8. Install throttle shaft cover (2) on fuel pump housing (1) by gently tapping two new drive pins (8) through throttle shaft cover (2) into holes.

9. Install throttle lever (3) on throttle shaft (9) with screw (7), washer (6), new lockwasher (5), and nut (4).

5-16. THROTTLE COVER AND SHAFT MAINTENANCE (Contd)

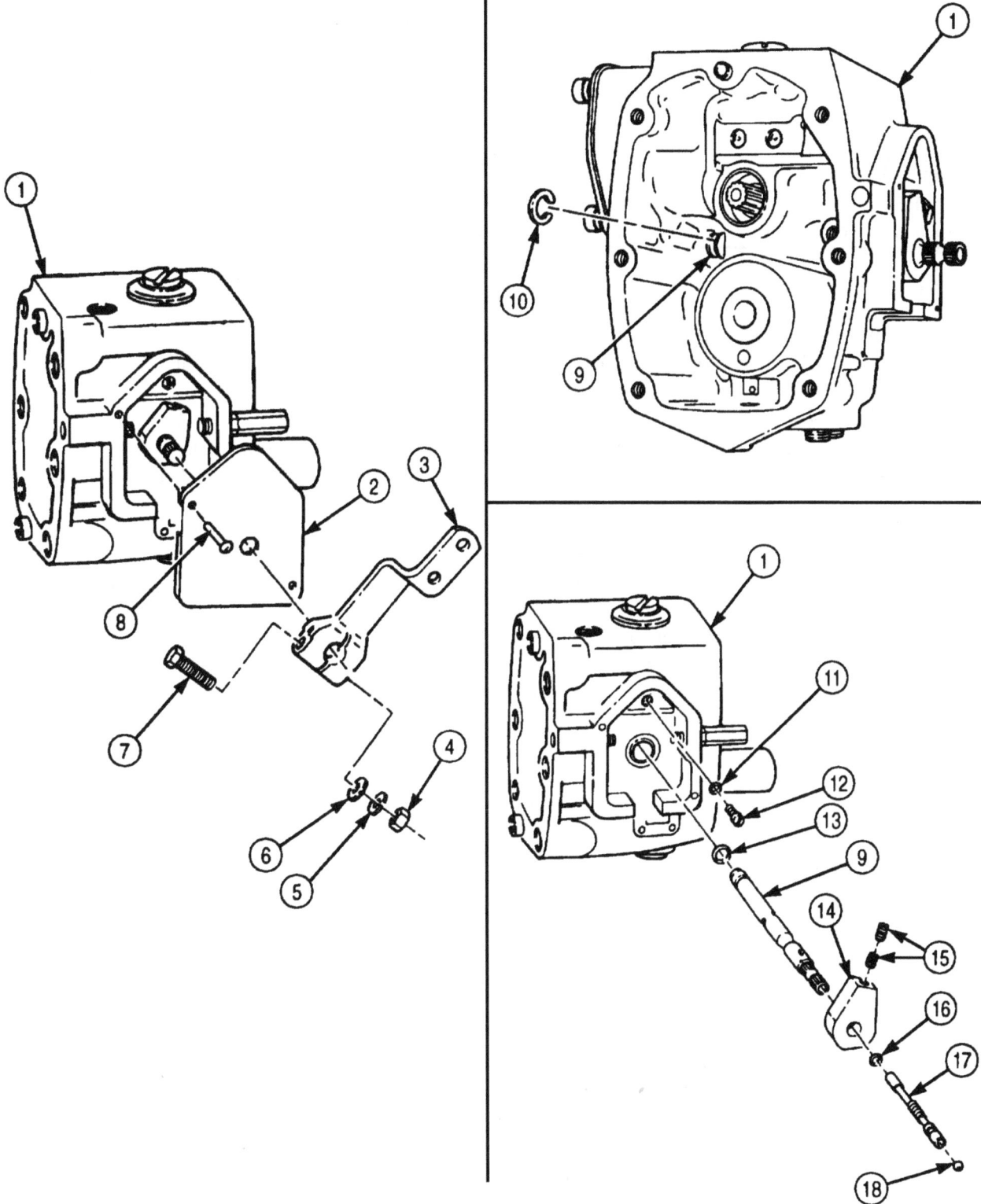

FOLLOW-ON TASK: Assemble fuel pump housing (para. 5-15).

5-17. GOVERNOR WEIGHT MAINTENANCE

THIS TASK COVERS:

a. Removal

b. Inspection

c. Installation

<u>INITIAL SETUP:</u>

<u>APPLICABLE MODELS</u>	<u>REFERENCES (TM)</u>
M939/A1	TM 9-2320-272-24P
<u>TOOLS</u>	<u>EQUIPMENT CONDITION</u>
General mechanic's tool kit (Appendix E, Item 1)	Throttle cover and shaft disassembled (para. 5-16).

a. Removal

1. Remove governor weight (6) and gear (2) from front drive cover (1).
2. Remove weight-assist plunger (5), spring (3), and shims (4) from governor weight (6).

b. Inspection

1. Inspect spring (3) for bent or broken coils. If bent or broken, replace spring (3).
2. Inspect governor weight gear (2) for cracked, broken, or pitted teeth. If cracked, broken, or pitted, replace governor weight (6).

c. Installation

1. Install shims (4), spring (3), and weight-assist plunger (5) in governor weight (6).

NOTE

Large end of weight-assist plunger is installed first.

2. Install governor weight gear (2) and governor weight (6) on front drive cover (1).

5-17. GOVERNOR WEIGHT MAINTENANCE (Contd)

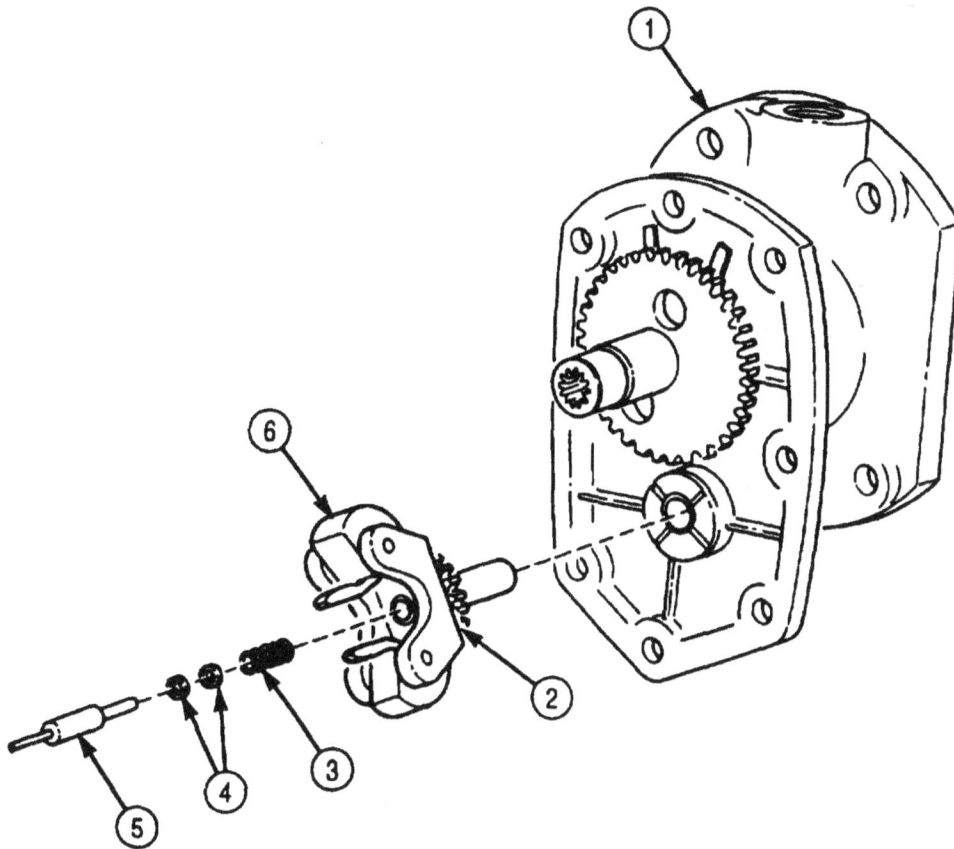

FOLLOW-ON TASKS • Install front cover (para. 5-15)
 • Install throttle shaft and cover (para. 5-16).

5-18. FUEL PUMP SETUP AND CALIBRATION

THIS TASK COVERS:

a. Throttle Shaft Cover Removal
b. Mounting Pump to Test Stand
c. Fuel Pump Run-In
d. Testing Pump Seals for Leaks
e. Testing Governor Cutoff RPM
f. Testing and Adjusting Throttle Leakage
g. Testing and Adjusting Idle Speed
h. Checking and Adjusting Throttle Lever Travel

i. Testing and Adjusting Pump Main Pressure
j. Testing and Adjusting Fuel Pressure
k. Testing and Adjusting Governor Fuel Pressure
l. Checking and Adjusting Governor Weight Pressure
m. Testing and Adjusting Idle Speed (VS Governor Only)
n. Shutdown and Removal from Test Stand

INITIAL SETUP:

APPLICABLE MODELS
M939/A1

SPECIAL TOOLS
Fuel injection tester (test stand) (Appendix E, Item 49)
Spring pack adjusting tool (Appendix E, Item 130)
Travel template (Appendix E, Item 150)
Level and angle indicator (Appendix E, Item 66)
Shaft installation tool (Appendix E, Item 120)
Gear pump block plate (Appendix E, Item 57)

TOOLS
General mechanic's tool kit (Appendix E, Item 1)
Straightedge

MATERIALS/PARTS
Two drive pins (Appendix D, Item 101)
Lockwasher (Appendix D, Item 400)
Throttle shaft soft ball bearing (Appendix D, Item 666)
Grease, GAA (Appendix C, Item 28)
Lubricating oil OE/HDO 30 (Appendix C, Item 50)
45A calibrating fluid (Appendix C, Item 13)
Sealing tape (Appendix C, Item 72)

REFERENCES (TM)
TM 9-2320-272-24P
TM 9-4910-571-12
TM 9-4910-571-12P

EQUIPMENT CONDITION
- Fuel pump removed from vehicle (para. 4-35).
- Fuel pump shutoff valve removed (para. 4-36 or 4-37).

a. Throttle Shaft Cover Removal

NOTE
Perform task a. only if throttle shaft cover has not been removed.

1. Center punch each of the two drive pins (8) connecting throttle shaft cover (2) to fuel pump housing (1).
2. Drill drive pins (8) out of throttle shaft cover (2). Discard drive pins (8).
3. Remove nut (4), lockwasher (5), washer (6), screw (7), and throttle lever (3) from throttle shaft cover (2). Discard lockwasher (5).
4. Remove throttle shaft cover (2) from fuel pump housing (1).
5. Install throttle lever (3) on fuel pump housing (1) with washer (6), new lockwasher (5), and nut (4).

5-18. FUEL PUMP SETUP AND CALIBRATION (Contd)

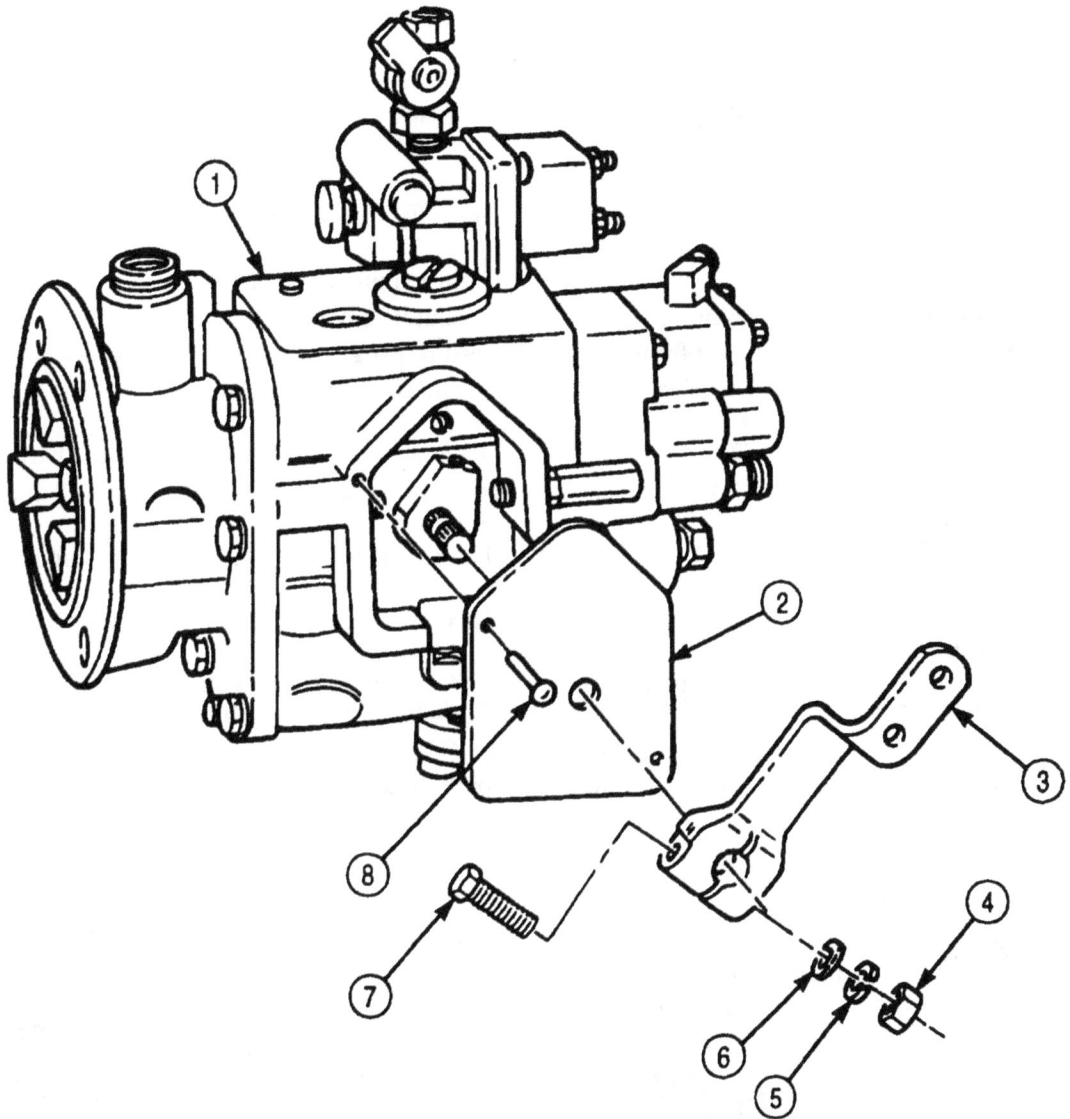

5-18. FUEL PUMP SETUP AND CALIBRATION (Contd)

b. Mounting Pump to Test Stand

1. Install adapter ring (7) on adapter bracket (8) with four washers (9) and screws (6). Ensure the word TOP or part number on adapter ring (7) faces up.

2. Install ring and adapter bracket (4) on mounting rails (16) of test stand with clamp bar (14). Tighten clamp bar (14) finger tight.

3. Mount fuel pump (11) to ring and adapter bracket (4) with four screws (12) and washers (13).

4. Place pump coupling insert (15) into test stand drive coupling (10).

5. Loosen clamp bar (14) and slide fuel pump (11) and ring and adapter bracket (4) forward to engage drive coupling (10).

NOTE

Clean all male pipe threads and wrap with sealing tape before installation.

6. Install inlet adapter (27) on pump elbow (26).

7. Connect 1/2 in. (12.7 mm) flexible hose (30) from test stand fuel pressure control valve (32) to adapter (27).

8. Connect 1/4 in. (6.35 mm) manifold hose (24) from manifold vacuum adapter (28) to adapter (21).

9. Install pump discharge fitting (18) in fuel pump shutoff solenoid valve (19).

10. Install fuel pressure hose (17) from test stand pressure gauge outlet (38) to pump discharge fitting (18).

11. Install 1/2 in. (12.7 mm) flexible hose (25) from test stand lube pressure (31) to test stand lube return (22).

12. Install fuel input hose (37) from test stand fuel input connector (36) to pump discharge fitting (18).

13. Install fuel outlet hose (33) from test stand fuel outlet connector (34) to test stand fuel return connector (23).

14. Install leakage accumulator hose (40) to No. 1 accumulator can (41) from test stand leak test connector (39).

15. Connect 1/4 in. (6.35 mm) flexible hose (29) to check valve fitting (20) on fuel pump (11) and to auxiliary return connector (35).

16. Install throttle lever position holding spring (1) from top of throttle shaft lever (3) to ring and adapter bracket (4). Spring (1) will hold throttle shaft lever (3) to full fuel position.

17. Remove governor spring pack housing pipe plug (2) from spring pack housing (5).

18. Install spring pack adjusting tool into spring pack housing (5).

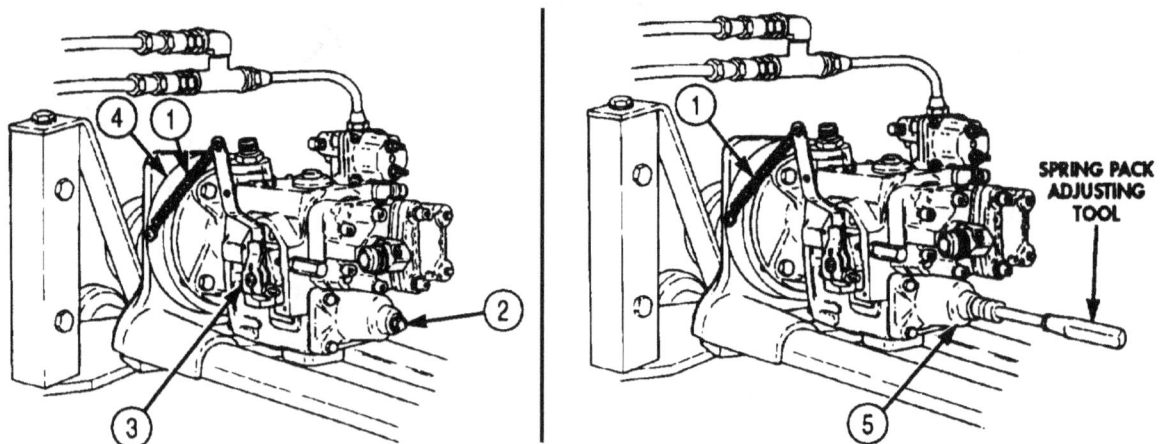

5-18. FUEL PUMP SETUP AND CALIBRATION (Contd)

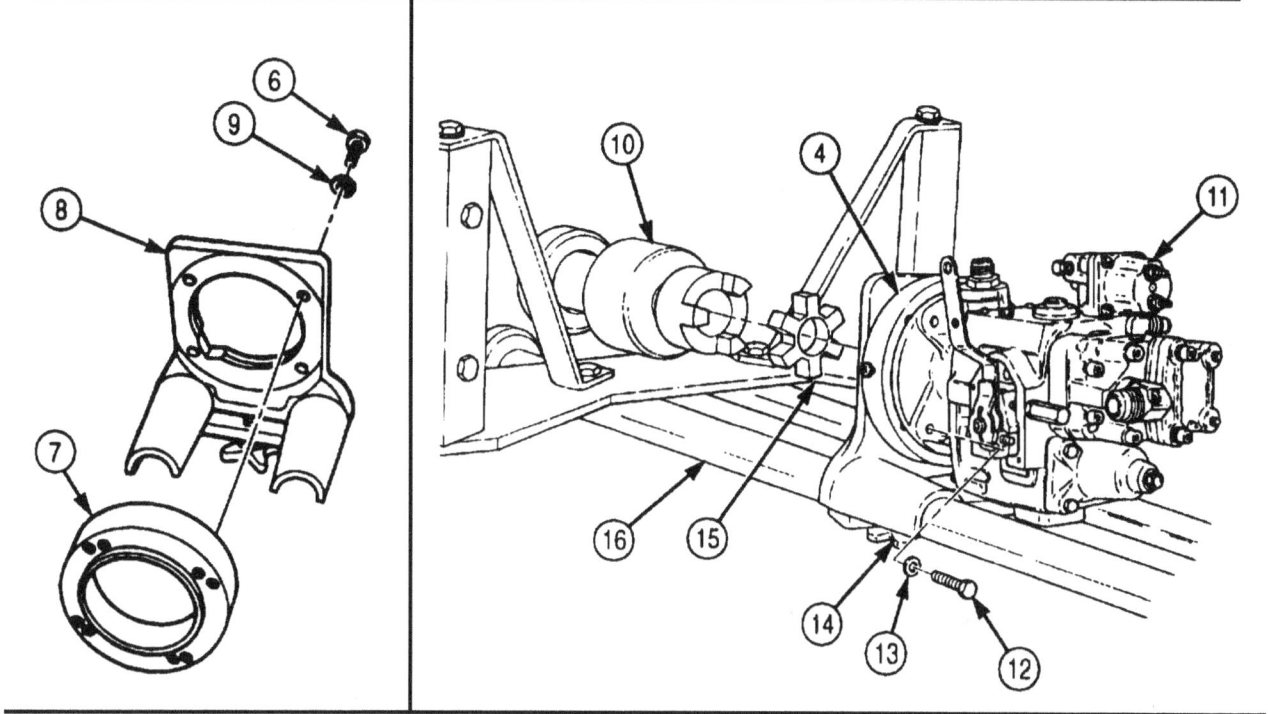

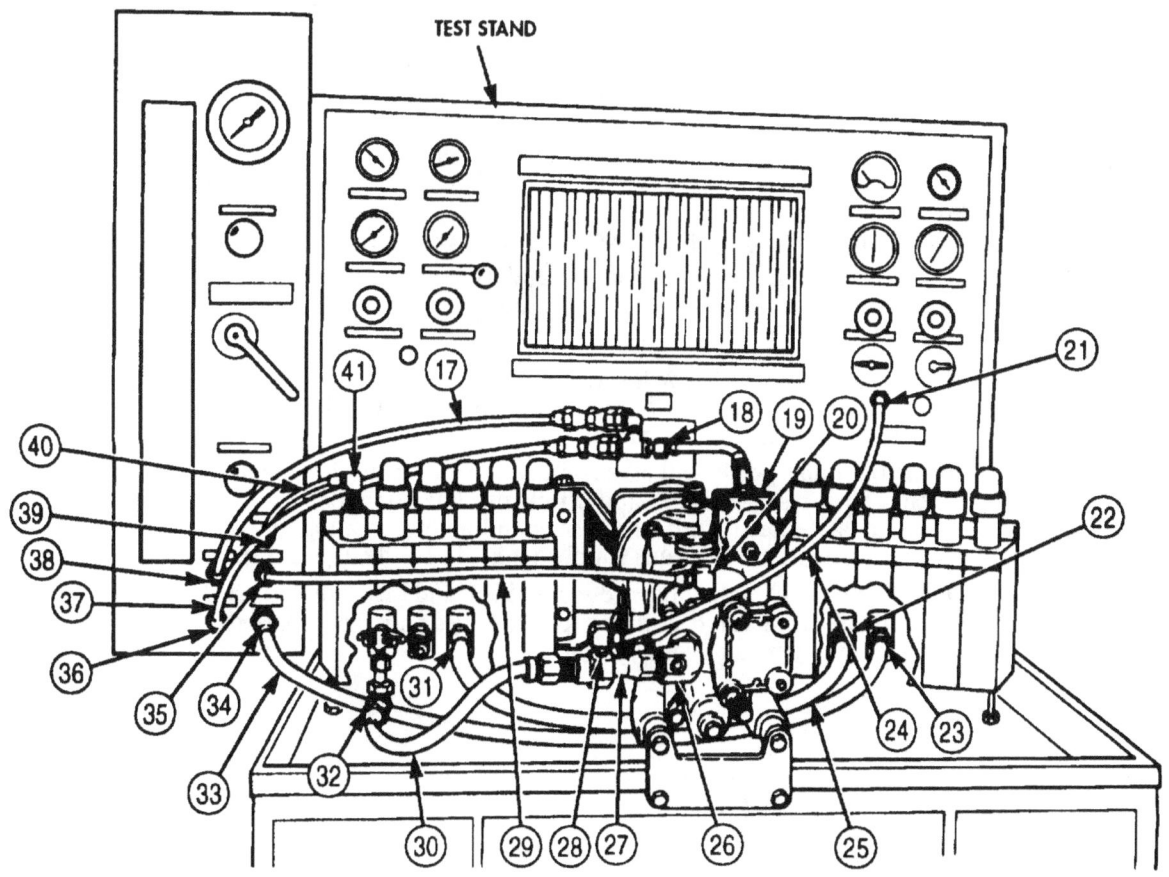

TEST STAND

5-18. FUEL PUMP SETUP AND CALIBRATION (Contd)

c. Fuel Pump Run-In

NOTE

Ensure all other test stand valves are in the CLOSED position to prevent leakage. Seat all other test stand valves by opening one quarter turn and reclosing.

1. Place test stand (1), bypass valve (10), fuel pressure valve (9), and flow control valve (12) in open position.

2. Open fuel pump (3) by turning fuel shutoff valve manual override knob (13) until seated.

CAUTION

Check tachometer drive for clockwise rotation. If rotation is not clockwise, reverse rotation of drive coupling.

3. Apply lubricating oil to tachometer drive seal (2).

4. Place test stand power switch (8) in ON position.

5. Place test stand fuel heat switch (4) in ON position. Observe that fuel temperature gauge reads between 90°-100°F (32°- 38°C) for diesel fuel.

6. Place selector valve (11) in ROTAMETER position.

7. Turn range crank (5) to HIGH range position.

NOTE

Perform steps 8 and 9 only on Variable Speed (VS) governor.

8. Back high-adjusting screw (14), low-adjusting screw (15), and throttle lever (16) out of VS governor (17) four turns.

9. Fasten VS governor throttle lever (17) in full fuel position.

CAUTION

Pump must pick up fuel at 500 rpm without priming. If no fluid pickup is indicated at ROTAMETER, check fuel filter for improper installation, motor switch for correct rotation, open suction valve and hose, and that gear pump connections are tight.

10. Start test stand (1) by depressing start button (8) and run up to 500 rpm. To maintain 500 rpm, depress and release FAST (7) rotate SLOW (6) button.

NOTE

- Check ROTAMETER for air in fuel flow. If air bubbles are present, work pump throttle from fuel full-open to idle several times to relieve entrapped air in pump.

- If air bubbling persists, it is an indication of an air leak in the system. Turn test stand off and check the line for loose connections between tank and test stand pump, mating of gear pump housing, and full fuel supply tank (TM 9-4910-571-12).

- If pump is new or has been disassembled and reassembled, run pump at 500 rpm for 5 minutes to allow bearings and seals to seat and to purge air from system.

5-18. FUEL PUMP SETUP AND CALIBRATION (Contd)

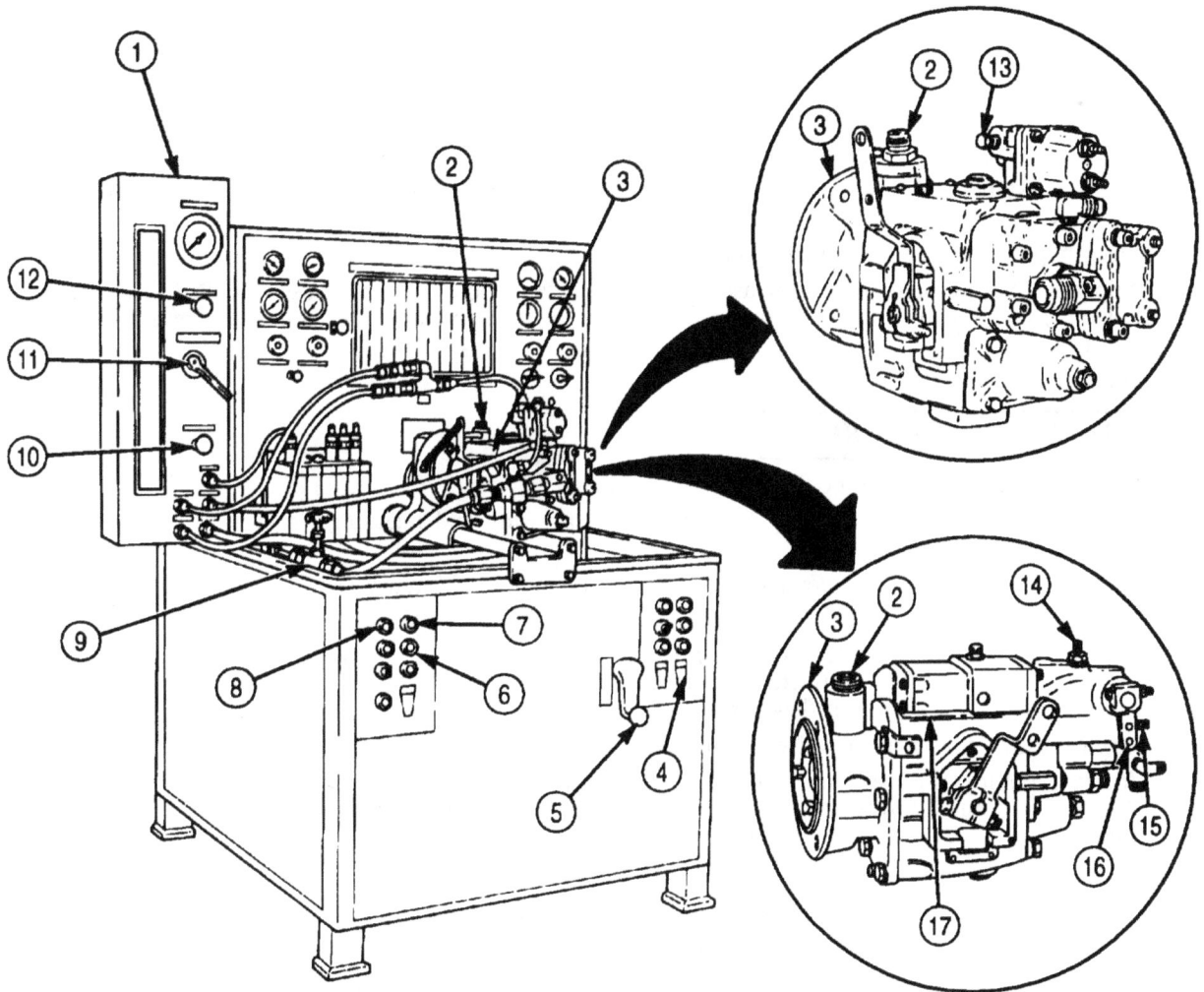

5-18. FUEL PUMP SETUP AND CALIBRATION (Contd)

d. Testing Pump Seals for Leaks

CAUTION

- Check tachometer drive for clockwise rotation. If rotation is not clockwise, reverse rotation of drive coupling.
- Do not leave fuel pressure valve closed more than five minutes because pump could overheat and be damaged.

NOTE

All steps must be completed before pump is considered calibrated.

1. With test stand (2) operating at 500 rpm, close fuel pressure valve (7).
2. Place test stand fuel flow control valve (1) to OPEN position. If 25 in. vacuum is not obtained, check all hose connections.
3. Place bypass valve (8) in closed position.
4. Apply a small amount of GM grease over vent of weep hole (12). If grease is pulled into weep hole (12) at 25 in. vacuum, fuel pump (10) oil seal is defective and fuel pump (10) must be replaced (para. 4-35).
5. Open fuel pressure valve (7).

e. Testing Governor Cutoff RPM

1. Open test stand fuel flow control valve (1) completely.
2. Increase fuel pump (10) speed to 2,100 rpm by depressing and releasing FAST (5) or SLOW (6) button.
3. Adjust test stand fuel flow control valve (1) to obtain 8 in. Mercury (HG) on vacuum gauge (4).

NOTE

Once 8 in. HG vacuum setting is obtained, do not change setting. Readings will fluctuate during other tests. Just note the increases or decreases as they occur.

4. Open the fuel flow control valve (1) and place the selector valve (9) in ROTAMETER position.
5. Increase pump (10) speed until the fuel pressure drops. Stand tachometer (3) reading should be 2,130-2,150 rpm. Depress and release FAST (5) or SLOW (6) button to increase or decrease rpm. The VALVES governor automotive governor portion is set 100 rpm higher.

CAUTION

Test stand must be shut off to change shims in spring pack.

NOTE

- Perform steps 6 and 7 only if spring pack shims are changed.
- Each .001 in. (0.25 mm) shim thickness will change speed approximately two rpm.
- Shims are available in 0.005, 0.010, and 0.020 in. (0.13, 0.25, and 0.51 mm) thicknesses.

6. If cutoff is too low, remove spring pack (11) and add shims. If cutoff is too high, remove shims (para. 5-14).
7. With fuel pump (10) at 500 rpm, move throttle lever (13) back and forth until ROTAMETER shows no air and recheck governor cutoff rpm, steps 1 through 6.

5-18. FUEL PUMP SETUP AND CALIBRATION (Contd)

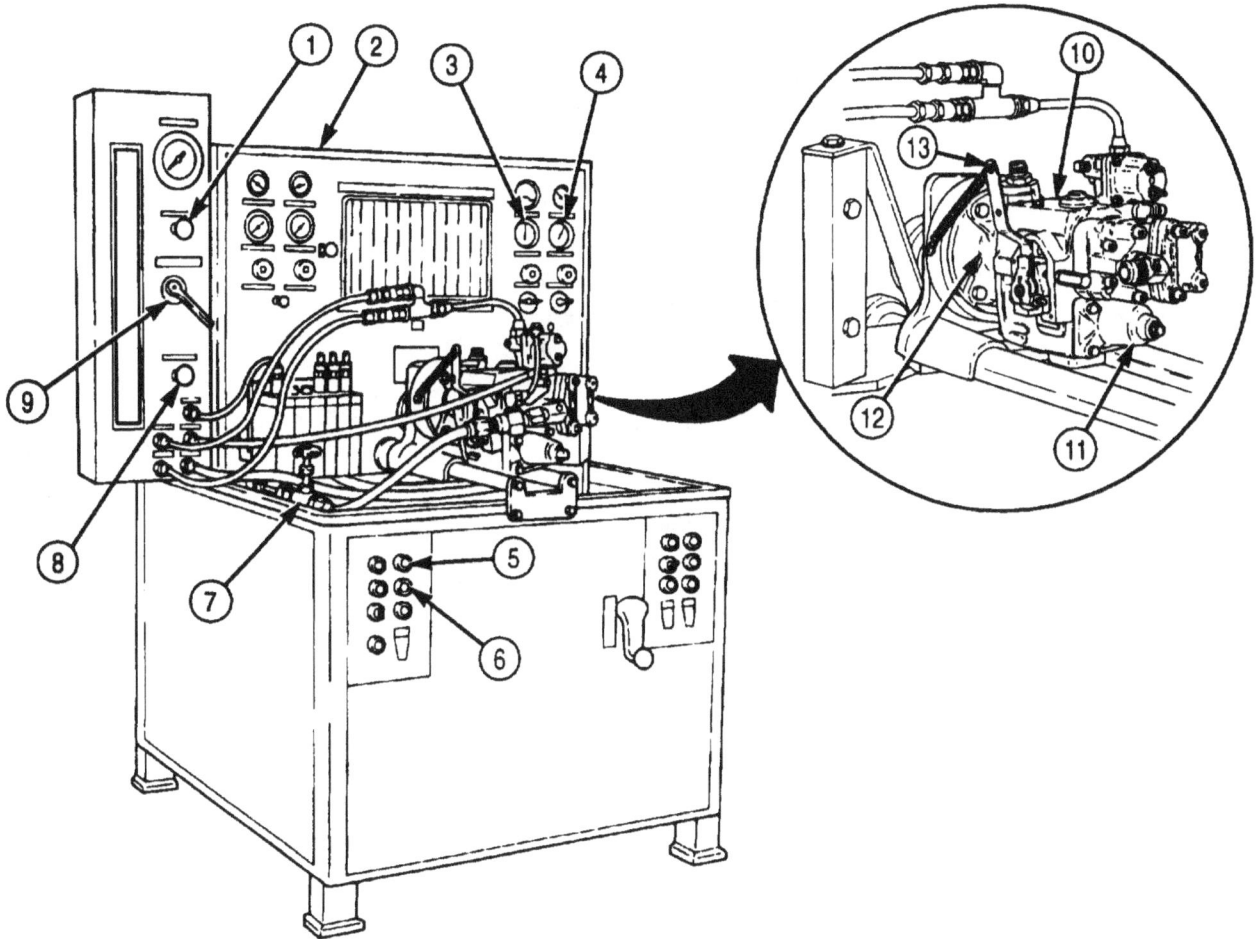

5-18. FUEL PUMP SETUP AND CALIBRATION (Contd)

f. Testing and Adjusting Throttle Leakage

1. Increase fuel pump (9) speed to 2,100 rpm with fuel pump throttle lever (13) and place selector valve (7) to ROTAMETER position to see if any air is in system.

NOTE

At 2,100 rpm fuel flow, ROTAMETER FLOAT should read 315 pph on the scale.

2. Set fuel flow control valve (8) for 315 pph reading.
3. Place selector valve (7) to LEAKAGE TEST position and place count selector switch (6) to the 1,000 position.
4. Pull out dumping lever (2) to retain fuel in No. 1 burette (3).

CAUTION

Do not hold throttle lever in idle position any longer than two minutes to complete test. Pump may overheat, since fuel flow is used to cool the pump.

5. Remove throttle spring (4) and manually position throttle lever (14) to idle position.
6. Depress pulse counter button (5) to fill No. 1 burette (3).
7. Push dumping lever (2) inward.

NOTE

- Ensure burette is cleared of fuel in order to prevent overflow at this time.
- A test cycle is one-half minute duration.
- For one half minute cycle, fuel delivery is 40-70 cc.

8. At the end of a cycle, read the amount of fuel in number one burette (3) on the scale.
9. If throttle leakage is not as specified, back rear throttle screw (12) out to decrease leakage and in to increase leakage.

g. Testing and Adjusting Idle Speed

1. Place selector valve (7) to IDLE position and increase fuel pump (9) rpm to 500.
2. Pull fuel pump throttle lever (13) to idle position.
3. If fuel pressure is not 26 psi (179.27 kPa), adjust idle screw (10) in governor spring pack housing (11) using adjusting tool.
4. If pressure is low and the adjusting screw bottoms, stop the test stand, add shims to the spring end of the adjusting screw (para. 5-14), and retest cutoff rpm and throttle leakage, tasks e. and f.

NOTE

Each time governor spring pack housing or adjusting tool is removed, run pump until purged of air.

5. After proper adjustment is made, stop the test stand (1) and remove adjusting tool.
6. Install 1/8 in. (3.17 mm) pipe plug (14) removed in steps 17 and 18, task b.
7. Purge the fuel pump (9) of air.

5-18. FUEL PUMP SETUP AND CALIBRATION (Contd)

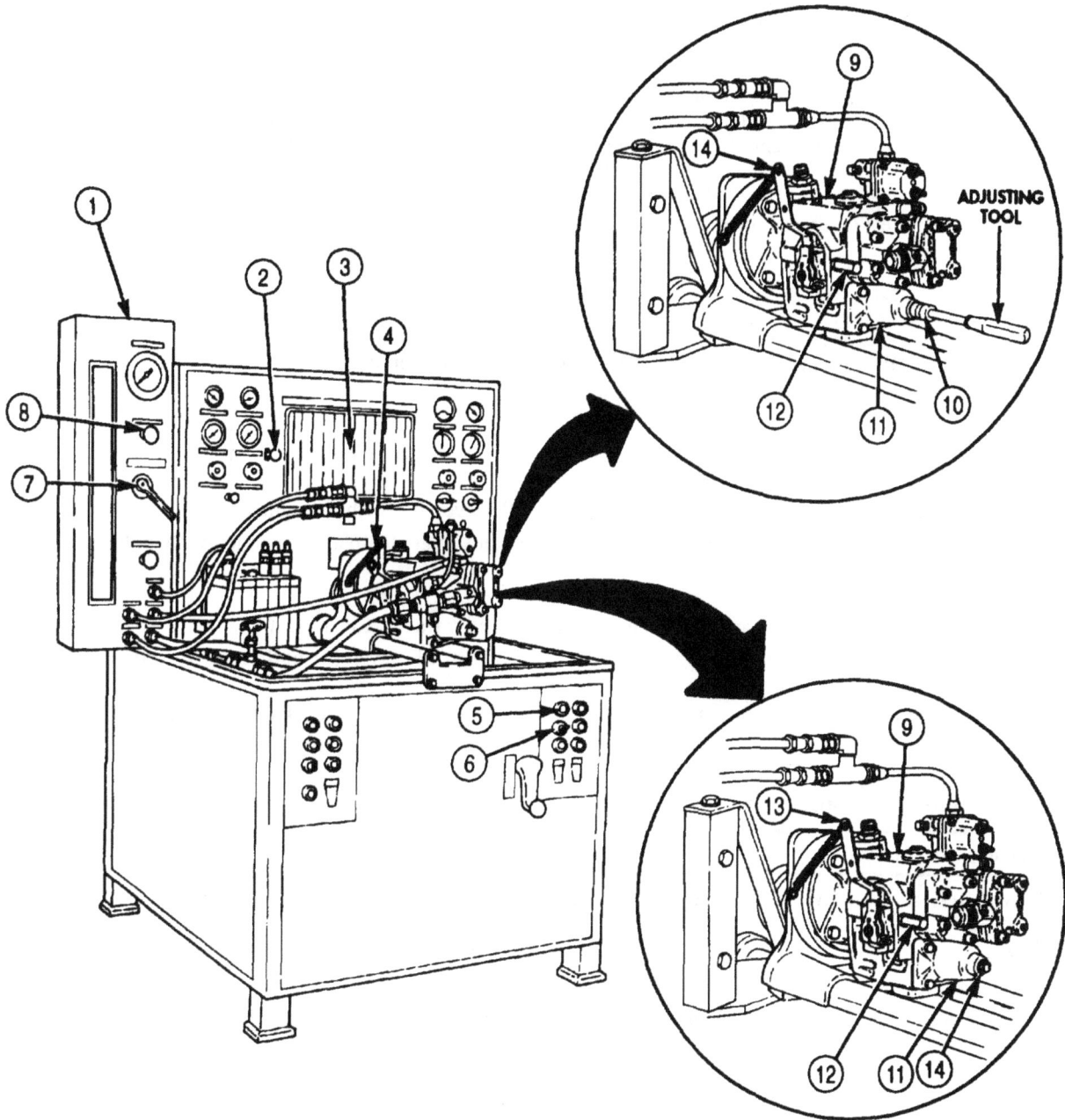

ADJUSTING TOOL

5-18. FUEL PUMP SETUP AND CALIBRATION (Contd)

h. Checking and Adjusting Throttle Lever Travel

NOTE

Travel template or indicator, level, and angle will be used to set pump throttle lever for travel adjustment. Ensure the combination of the first and third or second and fourth holes on the template are used. Any other combination will result in an inaccurate reading. Correct travel is 27-29.

1. Place template (7) against throttle housing (8) so inside flats are even on top and bottom.

CAUTION

DO NOT adjust rear throttle screw from valve set under throttle linkage. The rear throttle screw has already been set to provide the proper deceleration time for the engine, and any changes at this point will require recalibration of throttle linkage.

2. Move the throttle lever (6) to idle position.

NOTE

The throttle lever may be repositioned on shaft as required to line up the lever and template holes.

3. Line up template (7) idle hole and center of throttle lever (6) with straightedge.

4. Move the throttle lever (6) to full throttle position and align template (7) holes with hole in throttle lever (6).

NOTE

Steps 5 through 8 check throttle lever travel adjustment.

5. Place level and angle indicator against the bottom of throttle lever (6) and move throttle lever (6) to idle position. Note reading on scale of angle indicator.

6. Move throttle lever (6) to full throttle position. Note reading on scale of angle indicator.

7. Add readings taken in steps 5 and 6. If throttle lever (6) travel is not 27-29, adjust front throttle step screw (9) to obtain correct ravel.

8. After proper adjustment, the throttle lever (6) may be repositioned to accommodate throttle linkage.

i. Testing and Adjusting Pump Main Pressure

1. With test stand (2) vacuum set at 8 in. HG on vacuum gauge (5) and throttle wide open, adjust speed to 2,100 rpm.

2. Place selector valve (4) to ROTAMETER position.

3. Set fuel flow to 3.15 pph with fuel flow control valve (1).

4. If fuel pressure reading on pressure gauge (3) is not 172-178 psi (1186-1227 kPa), adjust fuel pressure.

5-18. FUEL PUMP SETUP AND CALIBRATION (Contd)

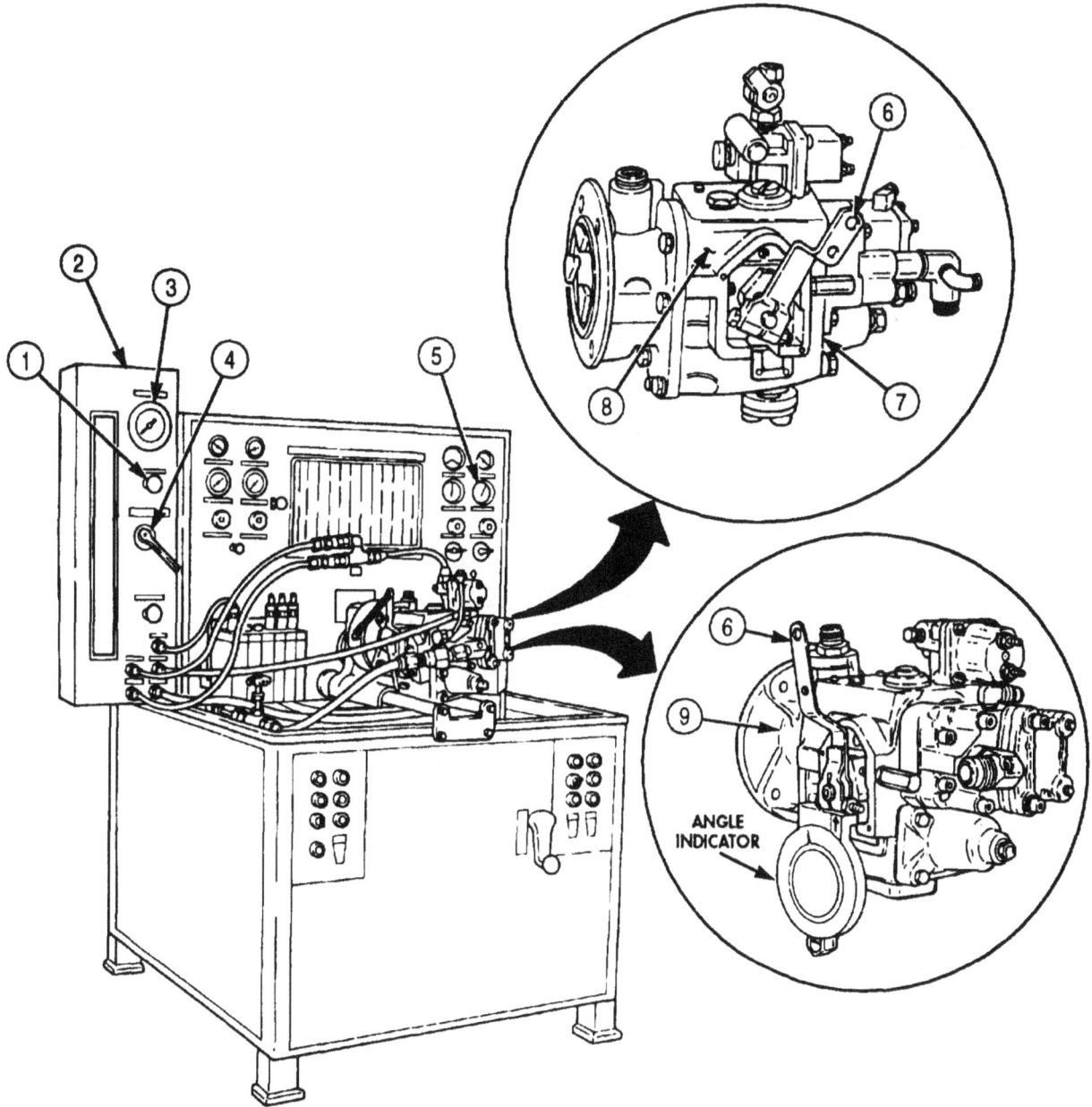

ANGLE INDICATOR

5-18. FUEL PUMP SETUP AND CALIBRATION (Contd)

j. Testing and Adjusting Fuel Pressure

NOTE

Throttle shaft internal adjusting screw is covered by a throttle shaft ball in the end of throttle shaft.

CAUTION

Be careful not to damage bore of throttle shaft when drilling out ball.

1. Center punch and drill ball (6) out of fuel pump throttle shaft (7) with 1/4 in. drill bit.
2. Screw internal fuel adjusting screw located in fuel pump throttle shaft (7) inward to increase fuel pressure and outward to decrease fuel pressure until it is set to 172-178 psi (1186-1227 kPa).
3. After fuel pressure is adjusted, insert new throttle shaft ball (6) with throttle shaft installation tool.

k. Testing and Adjusting Governor Fuel Pressure

1. Adjust fuel pump (3) speed to 1,500 rpm.
2. Place selector valve (4) on test stand (2) in ROTAMETER position.
3. Place fuel throttle lever (5) to wide-open position.
4. Set fuel flow to 2.30 pph with the fuel control valve (1).
5. If fuel pressure is not 100-106 psi (689-730 kPa), check governor cutoff rpm (task e).

l. Checking and Adjusting Governor Weight Pressure

1. Adjust fuel pump (3) speed to 1,000 rpm.
2. Place throttle lever (5) to wide-open position.
3. Place selector valve (4) in ROTAMETER position.
4. Set fuel flow to 150 pph with fuel flow control valve (1).

NOTE

Shims are available in 0.005 and 0.010 in. (0.13 and 0.25 mm) thickness. Do not change setting more than 0.020 in. (0.508 mm) from specification.

5. If fuel pressure is greater than 58 psi (399 kPa), decrease pressure by removing shims from behind weight plunger.
6. If fuel pressure is less than 50 psi (335 kPa), increase pressure by adding shims (para. 5-17).

5-18. FUEL PUMP SETUP AND CALIBRATION (Contd)

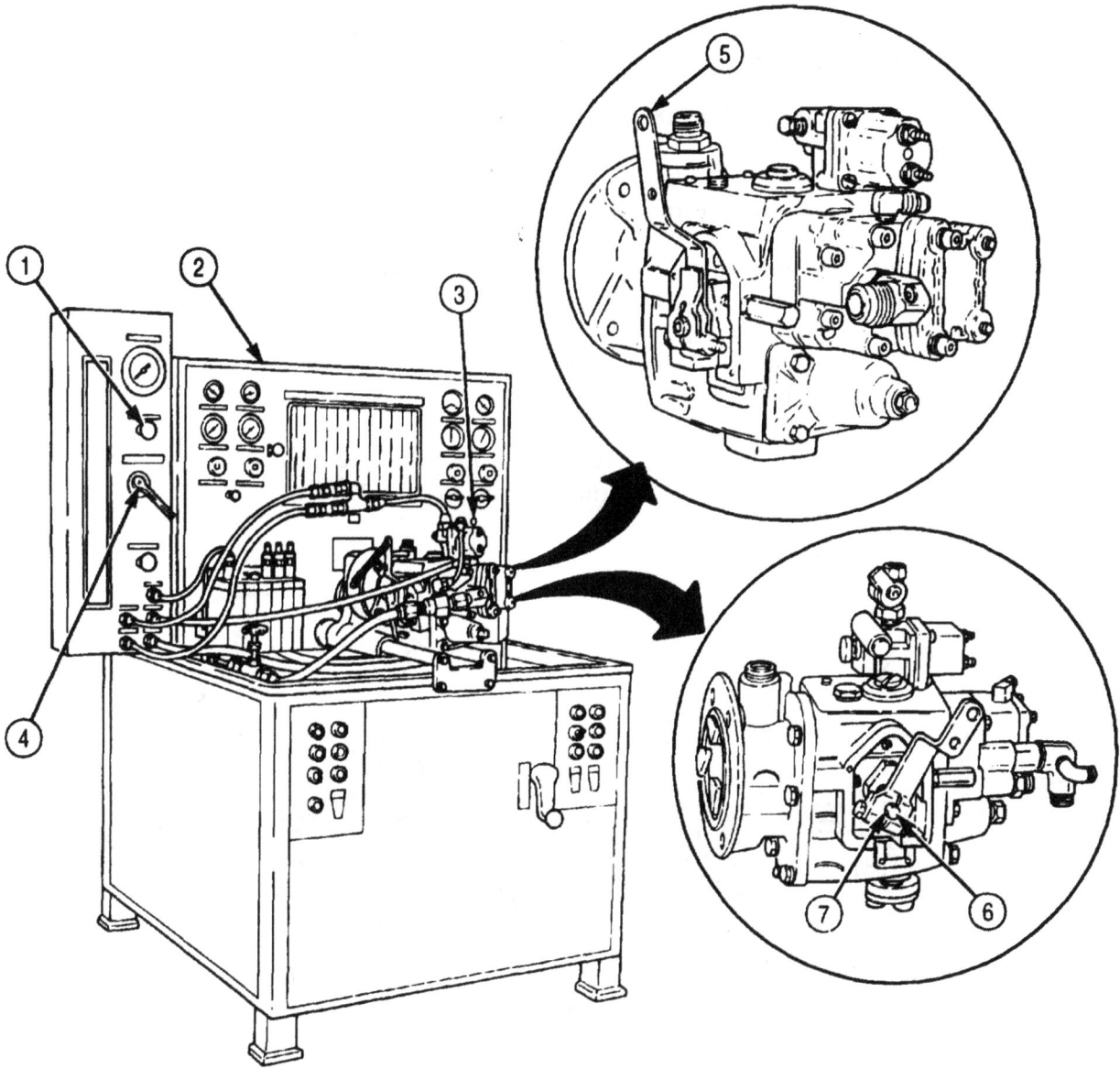

5-18. FUEL PUMP SETUP AND CALIBRATION (Contd)

m. Testing and Adjusting Idle Speed (VS Governor Only)

NOTE

To calibrate fuel pump equipped with Variable Speed (VS)
governor, the preceding steps are followed. In addition, task c,
steps 8 and 9 are to adjust VS governor only.

1. On test stand (2), place throttle lever (8) of fuel pump with VS governor (3) and VS governor lever (7) to full-fuel position.
2. Increase fuel pump (3) speed to 2,140 rpm.
3. Loosen locknut (5), turn HI-IDLE (top) screw (4) in until fuel pressure starts to drop. Tighten locknut (5).
4. Decrease fuel pump (3) speed to 2,100 rpm, and gradually increase until pressure starts to drop.
5. If fuel pump (3) speed is not 2,120-2,140 rpm, adjust VS governor HI-IDLE screw (4) until rpm is 2,120-2,140 rpm.

NOTE

The automotive governor cutoff must be set 100 rpm higher.

6. Decrease fuel pump (3) speed to 500 rpm.
7. Place VS governor lever (7) and stand selector valve (1) to IDLE position.
8. Adjust governor LO-IDLE screw (bottom) (6) to obtain 26 psi (179 kPa).

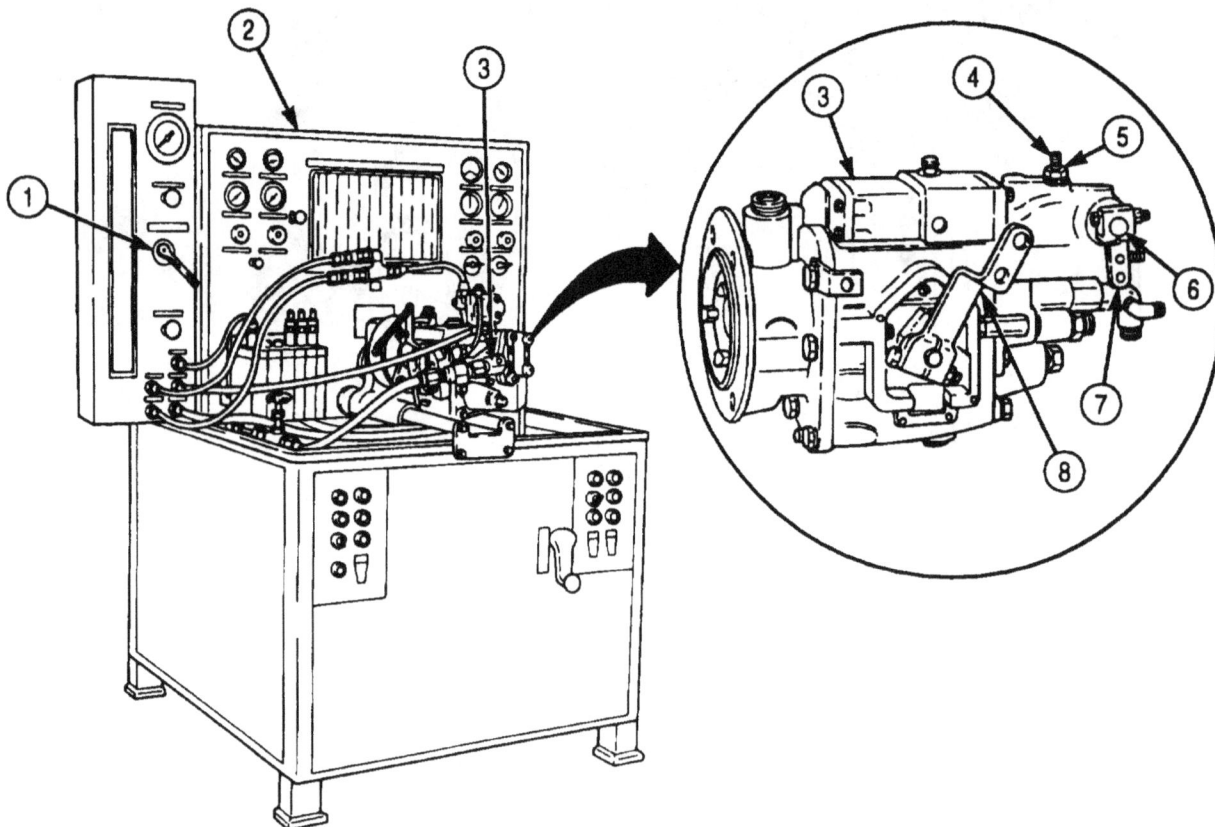

5-18. FUEL PUMP SETUP AND CALIBRATION (Contd)

n. Shutdown and Removal from Test Stand

1. Depress test stand stop button (11).
2. Place test stand fuel heat switch (10) in OFF position.
3. Close test stand fuel pressure control valve (12) and fuel flow control valve (17).
4. Remove 1/4 in. (6.35 mm) flexible hose (13) from small fitting (9) on fuel pump (3) and auxiliary return fitting (16) on test stand (2).
5. Remove leakage accumulator hose (15) from No. 1 accumulator can (14) and stand leak test connector (17).

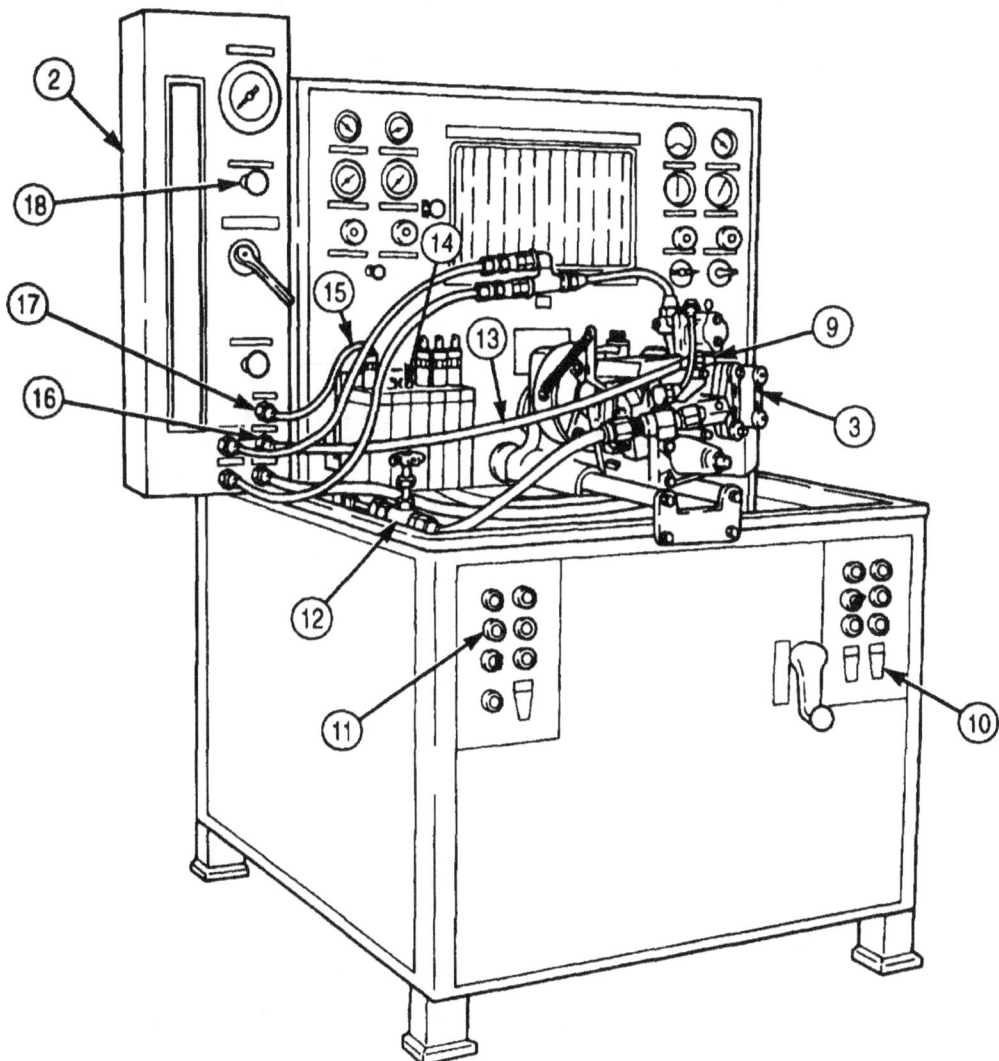

5-18. FUEL PUMP SETUP AND CALIBRATION (Contd)

6. Remove fuel outlet hose (10) from test stand fuel input connector (11) and fuel return connector (5).

7. Remove 1/2 in. (12.7 mm) flexible hose (8) from test stand lube pressure (9) and test stand lube return (6).

8. Remove fuel pressure hose (14) from test stand pressure gauge outlet (13) and pump discharge fitting (2).

9. Remove pump discharge fitting assembly (3) from pump fuel shutoff valve (4) and test stand fuel input connector (12).

10. Remove throttle lever spring (15) from fuel pump (7).

11. Remove 1/4 in. (6.35 mm) manifold hose (17) from test stand manifold vacuum gauge (16) and 1/4 in. (6.35 mm) adapter on inlet adapter assembly (18).

12. Remove 1/2 in. (12.7 mm) inner diameter flexible hose (22) from fuel pressure control valve (23) and adapter elbow on inlet adapter assembly (21).

13. Remove pump inlet adapter assembly (20) from pump inlet port (19).

14. Remove pump coupling insert (35) from test stand drive coupling (30).

15. Loosen bar clamp (34) and slide fuel pump (7) and ring and adapter bracket (31) back from test stand drive coupling (30).

16. Remove fuel pump (7) from ring and adapter bracket (31) by removing four screws (32) and washers (33).

17. Remove nut (26), lockwasher (27), washer (28), screw (29), and throttle lever (25) from throttle lever shaft (24).

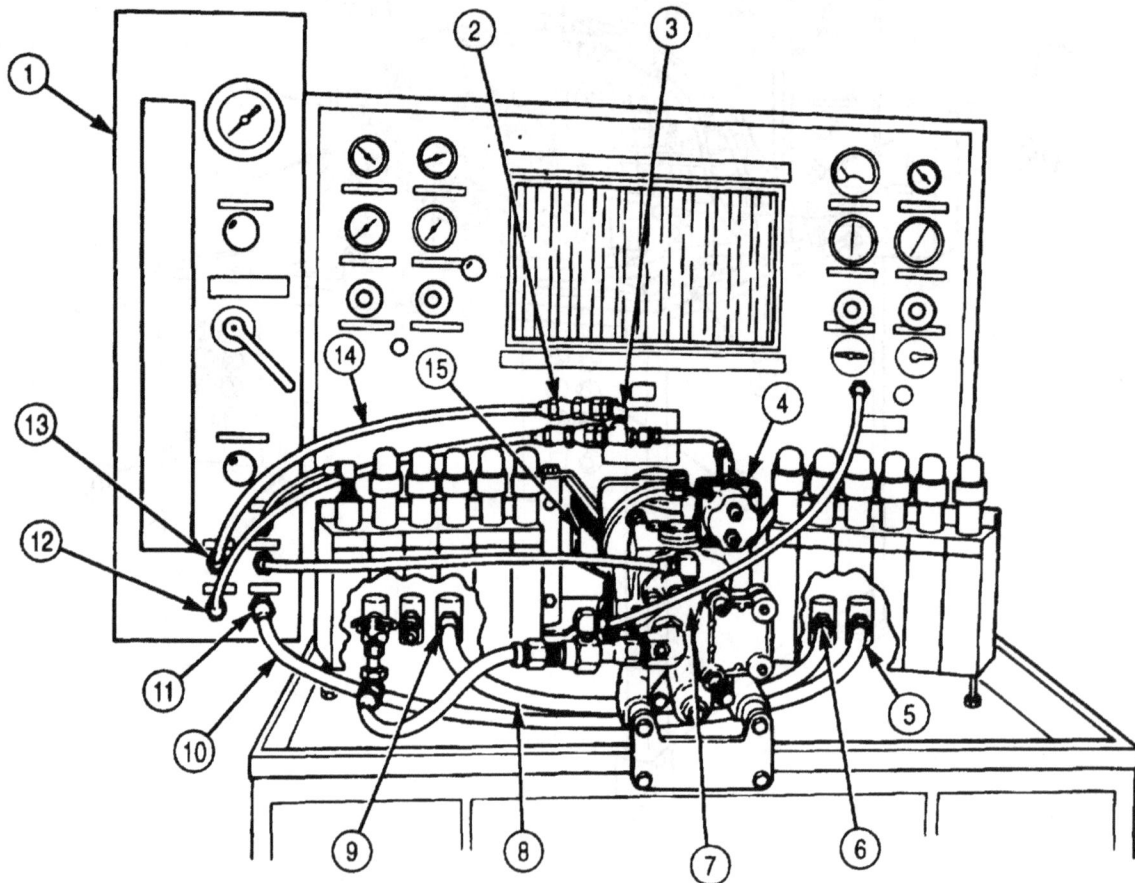

5-18. FUEL PUMP SETUP AND CALIBRATION (Contd)

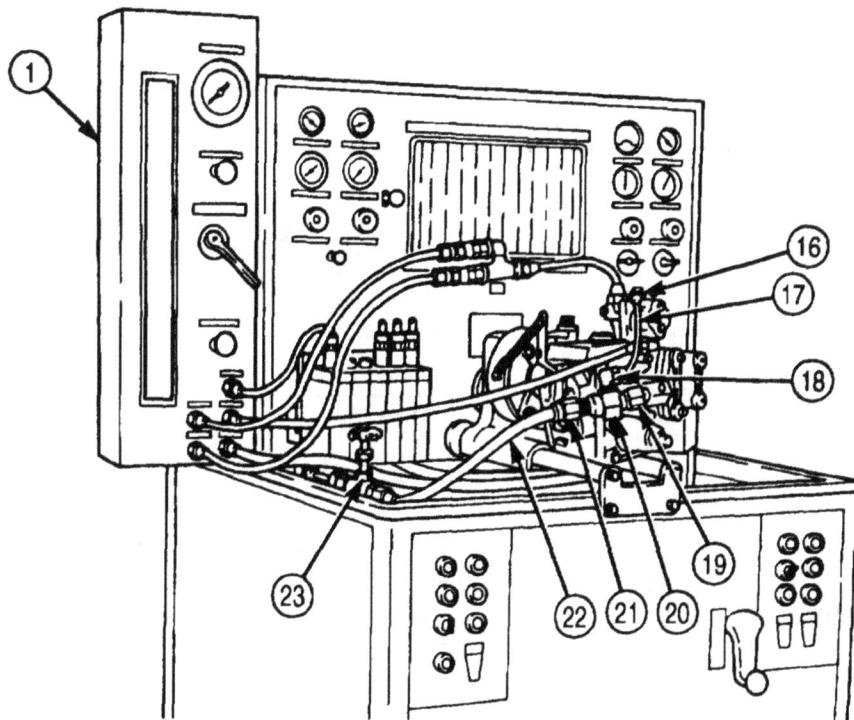

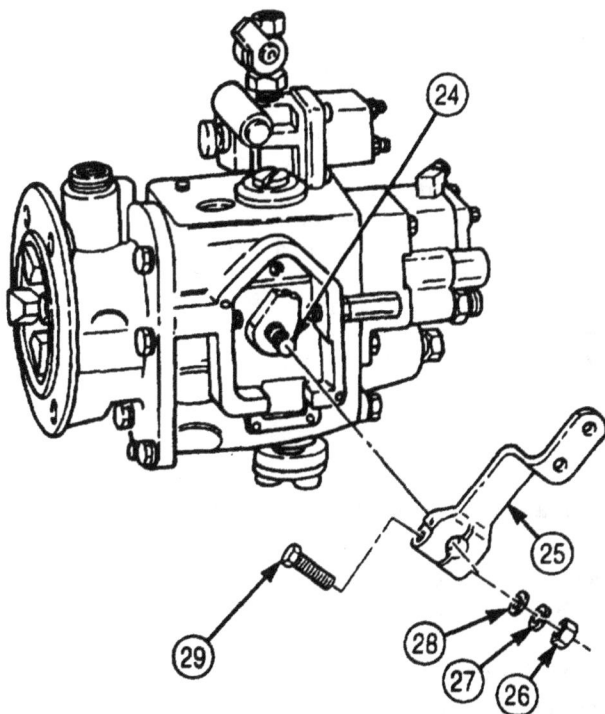

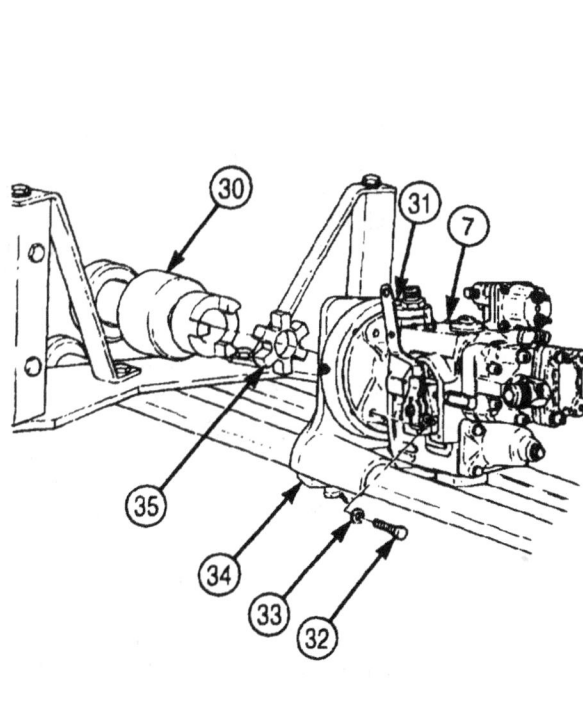

FOLLOW-ON TASKS: • Install throttle shaft cover (para. 5-16).
• Install fuel pump shutoff valve (para. 4-36 or 4-37).
• Install fuel pump on vehicle (para. 4-35).

5-19. FUEL INJECTOR REPAIR

THIS TASK COVERS:

a. Non-Top Stop Injector Disassembly
b. Top Stop Injector Disassembly
c. Inspection

d. Non-Top Stop Injector Assembly
e. Top Stop Injector Assembly

INITIAL SETUP:

APPLICABLE MODELS
M939/A1

SPECIAL TOOLS
Cup retainer wrench (Appendix E, Item 162)
Crowfoot injector wrench-(Appendix E, Item 164)
Adjusting wrench (Appendix E, Item 3)
Locknut wrench (Appendix E, Item 165)
Injector body wrench (Appendix E, Item 70)

TOOLS
General mechanic's tool kit (Appendix E, Item 1)
Torque wrench (Appendix E, Item 144)
Magnifying glass

MATERIALS/PARTS
Injector overhaul kit (Appendix D, Item 255)
Diesel fuel (Appendix C, Item 42)
Drycleaning solvent (Appendix C, Item 71)

REFERENCES (TM)
TM 9-2320-272-24P

EQUIPMENT CONDITION
Fuel injectors removed (para. 4-32).

GENERAL SAFETY INSTRUCTIONS
• Keep fire extinguisher nearby when using drycleaning solvent.
• Drycleaning solvent is flammable and toxic. Do not use near an open flame.

---

a. Non-Top Stop Injector Disassembly

1.   Remove injector link (1) from injector (8). Set injector link (1) aside to prevent damage.

WARNING

Drycleaning solvent is flammable and toxic. Do not use near open flame and always have a fire extinguisher nearby when solvents are used. Use only in well-ventilated places, wear protective clothing, and dispose of cleaning rags in approved container. Failure to do this may result in injury or death to personnel and/or damage to equipment.

CAUTION

Improper cleaning methods and use of unauthorized cleaning solvents can damage equipment.

2.   Clean exterior of injector (8) with drycleaning solvent (para. 2-14).

CAUTION

• Injector barrel and plunger are a matched pair (class fit). Do not interchange.
• Do not touch internal parts unless hands are clean and moistened with diesel fuel. Failure to do so may result in damage to internal parts.

3.   Remove plunger (2) and spring (3) from injector (8). Store plunger (2) by standing on end.

4.   Remove small O-ring (4) and two large O-rings (5) from injector (8). Discard O-rings (4) and (5).

5.   Remove screen retaining ring (6) and screen (7) from injector (8). Discard retaining ring (6) and screen (7).

5-19. FUEL INJECTOR REPAIR (Contd)

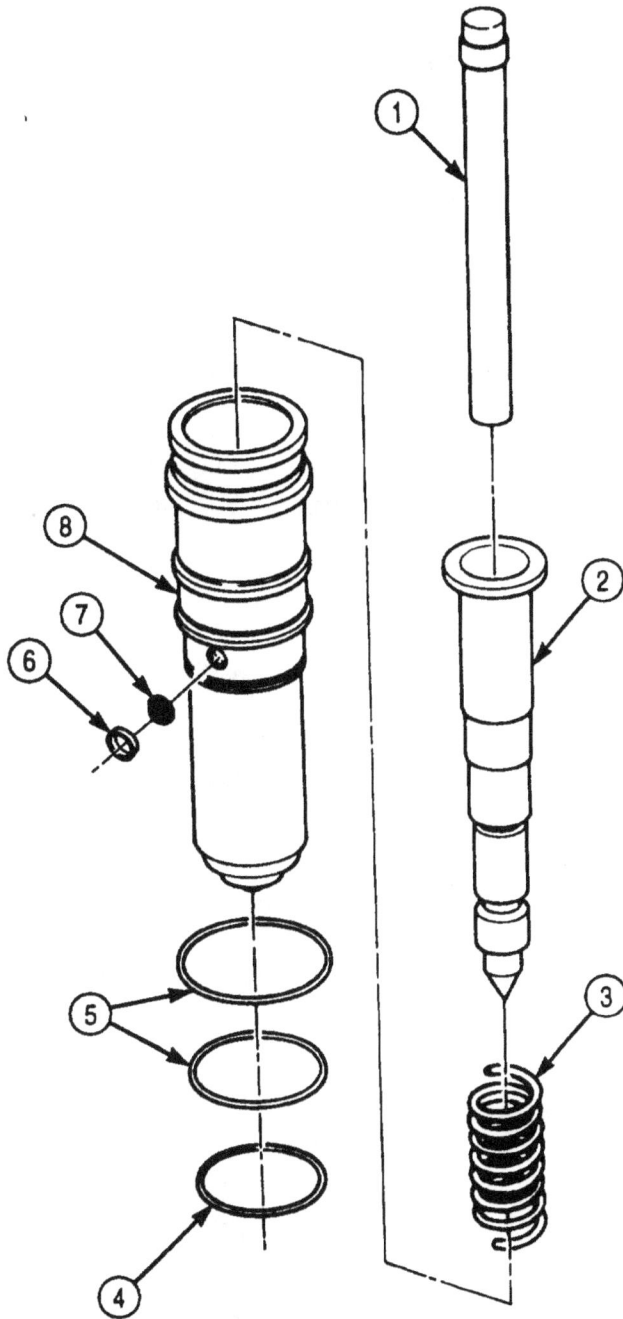

5-19. FUEL INJECTOR REPAIR (Contd)

6. Install injector body wrench in vise.
7. Slide flat machined areas of injector (6) into body wrench.
8. Loosen cup retainer (1) with cup retainer wrench.
9. Remove injector (6) and body wrench from vise.

CAUTION

When handling injector, use care not to drop or lose parts.

10. Set injector adapter (3) upright on flat surface and remove cup retainer (1) by lifting straight up.
11. Remove injector cup (2) from injector (6). Discard injector cup (2).

CAUTION

- Injector barrel and plunger are a matched pair (class fit). Do not interchange.
- Do not touch internal parts unless hands are clean and moistened with diesel fuel. Failure to do so may result in damage to internal parts.

12. Hold injector barrel (5) and adapter (3) together and set injector barrel (5) end upright on clean cloth. Lift adaptor (3) straight out of injector barrel (5) while holding injector barrel (5).
13. Lift injector barrel (5) up and tilt over hand until check ball (4) falls out.

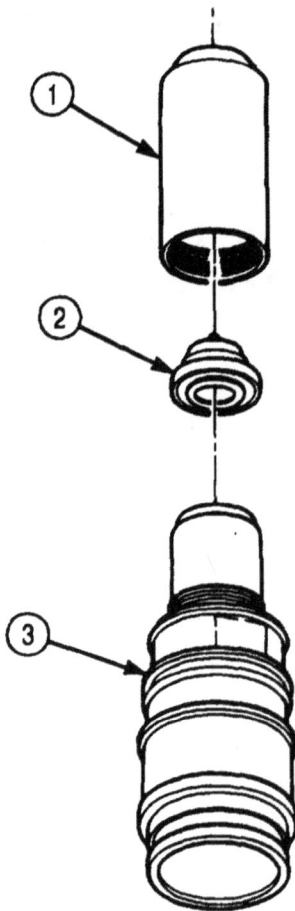

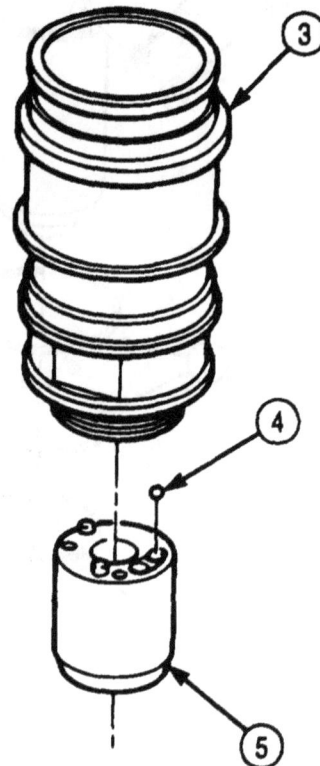

5-19. FUEL INJECTOR REPAIR (Contd)

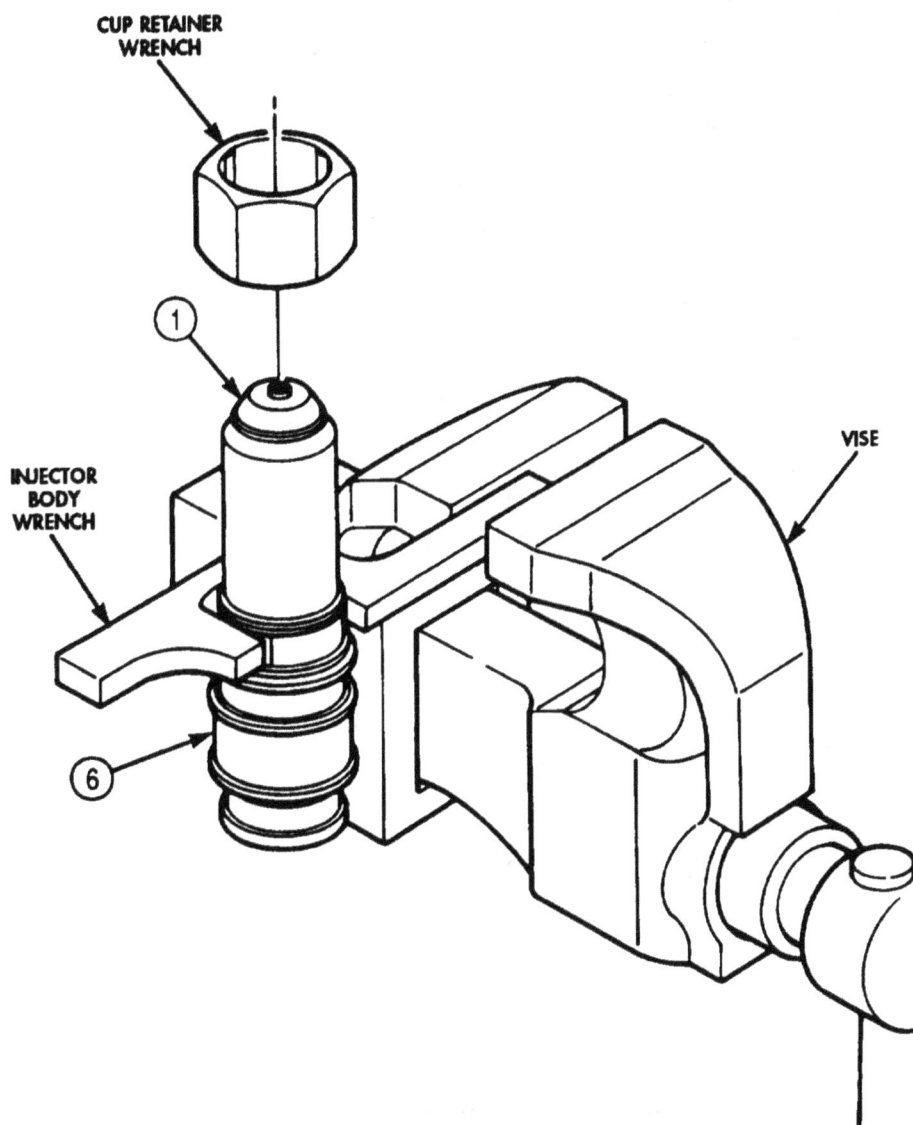

CUP RETAINER
WRENCH

VISE

1

INJECTOR
BODY
WRENCH

6

5-19. FUEL INJECTOR REPAIR (Contd)

b. Top Stop Injector Disassembly

WARNING

Drycleaning solvent is flammable and toxic. Do not use near open flame and always have a fire extinguisher nearby when solvents are used. Use only in well-ventilated places, wear protective clothing, and dispose of cleaning rags in approved container. Failure to do this may result in injury or death to personnel and/or damage to equipment.

CAUTION

- When handling injector, use care not to drop or lose parts.
- Improper cleaning methods and use of unauthorized cleaning solvents can damage equipment.

1. Clean exterior of injector (6) with drycleaning solvent (para. 2-14).
2. Install injector body wrench in vise.
3. Slide flat machined areas of injector (6) into injector body wrench.
4. Loosen locknut (4) with locknut wrench.
5. Loosen adjusting screw (3) with adjusting wrench.
6. Remove adjusting screw (3), locknut (4), plunger (2), spring retainer (1), and plunger spring (7) from injector (6).
7. Loosen cup retainer (5) using cup retainer wrench.
8. Remove injector (6) and body wrench from vise.
9. Set injector adapter (9) upright on flat surface and remove cup retainer (5) by lifting straight up.
10. Remove injector cup (8) from injector (6). Discard injector cup (8).

CAUTION

- Injector barrel and plunger are a matched pair (class fit). Do not interchange.
- Do not touch internal parts unless hands are clean and moistened with diesel fuel. Failure to do so may result in damage to internal parts.

11. Hold injector barrel (11) and adapter (9) together and set injector barrel (11) end upright on clean cloth. Lift adapter (9) straight up out of injector barrel (11) while holding injector barrel (11).
12. Lift injector barrel (11) up and tilt over hand until check ball (10) falls out. Discard check ball (10).
13. Remove small O-ring (13) and two large O-rings (12) from injector (6). Discard O-rings (13) and (12).
14. Remove screen retaining ring (14) and screen (15) from injector (6). Discard screen retaining ring (14) and screen (15).

5-19. FUEL INJECTOR REPAIR (Contd)

5-19. FUEL INJECTOR REPAIR (Contd)

c. Inspection

NOTE
- Bright spots or surface disruption at top of plunger-machined area, on opposite side at bottom or midpoint, are normal results of rocker lever action.
- This task pertains to the non-top stop and top stop injectors.

1. Inspect machined surfaces of plunger (1) for pitting, wear, cracks, and looseness. Narrow streaks running length of plunger (1) are normal. Plunger (1) should be one solid part. If plunger (1) is pitted or worn, or if cracks or looseness exist, replace plunger (1) and injector barrel (4).

2. Inspect injector barrel plunger bore (2) for scoring with strong magnifying glass. If scoring exists, replace injector barrel (4) and plunger (1).

3. Inspect injector barrel (4) surface at each end for burrs or scratches. If burrs or scratches exist, replace injector barrel (4) and plunger (1).

4. Inspect injector barrel check ball seat (3) for nicks or burrs. If nicks or burrs exists, replace injector barrel (4) and plunger (1).

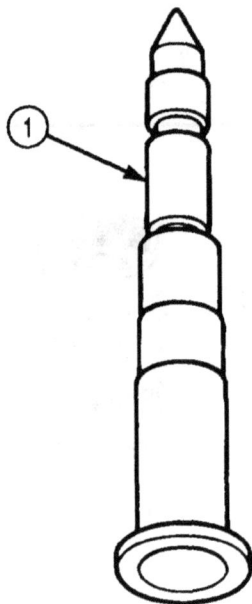

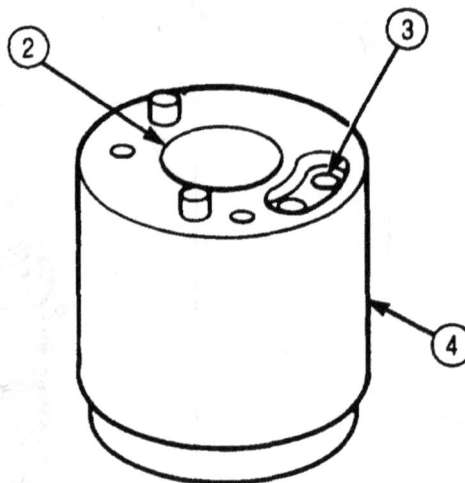

5-19. FUEL INJECTOR REPAIR (Contd)

d. Non-Top Stop Injector Assembly

CAUTION

- Injector barrel and plunger are a matched pair (class fit). Do not interchange.
- Do not touch internal parts unless hands are clean and moistened with diesel fuel. Failure to do so may result in damage to internal parts.

1. Place new check ball (6) in palm of hand and scoop up into injector barrel (4).
2. Place injector barrel (4) flat on clean cloth so mating surface faces upward and align spiral pins (7) with holes in adapter (5)
3. Place adapter (5) on injector barrel (4) and hold together. Set adapter (5) upright on clean cloth.
4. Position new injector cup (9) on top of injector barrel (4) and screw cup retainer (8) onto adapter (5). Finger tighten cup retainer (8), and back off 1/4 turn.

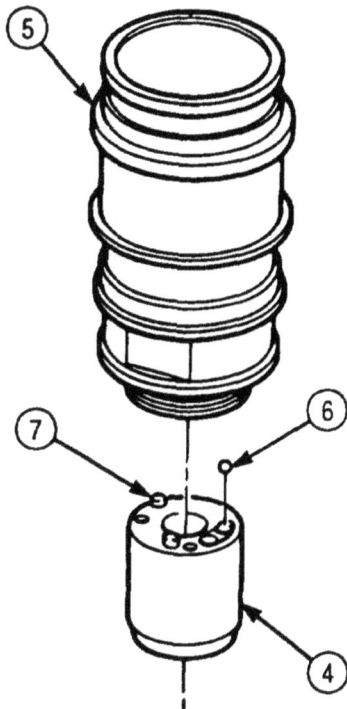

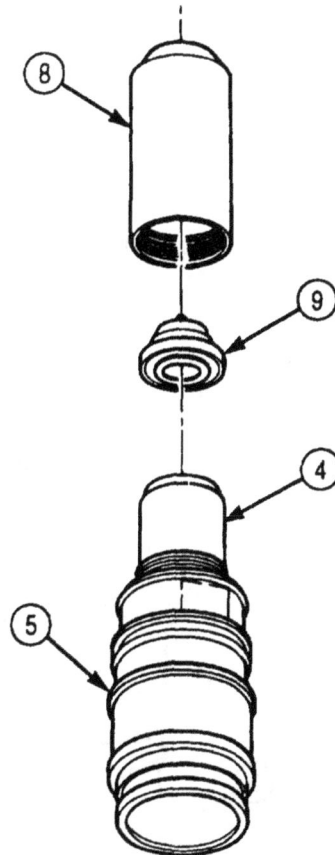

5-19. FUEL INJECTOR REPAIR (Contd)

5. Place injector cup retainer wrench over cup retainer (3).

CAUTION

- Injector barrel and plunger are a matched pair (class fit). Do not interchange.
- Do not touch internal parts unless hands are clean and moistened with diesel fuel. Failure to do so may result in damage to internal parts.

6. Apply clean diesel fuel to coat plunger (2) and insert plunger (2) into adapter (1).
7. Remove loading fixture stud (6) from loading fixture (4) and slide body wrench over injector (5) flats.
8. Position cup retainer wrench on cup retainer (3).
9. Install loading fixture stud (6) in loading fixture (4) and tighten 110 lb-in. (12 N·m).
10. Tighten cup retainer (3) 50 lb-ft (68 N·m) with torque wrench and crowfoot wrench.
11. Remove loading fixture stud (6) from loading fixture (4) and lift injector (5) out of loading fixture (4).
12. Remove plunger (2) from adapter (1).
13. Apply clean diesel fuel to coat plunger (2).
14. Insert plunger (2) into injector (5) barrel so plunger (2) remains 0.5 in. (12.7 mm) from edge of adapter (1).
15. Press plunger (2) into cup retainer (3) with palm of hand.

CAUTION

Do not allow plunger to fall out during step 16.

16. Turn cup retainer (3) and adapter (1) with one hand so cup retainer (3) faces upward. Plunger (2) should slide out immediately. If plunger (2) does not slide out immediately, the injector (5) is not aligned and must be reassembled as outlined in task b., steps 5 through 14.
17. Slide two new large O-rings (10) over end of injector (5) and into two upper grooves (15) on adapter.
18. Slide new small O-ring (9) over end of injector (5) onto lowest groove (13) on adapter (1).
19. Place new inlet fuel screen (11) over adapter orifice (14) and install with new retainer ring (12).
20. Place spring (8) on plunger (2) and slide plunger (2) into injector (5).
21. Install injector link (7) in plunger (2).

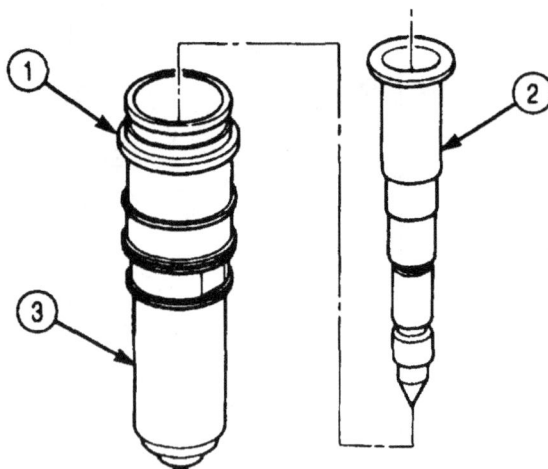

5-19. FUEL INJECTOR REPAIR (Contd)

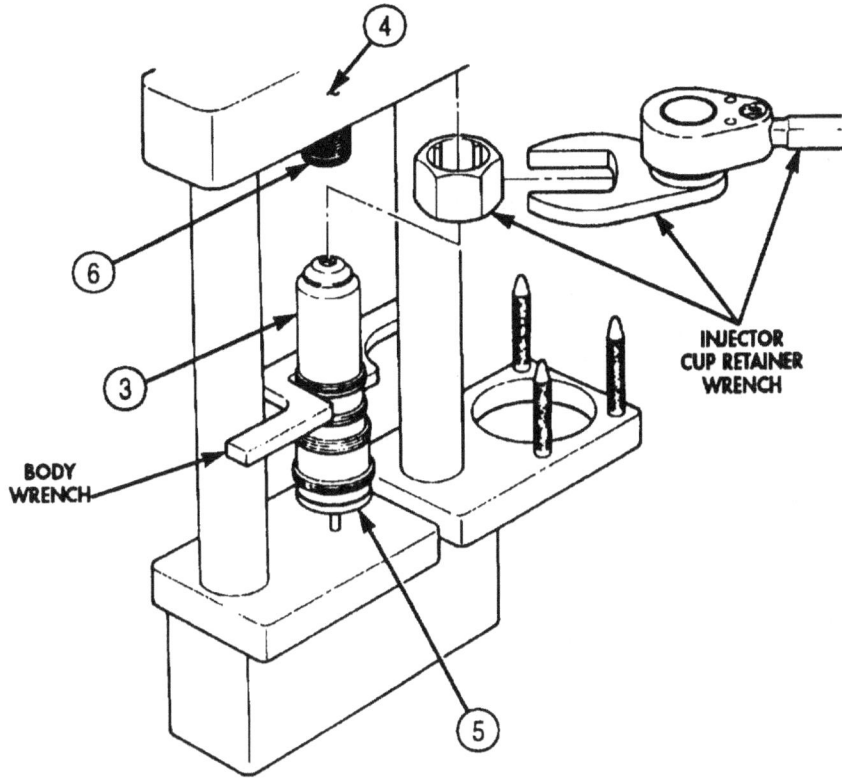

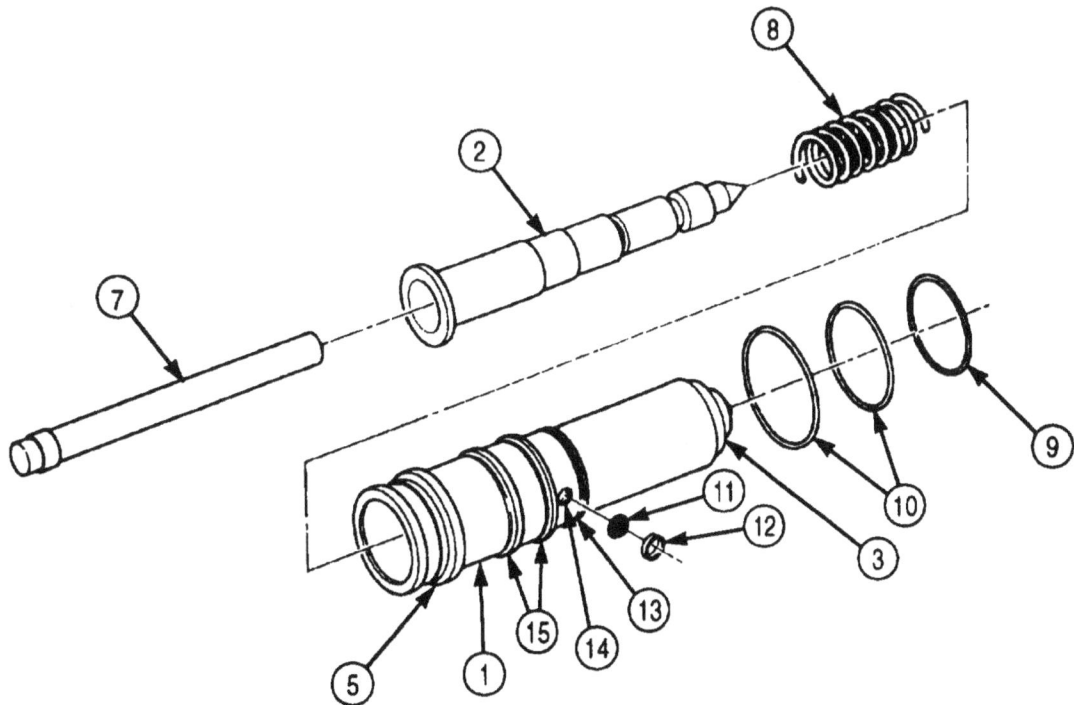

INJECTOR
CUP RETAINER
WRENCH

BODY
WRENCH

5-19. FUEL INJECTOR REPAIR (Contd)

e. Top Stop Injector Assembly

CAUTION

- Lubricate parts only with clean diesel fuel before assembly. Do not use lubricating oil. Oil can crystallize under excessive heat, causing damage to injector components.
- Do not touch internal parts unless hands are clean and moistened with diesel fuel. Failure to do so may result in damage to internal parts.

1. Place new check ball (3) in injector barrel (4).
2. Place injector barrel (4) flat on clean lint-free cloth so mating surface faces upward.
3. Place adapter (2) on injector barrel (4).
4. Hold injector barrel (4) and adapter (2) together and set adapter (2) upright on clean lint-free cloth.
5. Position new injector cup (6) on top of injector barrel (4).
6. Screw cup retainer (5) onto adapter (2), finger tighten, and back off 1/4 turn.

CAUTION

- Injector barrel and plunger are a matched pair (class fit). Do not interchange.
- Do not touch internal parts unless hands are clean and moistened with diesel fuel. Failure to do so may result in damage to internal parts.

7. Apply clean diesel fuel to coat plunger (1).
8. Insert plunger (1) into adapter (2).
9. Remove loading fixture stud (7) from loading fixture (8) and insert injector (9) into loading fixture (8).
10. Slide body wrench over flats on injector adapter (2).
11. Position cup retainer wrench on cup retainer (5).
12. Install loading fixture stud (7) in loading fixture (8) and tighten 75 lb-ft (8m N).
13. Tighten cup retainer (5) 50 lb-ft (68 N·m) with torque wrench and crowfoot wrench.

5-19. FUEL INJECTOR REPAIR (Contd)

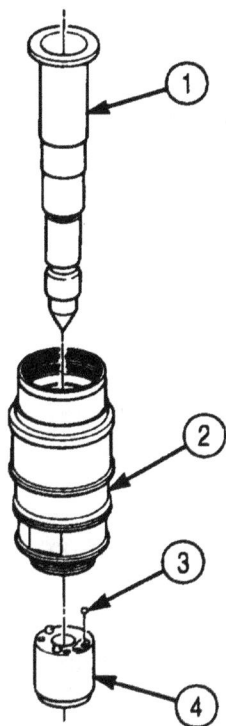

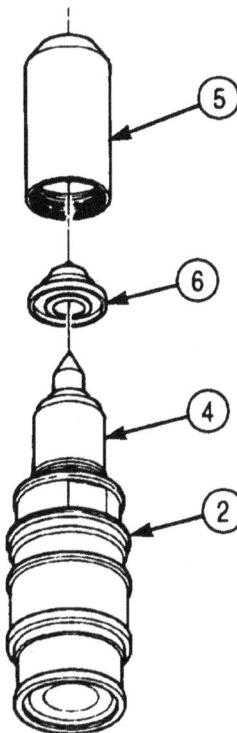

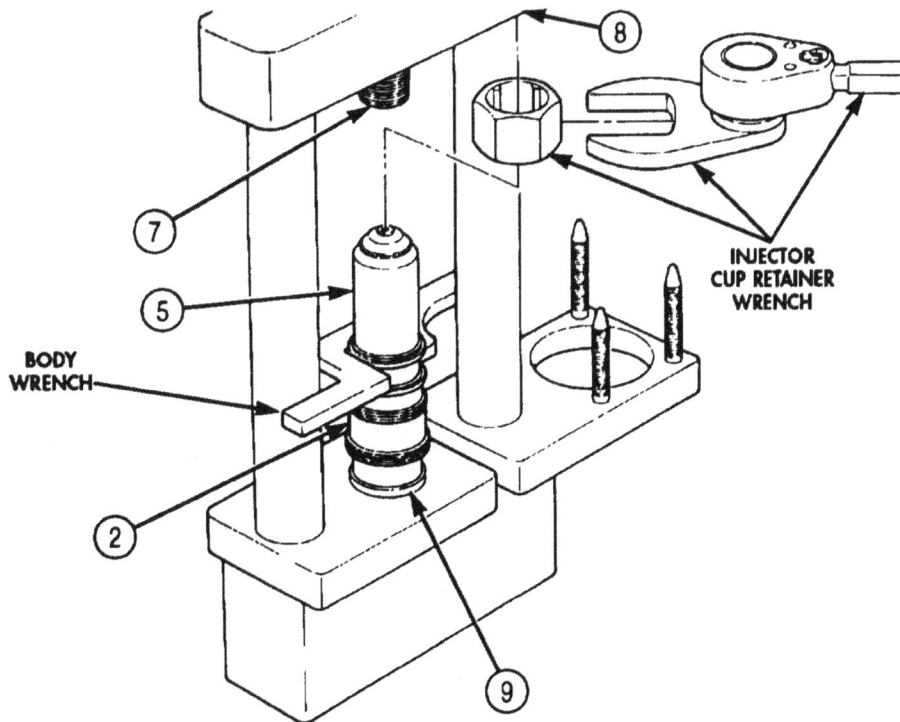

BODY
WRENCH

INJECTOR
CUP RETAINER
WRENCH

5-19. FUEL INJECTOR REPAIR (Contd)

14. Remove injector (4) and body wrench from loading fixture.
15. Remove plunger (2) from injector (4).
16. Apply clean diesel fuel to coat plunger (2).
17. Insert plunger (2) into injector (4) so plunger (2) remains 0.5 in. (12.7 mm) from edge of adapter (1).
18. Press plunger (2) into cup retainer (3) with palm of hand.

CAUTION

Do not allow plunger to fall out. Damage to plunger may result.

19. Turn cup retainer (3) and adapter (1) with one hand so cup retainer (3) faces upward. Plunger (2) should slide out immediately. If plunger (2) does not slide out immediately, the injector (4) is not aligned and must be reassembled as outlined in steps 1 through 18.
20. Install body wrench in vise.
21. Slide injector (4) into body wrench.
22. Place spring (9) and spring retainer (5) in injector (4).
23. Slide plunger (2) into injector (4).
24. Thread locknut (8) and adjusting screw (7) into injector (4) with locknut wrench and top-stop adjusting wrench (6) until locknut (8) contacts adapter (10). Tighten locknut (8) 55 lb-ft (75 N•m).
25. Remove injector (4) from body wrench.
26. Slide two new large O-rings (15) over end of injector (4) and into two upper grooves (11) and (12) of adapter (10).
27. Slide new small O-ring (14) over end of injector (4) into lowest groove (13) on adapter (10).
28. Place new inlet fuel screen (17) over adapter orifice (18) and install with new retainer ring (16).

5-19. FUEL INJECTOR REPAIR (Contd)

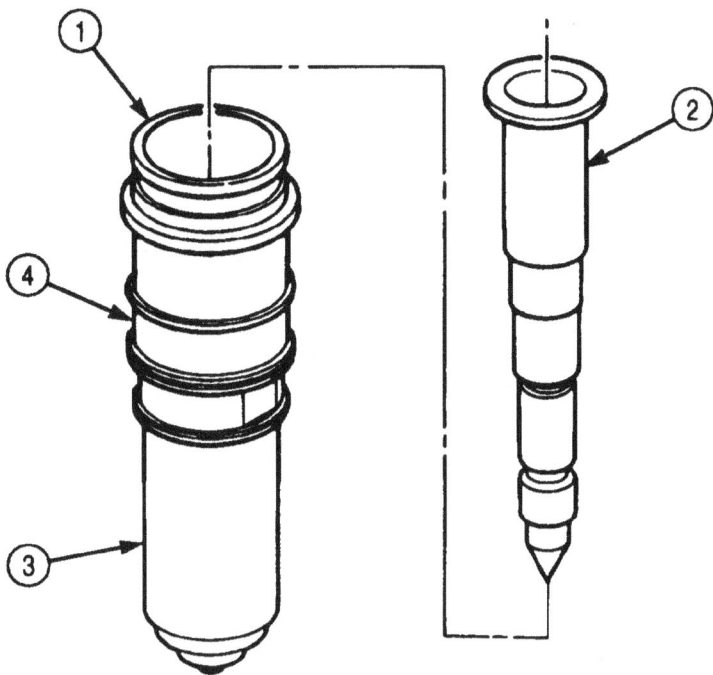

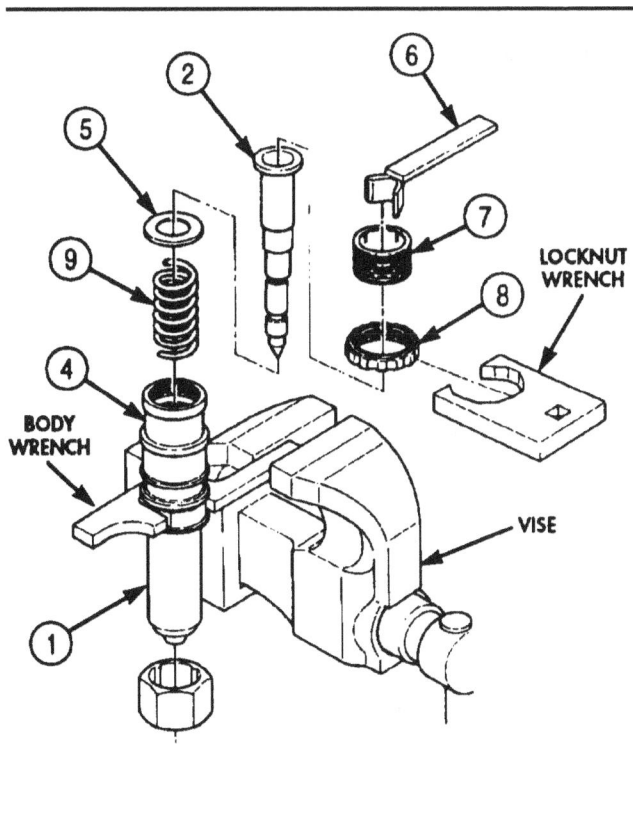

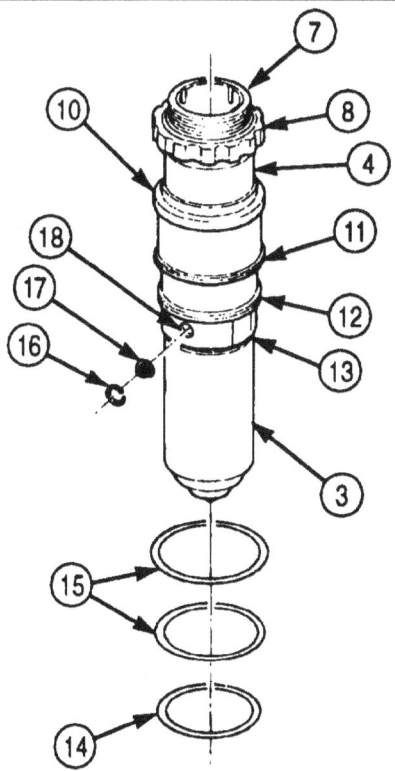

FOLLOW-ON TASK: Install fuel injectors (para. 4-32).

5-20. INJECTOR LEAK TEST

THIS TASK COVERS:

a. Setting Up Injector Leakage Detector   b. Testing Injector Plunger and Seat

---

INITIAL SETUP:

APPLICABLE MODELS
M939/A1

SPECIAL TOOLS
Injector leakage detector (Appendix E, Item 68)

TOOLS
General mechanic's tool kit (Appendix E, Item 1)
Feeler gauge

REFERENCES (TM)
TM 9-2320-272-24P

EQUIPMENT CONDITION
Fuel injectors removed (para. 4-32).

---

a. Setting Up Injector Leakage Detector

1. Set pressure regulator on leakage tester at 60 psi (414 kPa). Read pressure setting on pressure regulator gauge (1).

2. Operate retraction lever (3) on leakage tester three times, and reset pressure regulator gauge (1) at 60 psi (414 kPa).

NOTE
Air pressure must be held at 60 psi (414 kPa) during all tests.

3. Place leakage tester retraction lever (3) in position A.

4. Position load cell (2) on leakage tester.

5. Adjust knurled knob (4) on load cell (2) until load cell (2) reads 200 psi (1,379 kPa).

NOTE
Do not adjust locknut unless load cell is in use.

6. Place feeler gauge between knurled knob (4) and locknut (5) and adjust locknut (5).

7. Remove load cell (2) from leakage tester.

5-20. INJECTOR LEAK TEST (Contd)

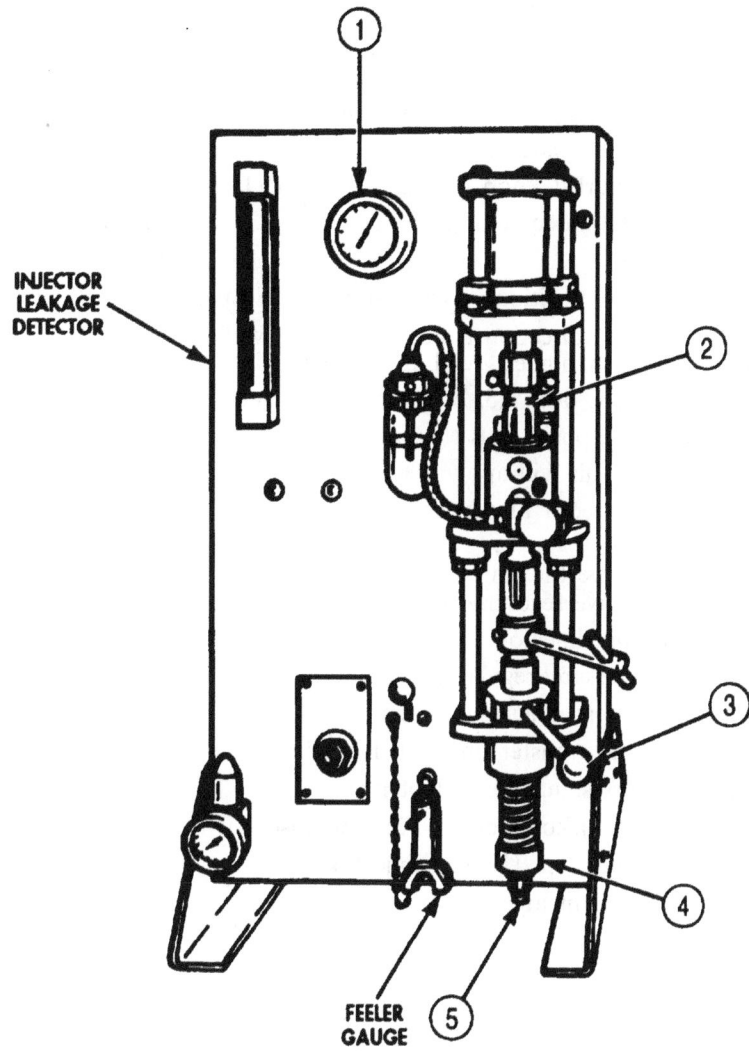

INJECTOR
LEAKAGE
DETECTOR

FEELER
GAUGE

5-20. INJECTOR LEAK TEST (Contd)

b. Testing Injector Plunger and Seat

1. Remove plunger (2) and spring (3) from injector (4).
2. Install injector link (1) in plunger (2) and place in injector (4).
3. Remove screen retaining ring (6) and inlet fuel screen (7) from injector (4).
4. Align injector delivery orifice (5) with burnishing tool hole (11) on burnishing tool adapter (10).
5. Insert locating screw (8) in locating screw hole (9) on burnishing tool adapter (10) and tighten.

NOTE
Support plate may be tilted.

6. Position injector (4) on support plate (14) and install in leakage tester.
7. Place feeler gauge between knurled knob (17) and locknut (18). Do not adjust locknut (18).
8. Install transfer line (20) and tighten in injector adapter (13) drain port.
9. Tighten T-handle clamp (15).
10. Shift retraction lever (16) from position A to position B. Ensure plunger (2) retracts.
11. Rotate plunger (2) in clockwise direction a little at a time while reading air flow meter (12). Stop rotating plunger (2) when highest reading is obtained.
12. If air flow meter (12) reading is over 4.5, overhaul injector (4) (para. 5-19).
13. Shift retraction lever (16) from position B to position A.
14. Loosen T-handle clamp (15). This will apply 200 lbs (91 kg) load to plunger (2).
15. While observing checker (21), see if any bubbles appear during the first ten seconds of testing. No bubbles should appear.
16. If a bubble does appear, observe the time it takes for the next one to appear. No more than one bubble can appear every five seconds. If bubbling is too high, overhaul injector (4) (para. 5-19).
17. Remove transfer line (20) from injector adapter (13) drain port.
18. Release air pressure from pressure regulator (19).
19. Remove injector adapter (13) from leakage tester.
20. Remove locating screw (8) and transfer line (20) from burnishing tool adapter (10).
21. Remove injector (4) from burnishing tool adapter (10).
22. Place inlet fuel screen (7) over injector orifice (5), and install with retainer ring (6).
23. Remove plunger (2) and injector link (1) from injector (4).
24. Place plunger (2) and spring (3) in injector (4).

5-20. INJECTOR LEAK TEST (Contd)

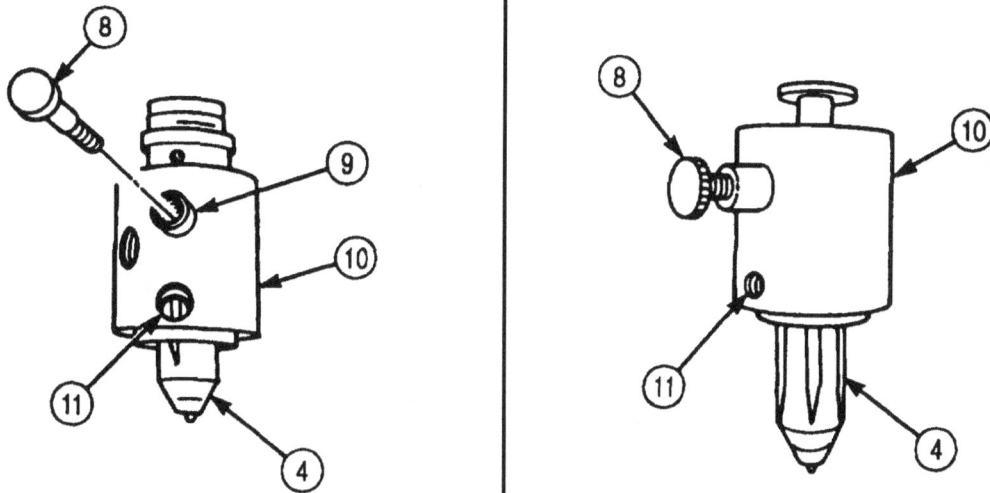

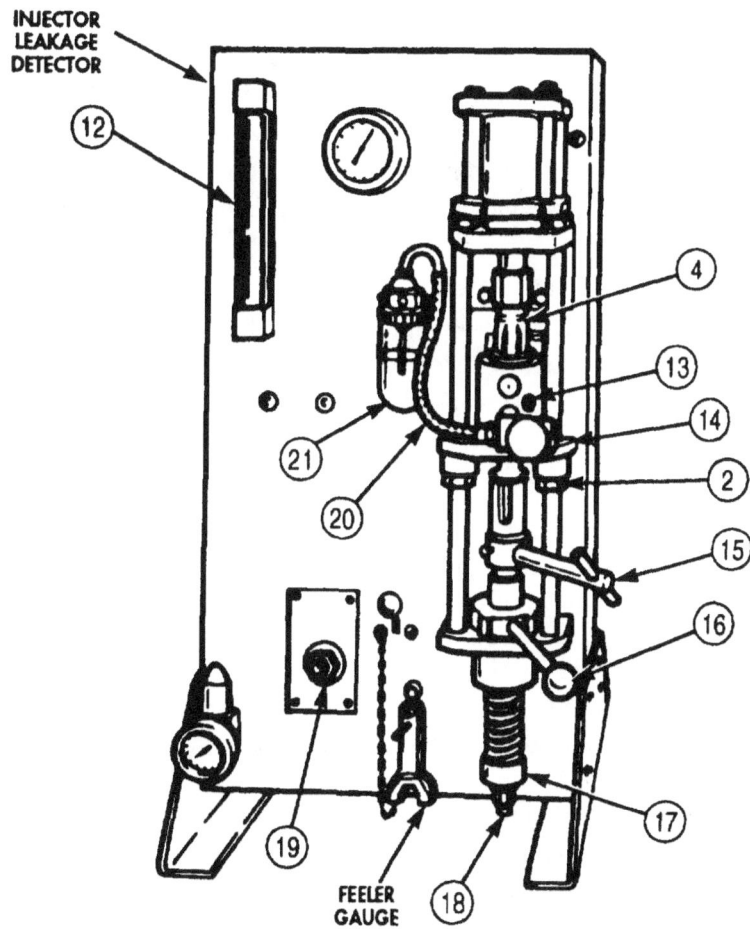

INJECTOR
LEAKAGE
DETECTOR

FEELER
GAUGE

FOLLOW-ON TASK: Install fuel injectors (para. 4-32).

5-21. INJECTOR SPRAY PATTERN TEST

THIS TASK COVERS:

a. Setting Up Spray Pattern Tester   b. Testing Spray Pattern

INITIAL SETUP:

APPLICABLE MODELS
M939/A1

SPECIAL TOOLS
Fuel injection tester (Appendix E, Item 49)
Spray angle tester (Appendix E, Item 138)

TOOLS
General mechanic's tool kit (Appendix E, Item 1)

REFERENCES (TM)
TM 9-2320-272-24P

EQUIPMENT CONDITION
Fuel injectors removed (para. 4-32).

GENERAL SAFETY INSTRUCTIONS
Whenever fuel is forced from injector spray holes, keep hands away from spray stream.

a. Setting Up Spray Pattern Tester

NOTE
Any source of 22 psi (152 kPa) constant fuel pressure will operate spray pattern tester.

1. Locate spray pattern tester near or on injector test stand.
2. Attach inlet hose (2) to adapter (8) from injector test stand.
3. Attach drain hose (4) to adapter base (5) and place loose end in test stand drain area.
4. Remove plunger (10) and spring (11) from injector (12) and store in safe place.
5. Remove screen retaining ring (15) and inlet fuel screen (14) from injector test stand.
6. Place injector (12) in adapter (8).
7. Install knurled plug (6) in adapter (8) drain opening.
8. Place injector (12) in spray pattern tester and tighten inlet hose connector (3).
9. Install injector (12) in spray patter tester holddown bracket (9) with knurled knob (1).
10. Install target ring (7) marked 8-.007 x ı̊7on spray pattern tester.

b. Testing Spray Pattern

WARNING
Keep hands away from spray stream when fuel is forced from injector spray holes. Failure to do this may result in injury to personnel.

1. Apply 22 psi (152 kPa) pressure to injector test stand.

NOTE
Each spray stream must hit a window in the target ring.

2. Shift target ring (7) so one spray stream hits on No. 1 or index window. If spray stream is 8f6f2 window, replace cup (para. 5-19).
3. Loosen knurled knob (1) on holddown bracket (9).
4. Remove injector (12) from spray pattern tester.
5. Remove knurled plug (6) and injector (12) from adapter (8).
6. Install spring (11) and plunger (10) in injector (12).
7. Install inlet fuel screen (14) over injector orifice (13) with screen retaining ring (15).

5-21. INJECTOR SPRAY PATTERN TEST (Contd)

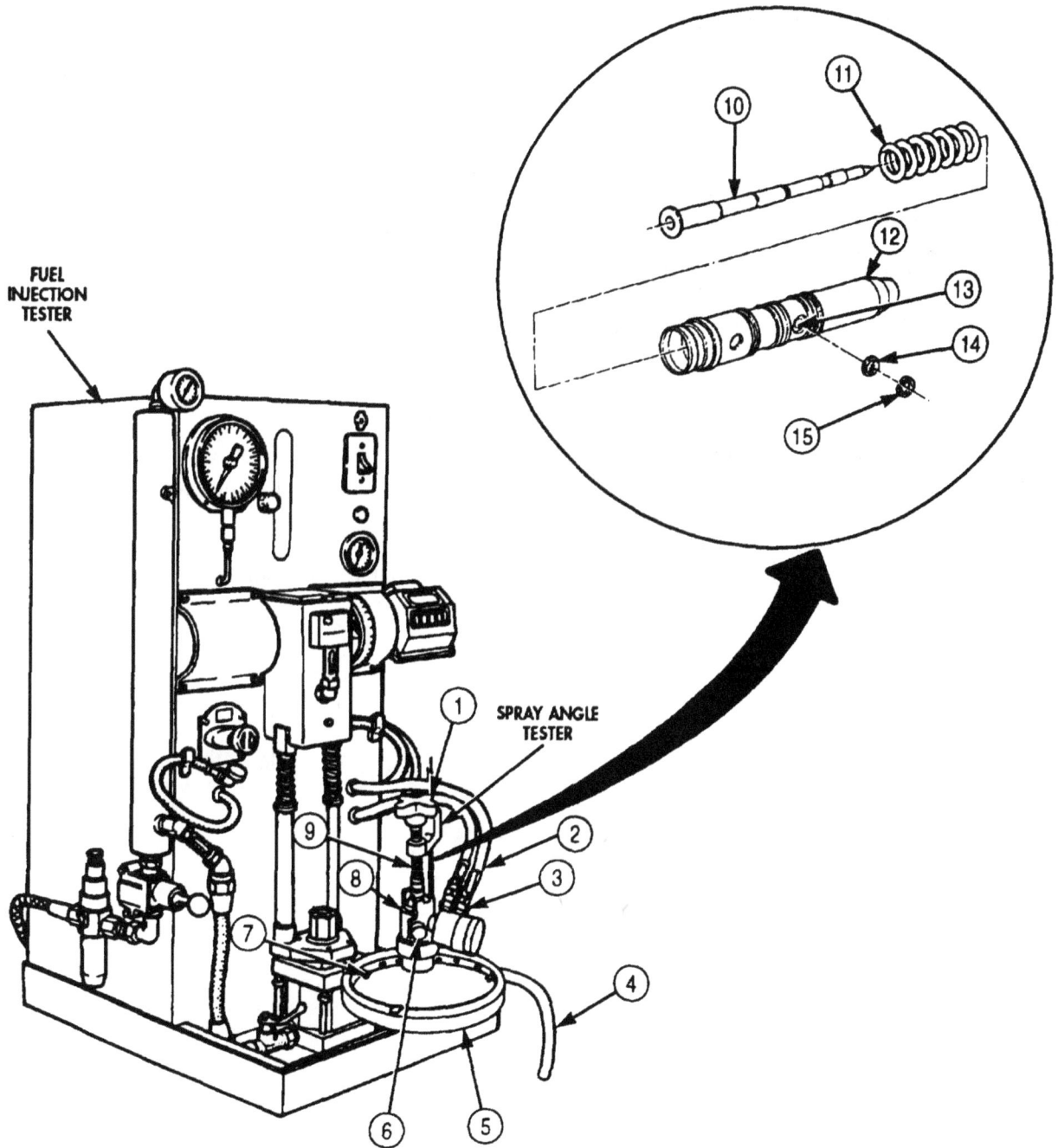

FOLLOW-ON TASK: Install fuel injectors (para. 4-32).

5-22. INJECTOR TEST STAND CALIBRATION

THIS TASK COVERS:

a. Setting Up Injection Tester   **b. Test Stand Calibration**

INITIAL SETUP:

APPLICABLE MODELS	TOOLS
M939/A1	General mechanic's tool kit (Appendix E, Item 1)

SPECIAL TOOLS	REFERENCES (TM)
Fuel injection tester (Appendix E, Item 49)	TM 9-2320-272-24P

a. Setting Up Injection Tester

1. Rotate counter wheel (2) on injection tester until wheel marks and pointer are aligned.
2. Open hydraulic valve (3).

CAUTION
Never operate test stand with load cell in position.

3. Position load cell tester on injection tester and secure by opening air valve (4).
4. Adjust air regulator (5) by turning knurled knob (7) until load cell tester registers within coded range markings. Record air gauge (1) pressure reading.
5. Lock air regulator (5) in place with locknut (6).
6. Place air valve (4) to center position and remove load cell tester from injection tester.

5-22. INJECTOR TEST STAND CALIBRATION (Contd)

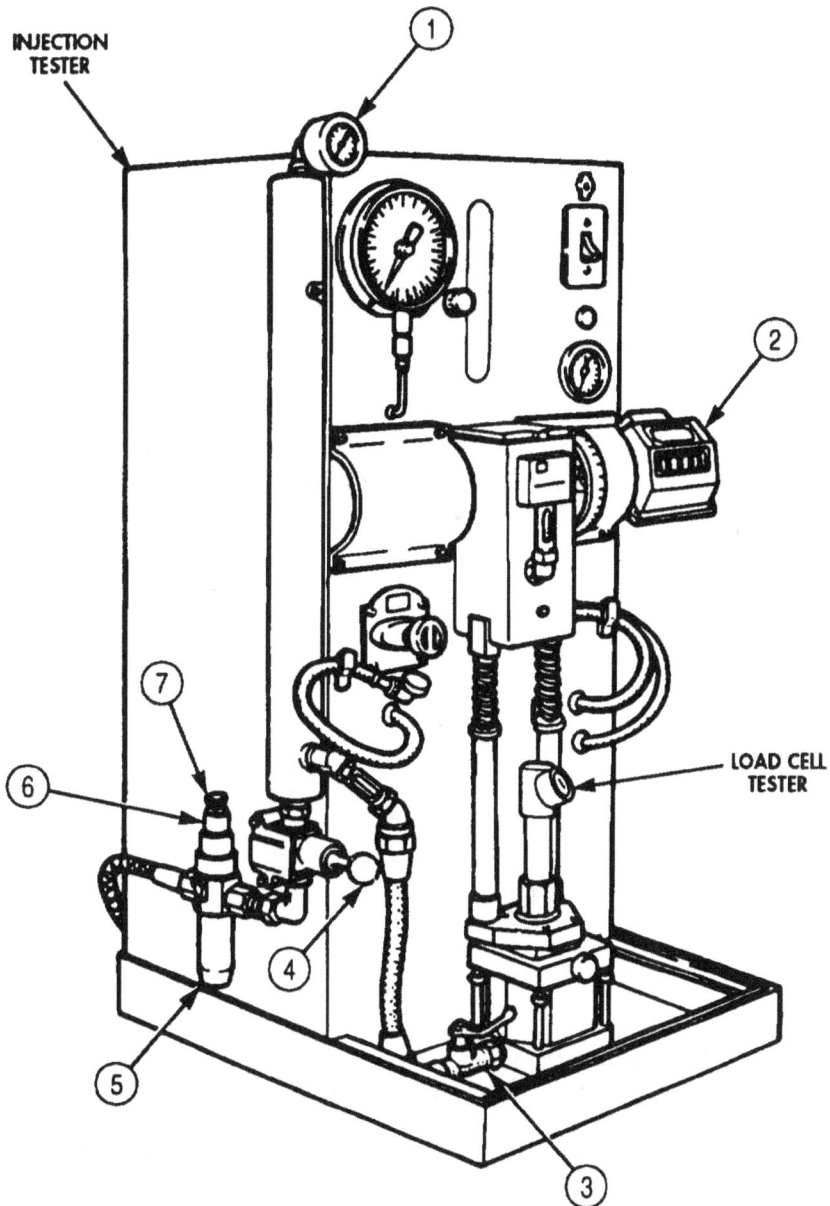

5-22. INJECTOR TEST STAND CALIBRATION (Contd)

7. When cylinder (11) is down, hydraulic fluid should show in sight bulb. Open air valve (13) when cylinder (11) is at the top of its travel; no air should show in sight glass (14). If air is indicated, tighten all line connections.

NOTE
The master injector is precalibrated and must never be tampered with.

8. Place adapter (20) on master injector (17). Ensure wheel marks and pointer are aligned (step 1).

NOTE
Ensure injector seat contains 0.20 in. (0.51 mm) restrictor orifice.

9. Position test stand link (15) over injector link (9) and place master injector (10) in injector seat (18).
10. Tip back until test stand link (15) is below test stand push rod (8) and not rubbing.
11. Open air valve (13) to clamp master injector (10) in place. Ensure test stand link (15) is aligned.
12. Close hydraulic valve (12) to lock master injector (10) in place.
13. Connect test stand fuel drain connector (19) to adapter (20).
14. Connect test stand fuel inlet connector (16) to adapter (20).

CAUTION
If temperature exceeds 135°F (57°C), drain and replace with new test oil.

15. Place injector test stand motor switch (6) in START position. Ensure temperature gauge (7) reads 90°-95°F (32-35°C).
16. Shift the silver colored counter (1) wheels (3) to the right, rotate counter (1) to indicate 1,020 strokes and release.
17. Clear counter (1) by rotating thumbscrew (4) one complete revolution. Ensure all white-colored counter wheels (2) read zero.

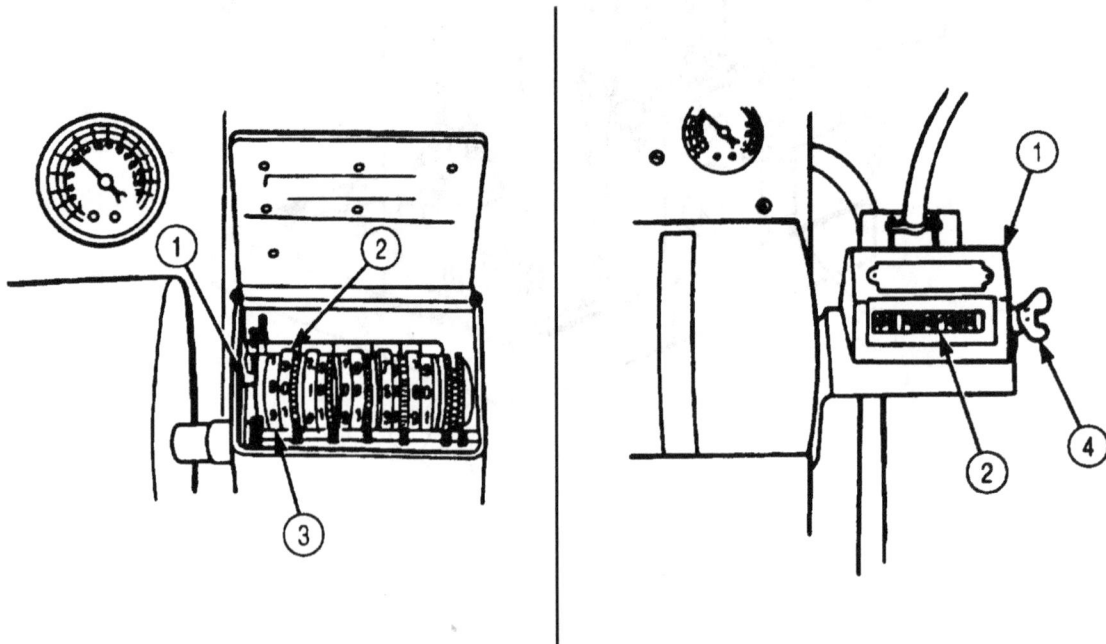

5-22. INJECTOR TEST STAND CALIBRATION (Contd)

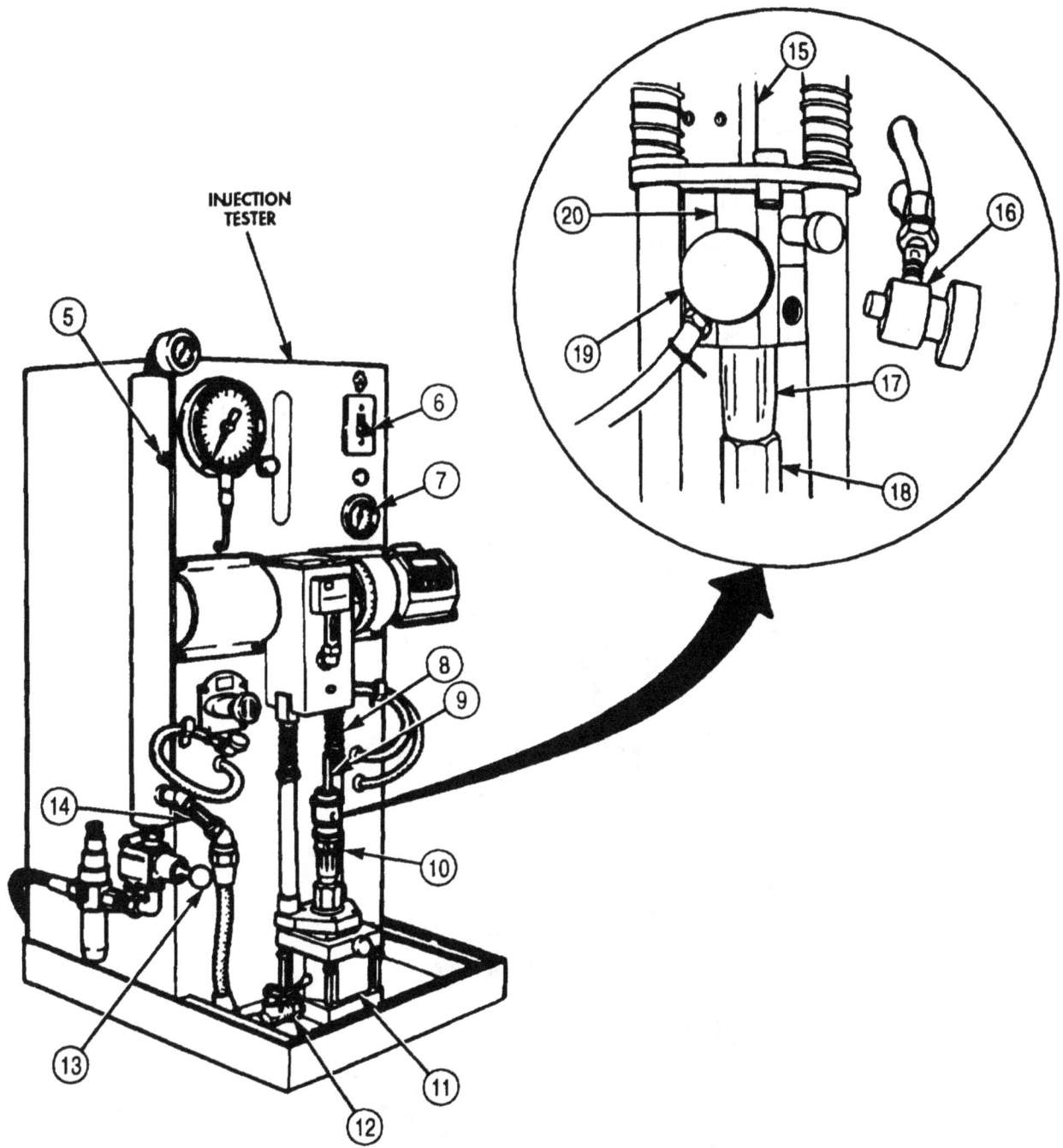

INJECTION
TESTER

5-22. INJECTOR TEST STAND CALIBRATION (Contd)

b. Test Stand Calibration

1. Adjust regulator knob (5) on injector test stand by turning until pressure gauge (1) reads 120 psi (827 kPa). Ensure pressure is maintained at 120 psi (827 kPa) during calibration.

2. Press red flow start switch (3) in and out until counter (4) reads 1,020 count strokes.

3. Look directly into vial (2) and observe reading.

NOTE

Reading of 132 counterclockwise at 120 psi (827 kPa) indicates test stand is in calibration. If reading is more than 132 cc, the test stand is not set up properly. Repeat setup steps. If reading is below 132 cc, continue.

4. Reset counter (4) on injector test stand to zero (task a., steps 16 and 17).

5. Set counter (4) up seven strokes for each cc less than 132 cc.

6. Dump fuel from vial (2).

NOTE

If counter is set beyond 1,050 strokes to obtain 132 cc, the test stand is not set up properly. Repeat setup steps. If counter reads 1,050 or below, the test stand is in calibration.

7. Press red flow start switch (3) in and out until counter (4) sets strokes to obtain 132 cc.

8. Place motor switch (6) in STOP position.

9. Remove test stand fuel drain connector (14) from adapter (15).

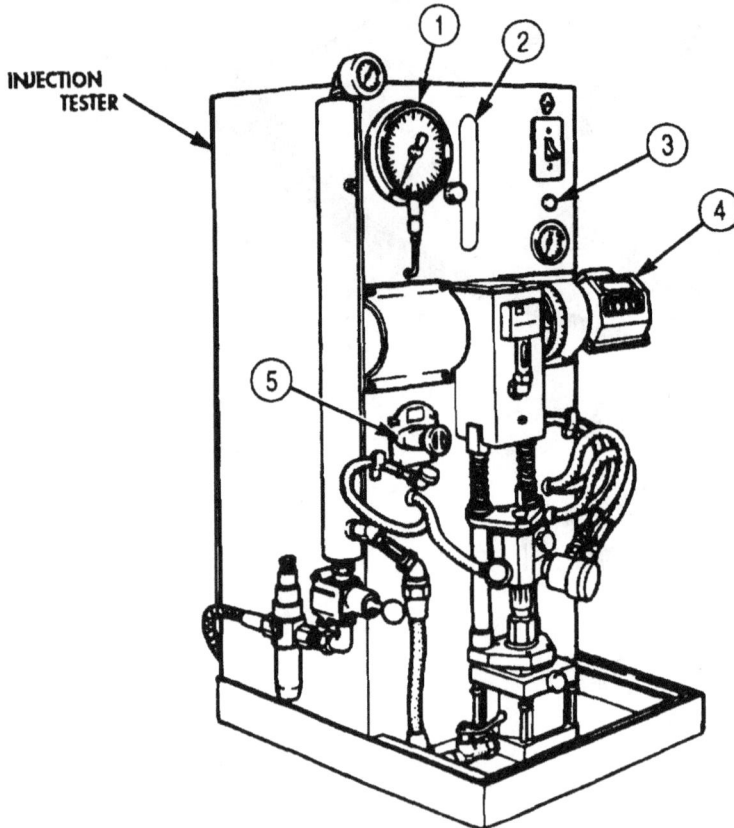

5-22. INJECTOR TEST STAND CALIBRATION (Contd)

10. Open hydraulic valve (9).
11. Close air valve (10).
12. Remove master injector (8) from injector seat (13).
13. Slide test stand link (11) off injector plunger link (7).
14. Remove test stand fuel inlet connector (12) from adapter (15).
15. Remove adapter (15) from master injector (8).

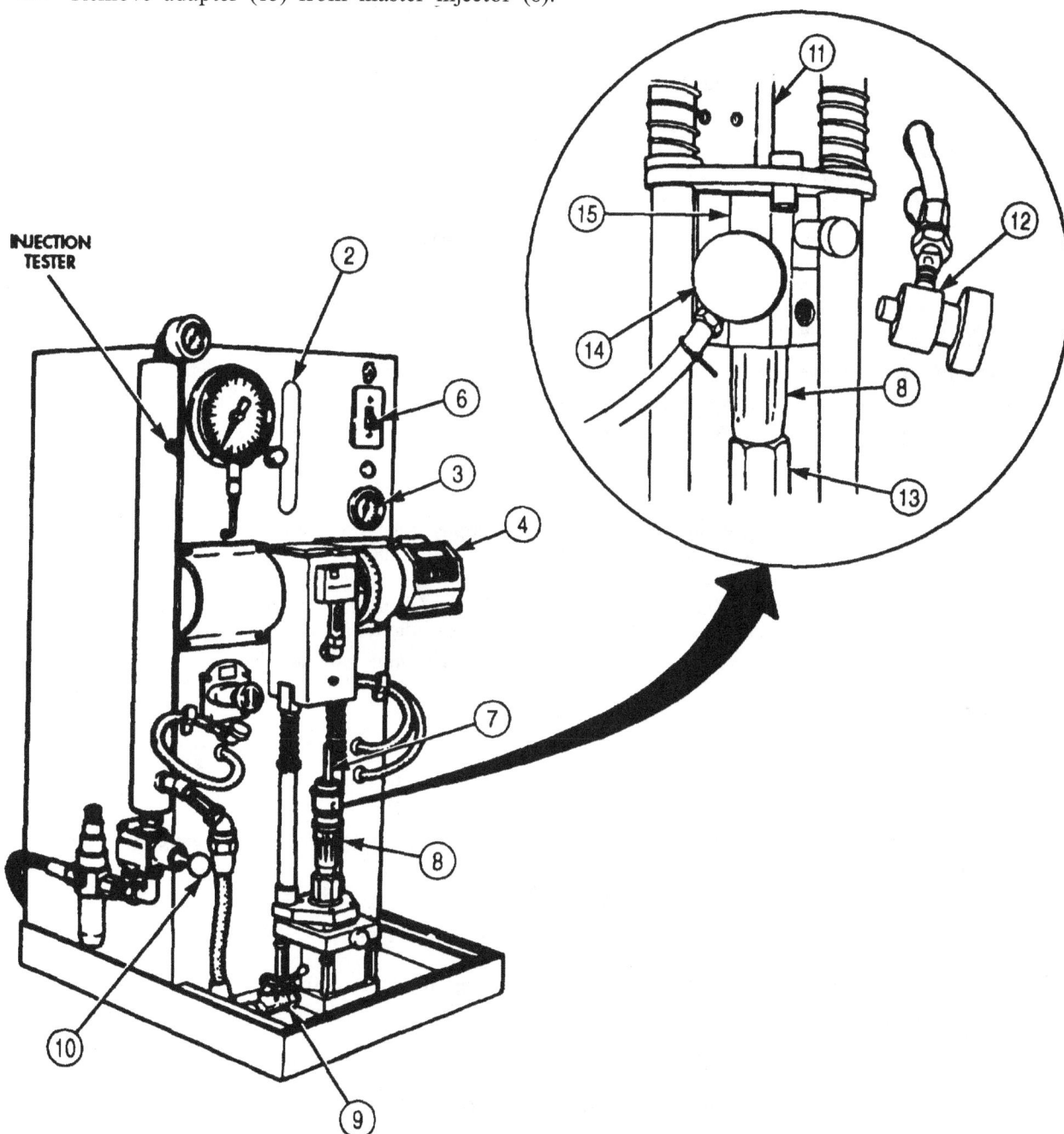

5-23. INJECTOR FLOW TEST

THIS TASK COVERS:

a. Setting Up Test Stand
b. Testing Check Ball Seating

c. Adjusting and Measuring Fuel Delivery

INITIAL SETUP:

APPLICABLE MODELS
M939/Al

SPECIAL TOOLS
Injection tester (Appendix E, Item 49)

TOOLS
General mechanic's tool kit (Appendix E, Item 1)
Burnishing tool adapter
Adjusting tool adapter

MATERIALS/PARTS
Gasket (Appendix D, Item 240)
Diesel fuel (Appendix C, Item 42)

REFERENCES (TM)
TM 9-2320-272-24P

EQUIPMENT CONDITION
• Fuel injector removed (para. 4-32).
• Injector test stand calibrated (para. 5-22).

SPECIAL ENVIRONMENTAL CONDITIONS
Work area clean and free from blowing dirt and dust.

a. Setting Up Test Stand

1. Remove screen retainer ring (5) and inlet fuel screen (4) from injector (3).
2. Remove plunger (1) and spring (2) from injector (3) and separate plunger (1) from spring (2).
3. Place plunger (1) in injector (3) without spring (2).
4. Apply diesel fuel to lubricate inside of burnishing tool adapter.

CAUTION

Injector inlet port and burnishing tool adapter inlet hole must be
aligned to prevent damage to the burnishing tool points when
installed later.

5. Position injector (3) in burnishing tool adapter so injector (3) inlet port and burnishing adapter inlet holes align. Tighten locating screw (12) on burnishing tool adapter to install injector (3).
6. Connect test stand inlet pressure line adapter (11) to burnishing tool adapter.

b. Testing Check Ball Seating

NOTE

Hold injector in hand; do not place in test stand holding device,

1. Hold plunger (1) down in injector (3).
2. Place motor switch (7) on injection tester in START position. Ensure temperature gauge (9) reads 90-95°F (32-35°C).

CAUTION

If temperature exceeds 135°F (57°C), drain and replace with new
test oil.

3. Adjust regulator knob (9) by turning until pressure gauge (6) reads 150 psi (1,034 kPa).
4. Check burnishing tool adapter installation hole (10) for leaks. A slight seepage is not harmful. If leakage is found, replace check ball (para. 5-19).
5. Place motor switch (7) in STOP position.
6. Disconnect test stand inlet pressure line adapter (11) from burnishing tool adapter.

5-23. INJECTOR FLOW TEST (Contd)

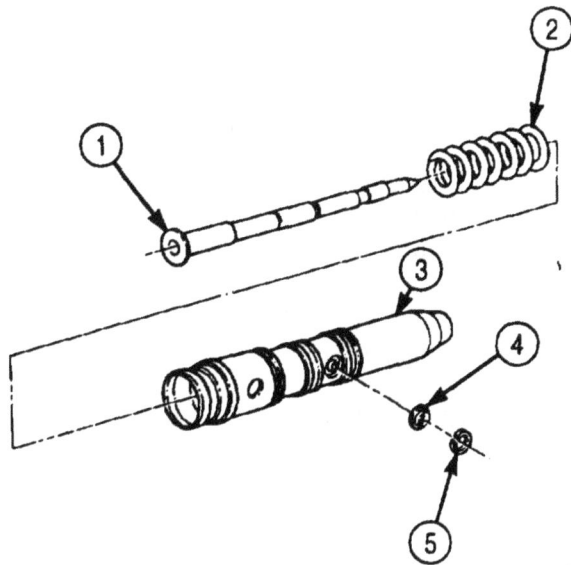

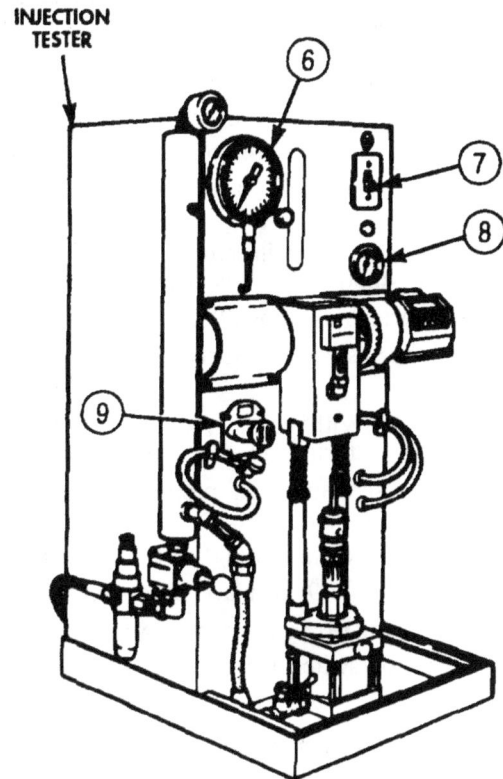

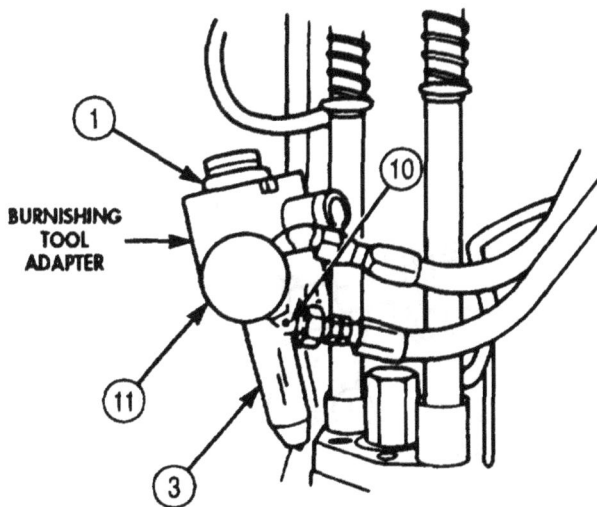

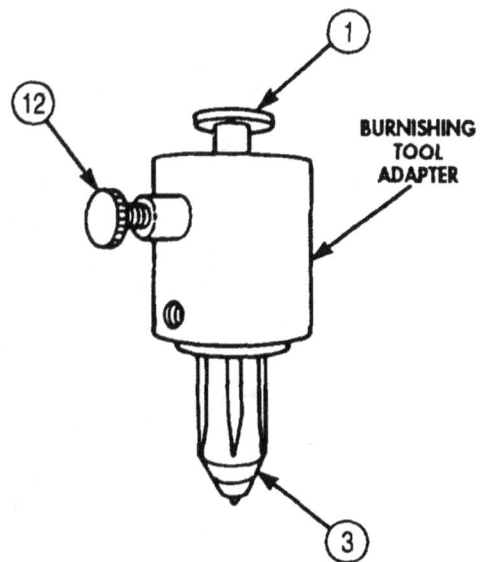

5-23. INJECTOR FLOW TEST (Contd)

c. Adjusting and Measuring Fuel Delivery

1. Remove plunger (11) from injector (9).
2. Slide spring (12) on plunger (11) and place in injector (9).
3. Place retainer plate (14) on burnishing tool adapter and secure with pins (13).
4. Place test stand link (1) in burnishing tool adapter. Test stand link (1) should be 6.5 in. (17 cm) long.
5. Place injector (9) in injector test stand (7) so injector (9) is in injector seat (8). Tip injector (9) back until test stand link (1) is below test stand push rod (16) and is not rubbing.
6. Install burnishing tool (2) in test stand inlet pressure line adapter (9).
7. Retract burnishing tool needle (4) by pulling small knob (3) out.

NOTE

With needle retracted, burnishing tool may be left in adapter during all test operations.

8. Connect test stand inlet pressure line adapter (6) to burnishing tool adapter inlet hole by screwing in large knob (5).
9. Install drain connector (10) in burnishing tool adapter.
10. Clamp injector (9) in place by opening air valve (18). Ensure test stand link (1) is aligned.
11. Lock injector (9) in place by closing hydraulic valve (17).
12. Place motor switch (15) in START position. Ensure temperature reads 90°-95° F (32°-35°C).

CAUTION

If temperature exceeds 135 °F (57°C), drain and replace with new test oil.

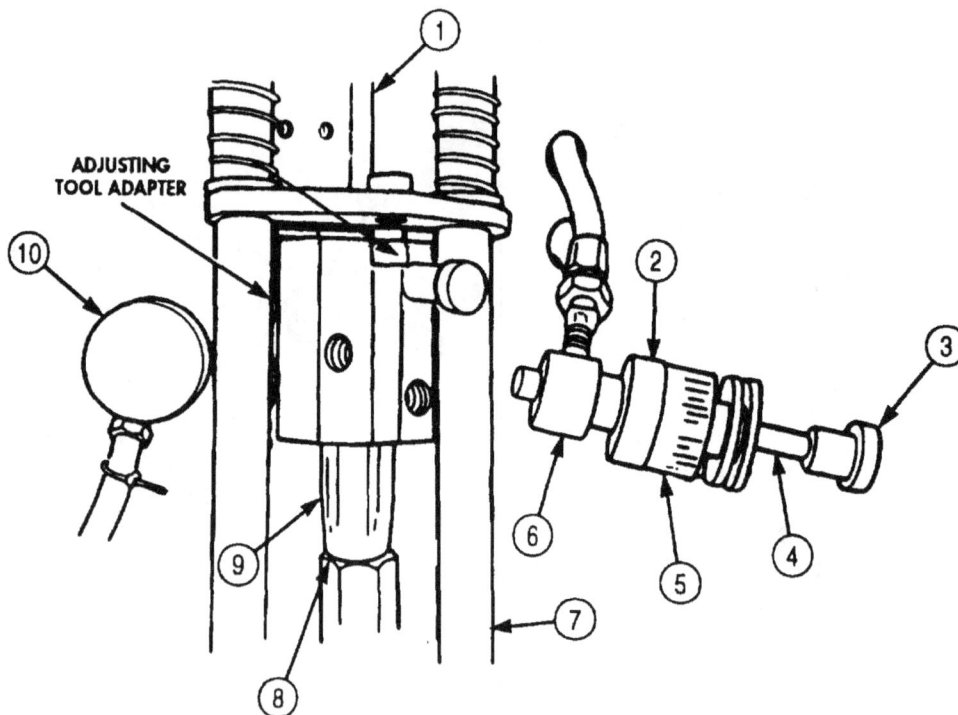

ADJUSTING TOOL ADAPTER

5-23. INJECTOR FLOW TEST (Contd)

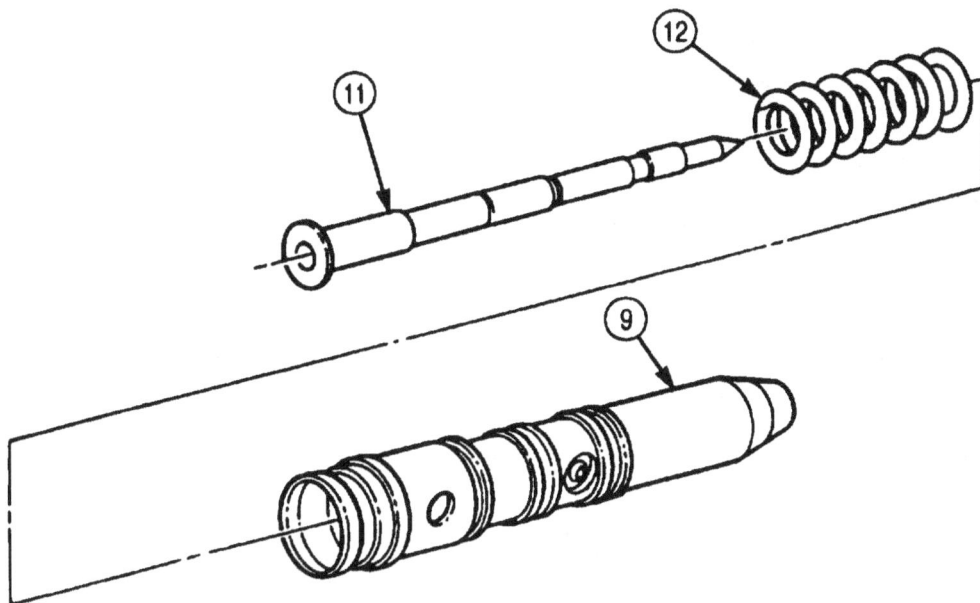

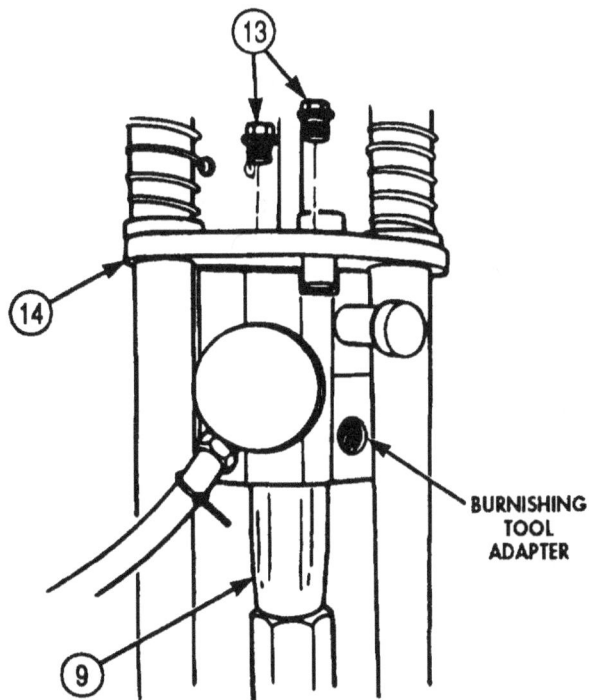

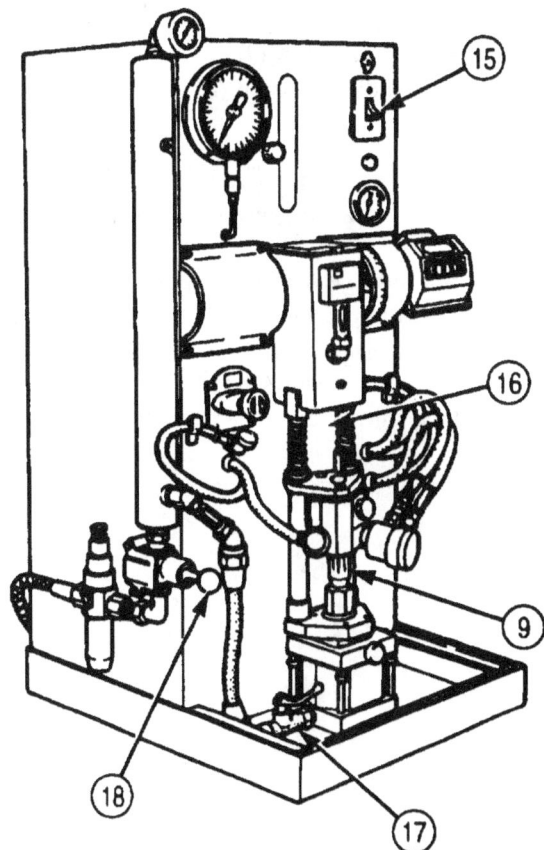

BURNISHING
TOOL
ADAPTER

5-23. INJECTOR FLOW TEST (Contd)

13. Adjust injector test stand regulator knob (6) by turning until pressure gauge (1) reads 120 psi (827 kPa).

14. Press flow start switch (4) in until counter (5) reads the same as master injector counter strokes.

15. Observe vial (2) reading, If reading is higher than 122 cc at 120 psi (827 kPa), perform steps 17 through 19 to install new orifice plug (16). If reading is lower than 121 cc at 120 psi (827 kPa), perform steps 20 through 24 to set fuel flow.

16. Remove test stand inlet pressure line adapter (11) and burnishing tool from burnishing tool adapter (13) by turning large knob (9) out until assembly is free.

17. Remove orifice plug (16) and gasket (15) from injector (10). Discard gasket (15).

NOTE

New orifice plug size .018-.019 in. (0.48-0.49 mm) is small enough so burnishing will increase fuel efficiency.

18. Install new gasket (15) and orifice plug (16) in injector orifice (14) and tighten orifice plug (16) 8-10 lb-in. (0.9-1.1N•   m).

19. Install test stand inlet pressure line adapter (11) and burnishing tool on burnishing tool adapter (13) by screwing large knob (9) into burnishing tool adapter (13) inlet hole until tight.

CAUTION

When seating burnishing tool, use care not to push small knob in too hard or overtighten indicator knob. When slight contact is made, stop. Damage can be caused to injector. Test stand must be running while burnishing.

20. Turn burnishing tool indicator knob (8) until spaced 3/8 in. (9.5 mm) from large knob (9).

21. Slowly push small burnishing tool knob (7) in until slight contact is made with injector (10).

22. Turn small burnishing tool knob (7) counterclockwise to lock large knob (9) and indicator knob (8).

23. Slowly turn indicator knob (8) in until slightly seated in injector (10). Index indicator knob (8) with mark on large knob (9).

24. Advance indicator knob (8) one mark, and back off until spaced 3/8 in. (9.5 mm).

NOTE

Perform steps 12 through 15 and recheck fuel delivery. If delivery is lower than 121-122 cc, repeat steps 20 through 24. If reading is higher, install new orifice plug (steps 17 through 19), and recheck fuel delivery.

25. Place injector test stand motor switch (3) in STOP position.

26. Remove drain connector (12) from burnishing tool adapter (13).

27. Screw out large knob (9) on burnishing tool and remove test stand inlet pressure line adapter (11) from burnishing tool adapter (13).

28. Remove burnishing tool from burnishing tool adapter (13). Store burnishing tool in clean place.

5-23. INJECTOR FLOW TEST (Contd)

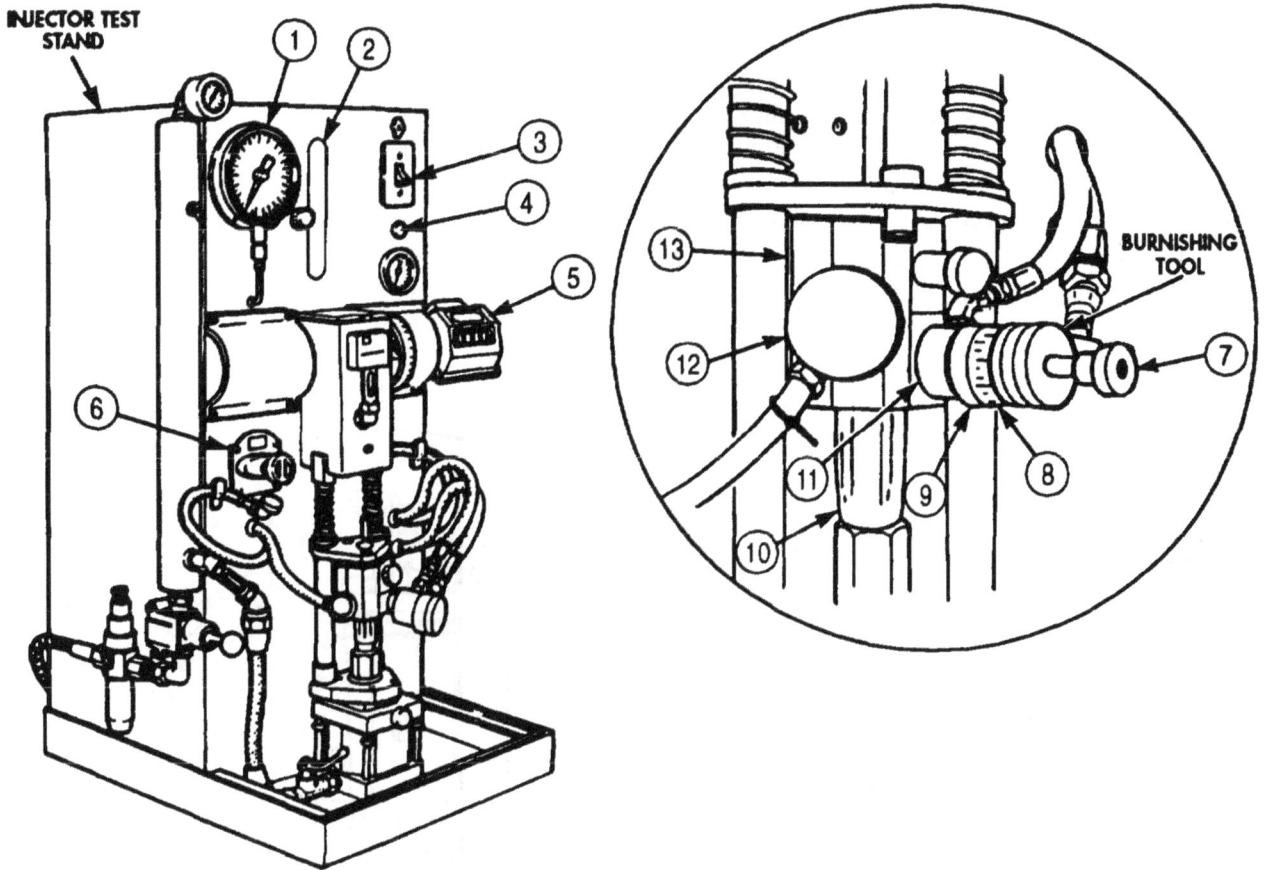

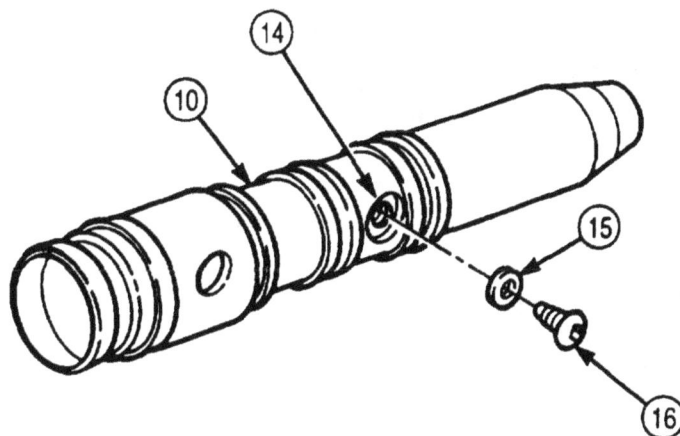

INJECTOR TEST STAND

BURNISHING TOOL

5-23. INJECTOR FLOW TEST (Contd)

29. Open hydraulic valve (2) on injector test stand.
30. Close air valve (3) on injector test stand.
31. Remove injector (1) from injector test stand.
32. Slide test stand link (7) out of burnishing tool adapter.
33. Remove two pins (4) and retainer plate (5) from burnishing tool adapter.
34. Remove burnishing tool adapter from injector (6).
35. Remove plunger (8) and spring (9) from injector (1).
36. Install screen (10) in injector orifice (12) with screen retainer (11).
37. Place spring (9) in plunger (8) and slide in injector (1).

NOTE
Store injectors in a clean place.

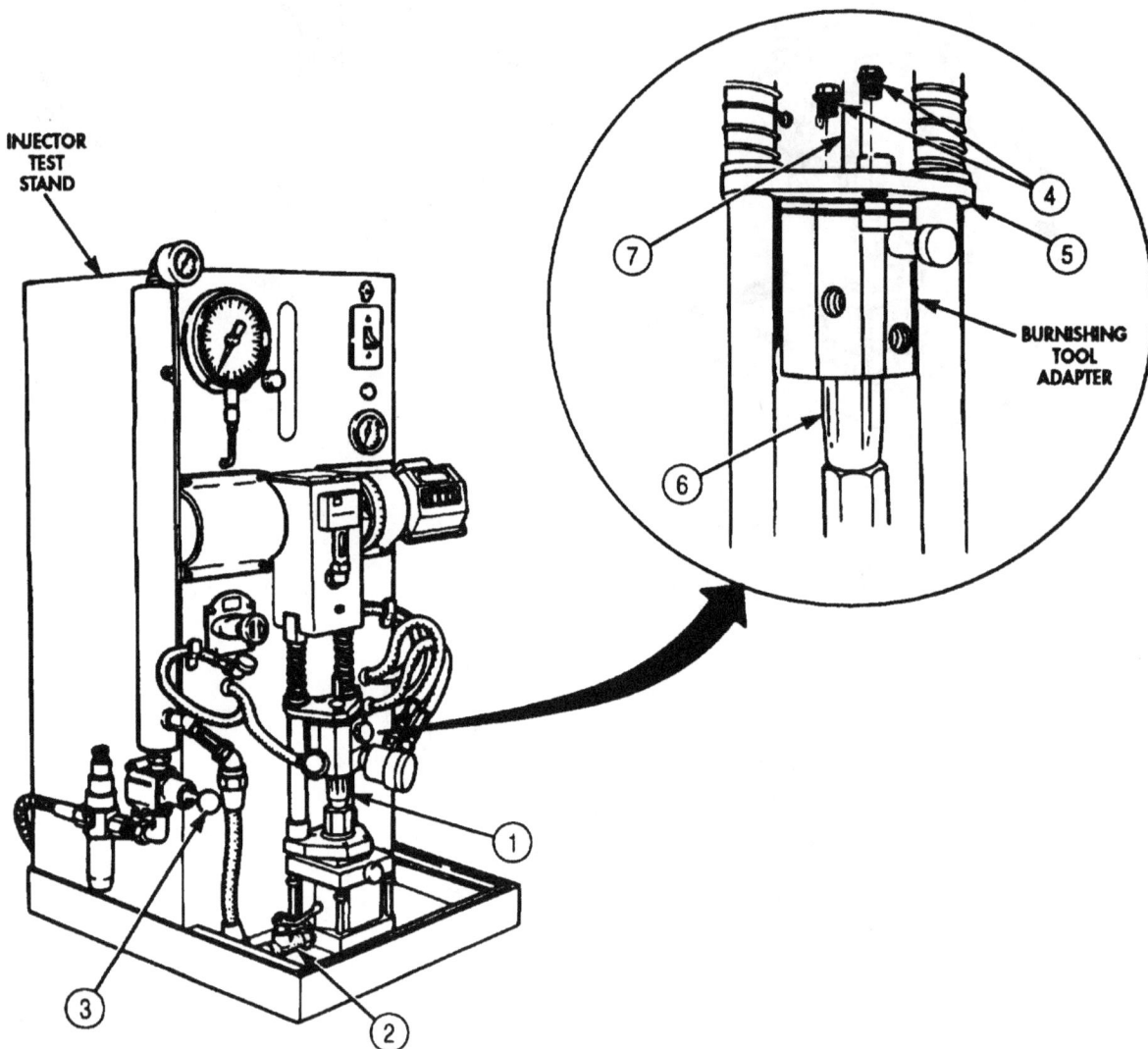

5-23. INJECTOR FLOW TEST (Contd)

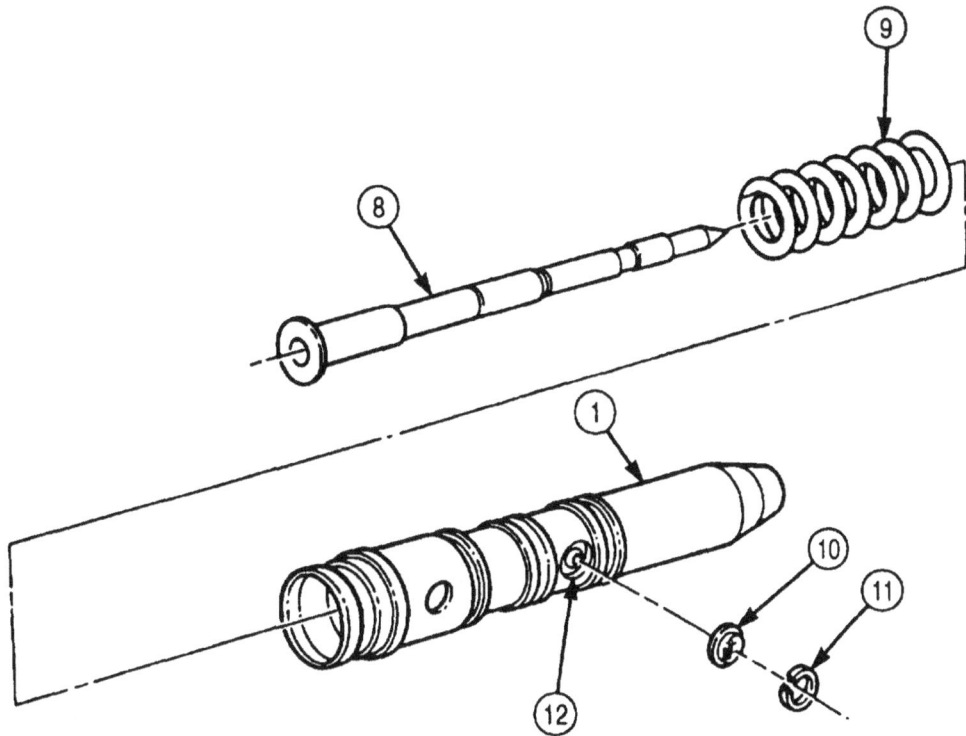

FOLLOW-ON TASK: Install fuel injector (para. 4-32).

Section II. ENGINE (M939A2) MAINTENANCE

5-24. ENGINE (M939A2) MAINTENANCE INDEX

5-25. CYLINDER HEAD REPAIR

THIS TASK COVERS:

a. Disassembly   d. Repair
b. Cleaning   e. Assembly
c. Inspection

INITIAL SETUP:

APPLICABLE MODELS

M939A2

TOOLS

General mechanic's tool kit (Appendix E, Item 1)
Spring compressor (Appendix E, Item 135)
Gauge block (Appendix E, Item 52)
Spring tester (Appendix E, Item 131))
Vernier caliper (Appendix E, Item 159)
Dial indicator (Appendix E, Item 36)
Feeler gauge
Valve refacing machine

MATERIALS/PARTS

Twenty-four collets (Appendix D, Item 253)
Twelve seals (Appendix D, Item 613)
Eight expansion plugs (Appendix D, Item 105)
Expansion plug (Appendix D, Item 106)
Two expansion plugs (Appendix D, Item 107)
Antiseize tape (Appendix C, Item 72)
Lapping compound (Appendix C, Item 37)
Lubricating oil (Appendix C, Item 50)
Pigment (Appendix C, Item 54)
Sealing compound (Appendix C, Item 64)

REFERENCES (TM)

TM 9-247
TM 9-2320-272-24P

EQUIPMENT CONDITION

Cylinder head removed (para. 4-41).

GENERAL SAFETY INSTRUCTIONS

Eyeshields must be worn when removing valves.

---

a. Disassembly

WARNING

Valve springs are under compression. Eyeshields must be worn when removing valves. Failure to do so may cause injury to personnel.

CAUTION

To remove intake and exhaust valves, position cylinder head on exhaust port face. Use wooden or protective surface work bench to prevent damage to cylinder head.

NOTE

- Mark location and position of valves for installation.
- All intake and exhaust valves are removed the same.

1. Using spring compressor, compress valve spring (3) and remove two collets (1) from valve (10) and cylinder head (9). Discard collets (1).

2. Release spring compressor and remove two spring retainers (2), valve spring (3), seal (4), and valve (10) from cylinder head (9). Discard seal (4).

3. Remove five expansion plugs (5) from top face of cylinder head (9). Discard expansion plugs (5).

4. Remove expansion plug (6) from rear face of cylinder head (9). Discard expansion plug (6).

5. Remove two expansion plugs (8) from front and rear faces of cylinder head (9). Discard expansion plugs (8).

6. Remove plug (7) from rear face of cylinder head (9).

5-25. CYLINDER HEAD REPAIR (Contd)

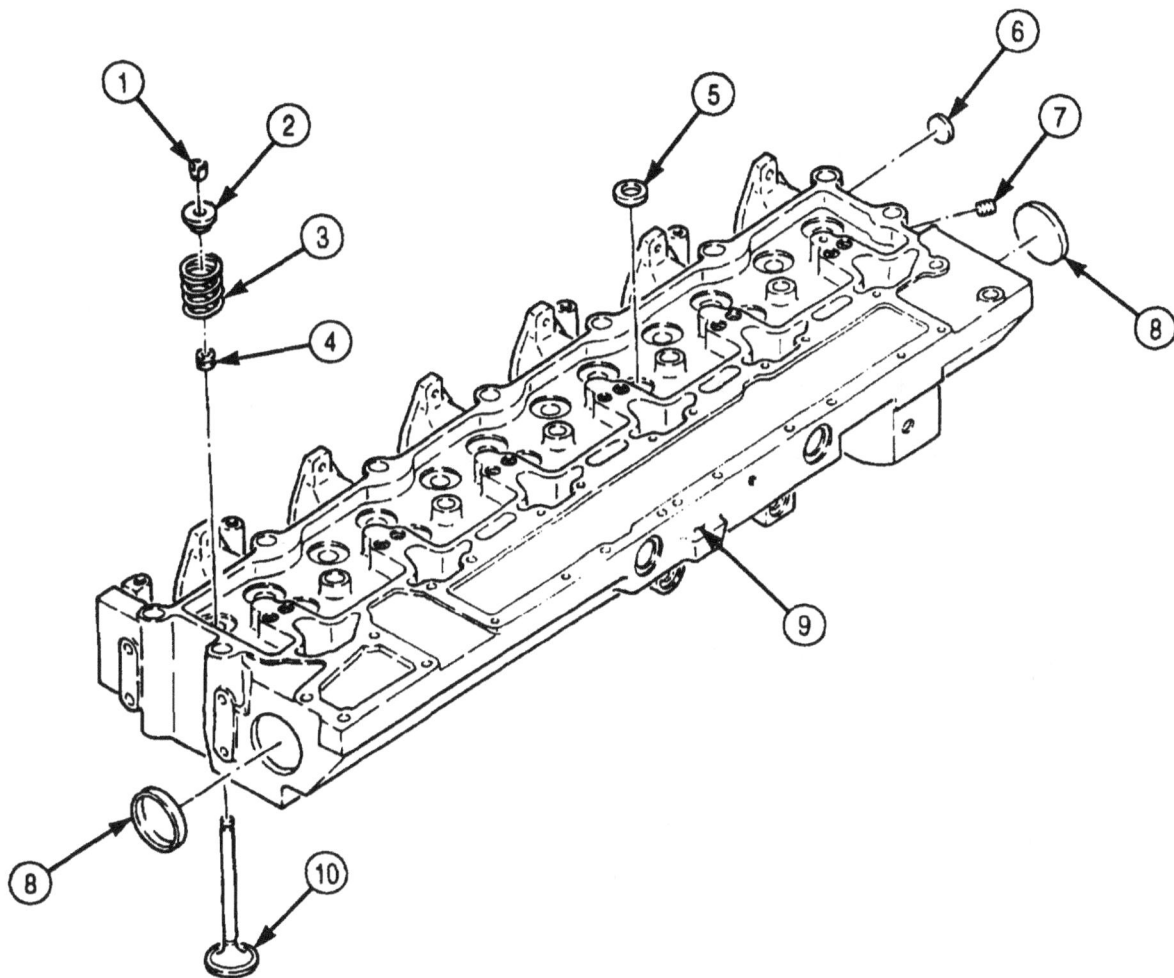

5-25. CYLINDER HEAD REPAIR (Contd)

b. Cleaning

Clean all cylinder head components (TM 9-247).

c. Inspection

1. Inspect cylinder head (1) for cracks, pits, and scratches.
    a. If pitted or scratched, remove with crocus cloth.
    b. If cracked, excessively pitted or scratched, replace cylinder head (1).
2. Inspect cylinder head (1) for warping.
    a. Using straight edge and feeler gauge, inspect gasket surface of cylinder head (1) from front to rear. Warping should not exceed 0.008 in. (0.20 mm).
    b. Inspect gasket surface of cylinder head (1) from side to side. Warping should not exceed 0.003 in. (0.076 mm).
    c. Using 2-in. (51-mm) straight edge and feeler gauge, inspect for warping across valve ports (2) and coolant passages (5). Warping should not exceed 0.001 in. (0.025 mm).

CAUTION

Machining of cylinder head surface is not to exceed 0.040 in. (1.02 mm) or damage to equipment will result.

    d. Replace or resurface cylinder head (1) if warping exceeds limits. Notify your supervisor.

NOTE

Perform steps 3 and 4 if cylinder head was resurfaced.

3. Install valves (3) in their respective valve ports (2).
4. Using gauge block, measure depth of valves (3). Mark valve seat(s) (6) for resurfacing if depth is less than 0.043 in. (1.09 mm).
5. Remove valves (3) from cylinder head (1).
6. Inspect valve springs (4) for distortion, cracks, and collapsed coils. Refer to table 5-9, Cylinder Head Wear Limits, for measurements. Replace valve springs (4) if damaged or worn.
7. Using spring tester, compress valve springs (4) to 2 in. (51 mm). Load measured should be 101.2-115 lb (45.9-52.2 kg). Replace valve spring(s) (4) if not within limits.
8. Inspect valve spring retainers (8) for wear and distortion. Replace valve spring retainer(s) (8) if worn or distorted.

Table 5-9. Cylinder Head Wear Limits.

ITEM NO.	ITEM/POINT OF MEASUREMENT	WEAR LIMITS/TOLERANCES	
		INCHES	MILLIMETERS
3	Intake and exhaust valves Rim thickness Valve stem diameter	0.060 0.373	1.52 9.47
14	Exhaust valve guide Valve guide length Valve guide bore, inner	0.896 0.376	22.76 9.55
15	Intake valve guide Valve guide length Valve guide bore, inner	0.893 0.376	22.68 9.55
4	Valve springs Free length	2.585	65.66

5-25. CYLINDER HEAD REPAIR (Contd)

9. Inspect valve seats (6) for wear, burrs, pits, and cracks. Replace cylinder head (1) if worn, pitted, cracked, or burred. Mark valve seat (6) for resurfacing.

10. Inspect exhaust (14) and intake (15) valve guides for chips, cracks, burrs, breaks, and wear. Replace cylinder head (1) if chipped, cracked, burred, broken, or worn.

11. Measure valve guide length (12) and inside diameter of valve guide bore (13). Refer to table 5-9, Cylinder Head Wear Limits, for measurements. Replace cylinder head (1) if not within limits.

12. Inspect valves (3) for cracks, bends, and wear. Replace valve(s) (3) if cracked, bent, or worn.

13. Measure outside diameter of valve stem (9). Refer to table 5-9, Cylinder Head Wear Limits, for measurements. Replace valve(s) (3) if not within limits.

14. Measure rim (10) thickness of valves (3). Refer to table 5-9, Cylinder Head Wear Limits, for measurement. Replace valve(s) (3) if not within limits.

15. Inspect collet grooves (11) for wear.

    a. Install new collets (7) in collet grooves (11).

    b. If collets (7) fit loosely, replace valve(s) (3).

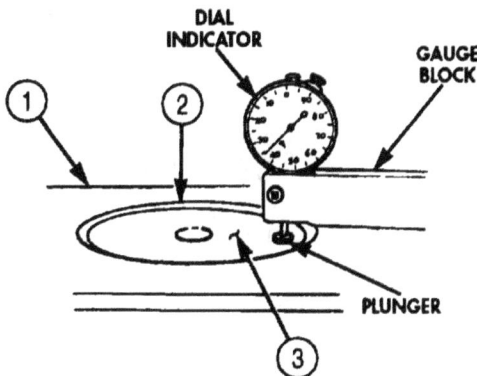
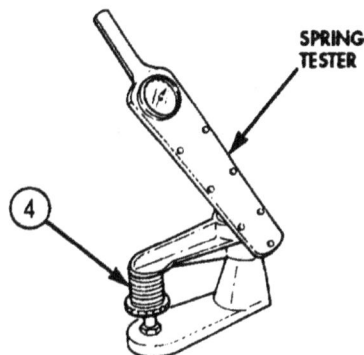
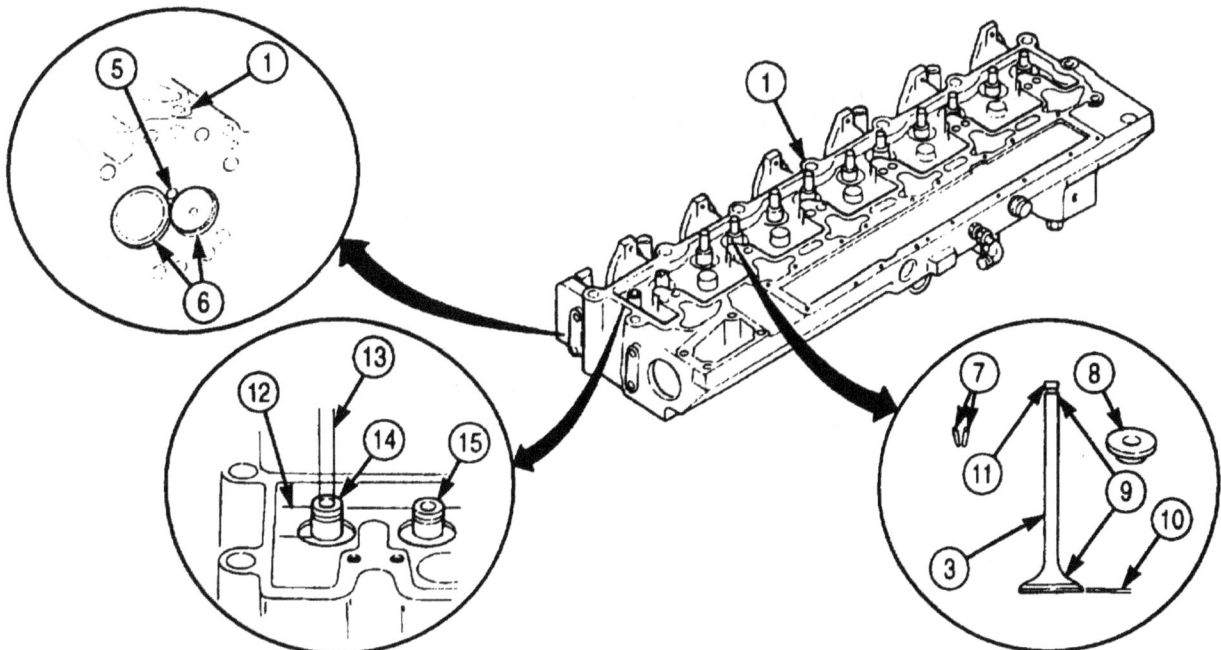

5-25. CYLINDER HEAD REPAIR (Contd)

d. Repair

1. Apply light coat of lapping compound to valve (2) and lap valve (2) to valve seat insert (1).
2. Clean lapping compound from valve (2) and valve seat insert (1).
3. Measure width of valve seat insert (1). Limits are 0.060-0.080 in. (1.52-2.03 mm). If beyond limits, replace cylinder head (4).
4. Check contact surface between valve (2) and valve seat insert (1) as follows:

   a. Apply light coat of pigment to valve face (3).

   b. Lower valve (2) into valve guide (5) and let valve (2) drop against valve seat insert (1).

   NOTE
   Do not rotate valve when pushing valve against valve seat insert.

   c. Push up on valve stem (6) until valve face (3) is about 1 in. (25.4 cm) above valve seat insert (1).

   d. While releasing valve stem (6), apply pressure to valve (2) until valve face (3) makes contact with valve seat insert (1). Repeat several times to get good imprint on pigment.

   NOTE
   Do not smear pigment from valve.

   e. Remove valve (2).

   f. Imprint on pigment should have an even contact surface all around center of valve face (3). If pigment imprint is uneven, mark valve (2) and valve seat insert (1) for grinding.

   NOTE
   Repeat step 4 for all valves to check contact surface between valve and valve seat insert.

5. Grind valve (2) and valve seat insert (1) if valve (2) is new or has an uneven contact surface.

   a. Use a 30° grinding wheel on valve seat insert (1) for intake port (7).

   b. Use a 45° grinding wheel on valve seat insert (1) for exhaust port (8).

   NOTE
   If contact surface is at bottom of valve face, valve seat insert will require more grinding with 15° wheel.

   c. To center contact surface on valve face (3), grind area (11) of valve seat insert (1) with a 60° wheel, and area (12) of valve seat insert (1) with a 15° wheel.

   d. Using valve refacing machine, grind face of intake valve (2) to a 30° angle or a 45° angle for exhaust valve.

6. If valve stem tip (10) is not flat, dress tip for flatness on grinding wheel.
7. Measure valve (2) depth from bottom of cylinder head (4) to surface when valve (2) is seated against valve seat insert (1).

   a. Position gauge block flat on surface of cylinder head (4) with plunger extended to cylinder head (4) surface. Adjust dial indicator to zero.

   b. Position gauge block flat on surface of cylinder head (4) with plunger extended to valve (2).

   c. Valve (2) depth must measure 0.035-0.056 in. (0.89-1.42 mm) below bottom surface of cylinder head (4).

   d. If measurement is not within limits, use new valve (2) and repeat steps a through c.

   e. If measure is beyond limit, replace cylinder head (4).

8. Using pigment, check contact surface (9) between face of valve (2) and valve seat insert (1) (step 4d.).
9. Clean cylinder head (4) and valves (2) after valves (2) and valve seat inserts (1) have been ground (para. 2-14).

5-25. CYLINDER HEAD REPAIR (Contd)

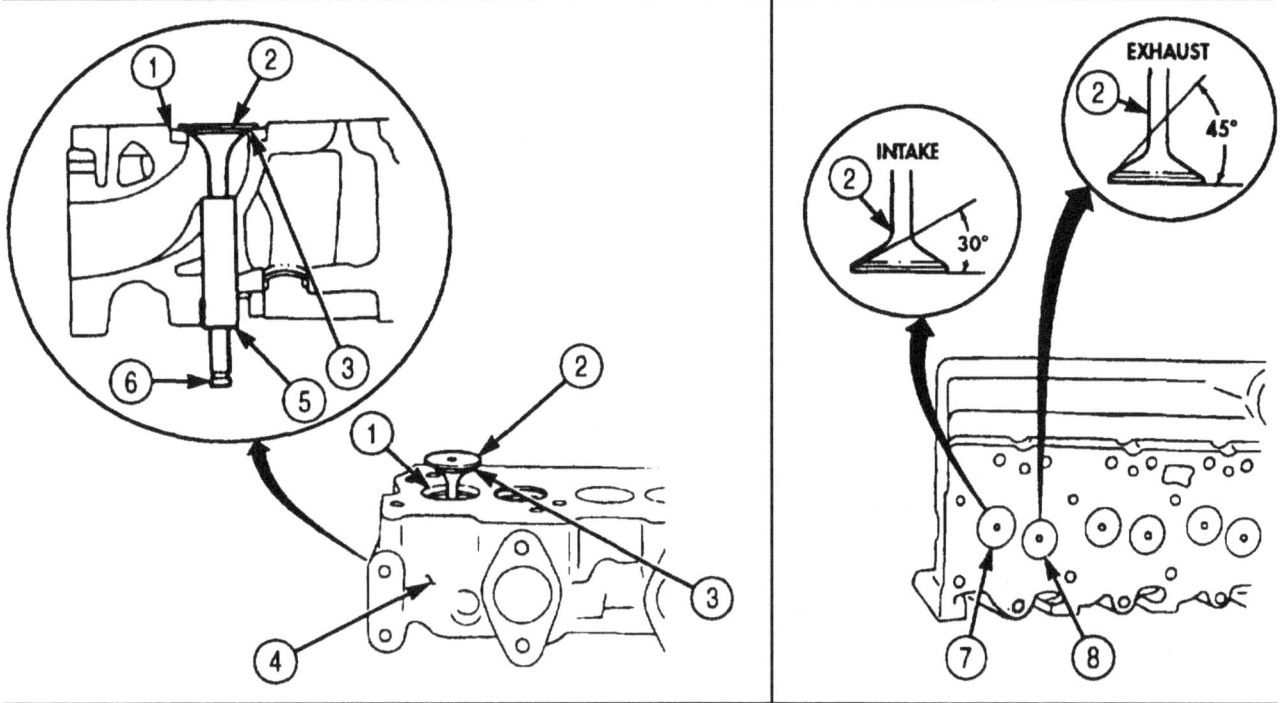

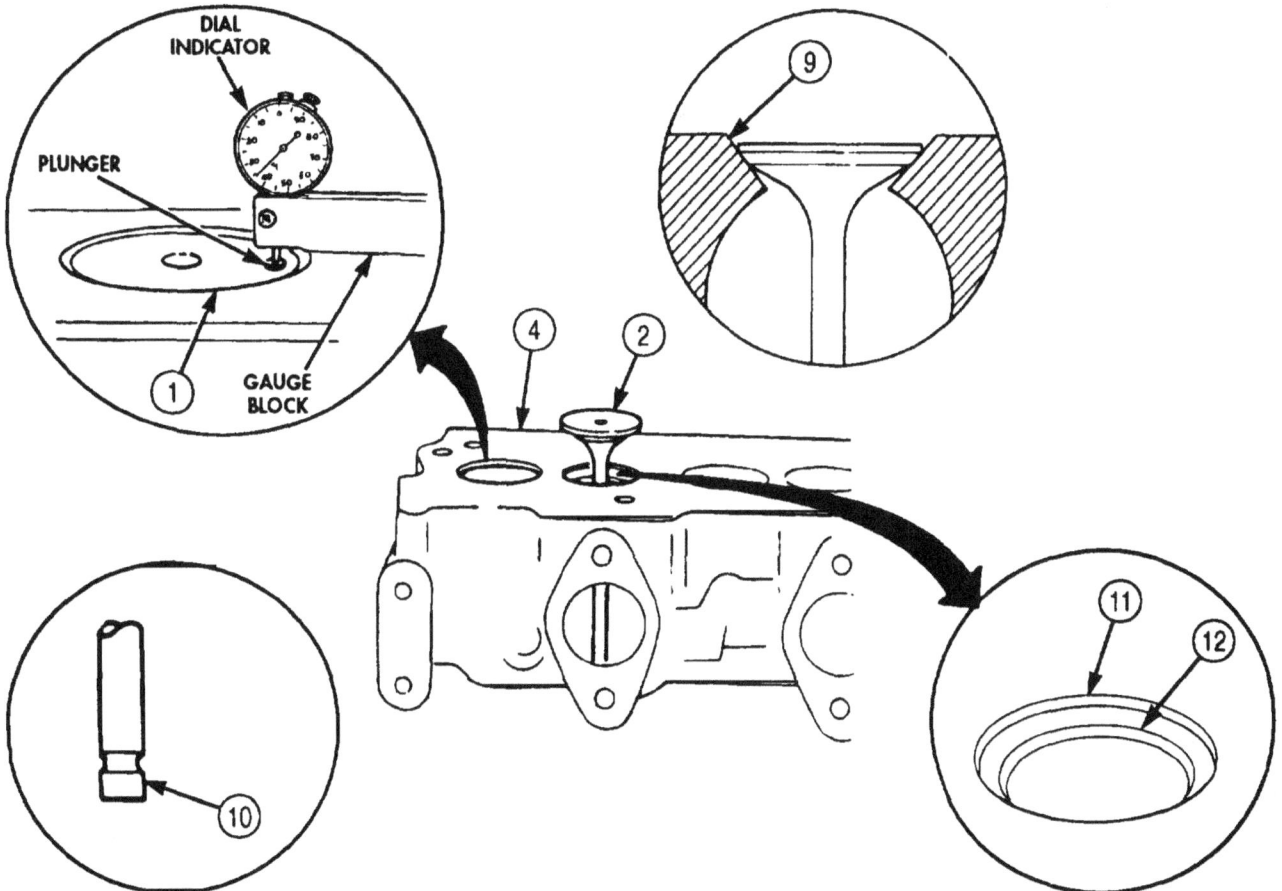

5-25. CYLINDER HEAD REPAIR (Contd)

e. Assembly

NOTE

• Wrap male pipe threads with antiseize tape before installation.

• Apply sealing compound to mating surfaces of expansion plugs before installation.

1. Install plug (8) on rear face of cylinder head (10).
2. Install two new expansion plugs (9) on front and rear faces of cylinder head (10).
3. Install new expansion plug (7) on rear face of cylinder head (10).
4. Install five new expansion plugs (6) on top face of cylinder head (10).

CAUTION

Position cylinder head on exhaust port face to install intake and exhaust valves. Use wooden or protective surface work bench to prevent damage to cylinder head. Bench must be clean.

NOTE

Ensure intake and exhaust valves are installed in original locations as previously noted.

5. Dip valve (11) stems in clean engine oil and install in valve guides (5) from face side of cylinder head (10).
6. Carefully position cylinder head (10) face down on bench.
7. Install twelve new seals (4), valve springs (3), and spring retainers (2) on valves (11).
8. Using spring compressor, compress valve springs (3) until collet grooves (12) on valves (11) are exposed.

NOTE

Ensure all valves are locked by collets.

9. Install twenty-four new collets (1) in collet grooves (12) and slowly release spring compressor.

5-25. CYLINDER HEAD REPAIR (Contd)

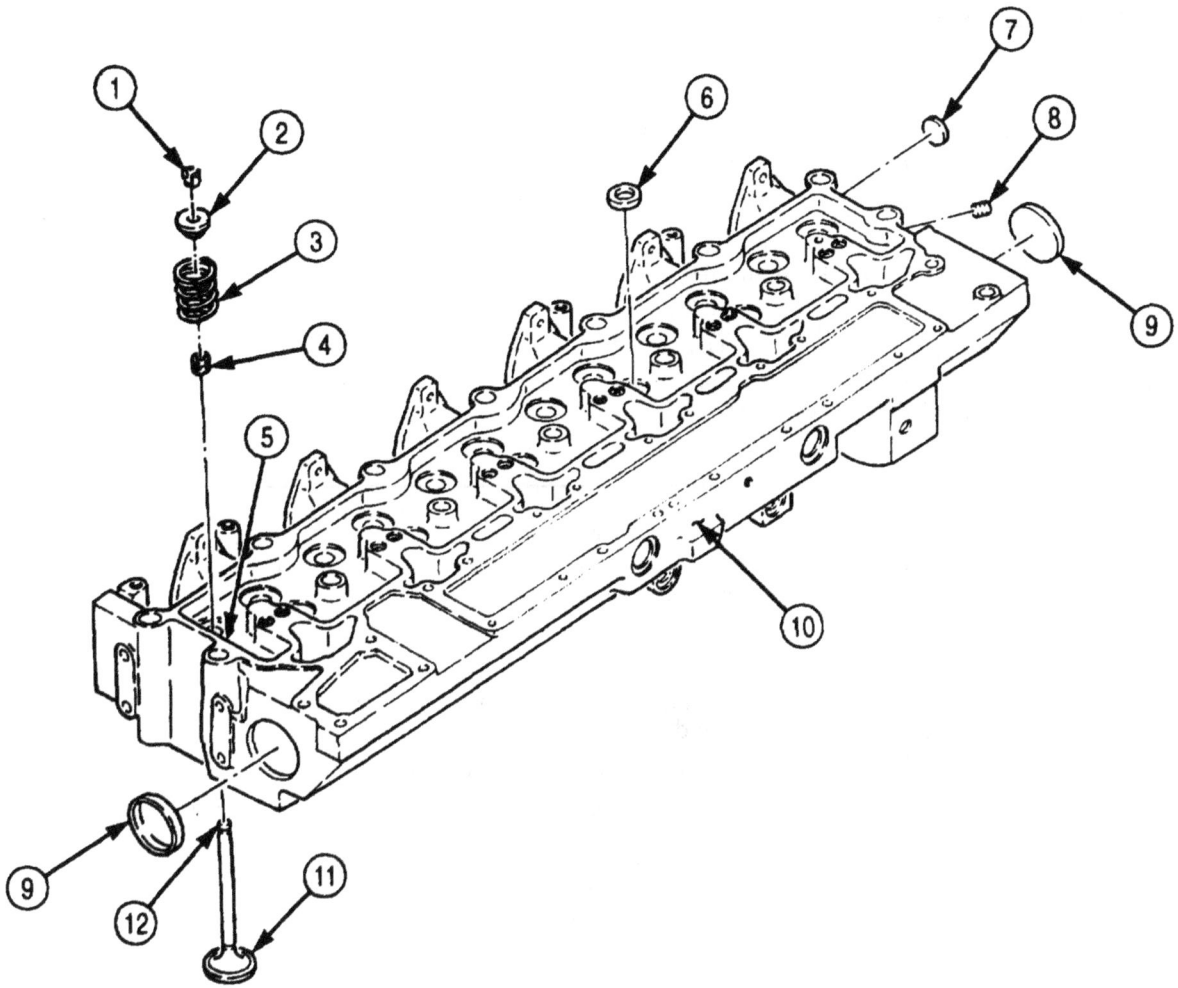

FOLLOW-ON TASK: Install cylinder head (para. 4-41).

5-26. PISTON AND CONNECTING ROD MAINTENANCE

THIS TASK COVERS:

a. Removal
b. Disassembly
c. Cleaning and Inspection

d. Assembly
e. Installation

INITIAL SETUP:

APPLICABLE MODELS
M939A2

SPECIAL TOOLS
Engine barring tool (Appendix E, Item 43)
Ring compressor (Appendix E, Item 32)

TOOLS
General mechanic's tool kit (Appendix E, Item 1)
Keystone gauge (Appendix E, Item 54)
Inside micrometer (Appendix E, Item 82)
Outside micrometer (Appendix E, Item 80)
Vernier calipers (Appendix E, Item 159)
Torque wrench (Appendix E, Item 144)
Depth gauge (Appendix E, Item 81)
Feeler gauge
Soft-headed hammer

MATERIALS/PARTS
Piston rings (Appendix D, Item 518)
Two retaining rings (Appendix D, Item 539)
Connecting rod bearing (Appendix D, Item 40)
Lubricating oil (Appendix C, Item 50)
Plastigage (Appendix C, Item 55)

REFERENCES (TM)
TM 9-2320-272-24P

EQUIPMENT CONDITION
• Cylinder head removed (para. 4-41).
• Oil pan and suction tube removed (para. 4-47).

NOTE

All piston assemblies are replaced the same way. This procedure
covers replacement of one piston assembly.

a. Removal

1. Using engine barring tool, rotate crankshaft (6) to lower pistons (1) enough to remove all carbon from upper inside wall of each cylinder (3).
2. Remove two nuts (8) from screws (5) and connecting rod cap (7).
3. Using soft-headed hammer, tap screws (5) and remove connecting rod cap (7) from crankshaft (6).

CAUTION

Use a tape-protected tool to push piston from cylinder block.
Failure to do so may cause damage to cylinder liners.

NOTE

Assistant will help with step 4.

4. Using tape-protected tool, push connecting rod (2) and piston (1) from engine block (4).

b. Disassembly

CAUTION

Pistons, connecting rods, and connecting rod caps must be kept together
as an assembly. Missing parts may cause damage to equipment.

1. Remove two retaining rings (12) from bore of piston (1). Discard retaining rings (12).
2. Remove piston pin (13) and connecting rod (2) from piston (1).
3. Remove ring (9), compression ring (10), and oil ring (11) from piston (1). Discard rings (9), (10), and (11).

5-26. PISTON AND CONNECTING ROD MAINTENANCE (Contd)

4. Remove two screws (5) from connecting rod (2).
5. Remove upper half bearing shell (14) from connecting rod (2). Discard upper half bearing shell (14).
6. Remove lower half bearing shell (15) from connecting rod cap (7). Discard lower half bearing shell (15).

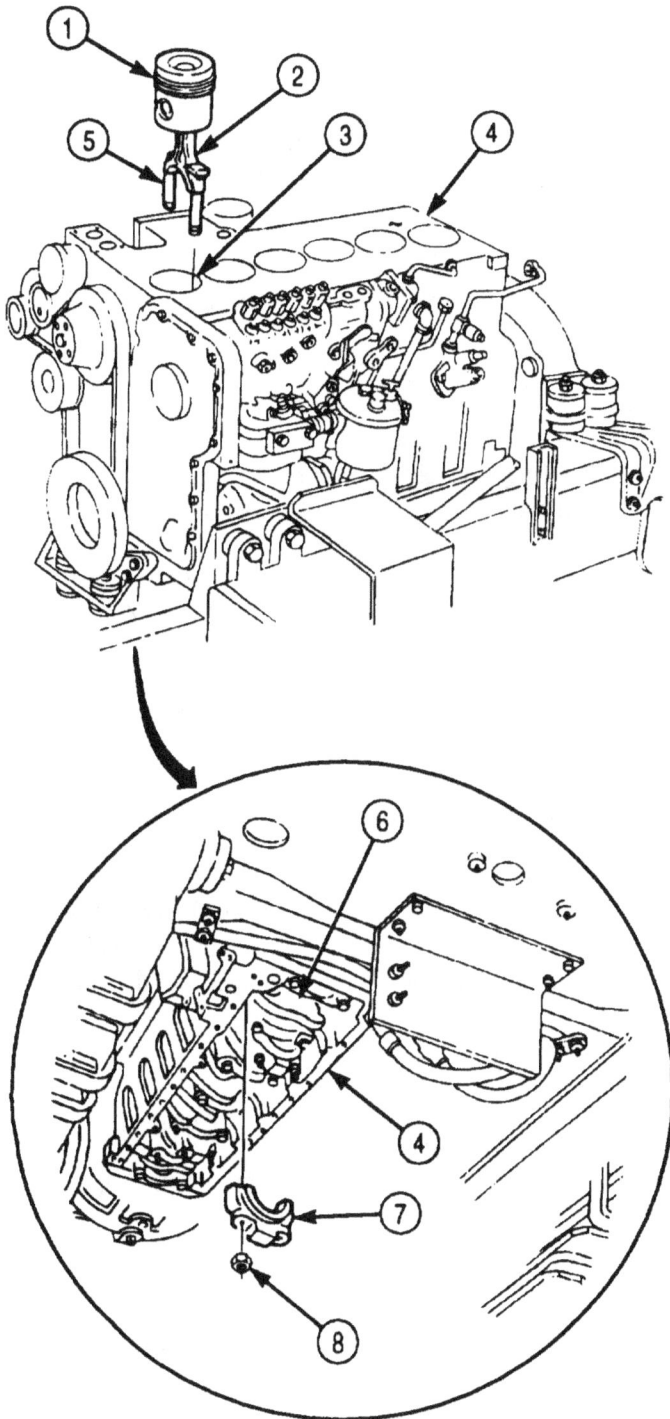

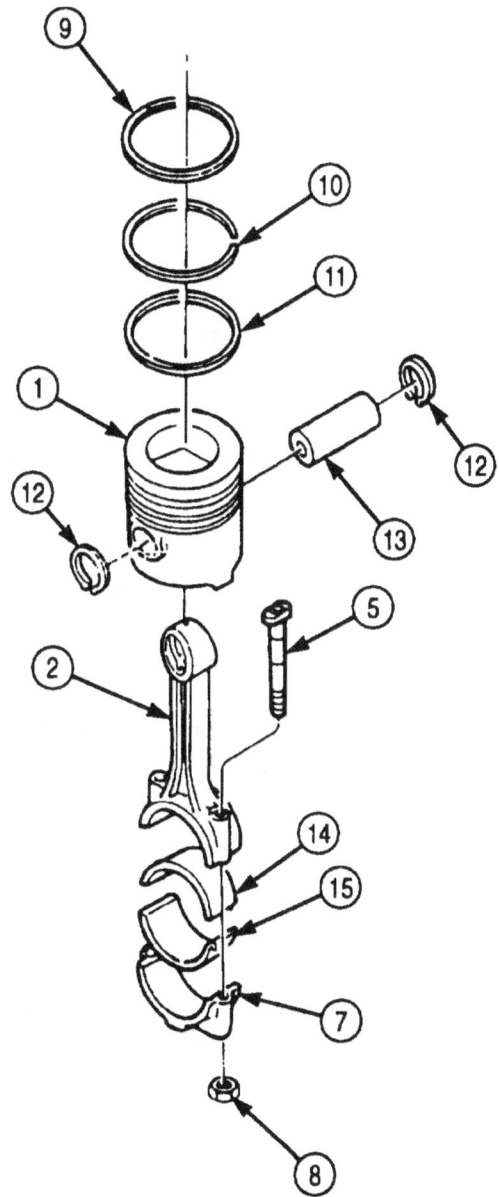

5-26. PISTON AND CONNECTING ROD MAINTENANCE (Contd)

c. Cleaning and Inspection

1. For general cleaning instructions, refer to para. 2-14.

2. For general inspection instructions, refer to para. 2-15.

3. Replace all parts failing inspection.

NOTE

Measurements to check piston skirt outside diameters are taken at right angle to piston pin bore.

4. Measure outside diameter of piston skirt (3) at several places. If diameter is less than 4.479 in. (113.8 mm), replace piston (4).

5. Check piston ring groove (5) for wear.

    a. Insert keystone gauge into ring groove (5).

    b. If shoulders of keystone gauge touch either end of ring groove (5), replace piston (4).

6. Check compression ring (6) and oil ring (7) grooves for wear.

    a. Insert new compression ring (1) into compression ring groove (6).

    b. Try to insert 0.006 in. (0.152 mm) feeler gauge between compression ring (1) and compression ring groove (6).

    c. If feeler gauge enters compression ring groove (6) without force, replace piston (4).

    d. Repeat steps a through c using new oil ring (2) and 0.005 in. (0.127 mm) feeler gauge to check oil ring groove (7) for wear.

7. Measure inside diameter of piston pin bore (9). If inside diameter exceeds 1.773 in. (45.03 mm), replace piston (4).

8. Measure outside diameter of piston pin (8). If outside diameter exceeds 1.771 in. (44.98 mm), replace piston pin (8).

9. Inspect I-beam section (15) of connecting rod (11) for nicks, dents, and gouges. If nicked, dented, or gouged, replace connecting rod (11).

10. Install two connecting rod screws (12) in connecting rod (11).

11. Install connecting rod cap (14) on connecting rod (11) with two nuts (13). Tighten nuts (13) to 73 lb-ft (99 N•m).

12. Check crankshaft journal bore (10) of connecting rod (11) for wear:

    a. Measure inside diameter of crankshaft journal bore (10) at several points 30° from each other.

    b. If inside diameter measurement exceeds 2.9926 in. (76.012 mm), replace connecting rod (11).

13. Measure inside diameter of piston pin bore (16). If inside diameter exceeds 1.773 in. (45.03 mm), replace connecting rod (11).

14. Remove two nuts (13) and connecting rod cap (14) from connecting rod (11).

15. Check connecting rod journals (18) on crankshaft (17) for wear.

    a. Measure outside diameter of six connecting rod journals (18) at A-A and B-B. If difference exceeds 0.0005 in. (0.013), replace crankshaft (17) (para. 5-29).

    b. Measure outside diameter of six connecting rod journals (18) at C-C and D-D. If difference exceeds 0.002 in. (0.05 mm), replace crankshaft (17) (para. 5-29).

5-26. PISTON AND CONNECTING ROD MAINTENANCE (Contd)

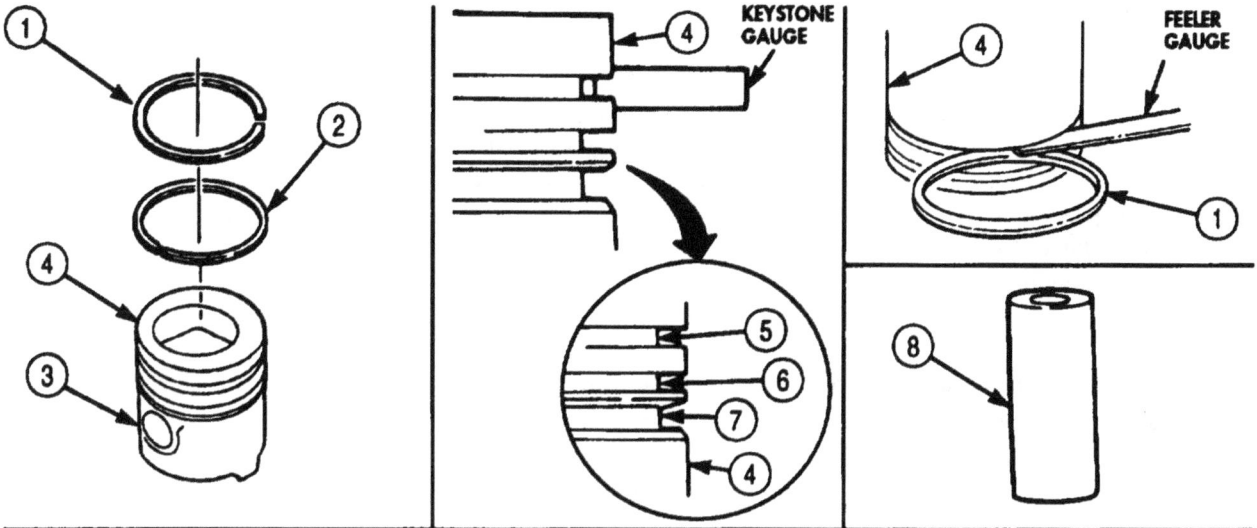

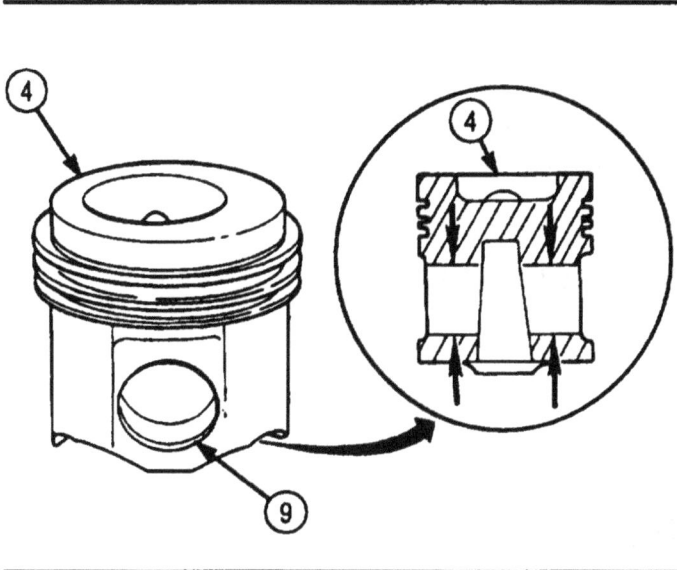

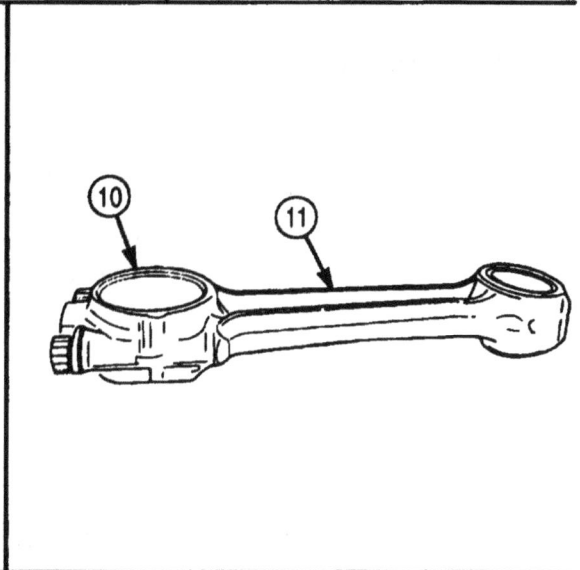

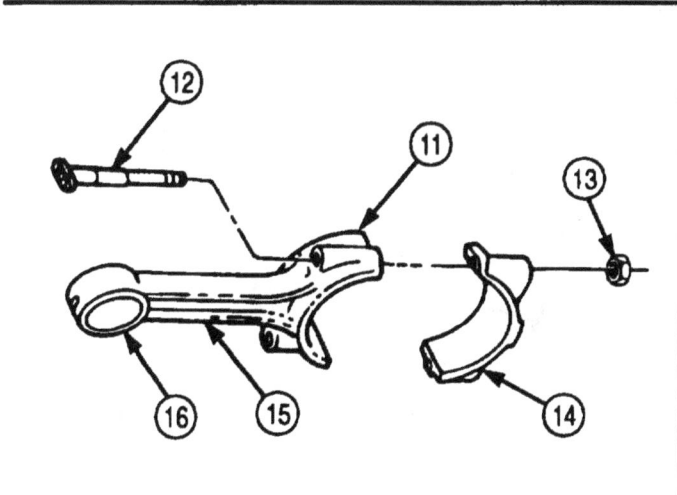

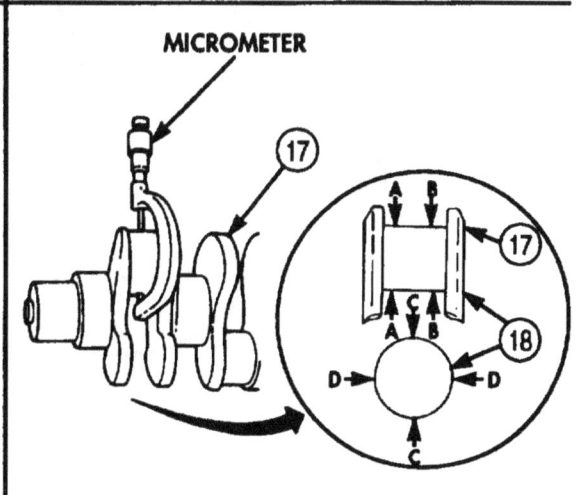

5-26. PISTON AND CONNECTING ROD MAINTENANCE (Contd)

d. Assembly

NOTE
- Ensure numbers on pistons, connecting rods, and connecting rod caps match numbers on the mating cylinders.
- Perform step 1 if assembling with new connecting rod.

1. Install two connecting rod screws (8) on connecting rod (2). Ensure heads of screws (8) are on flat of connecting rod (2).
2. Install retaining ring (13) in pin groove of piston (1).
3. Position connecting rod (2) in piston (1). Ensure connecting rod (2) matches FRONT mark on piston (1).
4. Apply clean engine oil to lubricate piston pin (7) and piston pin bore (12).
5. Install piston pin (7) through piston (1) and connecting rod (2).
6. Install retaining ring (6) in pin groove of piston (1).

NOTE
Piston rings must be installed on piston with the word TOP or supplier's mark facing upward.

7. Install new oil ring (5) in oil ring groove (14).
8. Using piston ring expander, install new compression ring (4) and new ring (3) on piston (1).

NOTE
Align bearing shell tabs with connecting rod and connecting rod cap slots when installing.

9. Install new upper half bearing shell (9) in connecting rod (2).
10. Install new lower half bearing shell (10) in connecting rod cap (11).

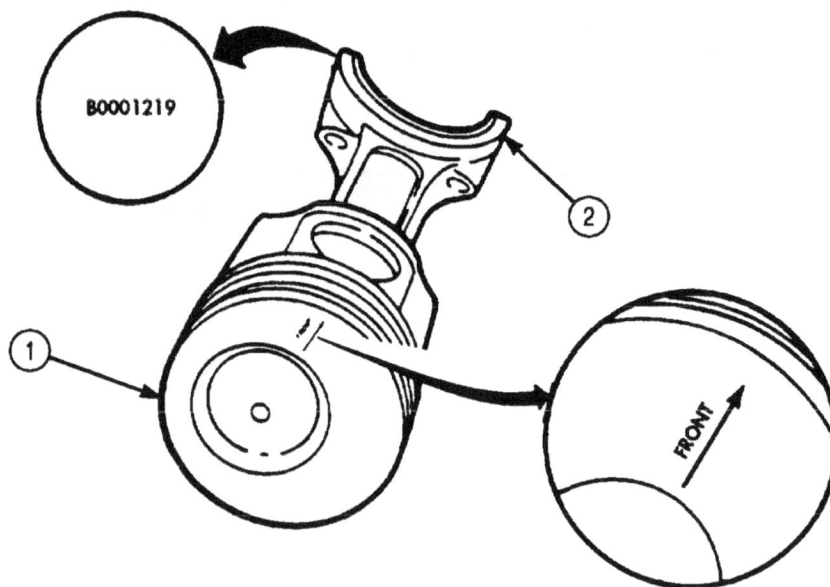

5-26. PISTON AND CONNECTING ROD MAINTENANCE (Contd)

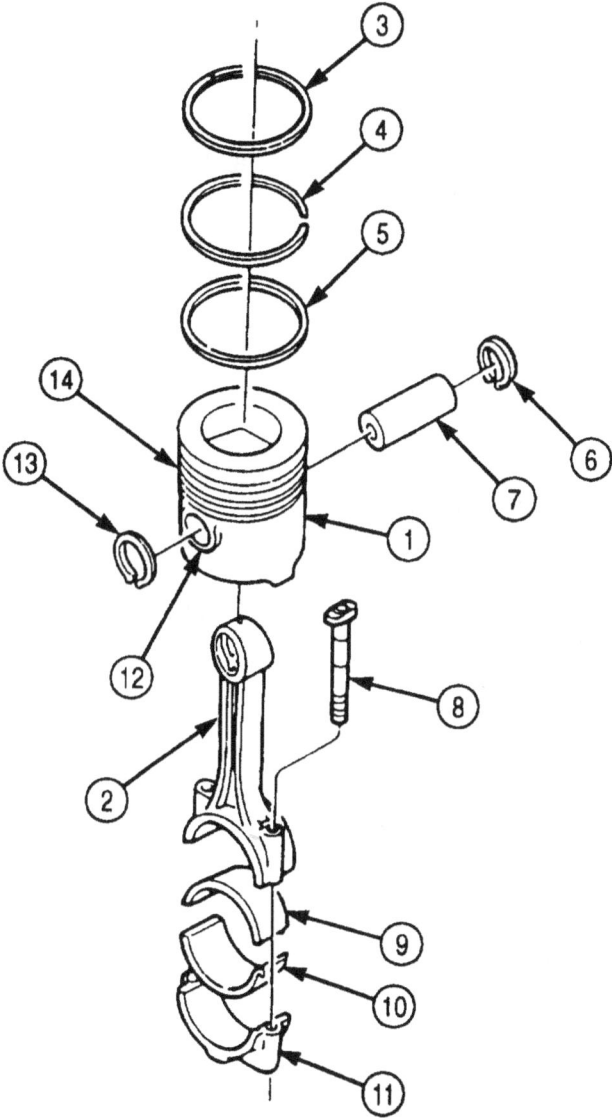

5-26. PISTON AND CONNECTING ROD MAINTENANCE (Contd)

e. Installation

1. Apply clean engine oil to lubricate piston (3) and cylinder bore (5).

NOTE

Ensure piston rings are staggered properly so they are not in line with each other or piston pin.

2. Lubricate ring compressor and install over piston rings (4) and piston (3).

3. Using engine barring tool, rotate crankshaft (91 until connecting rod journal (11) is at bottom dead center.

CAUTION

- Ensure piston assembly number matches that of mating cylinder. Failure to do so will cause poor engine performance.

- Ring compressor must be held firmly against engine block to prevent compressor from slipping and causing piston ring breakage when pushing piston into cylinder liner.

- Do not force piston assembly into liner. If piston does not install freely in liner, remove and check for broken rings.

NOTE

Ensure piston assemblies are installed with FRONT mark facing front of engine.

4. Insert connecting rod (7) in cylinder bore (5) and hold ring compressor tight and firmly seated against engine block (6).

5. Using rubber or wooden mallet handle, push piston (3) from ring compressor and into cylinder bore (5) until upper half bearing (8) seats against connecting rod journal (11).

6. Place a piece of plastic strip across lower half bearing shell (2).

7. Install connecting rod cap (1) on connecting rod (7) with two nuts (10). Ensure numbers on connecting rod cap (1) and connecting rod (7) are aligned.

8. Tighten two nuts (10) as follows:

   a. Tighten two nuts (10130 lb-ft (41 N•m).

   b. Tighten two nuts (10) 60 lb-ft (81 N•m).

   c. Tighten two nuts (10) 90 lb-ft (122 N•m).

9. Remove two nuts (10) and connecting rod cap (1) from connecting rod (9).

10. Using plastigage, measure width of plastic strip. Width should be 0.0013-0.0046 in. (0.033-0.117 mm). If width is not within limits, inspect connecting rod journal (11) (task c.).

11. Clean connecting rod cap (1) and install on connecting rod (7) with two nuts (10). Ensure numbers on connecting rod (7) and connecting rod cap (1) are aligned.

12. Tighten nuts (10) as follows:

    a. Tighten two nuts (10) 30 lb-ft (41 N•m).

    b. Tighten two nuts (10) 60 lb-ft (81 N•m).

    c. Tighten two nuts (10) 90 lb-ft (122 N•m).

13. Check connecting rod side clearance.

    a. Move connecting rod (7) back and forth on connecting rod journal (11).

    b. Using feeler gauge, measure connecting rod (7) side clearance. Clearance should be 0.004-0.012 in. (0.1-0.030 mm). If not within limits, remove connecting rod cap (1) and inspect for proper size and seating of lower half bearing shell (2), and for dirt or burrs.

    c. Repeat steps 12 and 13 to install connecting rod cap (1).

5-26. PISTON AND CONNECTING ROD MAINTENANCE (Contd)

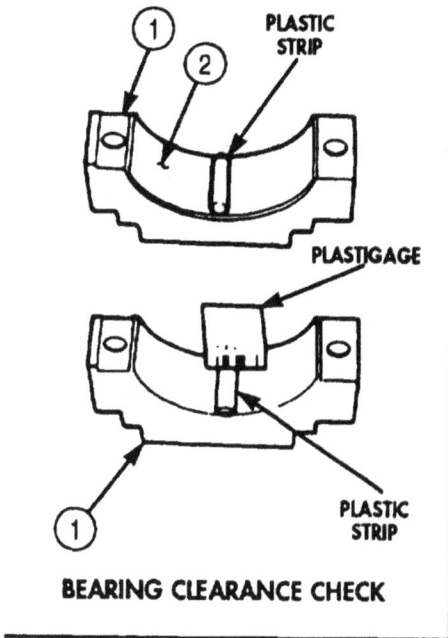

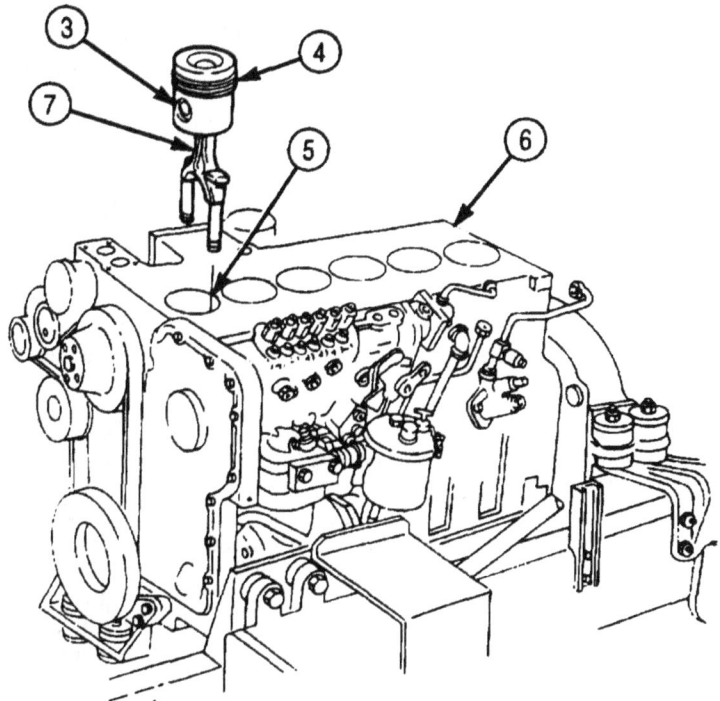

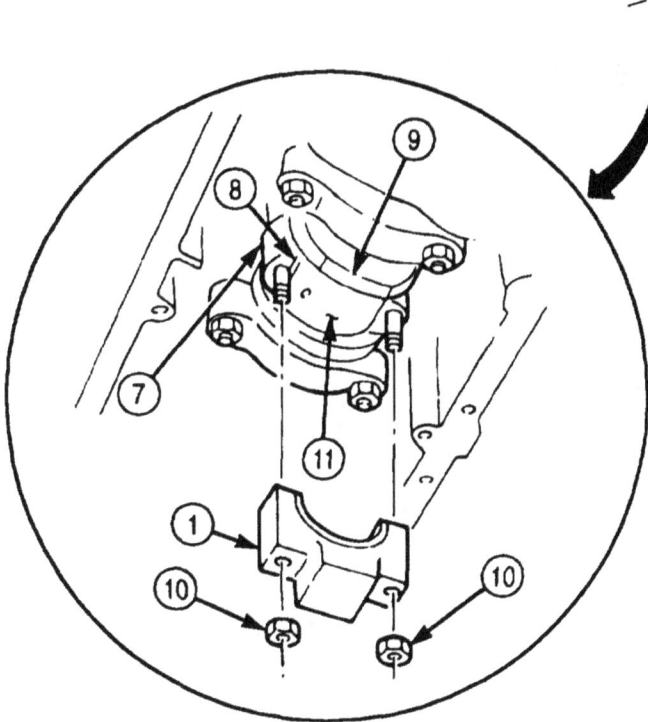

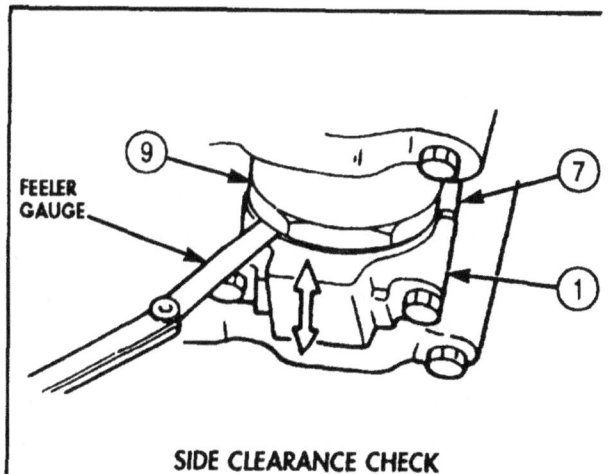

PLASTIC STRIP

PLASTIGAGE

PLASTIC STRIP

BEARING CLEARANCE CHECK

FEELER GAUGE

SIDE CLEARANCE CHECK

FOLLOW-ON TASKS: • Install cylinder head (para. 4-41).
• Install oil pan and suction tube (para. 4-47).

5-27. CAMSHAFT AND GEAR MAINTENANCE

THIS TASK COVERS:

a. Check Backlash and End Play
b. Removal
c. Disassembly

d. Cleaning and Inspection
e. Assembly
f. Installation

INITIAL SETUP:

APPLICABLE MODELS
M939A2

SPECIAL TOOLS
Cam bushing replacement tool
  (Appendix E, Item 26)

TOOLS
General mechanic's tool kit (Appendix E, Item 1)
Gear puller (Appendix E, Item 102)
Dial indicator (Appendix E, Item 36)
Inside micrometer (Appendix E, Item 82)
Outside micrometer (Appendix E, Item 80)
Vernier caliper (Appendix E, Item 159)
Torque wrench (Appendix E, Item 144)
Arbor press
Prybar

MATERIALS/PARTS
Dowel pin (Appendix D, Item 98)
Lockwasher (Appendix D, Item 396)
GAA grease (Appendix C, Item 28)
Adhesive sealant (Appendix C, Item 4)
Rags (Appendix C, Item 58)
Drycleaning solvent (Appendix C, Item 71)

REFERENCES (TM)
TM 9-2320-272-24P

EQUIPMENT CONDITION
• Engine mounted on repair stand (para. 4-9).
• Rocker levers and push rods removed (para. 4-41).
• Flywheel housing removed (para. 4-43).
• Oil pan and suction tube removed (para. 4-47).
• Front gearcase cover removed (para. 4-48).

GENERAL SAFETY INSTRUCTIONS
• Keep fire extinguisher nearby when using drycleaning solvent.
• Drycleaning solvent is flammable and toxic. Do not use near open flame.

a. Check Backlash and End Play

1. Rotate engine block (2) so bottom is facing upward.
2. Install dial indicator on engine block (2) and crankshaft gear (1).

NOTE
Prevent crankshaft from rotating when checking backlash.
Reading will be the total of both gears.

3. Rotate camshaft gear (3) clockwise until teeth contact teeth of crankshaft gear (1).
4. Set dial indicator to zero.
5. Rotate camshaft gear (3) counterclockwise until teeth contact crankshaft gear (1) teeth. Ensure crankshaft gear (1) does not turn.
6. Backlash of camshaft gear (3) should read 0.005-0.013 in. (0.13-0.33 mm). If reading is not within limits, replace camshaft (tasks b through f.).
7. Position dial indicator on face of camshaft gear (3). Press camshaft gear (3) towards rear of engine block (1).
8. Set dial indicator to zero.
9. Pull camshaft gear (3) to front of engine block (1).
10. End play of camshaft gear (3) should read 0.006-0.010 in. (0.152-0.254 mm). If reading is not within limits, replace camshaft support (tasks b through f).

5-27. CAMSHAFT AND GEAR MAINTENANCE (Contd)

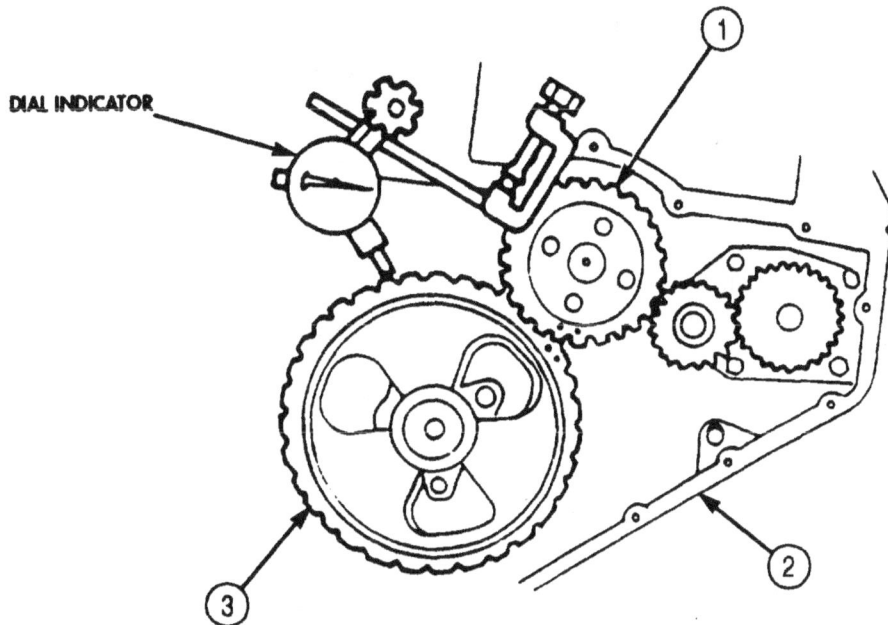

DIAL INDICATOR

CHECKING BACKLASH

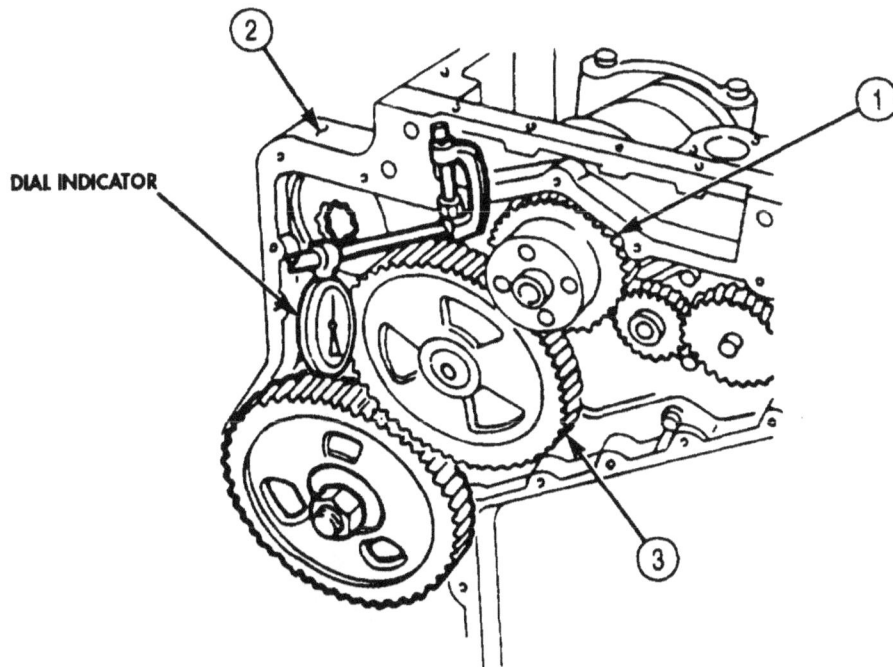

DIAL INDICATOR

CHECKING END PLAY

5-27. CAMSHAFT AND GEAR MAINTENANCE (Contd)

b. Removal

1. Install two vibration damper screws (1) on crankshaft (3).
2. Using prybar, prevent crankshaft (3) from rotating.
3. Remove nut (6) and lockwasher (7) from drive gear (8). Discard lockwasher (7).
4. Using gear puller, remove drive gear (8) from fuel injection pump shaft (5).
5. Rotate crankshaft (3) and align timing marks (14) on crankshaft gear (2) and camshaft gear (4).
6. Remove two screws (12) from camshaft support (11) and engine block (9).

CAUTION

Use care when removing camshaft. Damage to camshaft and bushings may result.

7. Remove camshaft (13) and camshaft support (11) from engine block (9).

NOTE

Tag all valve tappets for installation.

8. Rotate crankshaft (3) as necessary to remove twelve valve tappets (10) from engine block (9).

c. Disassembly

NOTE

Before removing camshaft gear from camshaft, perform task d of this procedure. If, as a result of inspection, the camshaft or camshaft gear requires replacement, perform steps 1 and 2.

1. Using arbor press, press camshaft gear (4) from camshaft (13).
2. Remove dowel pin (15) from camshaft (13). Discard dowel pin (15).

5-27. CAMSHAFT AND GEAR MAINTENANCE (Contd)

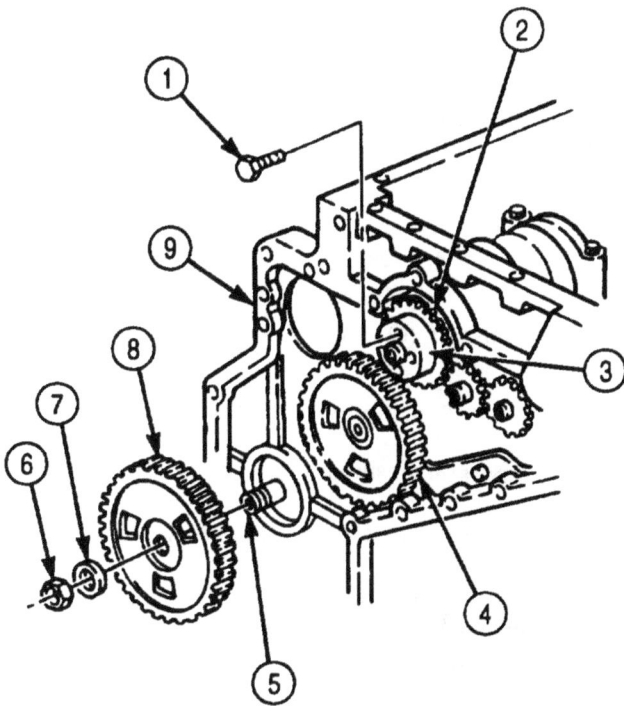

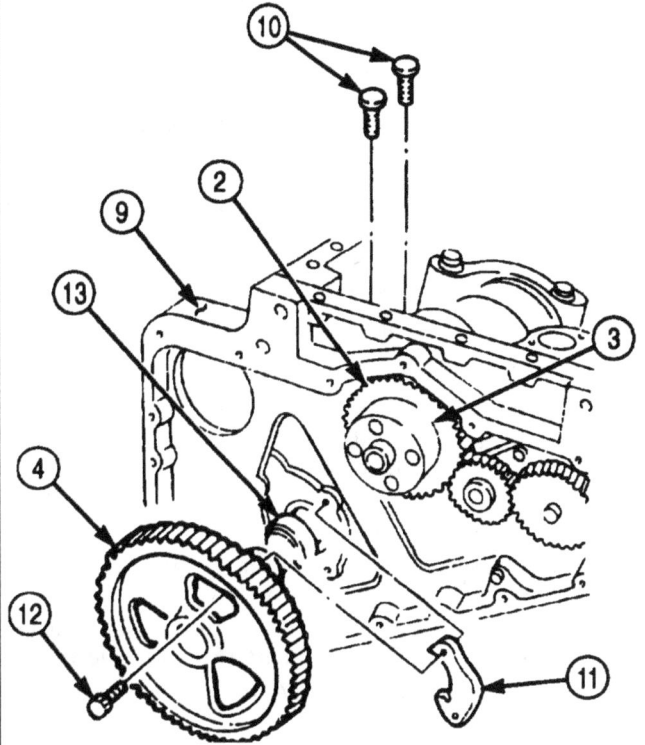

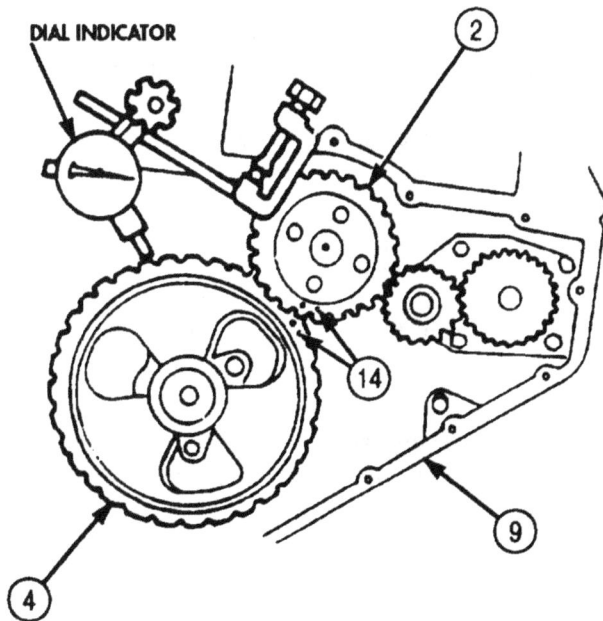

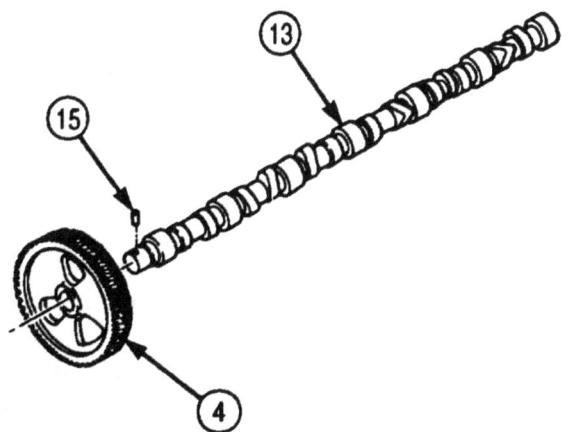

DIAL INDICATOR

5-27. CAMSHAFT AND GEAR MAINTENANCE (Contd)

d. Cleaning and Inspection

WARNING

Drycleaning solvent is flammable and toxic. Do not use near open flame and always have a fire extinguisher nearby when solvents are used. Use only in well-ventilated places, wear protective clothing, and dispose of cleaning rags in approved container. Failure to do this may result in injury or death to personnel and/or damage to equipment.

1. Clean all camshaft (6) components with drycleaning solvent and dry with clean rag.
2. Inspect camshaft support (1) for wear. Replace camshaft support (1) if excessively worn.
3. Inspect camshaft (6) for cracks, pits, and breaks. Replace camshaft (6) if cracked, pitted, or broken.
4. Inspect exhaust (4) and intake (2) camshaft lobes for wear.
    a. Measure outside diameter of exhaust camshaft lobe (4) at several positions. Replace camshaft (6) if outside diameter is less than 2.0313 in. (1.5950 mm).
    b. Measure outside diameter of intake camshaft lobe (2) at several positions. Replace camshaft (6) if outside diameter is less than 2.0383 in. (51.7728 mm).
    c. Measure outside diameter of lift pump lobe (5) at several positions. Replace camshaft (6) if outside diameter is less than 1.626 in. (59.9618 mm).
5. Inspect camshaft journals (3) for wear. Measure outside diameter of camshaft journals (3) at several positions. Replace camshaft (6) if outside diameter is less than 2.3067 in. (59.9618 mm).
6. Inspect camshaft gear (7) for cracks, pits, and breaks. Replace camshaft gear (7) if cracked, pitted, or broken.
7. Inspect camshaft gear teeth (8) for pitting and cracks at root of teeth (8). Replace camshaft gear (7) if pitted or cracked.
8. Inspect camshaft bushings (9) for wear. Measure inside diameter of camshaft bushing (9) at several positions. If inside diameter exceeds 2.367 in. (60.122 mm), mark camshaft bushing (9) for replacement.

NOTE
Perform steps 9 through 14 for replacing camshaft bushings.

9. Remove expansion plug (11) from rear of engine block (10). Discard expansion plug (11).
10. Using cam bushing replacement tool, remove camshaft bushing (9) from bushing bore (14). Discard camshaft bushing (9).
11. Measure inside diameter of camshaft bores (14). Replace engine block (10) if inside diameter exceeds 2.520 in. (64.008 mm).

CAUTION
When installing camshaft bushing, ensure oil hole and oil hole passages are aligned. Failure to do so will result in premature failure and damage to engine.

12. Mark engine block (10) and new camshaft bushing (9) to ensure oil holes (12) and (13) are aligned.
13. Using cam bushing replacement tool, install new camshaft bushing (9) in camshaft bushing bore (14).

NOTE
Perform step 14 if rear camshaft bushing was removed.

14. Apply adhesive sealant to mating surfaces of new expansion plug (11) and install on engine block (10).

5-27. CAMSHAFT AND GEAR MAINTENANCE (Contd)

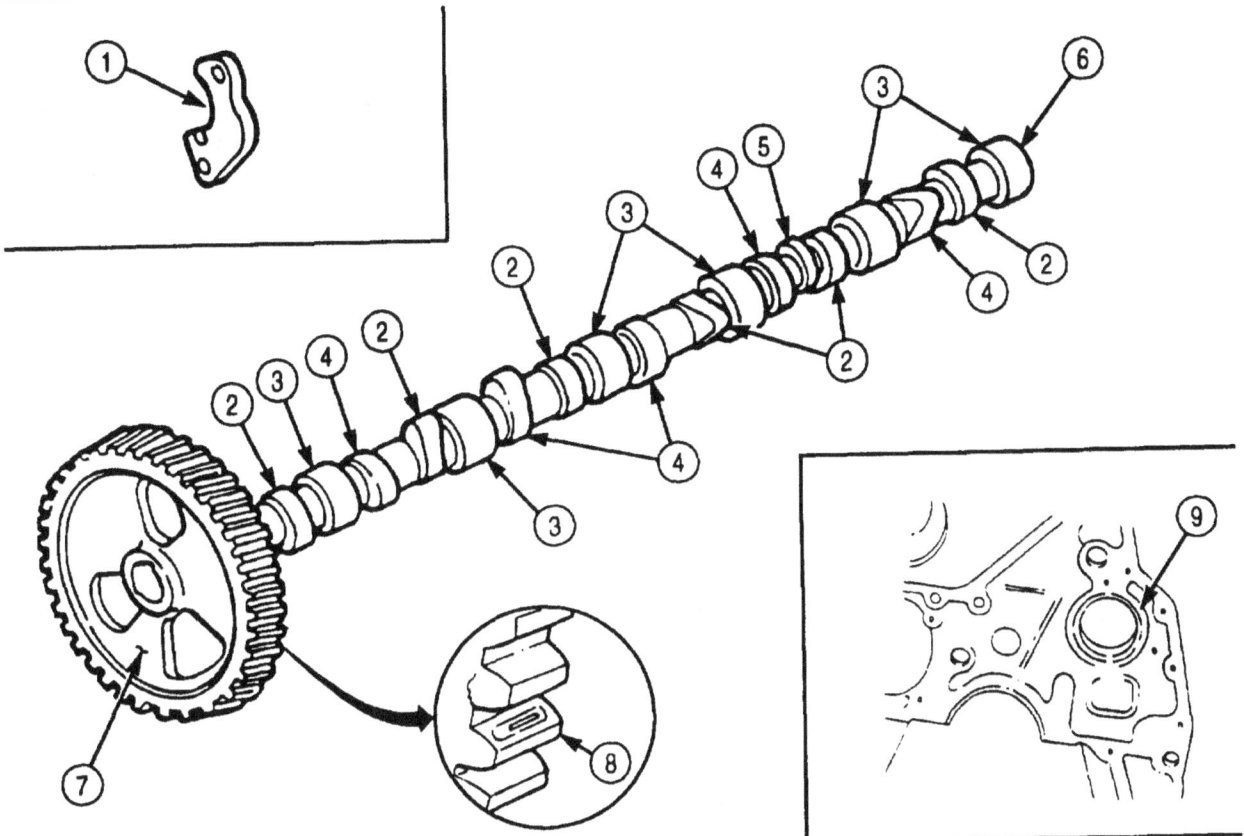

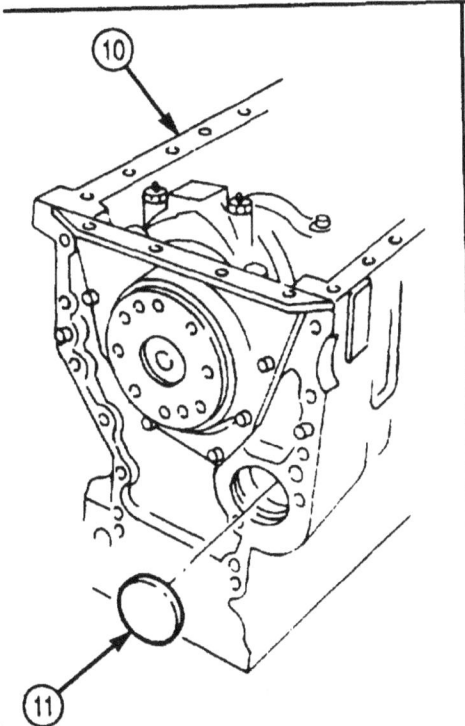

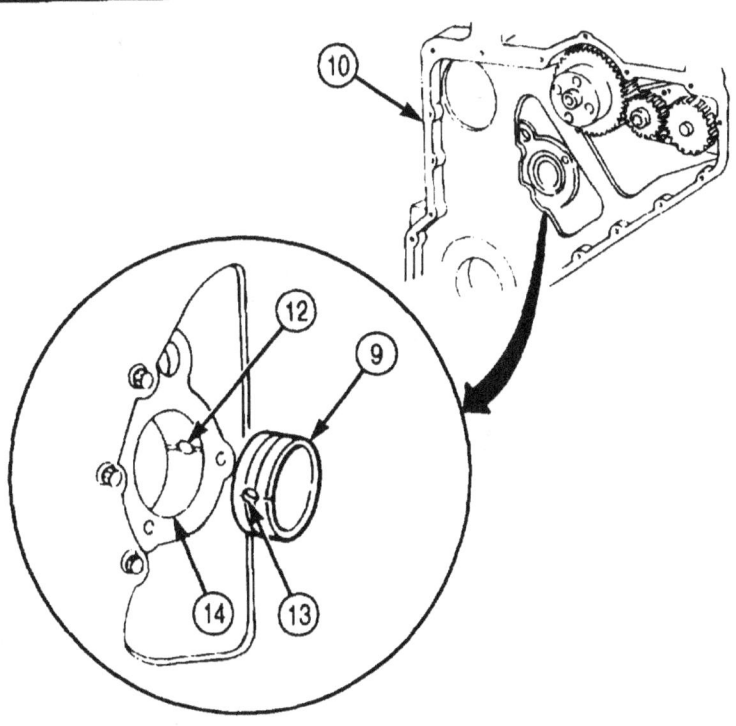

5-27. CAMSHAFT AND GEAR MAINTENANCE (Contd)

15. Inspect valve tappets (2) for nicks, burrs, and wear. Replace tappet(s) (2) if nicked, burred, or worn.
16. Measure outside diameter of tappet stem (1). Replace tappet(s) (2) if outside diameter is less than 0.628 in. (15.95 mm).
17. Inspect tappet bores (8) for scoring and wear. Replace engine block (7) if excessively scored or worn.
18. Measure inside diameter of tappet bore (8). Replace engine block (7) if inside diameter exceeds 0.632 in. (16.05 mm).

e. Assembly

NOTE
Perform steps 1 through 3 if camshaft gear was removed.

1. Install new dowel pin (3) in camshaft (4).
2. Apply GM grease to coat camshaft lobe (5).
3. Using arbor press, press camshaft gear (6) on camshaft (4).

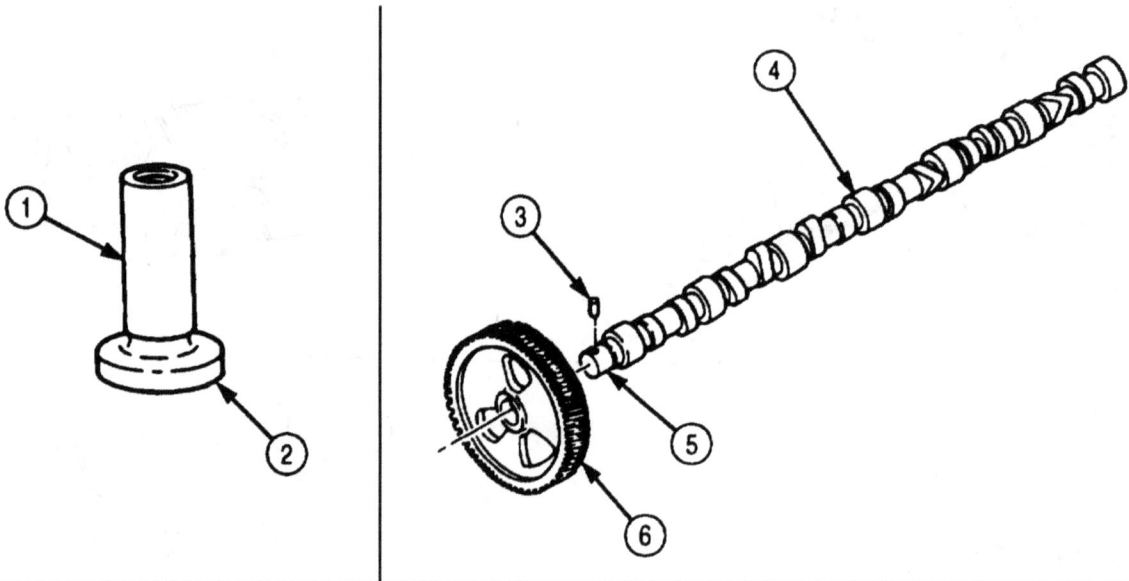

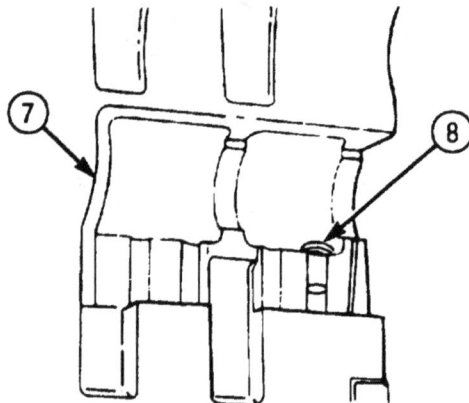

5-27. CAMSHAFT AND GEAR MAINTENANCE (Contd)

f. Installation

NOTE

- Used valve must be installed in mating tappet bores. If valve tappet position was not noted during removal, replace all valve tappets.
- Do not reuse old valve tappets if installing new camshaft.

1. Rotate crankshaft (10) as necessary to access tappet bores (8).
2. Coat twelve tappets (2) with GAA grease and install in tappet bores (8).

CAUTION

Use care when installing camshaft. Camshaft and bushings are easily damaged.

3. Apply coat of GM grease to camshaft (4) and install in engine block (7). Ensure timing marks (13) of camshaft gear (6) and crankshaft gear (9) are aligned.
4. Install camshaft support (11) on engine block (7) with two screws (12). Tighten screws (12) 18 lb-ft (24 N•m).
5. Check backlash and end play of camshaft (4) for proper seating (task a).

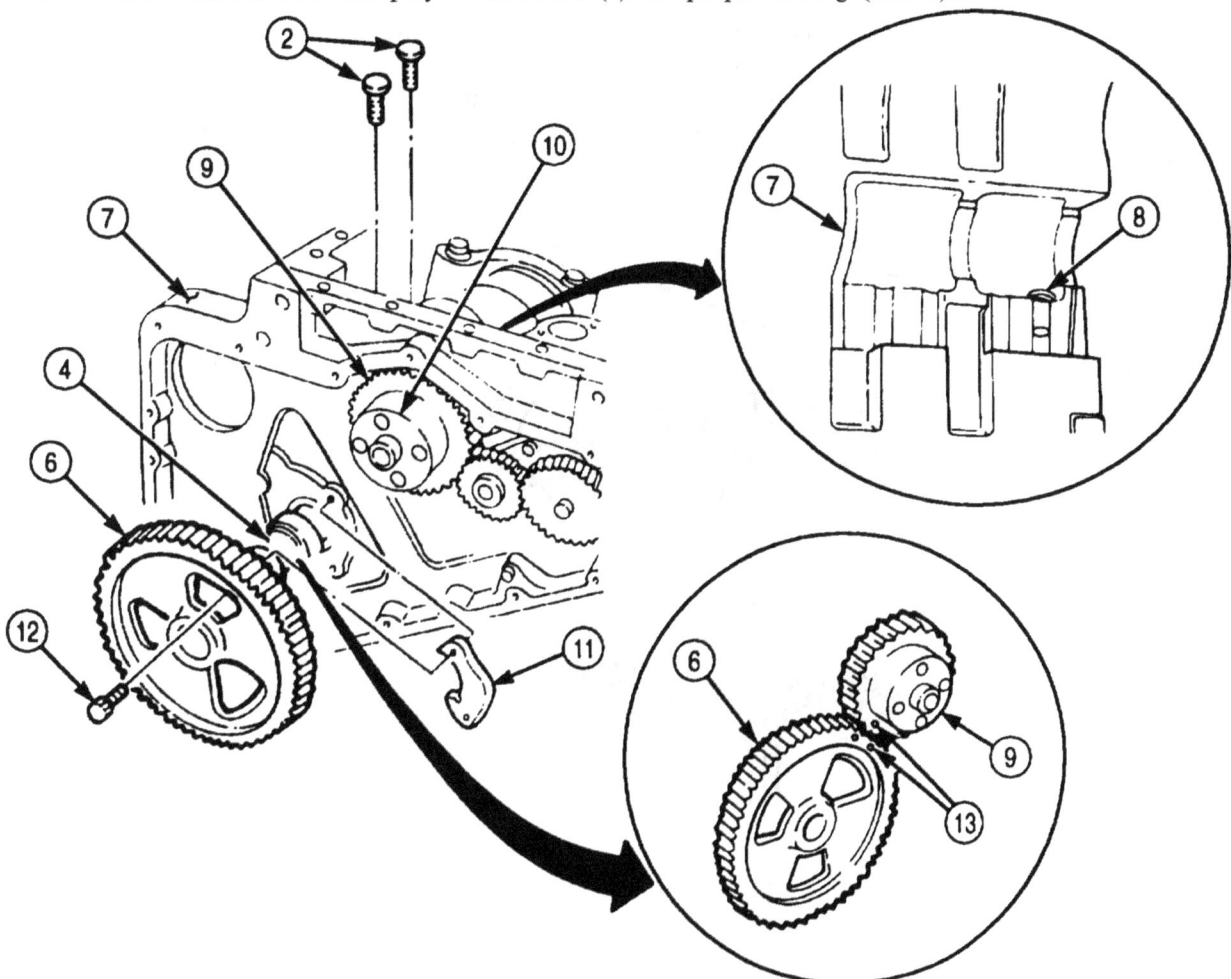

5-27. CAMSHAFT AND GEAR MAINTENANCE (Contd)

6. Using prybar between two vibration damper screws (1), rotate crankshaft (2).

7. Press in timing pin (15), while rotating crankshaft (2), until timing pin (15) engages camshaft gear (3).

8. Remove access plug (9), copper washer (8), and fuel injection pump timing pin (10) from fuel injection pump (11).

9. Install nut (4) on fuel injection pump shaft (12) and rotate until timing tooth (14) is visible in timing pin hole (13).

10. Install fuel injection pump timing pin (10) in hole (13). Ensure slot (7) of fuel injection pump timing pin (10) is positioned over timing tooth (14).

11. Install copper washer (8) and access plug (9) on fuel injection pump (11).

12. Remove nut (4) from fuel injection pump shaft (12).

NOTE

Do not exceed torque limit for fuel injection pump drive gear nut.
Torque is not the final limit.

13. Install drive gear (6) on fuel injection pump shaft (12) with new lockwasher (5) and nut (4). Tighten nut (4) 7-11 lb-ft (10-15 N•m).

14. Disengage timing pin (15) from camshaft gear (3).

15. Remove access plug (9), copper washer (8), and timing pin (10) from fuel injection pump (11).

16. Install timing pin (10) in fuel injection pump (11) with slot (7) facing outward from fuel injection pump (11).

17. Install copper washer (8) and access plug (9) on fuel injection pump (11).

18. Using prybar between two vibration damper screws (1), tighten nut (4) 80 lb-ft (109 N•m).

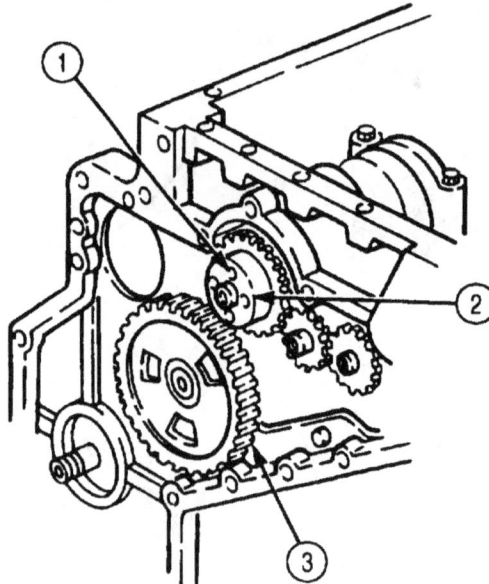

5-27. CAMSHAFT AND GEAR MAINTENANCE (Contd)

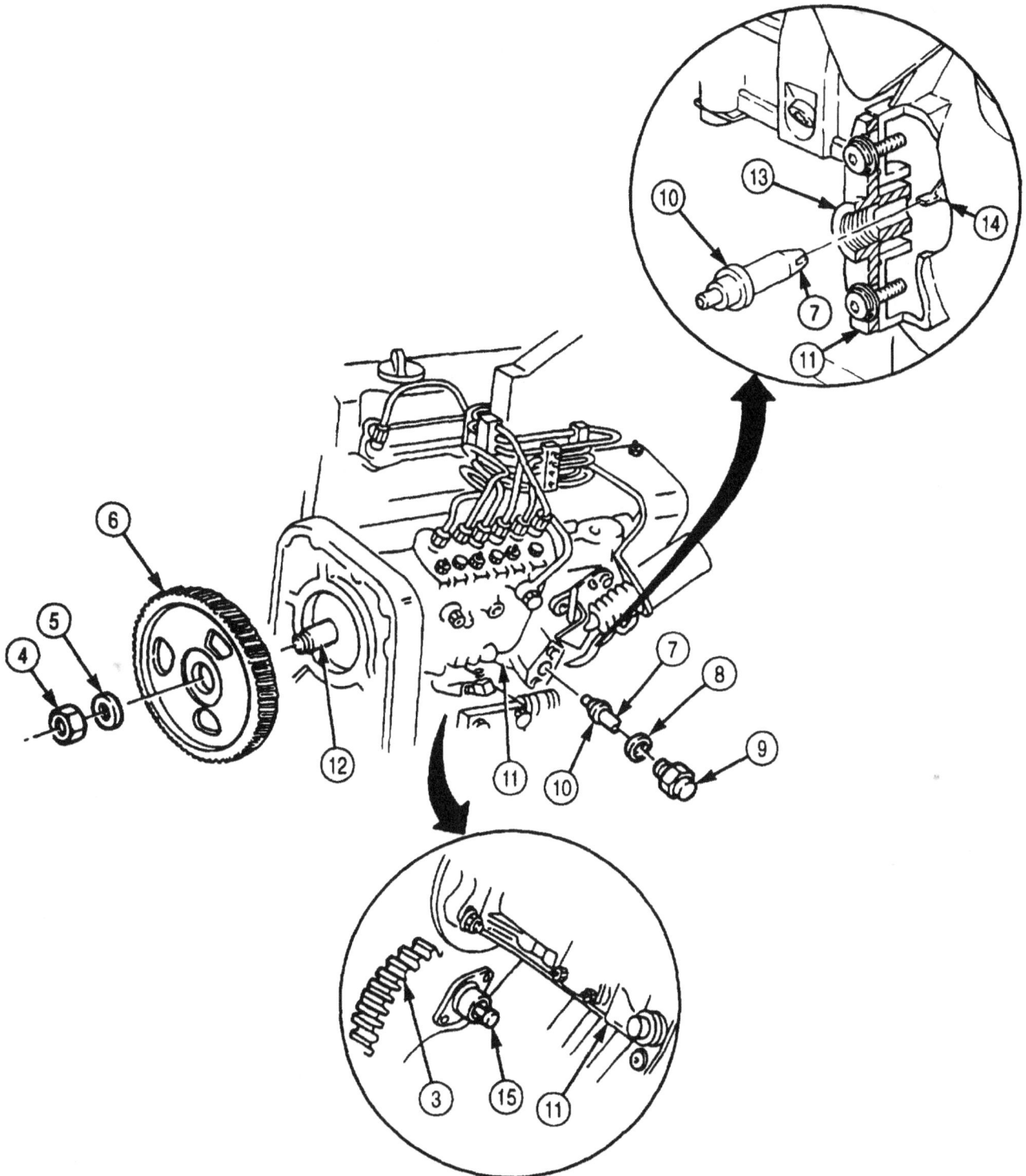

FOLLOW-ON TASKS: • Install oil pan and suction tube (para. 4-47).
• Install front gearcase cover (para. 4-48).
• Install flywheel housing (para. 4-43)
• Install rocker levers and push rods tpara. 4-41)
• Remove engine from repair stand (para. 4-9).

5-28. CYLINDER LINERS AND BLOCK MAINTENANCE

THIS TASK COVERS:

a. Cylinder Liner Inspection
b. Cylinder Liner Removal
c. Engine Block Disassembly

d. Cylinder Liner and Engine Block Cleaning
 and Inspection
e. Engine Block Assembly
f. Cylinder Liner Installation

INITIAL SETUP:

APPLICABLE MODELS
M939A2

REFERENCES (TM)
TM 9-2320-272-24P

SPECIAL TOOLS
Cylinder liner puller (Appendix E, Item 107)
Cylinder liner driver (Appendix E, Item 79)
Two cylinder liner clamps (Appendix E, Item 35)

EQUIPMENT CONDITION
• Connecting rods and pistons removed (para. 5-26).
• Camshaft removed (para. 5-27).
• Crankshaft removed (para. 5-29).

TOOLS
General mechanic's tool kit (Appendix E, Item 1)
Inside micrometer (Appendix E, Item 83)
Gauge block (Appendix E, Item 52)
Torque wrench (Appendix E, Item 144)
Dial indicator (Appendix E, Item 36)
Telescoping gauge (Appendix E, Item 136)
Feeler gauge

GENERAL SAFETY INSTRUCTIONS
• Keep fire extinguisher nearby when using drycleaning solvent.
• Drycleaning solvent is flammable and toxic.
• Eyeshields must be worn when cleaning with compressed air. Compressed air source will not exceed 30 psi (207 kPa).

MATERIALS/PARTS
Cylinder block kit (Appendix D, Item 93)
Cylinder liner seal (Appendix D, Item 95)
Drycleaning solvent (Appendix C, Item 71)
Rags (Appendix C, Item 58)
Crocus cloth (Appendix C, Item 20)
Detergent (Appendix C, Item 27)
Lubricating oil (Appendix C, Item 50)
Antiseize tape (Appendix C, Item 72)
Adhesive sealant (Appendix C, Item 4)

a. Cylinder Liner Inspection

CAUTION

Do not hone or deglaze cylinder liners. Honing or deglazing will prevent oil ring from sealing properly.

NOTE

Inspect all cylinder liners to determine need for replacement.

1. Inspect cylinder liner (1) for vertical scratches, grooves, pits, scuffing, scoring, and cracks. Tag cylinder liner (1) for replacement if scratched, grooved, pitted, scuffed, scored, or cracked.

2. Inspect for dirt or debris embedded in walls of cylinder liner (1). Tag cylinder liner (1) for replacement if debris is present.

3. Inspect walls of cylinder liner (1) for heavy polishing and wear. Heavy polishing exists when wall has a bright mirrored finish. Tag cylinder liner (1) for replacement if heavy polishing exists.

5-28. CYLINDER LINERS AND BLOCK MAINTENANCE (Contd)

4. Check cylinder liner (1) for wear:

   a. Measure inside diameter of cylinder liner (1) in two places offset 90° at top of piston travel.

   b. Record inside diameter measurements A and B.

   c. Measure inside diameter of cylinder liner (1) in two places offset 90° at bottom of piston travel.

   d. Record inside diameter measurements C and D.

   e. Compare inside diameter measurements A with B and C with D. Tag cylinder liner (1) for replacement if difference exceeds 0.0016 in. (0.041 mm).

   f. Compare inside diameter measurements A with D and B with C. Tag cylinder liner (1) for replacement if difference exceeds 0.0016 in. (0.041 mm).

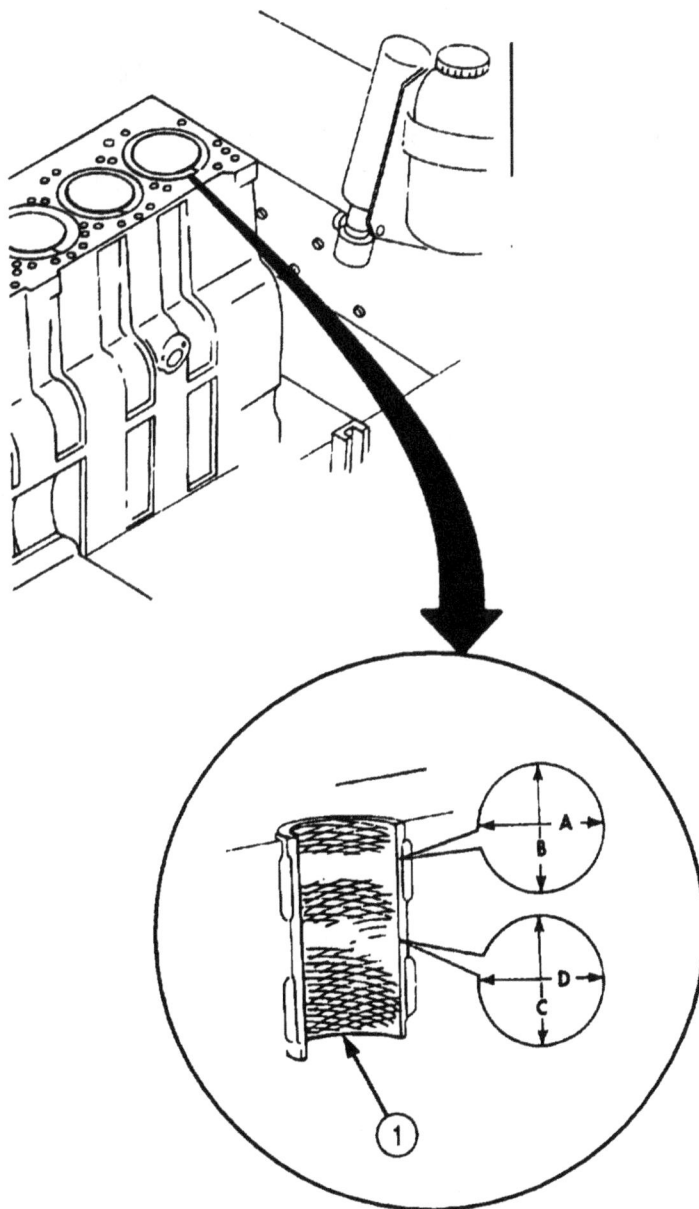

5-28. CYLINDER LINERS AND BLOCK MAINTENANCE (Contd)

b. Cylinder Liner Removal

NOTE

All cylinder liners are removed the same way. This procedure covers the removal of one cylinder liner.

1. Scribe a mark (3) across cylinder liner (2) and engine block (4) for installation.

CAUTION

Do not install cylinder liner puller on lower block ledges. Failure to comply may cause damage to engine block.

NOTE

Tag all cylinder liners with mating cylinder bore number for installation.

2. Install cylinder liner puller on engine block (4) and cylinder liner (2). Ensure cylinder liner puller does not catch on lower block ledges (5).

3. Remove cylinder liner (2) from engine block (4).

4. Remove cylinder liner seal (1) from cylinder liner (2). Discard cylinder liner seal (1).

5-28. CYLINDER LINERS AND BLOCK MAINTENANCE (Contd)

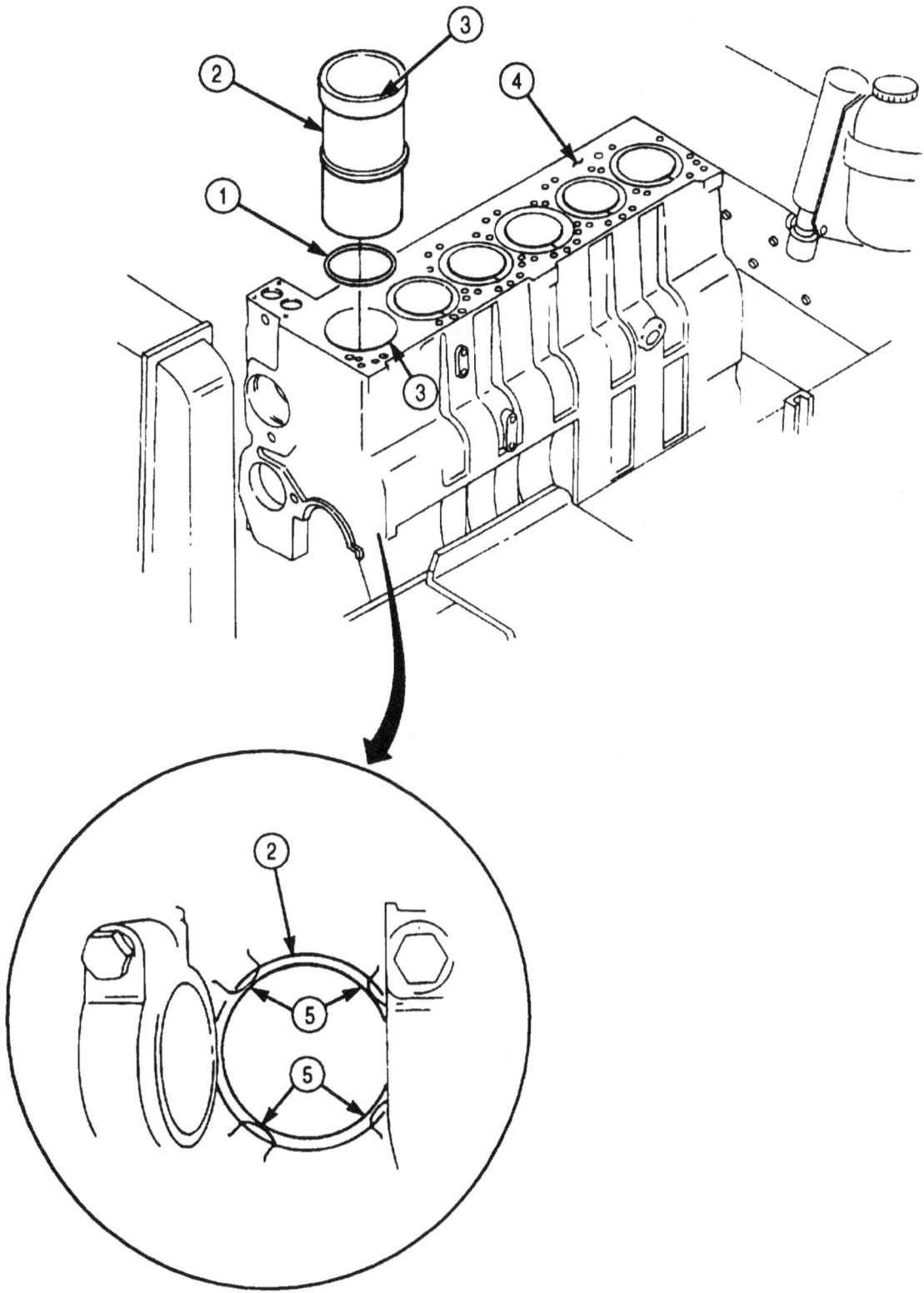

5-28. CYLINDER LINERS AND BLOCK MAINTENANCE (Contd)

c. Engine Block Disassembly

NOTE

Mark position and location of pipe and expansion plugs for installation.

1. Remove expansion plug (1) from engine block (2). Discard expansion plug (1).
2. Remove three expansion plugs (12) from engine block (2). Discard expansion plugs (12).
3. Remove two expansion plugs (11) from engine block (2). Discard expansion plugs (11).
4. Remove drainvalve (10) from engine block (2).
5. Remove expansion plug (4) from engine block (2). Discard expansion plug (4).
6. Remove pipe plugs (5) and (9) from engine block (2).
7. Remove expansion plug (6) from engine block (2). Discard expansion plug (6).
8. Remove expansion plug (7) from engine block (2). Discard expansion plug (7).
9. Remove expansion plug (8) from engine block (2). Discard expansion plug (8).

NOTE

Perform step 10 if dowel pins are damaged.

10. Remove four dowel pins (3) from engine block (2). Discard dowel pins (3).
11. Remove three pipe plugs (16) from engine block (2).
12. Remove expansion plugs (13) and (17) from engine block (2). Discard expansion plugs (13) and (17).
13. Remove expansion plug (15) from engine block (2). Discard expansion plug (15).

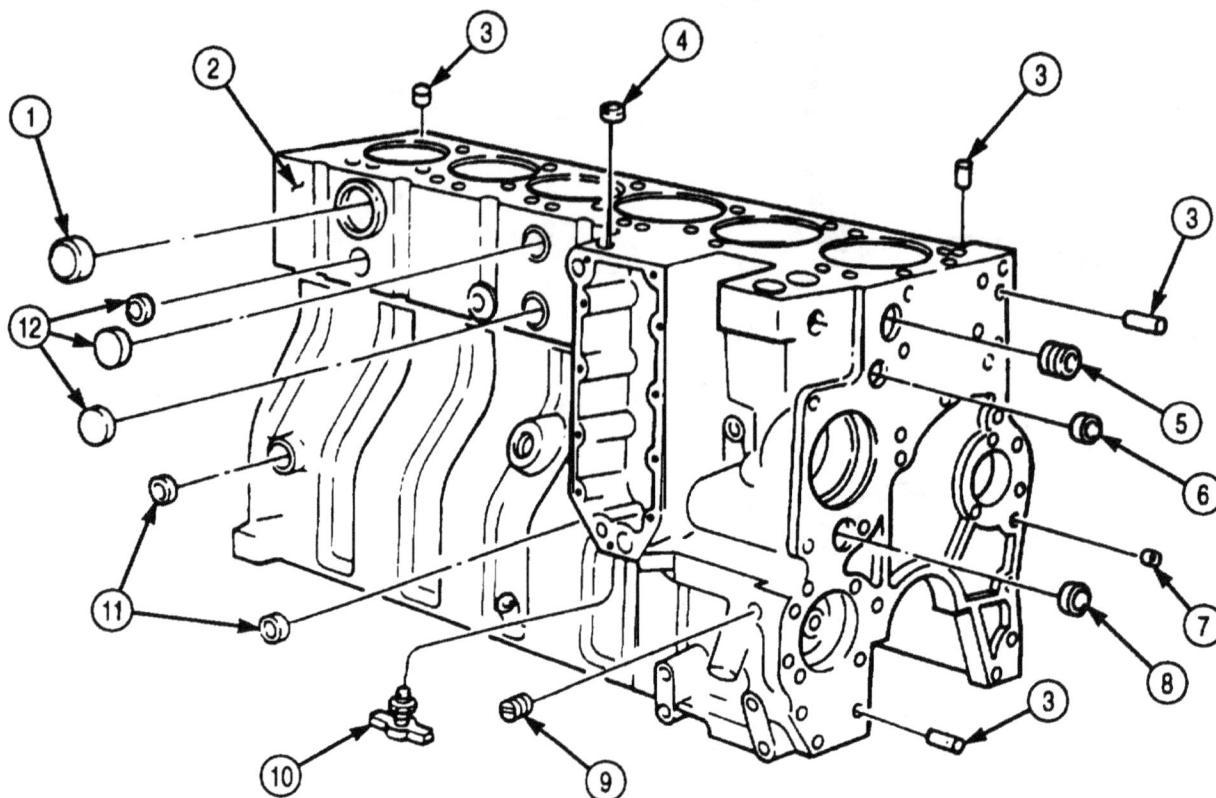

5-28. CYLINDER LINERS AND BLOCK MAINTENANCE (Contd)

NOTE

Perform step 14 if dowel pins are damaged.

14. Remove two dowel pins (14) from engine block (2). Discard dowel pins (14).

15. Remove six piston cooling nozzles (18) from engine block (2).

NOTE

Perform step 16 for late model engine.

16. Remove twelve piston cooling nozzles (18) from engine block (2).

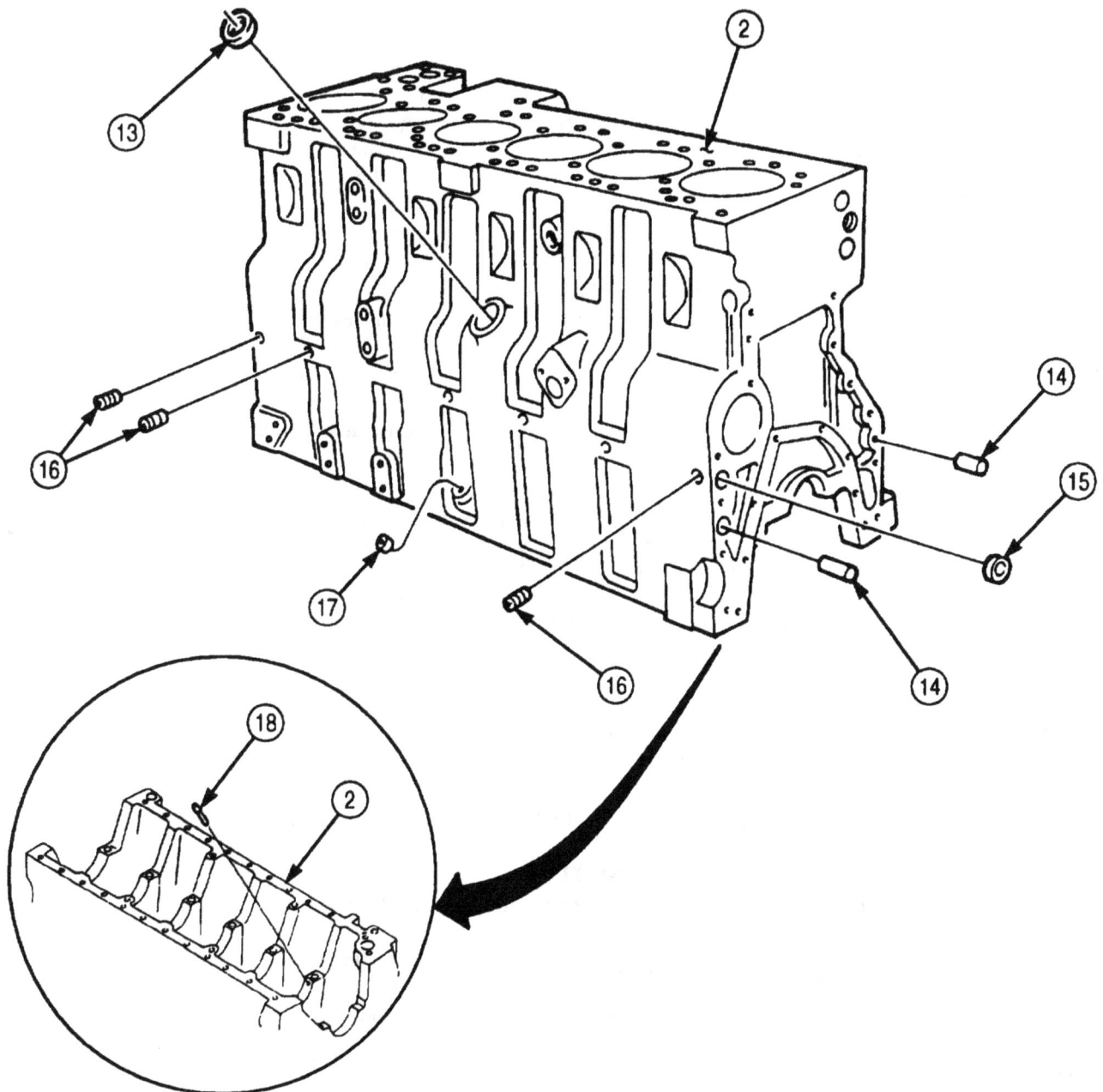

5-28. CYLINDER LINERS AND BLOCK MAINTENANCE (Contd)

d. Cylinder Liner and Engine Block Cleaning and Inspection

WARNING

- Drycleaning solvent is flammable and toxic. Do not use near an open flame and always have a fire extinguisher nearby when solvents are used. Use only in well-ventilated places, wear protective clothing, and dispose of cleaning rags in approved container. Failure to do this may result in injury or death to personnel and/or damage to equipment.

- Eyeshields must be worn when cleaning with compressed air. Compressed air source will not exceed 30 psi (207 kPa). Failure to do so may result in injury to personnel.

CAUTION

- Do not allow debris to fall into crankshaft oil holes. Damage to crankshaft bearings could result.

- Do not remove stock material from cylinder liner seating surface when cleaning. Machined tolerances are critical and premature engine failure will result if material is removed.

1. Cover crankshaft oil holes (6) with clean rags.

2. Clean all expansion plug holes of engine block (2) with drycleaning solvent. Dry with compressed air.

3. Clean sealing surfaces (3), (4), and (5) of engine block (2) with crocus cloth and drycleaning solvent.

4. Inspect sealing surfaces (3), (4), and (5) for cracks, pits, scores, and excessive wear. Replace engine block (2) if cracked, pitted, scored, or excessively worn.

5. Measure inside diameter of sealing surface (4). Replace engine block (2) if inside diameter exceeds 5.2342 in. (132.949 mm).

6. Clean all oil draining and water cooling passage bores with brush and drycleaning solvent. Dry with compressed air.

CAUTION

Engine block deck can be resurfaced twice in increments of 0.010 in. (0.254 mm). Do not exceed 0.020 in. (0.508 mm) when resurfacing or damage to equipment will result.

7. Using straight edge and feeler gauge, inspect engine block (2) deck for warping. Replace engine block (2) if warping exceeds 0.008 in. (0.203 mm). Notify your supervisor.

8. Install seven main bearing caps (8) on engine block (2) with fourteen screws (9). Tighten screws (9) 130 lb-ft (176 N•m).

9. Using bore gauge, measure inside diameter of main bearing bore (10). Record measurement. Replace main bearing (7) if inside diameter exceeds 3.8632 in. (98.125 mm).

10. Using bore gauge, measure inside diameter of camshaft bore (12). Replace bushing (13) if inside diameter exceeds 2.367 in. (60.12 mm).

11. Measure inside diameter of tappet bore (11). Replace engine block (2) if inside diameter exceeds 0.632 in. (16.05 mm).

NOTE

Cylinder liners should be inspected before cleaning so defects can be clearly noted.

12. Check cylinder liner (1) for cracks, scoring, or vertical grooves. Replace cylinder liner (1) if cracked, scored, or vertical grooves are present.

13. Check cylinder liner (1) for pits and eroded surfaces. Replace cylinder liner (1) if pitted or eroded.

5-28. CYLINDER LINERS AND BLOCK MAINTENANCE (Contd)

14. Check cylinder liner (1) for fretting and breaks in machined areas. Replace cylinder liner (1) if fretted or broken.

15. Check cylinder liner (1) for rust, scaling, and corrosion.

    a. Clean with steam or hot water and detergent.

    b. Replace if excessively rusted, scaled, or corroded.

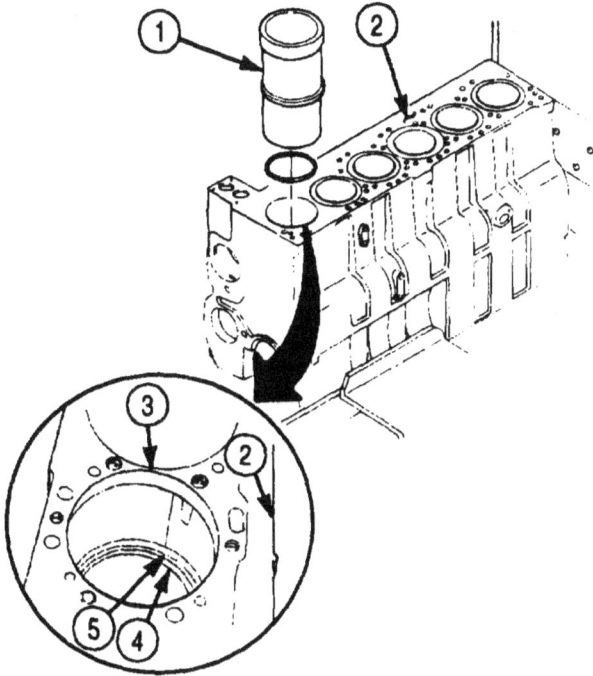

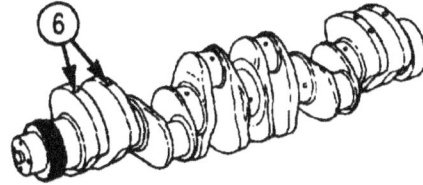

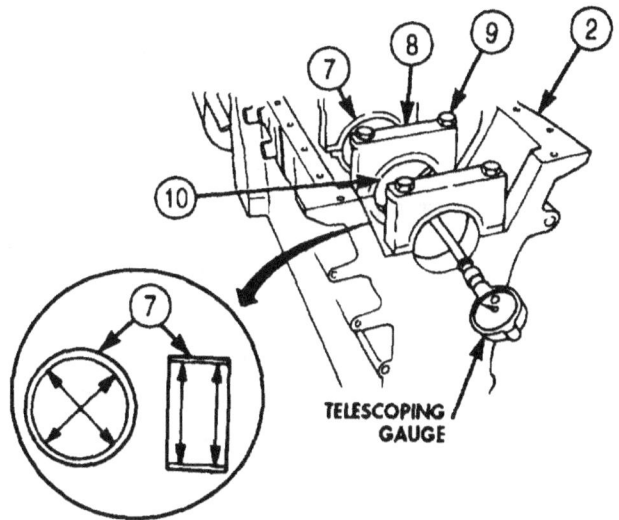

TELESCOPING GAUGE

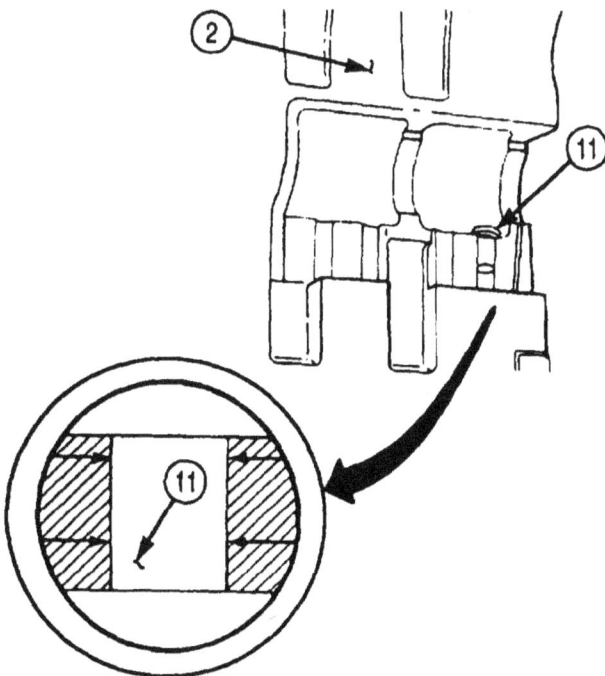

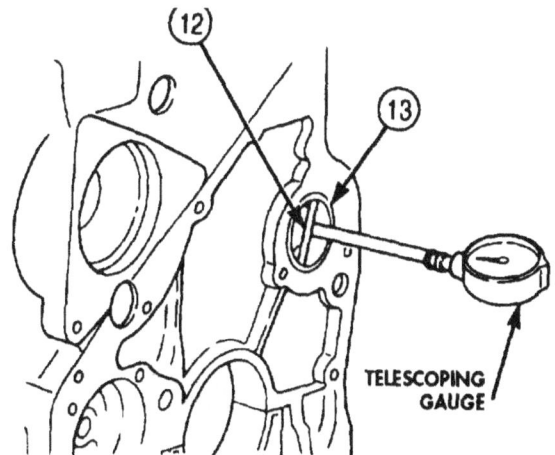

TELESCOPING GAUGE

5-28. CYLINDER LINERS AND BLOCK MAINTENANCE (Contd)

e. Engine Block Assembly

CAUTION

Ensure all engine block oil passages are free of obstructions.
Failure to do so may result in damage to engine.

NOTE
- Perform step 1 for late model engines, serial number 44487830 and after.
- Perform step 2 if dowel pins were removed.

1. Install twelve piston cooling nozzles (7) in engine block (2).

2. Install two new dowel pins (3) in engine block (2).

3. Install six piston cooling nozzles (7) in engine block (2).

NOTE
- Apply adhesive sealant to mating surfaces of all expansion plugs before installation.
- Wrap male pipe threads with antiseize tape before installation.

4. Install new expansion plug (4) in engine block (2).

5. Install new expansion plugs (6) and (1) on engine block (2).

6. Install three pipe plugs (5) on engine block (2). Tighten pipe plugs (5) 6 lb-ft (8 N•m).

NOTE
Perform step 7 if dowel pins were removed.

7. Install four new dowel pins (8) on engine block (2).

8. Install new expansion plug (13) on engine block (2).

9. Install new expansion plug (12) on engine block (2).

10. Install new expansion plug (11) on engine block (2).

11. Install pipe plugs (10) and (14) on engine block (2). Tighten pipe plug (14) 6 lb-ft (8 N•m). Tighten pipe plug (10) 27 lb-ft (37 N•m).

12. Install new expansion plug (9) on engine block (2).

13. Install new drainvalve (15) on engine block (2).

14. Install two new expansion plugs (16) on engine block (2).

15. Install three new expansion plugs (17) on engine block (2).

16. Install new expansion plug (18) on engine block (2).

5-28. CYLINDER LINERS AND BLOCK MAINTENANCE (Contd)

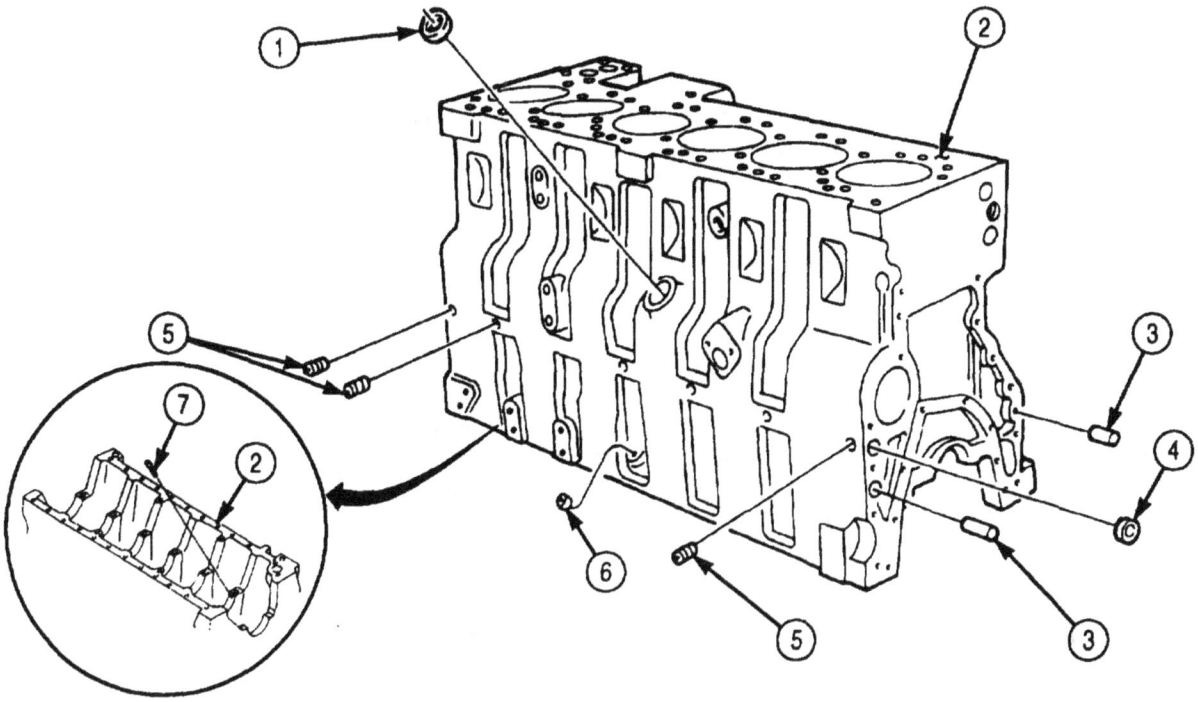

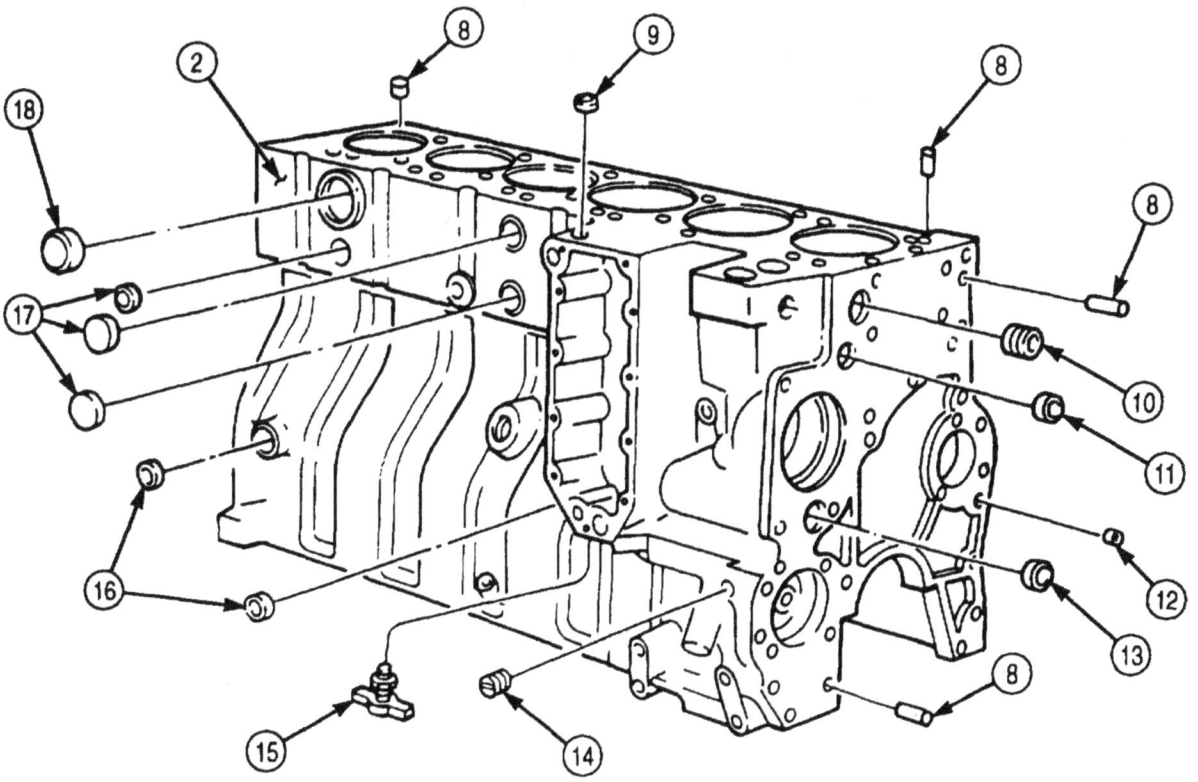

5-28. CYLINDER LINERS AND BLOCK MAINTENANCE (Contd)

f. Cylinder Liner Installation

NOTE

All cylinder liners are installed the same. This procedure covers
installation of one cylinder liner.

1. Apply clean engine oil to lubricate cylinder liner bore (4).

2. Install new cylinder liner seal (2) on cylinder liner (1).

3. Apply clean engine oil to lubricate sealing surfaces (5), (6), and (7), cylinder liner (1), and cylinder liner seal (2).

4. Install cylinder liner (1) in mating cylinder liner bore (4).

NOTE

Perform step 5 if installing old cylinder liner.

5. Rotate cylinder liner (1) 1/8 turn from position noted in removal. Using cylinder liner driver, drive cylinder liner (1) until it bottoms in cylinder liner bore (4).

6. Install two cylinder liner clamps on engine block (3) with two washers and screws. Tighten screws 50 lb-ft (68 N·m).

7. Remove two screws, washers, and cylinder liner clamps from engine block (3).

8. Using gauge block, measure cylinder liner protrusion:

   a. Position gauge block on engine block (3).

   b. Place plunger on engine block (2) and set dial indicator to zero.

   c. Position gauge block on engine block (3) and place plunger on cylinder liner (1).

   d. Cylinder liner protrusion should be 0.005-0.0043 in. (0.0127-0.1092 mm).

   e. Check cylinder liner (1) for proper installation if cylinder liner protrusion is not within limits.

   f. If installed correctly, counterbore of engine block (3) will require cutting. Notify your supervisor.

9. Check clearance between cylinder liner (1) and lower block ledges (8) using 0.009 in. (0.229 mm) feeler gauge. If feeler gauge does not pass between cylinder liner (1) and lower bottom ledges (8), remove cylinder liner (1) and reinstall.

5-28. CYLINDER LINERS AND BLOCK MAINTENANCE (Contd)

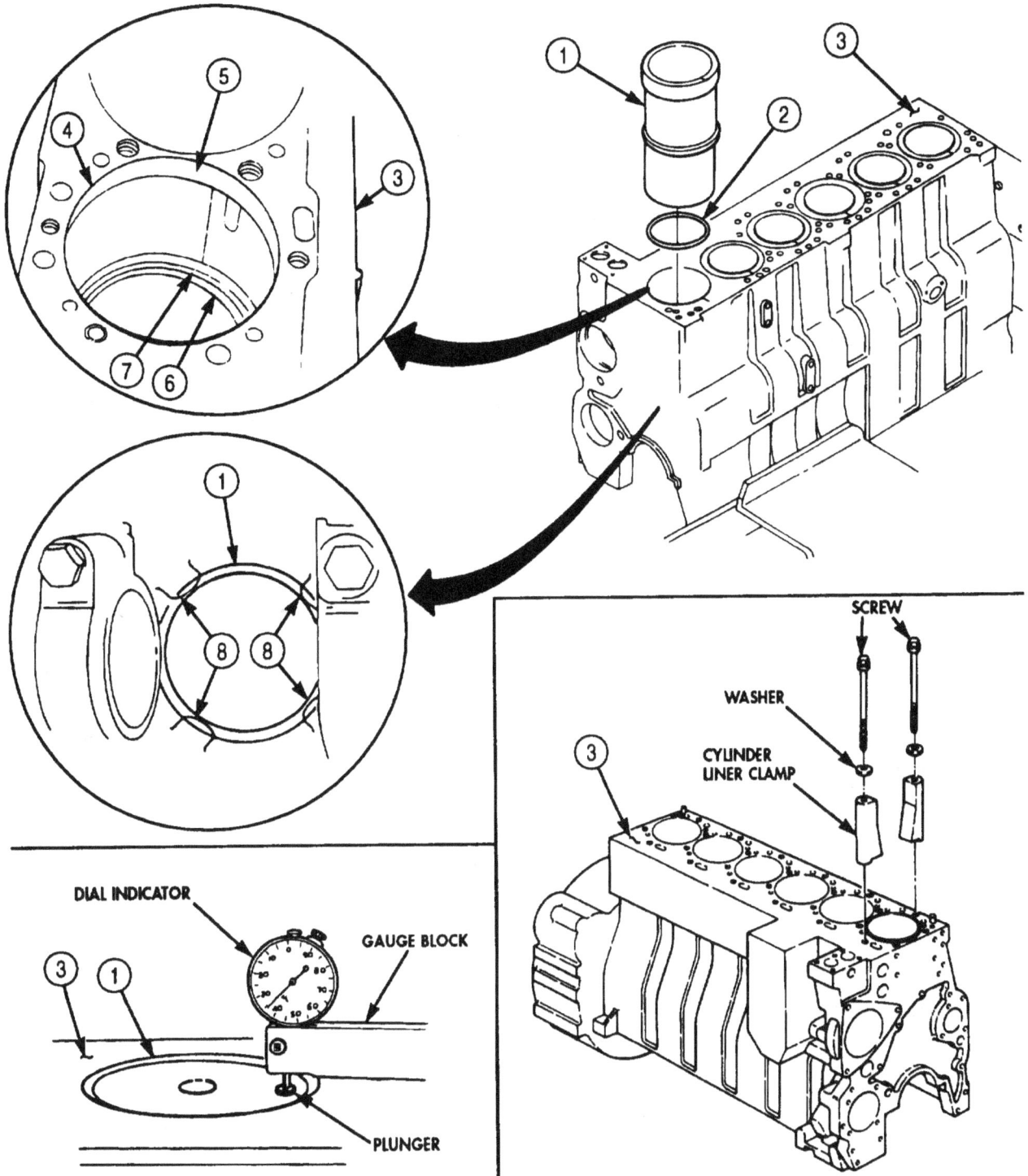

DIAL INDICATOR

GAUGE BLOCK

PLUNGER

SCREW

WASHER

CYLINDER
LINER CLAMP

FOLLOW-ON TASKS:
- Install connecting rods and pistons (para. 5-26).
- Install crankshaft (para. 5-29).
- Install camshaft (para. 5-27).

5-29. CRANKSHAFT AND GEAR MAINTENANCE

THIS TASK COVERS:

a. Check End Play
b. Removal
c. Disassembly

d. Cleaning and Inspection
e. Assembly
f. Installation

INITIAL SETUP:

APPLICABLE MODELS
M939A2

SPECIAL TOOLS
Torque angle gauge (Appendix E, Item 141)

TOOLS
General mechanic's tool kit (Appendix E, Item 1)
Heavy duty gear puller (Appendix E, Item 102)
Dial indicator (Appendix E, Item 36)
Torque wrench (Appendix E, Item 144)
Feeler gauge
Prybar
Soft-head hammer
Nylon brush

MATERIALS/PARTS
Dowel pin (Appendix D, Item 99)
Crankshaft main bearing set
  (Appendix D, Item 87)
Drycleaning solvent (Appendix C, Item 71)
GAA grease (Appendix C, Item 28)
Lubricating oil (Appendix C, Item 50)
Plastigage (Appendix C, Item 55)
Crocus cloth (Appendix C, Item 20)

REFERENCES (TM)
TM 9-2320-272-24P

EQUIPMENT CONDITION
• Fuel injectors removed (para. 4-55).
• Oil pan and suction tube removed (para. 4-47).
• Front gearcase housing removed (para. 4-46).
• Flexplate and flywheel housing removed (para. 4-43).
• Rear cover and oil seal removed (para. 4-44).

GENERAL SAFETY INSTRUCTIONS
• Drycleaning solvent is flammable and toxic. Do not use near open flame.
• Keep fire extinguisher nearby when using drycleaning solvent.
• Eyeshields must be worn when cleaning with compressed air. Compressed air source will not exceed 30 psi (207 kPa).

a. Check End Play

1. Position engine block (2) so crankshaft (1) is facing upward.
2. Install dial indicator on engine block (2) with plunger on face of crankshaft gear (3).
3. Move crankshaft (1) towards rear of engine block (2) using prybar.
4. Set dial indicator to zero.
5. Move crankshaft (1) toward front of engine block (2) using prybar. Record dial indicator reading.
6. Replace thrust bearing in subtasks b and f if crankshaft (1) end play is not between 0.006-0.013 in. (0.152-0.330 mm).

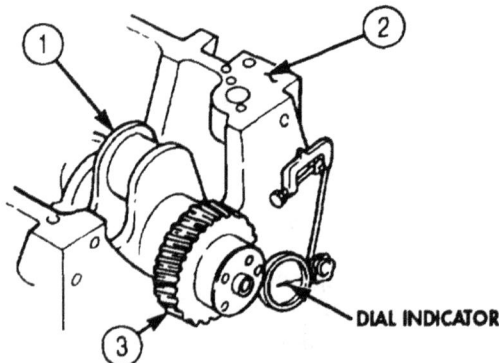

5-29. CRANKSHAFT AND GEAR MAINTENANCE (Contd)

b. Removal

NOTE
- Tag all connecting rod caps, main bearing caps, mating cylinders, and bearings for installation.
- All connecting rod caps and main bearing caps are removed the same.

1. Remove twelve nuts (5) from six connecting rod caps (4).
2. Using soft-head hammer, tap screws (6) and remove connecting rod caps (4) from crankshaft (1).
3. Remove fourteen screws (8) from seven main bearing caps (9).
4. Loosen main bearing caps (9) and remove from crankshaft (1) using screws (8).
5. Remove seven lower main bearings (7) from main bearing caps (9). Discard bearings (7).
6. Remove crankshaft (1) from engine block (2).

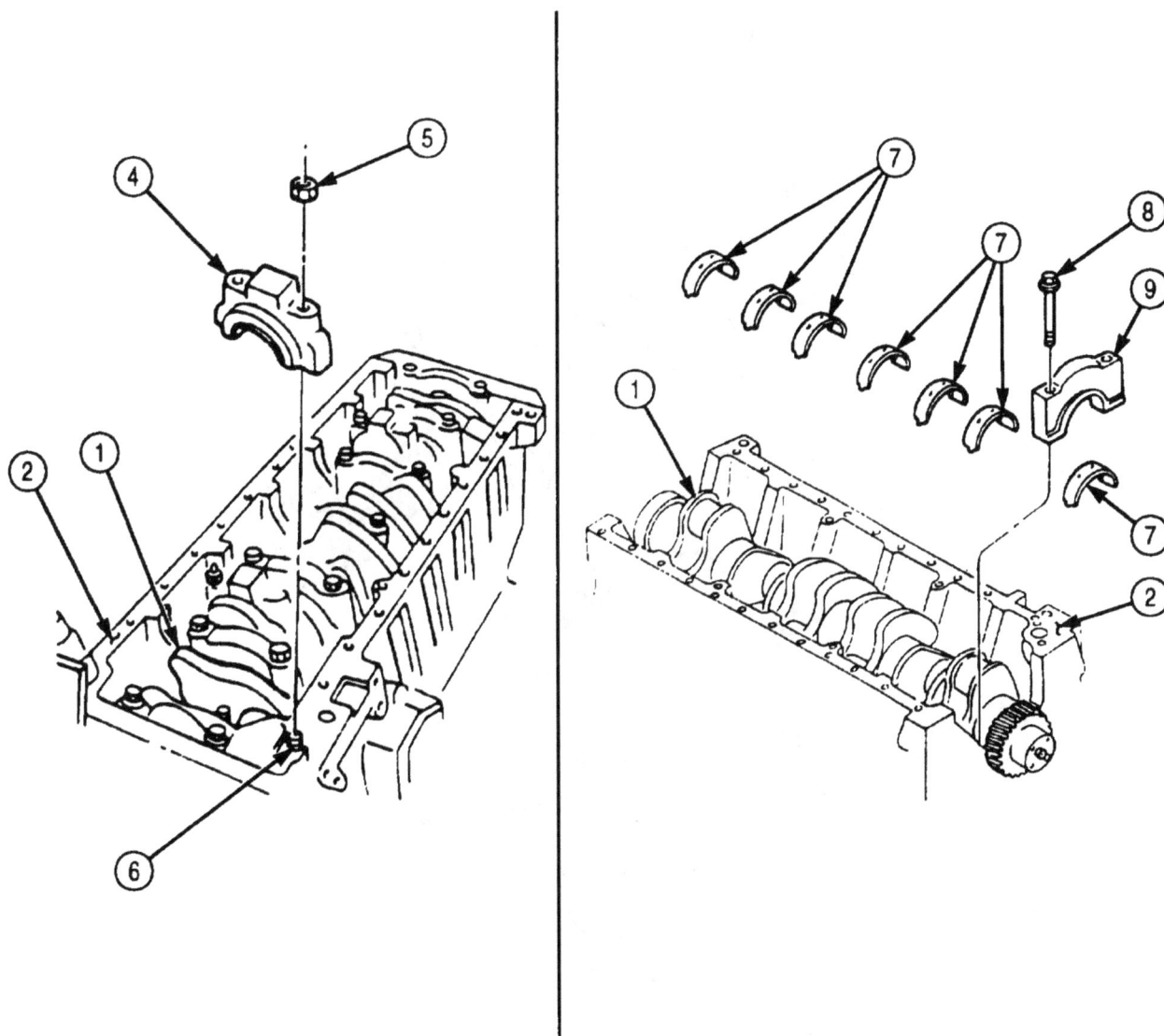

5-29. CRANKSHAFT AND GEAR MAINTENANCE (Contd)

7. Remove six upper main bearings (5) and thrust bearing (6) from engine block (4). Discard bearings (5) and (6).

NOTE
Perform step 8 for late model engines, serial number 44487830 and after.

8. Remove twelve piston cooling nozzles (7) from engine block (4).

9. Remove six piston cooling nozzles (7) from engine block (4).

c. Disassembly

NOTE
Before removing crankshaft gear from crankshaft, perform subtask d of this procedure. If, as a result of inspection, the crankshaft gear must be removed, perform steps 1 and 2.

1. Using gear puller, remove crankshaft gear (3) from crankshaft (1).

2. Remove dowel pin (2) from crankshaft (1). Discard dowel pin (2).

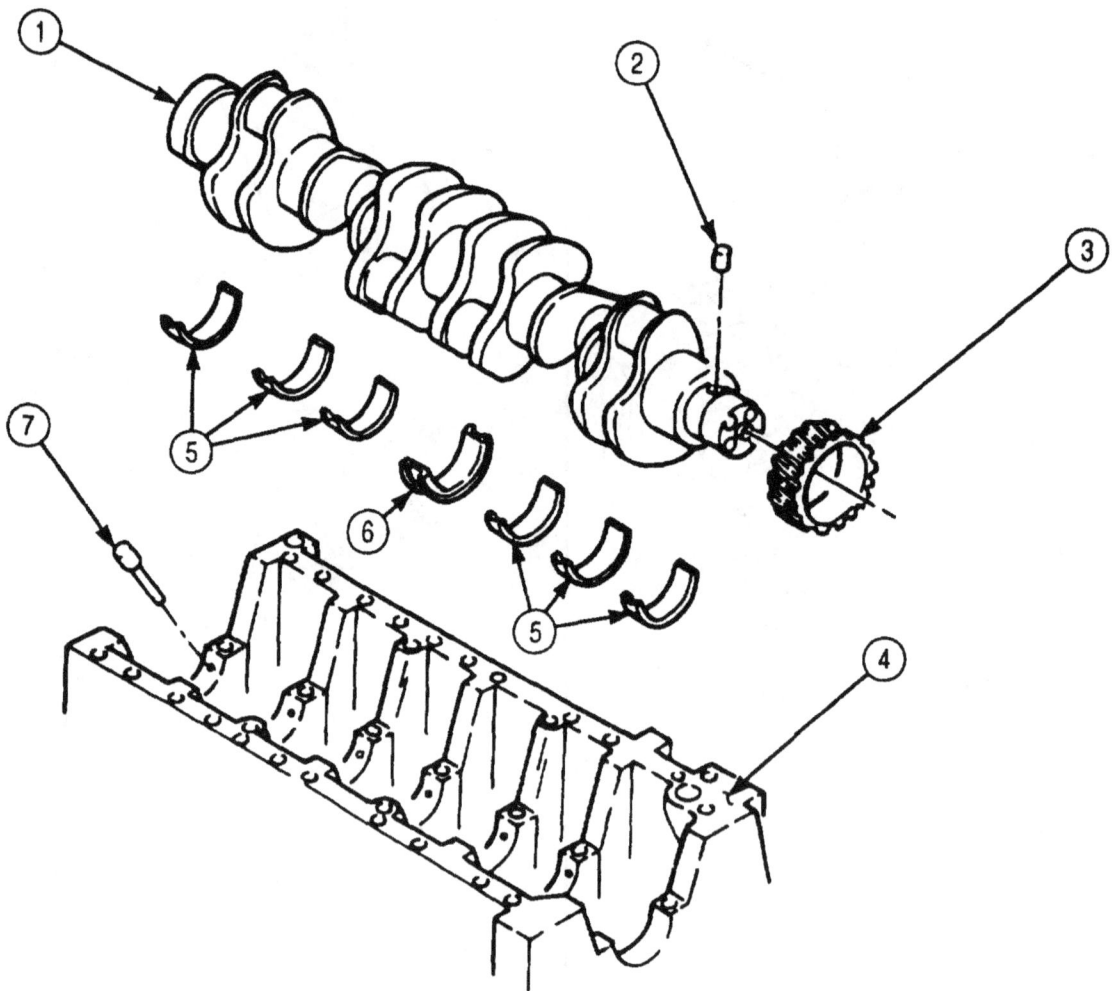

5-29. CRANKSHAFT AND GEAR MAINTENANCE (Contd)

d. Cleaning and Inspection

WARNING

- Drycleaning solvent is flammable and toxic. Do not use near open flame and always have a fire extinguisher nearby when solvents are used. Use only in well-ventilated places, wear protective clothing, and dispose of cleaning rags in approved container. Failure to do this may result in injury or death to personnel and/or damage to equipment.
- Eyeshields must be worn when cleaning with compressed air. Compressed air source will not exceed 30 psi (207 kPa). Failure to do so may result in injury to personnel.

1. Clean all parts with drycleaning solvent and dry with compressed air. Use nylon brush to clean all oil passages.

2. Inspect crankshaft (1) for cracks, gouges, pits, burrs, nicks, and scratches. Remove minor damage with fine mill file or crocus cloth. If excessively cracked, gouged, pitted, burred, nicked, or scratched, replace crankshaft (1).

3. Inspect crankshaft gear (3) for cracks, breaks, and damaged teeth. If cracked, broken, or teeth are damaged, replace crankshaft gear (3).

4. Inspect connecting rod journals (9) and main bearing journals (8) for scratches, grooves, excessive wear, and discoloring caused by overheating. Replace crankshaft (1) if scratched, grooved, worn, or discolored.

5-29. CRANKSHAFT AND GEAR MAINTENANCE (Contd)

5. Inspect connecting rod caps (14) for cracks, breaks, and pits. Replace connecting rod caps (14) if cracked, broken, or pitted.

6. Inspect main bearing caps (7) for cracks, breaks, and pits. Replace main bearing caps (7) if cracked, broken, or pitted.

7. Inspect engine block (9) for scratches, grooves, wear, and stripped threads. Replace or repair if engine block (9) is scratched, grooved, worn, or threads are stripped.

8. Measure outside diameter of main bearing journals (2) for out-of-roundness across points A and B. Replace crankshaft (1) if difference of measurements exceeds 0.002 in. (0.0051 mm).

9. Measure outside diameter of main bearing journals (2) for taper at points C and D. Replace crankshaft (1) if difference of measurements exceeds 0.0005 in. (0.013 mm).

10. Measure outside diameter of connecting rod journals (3) for out-of-roundness across points A and B. Replace crankshaft (1) if difference of measurements exceeds 0.002 in. (0.0050 mm).

11. Measure outside diameter of connecting rod journals (3) for taper at points C and D. Replace crankshaft (1) if difference of measurements exceeds 0.0005 in. (0.0127 mm).

12. Inspect front (4) and rear (5) crankshaft seal wear surfaces for scratches and grooves. Replace crankshaft (1) if scratched or grooved.

CAUTION

Ensure main bearings are installed in proper mating seats.
Failure to do so will cause engine to fail.

13. Install new thrust bearing (8) in No. 4 main bearing seat (11).

14. Install six new upper main bearings (7) in main bearing seats (10). Ensure tangs (12) of upper main bearings (7) are aligned with slots (13) of main bearings seats (10).

15. Install crankshaft (1) in engine block (9).

16. Install seven new lower main bearings (6) in main bearing caps (7).

17. Place plastic strips across width of seven main bearing journals (2).

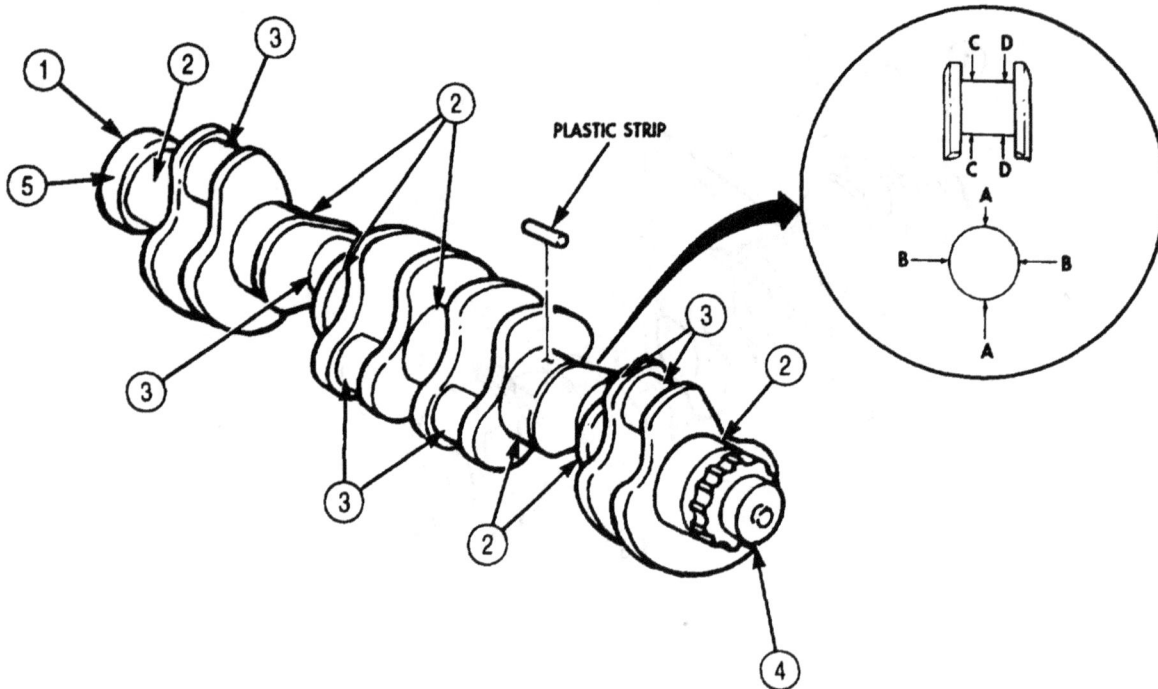

5-29. CRANKSHAFT AND GEAR MAINTENANCE (Contd)

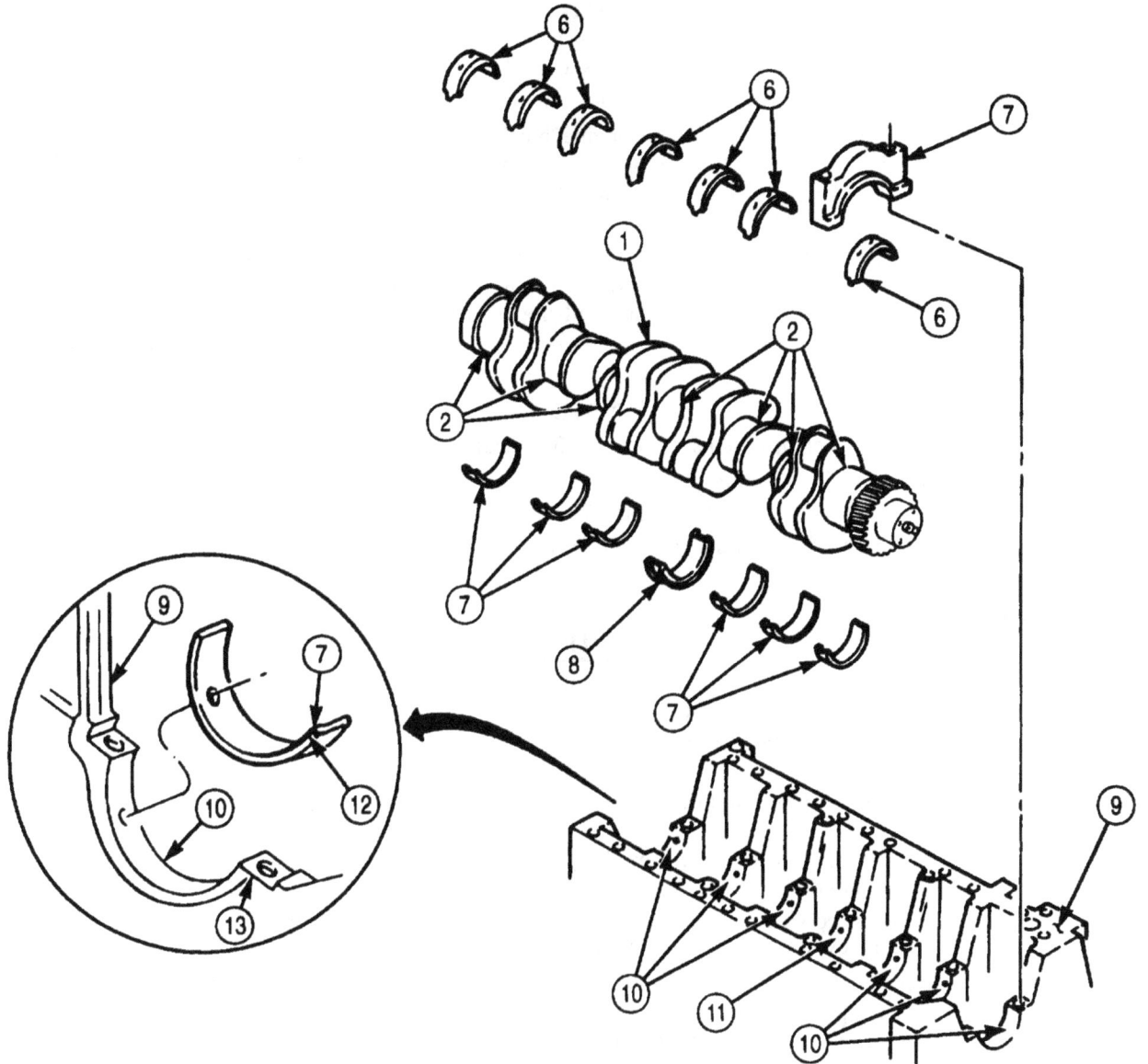

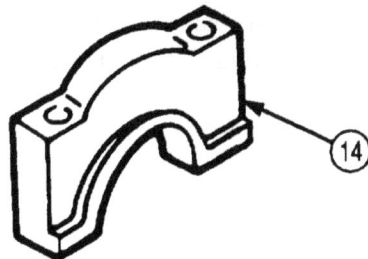

5-29. CRANKSHAFT AND GEAR MAINTENANCE (Contd)

NOTE

Ensure main bearing caps are installed with mating cylinders.
Number on main bearing caps should face camshaft.

18. Apply clean engine oil to coat threads of fourteen screws (2).
19. Position seven main bearing caps (3) on engine block (14) and crankshaft (7) and tap main bearing caps (3) using soft-head hammer until seated on main bearing journals (6).
20. Install main bearing caps (3) on engine block (14) with fourteen screws (2). Tighten screws (2) 125 lb-ft (170 N·m) in sequence shown.
21. Remove fourteen screws (2) and seven main bearing caps (3) from engine block (14).
22. Measure width of plastic strip at widest point using plastigage. If plastic strip does not measure 0.0026-0.0053 in. (0.660-0.1346 mm), replace all upper (10), lower (1), and thrust (11) bearings.

e. Assembly

NOTE

Perform steps 1 through 3 if gear was removed.

1. Install new dowel pin (5) in crankshaft (7).
2. Apply GAA grease to lobe (8) of crankshaft (7).
3. Press crankshaft gear (9) on crankshaft (7).

f. Installation

CAUTION

Late and early model engine main bearings and main bearing caps and components are not interchangeable. Mixing components will cause damage to engine.

NOTE

- Perform step 1 if inspection did not require replacement of main bearings.
- Perform step 3 for late model engines, serial number 44487830 and after.
- Piston cooling nozzles are located in all upper chain bearing seats, except for No. 3 cylinder.

1. Remove six upper main bearings (10) and thrust bearing (11) from engine block (14).
2. Install six piston cooling nozzles (15) in upper main bearing seats (12). Ensure piston cooling nozzles (15) seat below main bearing surface.
3. Install twelve piston cooling nozzles (15) in upper main bearing seats (12). Ensure piston cooling nozzles (15) seat below main bearing surface.

NOTE

Apply lubricating oil to all main bearings and bearing surfaces.

4. Align tangs (16) of six upper main bearings (10) with slots (17) of main bearing seats (12) and install on engine block (14).

5-29. CRANKSHAFT AND GEAR MAINTENANCE (Contd)

5. Install thrust bearing (11) in No. 4 main bearing seat (13).

6. Install crankshaft (7) in engine block (14).

NOTE

Perform step 7 if inspection required replacement of lower main bearings.

7. Align tangs (4) of seven lower main bearings (1) with slots of main bearing caps (3) and install seven lower main bearings (1) in main bearing caps (3).

8. Apply clean engine oil to main bearing journals (6) and install seven main bearing caps (3) on engine block (14).

9. Using soft-head hammer, tap main bearing caps (3) until seated against main bearing journals (6).

10. Apply clean engine oil to threads of fourteen screws (2) and install on main bearing caps (3).

11. Tighten screws (2) as follows:

    a. Tighten screws (2) 125 lb-ft (170 N·m) in sequence shown. Loosen screws (2).

    b. Tighten screws (2) to 70 lb-ft (95 N·m) in sequence shown.

    c. Using torque angle gauge, tighten screws (2) an additional 45-75°.

12. Rotate crankshaft (7) and check for freedom of movement.

13. Check crankshaft (7) end play (task a).

5-29. CRANKSHAFT AND GEAR MAINTENANCE (Contd)

NOTE
- All connecting rod caps are installed the same. This procedure covers the installation of one connecting rod cap.
- Rotate crankshaft as necessary to access connecting rod journals.

14. Place a piece of plastic strip across width of connecting rod cap (1).

15. Install connecting rod cap (1) on crankshaft (2) and connecting rod (3) with two nuts (4).

16. Tighten nuts (4) 30 lb-ft (41 N·m), then 60 lb-ft (81 N·m), then 90 lb-ft (122 N·m).

17. Remove two nuts (4) and connecting rod cap (1) from connecting rod (3) and crankshaft (2).

18. Using plastigage, measure width of plastic strip. If width is not 0.0013-0.0046 in. (0.033-0.117 mm), replace all connecting rod bearings (5) and (6).

19. Apply clean engine oil to lubricate connecting rod cap (1).

20. Install connecting rod cap (1) on crankshaft (2) and connecting rod (3) with two nuts (4).

21. Tighten nuts (4) 30 lb-ft (41 N·m), then 60 lb-ft (81 N·m), then 90 lb-ft (122 N·m).

22. Move connecting rod (3) back and forth on crankshaft (2).

23. Measure connecting rod (3) side clearance using feeler gauge. If clearance is not 0004-0.012 in. (0.1-0.30 mm), remove connecting rod cap (1) and inspect for proper size and seating of bearings (5) and (6).

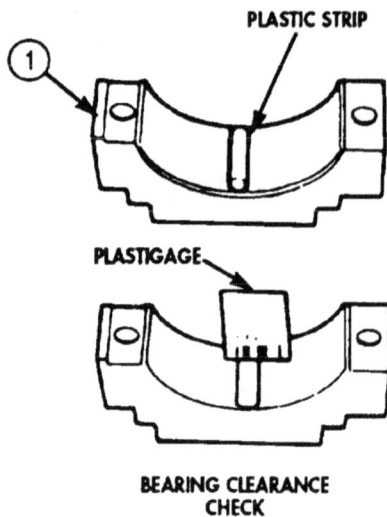

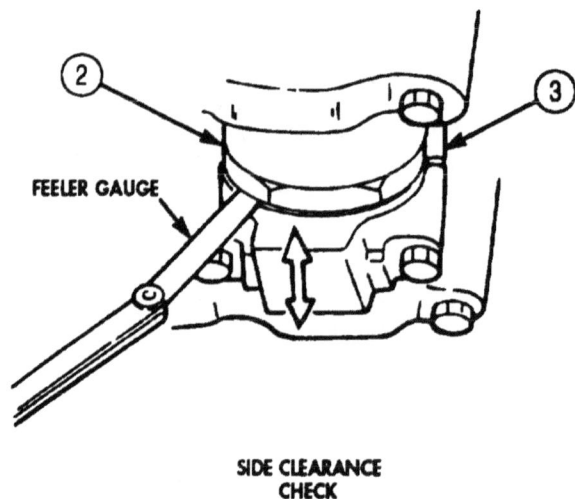

PLASTIC STRIP

PLASTIGAGE

BEARING CLEARANCE
CHECK

FEELER GAUGE

SIDE CLEARANCE
CHECK

5-29. CRANKSHAFT AND GEAR MAINTENANCE (Contd)

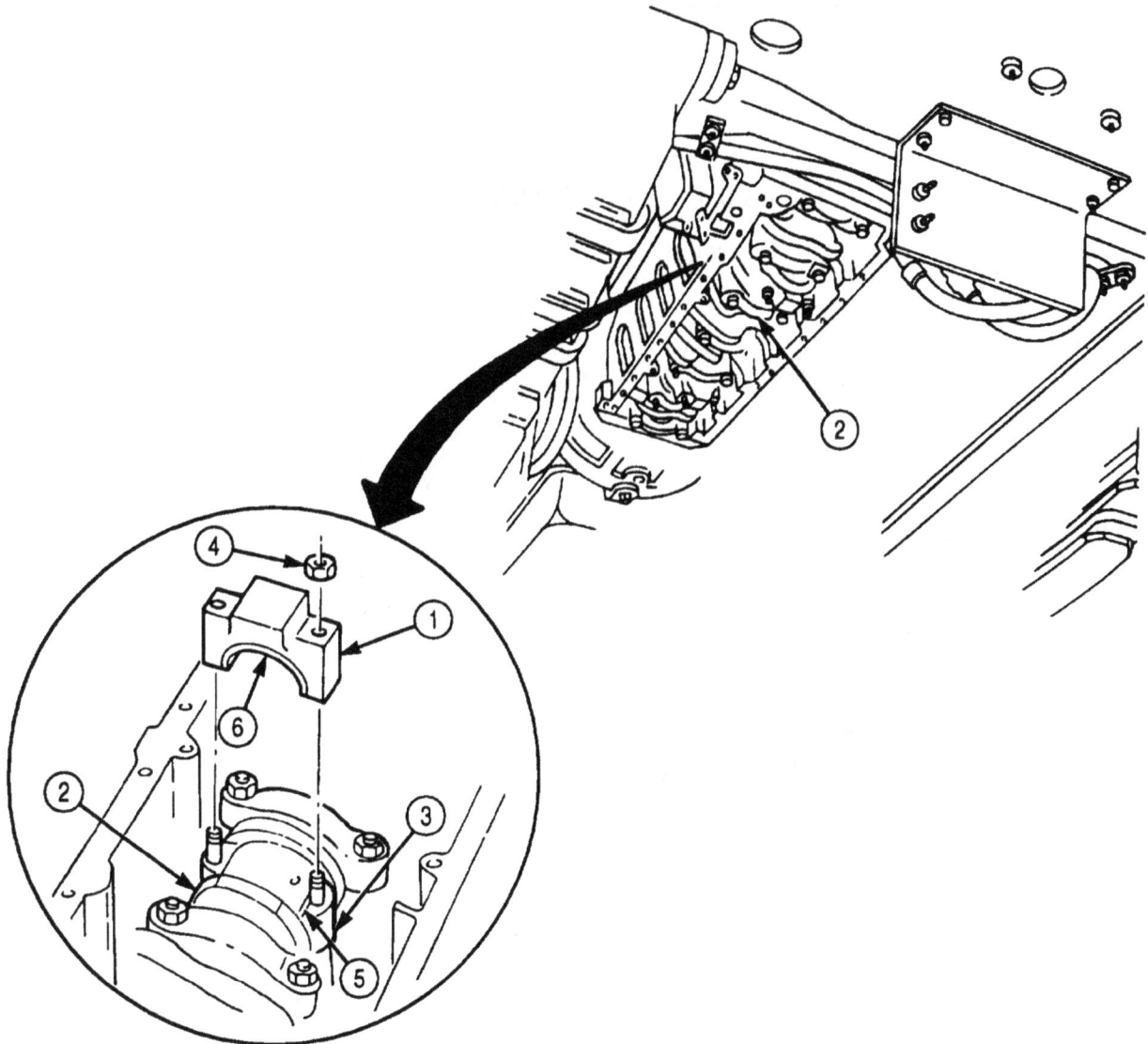

FOLLOW-ON TASKS:
- Install rear cover and oil seal (para. 4-44).
- Install flexplate and flywheel housing (para. 4-43).
- Install front gearcase housing (para. 4-46).
- Install oil pan and suction tube (para. 4-47).
- Install fuel injectors (para. 4-55).

5-30. ROCKER LEVERS AND PUSH RODS MAINTENANCE

THIS TASK COVERS:

a. Disassembly   c. Assembly
b. Cleaning and Inspection

INITIAL SETUP:

APPLICABLE MODELS
M939A2

TOOLS
General mechanic's tool kit (Appendix E, Item 1)
Inside micrometer (Appendix E, Item 82)
Outside micrometer (Appendix E, Item 80)
Torque wrench (Appendix E, Item 144)

MATERIALS/PARTS
Twelve locknuts (Appendix D, Item 326)
Drycleaning solvent (Appendix C, Item 71)
Lubricating oil (Appendix C, Item 50)

REFERENCES (TM)
TM 9-2320-272-10
TM 9-2320-272-24P

EQUIPMENT CONDITION
• Parking brake set (TM 9-2320-272-10).
• Rocker-levers and push rods removed (para. 4-41).

GENERAL SAFETY INSTRUCTIONS
• Keep fire extinguisher nearby when using drycleaning solvent.
• Drycleaning solvent is flammable and toxic.
• Eyeshields must be worn when cleaning with compressed air. Compressed air source will not exceed 30 psi (207 kPa).

a. Disassembly

NOTE
All rocker levers are disassembled the same way. This procedure covers one rocker lever assembly.

1. Remove two retaining rings (2), washers (3), spring washers (4), and rocker levers (1) from rocker shaft (7).
2. Remove locknut (5) and adjusting screw (6) from each rocker lever (1). Discard locknut (5).

b. Cleaning and Inspection

1. For general cleaning instructions, refer to para. 2-14
2. For general inspection instructions, refer to para. 2-15.
3. Replace all parts failing inspection.
4. Measure inside diameter of rocker lever bore (8) at several points, If inside diameter exceeds 0.878 in. (22.3 mm), replace rocker lever (1).
5. Measure outside diameter of rocker shaft (7) at several points. If outside diameter is less than 0.874 in. (22.2 mm), replace rocker shaft (7).

5-30. ROCKER LEVERS AND PUSH RODS MAINTENANCE (Contd)

5-30. ROCKER LEVERS AND PUSH RODS MAINTENANCE (Contd)

WARNING

- Drycleaning solvent is flammable and toxic. Do not use near open flame and always have a fire extinguisher nearby when solvents are used. Use only in well-ventilated places, wear protective clothing, and dispose of cleaning rags in approved container. Failure to do this may result in injury or death to personnel and/or damage to equipment.
- Eyeshields must be worn when cleaning with compressed air. Compressed air source will not exceed 30 psi (207 kPa). Failure to do so may result in injury to personnel.

6. Inspect oil manifold (2) for bends, cracks, and breaks. If bent, cracked, or broken, replace oil manifold (2).

CAUTION

Oil manifold screws are self-tapping and must be reinstalled in their proper positions to ensure proper alignment of screw flats or damage may occur to oil manifold.

7. Inspect oil manifold (2) for clogged oil ports. If oil ports are clogged, remove two screws (1), flush oil manifold (2) with drycleaning solvent. Dry with compressed air and replace two screws (1) in oil manifold (2).

d. Assembly

- All rocker levers are assembled the same way. This procedure is for one rocker lever assembly.
- Lightly oil all parts before assembling.
- Perform step 1 if assembling with new rocker shaft.

1. Install two plugs (8) on rocker shaft (13).
2. Install adjusting screw (12) and new locknut (7) on each rocker lever (6).
3. Install spring washer (5), washer (4), and retaining ring (3) on one side of rocker shaft (13).
4. Install two rocker levers (6) on rocker shaft (13).
5. Install spring washer (11), washer (10), and retaining ring (9) on rocker shaft (13).

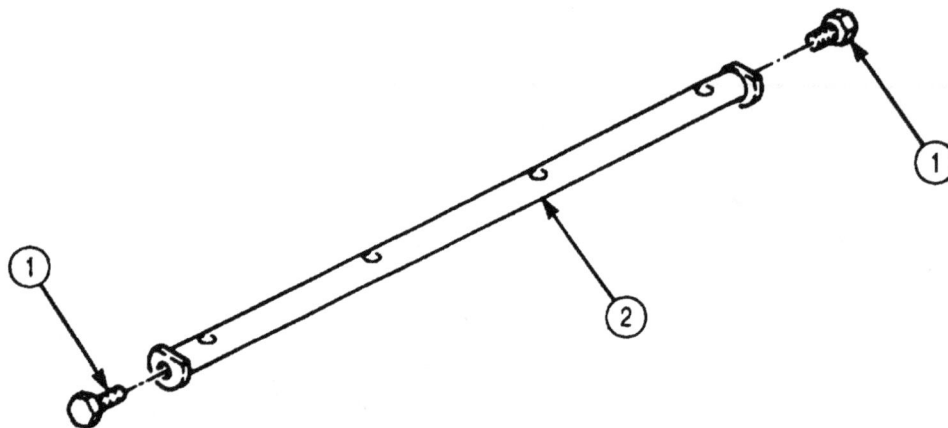

5-30. ROCKER LEVERS AND PUSH RODS MAINTENANCE (Contd)

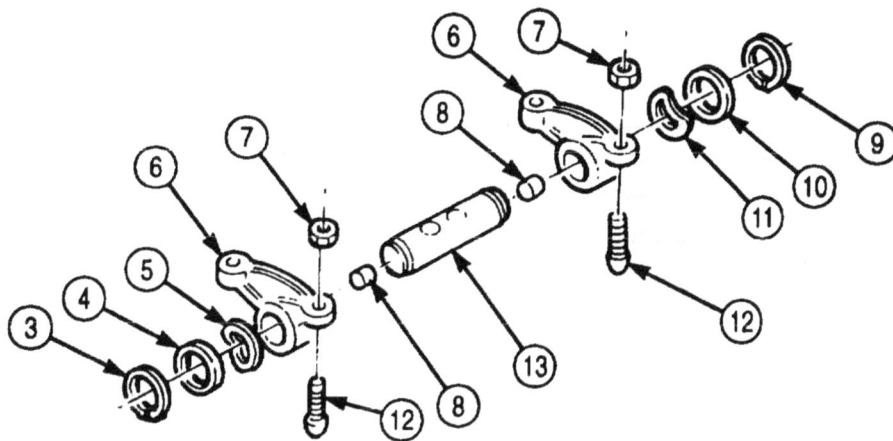

FOLLOW-ON TASK: Install rocker levers and push rods (para. 4-41).

5-31. MOUNTING FUEL INJECTION PUMP TO TEST STAND

THIS TASK COVERS:

a. Mounting Fuel Injection Pump to Test Stand

b. Removing Fuel Injection Pump from Test Stand

INITIAL SETUP:

APPLICABLE MODELS
M939A2

TOOLS
General mechanic's tool kit (Appendix E, Item 1)
Fuel injection pump test stand (Appendix E, Item 50)

MATERIALS/PARTS
Lockwasher (Appendix D, Item 383)
Two copper washers (Appendix D, Item 42)
Copper washer (Appendix D, Item 45)
Copper washer (Appendix D, Item 44)

REFERENCES (TM)
TM 9-2320-272-24P

EQUIPMENT CONDITION
Fuel injection pump removed (para. 4-57).

a. Mounting Fuel Injection Pump to Test Stand

1. Remove screw plug (11) and copper washer (12) from drive end of fuel pump (14). Discard copper washer (12).

2. Install adapter ring (23) on adapter bracket (6) with three washers (9) and screws (10).

3. Install adapter bracket (6) on test stand (20) with washer (7), screw (8), and clamp bar (25).

4. Install fuel pump (14) on adapter bracket (6) with four washers (26), screws (27), washers (22), and nuts (21).

5. Install woodruff key (13) and drive hub (3) on fuel pump (14) with lockwasher (2) and nut (1).

6. Slide fuel pump (14) forward so drive hub (3) engages with drive coupling (28) and secure fuel pump (14) in place with four lockwashers (4) and nuts (5).

7. Install rear support bracket (19) on rear of fuel pump (14) with two washers (17) and screws (18).

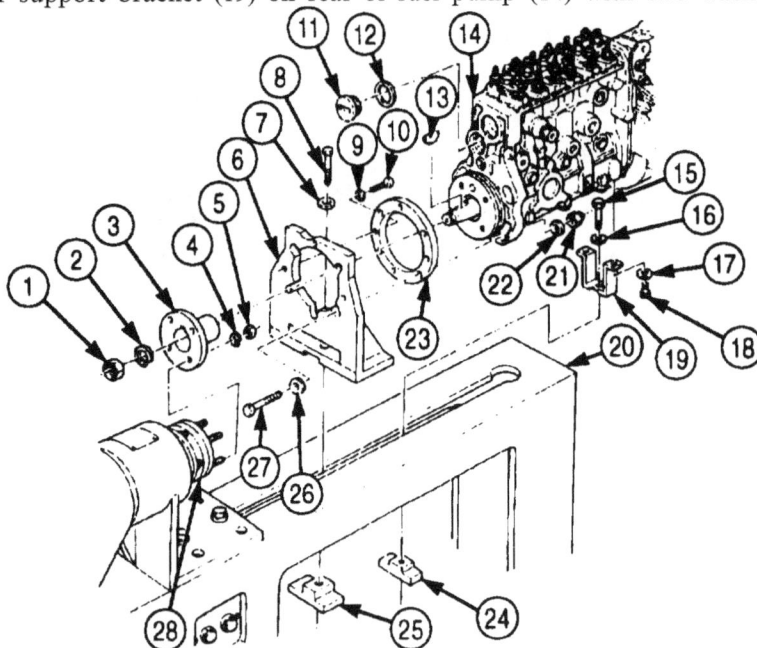

5-31. MOUNTING FUEL INJECTION PUMP TO TEST STAND (Contd)

8. Install rear support bracket (19) and clamp bar (24) on test stand (20) with two washers (16) and screws (15).

9. Connect six test stand fuel lines (30) to delivery valve holders (29) and test stand injectors.

10. Connect air line (32) to manifold pressure compensator (31).

11. Remove plug (36), two copper washers (35), and adapter (37) from fuel pump (14). Discard copper washers (35).

12. Install outlet elbow (34) on fuel pump (14).

13. Connect test stand supply line (33) to outlet elbow (34).

14. Remove plug (38) and copper washer (39) from side of fuel pump (14) for access to number one cylinder. Discard copper washer (39).

15. Refer to table 5-10, Test Stand Requirements, for further preliminary requirements.

Table 5-10. Test Stand Requirements.

ITEM	REQUIREMENTS
Calibration Fluid Temperature	100°-108°F (38°-42°C)
Inlet Pressure	22 psi (152 kPa)
Opening Pressure	3,000-3,046 psi (20,684-21,001 kPa)
Firing Order	1-5-3-6-2-4
Select Drive Rotation	1:2
Direction of Drive Rotation	Clockwise
Number of Injection Cylinders	Six

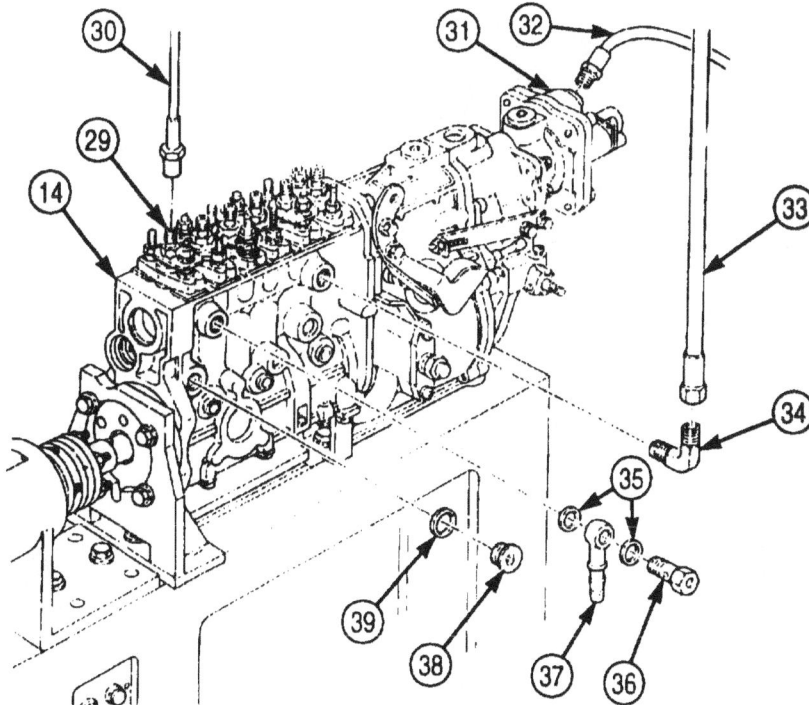

5-31. MOUNTING FUEL INJECTION PUMP TO TEST STAND (Contd)

b. Removing Fuel Injection Pump from Test Stand

1. Install new copper washer (10) and plug (9) on fuel pump (11).
2. Disconnect test stand supply line (4) from outlet elbow (5).
3. Remove outlet elbow (5) from fuel pump (11).
4. Install two new copper washers (6), adapter (8), and plug (7) on fuel pump (11).
5. Disconnect air line (3) from manifold pressure compensator (2).
6. Disconnect six test stand fuel lines (1) from delivery valve holders (12).
7. Remove two screws (29) and washers (28) from rear support bracket (30).
8. Remove four nuts (17) and lockwashers (16) from drive coupling (39). Discard lockwashers (16).

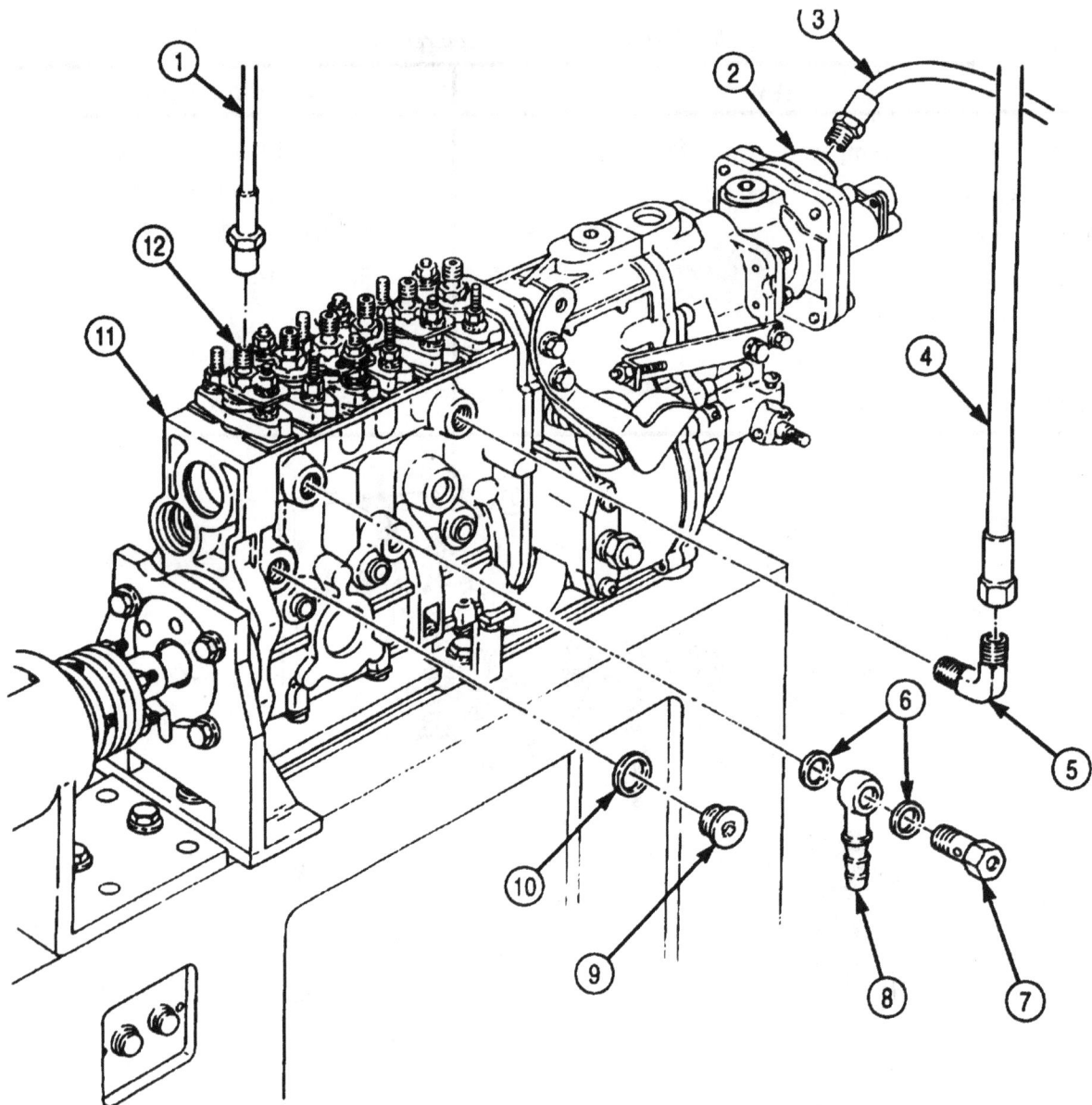

5-31. MOUNTING FUEL INJECTION PUMP TO TEST STAND (Contd)

9. Remove four nuts (32), washers (33), screws (38), and washers (37) from adapter bracket (18).

10. Loosen screw (20) and slide fuel pump (11) from drive coupling (39).

11. Remove nut (13), lockwasher (14), drive hub (15), and woodruff key (25) from fuel pump (11). Discard lockwasher (14).

12. Remove fuel pump (11) from adapter bracket (18) and test stand (31).

13. Install new copper washer (22) and screw plug (21) on fuel pump (11).

14. Remove three screws (24), washers (23), and adapter ring (34) from adapter bracket (18).

15. Remove screw (20), washer (19), adapter bracket (18), and clamp bar (36) from test stand (31).

16. Remove two screws (26), washers (27), rear support bracket (30), and clamp bar (35) from test stand (31).

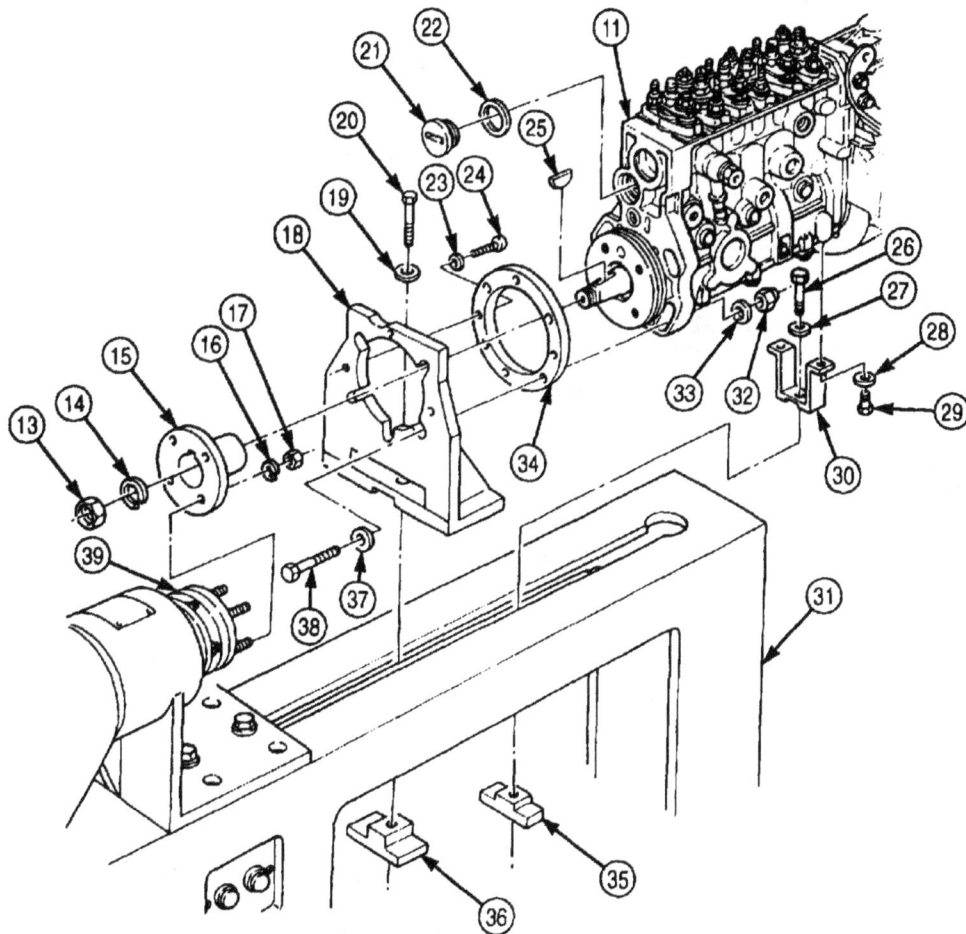

FOLLOW-ON TASK: Install fuel injection pump (para. 4-57).

5-32. FUEL INJECTION PUMP PRETEST

THIS TASK COVERS:

a. Prestroke and Rack Travel
b. Cylinder Phasing (Timing)
c. High-speed Fuel Delivery Test
d. Low-speed Fuel Delivery Test
e. Rated Speed Test
f. Full Load Delivery

g. Manifold Pressure Compensator Test
h. Start Cut Out Test
I. Fuel Delivery Test
j. Fuel Breakaway Test
k. Starting Fuel Delivery Test
l. Low Idle Fuel Delivery Test

INITIAL SETUP:

APPLICABLE MODELS
M939A2

SPECIAL TOOLS
Rack extension (Appendix E, Item 112)
Plunger lift device (Appendix E, Item 98)

TOOLS
General mechanic's tool kit (Appendix E, Item 1)
Dial Indicator (Appendix E, Item 36)

MATERIALS/PARTS
Two copper washers (Appendix D, Item 42)

REFERENCES (TM)
TM 9-2320-272-24P
TM 9-4910-778-14&P

EQUIPMENT CONDITION
Fuel injection pump mounted to test stand (para. 5-31).

NOTE

- Refer to TM 9-4910-778-14&P for operation of fuel injection pump test stand, DFP 156.
- If readings taken during pretest deviate from specifications listed below, fuel pump must be repaired.

a. Prestroke and Rack Travel

1. Mount dial indicator (10) and rack travel indicator holder (9) on drive end of fuel pump (13).
2. Install rack extension (11) on control rack (12).
3. Pull control lever (2) to full load position and, using rack extension (11), push control rack (12) all the way back.
4. Preload dial indicator (10) giving control rack (12) 0.004 in. (0.1 mm) of preload travel.
5. Install plug (3) in return line access hole (14).

NOTE

- Cylinder number one is located at drive end of fuel pump.
- Rotate test stand driveshaft until tappet reaches bottom dead center.

6. Install plunger lift device (5) on fuel pump (13) with finger of plunger lift device (5) resting on top of tappet located in cylinder number one.
7. Mount dial indicator (4) on plunger lift device (5) and zero dial indicator (4).
8. Manually rotate test stand driveshaft (8) until a reading is observed on dial indicator (4).
9. Manually reverse rotation test stand driveshaft (8) until dial indicator (4) reads zero.
10. Apply 435-464 psi (2,999-3,190 kPa) of calibration fluid pressure.

5-32. FUEL INJECTION PUMP PRETEST (Contd)

NOTE

Ensure a steady stream of calibration fluid flows from all bleed-off
valves. If not, increase calibration fluid pressure until a steady
stream of calibration fluid is obtained.

11. Manually rotate test stand driveshaft (8) in correct fuel pump rotation until calibration fluid stops flowing from bleed-off valve (1).
12. Check prestroke measurement. Measurement should be 0.360-0.480 in. (9.14-12.19 mm).
13. Remove plunger lift device (5) and dial indicator (4) from fuel pump (13).
14. Install copper washer (7) and plug (6) on fuel pump (13).

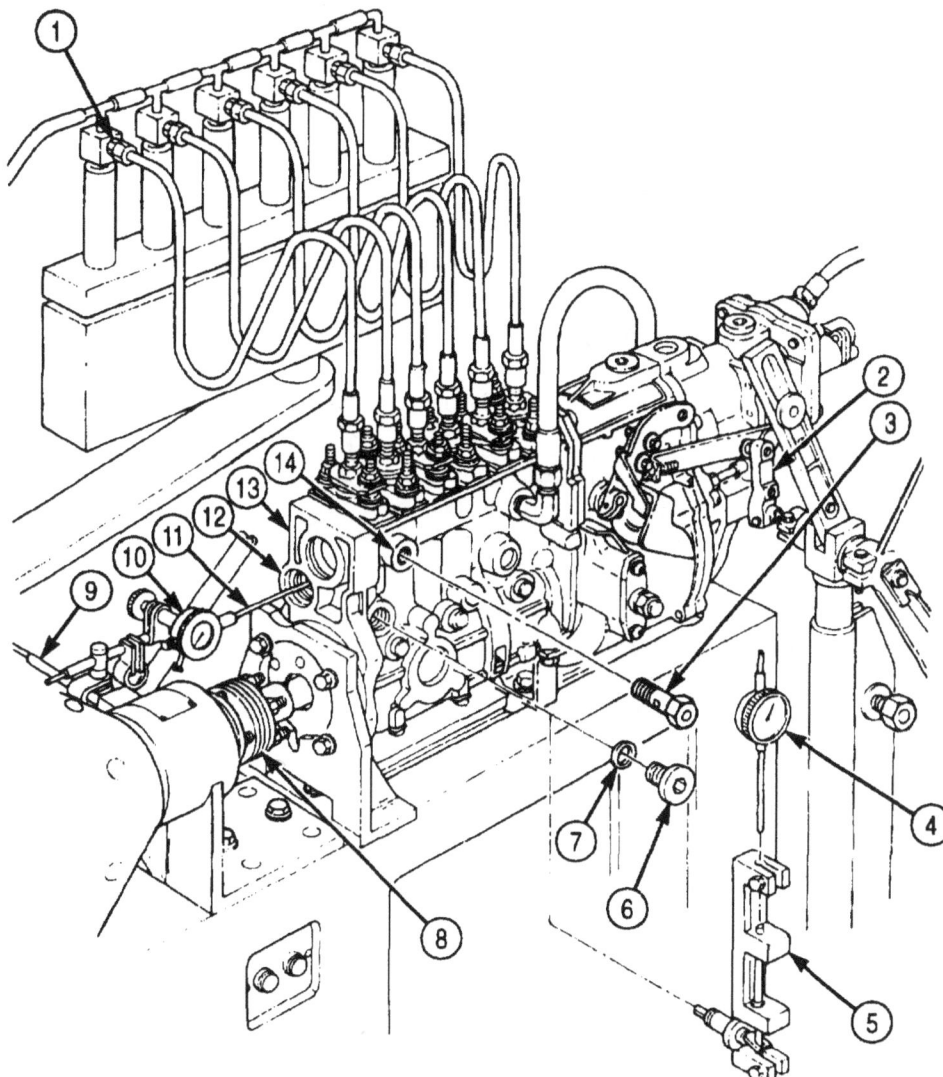

5-32. FUEL INJECTION PUMP PRETEST (Contd)

b. Cylinder Phasing (Timing)

NOTE

In order to perform this task, correct prestroke measurement must be obtained for cylinder number one (task a).

1. Set protractor (1) on drive end of test stand (6) to zero.
2. Remove plug (7) from fuel pump (4).
3. Install two new copper washers (5), plug (7), and adapter (8) on fuel pump (4).
4. Connect test stand return line (2) to fuel pump adapter (8).
5. Apply 435-464 psi (2,999-3,199 kPa) of calibration fluid pressure to fuel pump (4).
6. Manually rotate test stand driveshaft (3) in correct rotation of fuel pump (4) until calibration fluid stops flowing from number five bleed-off valve.
7. Observe reading on protractor (1). Correct reading should be 60°.

NOTE

- Following fuel pump firing order 1-5-3-6-2-4, repeat steps 6 and 7 until remaining cylinders are phased (timed).
- For each cylinder phased, test stand protractor will increase in 60° intervals.
- Perform steps 8 through 11 if 60° interval is not obtained.
- Perform steps 8 through 11 to adjust one cylinder. Adjustment of remaining five cylinders is performed basically the same way.

8. Remove fuel test line (9) from delivery valve holder (10).
9. Loosen three nuts (11) and remove two shims (12) between barrel (13) and fuel pump (4). Discard shims (12).

CAUTION

Shims of the same thickness must be used on both sides of barrel. Failure to do so will result in inaccurate prestroke measurement.

NOTE

Refer to TM 9-2320-272-24P to select shims.

10. Install two new shims (12) between barrel (13) and fuel pump (4).
11. Tighten three nuts (11) and install fuel test line (9) on delivery valve holder (10).

5-32. FUEL INJECTION PUMP PRETEST (Contd)

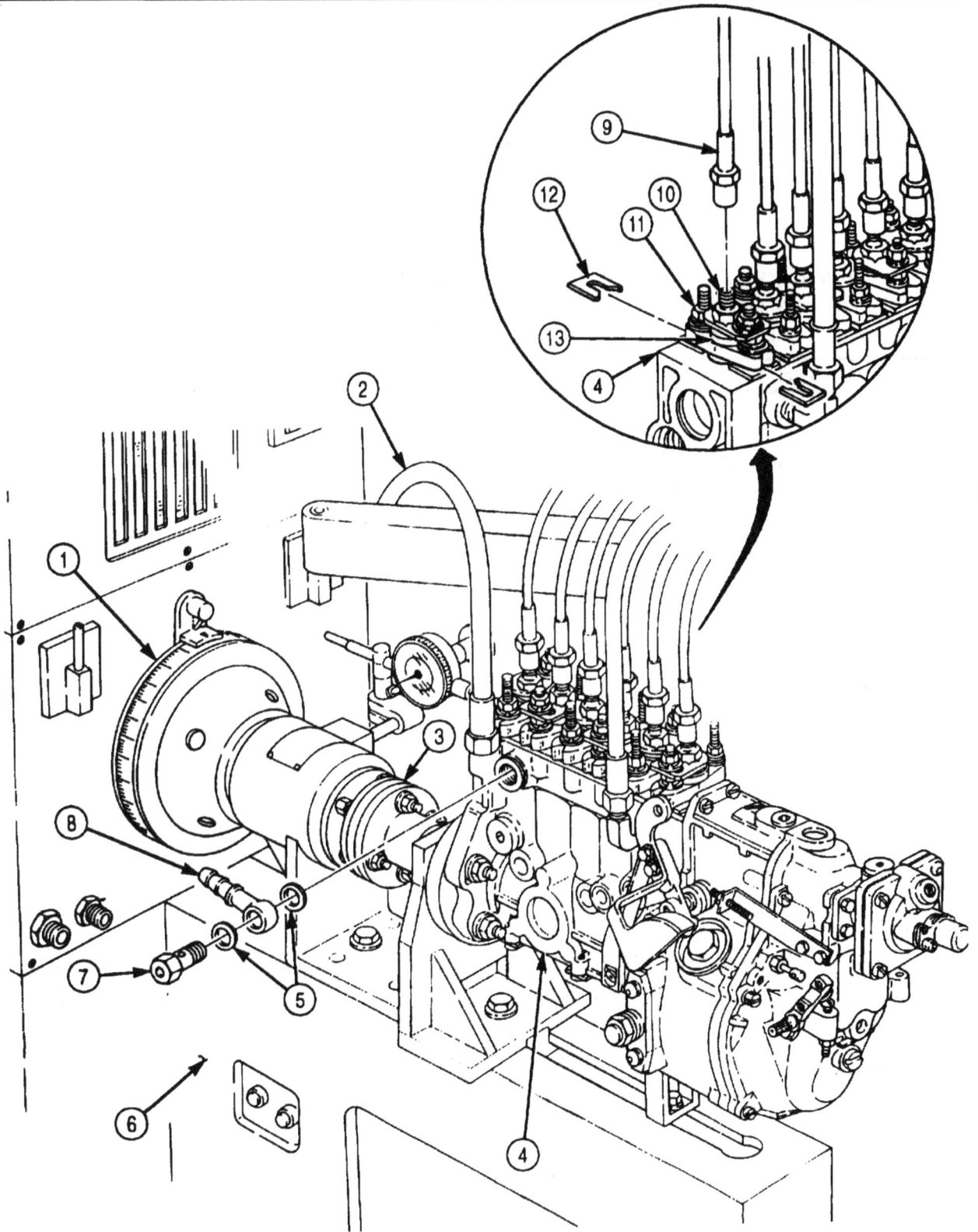

5-32. FUEL INJECTION PUMP PRETEST (Contd)

c. High-speed Fuel Delivery Test

NOTE

Maintain 21 psi (145 kPa) calibration fluid pressure throughout this task.

1. Set test stand (4) to measure fuel delivery at 1,000 strokes.
2. Set fuel pump (1) to 1,050 rpm.

NOTE

If needed, increase aneroid air pressure to obtain correct rack travel.

3. Move control lever (3) to obtain rack travel of 0.360-0.480 in. (9.14-12.19 mm).
4. Take fuel draw and check fuel delivery. Correct fuel delivery is 0.891-0.928 in^3 (14.6-15.2 cm^3).
5. Allowable reading spread is 0.037 in^3 (0.6 cm^3).

d. Low-speed Fuel Delivery Test

NOTE

Maintain 21 psi (145 kPa) calibration fluid pressure throughout this task.

1. Reduce fuel pump (1) to 300 rpm.
2. Set control rack travel between 0.308-0.316 in. (7.8-8.0 mm).
3. Take fuel draw and check fuel delivery. Correct fuel delivery 0.079-0.134 in^3 (1.3-2.2 cm^3)
4. Allowable reading spread is 0.030 in^3 (0.5 cm^3).

e. Rated Speed Test

1. Install protractor (2) on control lever (3).
2. Position control lever (3) between 42-50°
3. Increase rpm of fuel pump (1) until control rack travel of 0.464 in. (11.78 mm) is obtained.
4. Fuel pump (1) rpm should be between 1,090-1,100.
5. Increase rpm of fuel pump (1) to 1,185-1,215.
6. Check control rack travel. Correct control rack travel should be 0.016 in (0.40 mm).
7. Position control lever (3) between 10-18°.
8. Adjust fuel pump (1) to 300 rpm.
9. Check control rack travel. Correct control rack travel should be 0.312 in. (7.9 mm).
10. Decrease fuel pump (1) to 100 rpm.
11. Check control rack travel. Correct control rack travel should be 0.372 in. (9.45 mm).
12. Increase rpm of fuel pump (1) to 300. Check control rack travel. Correct control rack travel should be 0.308-0.316 in. (7.82-8.03 mm).

5-32. FUEL INJECTION PUMP PRETEST (Contd)

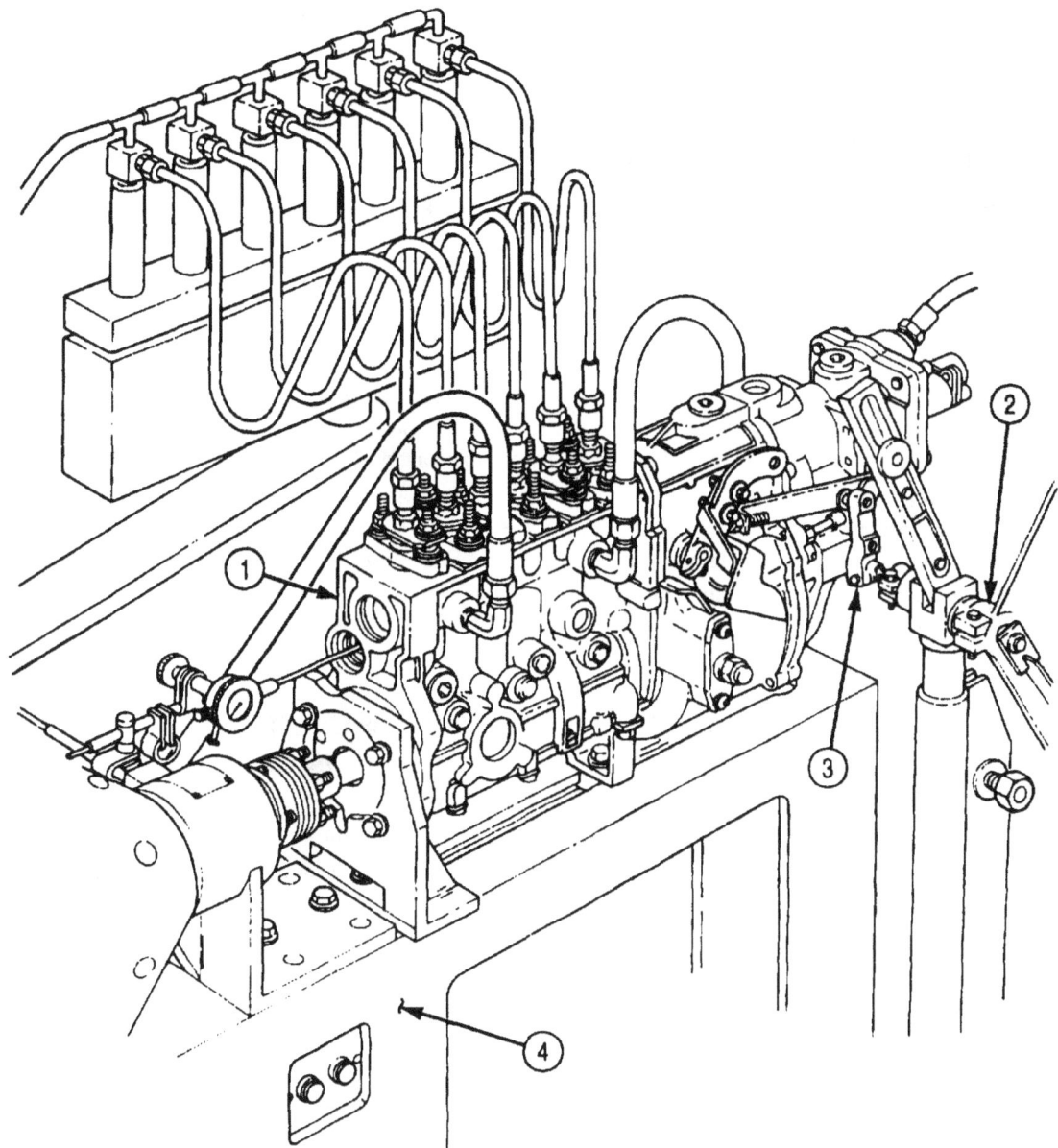

5-32. FUEL INJECTION PUMP PRETEST (Contd)

f. Full Load Delivery

1. Apply 131 psi (903 kPa) of air pressure to manifold pressure compensator (2).
2. Adjust fuel pump (1) to 1,050 rpm.
3. Position control lever (3) in full load position.
4. Take fuel draw and check delivery quantity. Correct fuel delivery quantity should be 8.90-9.28 in. (145.8-152.1 cm^3).
5. Allowable reading spread is 0.366 in. (5.998 cm^3).
6. Close air supply source to manifold pressure compensator (2).

g. Manifold Pressure Compensator Test

1. Set fuel pump (1) to 500 rpm and position control rack (5) at 0.038-0.039 in. (0.97-0.99 mm) control rack travel.
2. Apply 44 psi (303 kPa) of air pressure to manifold pressure compensator (2) and observe control rack travel. Correct control rack travel is 0.432-0.436 in. (10.97-11.07 mm).
3. Increased air pressure to 75 psi (517 kPa) and observe control rack travel. Correct control rack travel is 0.476-0.488 in. (12.09-12.39 mm).
4. Increase air pressure to 131 psi (903 kPa) and observe control rack travel. Correct control rack travel is 0.504-0.508 in. (12.80-12.90 mm).
5. Close air supply source to manifold pressure compensator (2).

h. Start Cut Out Test

1. Set fuel pump (1) to 240 rpm and run fuel pump (1) at this speed for approximately one minute.
2. Move throttle lever (4) to full load position and measure control rack travel. Control rack travel should be no more than 0.084 in. (2.13 mm).

i. Fuel Delivery Test

1. Apply 131 psi (903 kPa) of air pressure to manifold compensator (2).
2. Adjust fuel pump (1) to 700 rpm.
3. Set test stand to measure fuel delivery at 1,000 strokes.
4. Check fuel delivery quantity. Correct fuel delivery quantity should be 8.73-9.21 in^3 (143-150.9 cm^3).
5. Check fuel spread. Correct fuel spread should be 0.427 in. (6.99 cm).
6. Decrease fuel pump rpm to 500.
7. Check fuel delivery quantity. Correct fuel delivery quantity is 4.64-5.00 in^3 (76.0-81.9 cm^3).
8. Close air supply source to manifold pressure compensator (2).

5-32. FUEL INJECTION PUMP PRETEST (Contd)

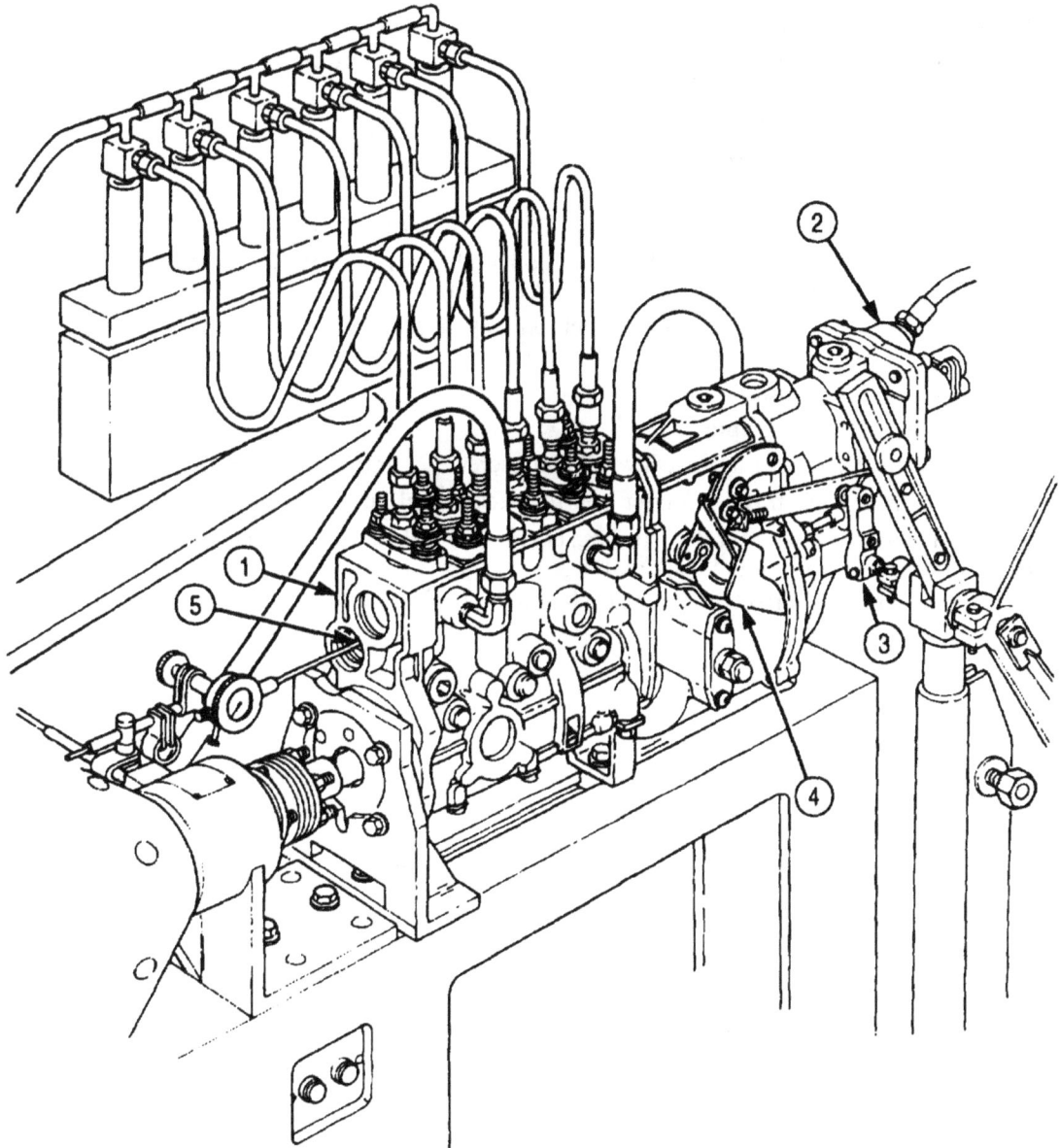

5-32. FUEL INJECTION PUMP PRETEST (Contd)

j. Fuel Breakaway Test

1. Move control lever (3) to full load position.
2. Gradually increase rpm of fuel pump (1) until control rack (6) moves 0.004 in. (0.1 mm) towards shut off.
3. Adjust fuel pump (1) to 1,090-1,100 rpm.
4. Move control lever (3) to full load position and observe control rack travel. Correct control rack travel should be 0.464 in. (11.79 mm).

k. Starting Fuel Delivery Test

1. Set fuel pump (1) to 100 rpm.
2. Set test stand (5) to measure calibration fluid at 1,000 strokes.
3. Position throttle lever (4) to full load position and observe control rack (6) travel. Correct control rack travel should be 0.076-0.084 in. (1.93-2.13 mm).
4. Check fuel delivery quantity. Correct fuel delivery quantity is 12.94-13.91 in^3 (212-227.9 cm^3).

l. Low Idle Fuel Delivery Test

1. Set fuel pump (1) to 300 rpm.
2. Position control lever (3) against low idle stopscrew (2) and take a fuel draw.
3. Observe control rack travel. Correct control rack travel is 0.308-0.316 in. (7.82-8.03 mm).
4. With test stand (5) set at 1,000 strokes, check fuel delivery quantity. Correct fuel delivery quantity should be 0.824-1.373 in^3 (13.5-22.5 cm^3).
5. Check fuel spread. Correct fuel spread should be 0.305 in^3 (4.90 cm^3).

5-32. FUEL INJECTION PUMP PRETEST (Contd)

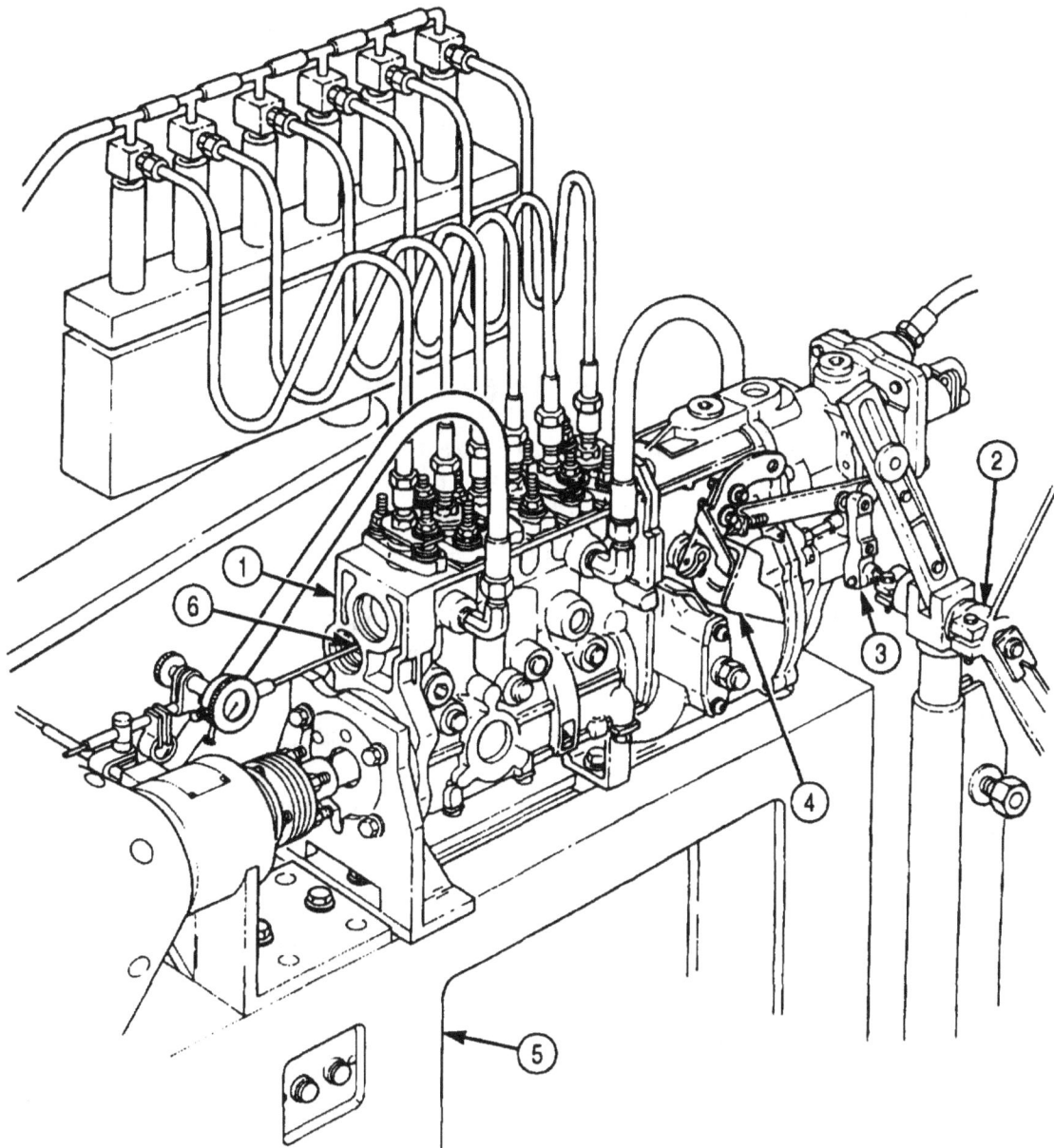

FOLLOW-ON TASK: Remove fuel injection pump from test stand (para. 5-31).

5-33. MANIFOLD PRESSURE COMPENSATOR MAINTENANCE

THIS TASK COVERS:

a. Removal   d. Assembly
b. Disassembly   e. Installation
c. Cleaning and Inspection

INITIAL SETUP:

APPLICABLE MODELS
M939A2

TOOLS
General mechanic's tool kit (Appendix E, Item 1)

MATERIALS/PARTS
Manifold pressure compensator maintenance
  kit (Appendix D, Item 423)
Silicone adhesive (Appendix C, Item 3)

REFERENCES (TM)
TM 9-2320-272-24P

EQUIPMENT CONDITION
Fuel injection pump pretest performed (para. 5-32).

a. Removal

1. Remove safety wire (1) from screw (24) and protective cap (2). Discard safety wire (1).

NOTE
It may be necessary to cut a slot on breakoff screws for removal.

2. Remove breakoff screw (22), screw (24), lockwashers (21) and (23), protective cap (2), and pull-stop (3) from pull-stop cover (16). Discard breakoff screw (22) and lockwashers (21) and (23).

3. Remove O-ring (4) from pull-stop (3). Discard O-ring (4).

4. Remove nuts (20) and (19) from stop bolt (12).

5. Remove three screws (18) and lockwashers (17) from pull-stop cover (16). Discard lockwashers (17).

6. Remove breakoff screw (5), lockwasher (6), and pull-stop cover (16) from manifold pressure compensator (7). Discard breakoff screw (5) and lockwasher (6).

7. Remove three screws (8), lockwashers (9), and washers (10) from manifold pressure compensator (7). Discard lockwashers (9).

8. Remove breakoff screw (15), lockwasher (14), washer (13), and manifold pressure compensator (7) from governor cover (11). Discard breakoff screw (15) and lockwasher (14).

b. Disassembly

1. Remove cotter pin (28), washer (27), and lever (26) from stop bolt (12). Discard cotter pin (28).

2. Remove stop bolt (12) and spring (29) from manifold pressure compensator (7).

3. Remove round nut (34), washer (33), washer plate (32), diaphragm (31), and washer plate (30) from stop bolt (12). Discard diaphragm (31).

4. Remove O-ring (25) from manifold pressure compensator (7). Discard O-ring (25).

5-33. MANIFOLD PRESSURE COMPENSATOR MAINTENANCE (Contd)

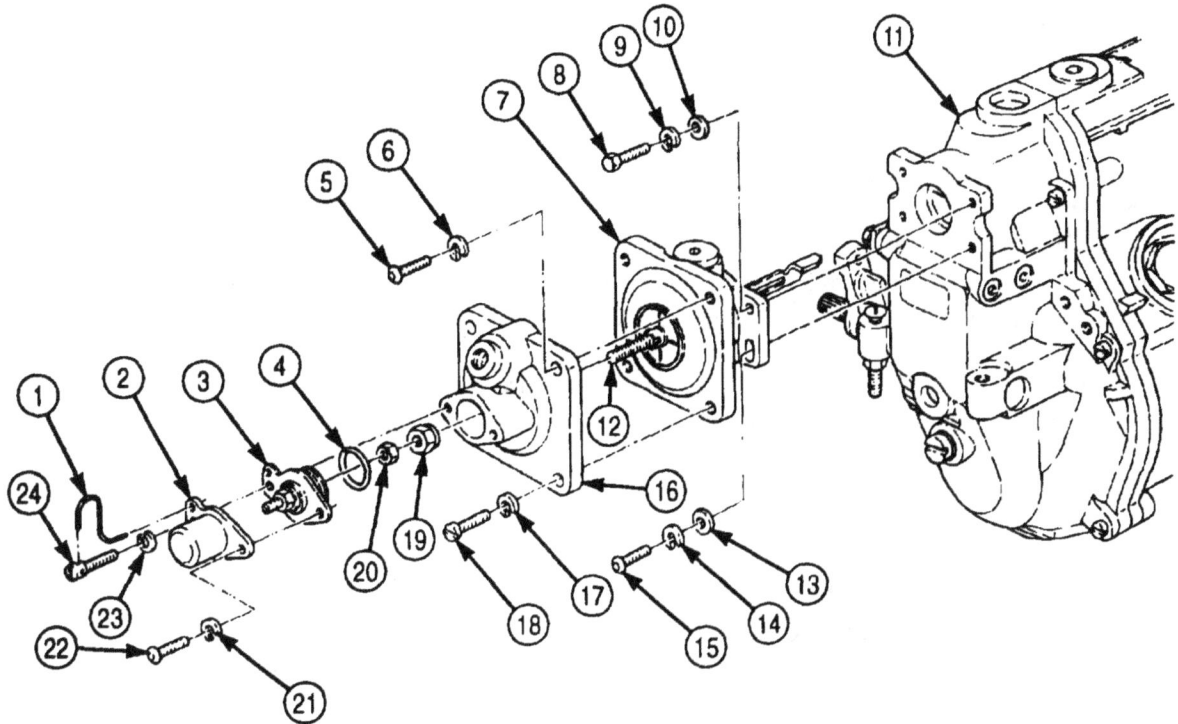

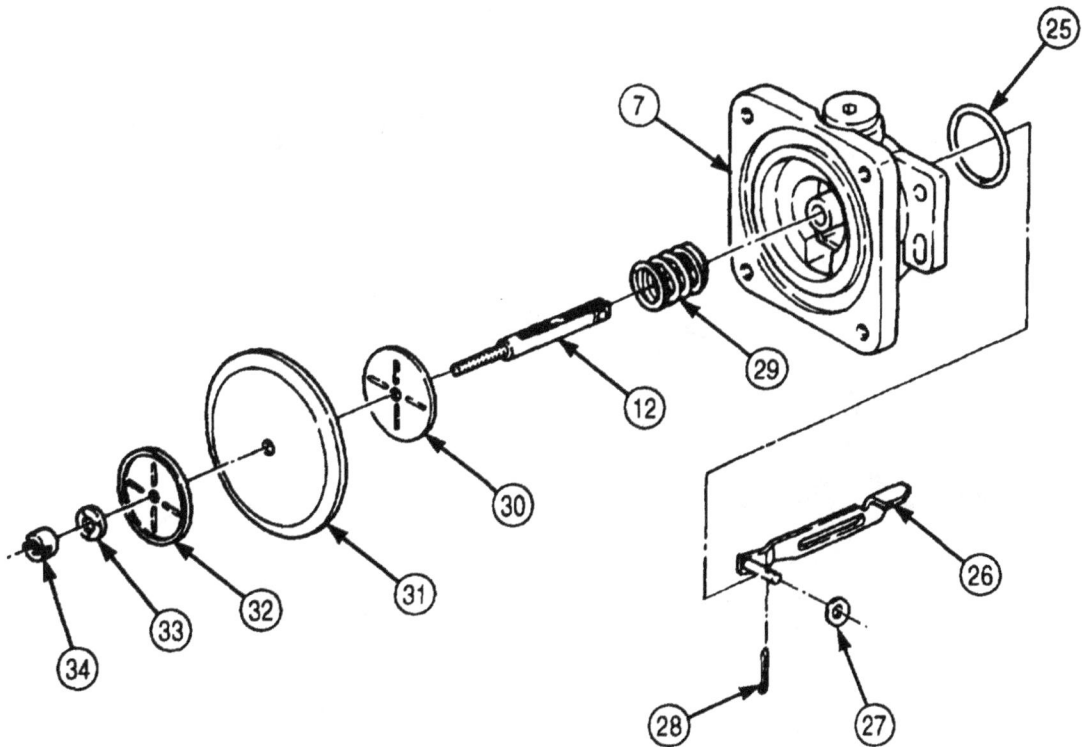

5-33. MANIFOLD PRESSURE COMPENSATOR MAINTENANCE (Contd)

c. Cleaning and Inspection

NOTE
Remove gasket sealant from mating surfaces.

1. For general cleaning instructions, refer to para. 2-14.
2. For general inspection instructions, refer to para. 2-15.

d. Assembly

1. Install new O-ring (2) on manifold pressure compensator (1).
2. Install washer plate (8), new diaphragm (9), washer plate (10), and washer (11) on stop bolt (7) with round nut (12).
3. Install spring (6) and stop bolt (7) in manifold pressure compensator (1).
4. Install lever (3) on stop bolt (7) with washer (4) and new cotter pin (5).

e. Installation

NOTE
Perform step 1 to ensure manifold pressure compensator lever is engaged to rocker arm in governor cover.

1. Remove plug (24) from governor housing (23).
2. Install manifold pressure compensator (1) on governor cover (22) with three washers (21), new lockwashers (20), and screws (19).
3. Install washer (25), new lockwasher (26), and new breakoff screw (27) on manifold pressure compensator (1).

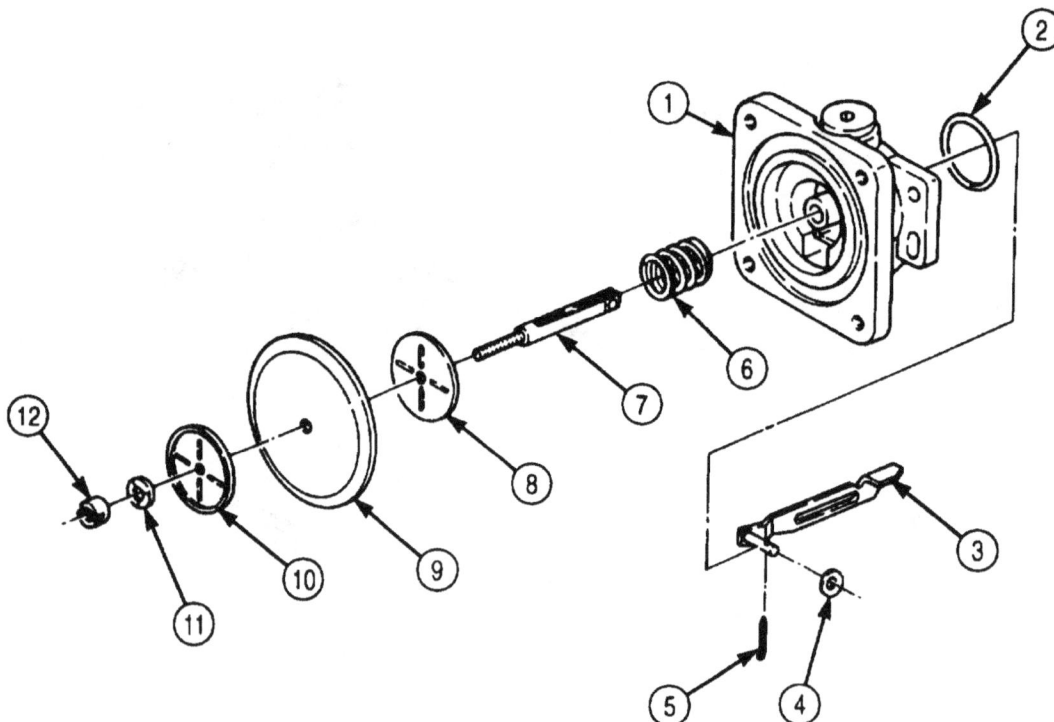

5-33. MANIFOLD PRESSURE COMPENSATOR MAINTENANCE (Contd)

NOTE

Apply silicone adhesive to mating surfaces of pull-stop cover and manifold pressure compensator.

4. Install pull-stop cover (28) on manifold pressure compensator (1) with three new lockwashers (29) and screws (30).

5. Install new lockwasher (18) and new breakoff screw (17) on pull-stop cover (28).

6. Install nuts (31) and (32) on stop bolt (7).

7. Install new O-ring (16) on pull-stop (15).

8. Place pull-stop (15) in pull-stop cover (28).

9. Install protective cap (33) on pull-stop cover (28) with new lockwashers (14) and (34), screw (13), and new breakoff screw (35).

10. Install plug (24) on governor housing (23).

NOTE

Safety wire will not be installed at this time. Safety wire will be installed after fuel pump is calibrated (para. 5-37).

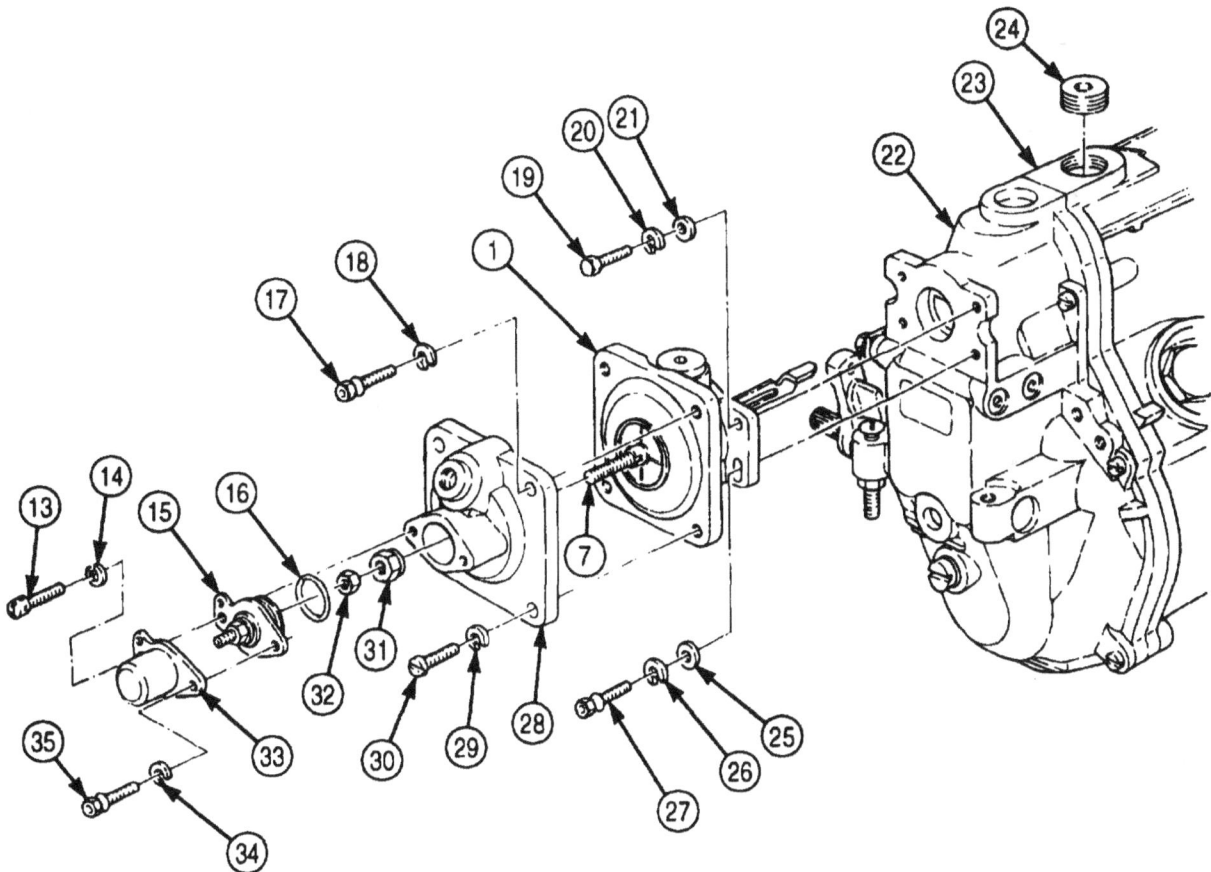

FOLLOW-ON TASK: Calibrate fuel injection pump (para. 5-37).

5-34. R.Q.V. GOVERNOR COVER MAINTENANCE

THIS TASK COVERS:

a. Removal   d. Assembly
b. Disassembly   e. Adjustment
c. Cleaning and Inspection   f. Installation

INITIAL SETUP:

APPLICABLE MODELS   REFERENCES (TM)
M939A2   TM 9-2320-272-24P

TOOLS   EQUIPMENT CONDITION
General mechanic's tool kit (Appendix E, Item 1) Manifold pressure compensator removed (para. 5-33).
Depth gauge (Appendix E, Item 81)

MATERIALS/PARTS
R.Q.V. governor cover maintenance kit
  (Appendix D, Item 562)
Sealing compound (Appendix C, Item 61)

a. Removal

1. Remove guide pin (8) and gasket (9) from governor cover (5). Discard gasket (9).
2. Remove five screws (10) from governor cover (5).
3. Remove two screws (12), lockwashers (13), governor stop (3), and two spacers (4) from governor cover (5). Discard lockwashers (13).

NOTE

When removing governor cover, move governor lever back and forth.

4. Remove safety wire (6) from screws (7) and (11). Discard safety wire (6).
5. Remove screw (7), governor cover (5), and gasket (2) from governor housing (1). Discard gasket (2).

b. Disassembly

1. Remove two drive pins (25) from linkage lever (23) and shaft (19). Discard drive pins (25).
2. Remove shaft (19) from governor cover (5).
3. Remove screw (14), lockwasher (15), shim (16), template (17), and shim (18) from governor cover (5). Discard lockwasher (15).
4. Remove linkage lever (23) and intermediate plates (22) and (24) from governor cover (5).
5. Remove rocker arm (21) and spring (20) from governor cover (5).

c. Cleaning and Inspection

1. For general cleaning instructions, refer to para. 2-14.
2. For general inspection instructions, refer to para. 2-15.

5-34. R.Q.V. GOVERNOR COVER MAINTENANCE (Contd)

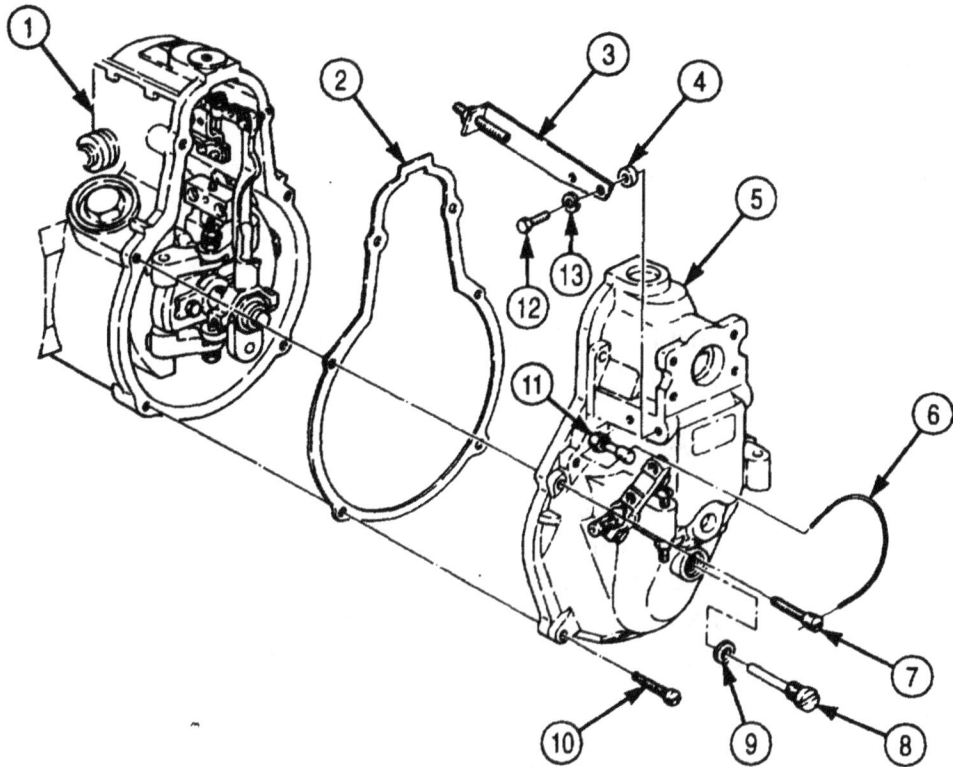

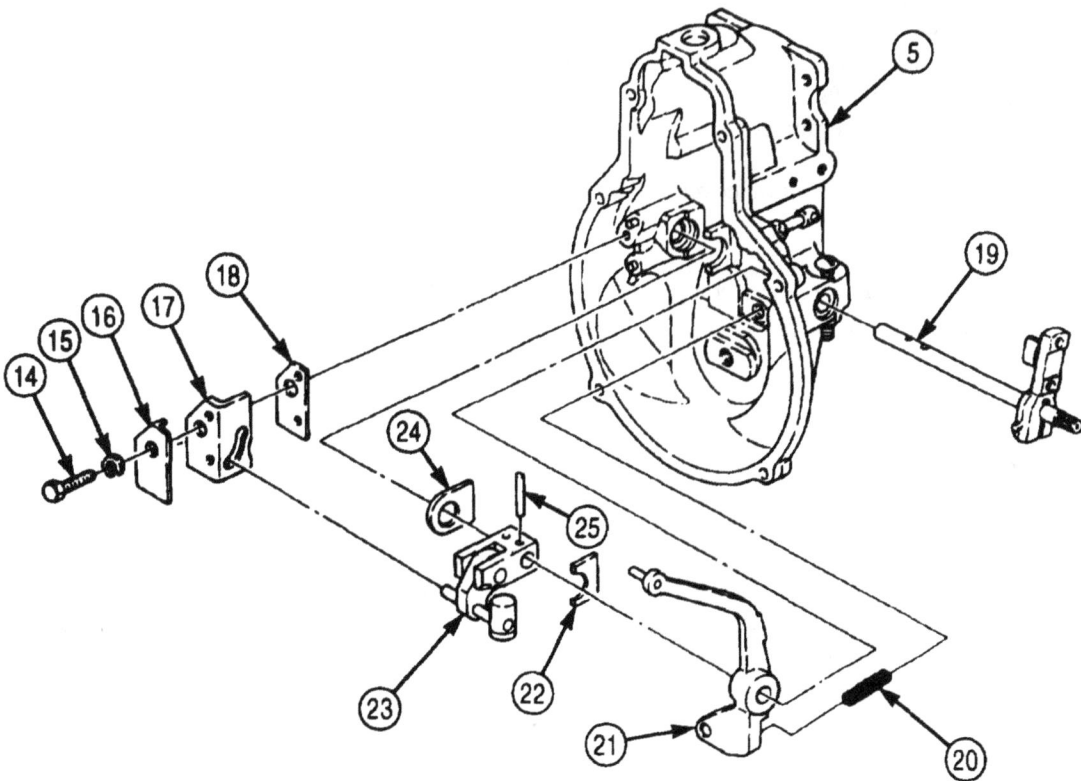

5-34. R.Q.V. GOVERNOR COVER MAINTENANCE (Contd)

d. Assembly

1. Install spring (12)and rocker arm (13) in governor cover (3).
2. Install intermediate plates (14) and (16) and linkage lever (15) in governor cover (3).
3. Install shaft (11) in governor cover (3).
4. Install shim (10), template (9), and shim (8) on governor cover (3) with new lockwasher (7) and screw (6).
5. Install two new drive pins (17) in linkage lever (15) and shaft (11).

e. Adjustment

1. Position control lever (4) in full load position so pilot (5) contacts bottom end of template slot (1).
2. Position new gasket (2) on governor cover (3).
3. Place straightedge on governor cover (3). Using depth gauge, measure from straightedge to pilot (5). Distance should be 0.098 in. (2.49 mm).

NOTE

Perform steps 4 through 7 if measurement is incorrect.

4. Remove screw (6), lockwasher (7), shim (8), template (9), and shim (10) from governor cover (3).
5. If distance is less than 0.098 in. (2.49 mm), use smaller shim (10).
6. If distance is more than 0.098 in. (2.49 mm), use larger shim (10).
7. Install shim (10), template (9), and shim (8) on governor cover (3) with lockwasher (7) and screw (6).

f. Installation

NOTE

- Safety wire will not be installed at this time. Safety wire will be installed after fuel pump is calibrated (para. 5-37).
- Linkage lever piston must fit into cylinder on lever located in governor housing.

1. Install new gasket (2) and governor cover (3) on governor housing (18) with five screws (24) and screw (21).
2. Install governor stop (19) on governor cover (3) with two spacers (20), new lockwashers (25), and screws (26).

NOTE

Apply sealing compound to male threads.

3. Install new gasket (23) and guide pin (22) in governor cover (3).

DEPTH GAUGE   STRAIGHTEDGE

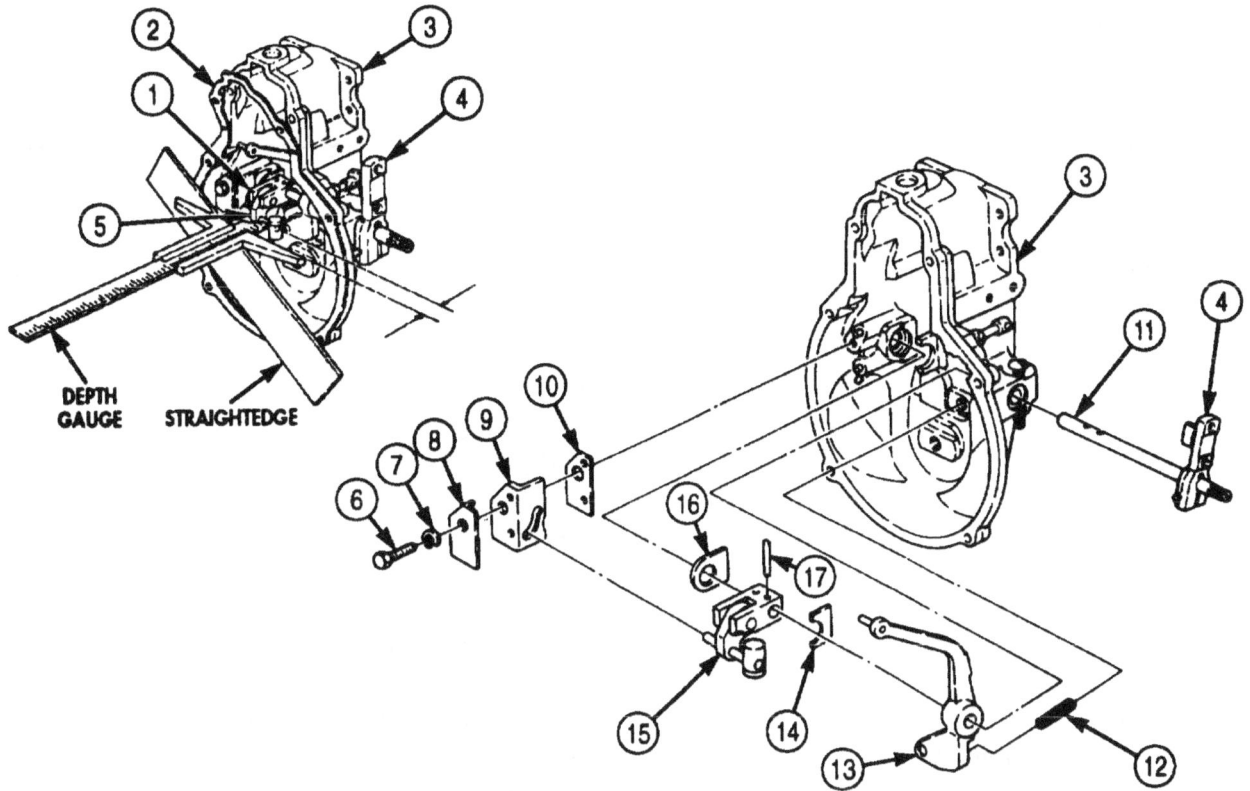

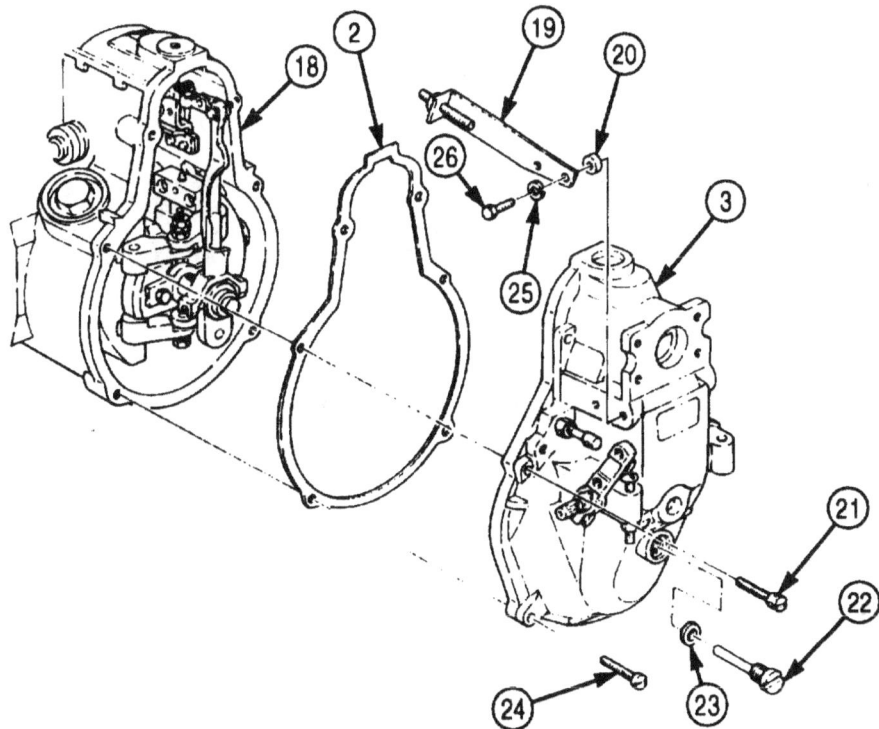

FOLLOW-ON TASK: Install manifold pressure compensator (para. 5-33).

5-35. R.Q.V. GOVERNOR HOUSING MAINTENANCE

THIS TASK COVERS:

a. Removal   d. Assembly
b. Disassembly   e. Installation
c. Cleaning and Inspection   f. Adjustment

<u>INITIAL SETUP:</u>

<u>APPLICABLE MODELS</u>
M939A2

<u>TOOLS</u>
General mechanic's tool kit (Appendix E, Item 1)
Depth gauge (Appendix E, Item 81)
Torque wrench (Appendix E, Item 144)

<u>MATERIALS/PARTS</u>
R.Q.V. governor housing maintenance kit
(Appendix D, Item 561)

<u>REFERENCES</u> (TM)
TM 9-2320-272-24P

<u>EQUIPMENT CONDITION</u>
R.Q.V. governor cover removed (para. 5-34).

GENERAL SAFETY INSTRUCTIONS
Wear eye protection when removing or installing springs under tension.

a. Removal

1. Remove retaining pin (12) from link pin (10). Discard retaining pin (12).
2. Remove link pin (10), slider (15), and lever (11) from link fork (9).
3. Remove retaining pin (5) and link fork (9) from connection plate (3). Discard retaining pin (5).
4. Bend tab washer (19) away from two nuts (20).
5. Remove nut (20). tab washer (19), nut (20), coupling pin (8), and washer (7) from flyweight assembly (6). Discard tab washer (19).
6. Remove bearing pin (16) from guide bushing (17).
7. Bend two tab washers (13) away from screws (14).
8. Remove two screws (14), tab washers (13), and guide bushing (17) from flyweight assembly (6). Discard tab washers (13).
9. Remove round nut (18) from camshaft (4) and governor flyweight assembly (6).
10. Remove governor flyweight assembly (6) from camshaft (4).
11. Remove capsule (21) and four rubber buffers (22) from driver (24). Discard rubber buffers (22).
12. Remove shim (23) and driver (24) from camshaft (4).

NOTE
Screws in steps 13 and 14 are torx head.

13. Remove four screws (28) and holding brackets (27) from governor housing (2).
14. Remove two screws (26) from governor housing (2).
15. Remove two screws (25), governor housing (2), and gasket (29) from fuel pump housing (1). Discard gasket (29).

5-35. R.Q.V. GOVERNOR HOUSING MAINTENANCE (Contd)

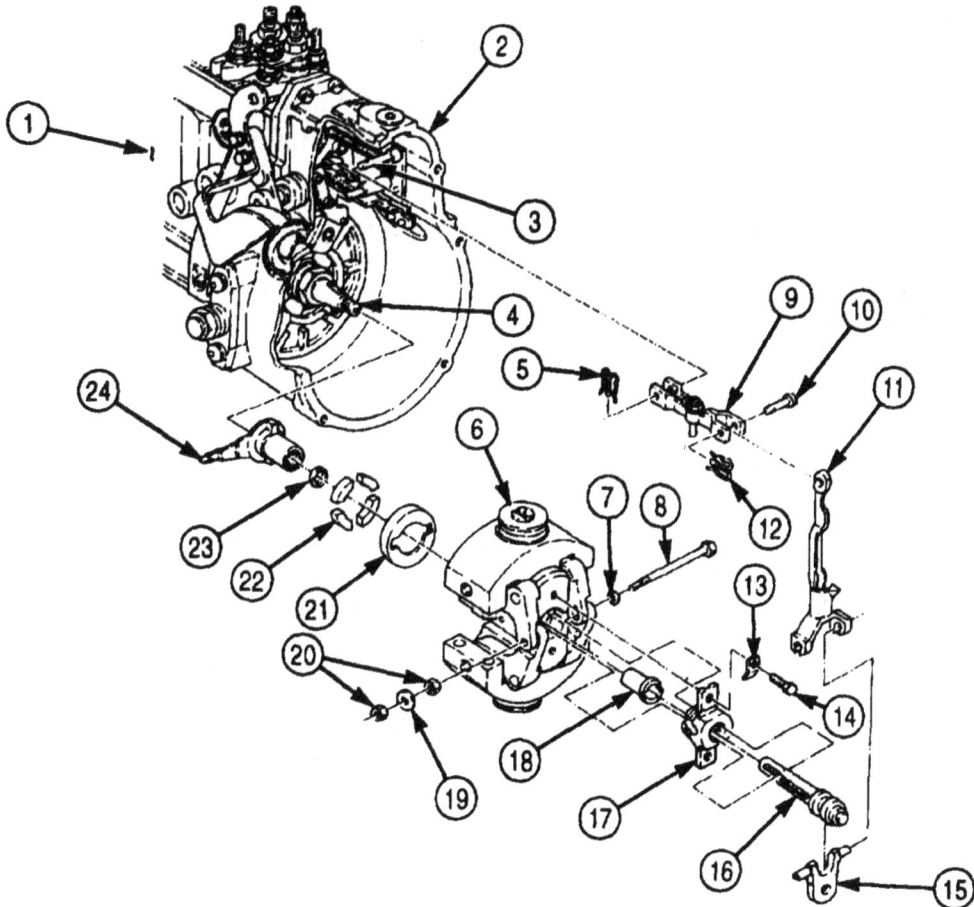

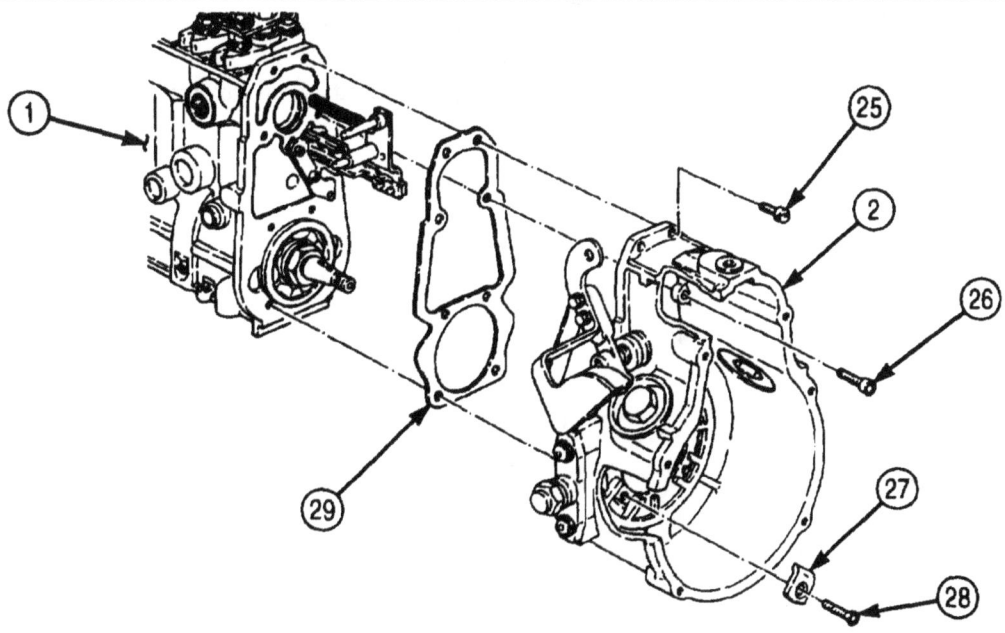

5-35. R.Q.V. GOVERNOR HOUSING MAINTENANCE (Contd)

b. Disassembly

NOTE

Relieve governor flyweight spring tension when performing step 1.

1. Press down on outer spring seat (3) and remove round nut (2) from threaded pin (9).

WARNING

Eye protection must be worn when removing springs under
tension. Failure to do so may result in injury to personnel.

2. Remove outer spring seat (3), springs (4), (5), and (6), shim (7), and lower spring seat (8) from governor flyweight assembly (12).

3. Remove shims (1) and (13) from governor flyweight assembly (12).

c. Cleaning and Inspection

1. For general cleaning instructions, refer to para. 2-14.

2. For general inspection instructions, refer to para. 2-15.

3. Inspect two governor flyweights (11) for nicks and scratches. If either flyweight (11) is nicked or scratched, replace governor flyweight assembly (12).

4. Inspect bell crank levers (10) for noticeable wear. Replace governor flyweight assembly (12) if either bell crank lever (10) is worn.

5. Inspect bearing pin (14) for scratches, scoring, nicks, and wear. Replace bearing pin (14) if scratched, scored, nicked, or worn.

6. Inspect guide bushing (15) for wear, scratches, or scores. Replace guide bushing (15) if worn, scratched, or scored.

d. Assembly

1. Install shims (13) and (1) in governor flyweight assembly (12).

WARNING

Eye protection must be worn when installing springs. Failure to
do so may result in injury to personnel.

2. Install lower spring seat (8), shim (7), and springs (6), (5), and (4) in governor flyweight assembly (12).

3. Install outer spring seat (3) on spring (4).

4. Install round nut (2) on threaded pin (9).

5-35. R.Q.V. GOVERNOR HOUSING MAINTENANCE (Contd)

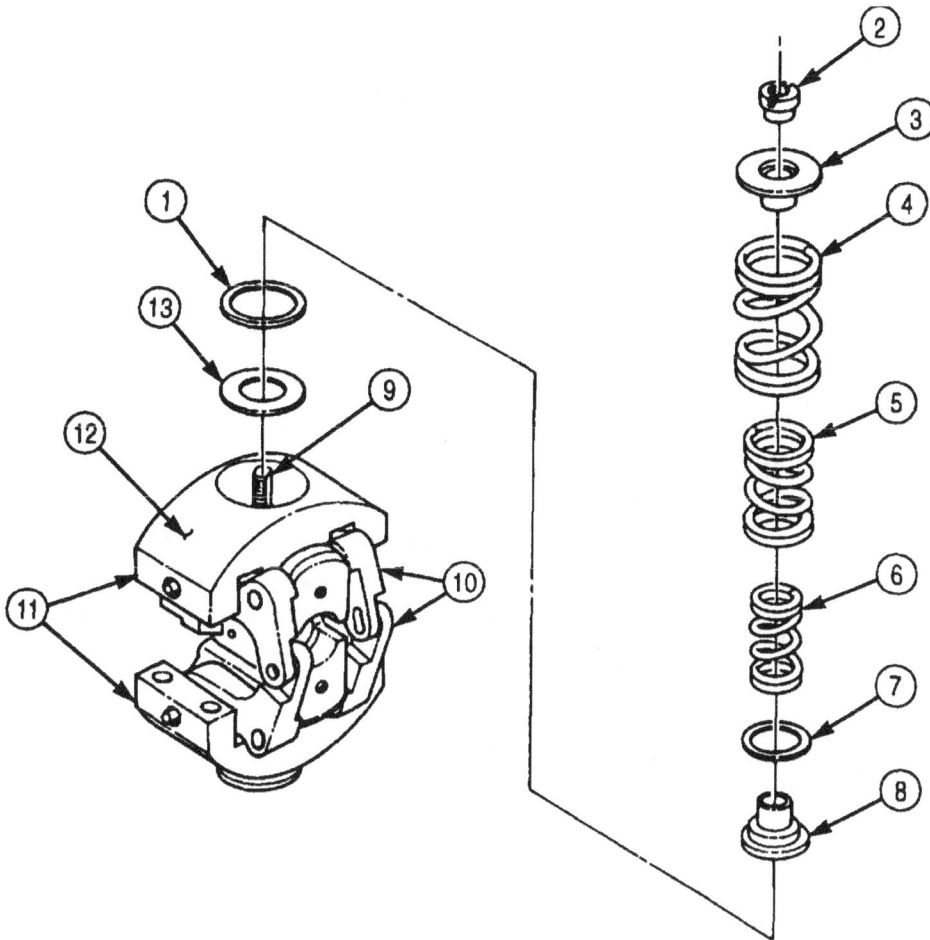

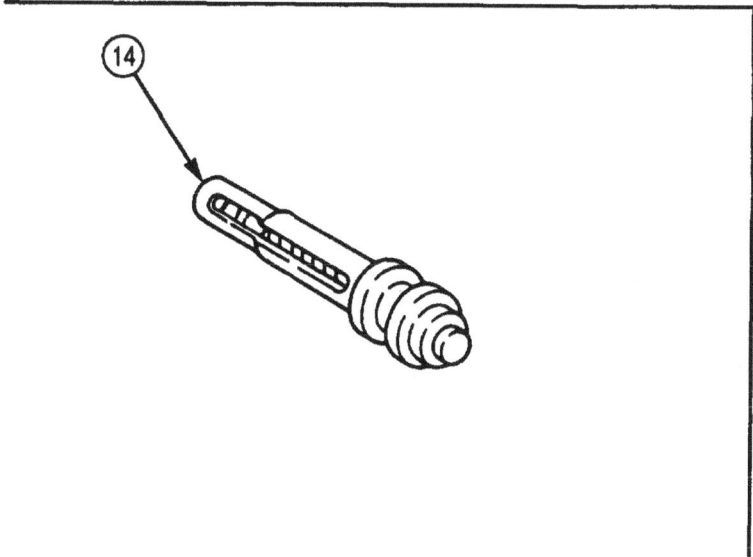

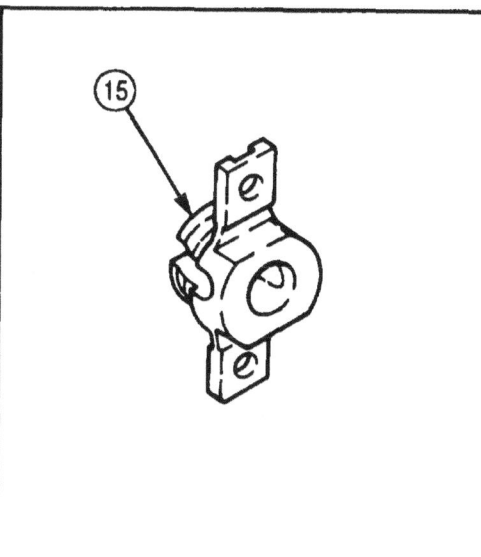

5-35. R.Q.V. GOVERNOR HOUSING MAINTENANCE (Contd)

e. Installation

1. Install new gasket (6) and governor housing (5) on fuel pump housing (7) with two screws (1) and screws (2).

2. Install four holding brackets (3) and screws (4) on governor housing (5).

3. Place shim (28) in driver (29) and install driver (29) on camshaft (9).

4. Install capsule (26) and four new rubber buffers (27) in back side of governor flyweight assembly (11).

5. Install governor flyweight assembly (11) on driver (29).

6. Install round nut (23) on camshaft (9). Tighten round nut (23) 30-35 lb-ft (41-47 N·m).

CAUTION

Failure to obtain proper axial play may cause damage to fuel pump.

7. Check governor flyweight assembly (11) for axial play:

   a. Normal range is 0.002-0.004 in. (0.05-0.1 mm).

   b. If axial play is less than 0.002 in. (0.05 mm), replace shim (28) with larger shim (28).

   c. If axial play is more than 0.004 in. (0.1 mm), replace shim (28) with smaller shim (28).

   d. When axial play is correctly adjusted, it is possible to turn governor flyweight assembly (11) without governor flyweight assembly (11) sticking.

NOTE

Do not bend tab washer until fuel pump is calibrated (para. 5-37).

8. Install guide bushing (22) on flyweight assembly (11) with two new tab washers (18) and screws (19). Tighten screws (19) 53-71 lb-in. (6-8 N·m).

9. Place bearing pin (21) in guide bushing (22).

NOTE

Adjust nuts to allow end play of 0.04-0.12 in. (1.0-3.0 mm).

10. Insert coupling pin (13) through washer (12), governor flyweight assembly (11), and bearing pin (21) and install nut (25), new tab washer (24), and nut (25) on coupling pin (13). Tighten nuts (25) 53-71 lb-in. (6-8 N·m).

11. Install slider (20) on bearing pin (21).

12. Install lever (16) on slider (20).

13. Install link fork (14) on connection plate (8) with new retaining pin (10).

14. Install lever (16) on link fork (14) with link pin (15) and new retaining pin (17).

5-35. R.Q.V. GOVERNOR HOUSING MAINTENANCE (Contd)

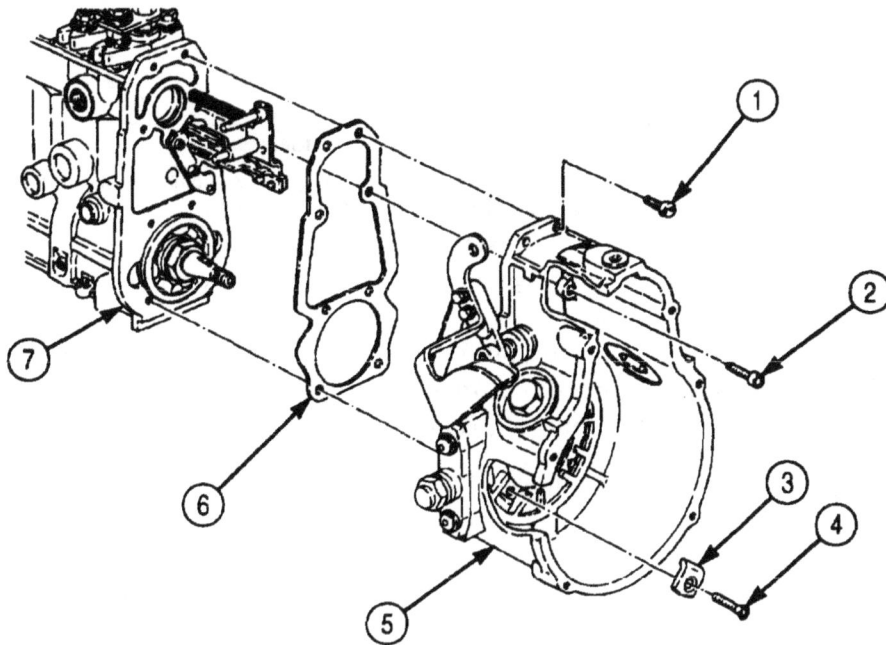

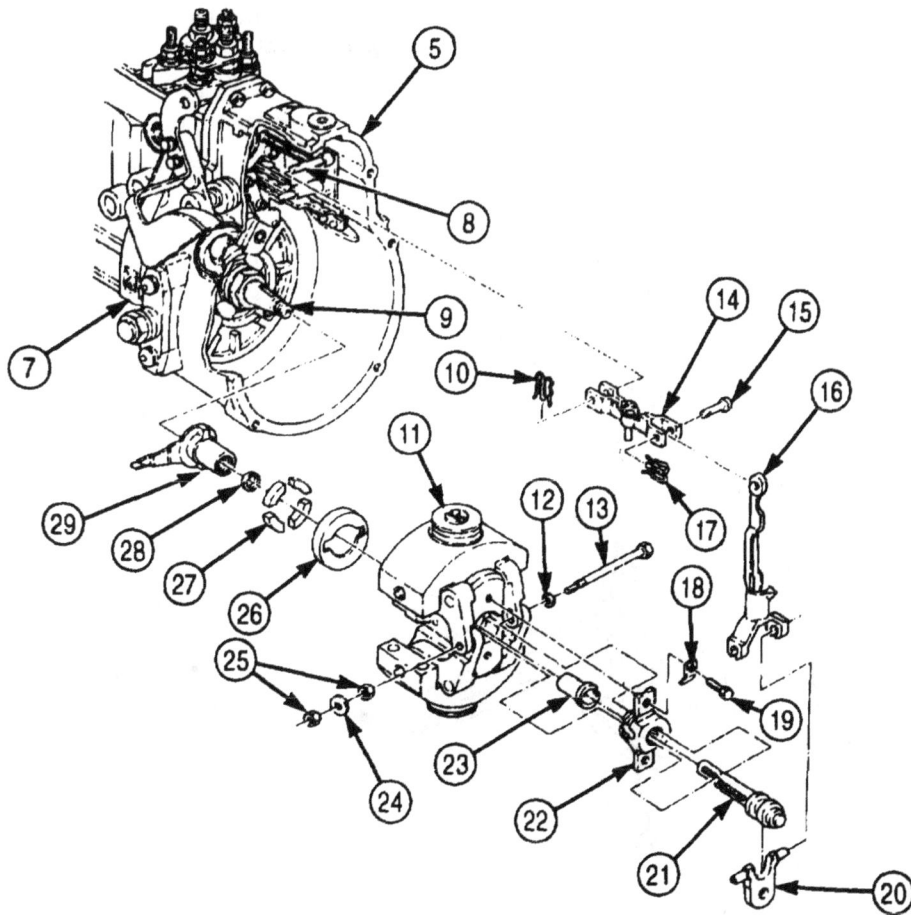

5-35. R.Q.V. GOVERNOR HOUSING MAINTENANCE (Contd)

f. Adjustment

NOTE

Distance is obtained by the measured distance plus the thickness
of straightedge minus half the thickness of slider.

1. Place straightedge on governor housing (1). Using depth gauge, measure distance between slider (2) and straightedge. Distance should be 1.39-1.40 in. (35.3-35.6 mm).

NOTE

Perform steps 2 through 15 if measurement is incorrect.

2. Remove retaining pin (7) from link pin (6).
3. Remove link pin (6). slider (2), and lever (8) from link fork (5) and bearing pin (11).
4. Remove retaining pin (4) and link fork (5) from connection plate (3).
5. Remove nut (14), tab washer (15), nut (14), coupling pin (9), and washer (10) from governor flyweight assembly (16).
6. Remove bearing pin (11) from guide bushing (13).
7. If slider measurement was less than 1.39 in. (35.3 mm), turn adjusting screw (12) in bearing pin (11) clockwise.
8. If slider measurement was more than 1.40 in. (35.6 mm), turn adjusting screw (12) in bearing pin (11) counterclockwise.
9. Place bearing pin (11) in guide bushing (13).
10. Install washer (10) on coupling pin (9) and insert coupling pin (9) through governor flyweight assembly (16) and bearing pin (11) and secure with nut (14), tab washer (15), and nut (14) on coupling pin (9). Tighten nuts (14) 53-71 lb-in. (6-8 N·m).
11. Bend tab washer (15) on nuts (14).
12. Install slider (2) on bearing pin (11).
13. Install lever (8) on slider (2).
14. Install link fork (5) on connection plate (3) with retaining pin (4).
15. Install lever (8) on link fork (5) with link pin (6) and retaining pin (7).
16. Check distance between slider (2) and straightedge (step 1).

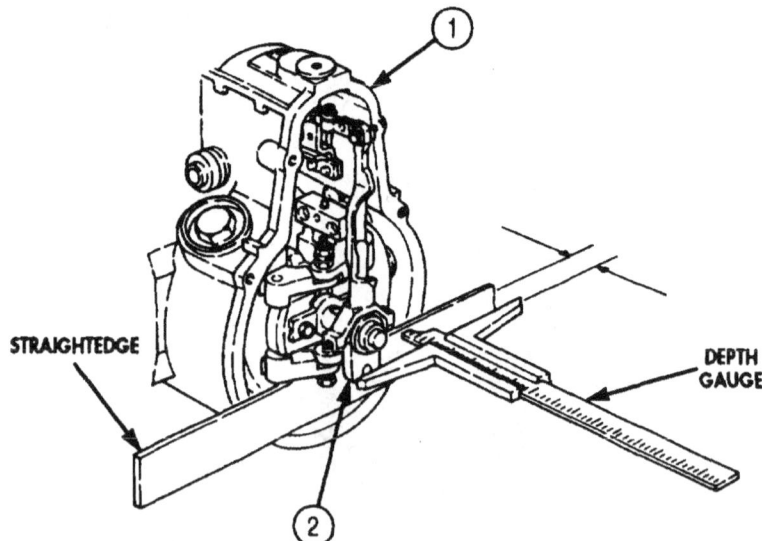

STRAIGHTEDGE   DEPTH GAUGE

5-35. R.Q.V. GOVERNOR HOUSING MAINTENANCE (Contd)

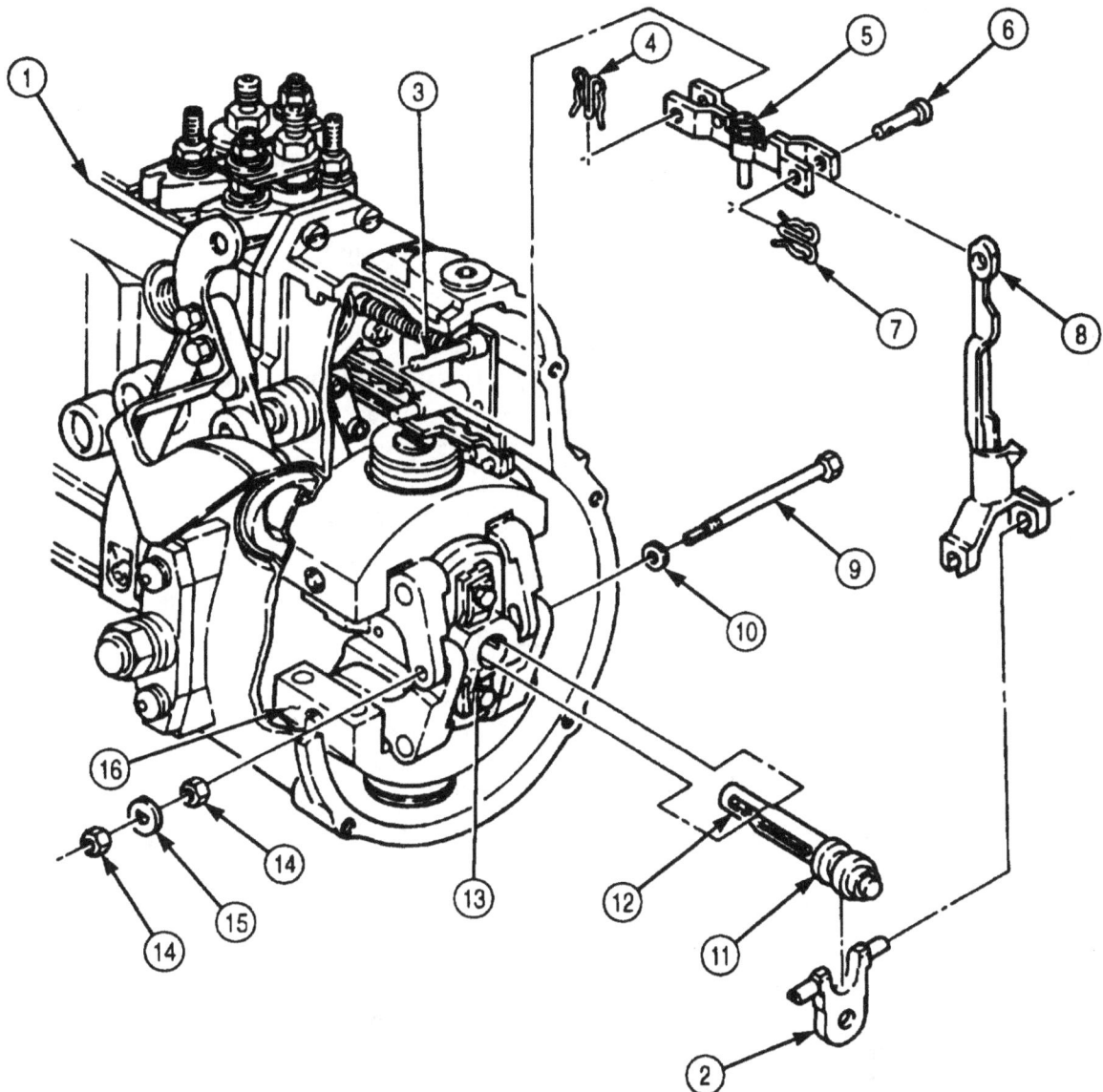

FOLLOW-ON TASK: Install R.Q.V. governor cover (5-34).

```
5-36.  FUEL INJECTION PUMP MAINTENANCE
```

THIS TASK COVERS:

a. Camshaft Removal
b. Tappet and Plunger Removal
c. Barrel and Control Rack Removal
d. Cleaning and Inspection
e. Leakage Test
f. Delivery Valve and Barrel Installation
g. Plunger and Control Rack Installation
h. Tappet and Spring Installation
i. Camshaft Installation

INITIAL SETUP:

APPLICABLE MODELS
M939A2

SPECIAL TOOLS
Two separation tubes (Appendix E, Item 119)
Side plug puller (Appendix E, Item 111)
Three tappet holders (Appendix E, Item 134)
Tappet spring compressor (Appendix E, Item 135)

TOOLS
General mechanic's tool kit (Appendix E, Item 1)
Torque wrench (Appendix E, Item 144)
Arbor press

MATERIALS/PARTS
Fuel injection pump repair kit (Appendix D, Item 137)
Fuel injection pump maintenance kit (Appendix D, Item 136)

MATERIALS/PARTS (Contd)
Sealing compound (Appendix C, Item 61)
GAA grease (Appendix C, Item 28)
Drycleaning solvent (Appendix C, Item 71)
Lubricating oil (Appendix C, Item 50)

REFERENCES (TM)
TM 9-2320-272-24P

EQUIPMENT CONDITION
R.Q.V. governor housing removed (para. 5-35).

GENERAL SAFETY INSTRUCTIONS
• Diesel fuel is flammable. Do not perform this task near open flames.
• Keep fire extinguisher nearby when using drycleaning solvent.
• Drycleaning solvent is flammable and toxic. Do not use near open flame.

```
a.  Camshaft Removal
```

WARNING

• Diesel fuel is flammable. Do not perform fuel system procedure near open flame. Injury to personnel may result.
• Drycleaning solvent is flammable and toxic. Do not use near open flame and always have a fire extinguisher nearby when solvents are used. Use only in well-ventilated places, wear protective clothing, and dispose of cleaning rags in approved container. Failure to do this may result in injury or death to personnel and/or damage to equipment.

CAUTION

Improper cleaning methods and use of unauthorized cleaning solvents can damage equipment.

1. Clean exterior of fuel pump housing (10) with drycleaning solvent.
2. Using side plug puller, remove three side plugs (7) from fuel pump housing (10). Discard side plugs (7).
3. Remove three spacer blocks (8) from side holes (9).

CAUTION

Keep cylinder parts together for replacement in their original cylinder. Failure to do so may result in damage to fuel pump.

4. Remove six nuts (4), lockwashers (3), spacer rings (2), and locking tabs (1) from studs (5) and delivery valve holders (13). Discard lockwashers (3).

5-36. FUEL INJECTION PUMP MAINTENANCE (Contd)

NOTE

Perform step 5 to aid in insertion of tappet holder.

5. Loosen twelve barrel flange nuts (12) and remove twelve shims (6) from between barrel (11) and fuel pump housing (10).

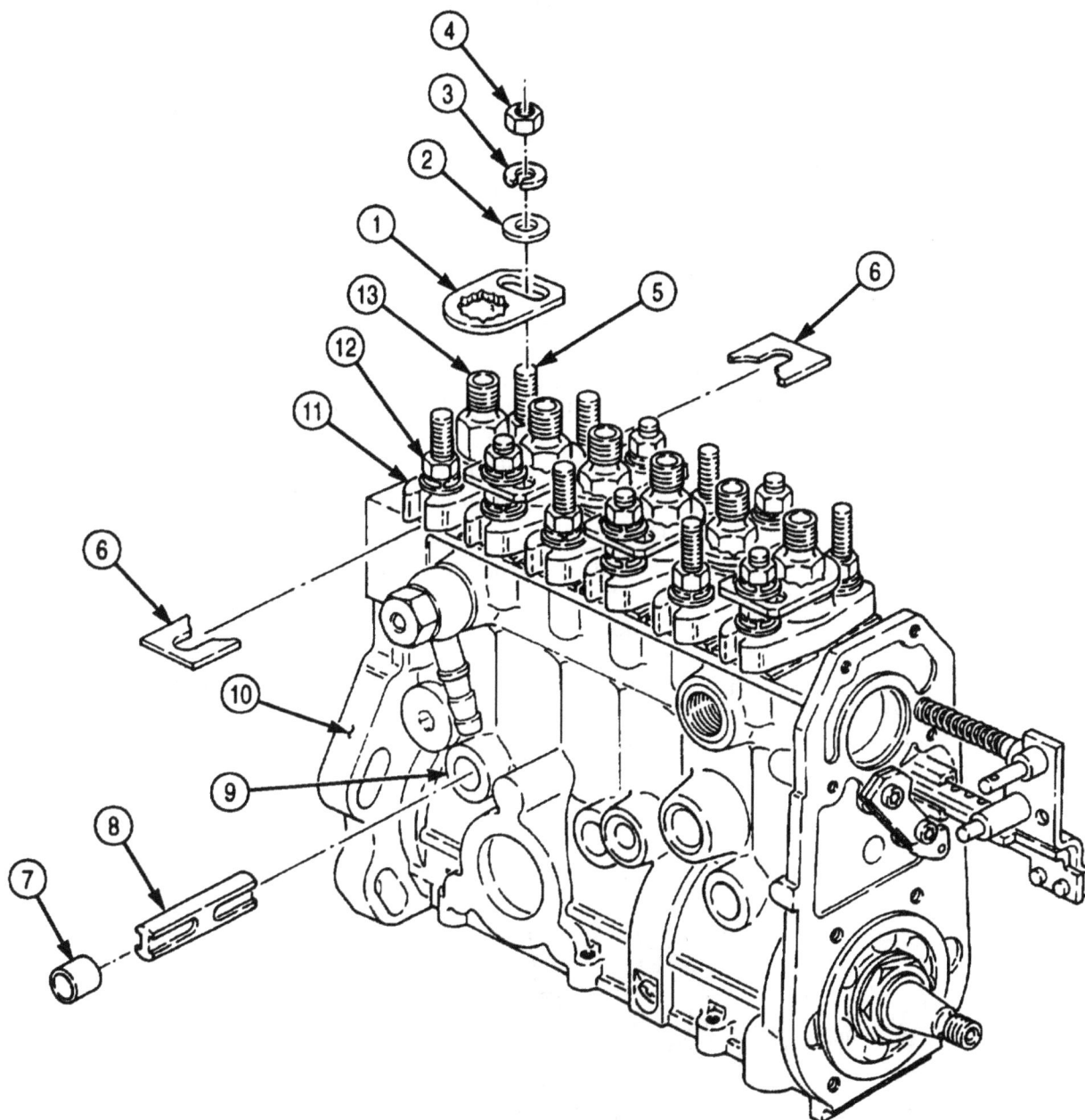

5-36. FUEL INJECTION PUMP MAINTENANCE (Contd)

6. Reverse position of fuel pump housing (1) and mount in vise.

7. Remove six screws (9), holding brackets (19), cover (18), and seal ring (17) from fuel pump housing (1). Discard seal ring (17).

NOTE
Cylinder number one is located at drive end of fuel pump.

8. Rotate camshaft (16) until cylinder number one is at Top Dead Center (TDC).

9. Turn knurled handle (14) on tappet holder (15) fully counterclockwise.

10. Apply GM grease to ramp (12) and guide (11) on tappet holder (15).

NOTE
Tappet holder will insert only part way in when lifting tappet number one off camshaft.

11. Hold tappet holder (15) with ramp (12) up and insert tappet holder (15) into side hole (10) of fuel pump housing (1).

NOTE
- Pressure must be maintained on handle of tappet holder to prevent tappet holder from popping out.
- When all six tappets are lifted off the camshaft, camshaft will rotate with little resistance.

12. Rotate camshaft (16) until cylinder number two is at TDC and push tappet holder (15) until tapered edge (13) on tappet holder (15) contacts fuel pump housing (1).

NOTE
Repeat steps 8 through 12 to install remaining two tappet holders.

13. Remove two screws (21) and seal rings (22) from fuel pump housing (1). Discard seal rings (22).

CAUTION
If arbor press is used, care should be taken to prevent damage to control rack guide pin.

14. Remove camshaft (16), with bearing (23) and bearing shell (20), from governor end of fuel pump housing (1).

15. Remove four screws (5), lockwashers (4), end plate (3), and gasket (2) from fuel pump housing (1). Discard gasket (2) and lockwashers (4).

16. Remove O-ring (6) and oil seal (7) from end plate (3). Discard O-ring (6) and oil seal (7).

17. Remove O-ring (8) from fuel pump housing (1). Discard O-ring (8).

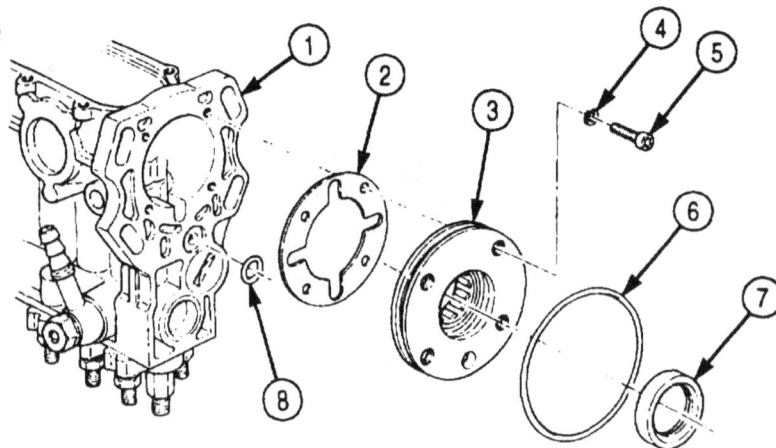

5-36. FUEL INJECTION PUMP MAINTENANCE (Contd)

5-36. FUEL INJECTION PUMP MAINTENANCE (Contd)

b. Tappet and Plunger Removal

NOTE

Camshaft must be removed to perform this task.

1. Position fuel pump housing (14) bottom side up in vise.

NOTE

- Cylinder number one is located at drive end of fuel pump.
- Step 2 covers the removal of tappet holder from cylinders one and two. Remaining tappet holders are removed the same.

2. Install spring compressor clamp (1) on governor side of fuel pump hosing (14) with two washers (2) and screws (3).
3. Place spring compressor adapter (6) on shaft (5).
4. Position shaft (5) and adapter (6) on the first two tappets (7).
5. Position lever (4) on spring compressor clamp (1) and shaft (5).
6. Depress lever (4), compressing both tappets (7).
7. Turn knurled knob (11) on tappet holder (10) fully clockwise and apply GAA grease to ramp (12) and guide (9) on tappet holder (10).
8. While both tappets (7) are compressed, remove tappet holder (10) from side hole (8) of cylinders one and two.

NOTE

Repeat steps 2 through 8 until all tappet holders are removed.

9. Remove two screws (3), washers (2), spring compressor clamp (1), lever (4), shaft (5), and adapter (6) from fuel pump housing (14).

CAUTION

Keep cylinder parts together for replacement in their original cylinder. Failure to do so may result in damage to fuel pump.

10. Remove six tappets (7), lower spring seats (15), plungers (16), and springs (20) from fuel pump housing (14).
11. Position control rack (13) to align control sleeve ball (17) with notch (21) on fuel pump housing (14).
12. Remove six upper control sleeves (18) and upper spring seats (19) from fuel pump housing (14).

5-36. FUEL INJECTION PUMP MAINTENANCE (Contd)

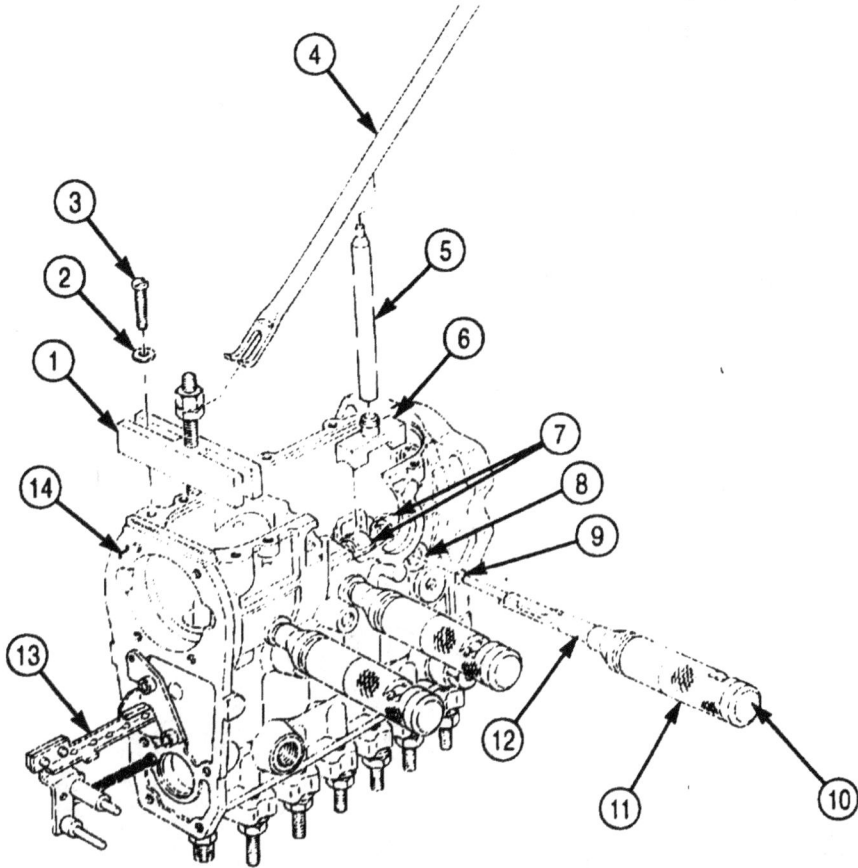

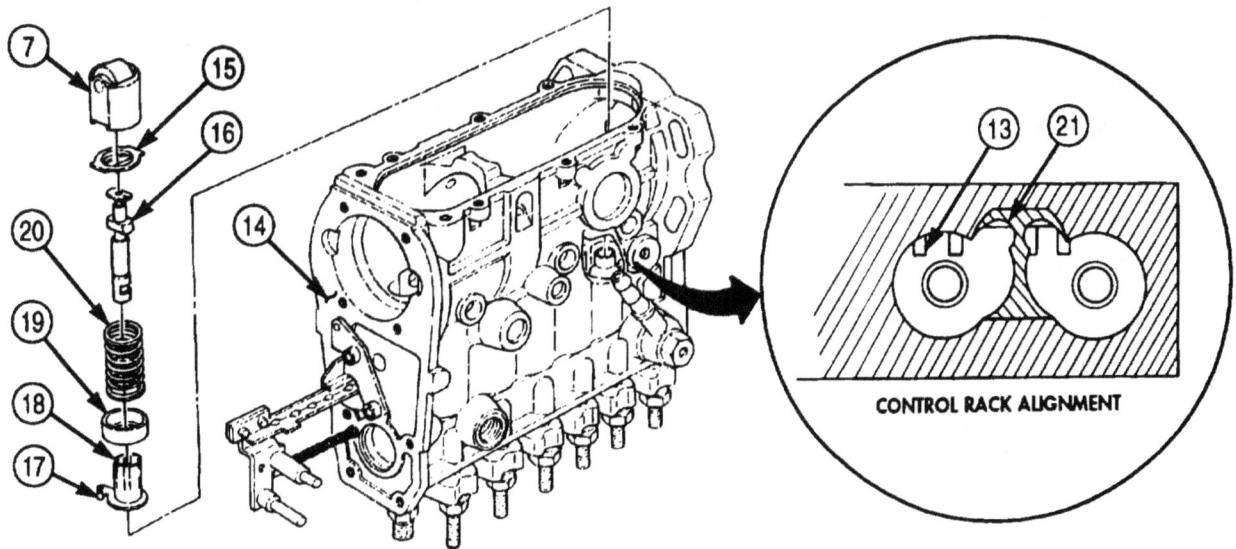

CONTROL RACK ALIGNMENT

5-36. FUEL INJECTION PUMP MAINTENANCE (Contd)

c. Barrel and Control Rack Removal

1. Position fuel pump housing (5) top side up in vise.
2. Loosen six delivery valve holders (1) from barrels (9).
3. Remove twelve barrel flange nuts (2), lockwashers (3), and spacer rings (4) from fuel pump housing (5). Discard lockwashers (3).
4. Remove six barrels (9) from fuel pump housing (5).
5. Remove six O-rings (8) from barrels (9). Discard O-rings (8).
6. Remove six retainers (6) and capsules (7) from barrels (9). Discard retainers (6).
7. Remove six delivery valve holders (1), washers (12), springs (11), and delivery valves (10) from barrels (9).
8. Remove six O-rings (13) from delivery valve holders (1). Discard O-rings (13).

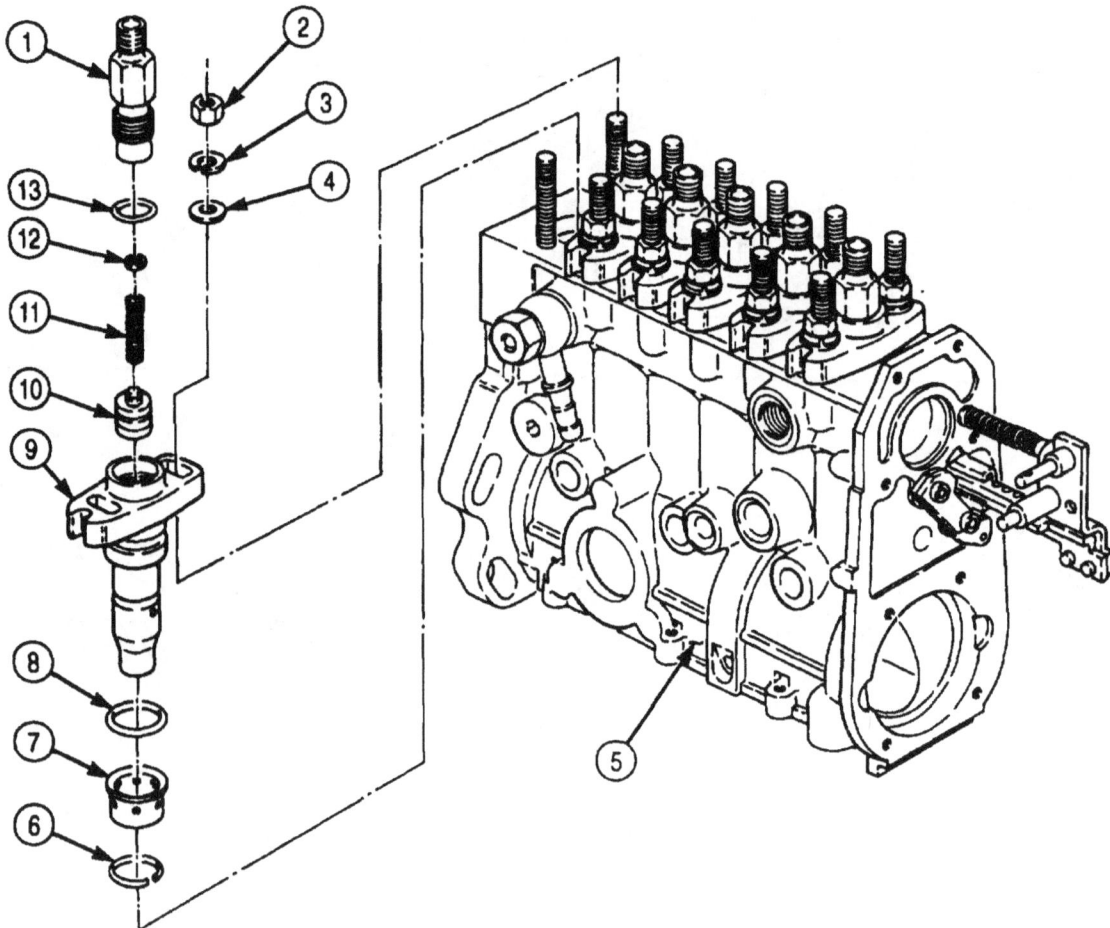

5-36. FUEL INJECTION PUMP MAINTENANCE (Contd)

9. Remove two screws (19) and retaining plate (20) from control rack (18) and fuel pump housing (5).

10. Remove control rack (18), spring retainer (17), spring (16), and washer (15) from fuel pump housing (5).

NOTE
Perform step 11 if control rod is damaged.

11. Remove control rod (14) from fuel pump housing (5).

12. Remove fuel pump housing (5) from vise.

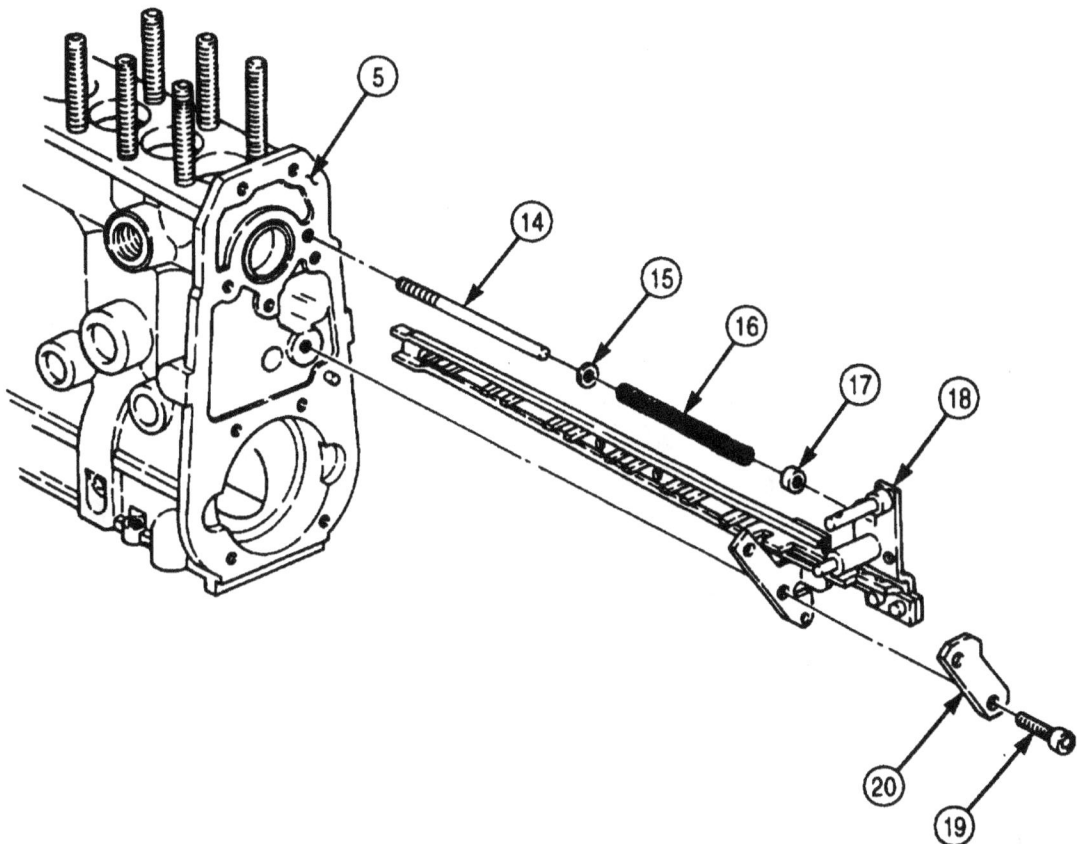

5-36. FUEL INJECTION PUMP MAINTENANCE (Contd)

d. Cleaning and Inspection

1. For general cleaning instructions, refer to para. 2-14.

2. Inspect plunger (1) for scratches and scoring. Replace plunger (1) and barrel (5) if plunger (1) is scratched or scored.

3. Inspect delivery valve (4) for wear and scoring. If worn or scored, replace delivery valve (4).

NOTE

Replace all camshaft bearings if camshaft requires replacement.

4. Inspect camshaft (7) for scratches and gouged lobes (6). If lobes (6) are scratched or gouged, replace camshaft (7).

NOTE

Perform steps 5 through 7 if camshaft and endplate bearings require replacement.

5. Press roller bearing (2) from end plate (3).

6. Remove nut (10) and lockwasher (9) from camshaft (7). Discard lockwasher (9).

7. Press bearing (8) from camshaft (7).

8. Inspect tappets (11) for wear or scratches. If worn or scratched, replace tappets (11).

9. Inspect six spacers (14) in fuel pump housing (13) for burrs, scratches, and nicks. If burred, scratched or nicked, replace spacer (14).

NOTE

Perform steps 10 through 12 if spacer requires replacement.

10. Bend edges of spacers (14) in towards center of port (12).

11. Drive six spacers (14) out through top of fuel pump housing (13). Discard spacers (14).

12. Remove six O-rings (15) from fuel pump housing (13). Discard O-rings (15).

13. For general inspection instructions, refer to para. 2-15.

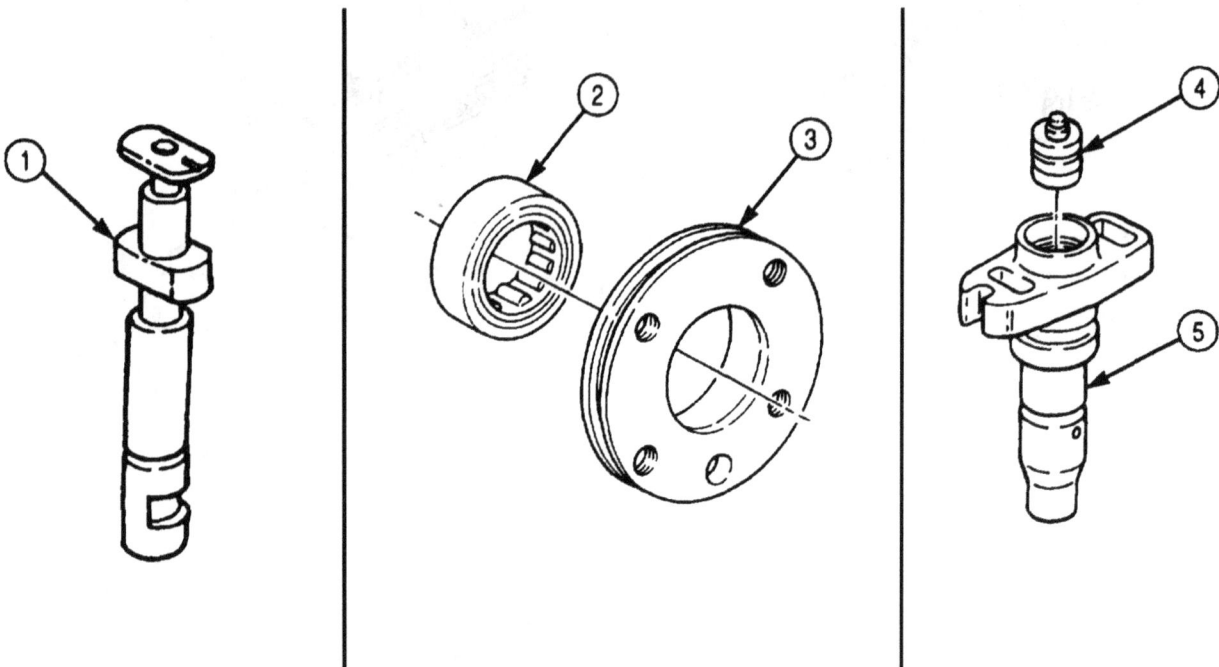

5-36. FUEL INJECTION PUMP MAINTENANCE (Contd)

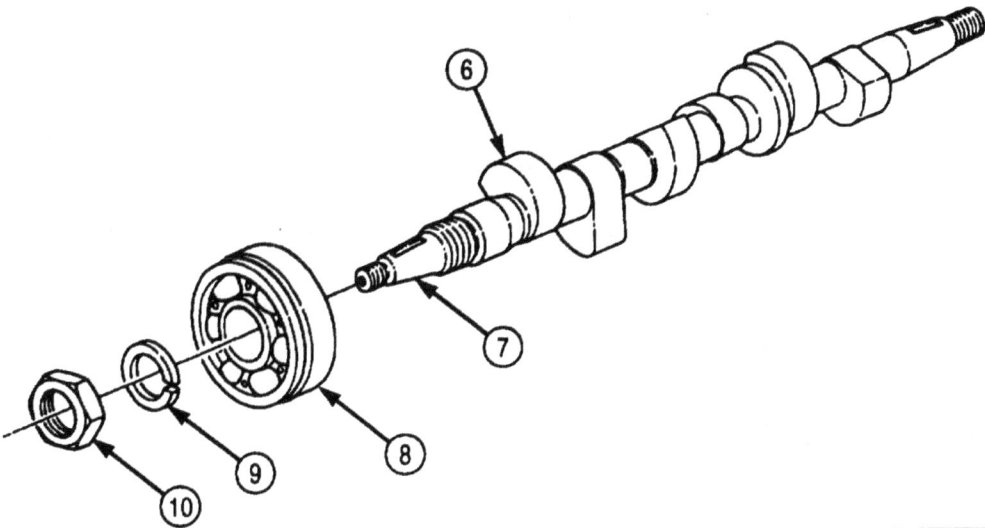

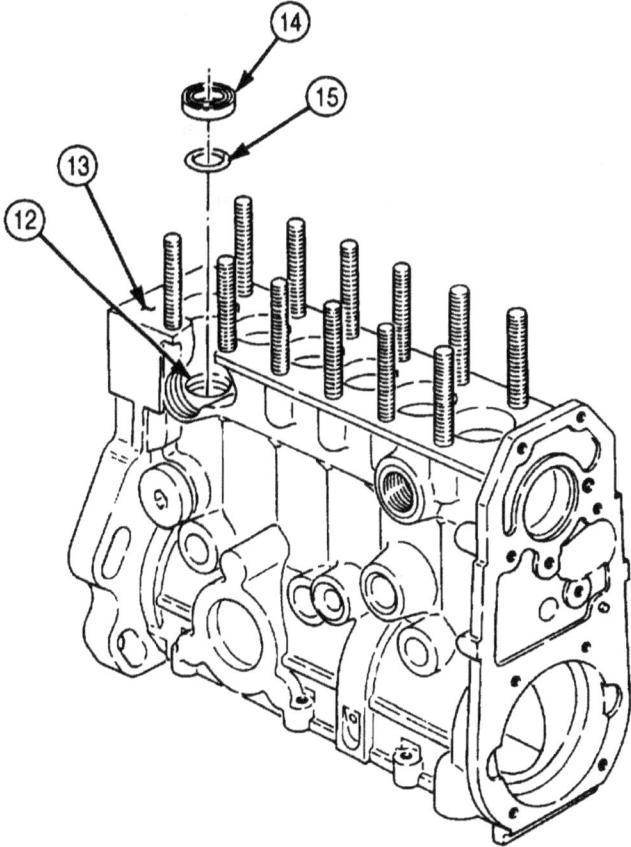

5-36. FUEL INJECTION PUMP MAINTENANCE (Contd)

e. Leakage Test

CAUTION

Plungers and barrels are matched sets. Do not interchange parts
or damage to fuel pump may result.

NOTE

Barrels must be installed (subtask f) to perform this task.

1. Position fuel pump housing (1) bottom side up in vise.

2. Install six plungers (4) in mating barrels (7). Position plunger (4) so groove (5) on plunger (4) points away from rack.

3. Install plunger retainer tubes (2) on fuel pump housing (1) with two bottom cover screws (3).

4. Remove fuel pump housing (1) from vise.

5. Install screw plug (8) in fuel gallery outlet (9).

6. Connect air hose to fuel gallery inlet (6).

7. Submerge fuel pump housing (1) in clean test oil bath and pressurize fuel pump housing (1) to 75 psi (517 kPa).

8. Check for bubbles at top and bottom of fuel pump housing (1). Any bubbles, other than small bubbles from inside of plunger retainer tubes (2), indicate a leak.

9. If leaks are present, replace fuel pump housing (1).

10. Disconnect air hose from gallery inlet (6).

11. Remove screw plug (8) from gallery outlet (9).

12. Remove two bottom cover screws (3) and plunger retainer tubes (2) from fuel pump housing (1).

13. Remove six plungers (4) from barrels (7).

14. Remove fuel pump housing (1) from vise.

5-36. FUEL INJECTION PUMP MAINTENANCE (Contd)

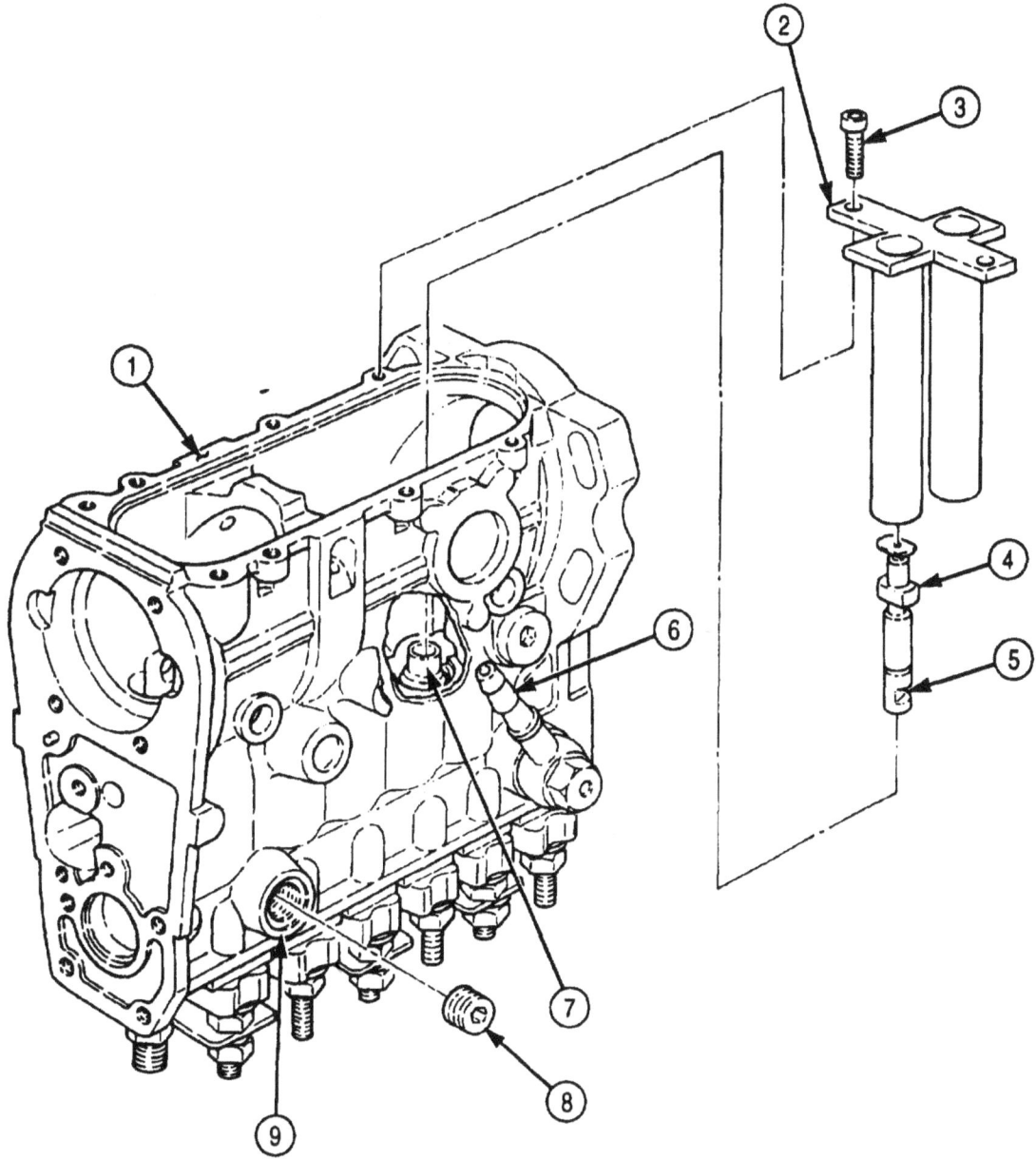

5-36. FUEL INJECTION PUMP MAINTENANCE (Contd)

f. Delivery Valve and Barrel Installation

NOTE

- The following subtask covers installation of one barrel. Remaining barrels are installed the same.
- Coat all parts with oil before installation.
- Coat all O-rings with GAA grease.

NOTE

Perform steps 1 and 2 if spacer was removed.

1. Install new O-ring (13) in fuel pump housing (11).
2. Using arbor press, press new spacer (14) through top of fuel pump housing (11).
3. Position fuel pump housing (11) top side up in vise.
4. Install new O-ring (22) on delivery valve holder (1).
5. Install delivery valve (191, spring (20), washer (21), and delivery valve holder (1) in corresponding barrel (18).
6. Install new O-ring (17) on barrel (18).
7. Install capsule (16) on barrel (18) with new retainer (15).
8. Position barrel (18) in corresponding cylinder (12) and install with two spacer rings (8), new lockwashers (7), and barrel flange nuts (6). Tighten nuts (6) finger tight.
9. Place two shims (9) between barrel (18) and fuel pump housing (11).
10. Tighten nuts (6) 15-20 lb-ft (20-27 N•m).
11. Tighten delivery valve holder (1) 35-45 lb-ft (47-61 N•m).
12. Install locking tab (5), spacer ring (4), new lockwasher (2), and nut (3) on stud (10) and delivery valve holder (1). Tighten nut (3) 15-20 lb-ft (20-27 N•m).
13. Remove fuel pump housing (11) from vise.

5-36. FUEL INJECTION PUMP MAINTENANCE (Contd)

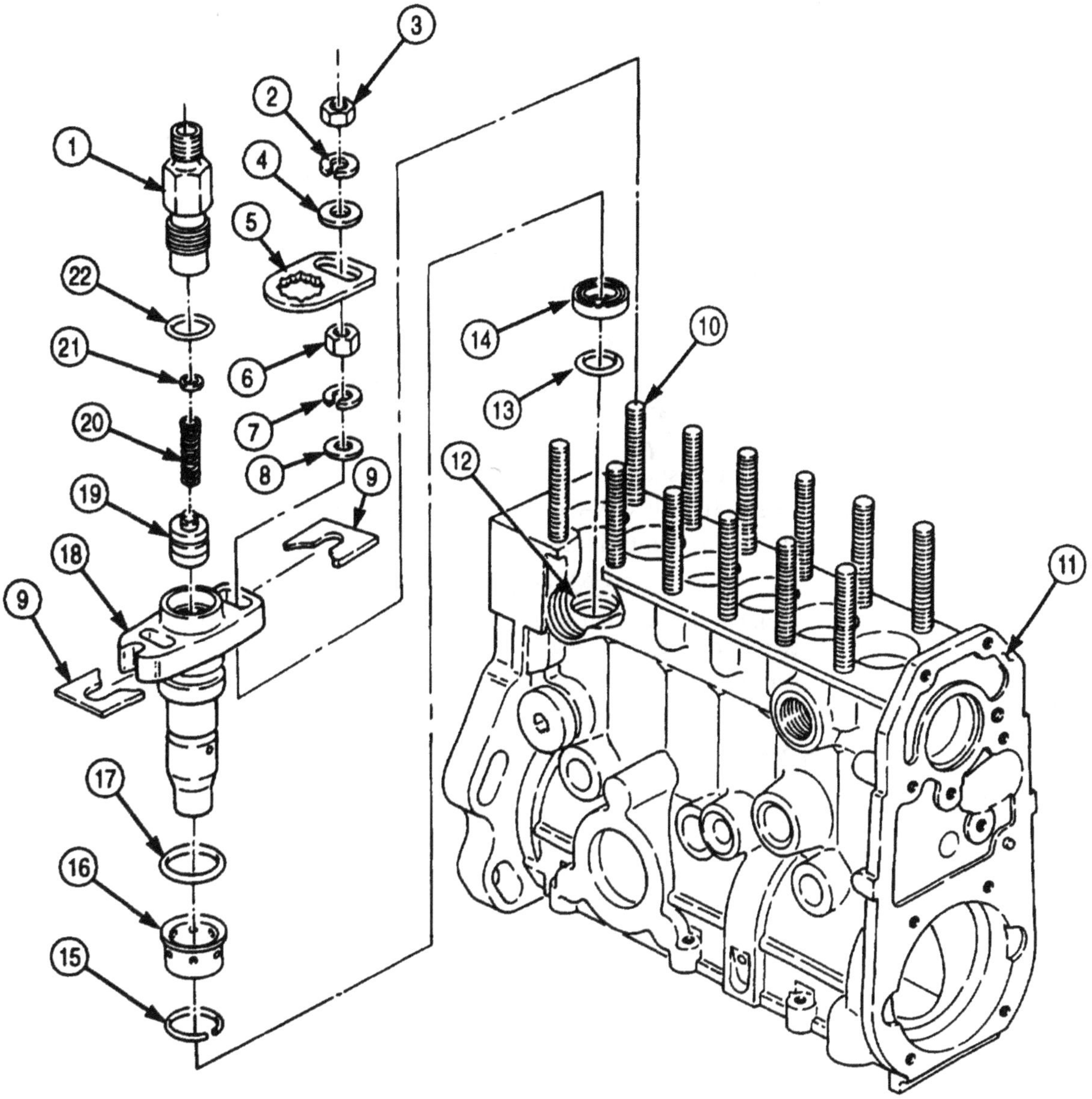

5-36. FUEL INJECTION PUMP MAINTENANCE (Contd)

g. Plunger and Control Rack Installation

NOTE

Perform step 1 if control rod was removed.

1. Install new control rod (1) on fuel pump housing (9).
2. Install washer (2), spring (3), and cap (4) on control rod (1).
3. Install control rack (5) and retaining plate (6) on fuel pump housing (9) with two screws (7).
4. Mount fuel pump housing (9) bottom side up in vise.
5. Position control rack (5) so guide slits (8) align with notch (18) on fuel pump housing (9).

NOTE

Control sleeve balls engage guide slit on control rack.

6. Install six upper control sleeves (15) into mating barrels (16).
7. After all control sleeves (15) are installed, move control rack (5) back and forth to ensure control sleeves (14) move freely.

CAUTION

Once correct alignment is obtained, secure control rack in this position until tappets are installed. Failure to do so may result in damage to control sleeves.

8. Position control rack (5) so guide slits (8) face adjusting groove (17) on barrel (16).
9. Coat six upper spring seats (14) with GAA grease and position on control sleeves (15).
10. Install six springs (13) on mating control sleeves (15).

CAUTION

Notch on plungers must align exactly with groove on barrel. If alignment is not obtained, damage to fuel pump may result.

11. Install six plungers (12) into mating barrels (16) so notched mark (11) on plunger base points toward adjusting groove (17) on barrel (16).
12. Install six lower spring seats (10) on plungers (12) and rotate lower spring seats (10) 90°.

5-36. FUEL INJECTION PUMP MAINTENANCE (Contd)

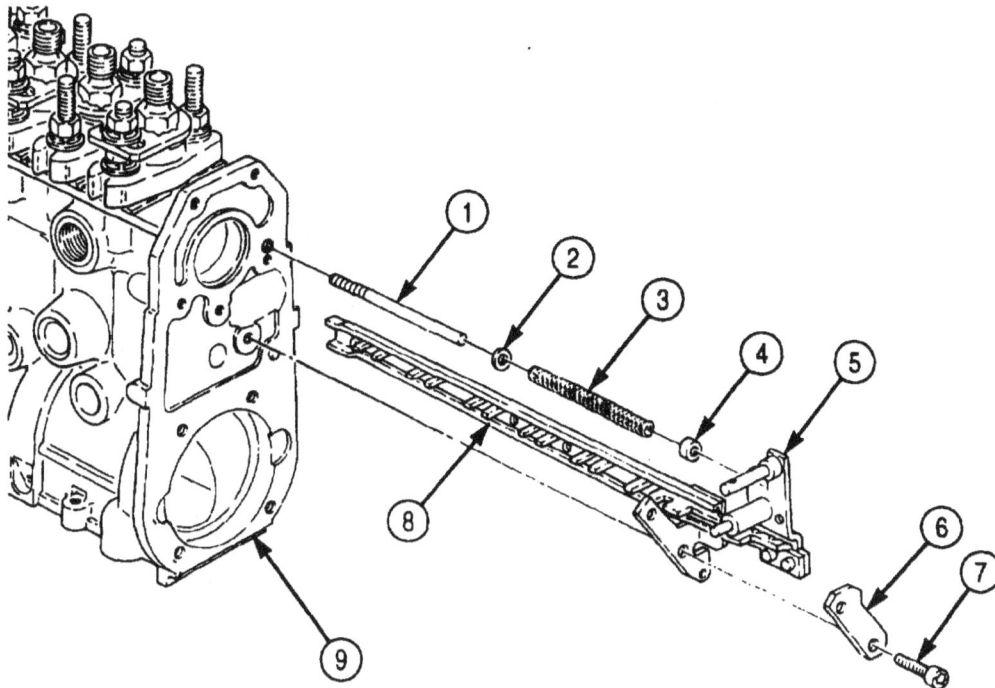

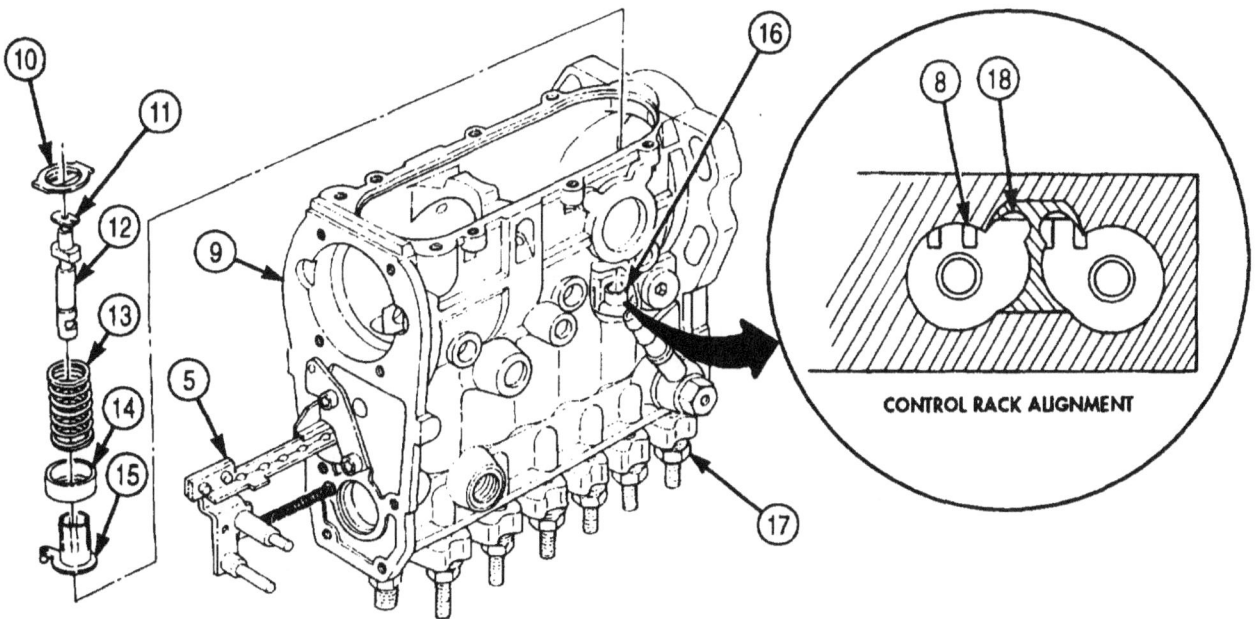

CONTROL RACK ALIGNMENT

5-36. FUEL INJECTION PUMP MAINTENANCE (Contd)

h. Tappet and Spring Installation

CAUTION

When performing step 1, ensure flats on tappet align with recess on the lower spring seat. Failure to do so may result in damage to fuel pump housing.

NOTE

This task covers installation of two tappets for cylinders number one and two. Remaining tappets are installed the same.

1. Install two tappets (1) in mating cylinders.
2. Install spring compressor clamp (3) on governor side of fuel pump housing (2) with two washers (4) and screws (5).
3. Place spring compressor adapter (8) on shaft (7).
4. Position shaft (7) and adapter (8) on the first two tappets (1).
5. Position lever (6) on spring compressor clamp (3) and shaft (7).
6. Depress lever (6), compressing both tappets (1).
7. Turn knurled knob (12) on tappet holder (13) fully counterclockwise and apply GM grease to ramp (11) and guide (10) on tappet holder (13).
8. While both tappets (1) are compressed, insert tappet holder (13) into side hole (9) of cylinders one and two.

NOTE

Repeat steps 1 through 8 until all tappets and tappet holders are installed.

9. Remove two screws (5), washers, (4), spring compressor clamp (3), lever (6), shaft (7), and compressor adapter (8) from fuel pump housing (2).
10. Remove fuel pump housing (2) from vise.

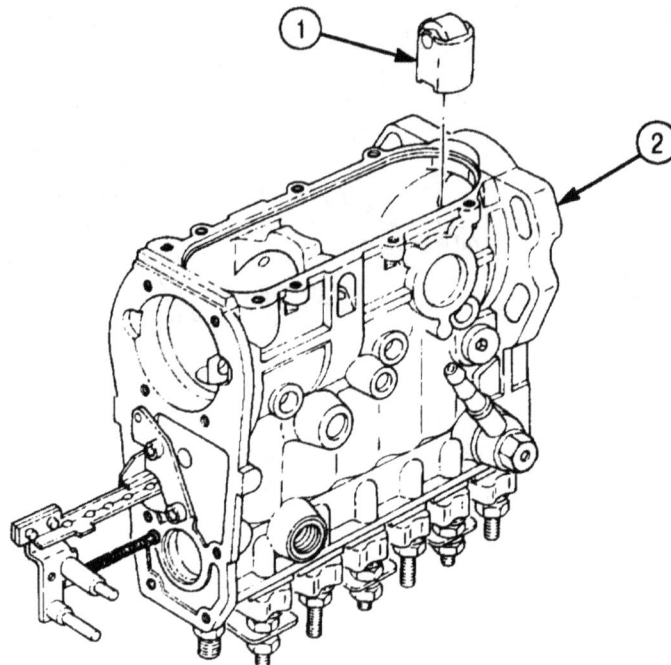

5-36. FUEL INJECTION PUMP MAINTENANCE (Contd)

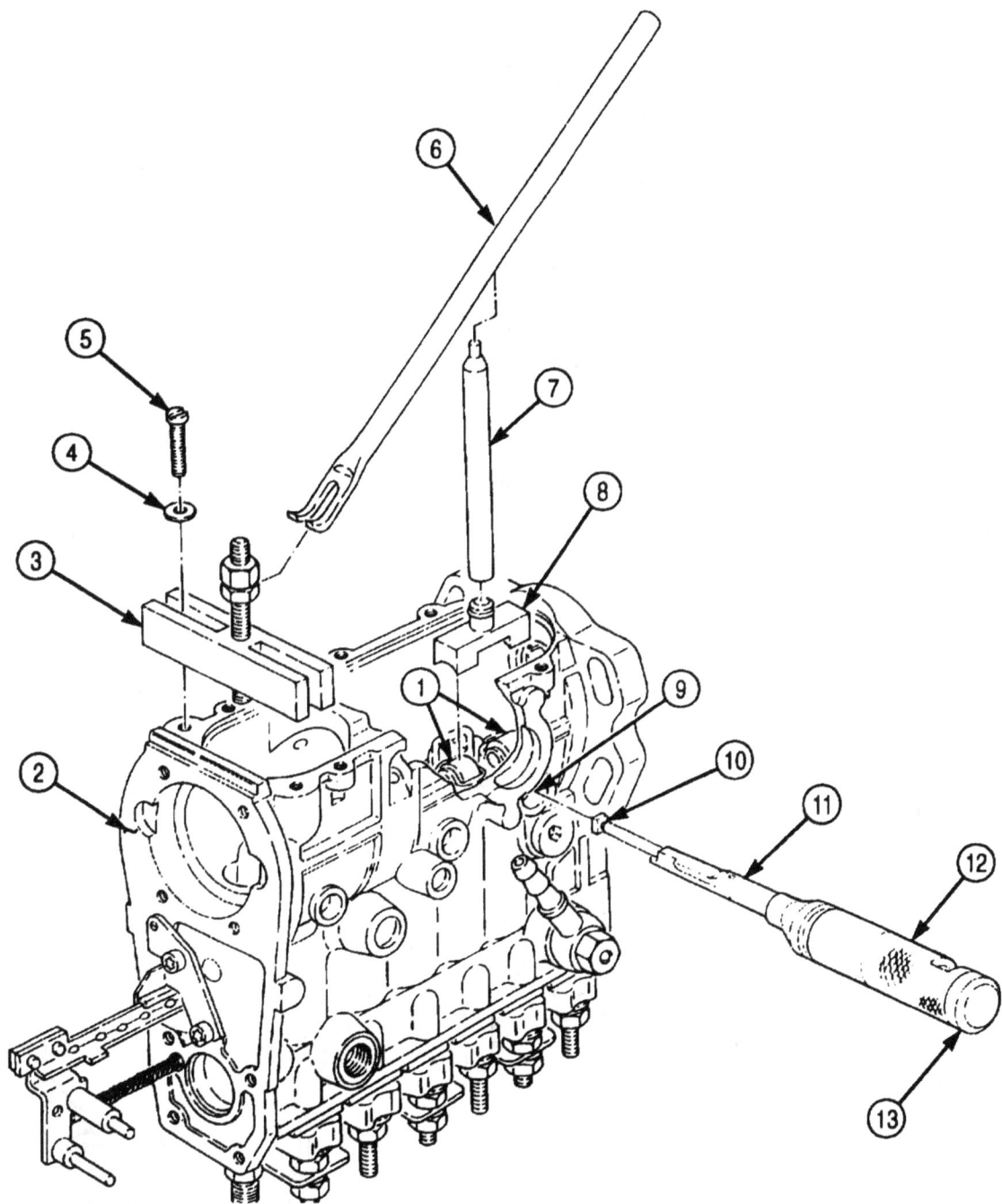

5-36. FUEL INJECTION PUMP MAINTENANCE (Contd)

i. Camshaft Installation

1. Position fuel pump housing (1) bottom side up in vise.

NOTE

Perform step 2 if roller bearing was removed.

2. Press roller bearing (8) on end plate (3).
3. Install new O-ring (6) and oil seal (7) on end plate (3).
4. Position new gasket (2) and end plate (3) on fuel pump housing (1). Ensure oil return holes are located at top of fuel pump housing (1).
5. Install end plate (3) on fuel pump housing (1) with four new lockwashers (4) and screws (5).
6. If bearing (16) was removed, press bearing (16) on camshaft (10) and secure bearing (16) on camshaft (10) with new lockwasher (15) and nut (14).
7. Install new O-ring (9) on fuel pump housing (1).

CAUTION

When installing camshaft into fuel pump housing, tap only on camshaft outer nut. Failure to do so may result in damage to fuel pump.

8. Insert camshaft (10) and bearing shell (13) into fuel pump housing (1) through governor end of fuel pump housing (1).
9. Install bearing shell (13) on fuel pump housing (1) with two new seal rings (12) and screws (11).

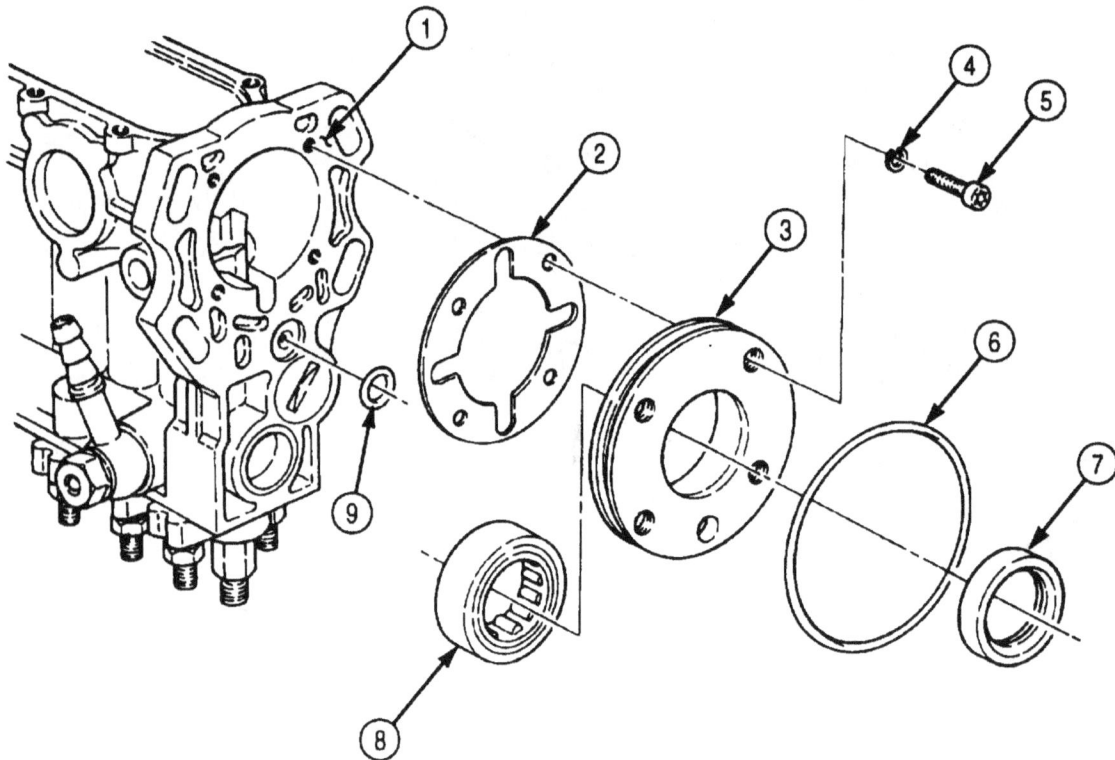

5-36. FUEL INJECTION PUMP MAINTENANCE (Contd)

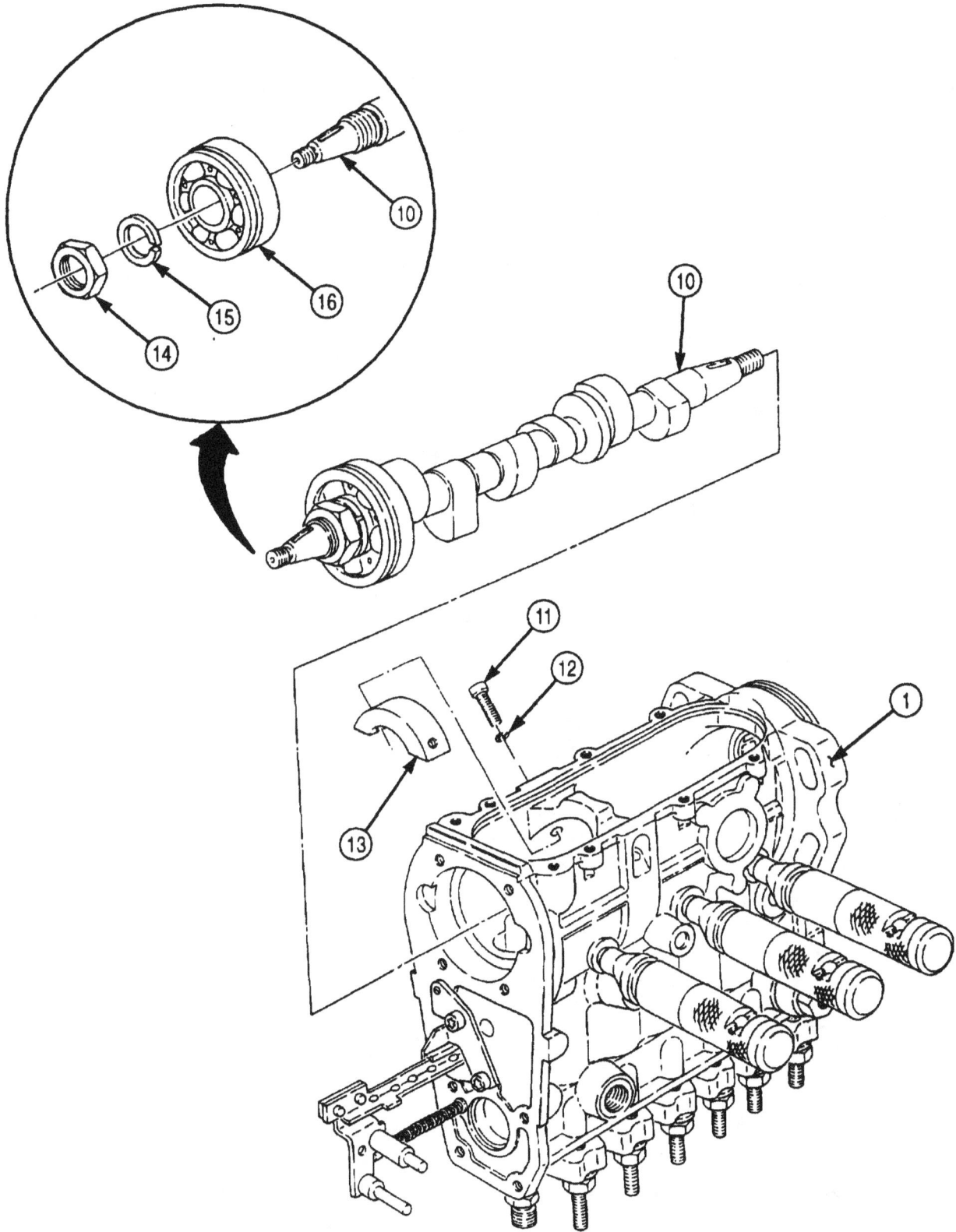

5-36. FUEL INJECTION PUMP MAINTENANCE (Contd)

NOTE

Steps 10 through 12 cover removal of one tappet holder. Remaining tappet holders are removed the same.

10. Turn knurled handle (7) on tappet holder (8) fully clockwise.

NOTE

Cylinder number one is located at drive end of fuel injection pump.

11. Hold tappet holder (8) and rotate camshaft (9) until cylinder number one is at TDC and pull tappet holder (8) from side hole (6).

12. Rotate camshaft (9) until cylinder number two is at TDC and remove tappet holder (8) from side hole (6).

13. Install new seal ring (4) and bottom cover (3) on fuel pump housing (5) with six holding brackets (1) and screws (2).

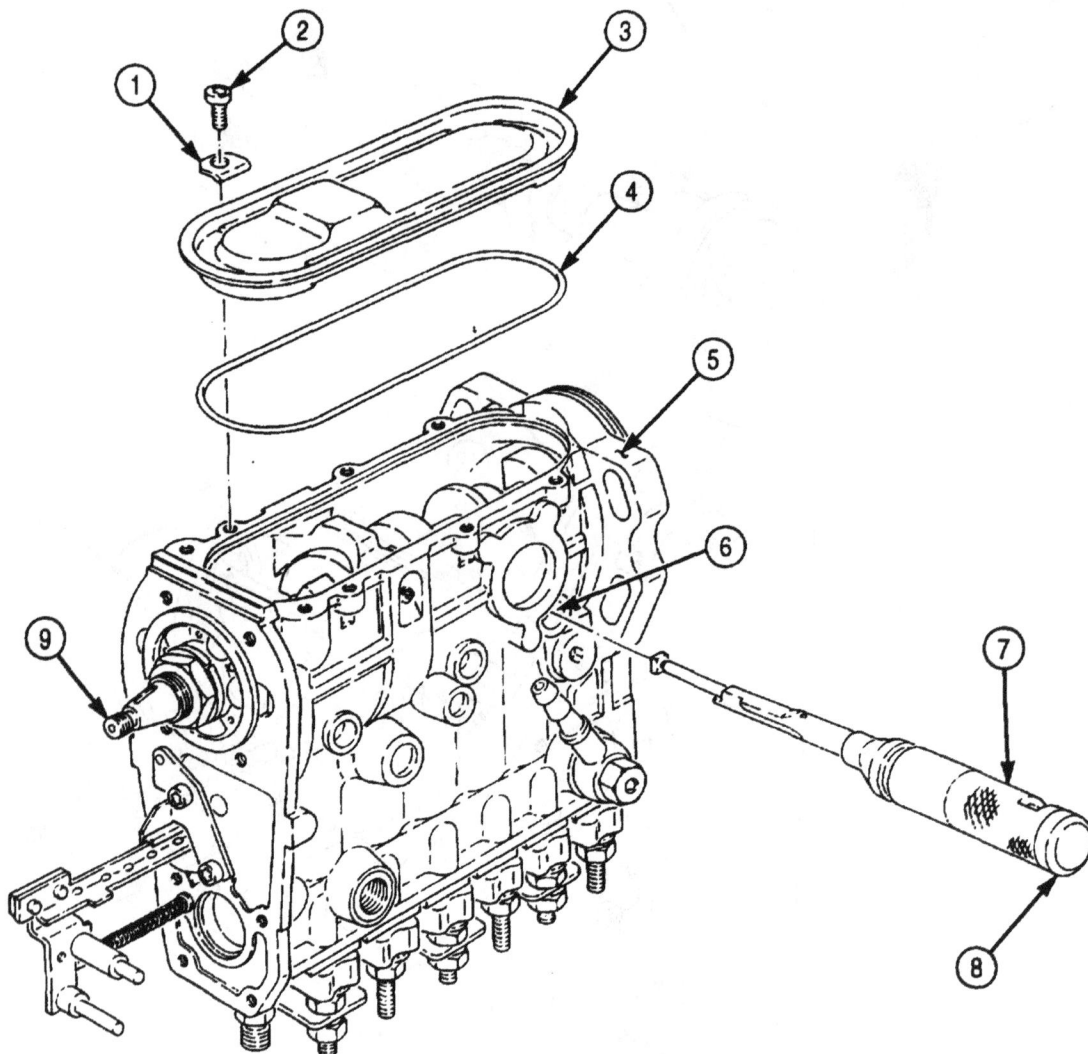

5-36. FUEL INJECTION PUMP MAINTENANCE (Contd)

14. Install three spacer blocks (11) in fuel pump housing (5) through side holes (12).

15. Apply sealing compound to new side plugs (10) and install side plugs (10) in fuel pump housing (5).

16. Remove fuel pump housing (5) from vise.

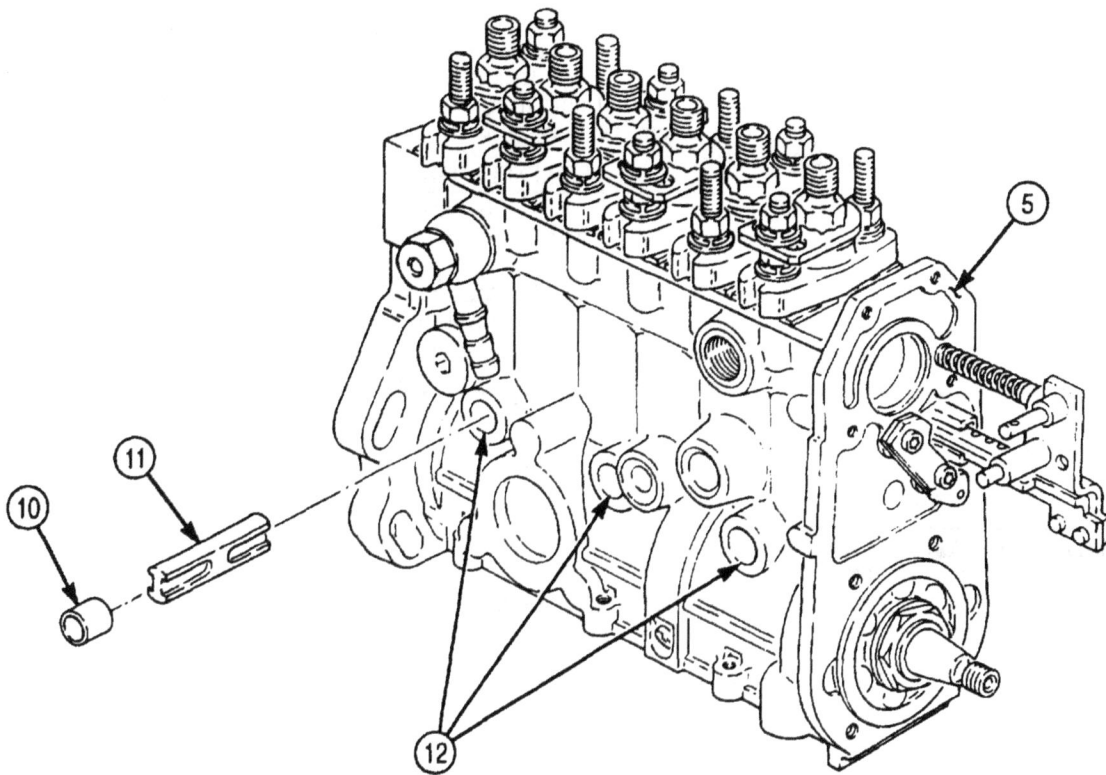

FOLLOW-ON TASK: Install R.Q.V. governor housing (para. 5-35).

5-37. FUEL INJECTION PUMP CALIBRATION

THIS TASK COVERS:

a. Prestroke and Back Travel
b. Cylinder Phasing (Timing)
c. Lock Timing
d. High-speed Fuel Delivery Test
e. Low-speed Fuel Delivery Test
f. Bated Speed Test
g. Full Load Delivery

h. Manifold Pressure Compensator Test
i. Fuel Delivery Test
j. Fuel Breakaway Test
k. Starting Fuel Delivery Test
l. Low Idle Fuel Delivery Test
m. External Stop Setting

<u>INITIAL SETUP:</u>

<u>APPLICABLE MODELS</u>
M939A2

<u>SPECIAL TOOLS</u>
Plunger lift device (Appendix E, Item 98)
Rack extension (Appendix E, Item 112)

<u>TOOLS</u>
General mechanic's tool kit (Appendix E, Item 1)
Dial indicator (Appendix E, Item 36)

<u>MATERIALS/PARTS</u>
Two copper washers (Appendix D, Item 44)
Two break off screws (Appendix D, Item 19)
Two lockwashers (Appendix D, Item 397)
Copper washer (Appendix D, Item 42)
Two breakoff screws (Appendix D, Item 20)
Four lockwashers (Appendix D, Item 398)
Lubricating oil (Appendix C, Item 50)
Safety wire (Appendix C, Item 79)

<u>REFERENCES (TM)</u>
TM 9-2320-272-24P
TM 9-4910-778-14&P

<u>EQUIPMENT CONDITION</u>
• Fuel injection pump repaired (para. 5-36).
• Fuel injection pump mounted to test stand (para. 5-31).

NOTE
Refer to TM 9-4910-778-14&P for operation of fuel injection pump test stand, DFP 156.

a. Prestroke and Rack Travel

1. Mount dial indicator (2) and rack travel indicator holder (1) on drive end of fuel pump (5).

NOTE
Cylinder number one is located at drive end of fuel injection pump.

2. Install plunger lift device (8) on fuel pump (5) with finger of lifting device (8) resting on top of tappet located in cylinder number one.

3. Mount dial indicator (7) on lifting device (8) and zero dial indicator (7).

4. Pull control lever (6) to full load position and using dial indicator shaft (3), push control rack (4) all the way back.

5. Preload dial indicator (2) giving control rack (4) 0.004 in. (0.1 mm) of preload travel.

6. Manually rotate test stand driveshaft (9) until a reading is observed on dial indicator (7).

7. Manually reverse rotation test stand driveshaft (9) until dial indicator (7) reads zero.

8. Manually rotate test stand driveshaft (9) clockwise and take prestroke measurement from dial indicator (7). Correct prestroke measurement should be 0.120-0.128 in. (3.05-3.25 mm).

9. If correct prestroke measurement is not obtained, perform steps 10 through 13.

5-37. FUEL INJECTION PUMP CALIBRATION (Contd)

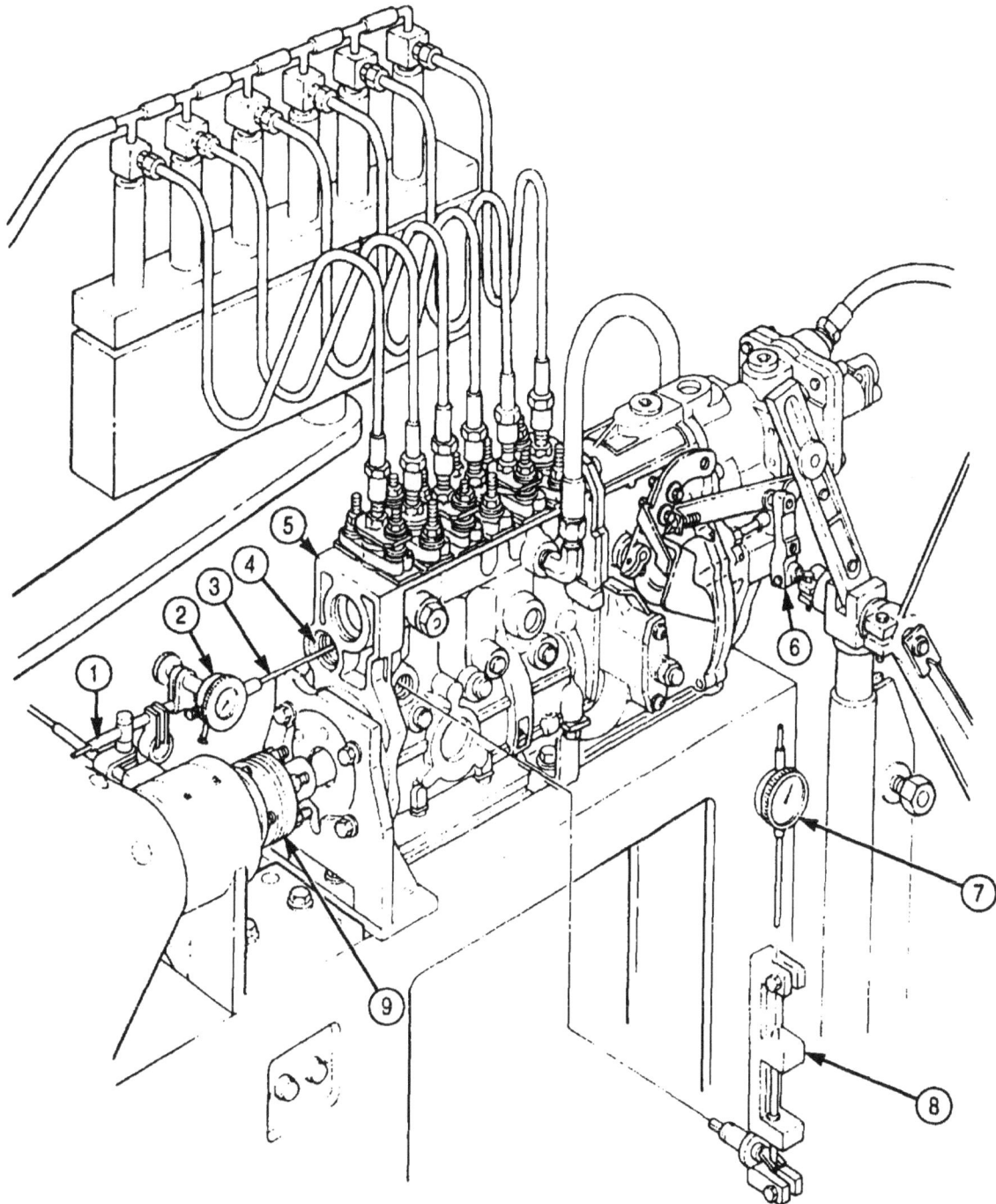

5-37. FUEL INJECTION PUMP CALIBRATION (Contd)

NOTE

Steps 10 through 13 pertain to adjusting one cylinder. Remaining five cylinders are adjusted the same.

10. Remove fuel test line (12) from delivery valve holder (11).

11. Loosen three nuts (10) and remove two shims (2) between barrel (9) and fuel pump (8). Discard shims (2).

CAUTION

Adjusting shims of the same thickness must be used on both sides of barrel. Failure to do so may result in inaccurate prestroke measurement.

NOTE

Refer to TM 9-2320-272-24P to select shims.

12. Install two new shims (2) between barrel (9) and fuel pump (8).

13. Tighten three nuts (10) and install fuel test line (12) on delivery valve holder (11).

NOTE

Cylinder number one is located at drive end of fuel pump.

14. Close all bleedoff valves (1) except for cylinder number one.

15. Apply 435-464 psi (2,999-3,199 kPa) of calibration fluid pressure.

NOTE

Ensure a steady stream of calibration fluid flows from number one bleedoff valve. If not, increase calibration fluid pressure until a steady stream of calibration fluid is obtained.

16. Manually rotate test stand driveshaft (7) in correct fuel pump (8) rotation, until calibration fluid starts to drip from bleedoff valve (1).

17. Check prestroke measurement. Correct measurement should be 0.360-0.480 in. (9.14-12.19 mm).

18. Remove dial indicator (3) and plunger lift device (4) from fuel pump (8).

19. Install new copper washer (6) and plug (5) on fuel pump (8).

5-37. FUEL INJECTION PUMP CALIBRATION (Contd)

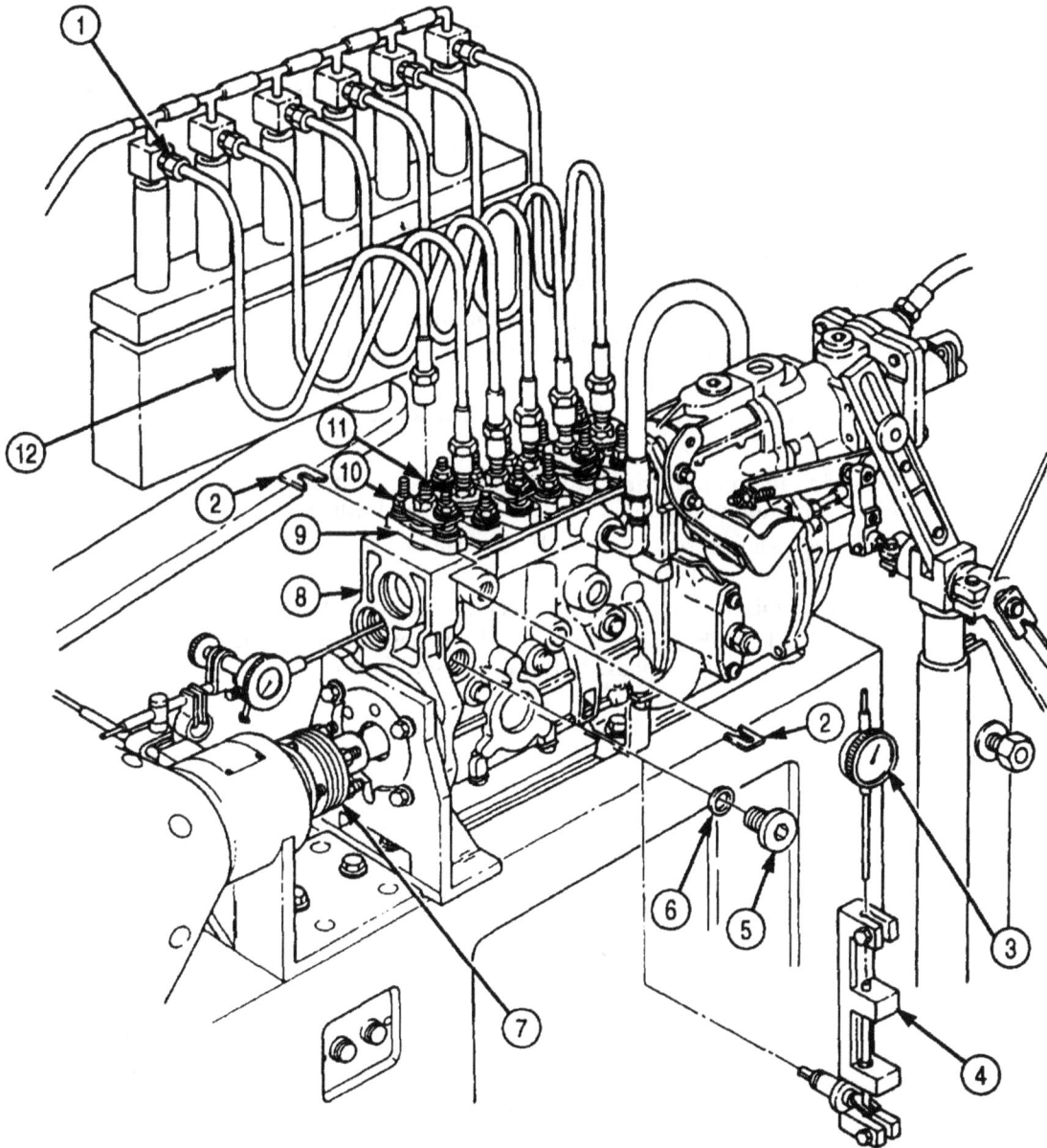

5-37. FUEL INJECTION PUMP CALIBRATION (Contd)

b. Cylinder Phasing (Timing)

NOTE

- When phasing (timing) cylinders, test stand calibration fluid pressure must be set at 435-464 psi (2,999-3,199 kPa).

- When performing this task, correct prestroke measurement must be obtained for cylinder number one. Refer to task a.

1. Zero test stand protractor (6) on drive end of test stand (5).

2. Close all bleedoff valves (2) except for cylinder number five.

NOTE

When bleedoff valve is opened, a stream of calibration fluid will flow from bleedoff valve. If not, increase fluid pressure until a stream of calibration fluid is obtained.

3. Open bleedoff valve (2) on test nozzle (3) for cylinder number five.

4. Manually rotate test stand driveshaft (7) in correct rotation of fuel pump (4) until calibration fluid starts to drip from number five bleedoff valve (2).

5. Observe reading on test stand protractor (6). Correct reading is 60°.

NOTE

- Following fuel pump firing order 1-5-3-6-2-4, repeat steps 2 through 5 until remaining cylinders are phased (timed).

- For each cylinder phased (timed), test stand protractor should increase in 60° intervals.

6. If 60° intervals are not obtained, adjust shims under barrel (1) (task a).

7. If any adjustments are made to bring the remaining cylinders to specification, repeat steps 1 through 6 to ensure all cylinders are phased (timed) to specifications.

5-37. FUEL INJECTION PUMP CALIBRATION (Contd)

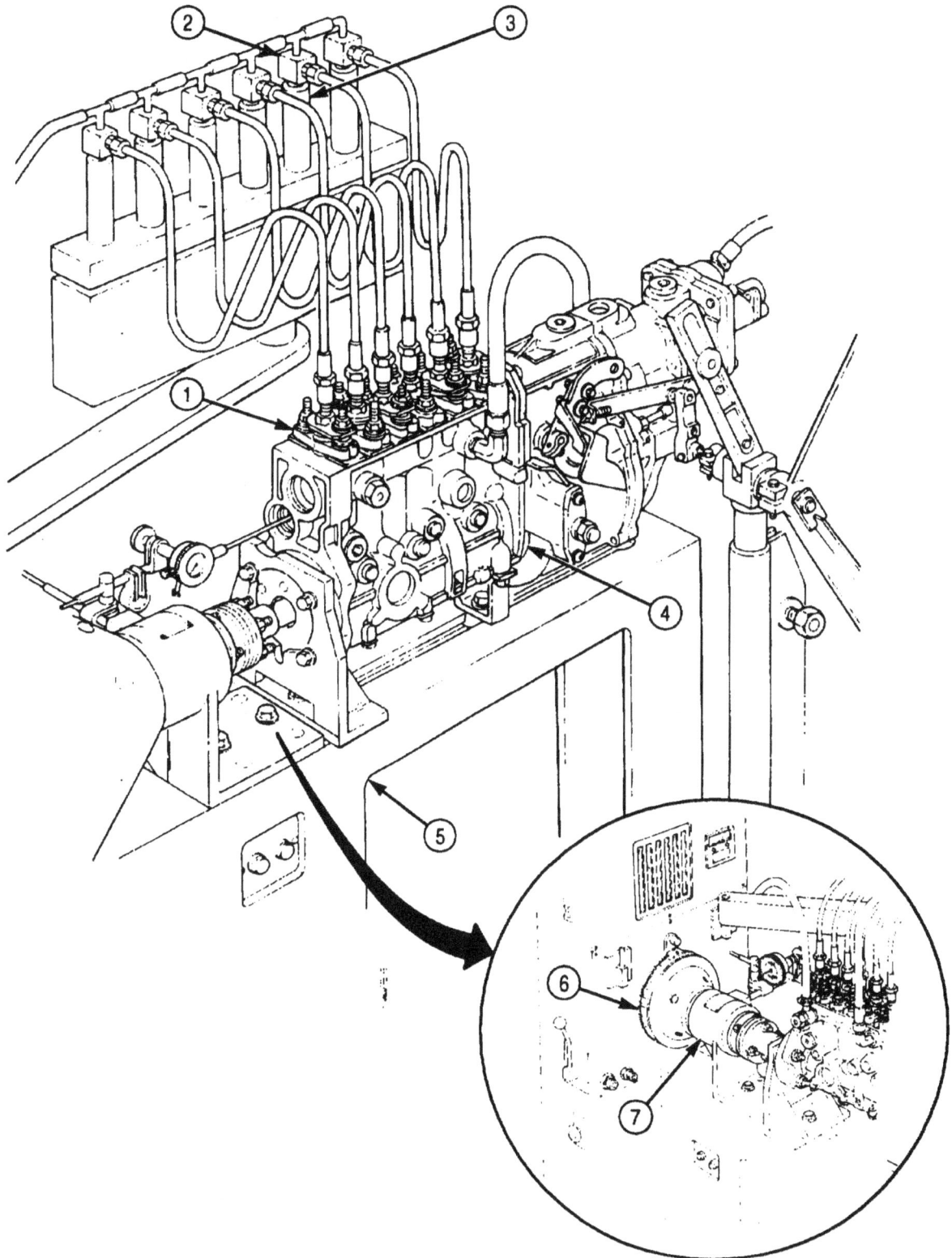

5-37. FUEL INJECTION PUMP CALIBRATION (Contd)

c. Lock Timing

NOTE

Fuel pump must be phased (timed) before performing this task.
Refer to task b.

1. Remove access plug (7), washer (8), and timing pin (3) from timing plate (2).
2. Rotate test stand drive (9) until number one cylinder is at port closure.
3. Rotate test stand drive (9) in correct pump rotation 8° from cylinder one port closure.
4. Timing tooth (11) should be aligned with timing pin access hole (10).

NOTE

Perform steps 5 through 11 if drive pointer is not aligned with
timing pin access hole.

5. Remove governor cover from governor housing (para. 5-34).
6. Remove two breakoff screws (6), lockwashers (5), washers (4), and timing plate (2) from governor housing (1). Discard lockwashers (5) and breakoff screws (6).
7. Remove governor flyweight and driver from fuel pump (para. 5-35).
8. Install timing plate (2) on governor housing (1) with two washers (4), new lockwashers (5), and new breakoff screws (6).

NOTE

- Align driver so timing tooth aligns with timing pin access hole.
- Timing plate can be adjusted up or down for final alignment.

9. Install governor flyweight and driver on fuel pump (para. 5-35).
10. Install timing pin (3) in timing plate (2) with slot of timing pin (3) facing timing tooth (11).
11. Install washer (8) and access plug (7) on timing plate (2).
12. Tighten breakoff screws (6).
13. Install governor cover on governor housing (para. 5-34).

5-37. FUEL INJECTION PUMP CALIBRATION (Contd)

5-37. FUEL INJECTION PUMP CALIBRATION (Contd)

d. High-speed Fuel Delivery Test

NOTE

- Maintain 22 psi (152 kPa) calibration fluid pressure throughout this task.
- Close all bleedoff valves.

1. Set test stand (12) to measure fuel delivery at 1,000 strokes.
2. Remove plug (4) from top of governor housing (5) and pour two pints of lubricating oil in governor housing (5).
3. Remove plug (13) and washer (10) from fuel pump (9).
4. Install new copper washer (10), adapter (11), new copper washer (10), and plug (13) on fuel pump (9).
5. Connect return line (1) to adapter (11).
6. Adjust fuel pump (9) to 1,050 rpm.

NOTE

Adjust aneroid air pressure to help obtain correct rack travel.

7. Adjust control lever (8) to obtain rack travel of 0.360-0.480 in. (9.14-12.19 mm).

NOTE

Perform steps 8 and 9 if correct rack travel is not obtained.

8. Remove breakoff screw (19), screw (21), two lockwashers (18), protective cap (17), and stop (16) from manifold pressure compensator (6). Discard breakoff screw (19) and two lockwashers (18).
9. Turn nuts (15) and (14) counterclockwise to increase rack travel; or clockwise to decrease rack travel.
10. Position control lever (8) to full load and take fuel draw.
11. Check fuel delivery. Correct fuel delivery should be 0.891-0.928 in^3(14.6-15.2 cm^3).
12. Allowable spread reading is 0.037 in.^{3}(0.6 cm^3).

NOTE

- Perform step 13 if fuel delivery deviates from specifications.
- Step 13 pertains to one cylinder only.

13. Loosen three nuts (2) and rotate barrel (3) clockwise to increase fuel delivery; counterclockwise to decrease fuel delivery.
14. Repeat step 13 for remaining cylinders.
15. Install stop (16) and protective cap (17) on manifold pressure compensator (6) with two new lockwashers (18), screw (21), and new breakoff screw (20).
16. Install new safety wire (22) on screw (21) and protective cap (17).

e. Low-speed Fuel Delivery Test

NOTE

Maintain 22 psi (152 kPa) calibration fluid pressure throughout this task.

1. Reduce fuel pump (9) to 300 rpm.
2. Set control rack travel at 0.308-0.316 in. (7.8-8.0 mm).
3. Check fuel delivery. Correct fuel delivery should be 0.079-0.134 in^3(1.3-2.2 cm^3).
4. Allowable spread reading is 0.030 in.^{3}(0.5 cm^3).
5. If low-speed fuel delivery deviates from specifications given, adjust low idle stopscrew (7) until correct fuel delivery is obtained.

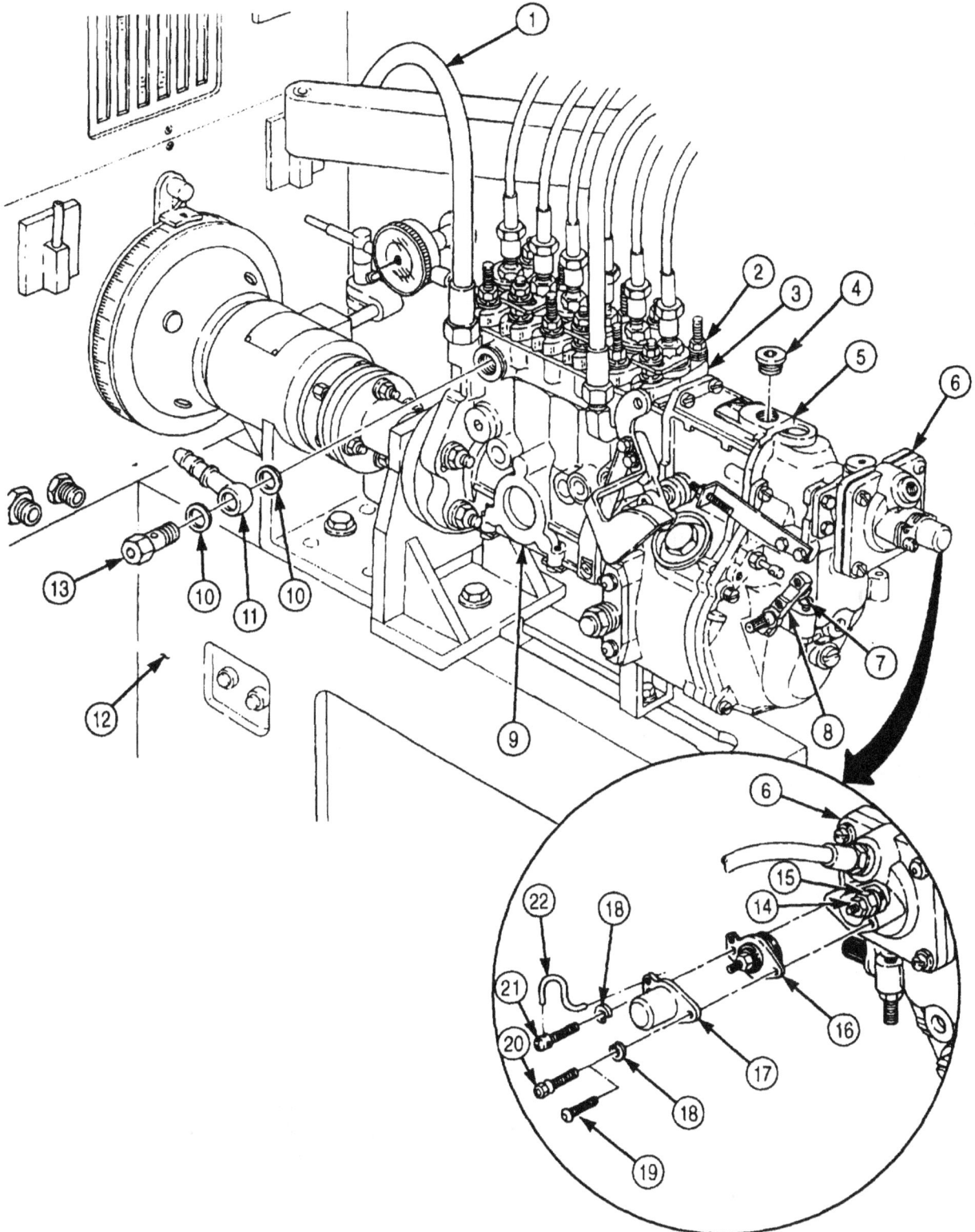

5-37. FUEL INJECTION PUMP CALIBRATION (Contd)

f. Rated Speed Test

1. Loosen nut (7) on low idle stopscrew (5) and turn low idle stopscrew (5) until flush with governor cover (6).
2. Position control level (4) until dial indicator (1) on drive end of fuel pump (2) reads zero.
3. Slowly move control lever (4) towards full load, until dial indicator (1) receives reading.
4. Set protractor (3) at zero.
5. Position control lever (4) to read 42-50° on protractor (3) by adjusting low idle stopscrew (5).
6. Increase rpm of fuel pump (2) until a rack travel of 0.464 in. (11.79 mm) is obtained.
7. Rpm of fuel pump (2) should be at 1,090-1,100 rpm.
8. Adjust rpm of fuel pump (2) until rack travel of 0.312 in. (7.9 mm) is obtained.
9. Rpm of fuel pump (2) should be at 300 rpm.
10. Decrease rpm of fuel pump (2) to 100 rpm. Rack travel should be 0.372 in. (9.45 mm).

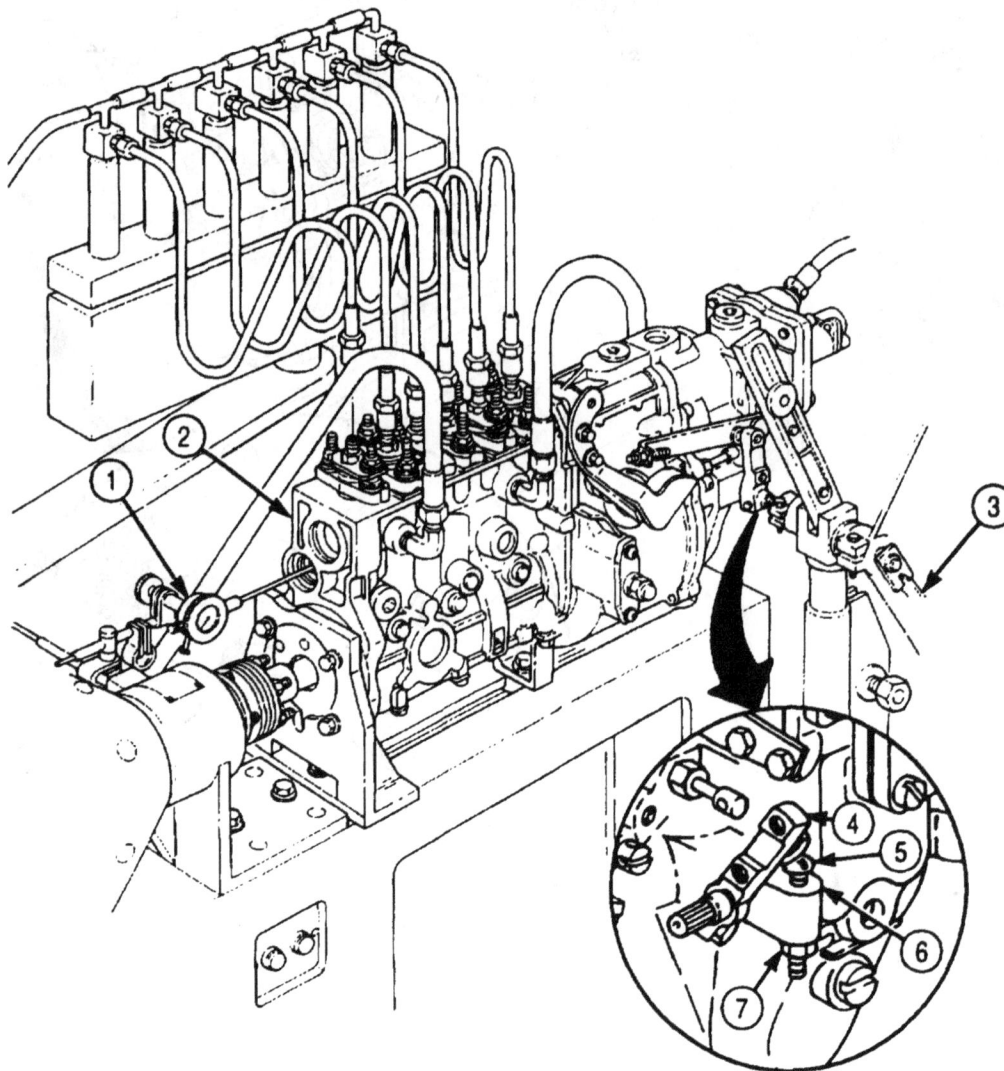

5-37. FUEL INJECTION PUMP CALIBRATION (Contd)

NOTE

Perform steps 11 through 17 if proper control rack travel was not obtained.

11. Remove safety wire (10) from governor housing (8).

12. Remove plug (9) from governor housing (8), giving access to governor springs.

NOTE

Repeat steps 13 and 14 until both governor springs are adjusted.

13. If control rack travel reading is too low at a given speed, increase flyweight spring tension by turning round nut (11) clockwise.

14. If control rack travel reading is too high at a given speed, decrease flyweight spring tension by turning round nut (11) counterclockwise.

15. Recheck control rack travel. If control rack travel is incorrect, repeat step 13 or 14. If rack travel is correct, proceed to step 16.

16. Install plug (9) on governor housing (5).

17. Install new safety wire (10) on governor housing (8).

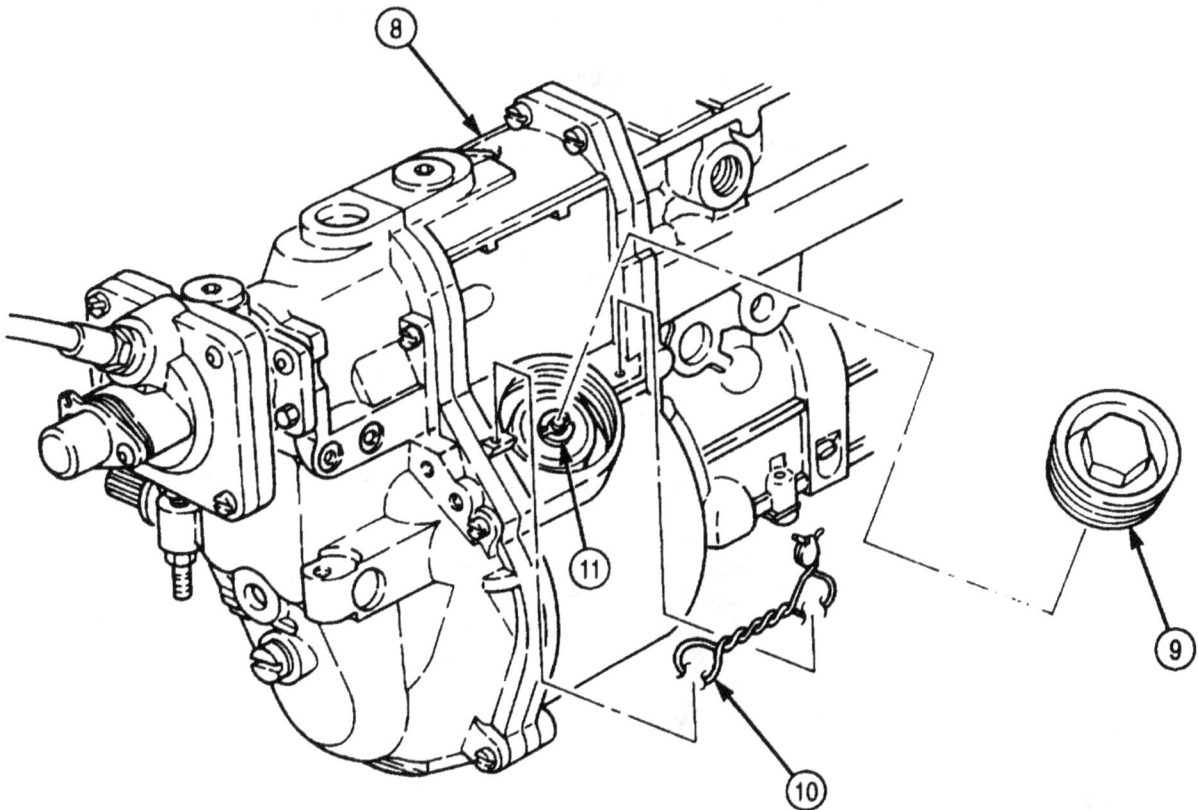

5-37. FUEL INJECTION PUMP CALIBRATION (Contd)

g. Full Load Delivery

1. Apply 13.05 psi (90 kPa) of air to manifold pressure compensator (6).
2. Adjust fuel pump (7) to 1,050 rpm.
3. Set fuel test stand to measure fuel delivery at 1,000 strokes.
4. Reduce fuel pump (7) to 700 rpm.
5. Position control lever (8) to full load position.
6. Take fuel draw and observe delivery quantity. Correct delivery quantity should be 9.1-9.2 in.^{3} (149.1-150.8 cm^3).
7. Check spread reading. Correct spread reading is 0.366 in.^{3} (6.00 cm^3).
8. Close air supply source to manifold pressure compensator (6).
9. Reduce fuel pump (7) to 500 rpm.
10. Position control lever (8) to full load position.
11. Take fuel draw and observe delivery quantity. Correct delivery quantity is 4.8-4.9 in.^{3} (78.7-80.3 cm^3).

NOTE
Perform steps 12 through 14 if fuel delivery quantity deviates from specifications.

12. Remove breakoff screw (10), screw (1), two lockwashers (3), protective cap (4), and stop (5) from manifold pressure compensator (6). Discard two lockwashers (3) and breakoff screw (10).
13. Turn nut (9) clockwise to increase fuel delivery; counterclockwise to decrease fuel delivery.
14. Install stop (5) and protective cap (4) on manifold pressure compensator (6) with two new lockwashers (3), new breakoff screw (11), and screw (1).
15. Install new safety wire (2) on screw (1) and protective cap (4).

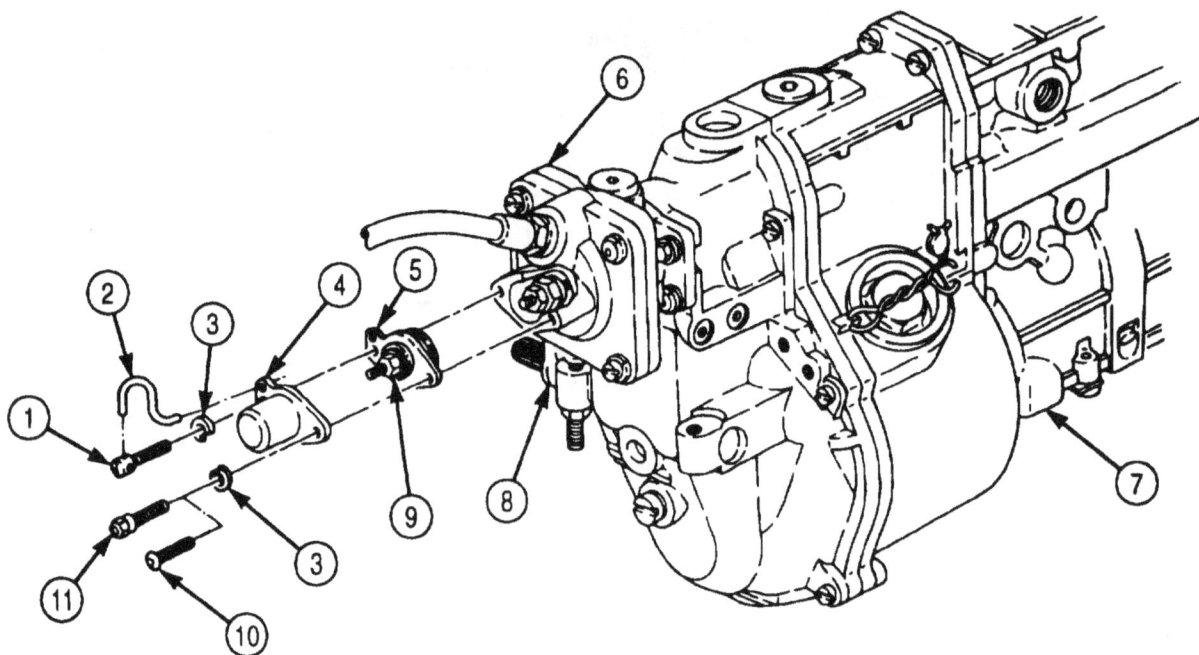

5-37. FUEL INJECTION PUMP CALIBRATION (Contd)

h. Manifold Pressure Compensator Test

1. Set fuel pump (7) to 500 rpm and position control rack (12) between 0.041-0.042 in. (1.040-1.060 mm) rack travel.
2. Apply 3.26 psi (22.5 kPa) of air pressure to manifold pressure compensator (6) and observe control rack (12) travel. Correct control rack (12) travel is 0.429-04.36 in. (11.90-12.30 mm).
3. Increase air pressure of manifold pressure compensator (6) to 6.53 psi (45.0 kPa) and observe control rack (12) travel. Correct control rack (12) travel is 0.468-0.484 in. (11.90-12.30 mm).

NOTE
Perform steps 5 through 8 if control rack travel deviates from specifications.

4. Remove plug (14) from top of governor housing (13).
5. If control rack (12) travel is less than required specifications, turn screw (15) clockwise.
6. If control rack (12) travel is more than required specifications, turn screw (15) counterclockwise.
7. Recheck control rack (12) travel. If control rack (12) travel is incorrect, repeat step 6 or 7. If control rack (12) travel is correct, proceed to step 9.
8. Install plug (14) on governor housing (13).

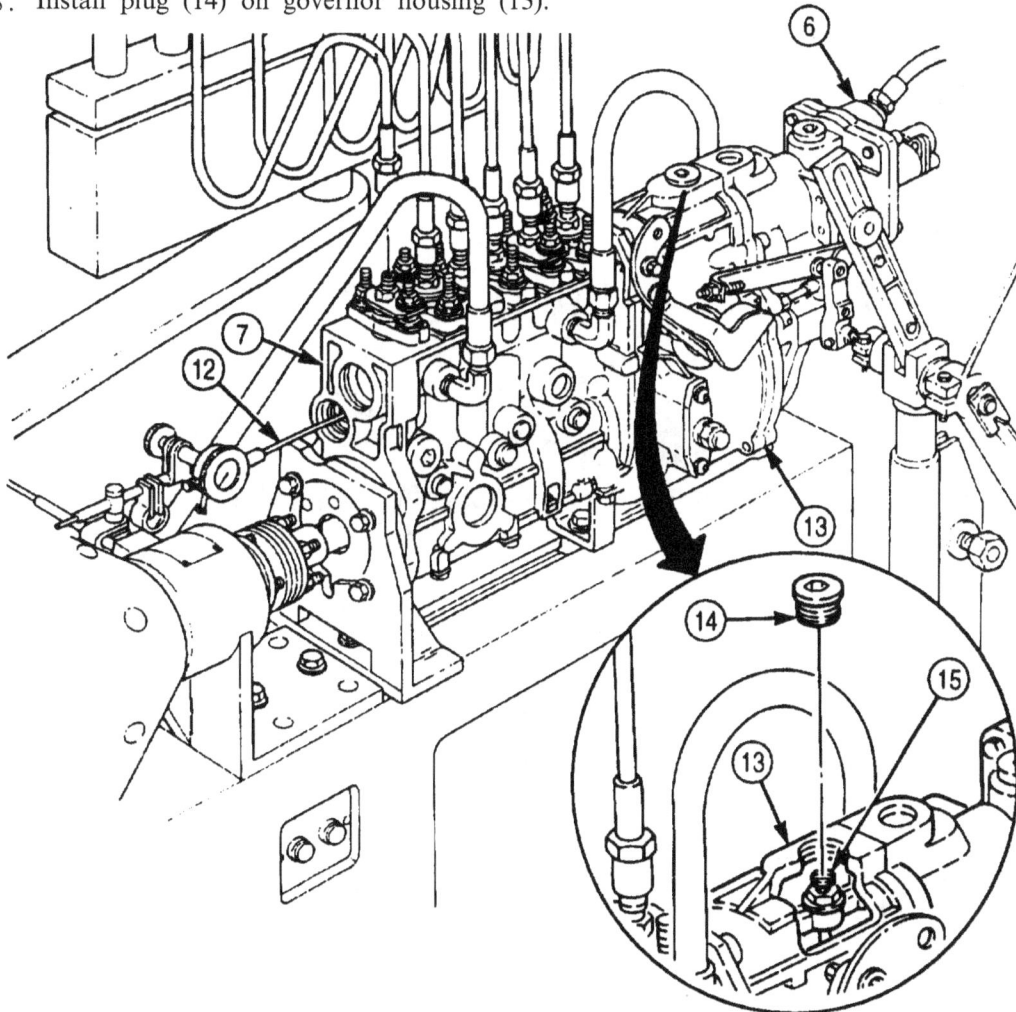

5-37. FUEL INJECTION PUMP CALIBRATION (Contd)

i. Fuel Delivery Test

1. Apply 13.1 psi (90.3 kPa) to manifold pressure compensator (2).
2. Adjust fuel pump (1) to 700 rpm.
3. Set test stand to measure fuel delivery at 1,000 strokes.
4. Set control lever (3) to full load position.
5. Check fuel delivery quantity. Correct fuel delivery quantity should be 8.73-9.21^3 in. (143.0-151.0 cm^3).
6. Check fuel spread. Fuel spread should be 0.427 in (7.0 cm^3).
7. Decrease fuel pump speed (1) to 500 rpm.
8. Check fuel delivery quantity. Correct fuel delivery quantity is 5.61-5.74^3 in (92.0-94.0 cm^3).
9. Adjust fuel delivery quantity by turning each barrel assembly (4) counterclockwise or clockwise until correct fuel delivery quantity is obtained.

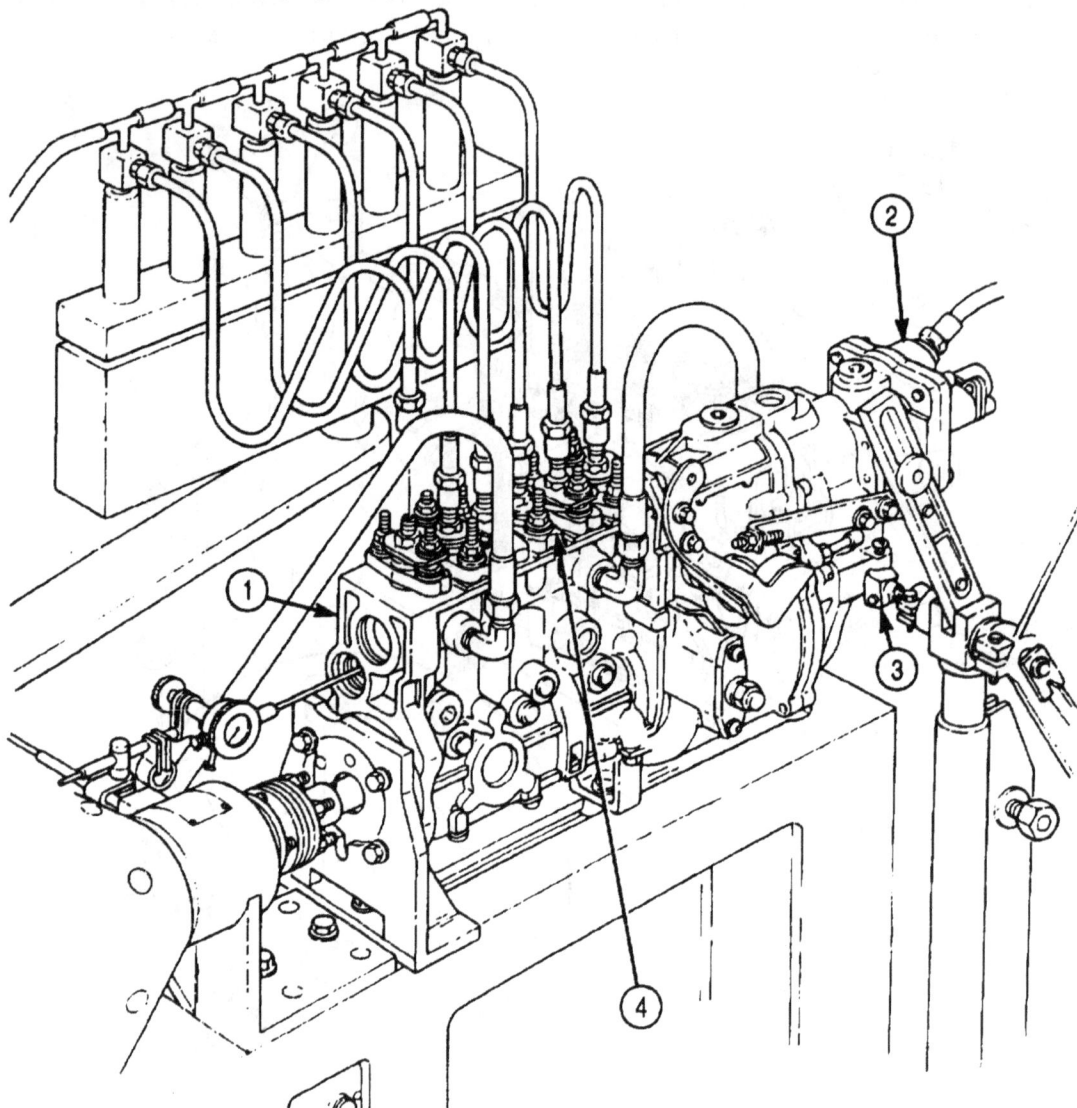

5-37. FUEL INJECTION PUMP CALIBRATION. (Contd)

j. Fuel Breakaway Test

1. Move control lever (3) to full load position.
2. Gradually increase rpm of fuel pump (1) rpm until control rack (5) moves 0.004 in. (0.1 mm) towards shutoff.
3. Rpm of fuel pump (1) should be 1,090-1,100 rpm. Perform step 5 if correct rpm is not obtained.
4. Move control lever (3) to full load position and observe travel of control rack. Correct control rack travel should be 0.464 in. (11.79 mm).
5. If breakaway is not as specified rpm, adjust high-speed stopscrew (7) until correct breakaway is obtained.

k. Starting Fuel Delivery Test

1. Set fuel pump (1) to 100 rpm.
2. Set test stand to measure calibration fluid at 1,000 strokes.
3. Position throttle lever (6) so control rack (5) is at full load 0.827 in. (21 mm).
4. Take fuel draw and check delivery quantity. Correct delivery quantity is 12.94-13.91^3 in. (212.0-228.0 cm^3).

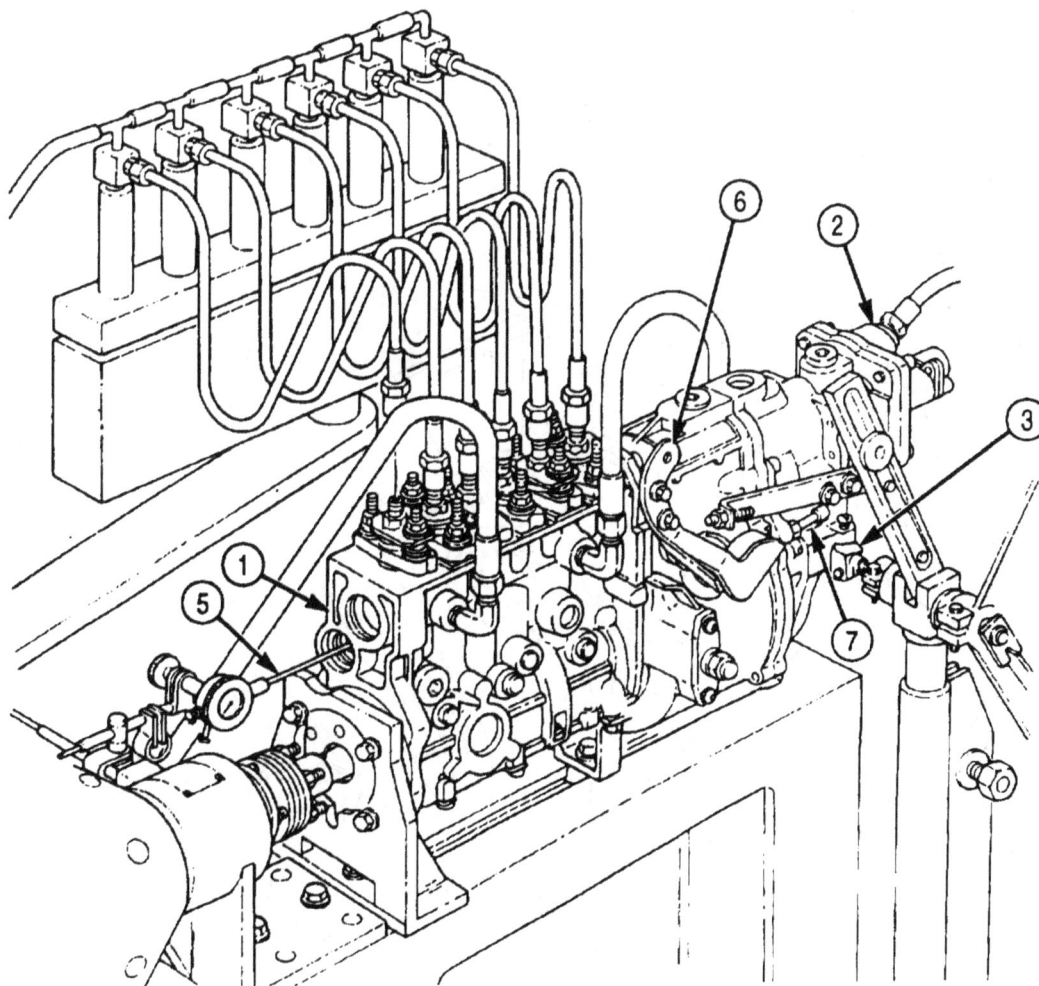

5-37. FUEL INJECTION PUMP CALIBRATION (Contd)

l. Low Idle Fuel Delivery Test

1. Set fuel pump (1) to 300 rpm.
2. Position control lever (6) against idle stopscrew (5) and take a fuel draw.
3. Observe control rack travel. Correct control rack travel should be 0.308-0.316 in. (7.80-8.03 mm).
4. With test stand set at 1,000 strokes, take fuel draw and check delivery quantity. Correct delivery quantity is 0.824-1.373 in.^{3} (13.5-22.5 cm^3).
5. If fuel delivery deviates from specifications, adjust idle stopscrew (5) until correct fuel delivery is obtained.

m. External Stop Setting

CAUTION

Perform this task so control rack has a 0.039 in. (1 mm) preload.
Failure to do so may result in damage to fuel pump.

1. Loosen adjusting nut (3) on throttle stopscrew (4).
2. Position throttle lever (2) against throttle stopscrew (4).
3. Adjust throttle stopscrew (4) until 0.039 in. (1 mm) preload is observed on dial indicator (7) and tighten adjusting nut (3).

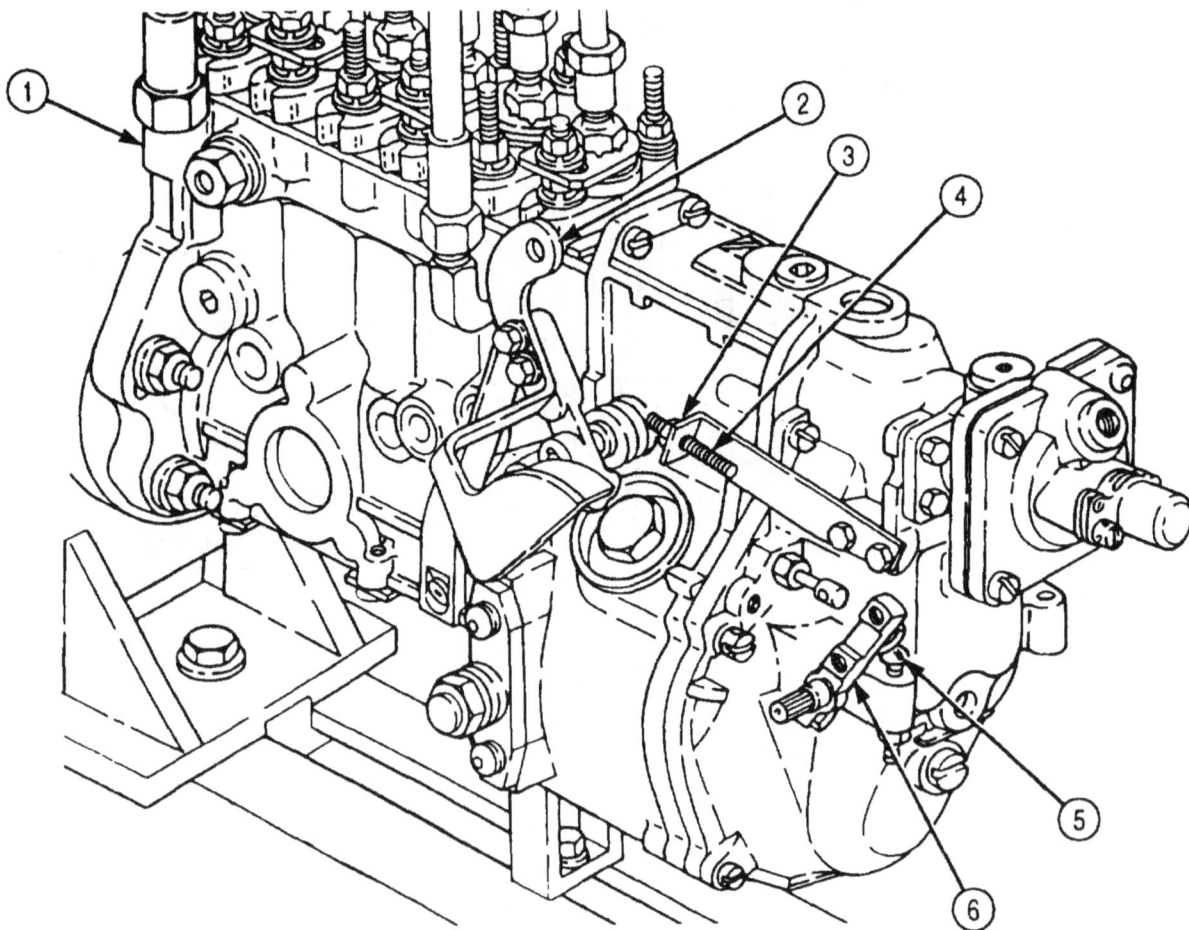

5-37. FUEL INJECTION PUMP CALIBRATION (Contd)

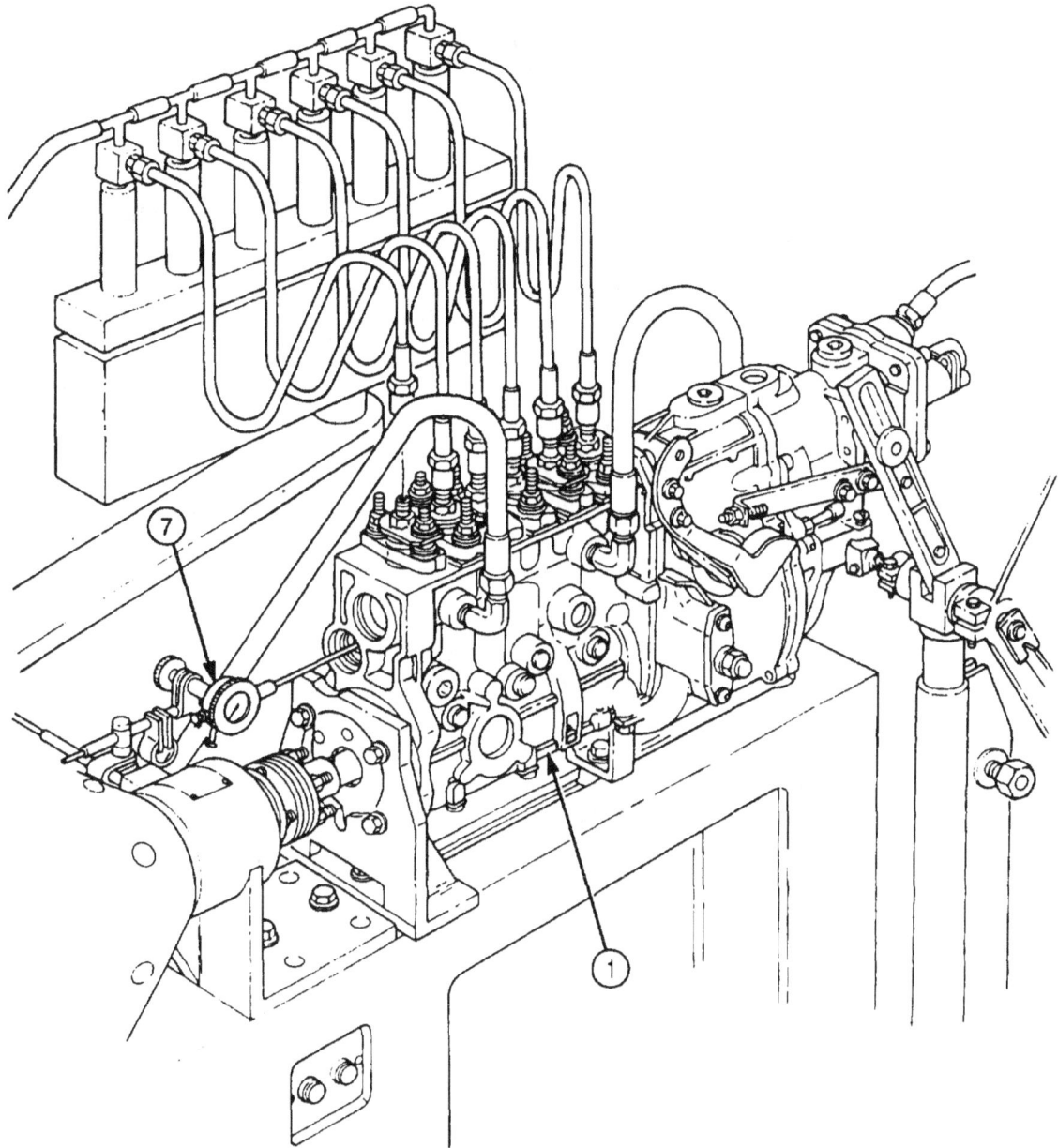

FOLLOW-ON TASK: Remove fuel injection pump from test stand (para. 5-31).

Section III. TRANSMISSION MAINTENANCE

5-38. TRANSMISSION MAINTENANCE INDEX

5-39. GENERAL MAINTENANCE INSTRUCTIONS

WARNING

- Drycleaning solvent is flammable and will not be used near open flame. Use only in well-ventilated places. Failure to do so may result in injury to personnel.

- Compressed air source will not exceed 30 psi (20 kPa). When cleaning with compressed air, eyeshields must be worn. Failure to wear eyeshields may result in injury to personnel.

CAUTION

- When converter pump hub, front support (including ground sleeve), or oil pump is defective, install wide oil pump kit. This kit converts an early model transmission (P/N 6885292) into a late model transmission (P/N 23040127). Failure to simultaneously install all parts of kit may cause damage to transmission during assembly or may cause transmission malfunction.

NOTE

- Refer to para. 2-14 for cleaning all transmission components.

- Make sure location markings are not removed when cleaning.

- All transmission parts must be lubricated with clean OE/HDO-10 lubricating oil before assembly.

5-40. TRANSMISSION MOUNTING TO HOLDING FIXTURE

THIS TASK COVERS
a. Installation   b. Removal

INITIAL SETUP:

APPLICABLE MODELS
All

SPECIAL TOOLS
Holding fixture adapter set (Appendix E, Item 64)
Holding fixture (Appendix E, Item 63)
Holding fixture base (Appendix E, Item 65)

TOOLS
General mechanic's tool kit (Appendix E, Item 1)
Lifting device
Chain

MATERIALS/PARTS
Gasket (Appendix D, Item 224)

REFERENCES (TM)
TM 9-2320-272-24P

EQUIPMENT CONDITION
Transmission removed from vehicle (para. 4-71) or transmission removed from engine (para. 4-72).

SPECIAL ENVIRONMENTAL CONDITIONS
Work area clean and free from blowing dirt and dust.

GENERAL SAFETY INSTRUCTIONS
• All personnel must stand clear during lifting operations.
• Do not remove hoist chain from transmission until transmission is stable. Injury to personnel may result.

a. Installation

1. Remove six screws (4), power takeoff cover (3), and gasket (2) from transmission (1). Discard gasket (2).
2. Install holding plate (6) and holding fixture (5) on right side of transmission (1) with six washers (7) and screws (8).

NOTE
Transmission may be mounted on any overhaul stand or holding fixture. However, the front, rear, and bottom of transmission must be freely accessible for removal and installation of components.

3. Install holding fixture base (10) on overhaul stand (11) with four screws (14), washers (13), and nuts (12).

WARNING
• All personnel must stand clear during lifting operations. A snapped cable, or swinging or shifting load, may cause injury to personnel.
• Do not remove hoist chain from transmission until transmission is stable on holding fixture base, or injury to personnel may result.

4. Using lifting device and chain, lift transmission (1) and holding fixture (5) into position and install in holding fixture base (10) with pin (15) and setscrew (9).
5. Remove chain and lifting device from transmission (1)

5-40. TRANSMISSION MOUNTING TO HOLDING FIXTURE (Contd)

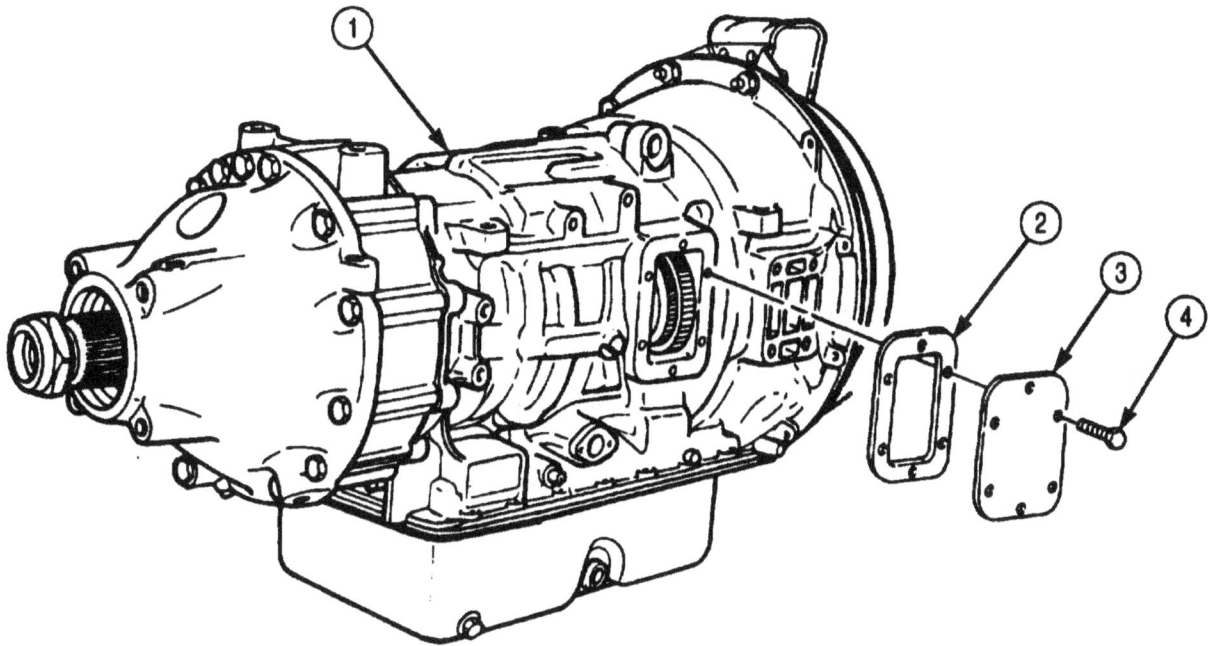

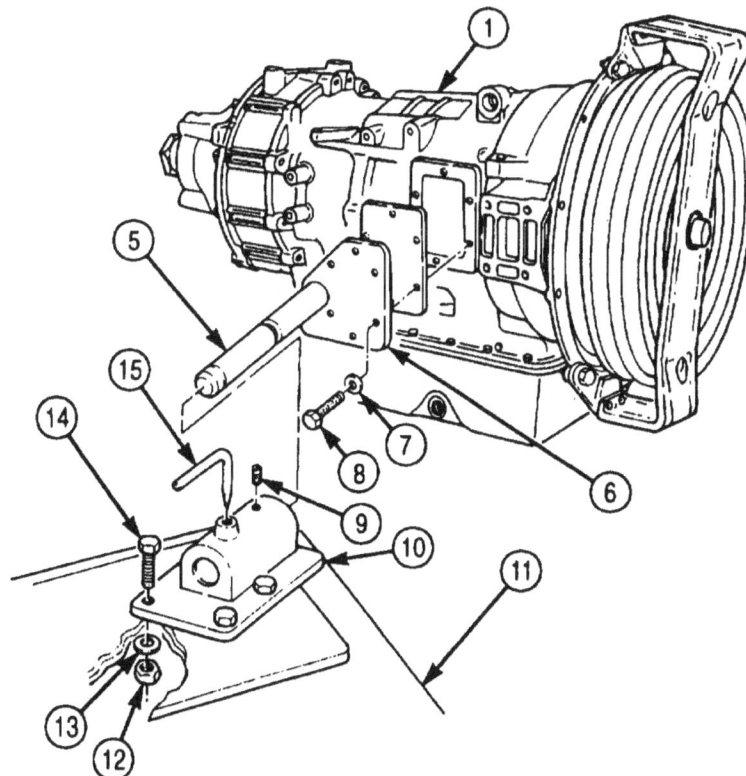

5-40. TRANSMISSION MOUNTING TO HOLDING FIXTURE (Contd)

b. Removal

WARNING

All personnel must stand clear during lifting operations.
A snapped cable, or swinging or shifting load, may cause
injury to personnel.

NOTE

Assistant will help with step 2.

1. Attach lifting device and chain to transmission (1).
2. Remove pin (7) from holding fixture base (6) and loosen setscrew (5).
3. Support transmission (1) with chain and lifting device and remove transmission (1) and holding fixture (8) from holding fixture base (6).
4. Remove six screws (4), washers (3), holding fixture (8), and holding plate (2) from transmission (1) and holding fixture (8).
5. Install new gasket (9) and power takeoff cover (10) on transmission (1) with six screws (11).
6. Remove lifting device and chain from transmission (1).

5-40. TRANSMISSION MOUNTING TO HOLDING FIXTURE (Contd)

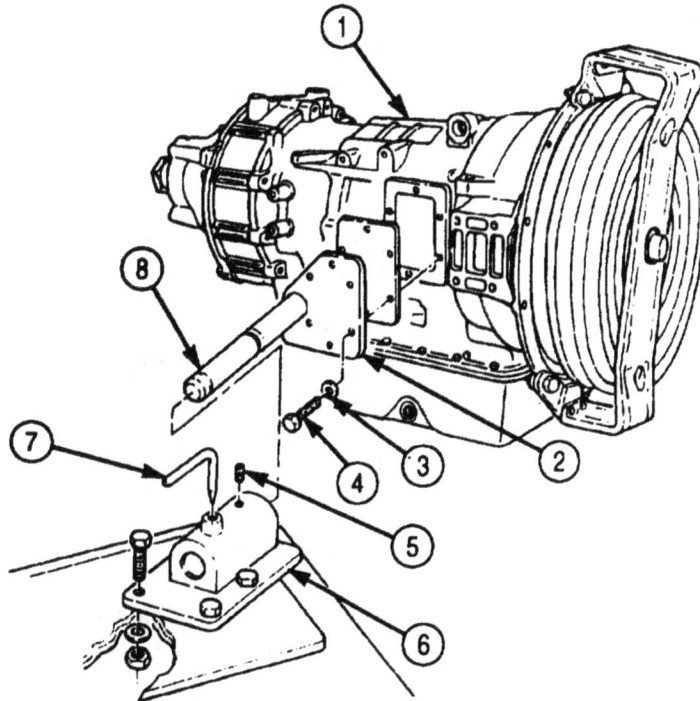

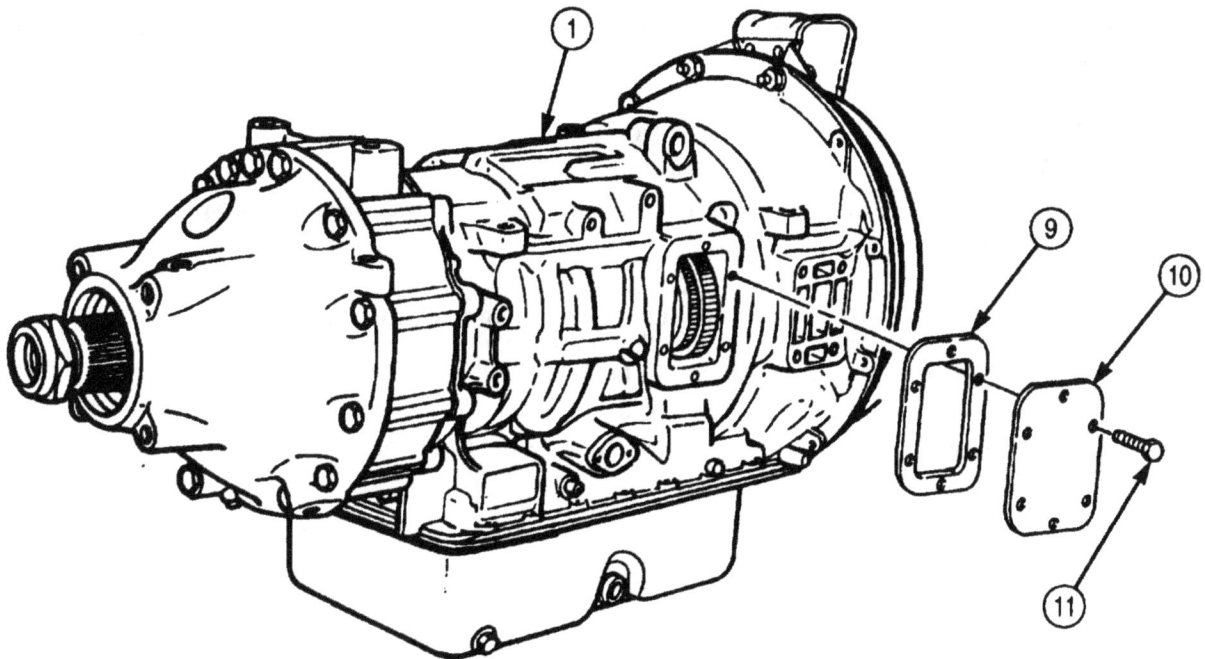

FOLLOW-ON TASK: Install transmission on engine (para. 4-72) or install transmission in vehicle (para. 4-71).

5-41. TRANSMISSION TORQUE CONVERTER MAINTENANCE

THIS TASK COVERS

a. Removal
b. Disassembly
c. Cleaning and Inspection

d. Assembly
e. Installation

<u>INITIAL SETUP:</u>

<u>APPLICABLE MODELS</u>
All

<u>SPECIAL TOOLS</u>
Bearing puller set (Appendix E. Item 105)
Drive handle (Appendix E, Item 38)
Bearing replacer (Appendix E, Item 15)
Bushing replacer (Appendix E, Item 25)

<u>TOOLS</u>
Arbor press
General mechanic's tool kit (Appendix E, Item 1)
Vernier caliper (Appendix E, Item 159)
Torque wrench (Appendix E, Item 144)

<u>MATERIALS/PARTS</u>
Converter pump seal ring (Appendix D, Item 636)
Flywheel seal ring (Appendix D, Item 638)
Hook-type seal (Appendix D, Item 637)
Piston seal ring (Appendix D, Item 520)
Pump hub gasket (Appendix D, Item 225)
Six lockstraps (Appendix D, Item 342)
Thirty self-locking screws (Appendix D, Item 648)
Lubricating oil (Appendix C, Item 49)
Oil-soluble grease (Appendix C, Item 53)

<u>REFERENCES (TM)</u>
TM 9-2320-272-24P

<u>EQUIPMENT CONDITION</u>
Transmission mounted to holding fixture (para. 5-40).

<u>SPECIAL ENVIRONMENTAL CONDITIONS</u>
Work area clean and free from blowing dirt and dust.

a. Removal

NOTE
Assistant will help with steps 1 and 2.

1. Remove four nuts (1), washers (2), and screws (4) from converter retaining strap (3) and transmission (7).

CAUTION
Torque converter must be pulled straight out of transmission and not moved from side to side. Side movement will damage hook-type seal ring on torque converter pump hub.

2. Pull converter retaining strap (3) with torque converter (5) straight out of transmission (7).
3. Remove two screws (6) and converter retaining strap (3) from torque converter (5).

5-41. TRANSMISSION TORQUE CONVERTER MAINTENANCE (Contd)

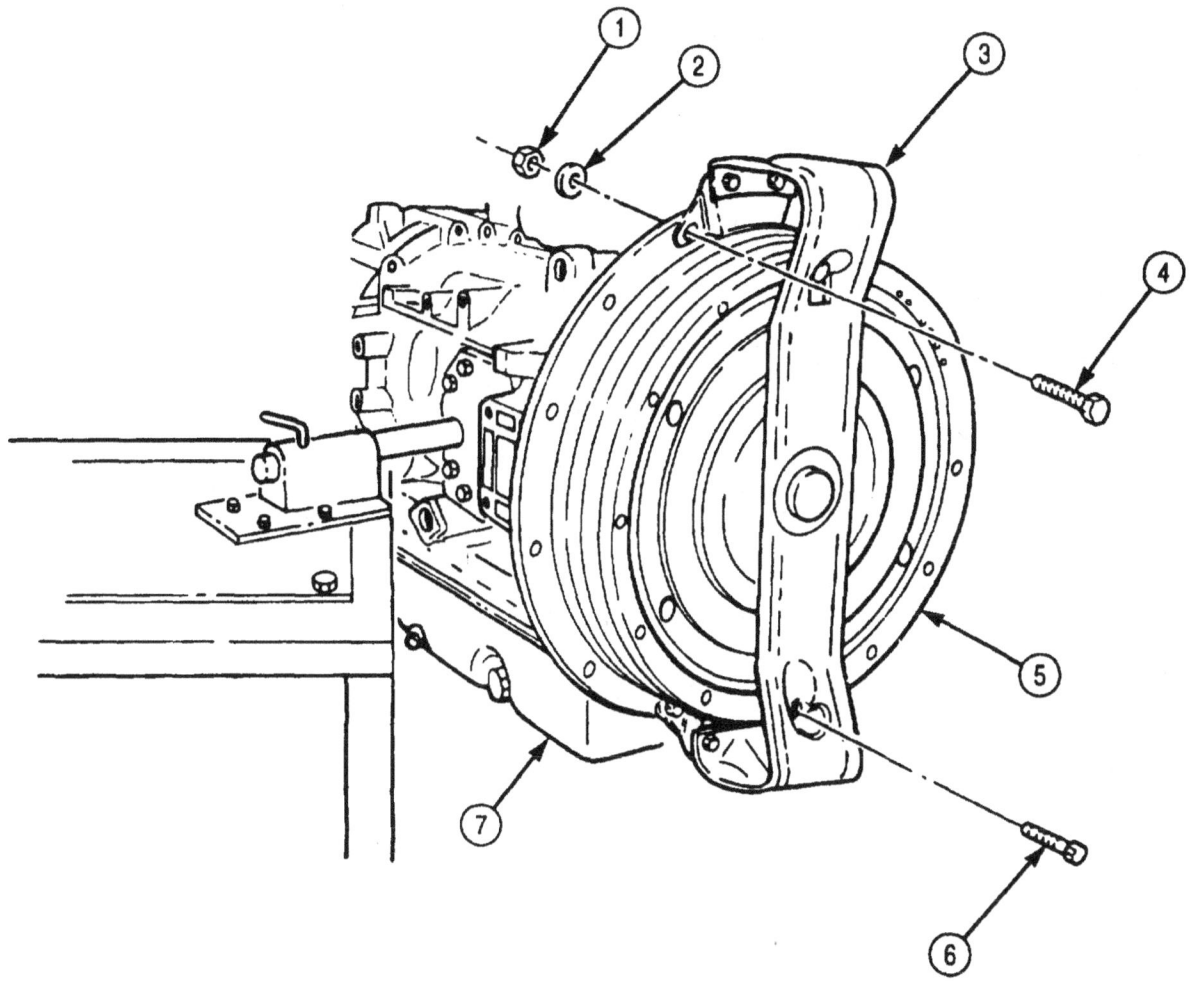

5-41. TRANSMISSION TORQUE CONVERTER MAINTENANCE (Contd)

b. Disassembly

NOTE

Hook-type seal ring used on early models.

1. Remove hook-type seal ring (12) from torque converter pump hub (11). Discard hook-type seal ring (12).

NOTE

- Have drainage container ready to catch transmission oil.
- Assistant will help with step 2.

2. Remove thirty self-locking screws (13) and washers (14) from converter pump (7) and converter flywheel (1), and drain transmission oil. Discard self-locking screws (13).

3. Separate converter pump (7) from converter flywheel (1).

4. Remove seal ring (6) from converter pump (7). Discard seal ring (6).

5. Remove roller thrust bearing (2) and bearing race (3) from converter flywheel (1) or converter pump (7).

6. Bend corners of six lockstraps (5) away from twelve screws (4) and remove twelve screws (4), six lockstraps (5), and retainer (15) from converter pump (7). Discard lockstraps (5).

7. Remove torque converter pump hub (11) and gasket (10) from converter pump (7). Discard gasket (10).

NOTE

- Early model transmission converter pump has a snapring and roller bearing inside front bore. Late model converter pump hub has no snapring or roller bearing, but does have a nose with flats.
- Perform step 8 for early model transmission.

8. Using bearing puller set, remove snapring (8) and roller bearing (9) from torque converter pump hub (11).

9. Remove stator (26) from converter turbine (25).

10. Turn roller race (31) clockwise and lift off stator (26).

11. Remove ten rollers (27) and springs (28) from stator (26).

12. Remove thrust bearing (30) and race (29) from stator (26).

NOTE

Converter turbine must be pulled straight off flywheel, or it will bind.

13. Remove converter turbine (25) from converter flywheel (1).

14. Using bearing puller, remove ball bearing (22) and spacer (23) from turbine hub (24).

15. Remove lockup clutch plate (19), backplate (20), and two backplate keys (21) from converter flywheel (1).

NOTE

Scribe mark on converter flywheel and lockup clutch piston for installation.

16. Place converter flywheel (1) upright and remove lockup clutch piston (17). If lockup clutch piston (17) sticks, tap flywheel (1) lightly with soft-faced hammer.

17. Remove piston seal ring (18) from piston (17). Discard piston seal ring (18).

18. Remove flywheel seal ring (16) from converter flywheel (1). Discard flywheel seal ring (16).

5-41. TRANSMISSION TORQUE CONVERTER MAINTENANCE (Contd)

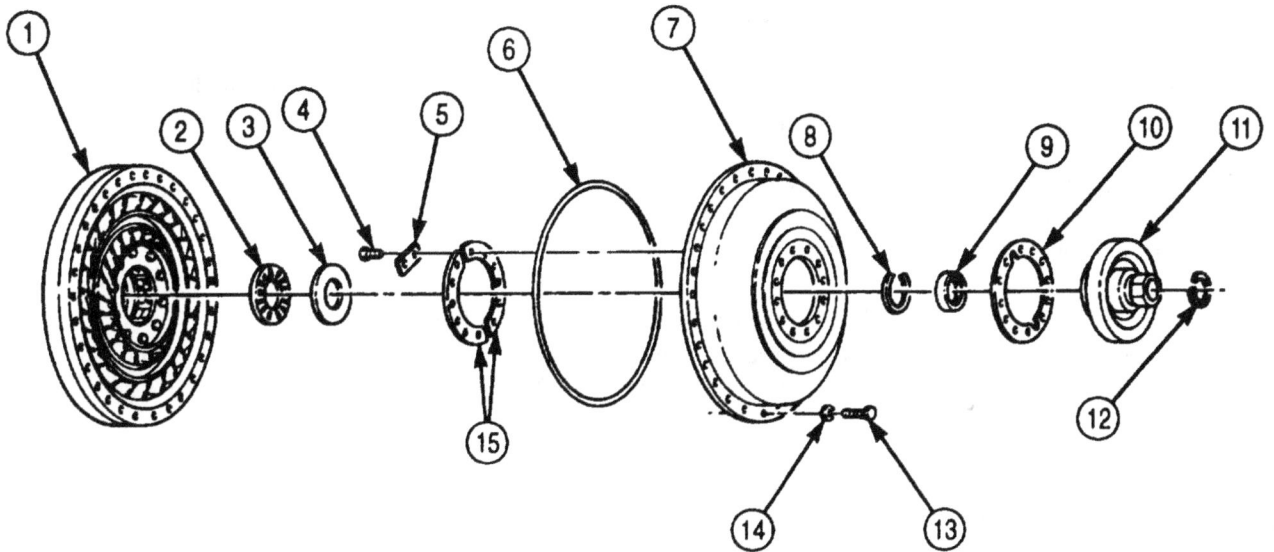

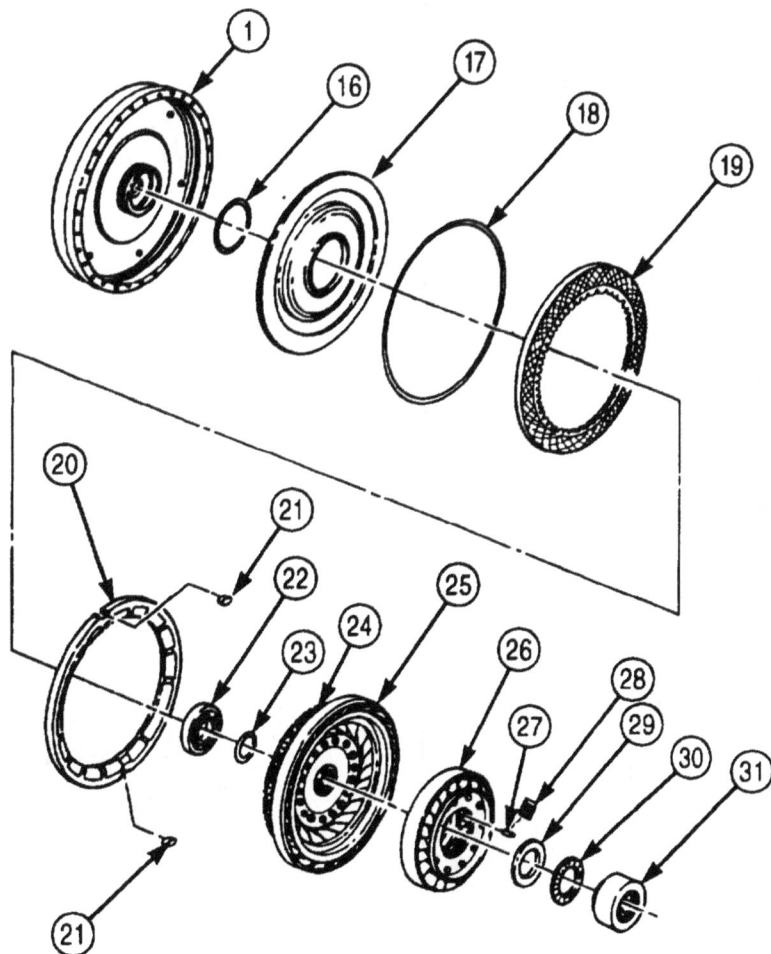

5-41. TRANSMISSION TORQUE CONVERTER MAINTENANCE (Contd)

c. Cleaning and Inspection

1. For general cleaning instructions, refer to para. 2-14.
2. For general inspection instructions, refer to para. 2-15.
3. Inspect lockup clutch disc (1) for burned surfaces. If burned, discard lockup clutch disc (1).
4. Measure thickness of lockup clutch disk (1). If thickness is less than 0.190 in. (4.32 mm), discard lockup clutch disc (1).
5. Measure thickness of thrust bearing race (2). If thickness is less than 0.029 in. (0.74 mm), discard thrust bearing race (2).
6. Inspect roller race (4) for scoring. If scored, discard roller race (4).

NOTE

Step 7 refers to early model pump hub.

7. Inspect pump hub (3) for cracks, breaks, burred flats, and scored seal surface and bearing bore. If damage is more than minor scoring, discard pump hub (3).

NOTE

Step 8 refers to late model pump hub.

8. Inspect pump hub (3) for cracks, breaks, burred flats, and scored seal surface and bushing journal. If damage is more than minor scoring, discard pump hub (3).

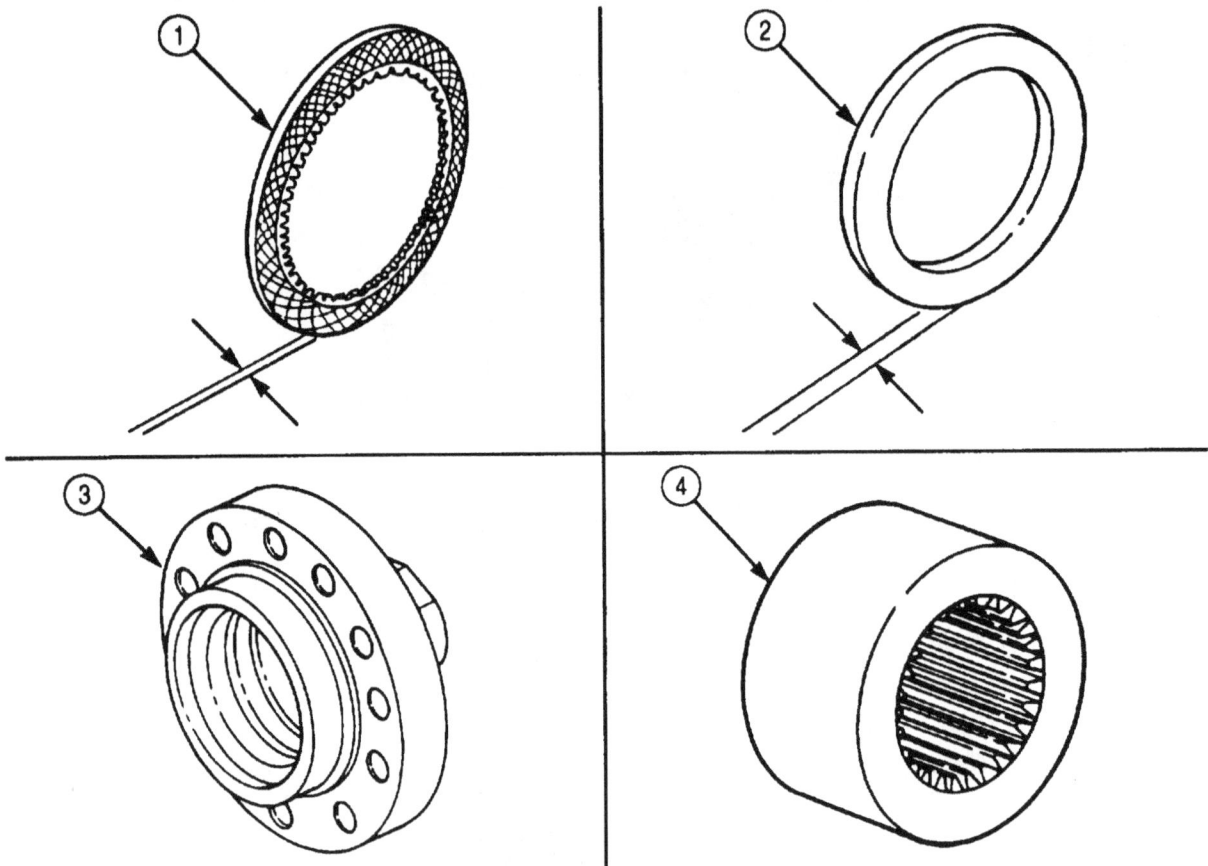

5-41. TRANSMISSION TORQUE CONVERTER MAINTENANCE (Contd)

d. Assembly

NOTE

- Steps 2 and 14 obtain measurements to select proper spacer thickness. These measuring steps are treated as assembly steps.
- Steps 2 and 3 are performed only if new bearing is being installed.

1. Install ball bearing (5) in flywheel bore (7). Ensure ball bearing (5) seats against shoulder of flywheel bore (7).

2. Place straightedge (6) across flywheel bore (7) and measure the distance from inner race of ball bearing (5) to straightedge (6). Record this measurement for use in step 14d.

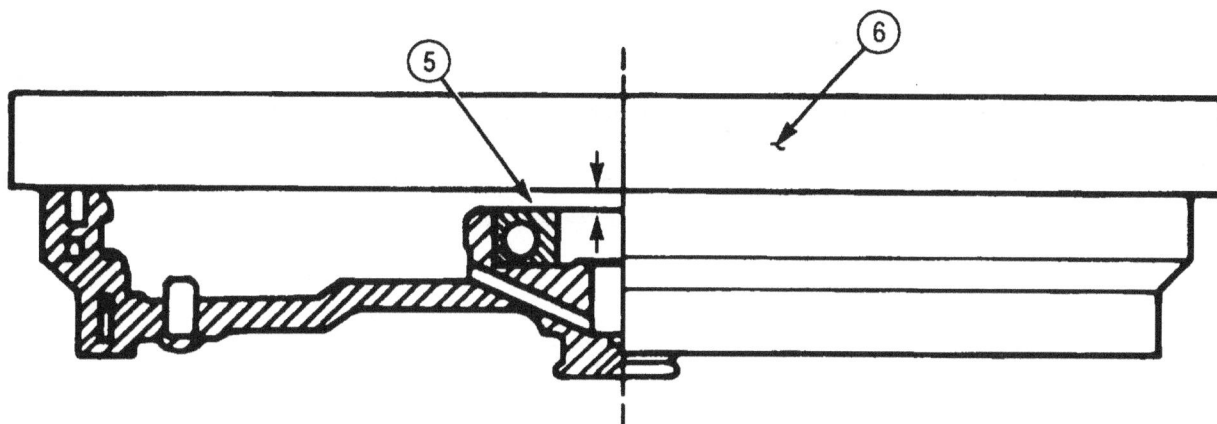

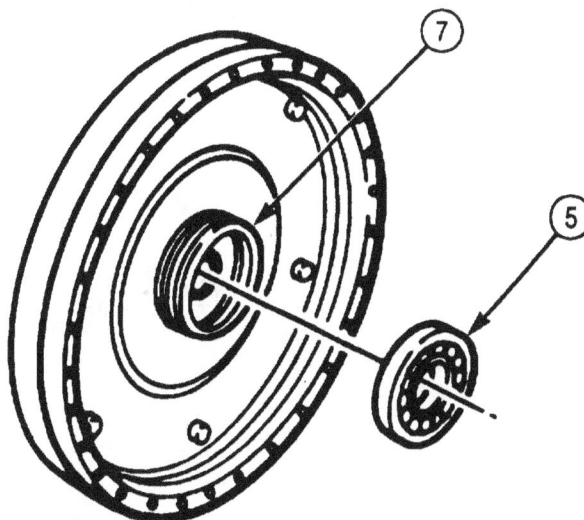

5-41. TRANSMISSION TORQUE CONVERTER MAINTENANCE (Contd)

3. Remove ball bearing (8) from flywheel (1).

4. Apply lubricating oil to new flywheel seal ring (2) and install on flywheel (1).

5. Apply lubricating oil to new piston seal ring (4) and install on lockup clutch piston (3).

6. Align previously scribed marks on lockup clutch piston (3) and install, cupped side first, on flywheel (1).

7. Soak lockup clutch plate (5) in clean lubricating oil for at least two minutes. Install lockup clutch plate (5) on clutch piston (3).

8. Position two backplate keys (7) in flywheel (1) recesses.

NOTE

Backplate and backplate keys must be level for the next step.

9. Install lockup clutch backplate (6).

a. Position lockup clutch backplate (6) in flywheel (1), engaging notch in lockup clutch backplate (6) with two backplate keys (7).

CAUTION

If backplate is not approximately level with the key, the lockup clutch piston is not properly engaged with the drive pins in the flywheel. Rotate the piston until it drops into place on the pins. Damage to backplate will result if piston is not engaged properly.

NOTE

Perform step b only if backplate is not level with the keys.

b. Rotate piston (3) until pin holes and flywheel (1) pins are aligned.

5-41. TRANSMISSION TORQUE CONVERTER MAINTENANCE (Contd)

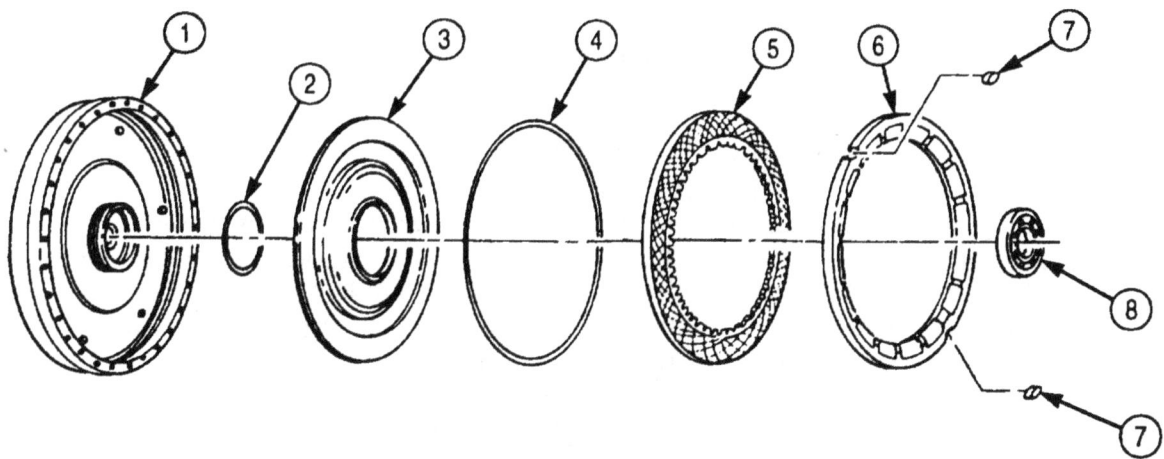

5-41. TRANSMISSION TORQUE CONVERTER MAINTENANCE (Contd)

NOTE

- Late model transmission does not have roller bearing or snapring in front bore of torque converter pump hub.
- Perform step 10 if bearing is to be installed in early model transmission.

10. Using an arbor press and bearing remover/installer, press roller bearing (16) in torque converter pump hub (18) and install with snapring (15).

CAUTION

Identify transmission model. Late model converter pump hub has longer nose and has 1 in. (25.4 mm) long flats that fit into oil pump (para. 5-39). Do not intermix parts from early and late model transmissions. Damage to transmission will result.

11. Install new gasket (17) and torque converter pump hub (18) on converter pump (14) with retainer (13), six new lockstraps (12), and twelve screws (11). Tighten screws (11) 33-40 lb-ft (45-54 N•m). Bend comers of lockstraps (12) against head of each screw (11).

12. Install bearing race (10) and roller thrust bearing (9) in converter pump (14).

13. Install bearing race (25), thrust bearing (26), and freewheel roller race (27) in stator (24) and place stator (24) and converter turbine (21) in converter pump (14). Center stator (24) and converter turbine (21) carefully.

NOTE

Step 14 is performed only if new bearing was installed in step 1.

14. Select spacer (23) size:

a. Place gauge blocks (22) of equal height on edge of converter pump (14) as shown

b. Place straightedge (19) across gauge blocks (22) and measure distance from straightedge (19) to shoulder (20) adjacent to hub of turbine (21). Record this measurement.

c. Subtract measurement obtained in step 14b. from gauge block height. Record this measurement.

d. Subtract measurement obtained in step 14c. from measurement obtained in step 2. Using this measurement, refer to table 5-11 to select correct size spacer (23).

Table 5-11. Torque Converter Spacer Sizes.

Measurements	Spacer
0.018-0.032 in. (0.457-0.813 mm)	Gold
0.032-0.044 in. (0.813-1.117 mm)	Silver
0.044-0.062 in. (1.117-1.575 mm)	Plain
0.062-0.077 in. (1.575-1.956 mm)	Black
0.077-0.096 in. (1.956-2.438 mm)	Copper

15. Remove converter turbine (21) from converter pump (14).

5-41. TRANSMISSION TORQUE CONVERTER MAINTENANCE (Contd)

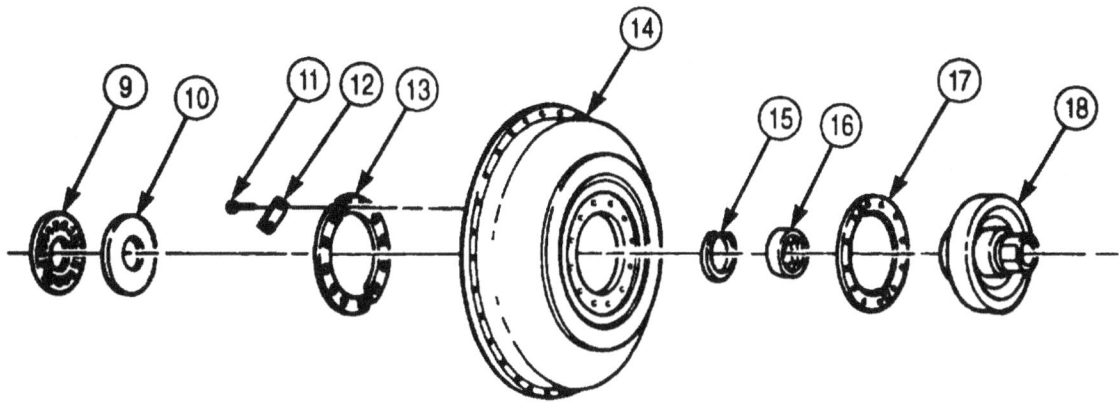

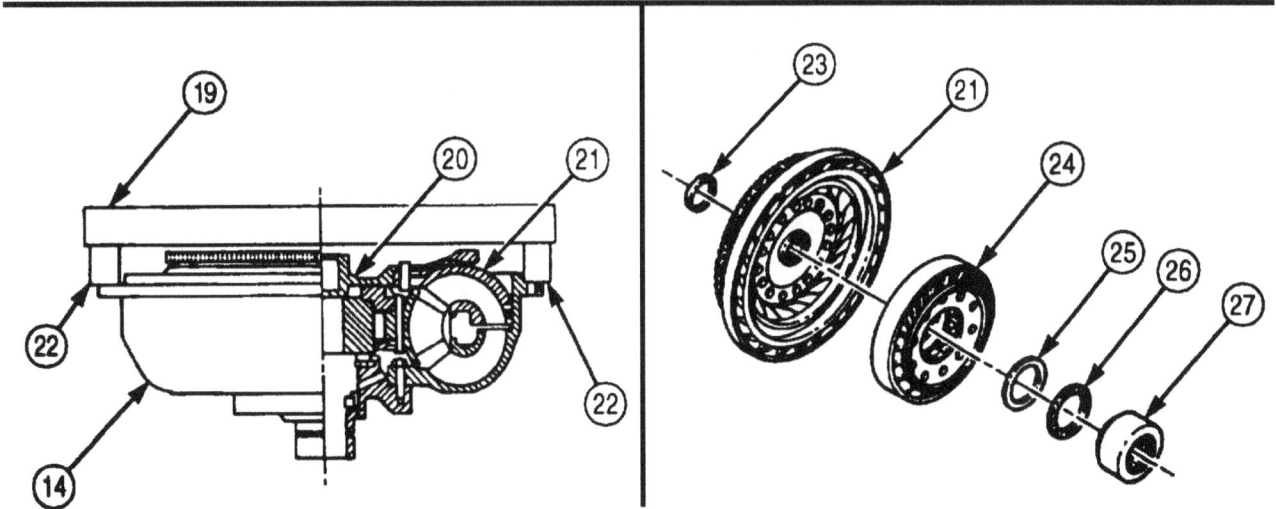

5-41. TRANSMISSION TORQUE CONVERTER MAINTENANCE (Contd)

16. Install spacer (3) on converter turbine hub (4).
17. Install ball bearing (2) on converter turbine hub (4) until seated against spacer (3).
18. Install converter turbine (5) in flywheel (1).
19. Remove stator (6) from converter pump (12).
20. Remove freewheel roller race (10) and thrust bearing (9) from stator (6).

NOTE
- One of three different types of roller springs will be present in early stators as shown. Either the later or latest type should be used, but all springs must be of the same type.
- The latest type roller springs are installed in a different manner. The old tab may be either up or down.
- Oil-soluble grease will hold springs and rollers in place.

21. Pack stator cam pockets (17) with oil-soluble grease and install springs (8) and rollers (7) in stator cam pockets (17).
22. Apply a small amount of oil-soluble grease to thrust bearing (9) and position thrust bearing (9) to freewheel roller race (10).
23. Position freewheel roller race (10) and thrust bearing (9) in stator (6) and turn clockwise until seated in stator (6).
24. Position stator (6) on converter turbine (5).
25. Apply lubricating oil to new seal ring (11) and install on converter pump (12).
26. Install converter pump (12) on flywheel (1) with thirty new self-locking washers (16) and new self-locking screws (15). Tighten screws (15) 41-49 lb-ft (56-66 N•m).

NOTE
Hook-type seal ring used on early models.

27. Install new hook-type seal ring (13) on torque converter pump hub (14).
28. Install converter retaining strap (19) on torque converter (18) with two screws (20).

5-41. TRANSMISSION TORQUE CONVERTER MAINTENANCE (Contd)

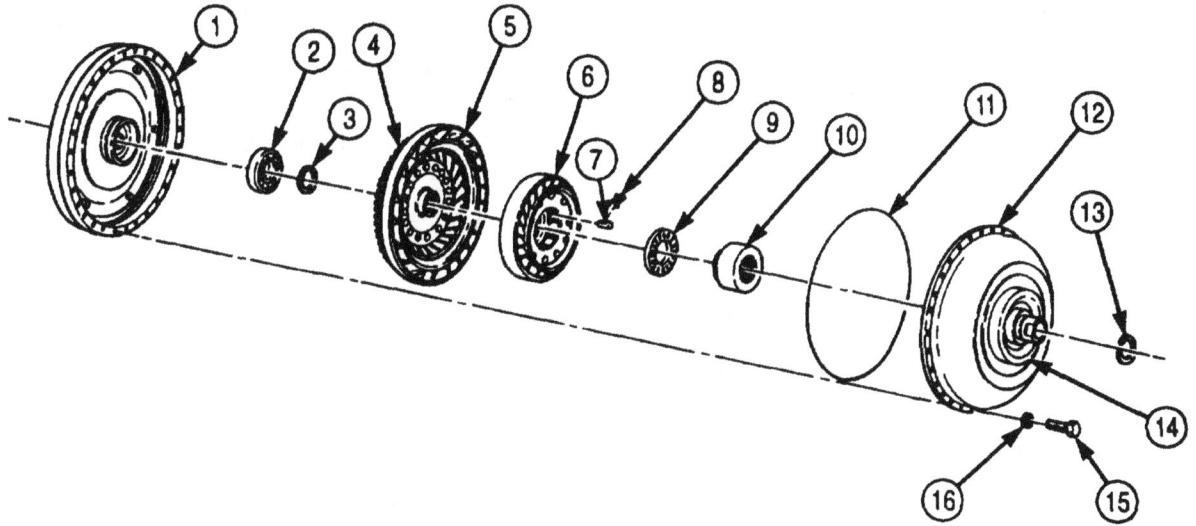

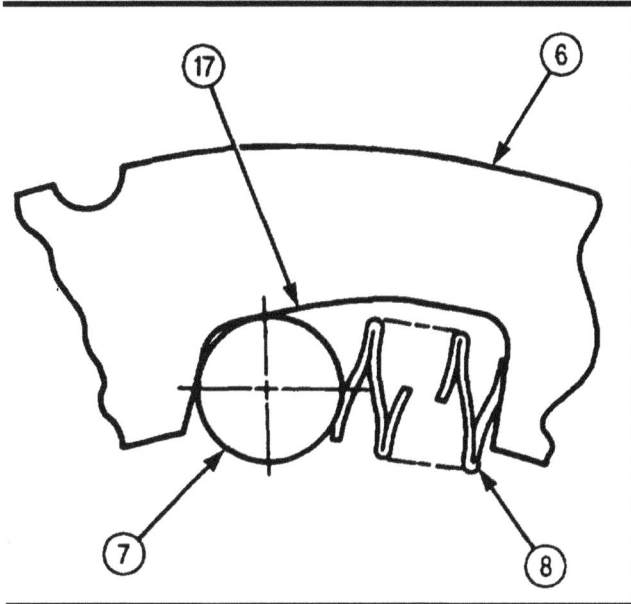

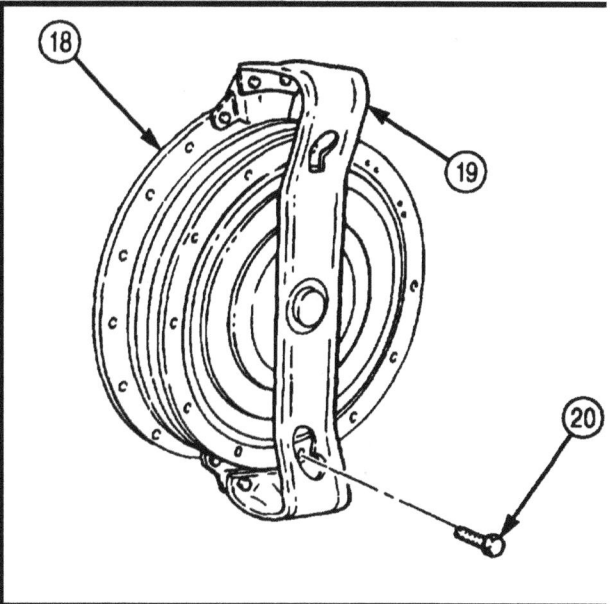

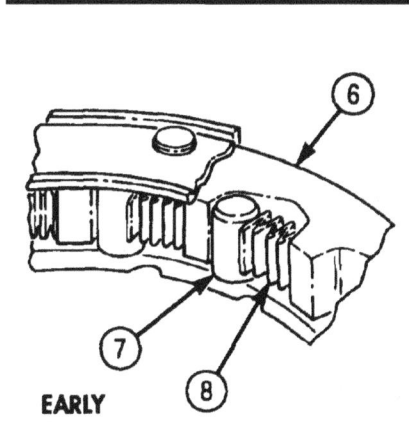

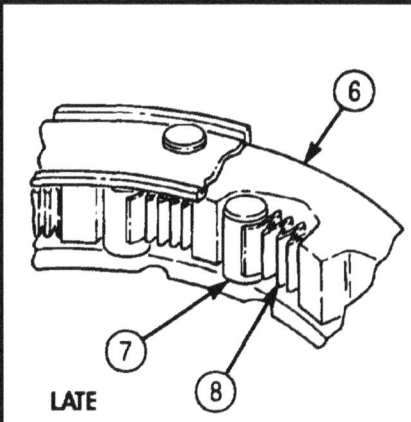

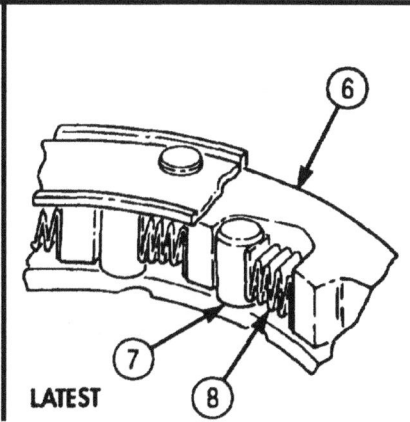

EARLY

LATE

LATEST

5-41. TRANSMISSION TORQUE CONVERTER MAINTENANCE (Contd)

e. Installation

1. Tilt transmission (7) to horizontal position, with oil pan (6) facing downward.

CAUTION

Torque converter must be installed straight on transmission and not moved from side to side. Side movement will damage hook-type seal ring on turbine shaft.

NOTE

Assistant will help with step 2.

2. Install torque converter (5) in transmission (7). Ensure seal ring is properly positioned. Torque converter (5) may be rotated until flat sides of pump hub engage flats in oil pump drive gear.

3. Install converter retaining strap (3) on transmission (7) with four screws (4), washers (2), and nuts (1).

5-41. TRANSMISSION TORQUE CONVERTER MAINTENANCE (Contd)

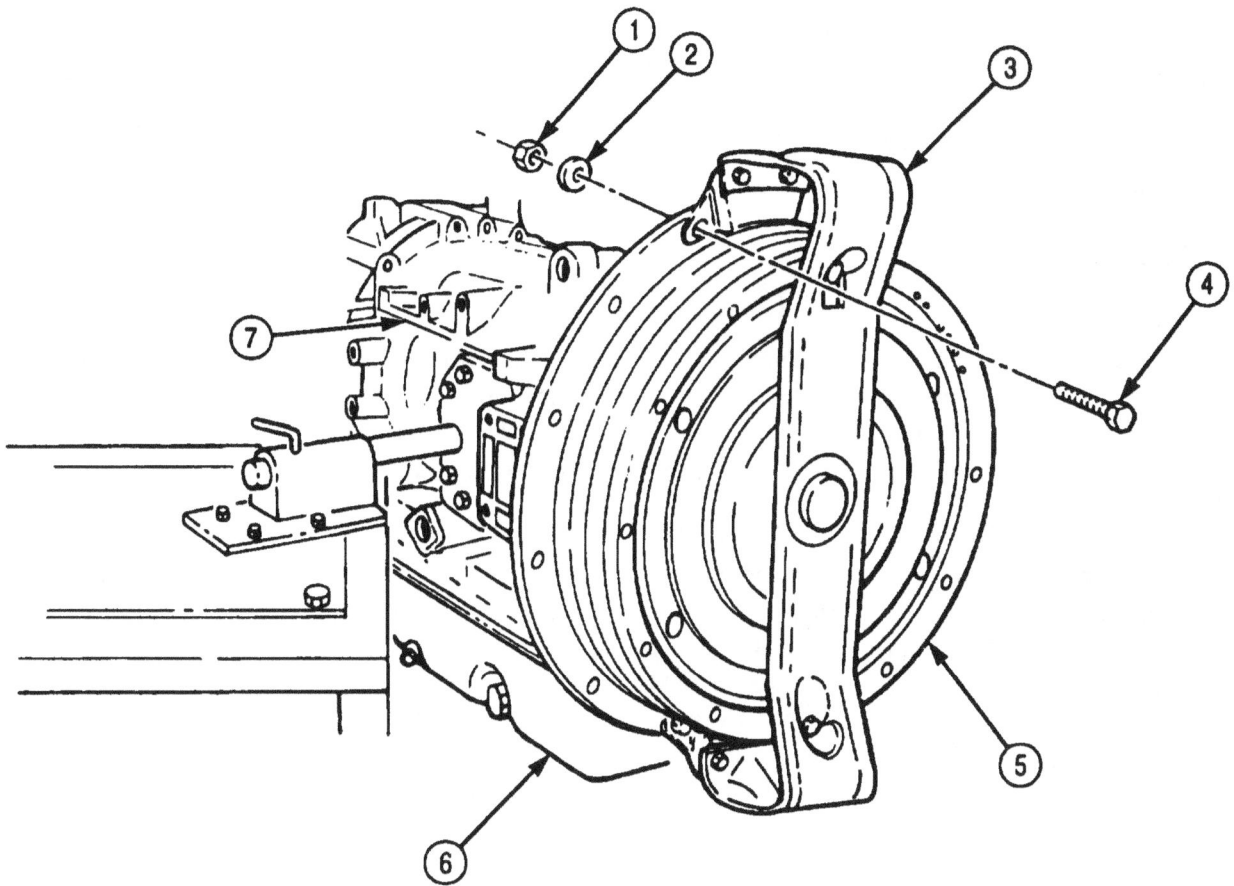

FOLLOW-ON TASK: Remove transmission from holding fixture (para. 5-40).

5-42. TRANSMISSION OIL PAN AND FILTER MAINTENANCE

THIS TASK COVERS:

a. Removal
b. Cleaning and Inspection

c. Installation

INITIAL SETUP:

APPLICABLE MODELS
All

TOOLS
General mechanic's tool kit (Appendix E, Item 1)
Torque wrench (Appendix E, Item 146)

MATERIALS/PARTS
O-ring (Appendix D, Item 429)
Transmission oil filter (Appendix D, Item 699)
Gasket (Appendix D, Item 159)

REFERENCES (TM)
TM 9-2320-272-24P

EQUIPMENT CONDITION
Transmission torque converter removed (para. 5-41).

SPECIAL ENVIRONMENTAL CONDITIONS
Work area clean and free from blowing dirt and dust.

a. Removal

1. Remove twenty-one screws (l), oil pan (2), and gasket (3) from transmission (4). Discard gasket (3).
2. Rotate transmission (4) in stand until oil pan (2) mounting surface is horizontal and facing upward.
3. Remove oil filter screw (6), oil filter (7), and filter suction tube (8) from transmission (4).
4. Remove filter suction tube (8) from oil filter (7). Discard oil filter (7).
5. Remove O-ring (9) from filter suction tube (8). Discard O-ring (9).

b. Cleaning and Inspection

1. For general cleaning instructions, refer to para. 2-14.
2. For general inspection instructions, refer to para. 2-15.
3. Inspect oil pan (2) for cracks. If cracked, replace oil pan (2).

NOTE
- New oil pan has a plug in left side fill port.
- If oil pan is defective, remove plug if present.

4. Remove oil plug (5). Retain for installation in new oil pan (2).

5-42. TRANSMISSION OIL PAN AND FILTER MAINTENANCE (Contd)

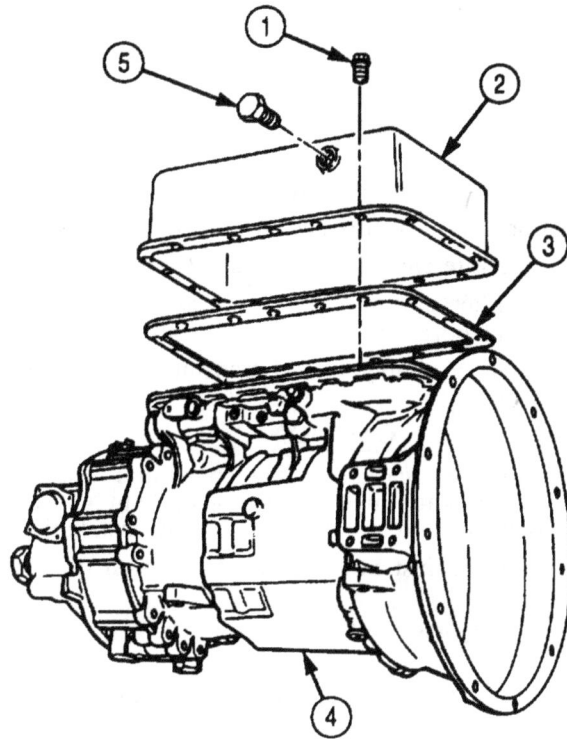

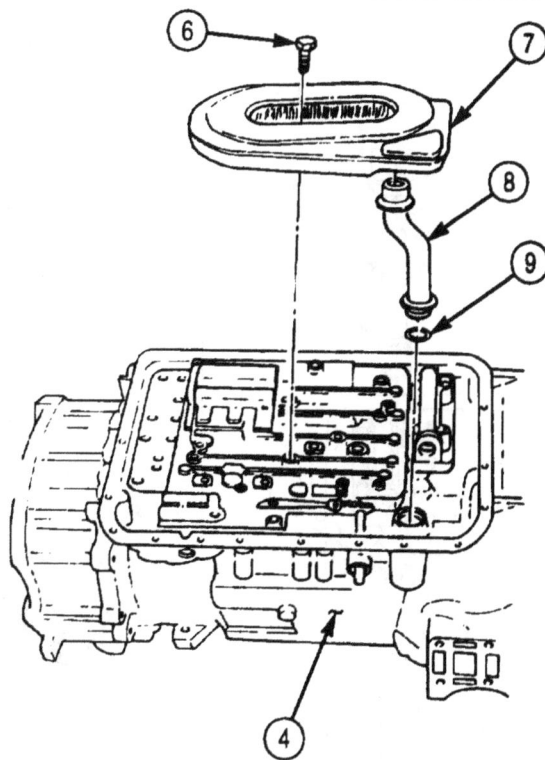

5-42. TRANSMISSION OIL PAN AND FILTER MAINTENANCE (Contd)

c. Installation

NOTE
- Keep oil pan mounting surface in horizontal position and facing upward.
- Filter suction tube ends are interchangeable.

1. Insert one end of filter suction tube (3) into new transmission oil filter assembly (2).

2. Install new O-ring (4) onto opposite end of filter suction tube (3).

3. Position new transmission oil filter assembly (2) on transmission (6). Ensure filter suction tube (3) is inserted into oil input port (5).

4. Install new transmission oil filter assembly (2) on transmission (6) with screw (1). Tighten screw (1) 10-15 lb-ft (14-20 N•m).

CAUTION

Do not use silicone-type gasket sealing compound when installing gasket, because oil leakage may result. Oil or light grease coating may be used to hold gasket in position during installation.

5. Position new oil pan gasket (10) against housing of transmission (6) and align with holes of transmission (6).

6. Position transmission oil pan (7) against oil pan gasket (10) and install with twenty-one screws (9). Tighten screws (9) finger-tight.

7. Tighten twenty-one screws (9) 10-15 lb-ft (14-20 N•m) following torque sequence shown.

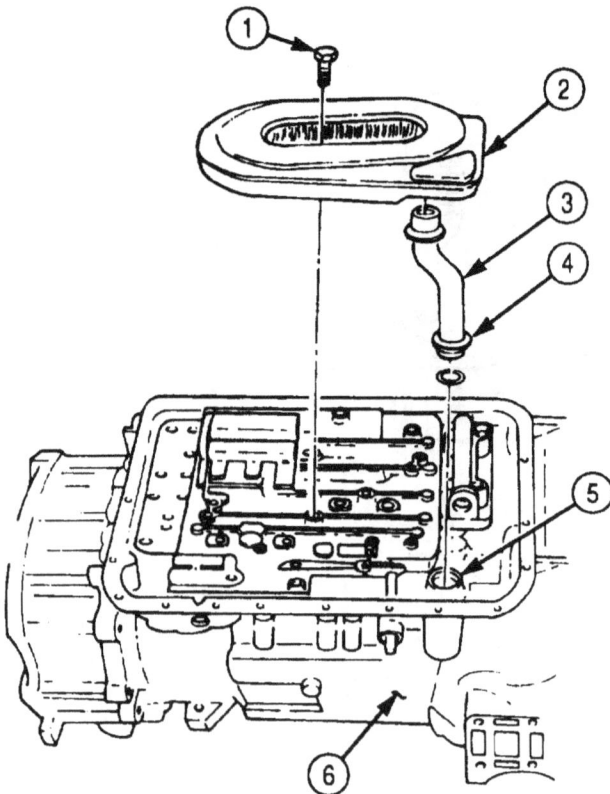

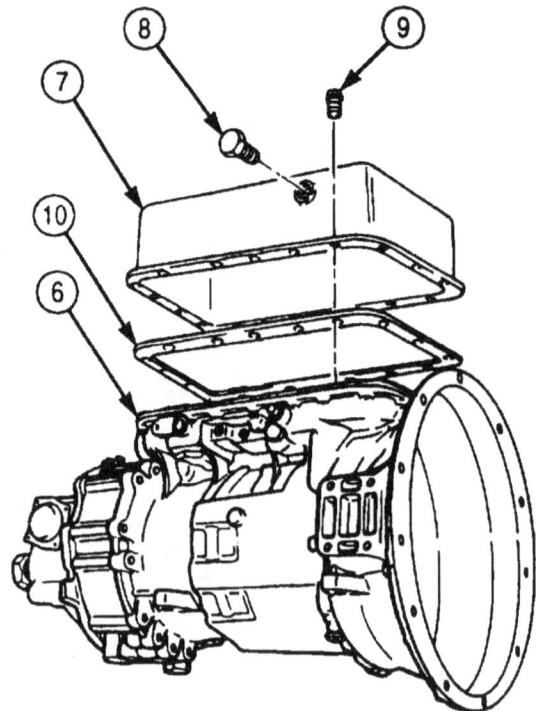

5-42. TRANSMISSION OIL PAN AND FILTER MAINTENANCE (Contd)

NOTE

Due to gasket compression, torque values will be lost and screws
must be retightened.

8. Tighten twenty-one screws (9) 15-20 lb-ft (20-27 N•m) following torque sequence shown to achieve final torque.

NOTE

Perform step 9 only if new oil pan was installed.

9. Install plug (8) in oil pan (7). Ensure plug (8) is installed opposite hole for dipstick.

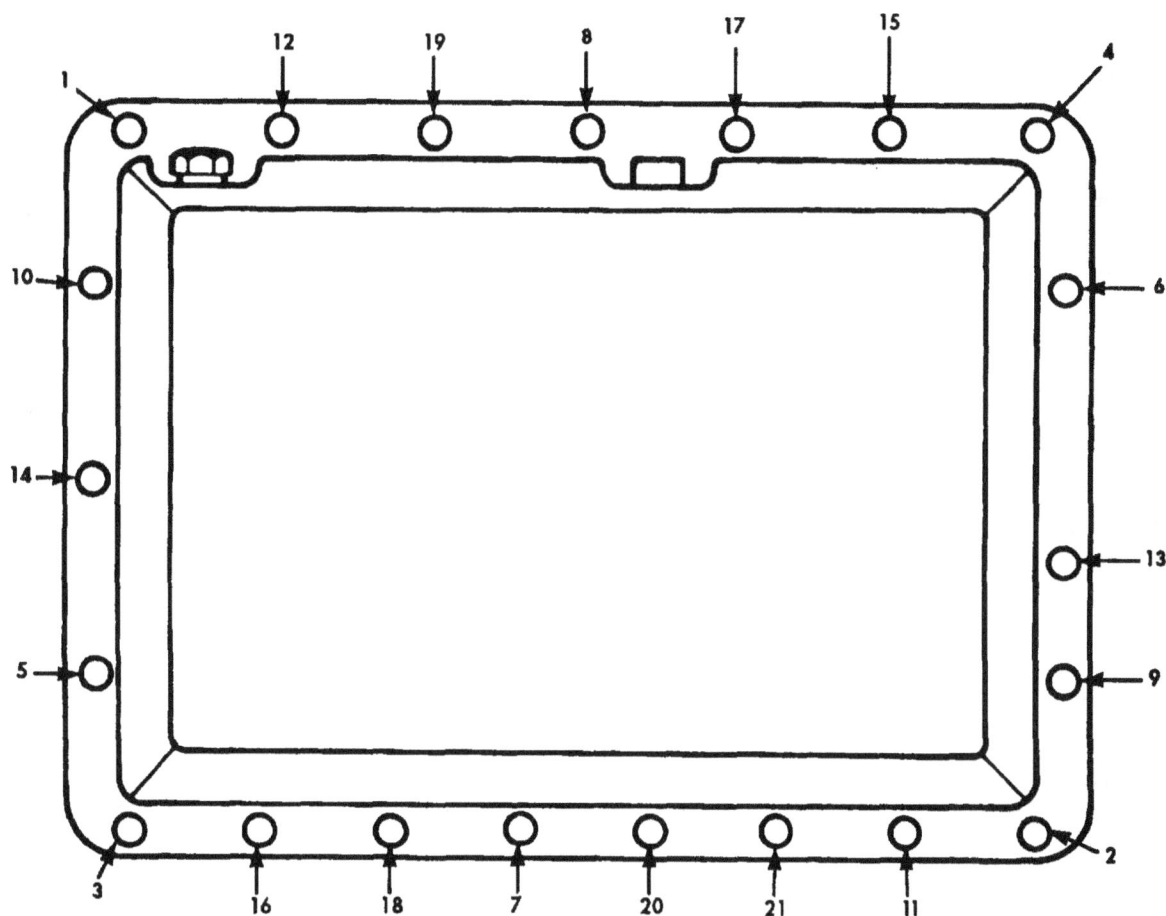

**OIL PAN SCREW
TORQUE SEQUENCE**

FOLLOW-ON TASK: Install transmission torque converter (para. 5-41).

5-43. MODULATE LOCKUP VALVE, LOW TRIMMER VALVE, AND LOW SHIFT VALVE REPLACEMENT

THIS TASK COVERS:

a. Removal   b. Installation

INITIAL SETUP:

APPLICABLE MODELS
All

TOOLS
General mechanic's tool kit (Appendix E, Item 1)
Torque wrench (Appendix E, Item 146)

REFERENCES (TM)
TM 9-2320-272-24P

EQUIPMENT CONDITION
Transmission oil pan and filter removed (para. 5-42).

SPECIAL ENVIRONMENTAL CONDITIONS
Work area clean and free from blowing dirt and dust.

a. Removal

1. Remove three screws (5) and modulated lockup valve (6) from oil transfer plate (4).
2. Remove six screws (1) and low trimmer valve (2) from low shift valve (3).
3. Remove two screws (7) and low shift valve (3) from oil transfer plate (4).

b. Installation

NOTE
Keep oil pan mounting surface in horizontal position and facing upward.

1. Install low shift valve (3) on oil transfer plate (4) with two screws (7). Finger-tighten screws (7).
2. Install low trimmer valve (2) on low shift valve (3) with six screws (1). Tighten screws (1) and (7) 9-11 lb-ft (12-15 N•m) in sequence shown.
3. Install modulated lockup valve (6) on oil transfer plate (4) with three screws (5). Tighten screws (5) 9-11 lb-ft (12-15 N•m) in sequence shown.

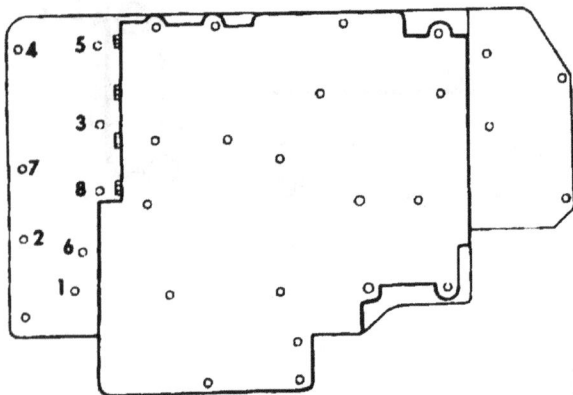

LOW TRIMMER VALVE INSTALLATION TORQUE SEQUENCE

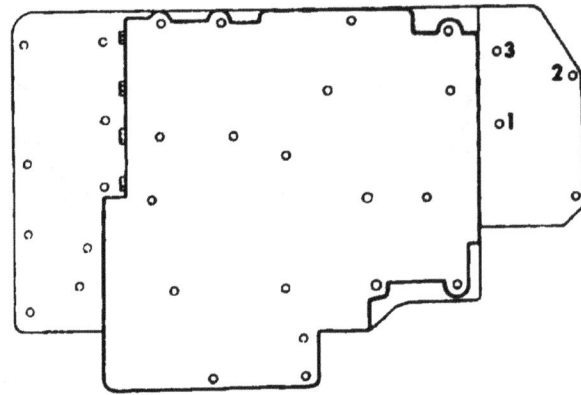

MODULATED LOCKUP VALVE INSTALLATION TORQUE SEQUENCE

5-43. MODULATE LOCKUP VALVE, LOW TRIMMER VALVE, AND LOW SHIFT VALVE REPLACEMENT (Contd)

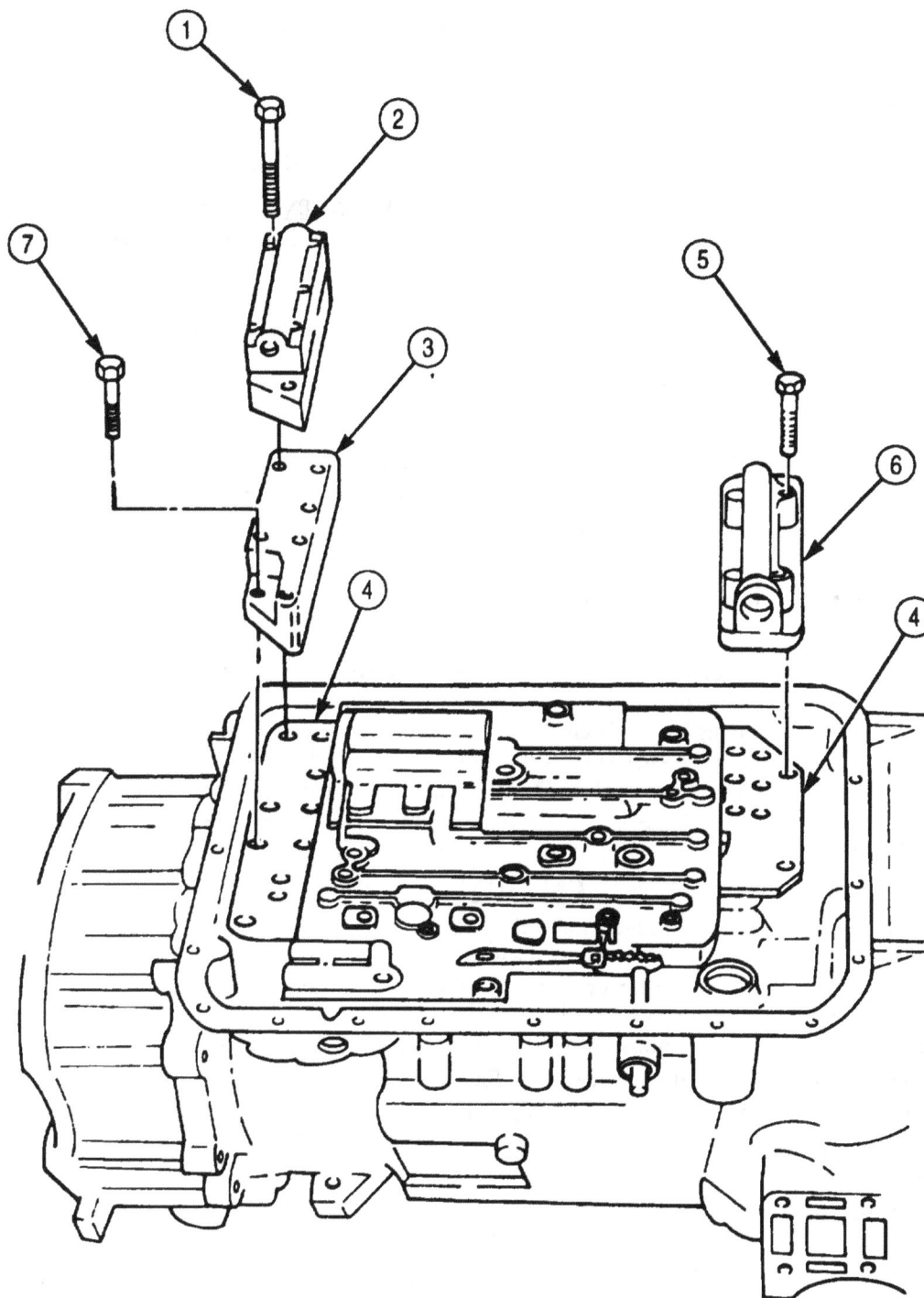

FOLLOW-ON TASK: Install transmission oil pan and filter (para. 5-42).

5-44. TRANSMISSION CONTROL VALVE REPLACEMENT

THIS TASK COVERS:

a. Removal   b. Installation

INITIAL SETUP:

APPLICABLE MODELS

All

TOOLS

General mechanic's tool kit (Appendix E, Item 1)
Torque wrench (Appendix E, Item 146)

REFERENCES (TM)

TM 9-2320-272-24P

EQUIPMENT CONDITION

Low shift valve, low trimmer valve, and modulate lockup valve removed (para. 5-43).

SPECIAL ENVIRONMENTAL CONDITIONS

Work area clean and free from blowing dirt and dust.

---

a. Removal

NOTE

Separate and tag screws for installation.

1. Remove screw (4) and detent spring and roller (5) from control valve (2).
2. Remove three screws (3) from control valve (2).
3. Remove screw (1) from control valve (2).
4. Remove fifteen screws (6) from control valve (2).

CAUTION

- Do not tilt control valve to allow selector valve to drop out.
  Selector valve may be damaged.

- Remove items as an assembly and do not separate. Loose parts
  inside will fall free, resulting in damage to control valve.

5. Remove control valve (2), separator plate (7), and oil transfer plate (8) from transmission (11).

b. Installation

CAUTION

Do not tilt control valve to allow selector valve to drop out.
Damage to selector valve may result.

1. Position oil transfer plate (8), separator plate (7), and control valve (2) on transmission (11) so pin on detent lever (10) aligns with slot on selector valve (9).
2. Install control valve (2) on transmission (11) with fifteen screws (6). Finger-tighten screws (6).
3. Install control valve (2) with three screws (3). Finger-tighten screws (3).
4. Install control valve (2) with screw (1). Finger-tighten screw (1).
5. Position detent spring and roller (5) on control valve (2) with roller in notch of detent lever (10), and install with screw (4).
6. Tighten twenty screws (1), (3), (4), and (6) 9-11 lb-ft (12-15 N•m) in sequence shown.

5-44. TRANSMISSION CONTROL VALVE REPLACEMENT (Contd)

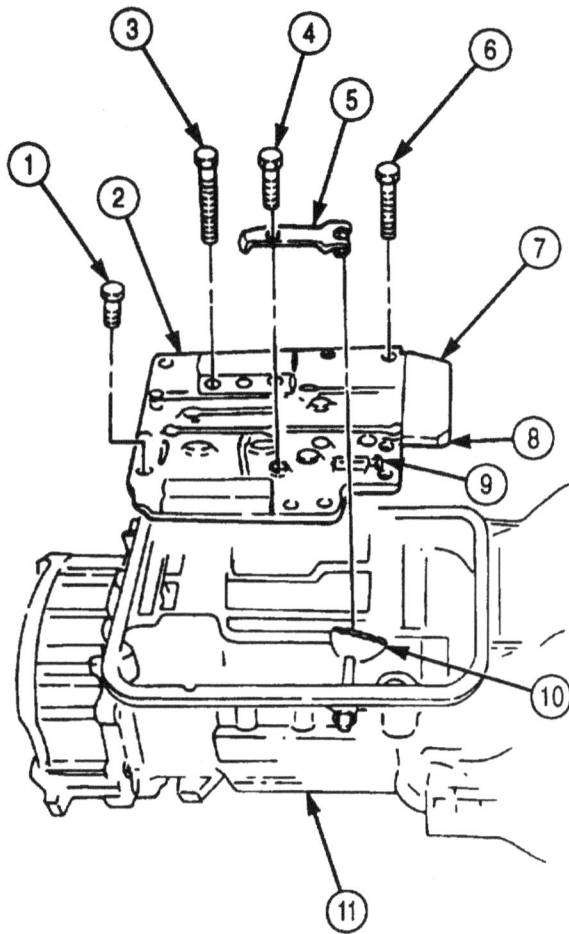

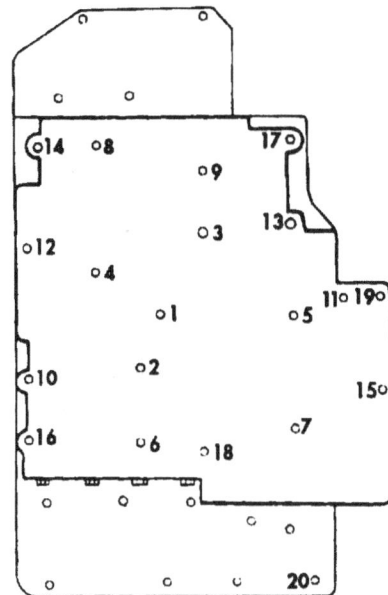

FOLLOW-ON TASK: Install low shift valve, low trimmer valve, and modulate lockup valve (para. 5-43).

5-45. TRANSMISSION MANUAL SELECTOR SHAFT REPLACEMENT

THIS TASK COVERS:

a. Removal

b. Installation

INITIAL SETUP:

APPLICABLE MODELS
All

TOOLS
General mechanic's tool kit (Appendix E, Item 1)
Torque wrench (Appendix E, Item 146)

REFERENCES (TM)
TM 9-2320-272-24P

EQUIPMENT CONDITION
Transmission selector shaft oil seal removed (para. 4-76).
Transmission control valve removed (para. 5-44).

SPECIAL ENVIRONMENTAL CONDITIONS
Work area clean and free from blowing dirt and dust.

a. Removal

1. Remove shaft retainer pin (3) from transmission housing (4).
2. Remove nut (2) and detent lever (1) from transmission manual selector shaft (5).
3. Remove transmission manual selector shaft (5) from transmission housing (4).

1. Tilt transmission housing (4) to horizontal position with the bottom facing upward.
2. Install manual selector shaft (5) and detent lever (1) on transmission housing (4) with nut (2). Tighten nut 15-20 lb-ft (20-27 N•m).
3. Secure manual selector shaft (5) in transmission housing (4) with shaft retainer pin (3).

5-45. TRANSMISSION MANUAL SELECTOR SHAFT REPLACEMENT (Contd)

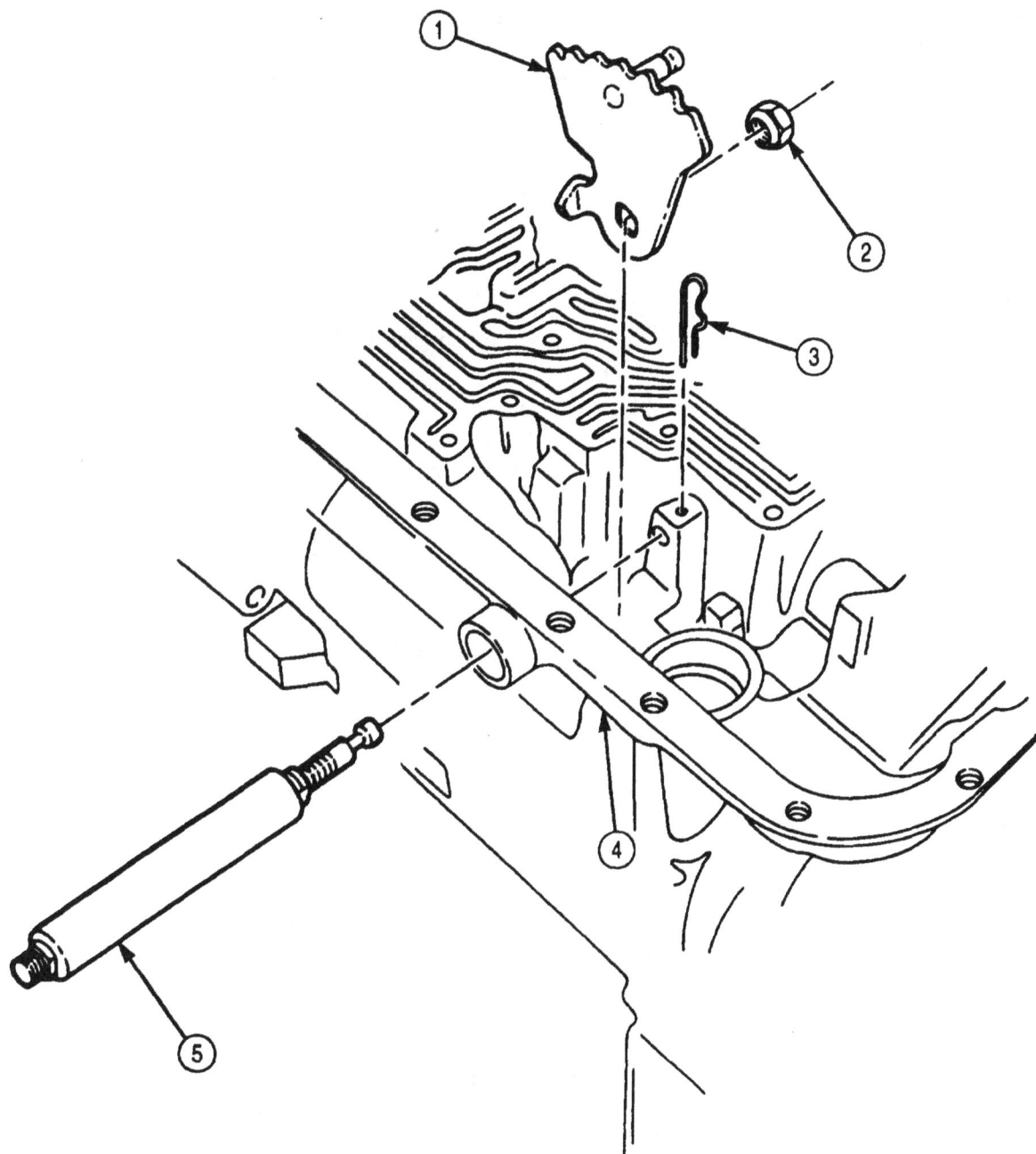

FOLLOW-ON TASK: • Install transmission selector shaft oil seal (para. 4-76).
• Install transmission control valve (para. 5-44).

5-46. TRANSMISSION OIL PUMP AND FRONT SUPPORT MAINTENANCE

THIS TASK COVERS:

a. Removal   d. Assembly
b. Disassembly   e. Installation
c. Cleaning and Inspection

INITIAL SETUP:

APPLICABLE MODELS
All

SPECIAL TOOLS
Two guide pins (Appendix E, Item 60)
Front support lifter (Appendix E, Item 48)
Bearing installer (Appendix E, Item 11)
Driver handle (Appendix E, Item 38)
Valve pin remover (Appendix E, Item 155)
Slide hammer (Appendix E, Item 122)
Main regulator and lockup spring compressor
   (Appendix E, Item 129)
Adapters (Appendix E, Item 2)
Centering band (Appendix E, Item 90)
Bushing replacer (Appendix E, Item 25)

TOOLS
General mechanic's tool kit (Appendix E, Item 1)
Depth micrometer (Appendix E, Item 81)
Torque wrench (Appendix E, Item 146)
Arbor press

MATERIALS/PARTS
Oil seal (Appendix D, Item 502)
Two self-locking screws (Appendix D, Item 649)
Twelve self-locking screws (Appendix D, Item 650)
Two hook-type seal rings (Appendix D, Item 638)
Twelve washers (Appendix D, Item 713)
Oil pump seal ring (Appendix D, Item 226)
Oil pump gasket (Appendix D, Item 227)
Valve guide pin (Appendix D, Item 705)
Lubricating oil (Appendix C, Item 47)
Oil-soluble grease (Appendix C, Item 53)
Sealing compound (Appendix C, Item 62)

REFERENCES (TM)
TM 9-2320-272-24P

EQUIPMENT CONDITION
Transmission manual selector shaft removed
(para. 5-45).

SPECIAL ENVIRONMENTAL CONDITIONS
Work area clean and free from blowing dirt and dust.

GENERAL SAFETY INSTRUCTIONS
Use care when removing snaprings and retaining rings.

---

a. Removal

1. Position transmission (5) front upward.
2. Remove twelve screws (2) and washers (1) connecting oil pump and front support (4) to transmission (5). Discard washers (1).
3. Attach front support lifter to converter ground sleeve (3) on oil pump and front support (4).

NOTE
Assistant will help with step 4.

4. Remove oil pump and front support (4) from transmission (5) by lifting straight up. Assistant may tap surface with a rubber hammer to ease removal.
5. Remove oil pump gasket (6). Discard oil pump gasket (6).

5-46. TRANSMISSION OIL PUMP AND FRONT SUPPORT MAINTENANCE (Contd)

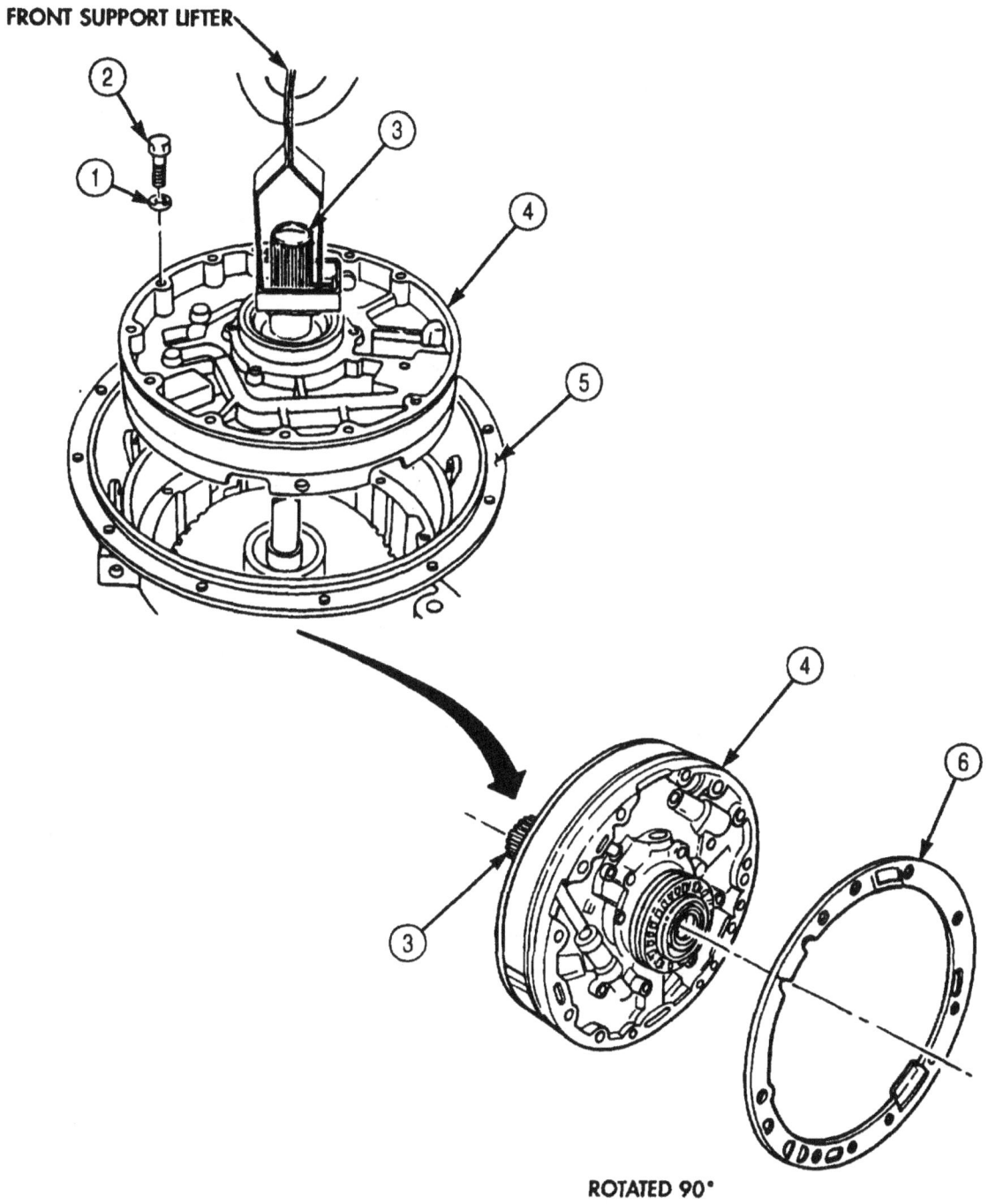

FRONT SUPPORT LIFTER

ROTATED 90°

5-46. TRANSMISSION OIL PUMP AND FRONT SUPPORT MAINTENANCE (Contd)

b. Disassembly

1. Remove two hook-type seal rings (8) from oil pump (19) and front support (5). Discard seal rings (8).
2. Remove bearing and race (7) from oil pump (19) and front support (5) or transmission.
3. Remove oil pump seal ring (1) from oil pump (19) and front support (5). Discard oil pump seal ring (1).
4. If replacement is necessary, remove needle bearing (6) from front support hub ground sleeve (9).
5. Install valve pin remover tool on slide hammer on oil pump (19) and front support (5).
6. Attach valve pin remover tool between coils of valve spring (3) and valve guide pin (2).

NOTE
Assistant will help with step 7.

7. Remove valve guide pin (2), valve spring (3), and converter pressure regulator valve (4). Discard valve guide pin (2). Tag valve spring (3) for installation.

WARNING
Regulator valve spring is under approximately 65 lb (29 kg) compression. Do not remove retaining snapring until compressor is in place, or regulator valve spring may fly out causing injury.

NOTE
For transmissions with serial numbers 21628 and higher, use two adapters with spring compressor.

8. Attach main regulator and lockup spring compressor to front support (5) and tighten two screws (23) to relieve spring compression.

NOTE
Perform steps 9 and 10 for transmissions with serial numbers 21628 and higher.

9. Remove twelve self-locking screws (25) and two self-locking screws (24) and separate oil pump (19) from front support (5). Mark location for installation. Discard self-locking screws (24) and (25).
10. Remove two cross pins (26) from front support (5).

NOTE
Perform step 11 for transmissions with serial numbers prior to 21628.

11. Remove snaprings (10) and (11) from oil pump (19) and front support (5).
12. Carefully remove and detach main regulator lockup spring compressor, valve stops (12) and (17), valve springs (13) and (16), regulator valve (15), and lockup valve (14) from front support (5). Mark locations for installation. Tag valve springs (13) and (16) for installation.

NOTE
Perform step 13 for transmissions with serial numbers prior to 21628.

13. Remove twelve self-locking screws (25) and two self-locking screws (24) connecting oil pump (19) to front support (5). Separate oil pump (19) from front support (5). Mark location for installation and discard self-locking screws (25) and (24).

5-46. TRANSMISSION OIL PUMP AND FRONT SUPPORT MAINTENANCE (Contd)

NOTE

Do not perform steps 14 and 15 unless relative movement is apparent between ground sleeve and front support.

14. If plug (22) is loose or damaged, remove plug (22) from front support (5).

15. Remove oil pump gears (20) and (21) and oil pump body oil seal (18) from oil pump (19). Discard body oil seal (18).

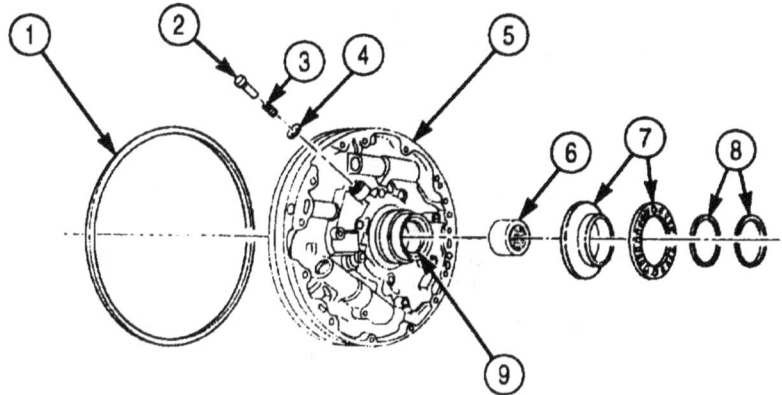

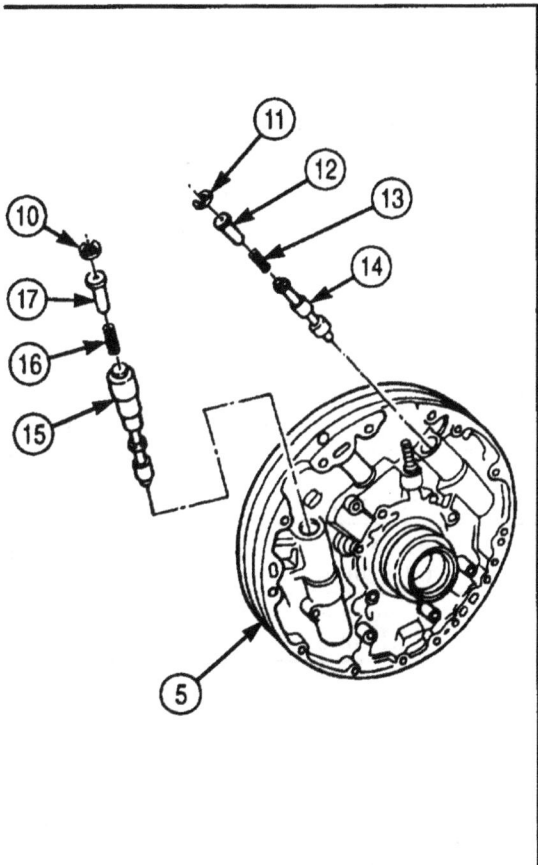

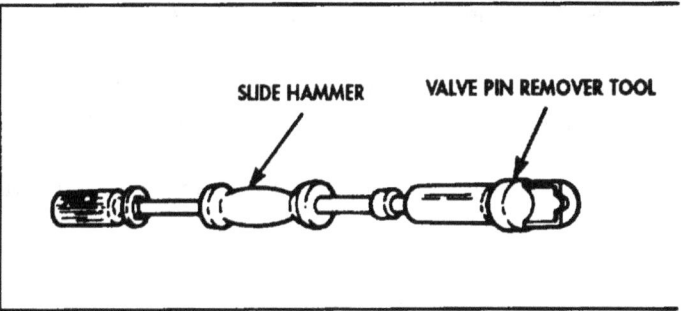

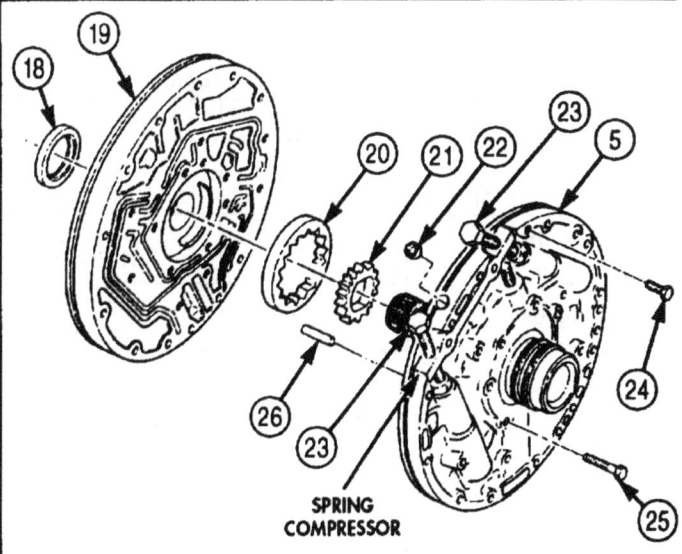

SLIDE HAMMER   VALVE PIN REMOVER TOOL

SPRING COMPRESSOR

5-46. TRANSMISSION OIL PUMP AND FRONT SUPPORT MAINTENANCE (Contd)

c. Cleaning and Inspection

CAUTION
Early model and late model converter pump hub, Front support seals, and oil pump are not interchangeable. Do not intermix parts from early and late transmissions. Damage to transmission will result.

1. For general cleaning instructions, refer to para. 2-14.
2. For general inspection instructions, refer to para. 2-15.
3. Inspect oil pump gears (3) and (4) for broken teeth. Discard gears if broken.
4. Install oil pump gears (3) and (4) in pump body (2).
5. Place straightedge (1) across face of oil pump (2).
6. Measure clearance between straightedge (1) and gear (4). If clearance is more than 0.0020 in. (0.051 mm), replace with thicker gear (4). Repeat steps 4 through 6.

NOTE
Perform step 7 for late model transmissions only. Early model transmissions do not have bushing.

7. Measure inside diameter of bushing (5) on oil pump (2) at 5 o'clock position when viewed from the front. Maximum wear limit is 2.257 in. (57.33 mm). Brass backing showing through is acceptable if wear limit is not exceeded.
8. Inspect bushing (5) on oil pump (2) for scoring. Replace bushing (5) if scoring can be felt. Use installer/remover tool and arbor press.
9. Inspect bushing (5) on oil pump (2) for discoloration due to overheating. Replace if discolored.
10. Inspect all springs for discoloration due to overheating. Discard if discolored.
11. Inspect all springs for broken coils or coils distorted due to wear. Discard if broken or distorted.
12. Using spring tester, inspect for serviceability by checking load when spring is compressed to the correct length. Discard spring if spring does not give the correct load (table 5-12).

Table 5-12. Spring Data.

SPRING	COLOR	FREE LENGTH	COMPRESSED LENGTH	UNDER LOAD
Main pressure regulator valve spring	Green	3.57 in. (90.7 mm)	2.01 in. (51.0 mm)	70.6-76.6 lb (314-341 N)
Converter pressure regulator valve spring	Blue	1.24 in. (31.5 mm)	1.05 in. (26.7 mm)	33.1-40.5 lb (147-180 N)
Lockup valve spring	Yellow	2.84 in. (72.1 mm)	1.46 in. (37.1 mm)	26.1-28.9 lb (116-129 N)

5-46. TRANSMISSION OIL PUMP AND FRONT SUPPORT MAINTENANCE (Contd)

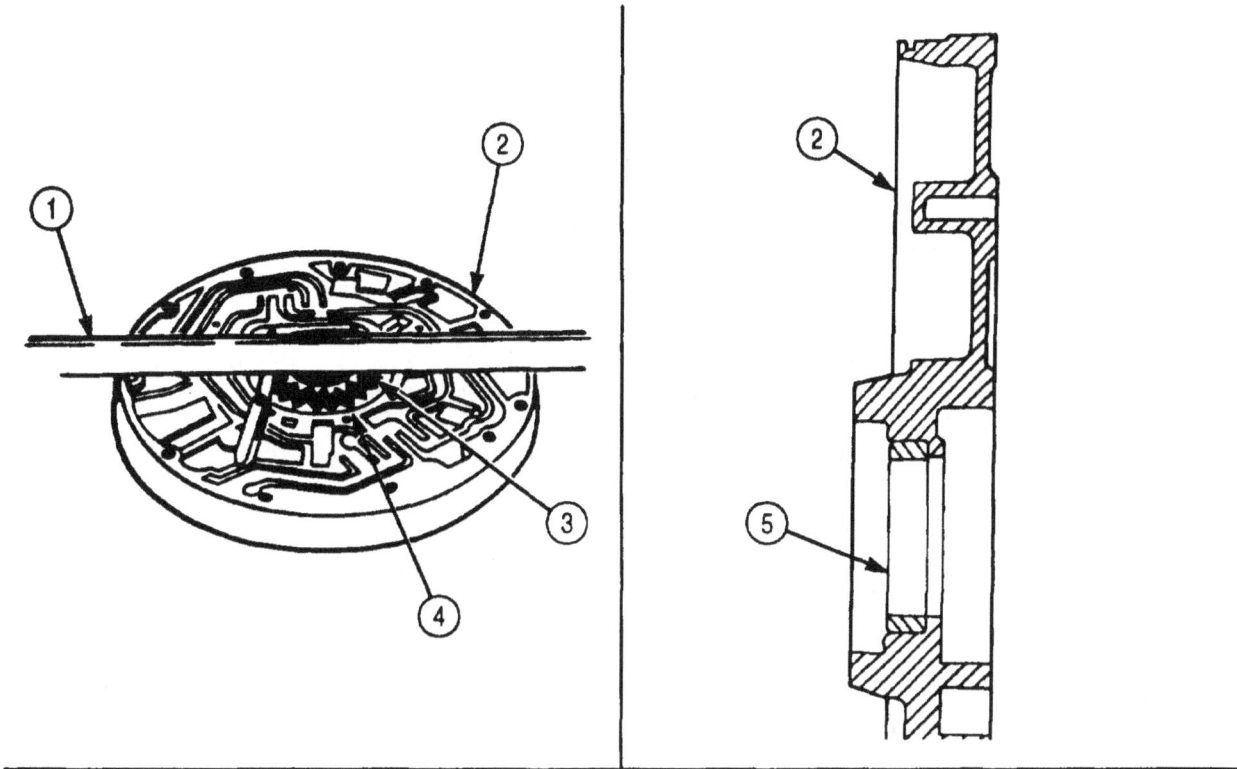

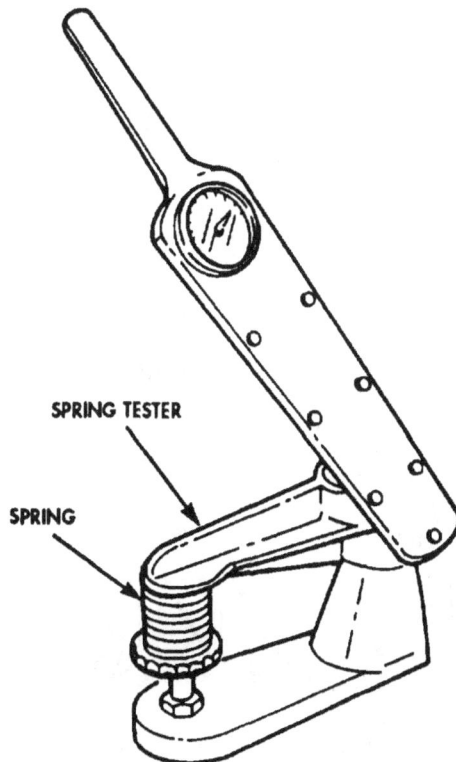

SPRING TESTER

SPRING

5-46. TRANSMISSION OIL PUMP AND FRONT SUPPORT MAINTENANCE (Contd)

d. Assembly

CAUTION

Oil seals for early and late model oil pumps are not
interchangeable (para. 5-39).

NOTE

- Coat all internal parts, except oil pump, with lubricating oil
  OE/HDO-10 prior to assembly.
- Perform steps 1 through 3 only if gears and oil seal were
  removed.

1. Using oil-soluble grease, coat outside diameter of new oil pump seal ring (1)

2. Install new oil pump seal ring (1) in oil pump body (2) 0.030-0.050 in. (1.76-1.27 mm) below outer
   edge of bore. Install spring-loaded lip first. Use depth micrometer for measurement.

3. Using oil-soluble grease, coat oil pump body seal (1) lip.

4. Install oil pump gears (3) and (4).

NOTE

For transmissions with serial numbers 21628 and higher, use two
adapters with spring compressor.

5. Install regulator valve (11), lockup valve (9), valve springs (12) and (8), and valve stops (13) and (7) in
   front support (10) at location marked during disassembly.

6. Attach main regulator and lockup spring compressor to front support (10) and tighten screws (14)
   until snapring grooves or crosspin holes are visible in front of valve stops.

NOTE

Perform step 7 for transmissions with serial numbers prior to
21628.

7. Install snaprings (5) and (6) over valve stops (13) and (7). Tighten screws (14) until snapring
   grooves are visible in front of valve stops (7) and (13).

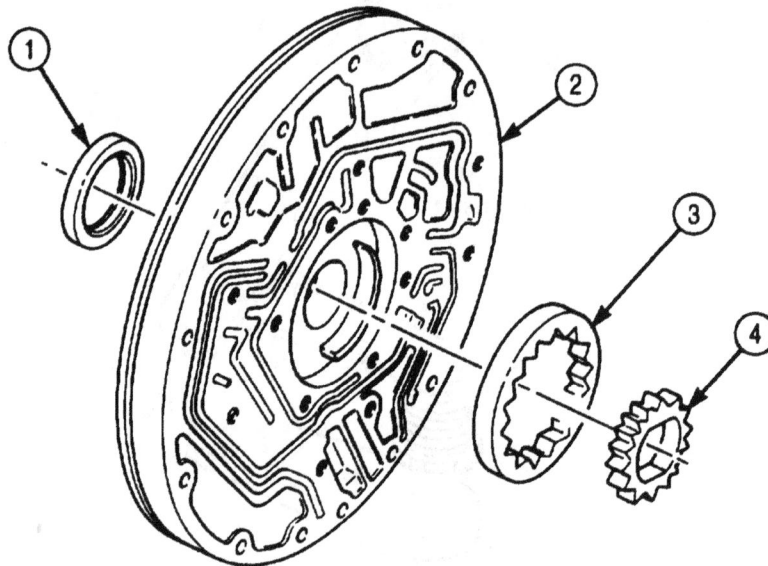

5-46. TRANSMISSION OIL PUMP AND FRONT SUPPORT MAINTENANCE (Contd)

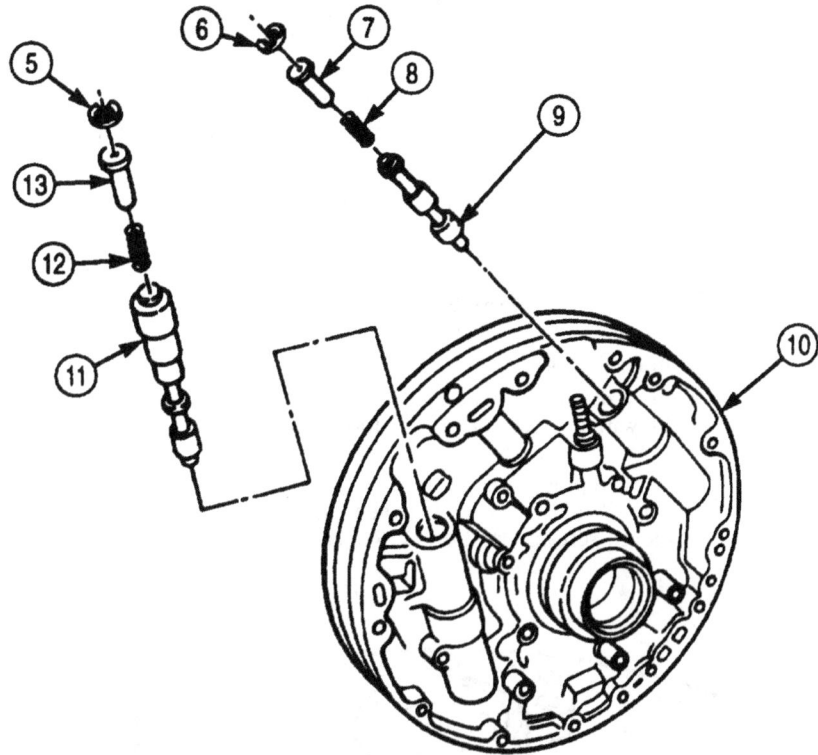

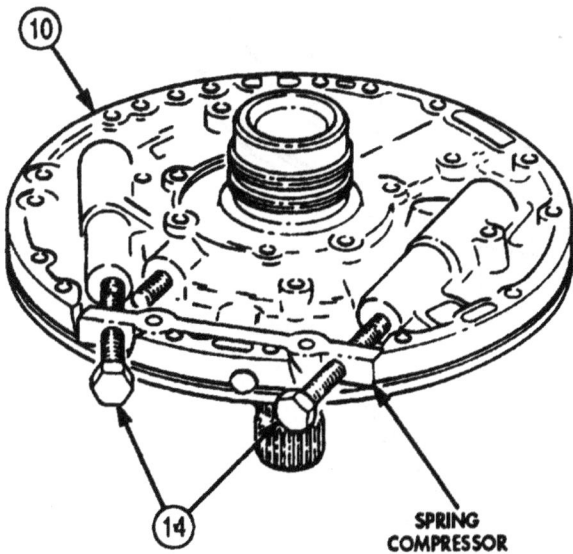

TRANSMISSION PRIOR TO
SERIAL NO. 21628

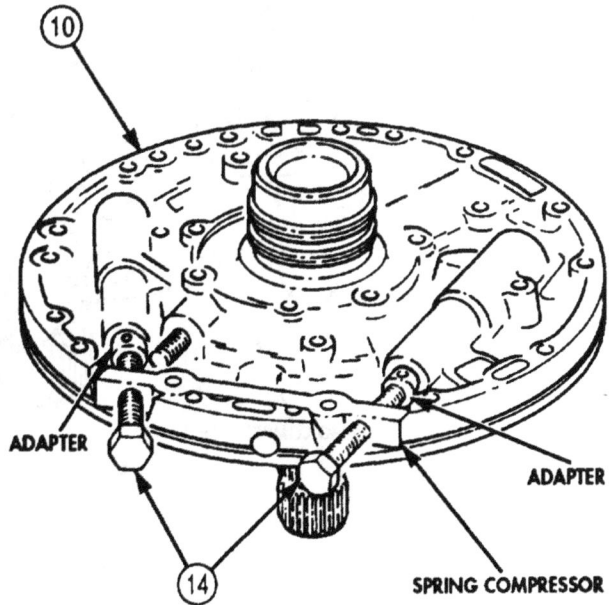

TRANSMISSION
SERIAL NOS. 21628 AND HIGHER

5-46. TRANSMISSION OIL PUMP AND FRONT SUPPORT MAINTENANCE (Contd)

NOTE

Perform step 8 for transmissions with serial numbers 21628 and higher.

8. Install two cross pins (2) in pin holes of front support (1).

9. Detach and remove main regulator and spring compressor from front support (1).

NOTE

Perform step 10 only if plug was previously removed.

10. Install plug (3) in front support (1).

11. Position oil pump (6) to front support (1) and align holes for screws (4) and (5). Install oil pump (6) to front support (1) with two new self-locking screws (4) positioned 180° apart. Finger-tighten screws (4).

12. Install centering band around oil pump (6).

13. Install oil pump (6) on front support (1) with remaining ten new self-locking screws (4) and two new self-locking screws (5). Tighten screws (4) 17-20 lb-ft (23-27 N•m). Tighten screws (5) 36-42 lb-ft (23-27 N•m).

14. Remove centering band from oil pump (6) to front support (1) and check separation line between oil pump (6) and front support (1). If mating point between oil pump (6) and front support (1) is not perfectly smooth by touch, loosen screws (4) and (5). Repeat steps 12 through 14.

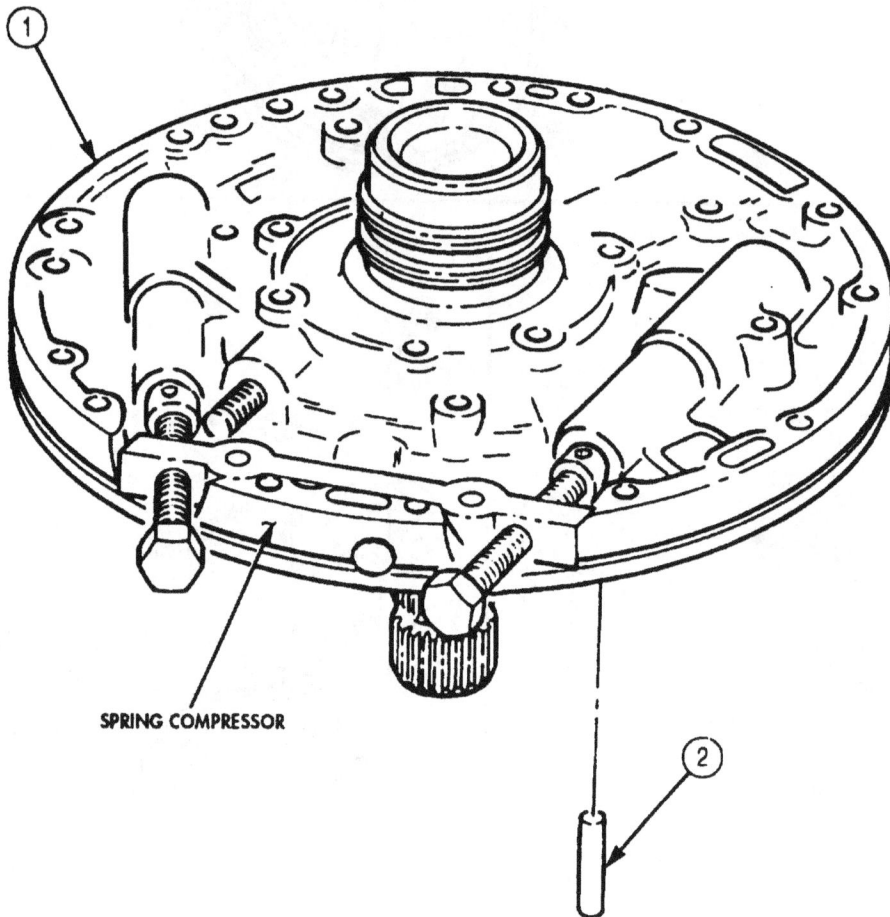

SPRING COMPRESSOR

5-46. TRANSMISSION OIL PUMP AND FRONT SUPPORT MAINTENANCE (Contd)

NOTE

Perform step 15 only if needle bearing was previously removed.

15. Position needle bearing (8) on converter ground sleeve (9). Ensure numbered end of needle bearing (8) faces outward.

16. Using bearing installer and driver handle, press needle bearing (8) into converter ground sleeve (9) 1.240-1.260 in. (31.50-32.00 mm) below outer edge of bore.

NOTE

Perform steps 17 and 18 only if converter pressure regulator valve was previously removed.

17. Place valve spring (11) and converter pressure valve (10) on new valve guide pin (7).

18. Press new valve guide pin (7) into front support (1) until new value guide pin (7) extends 1.16-1.20 in. (29.46-30.48 mm) above finished surface.

NOTE

Two seal rings for front support hub are installed when oil pump and front support are installed in transmission.

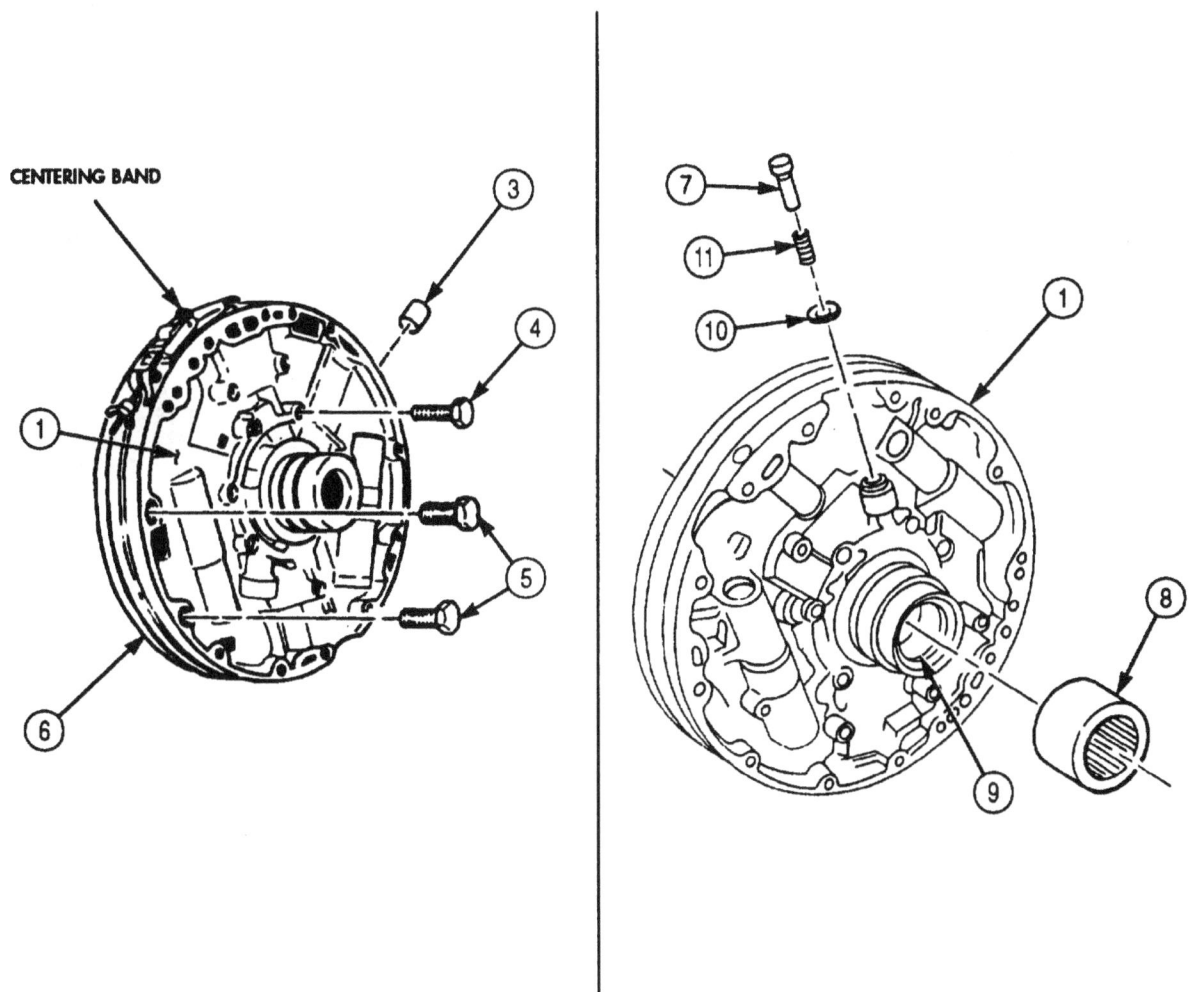

CENTERING BAND

5-46. TRANSMISSION OIL PUMP AND FRONT SUPPORT MAINTENANCE (Contd)

e. Installation

CAUTION

Identify transmission model before assembly of oil pump and front support. If early model converter hub was replaced with late model converter hub, a late model oil pump and front support must be installed, or transmission may be damaged during assembly (para. 5-39).

NOTE

- Late model oil pump is thicker than early model oil pump.
- Late model front support has a bushing and a thicker seal in oil pump hub. Early model oil pump hub has a thin seal and the torque converter hub has a roller bearing.

1. Install bearing and race (2) on turbine shaft (3).

NOTE

Guide pins maintain gasket alignment.

2. Install two guide pins (5) and new oil pump gasket (1) into transmission (4).

3. Apply oil-soluble grease sparingly to two new seal rings (10) to hold in place, and install seal rings (10) on hub (9) of oil pump (7) and front support (8).

4. Apply oil-soluble grease to new oil pump and ring coil, and install new oil pump seal ring (6) on oil pump (7) and front support (8).

5. Attach front support lifter to converter ground sleeve (12).

6. Align front support holes (13) with corresponding holes in transmission (4), and carefully install front support (8) in transmission (4) over guide pins (5).

7. Remove two guide pins (5) and front support lifter from transmission (4).

8. Install front support (8) on transmission (4) with twelve new rubber-covered washers (14) and screws (11). Tighten screws (11) as follows:

   a. Tighten alternately 180° apart. Tighten screws (11) 15 lb-ft (20 N•m).

   b. Repeat tightening sequence 180° apart to achieve final torque. Tighten screws (11) 24-32 lb-ft (33-43 N•m).

5-46. TRANSMISSION OIL PUMP AND FRONT SUPPORT MAINTENANCE (Contd)

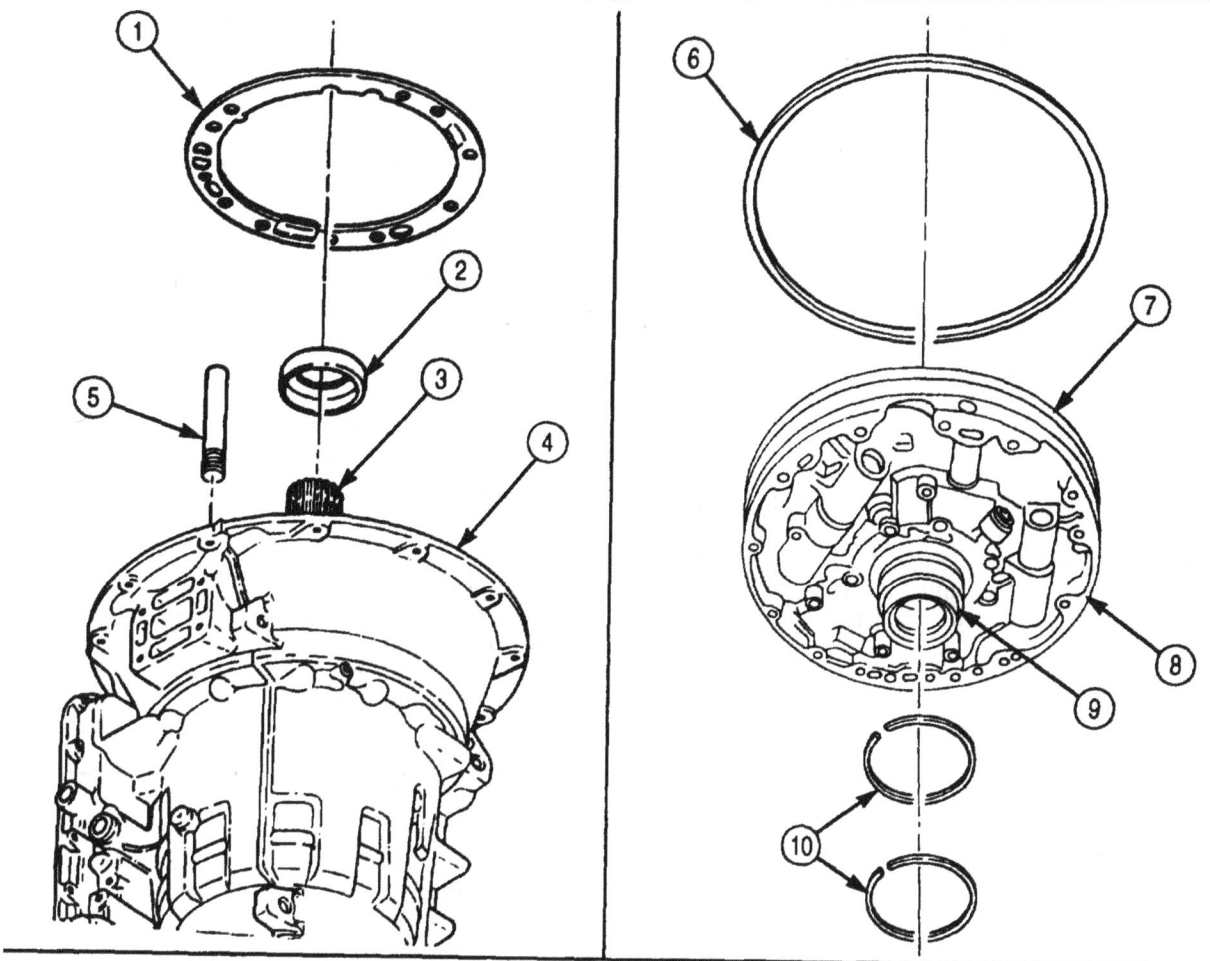

FRONT SUPPORT LIFTER

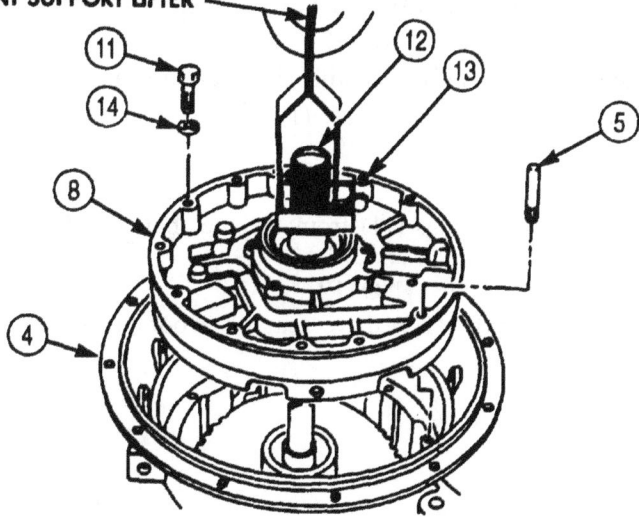

FOLLOW-ON TASK: Install transmission manual selector shaft (para. 5-45).

5-47. FORWARD CLUTCH MAINTENANCE

THIS TASK COVERS:

a. Removal   d. Assembly
b. Disassembly   e. Installation
c. Cleaning and Inspection

INITIAL SETUP:

APPLICABLE MODELS

All

SPECIAL TOOLS

Spring compressor (Appendix E, Item 95)
Forward clutch clearance gauge (Appendix E, Item 46)

TOOLS

General mechanic's tool kit (Appendix E, Item 1)
Spring tester (Appendix E, Item 131)
Outside micrometer (Appendix E, Item 80)
Vernier caliper (Appendix E, Item 159)
Depth gauge (Appendix E, Item 81)
Arbor press

MATERIALS/PARTS

Two turbine shaft seal rings (Appendix D, Item 640)
Turbine shaft seal ring (Appendix D, Item 639)
Piston inner seal ring (Appendix D, Item 515)
Piston outer seal ring (Appendix D, Item 516)
Oil-soluble grease (Appendix C, Item 53)
Lubricating oil (Appendix C, Item 47)

REFERENCES (TM)

TM 9-2320-272-24P

EQUIPMENT CONDITION

Transmission oil pump and front support removed (para. 5-46).

SPECIAL ENVIRONMENTAL CONDITIONS

Work area clean and free from blowing dirt and dust.

---

a. Removal

1. Lift forward clutch housing (9) and turbine shaft (1) straight out from transmission (5).

NOTE

Perform step 2 for transmissions with three-piece bearing.

2. Remove bearing race (2), roller bearing (3), and bearing race (4) from turbine shaft (1).

NOTE

Perform step 3 for transmissions with single-piece bearing assembly.

3. Remove bearing assembly (6) from turbine shaft (1).

b. Disassembly

1. Remove seal ring (10) and two seal rings (11) from turbine shaft (1). Discard seal ring (10) and two seal rings (11).

NOTE

Support Power Takeoff (PTO) drive gear on arbor press for steps 2 through 6.

5-47. FORWARD CLUTCH MAINTENANCE (Contd)

2. Locate opening (missing spline) in forward clutch housing (9) nearest snapring (12) gap.

3. Press snapring (12) into groove in clutch housing (9) and insert one piece of shim stock (8) through opening between snapring (12), Power Takeoff (PTO) gear (7), and teeth of clutch housing (9).

4. Repeat step 3, inserting remaining pieces of shim stock (8) into PTO gear approximately 5 in. (127 mm) apart.

5. Lift PTO gear (7) off clutch housing (9).

6. Remove snapring (12) and shim stock (8) from clutch housing (9).

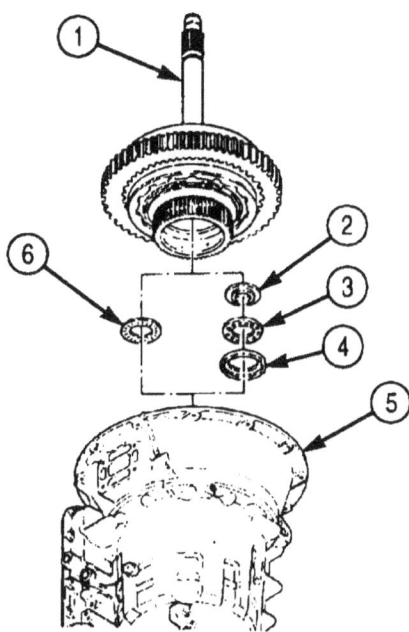

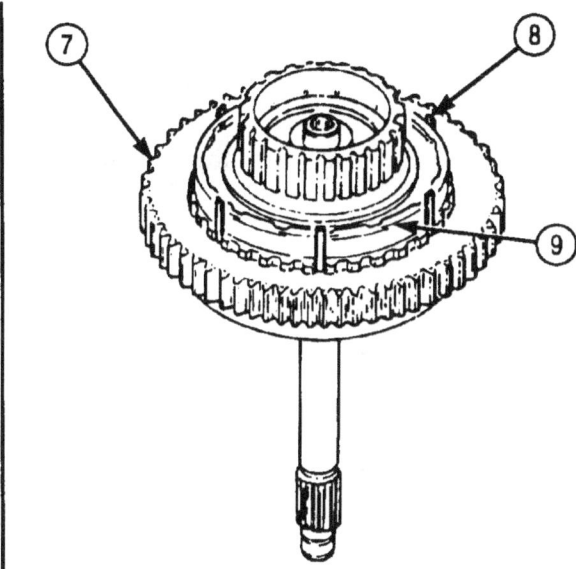

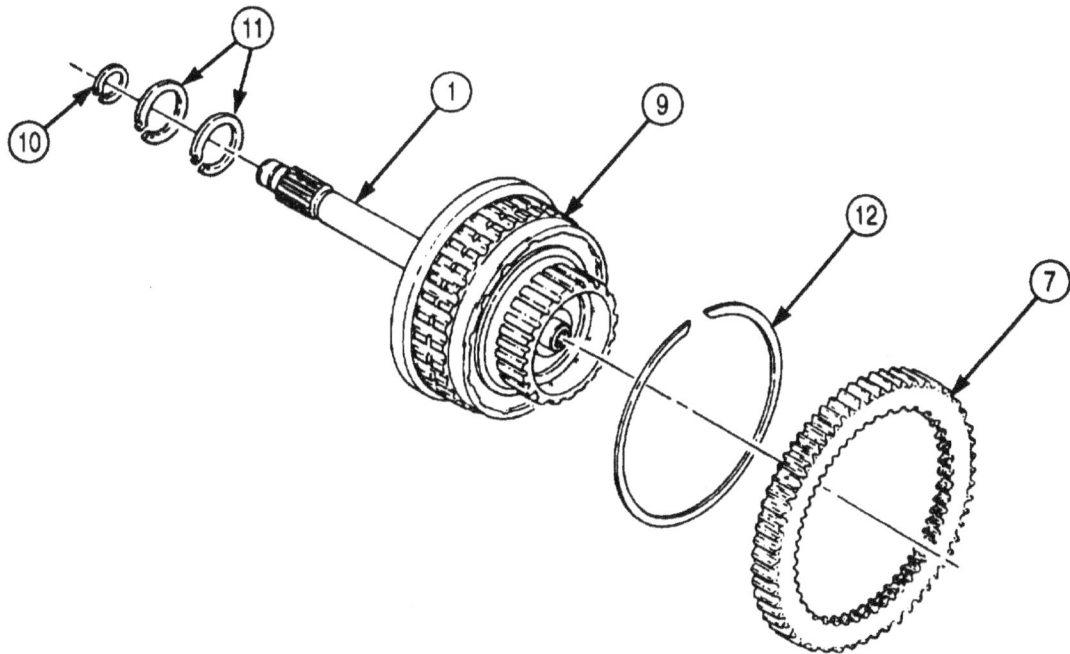

5-47. FORWARD CLUTCH MAINTENANCE (Contd)

7. Remove snapring (6) from forward clutch housing (16).

8. Remove fourth clutch drive hub (5) and forward clutch hub (4) from forward clutch housing (16).

NOTE

Perform step 9 for transmission with three-piece bearing.

9. Remove bearing race (7), bearing (8), and bearing race (9) from forward clutch housing (16) or fourth clutch hub (5).

NOTE

Perform step 10 for transmissions with single-piece bearing assembly.

10. Remove bearing assembly (3) from forward clutch housing (16).

CAUTION

Keep all parts together. Intermixing forward clutch parts with any other clutch pack will cause transmission damage.

NOTE

Tag all clutch parts for reassembly.

11. Remove six clutch plates (1) and six clutch discs (2) from forward clutch housing (16).

NOTE

Use arbor press with compression tool for step 12.

12. Position compressor tool on spring retainer (11) and place on arbor press.

13. Apply pressure on spring retainer (11) and remove snapring (10) from forward clutch housing (16).

14. Remove compressor tool from spring retainer (11) and arbor press.

15. Remove spring retainer (11), piston return spring (12), piston (13), and piston inner seal ring (15) from forward clutch housing (16). Discard piston inner seal ring (15).

16. Remove piston outer seal ring (14) from piston (13). Discard piston outer seal ring (14).

17. Remove retainer pin (20), valve plug (19), valve spring (18), and centrifugal valve (17) from forward clutch housing (16).

5-47. FORWARD CLUTCH MAINTENANCE (Contd)

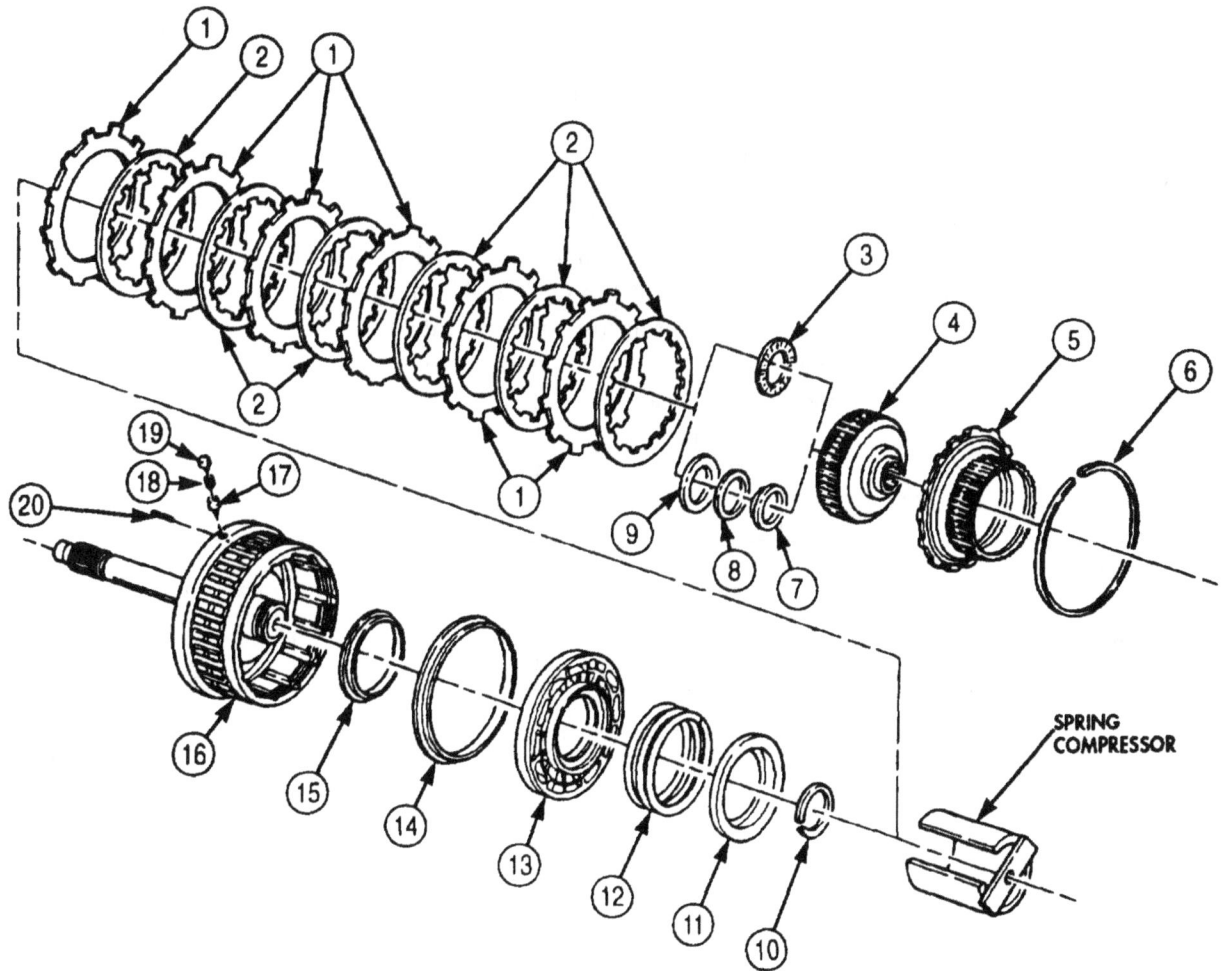

SPRING
COMPRESSOR

5-47. FORWARD CLUTCH MAINTENANCE (Contd)

c. Cleaning and Inspection

1. For general cleaning instructions, refer to para. 2-14.
2. For general inspection instructions, refer to para. 2-15.
3. Inspect turbine shaft and forward clutch housing (1) for scoring in area where seals contact turbine shaft (1). If scored, discard turbine shaft and forward clutch housing (1).
4. Inspect clutch plates (2) for burned surfaces. If burned, discard clutch plates (2).
5. Inspect clutch discs (3) for burned surfaces. If burned, discard clutch discs (3).
6. Measure thickness of fourth clutch drive hub (4) at clutch disc (3) contact area. If thickness is less than 0.248 in. (6.30 mm), discard fourth clutch drive hub (4).
7. Inspect valve spring (5) and piston return spring (6) for discoloration due to overheating. If discolored, discard.
8. Inspect valve spring (5) and piston return spring (6) for broken coils or coils distorted due to wear. If broken or distorted, discard.
9. Using spring tester, inspect valve spring (5) for serviceability.
   a. Compress valve spring (5) to 0.61 in. (15.5 mm) and check load (table 5-13).
   b. If valve spring (5) does not give load of 5.7-6.3 lb (25.3-28 N), discard.
10. Using spring tester, inspect piston return spring (6) for serviceability.
   a. Compress piston return spring (6) to 1.28 in. (32.5 mm) (table 5-13).
   b. If piston return spring (6) does not give load of 158-178 lb (707-791 N), discard.

Table 5-13. Spring Data.

SPRING	COLOR	FREE LENGTH	COMPRESSED LENGTH	UNDER LOAD
Valve spring	Green	0.76 in. (19.3 mm)	0.61 in. (15.5 mm)	5.7-6.3 lb (25.3-28 N)
Piston return spring	None	3.22 in. (81.8 mm)	1.28 in. (32.5 mm)	158-178 lb (702-791 N)

5-47. FORWARD CLUTCH MAINTENANCE (Contd)

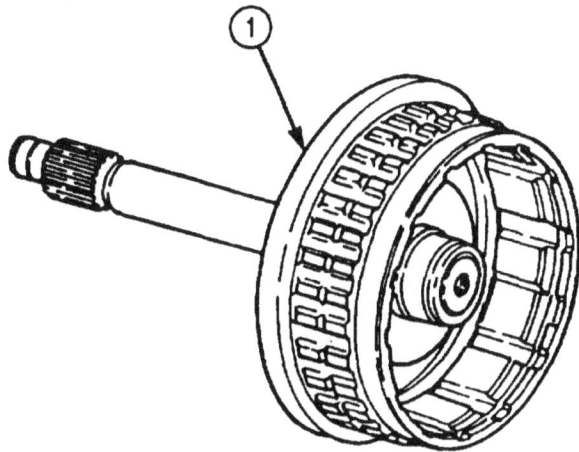

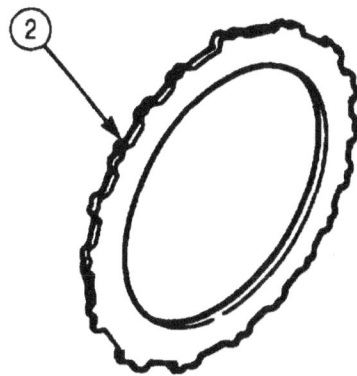

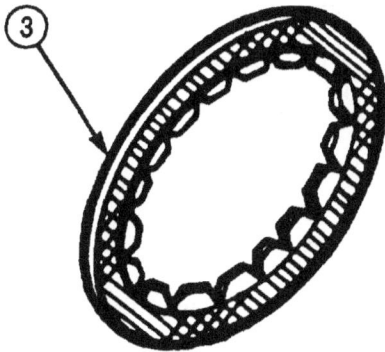

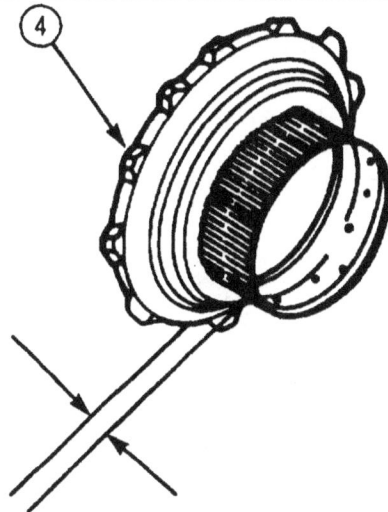

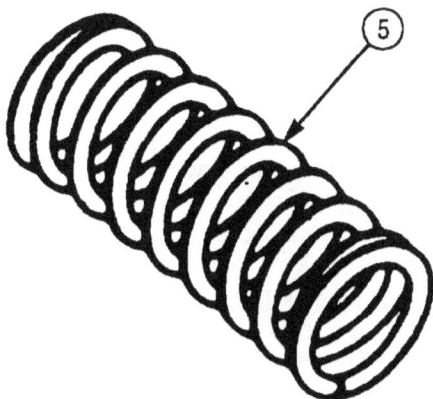

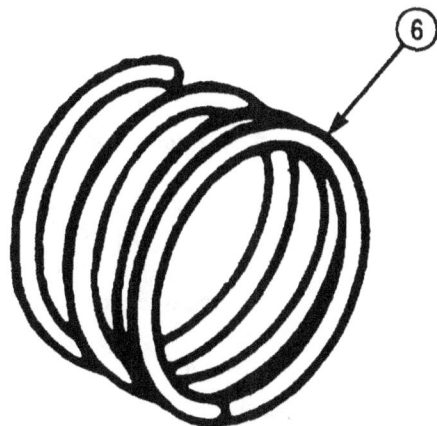

5-47. FORWARD CLUTCH MAINTENANCE (Contd)

d. Assembly

NOTE

If new centrifugal valve is being installed, all components must have same color code as old valve.

1. Install centrifugal valve (3) and valve spring (2) in forward clutch housing (16). Ensure pointed end of centrifugal valve (3) is installed first.

2. Push valve plug (1) into forward clutch housing (16) and install with retainer pin (17). Retainer pin (17) must protrude 0.080-0.100 in. (2.03-2.54 mm) above front surface of forward clutch housing (16).

3. Apply oil-soluble grease to new piston outer seal ring (12) and install new piston outer seal ring (12) on piston (10). Ensure piston outer seal ring (12) lips face toward oil pressure side of piston (10).

4. Apply oil-soluble grease to new piston inner seal ring (15) and install new piston inner seal ring (15) on forward clutch housing (16) hub. Ensure piston inner seal ring (15) lips face toward oil pressure side of piston (10).

5. Install piston (10) in forward clutch housing (16).

NOTE

Steps 6 through 12 obtain clutch running clearance.

6. Starting with clutch plate (5), alternately install six clutch plates (5) and six clutch discs (4) in forward clutch housing (16).

7. Position fourth clutch drive hub (6) in forward clutch housing (16) and install with snapring (7).

8. Measure clearance between fourth clutch drive hub (6) and clutch disc (4) with forward clutch clearance gauge.

9. Clutch running clearance should be 0.094-0.148 in. (2.387-3.759 mm).

NOTE

Perform steps 10 through 12 only if clutch running clearance is incorrect.

10. If clearance exceeds 0.148 in. (3.759 mm), replace clutch plates (5) and clutch discs (4) with new, thicker clutch plates (5) and clutch discs (4).

11. If clearance still exceeds 0.148 in. (3.759 mm) after completing step 10, replace piston (10) with new, thicker piston (10).

12. If clearance is less than 0.094 in. (2.387 mm), replace piston (10) with a new, thinner piston (10).

13. Remove snapring (7), fourth clutch drive hub (6), six clutch discs (4), and six clutch plates (5) from forward clutch housing (16).

14. Place piston return spring (14) and spring retainer (13) in forward clutch housing (16).

15. Place compressor tool on spring retainer (13) and arbor press.

16. Compress piston return spring (14) with compressor tool and arbor press and install spring retainer (13) with snapring (11).

17. Remove compressor tool from spring retainer (13).

NOTE

- Use oil-soluble grease to hold bearing race and bearing in place.
- Soak six clutch discs in clean OE/HDO-10 lubricating transmission oil for at least two minutes before installing in forward clutch housing.
- Perform step 18 for transmissions with single-piece bearing assembly.

18. Install bearing assembly (9) in forward clutch housing (16). Ensure lube scallops of bearing assembly (9) face down.

5-47. FORWARD CLUTCH MAINTENANCE (Contd)

NOTE

Perform steps 19 through 21 for transmission with three-piece bearing.

19. Install bearing race (8) in forward clutch housing (16). Ensure outer lip of bearing race (8) is installed first.

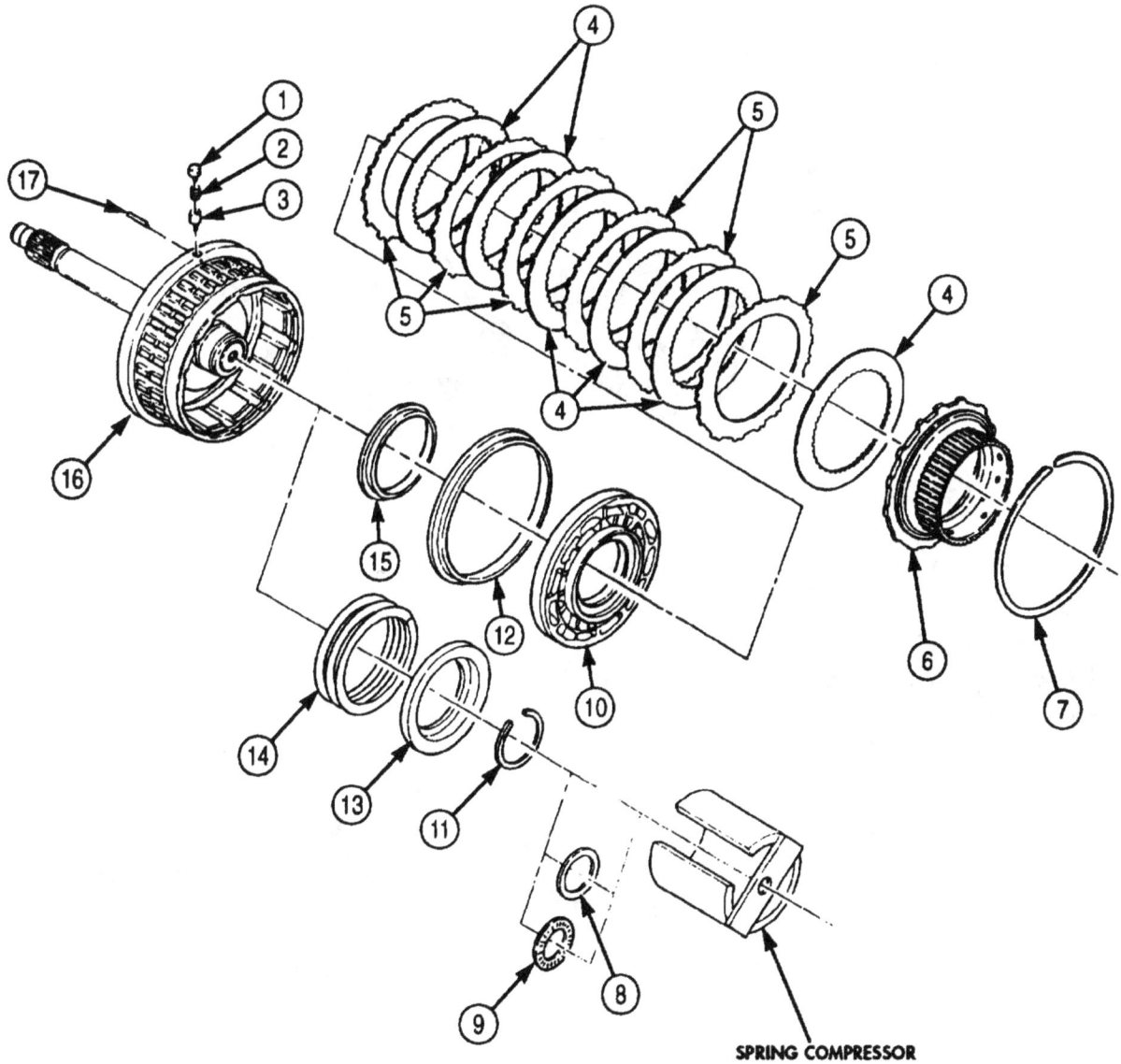

SPRING COMPRESSOR

5-47. FORWARD CLUTCH MAINTENANCE (Contd)

20. Install bearing race (3) on forward clutch hub (4). Ensure flat side of bearing race (3) is installed first.
21. Apply liberal amount of oil-soluble grease on bearing (2) and install bearing (2) on forward clutch hub (4).
22. Install forward clutch hub (4) in forward clutch housing (1). Ensure splined edge is installed first.
23. Starting with clutch plate (5), alternately install six clutch plates (5) and six clutch discs (6) in forward clutch housing (1).
24. Install fourth clutch drive hub (7) in forward clutch housing (1) with snapring (8).

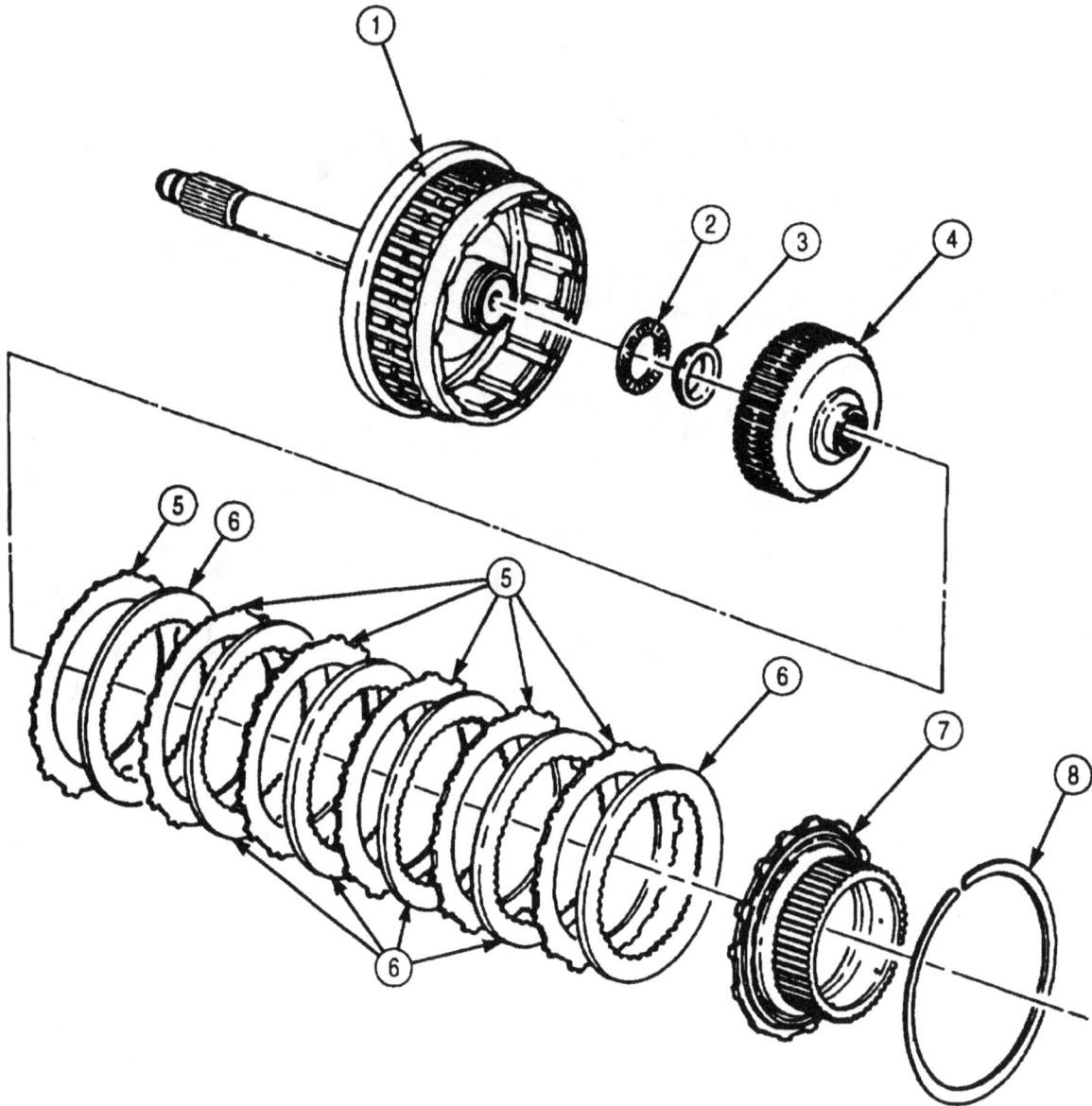

5-47. FORWARD CLUTCH MAINTENANCE (Contd)

25. Install PTO gear snapring (12) on forward clutch housing (1).

26. Slide PTO gear (13), chamfered end first, over PTO gear snapring (12) on forward clutch housing (1). Ensure PTO gear snapring (12) springs outward into internal groove of PTO gear (13).

NOTE
Perform step 27 for transmissions with single-piece bearing assembly.

27. Install roller bearing (17) on forward clutch hub (4).

NOTE
- Perform steps 28 and 29 for transmission with three-piece bearing.
- Use oil-soluble grease to hold bearing race, bearing, and seal rings in place.

28. Install bearing race (14), flat side first, on forward clutch hub (4).

29. Install bearing (15) and bearing race (16) on forward clutch hub (4).

30. Install two new seal rings (10) on turbine shaft (11).

31. Install new seal ring (9) on turbine shaft (11).

e. Installation

Install forward clutch housing (1) and turbine shaft (11) in transmission housing while engaging fourth clutch drive hub (7) with internal-splined clutch discs of fourth clutch. Ensure the fourth clutch housing (1) is properly seated.

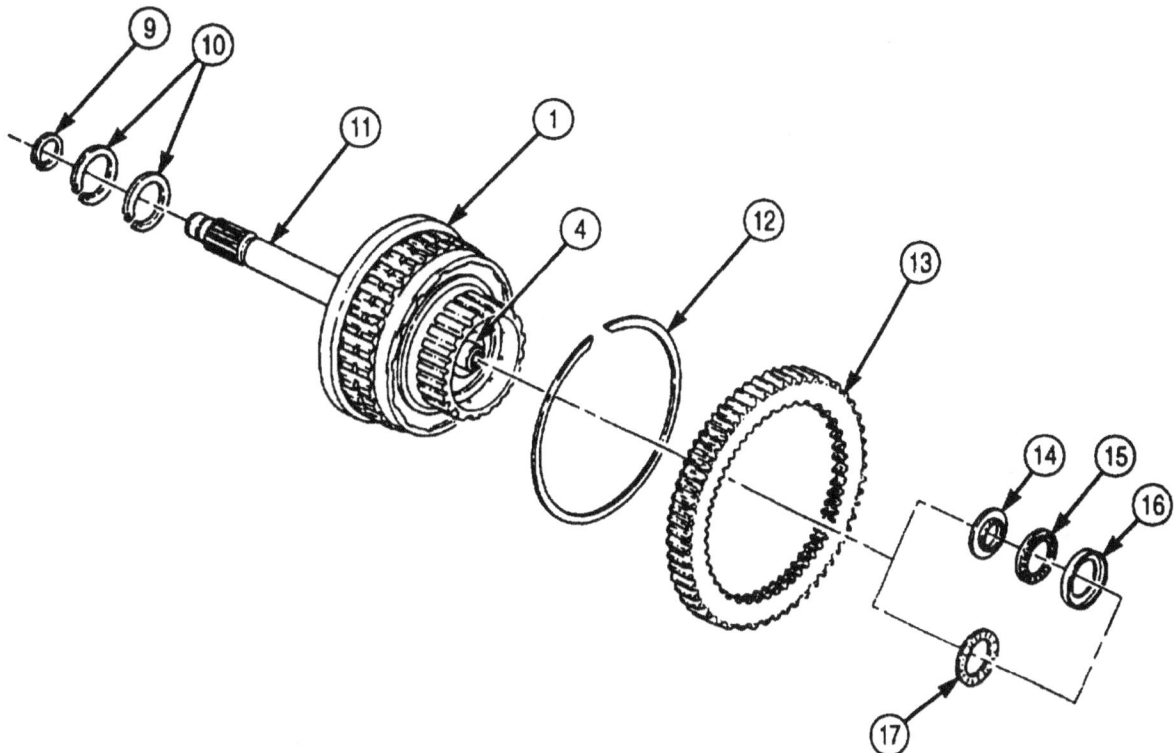

FOLLOW-ON TASK: Install transmission oil pump and front support (para. 5-46).

5-48. FOURTH CLUTCH MAINTENANCE

THIS TASK COVERS:

a. Removal
b. Disassembly
c. Cleaning and Inspection

d. Assembly
e. Installation

INITIAL SETUP:

APPLICABLE MODELS

All

SPECIAL TOOLS

Spring compressor (Appendix E, Item 95)
Fourth clutch clearance gauge (Appendix E,
Item 47)

TOOLS

General mechanic's tool kit (Appendix E, Item 1)
Vernier caliper (Appendix E, Item 159)
Depth gauge (Appendix E, Item 81)
Spring tester (Appendix E, Item 131)
Arbor press

MATERIALS/PARTS

Piston inner seal ring (Appendix D, Item 515)
Piston outer seal ring (Appendix D, Item 516)
Lubricating oil (Appendix C, Item 47)
Oil-soluble grease (Appendix C, Item 53)

REFERENCES (TM)

TM 9-2320-272-24P

EQUIPMENT CONDITION

Forward clutch removed (para. 5-47).

SPECIAL ENVIRONMENTAL CONDITIONS

Work area clean and free from blowing dirt and
dust.

GENERAL SAFETY INSTRUCTIONS

Use care when removing piston return spring.
Spring is under great compression.

---

a. Removal

1. Lift fourth clutch (1) straight out of transmission (4).

NOTE

Perform step 2 for transmissions with two-piece bearing.

2. Remove bearing race (2) and bearing assembly (3) from fourth clutch (1) or transmission (4).

NOTE

Perform step 3 for transmissions with single-piece bearing
assembly.

3. Remove bearing assembly (3) from fourth clutch (1) or transmission (4).

b. Disassembly

1. Position piston return spring compressor on spring retainer (5) and place on arbor press.

WARNING

Piston return spring is under great compression. Do not remove
snapring until pressure is applied to spring retainer. If not, piston
return spring may fly out, causing injury to personnel.

2. Apply pressure to spring retainer (5) and remove snapring (6) from fourth clutch (1).

3. Carefully remove piston return spring compressor.

4. Remove spring retainer (5), piston return spring (7), and optional spacer (8) from fourth clutch
housing (1).

5-48. FOURTH CLUTCH MAINTENANCE (Contd)

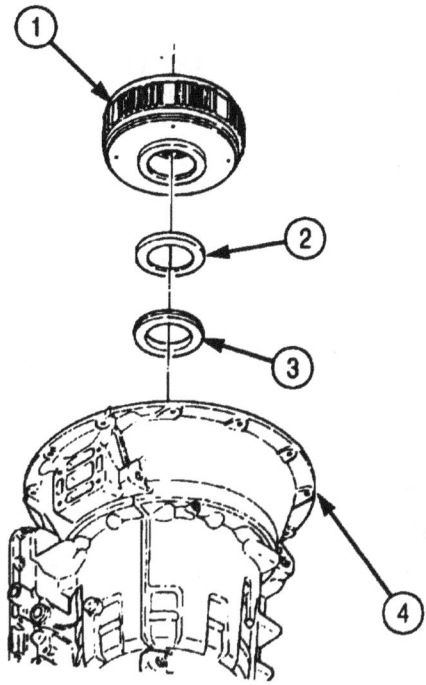

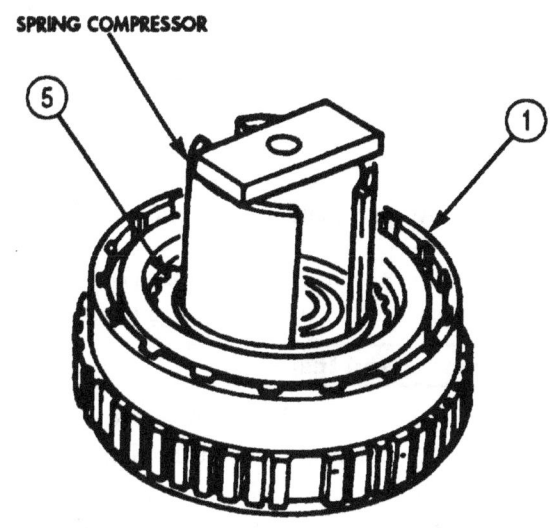

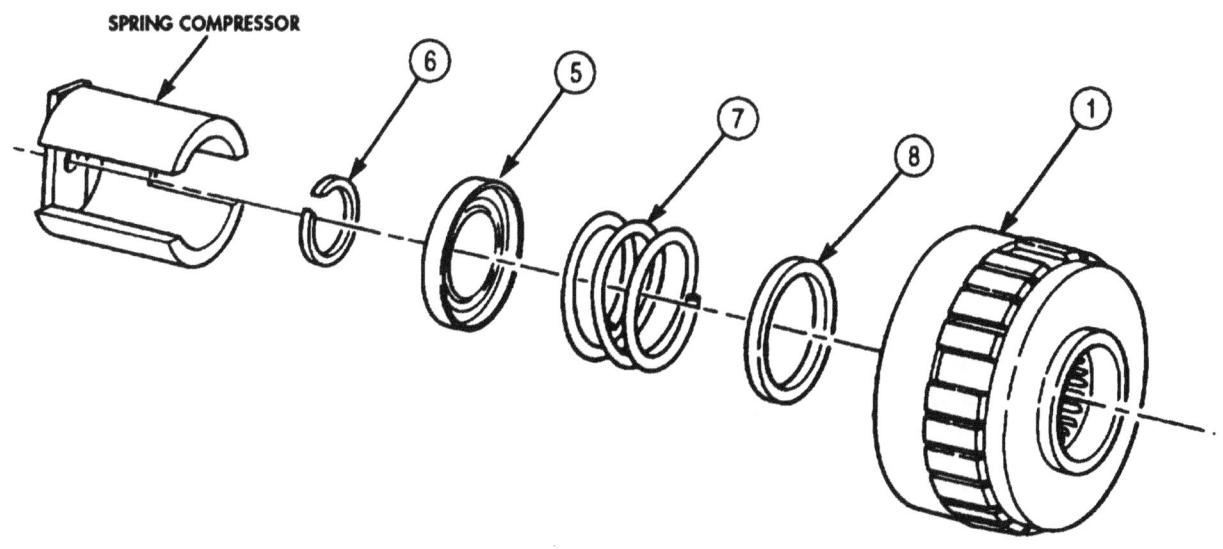

SPRING COMPRESSOR

SPRING COMPRESSOR

5-48. FOURTH CLUTCH MAINTENANCE (Contd)

5. Remove backplate snapring (1) from fourth clutch housing (5).

CAUTION

The fourth clutch consists of a backplate, five clutch plates, and five clutch discs. Keep all clutch parts together. Intermixing fourth clutch parts with any other clutch part will cause transmission damage.

NOTE

Tag all fourth clutch parts for reassembly.

6. Remove backplate (2), five clutch plates (4), five clutch discs (3), and piston (8) from fourth clutch housing (5).

7. Remove piston outer seal ring (7) from piston (8). Discard seal ring (7).

8. Remove piston inner seal ring (6) from fourth clutch housing (5). Discard seal ring (6).

c. Cleaning and Inspection

1. For general cleaning instructions, refer to para. 2-14.

2. For general inspection instructions, refer to para. 2-15.

3. Inspect five clutch plates (4) for burned surfaces. If burned, discard clutch plates (4).

4. Inspect five clutch discs (3) for burned surfaces. If surface of clutch disc (3) is burned, discard clutch disc (3).

5. Inspect piston return spring (9) for discoloration due to overheating. If discolored, discard piston return spring (9).

6. Inspect piston return spring (9) for broken coils or coils distorted due to wear. If coils are broken or distorted, discard piston return spring (9).

NOTE

Free length of piston return spring is 3.22 in. (81.8 mm).

7. Using spring tester, inspect piston return spring (9) for serviceability.

    a. Compress piston return spring (9) to 1.28 in. (32.5 mm).

    b. If piston return spring (9) does not give load of 158-178 lb (703-791 N), discard.

8. Measure backplate (2) thickness. If backplate (2) thickness is less than .248 in. (6.30 mm), discard.

5-48. FOURTH CLUTCH MAINTENANCE (Contd)

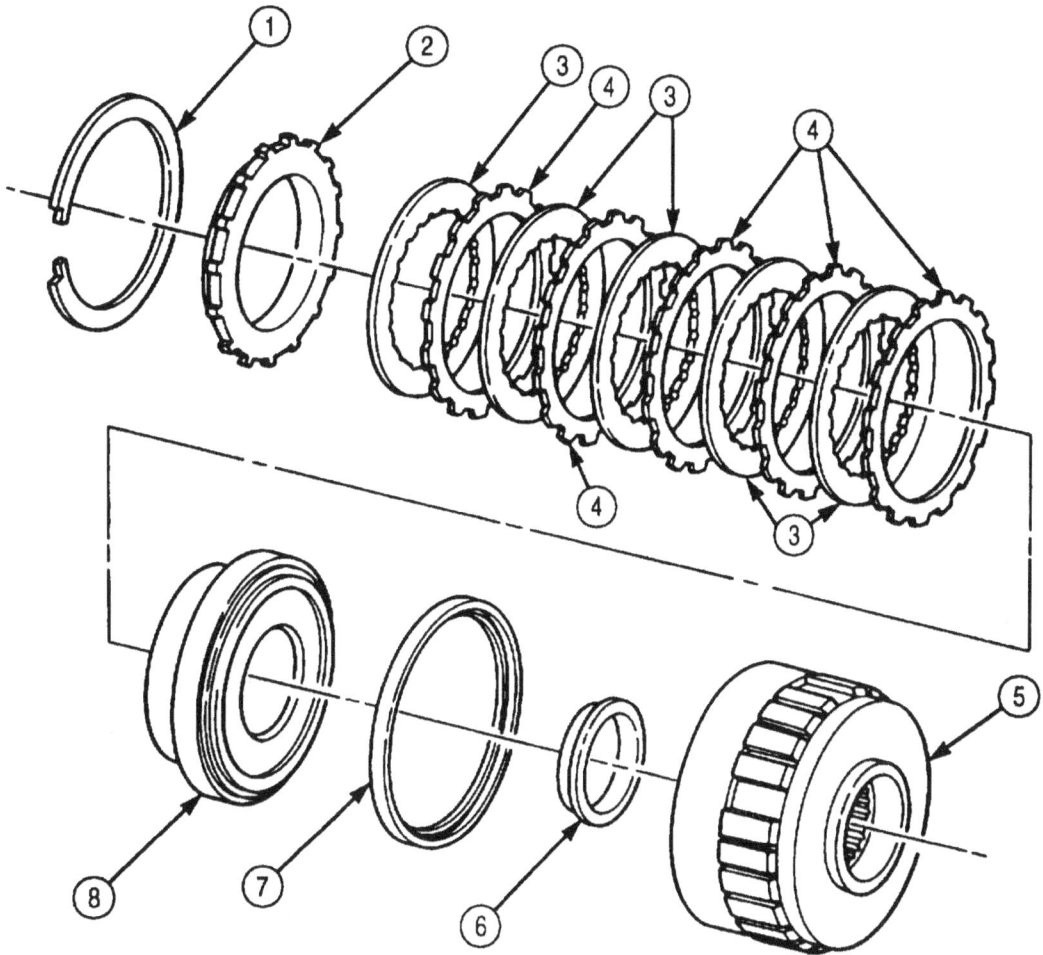

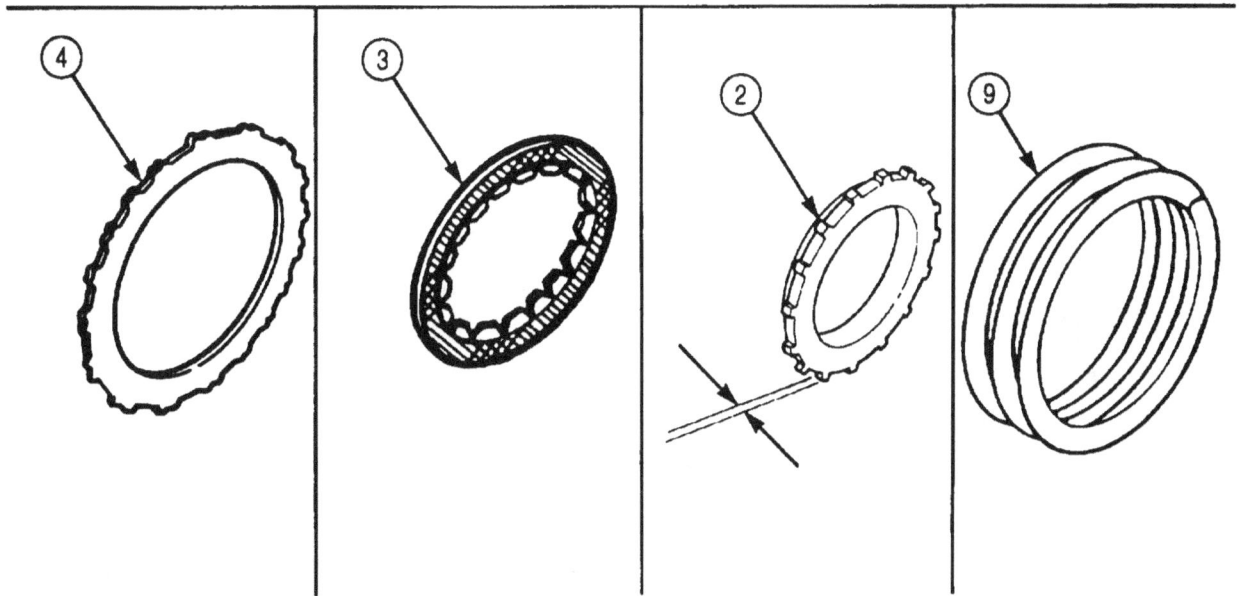

5-48. FOURTH CLUTCH MAINTENANCE (Contd)

d. Assembly

NOTE

Ensure new seal ring lips face toward oil pressure side of piston.

1. Apply oil-soluble grease to new piston outer seal ring (11) and install new piston outer seal ring (12) on piston (10).

2. Apply oil-soluble grease to new piston inner seal ring (12) and install new piston inner seal ring (12) in fourth clutch housing (5).

3. Install piston (10) in fourth clutch housing (5).

NOTE

Steps 4 through 9 obtain clutch running clearance.

4. Starting with clutch plate (9), alternately install five clutch plates (9) and five clutch discs (8) in fourth clutch housing (5).

5. Position backplate (7) in fourth clutch housing (5) and install with snapring (6).

6. Measure clearance between backplate (7) and clutch disc (8). Clearance should be 0.68-0.127 in. (17.27-3.226 mm).

7. If clearance exceeds 0.127 in. (3.226 mm), replace clutch discs (8) and clutch plates (9) with new, thicker clutch discs (8) and plates (9).

8. If clearance still exceeds 0.127 in. (3.226 mm) after completing step 7, replace piston (10) with new, thicker piston (10).

9. If clearance is less than 0.68 in. (17.27 mm), replace piston (10) with new, thinner piston (10).

10. Remove snapring (6), backplate (7), five clutch discs (8), and five clutch plates (9) from fourth clutch housing (5).

NOTE

Soak five clutch discs in clean OE/HDO-10 lubricating oil for at least two minutes before installing in fourth clutch housing.

11. Starting with clutch plate (9), alternately install five clutch plates (9) and clutch discs (8) in fourth clutch housing (5).

12. Place optional spacer (4), piston return spring (3), and spring retainer (2) in fourth clutch housing (5).

13. Position piston return spring compressor on spring retainer (2) and arbor press, and compress piston return spring (3).

14. Install spring retainer (2) in fourth clutch housing (5) with snapring (1).

15. Remove piston return spring compressor from spring retainer (2).

e. Installation

NOTE

Perform step 1 for transmission with single-piece bearing assembly.

1. Placing race section of bearing assembly (14) into transmission housing (15) first, install bearing assembly (14) in transmission housing (15).

NOTE

• Perform step 2 for transmission with two-piece bearing.

• Use oil-soluble grease to hold bearing race in place.

2. Install bearing race (13) and bearing assembly (14) on fourth clutch housing (5). Ensure bearing race (13) is installed outer lip first.

3. Install fourth clutch housing (5) in transmission housing (15).

5-48. FOURTH CLUTCH MAINTENANCE (Contd)

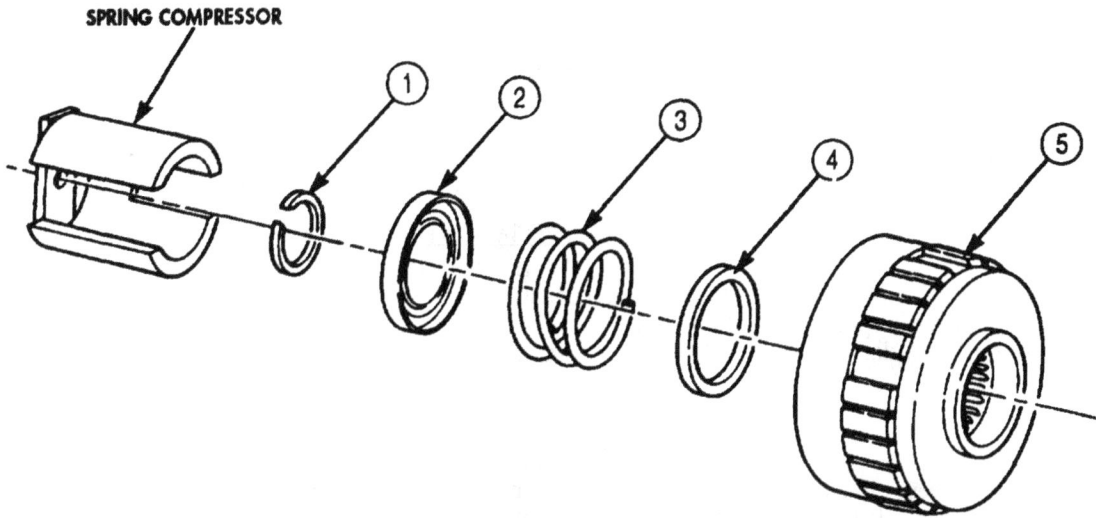

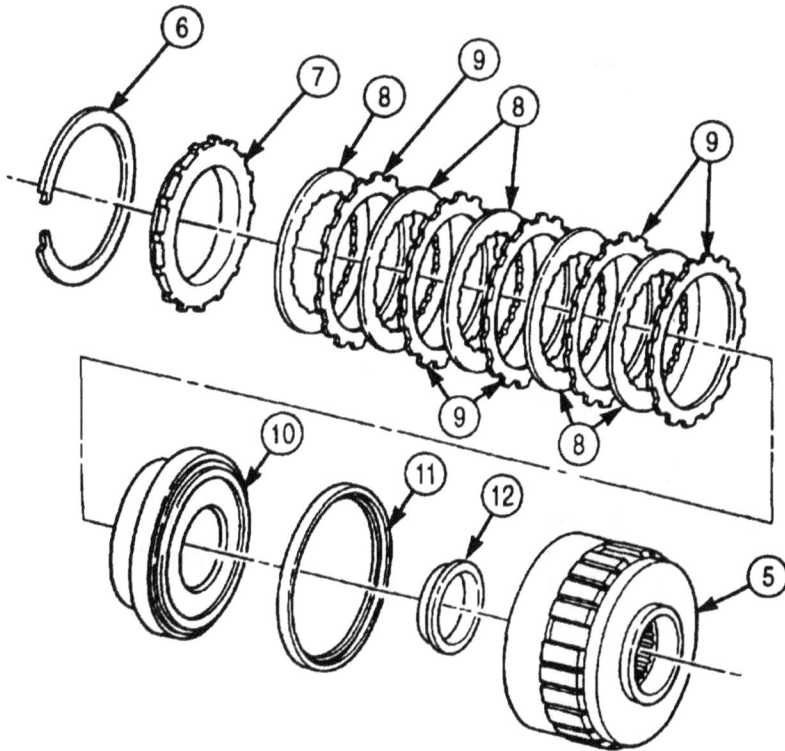

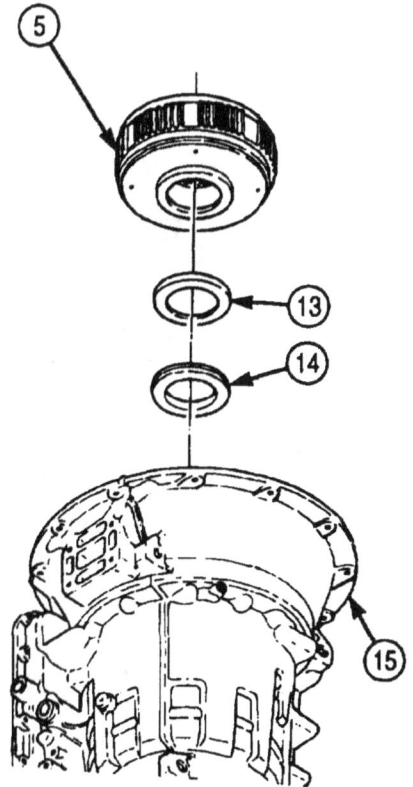

FOLLOW-ON TASK: Install forward clutch. (para. 5-47).

5-49. THIRD CLUTCH MAINTENANCE

THIS TASK COVERS:

a. Removal   c. Installation
b. Cleaning and Inspection

INITIAL SETUP:

APPLICABLE MODELS
All

SPECIAL TOOLS
Third clutch clearance gauge (Appendix E, Item 139)

TOOLS
General mechanic's tool kit (Appendix E, Item 1)
Outside micrometer (Appendix E, Item 80)
Spring tester (Appendix E, Item 131)

REFERENCES (TM)
TM 9-2320-272-24P

EQUIPMENT CONDITION
Fourth clutch removed (para. 5-48).

SPECIAL ENVIRONMENTAL CONDITIONS
Work area clean and free from blowing dirt and dust.

CAUTION

Keep all clutch parts together. Intermixing third clutch parts with any other clutch pack will cause transmission damage.

a. Removal

NOTE

- Tag all clutch parts for reassembly.
- Mark position of all clutch plates in transmission housing for reassembly.

1. Remove snapring (1), backplate (2), three clutch discs (3), and clutch plates (5) from transmission housing (4).

b. Cleaning and Inspection

1. For general cleaning instructions, refer to para. 2-14.
2. For general inspection instructions, refer to para. 2-15.
3. Inspect clutch discs (3) for burned surfaces. If burned, discard clutch discs (3).
4. Inspect clutch plates (5) for burned surfaces. If burned, discard clutch plates (5).
5. Measure backplate (2) thickness.
   a. If thickness of backplate (2) marked No. 7 is less than 0.476 in. (12.09 mm), discard backplate (2).
   b. If thickness of backplate (2) marked No. 8 is less than 0.450 in. (11.43 mm), discard backplate (2).
   c. If thickness of backplate (2) marked No. 9 is less than 0.463 in. (11.76 mm), discard backplate (2).

5-49. THIRD CLUTCH MAINTENANCE (Contd)

c. Installation

1. Starting with clutch plate (5), alternately install three clutch plates (5) and three clutch discs (3) in transmission housing (4).

2. Install backplate (2) and snapring (1) in transmission housing (4).

3. Measure clearance between backplate (2) and snapring (1) with third clutch clearance gauge. Third clutch clearance should be 0.050-0.114 in. (1.270-2.896 mm).

4. If third clutch clearance exceeds 0.114 in. (2.896 mm), replace clutch discs (3) with new, thicker discs (3).

5. If third clutch clearance still exceeds 0.114 in. (2.896 mm) after completing step 4, replace backplate (2) with new, thicker backplate (2).

6. If third clutch clearance is less than 0.050 in. (1.270 mm), replace backplate (2) with new, thinner backplate (2).

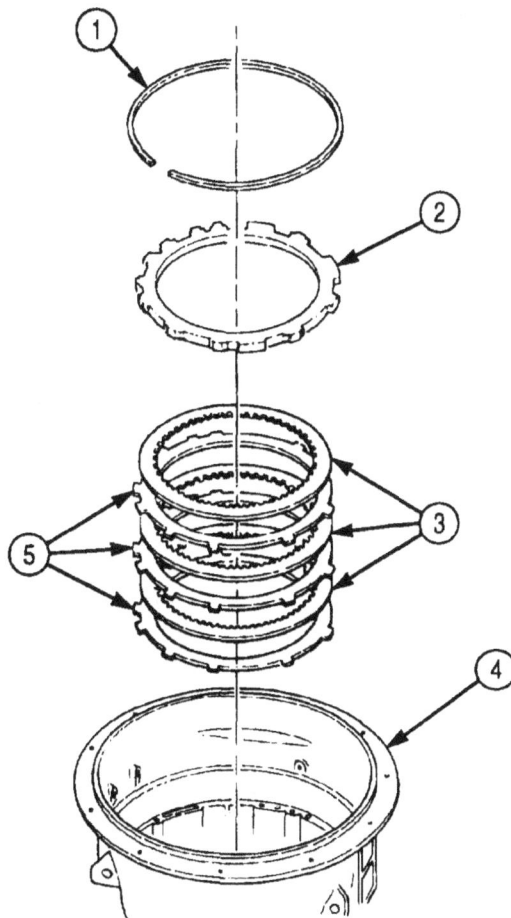

FOLLOW-ON TASK: Install fourth clutch (para. 5-48).

5-339

5-50. CENTER SUPPORT MAINTENANCE

THIS TASK COVERS:

a. Removal
b. Disassembly
c. Cleaning and Inspection

d. Assembly
e. Installation

INITIAL SETUP:

APPLICABLE MODELS
All

SPECIAL TOOLS
Bushing installer (Appendix E, Item 24)
Retainer ring depth tool (Appendix E, Item 114)
Center support lifter (Appendix E, Item 27)
Compressor bar and screw (Appendix E, Item 29)
Compressor base (Appendix E, Item 30)
Snapring gauge (Appendix E, Item 124)

TOOLS
General mechanic's tool kit (Appendix E, Item 1)
Spring tester (Appendix E, Item 131)
Depth gauge (Appendix E, Item 81)
Torque wrench (Appendix D, Item 144)
Arbor press
Hoist

MATERIALS/PARTS
Seal ring (Appendix D, Item 641)
Bushing (Appendix D, Item 27)
Eight self-locking retainers (Appendix D, Item 647)
Filter screen (Appendix D, Item 126)
Two piston inner seal rings (Appendix D, Item 258)
Two piston outer seal rings (Appendix D, Item 509)
Checkball (Appendix D, Item 37)
Two step-joint seal rings (Appendix D, Item 638)
Chamfered washer (Appendix D, Item 35)
Anchor bolt (Appendix D, Item 5)
Oil-soluble grease (Appendix C, Item 53)

REFERENCES (TM)
TM 9-2320-272-24P

EQUIPMENT CONDITION
Third clutch removed (para. 5-49).

SPECIAL ENVIRONMENTAL CONDITIONS
Work area clean and free from blowing dirt and dust.

a. Removal

1. Remove center support anchor bolt (3) or (4) and washer (5) or (2) from transmission housing (1). Discard anchor bolt (3) or (4) and washer (5) or (2).

NOTE

If old style hex-head bolt and washer were present, center support must be rethreaded. Refer to task c.

2. Remove snapring (6) from transmission housing (1).

3. Attach center support lifter to center support (7) and remove center support (7) from transmission housing (1).

NOTE

Record location of any shims present.

5-50. CENTER SUPPORT MAINTENANCE (Contd)

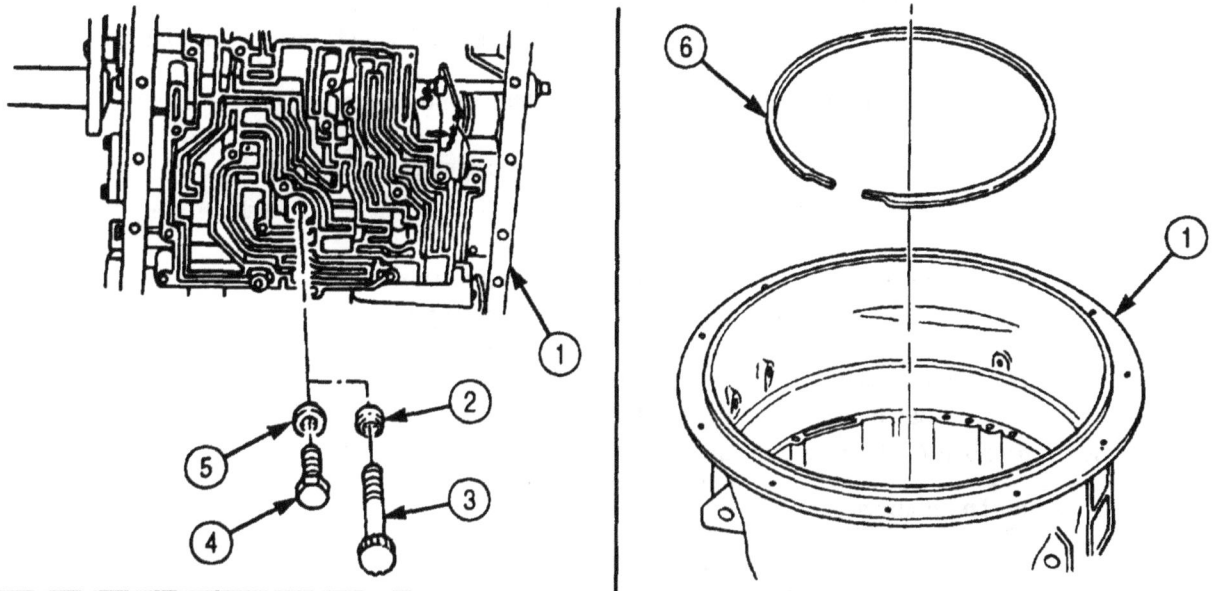

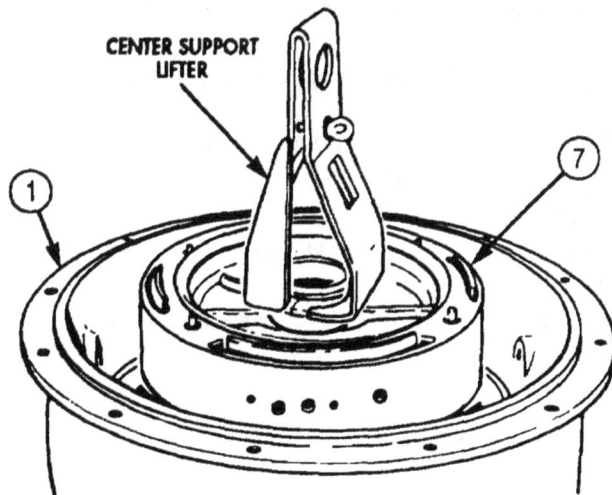

CENTER SUPPORT LIFTER

5-50. CENTER SUPPORT MAINTENANCE (Contd)

b. Disassembly

1. Remove filter screen (14) and seal ring (13) from center support (10). Discard filter screen (14) and seal ring (13).

NOTE

The second and third clutch pistons are disassembled the same way. This procedure covers the second clutch piston.

2. Remove second clutch piston (11) from center support (10).
3. Remove inner seal ring (5) and outer seal ring (6) from second clutch piston (11). Discard seal rings (5) and (6).
4. Remove four self-locking retainers (1) from pins (12) on second clutch piston (11). Discard self-locking retainers (1).
5. Remove retainer (2) and twenty springs (3) from second clutch piston (11).
6. Remove two step-joint seal rings (7) from center support (10). Discard step-joint seal rings (7).

NOTE

- Bushing is removed only if it fails inspection (task c).
- Mark location of bushing notch on center support for reassembly.
- Replace checkball if bushing is replaced.

7. Press bushing (8) out of center support (10) with arbor press and mandrel. Discard bushing (8).
8. Remove checkball (9) from center support (10). Discard checkball (9).

c. Cleaning and Inspection

1. For general cleaning instructions, refer to para. 2-14.
2. For general inspection instructions, refer to para. 2-15.
3. Inspect center support (10) cavities for obstruction or foreign material. Remove obstruction or foreign material.
4. Measure depth of anchor bolt hole in center support (10). If hole is not 1.610 in. (40.89 mm) deep, rework center support (10) as follows:

    a. Drill 1.610 in. (40.89 mm) hole in center support (10).

    b. Counterbore 0.469 in. (11.91 mm) diameter 0.66-0.70 in. (16.8-17.8 mm) deep.

    c. Tap hole using 3/8 in. (9.53 mm) tap UNC2B min. Full thread to depth of dim. A.

5. Inspect four pins (12) on each piston (4) and (11) for bends and breaks. If any pin (12) is damaged, replace piston (4) or (11).
6. Inspect all springs (3) for discoloration due to overheating. If discoloration is present, discard springs (3).

5-50. CENTER SUPPORT MAINTENANCE (Contd)

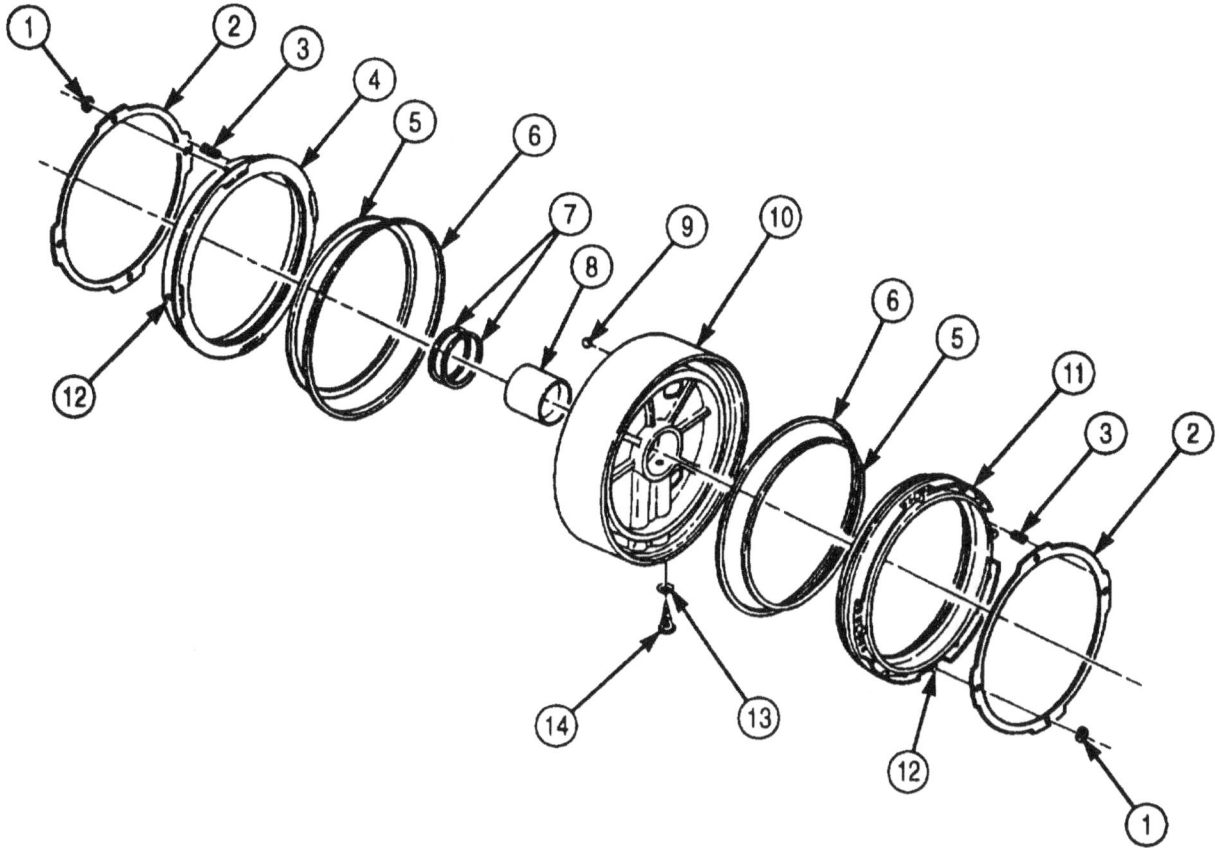

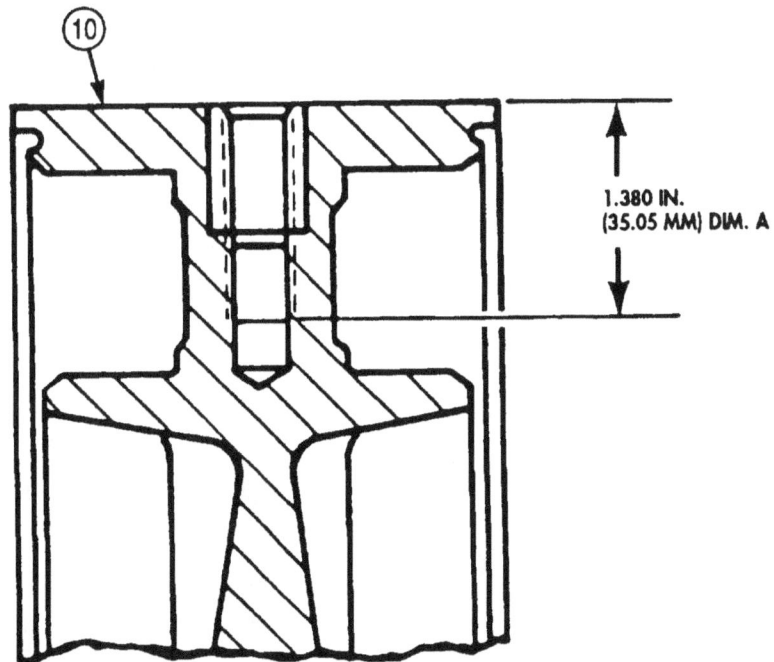

1.380 IN.
(35.05 MM) DIM. A

5-50. CENTER SUPPORT MAINTENANCE (Contd)

7. Inspect all springs (6) for broken coils or coils distorted due to wear. If spring (6) is broken or distorted, discard.

8. Using spring tester, inspect all springs (6) for serviceability.

   a. Compress spring (6) to 0.81 in. (20.7 mm) and check load.

   b. If spring (6) does not give load of 4.30-5.70 lb (19.1.-25.4 N), discard.

d. Assembly

NOTE
- Perform steps 1 and 2 only if bushing was removed.
- Checkball will be installed when new bushing is pressed in.

1. Position checkball (2) in center support (3) and align notch on new bushing (1) with marked location on center support (3).

2. Press new bushing (1) in center support (3) with bushing installer and arbor press until flush.

NOTE
The second and third clutch pistons are assembled the same way.
This procedure covers the second clutch piston.

3. Place second clutch piston (10) in center support (3).

4. Place twenty springs (6) in pockets on piston (11).

5. Align spring retainer (5) with pins (11) on piston (10) with retainer ring depth tool and install four new self-locking retainer rings (4)

6. Remove second clutch piston (10) from center support (3).

7. Apply oil-soluble grease to new inner seal ring (8) and new outer seal ring (9).

8. Install new inner seal ring (8) and new outer seal ring (9) on second clutch piston (10). Ensure lips of seal rings (8) and (9) face toward center support (3) when second clutch piston (10) is installed.

9. Repeat steps 3 through 8 to assemble third clutch piston (7).

NOTE
Do not install third clutch piston in center support at this time.

10. Install second clutch piston (11) in center support (3).

NOTE

11. Install new filter screen (13) and new seal ring (12) on center support (3).

5-50. CENTER SUPPORT MAINTENANCE (Contd)

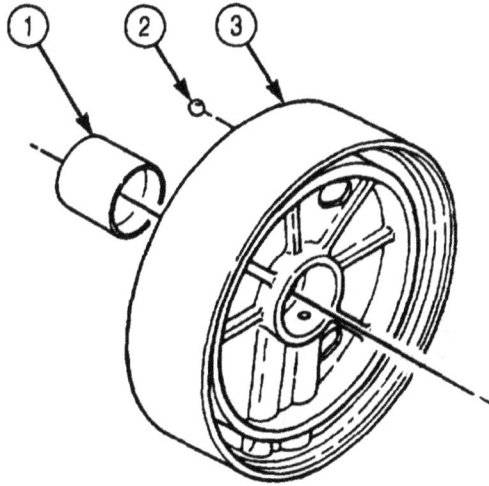

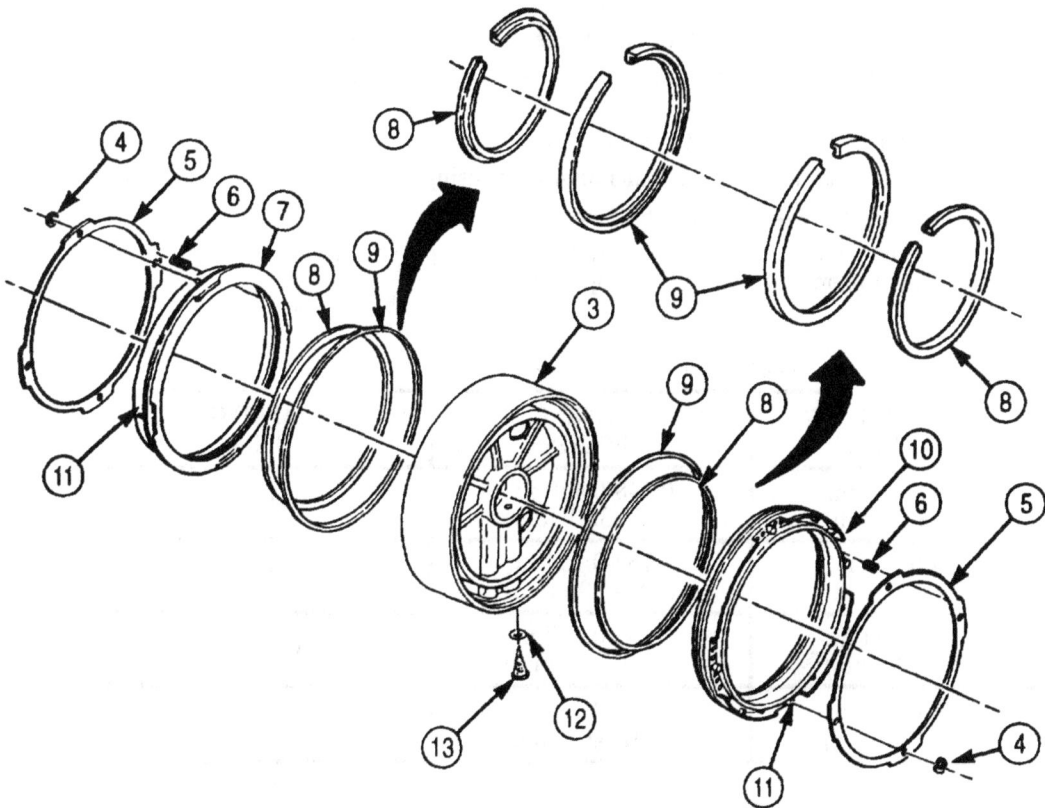

5-50. CENTER SUPPORT MAINTENANCE (Contd)

e. Installation

NOTE

Tilt front end of transmission housing upward before beginning installation.

1. Attach center support lifter to center support (4).

NOTE

- Ensure center support is seated firmly against second clutch snapring.
- Do not use old-style washer or hex-head anchor bolt.

2. Carefully position center support (4) in transmission housing (5) with the anchor bolt hole (6) on center support (4) and transmission housing center support hole (14) aligned.

3. Install center support (4) in transmission housing (5) with new chamfered washer (13) and anchor bolt (12). Tighten anchor bolt (13) finger-tight.

4. Remove center support lifter from center support (4).

5. Position compressor base (9) on center support (4).

6. Install compressor bar (10) and screw (8) on transmission housing (5) with two screws (11).

7. Tighten compressor screw (8) 5 lb-ft (7 N•m).

8. Measure clearance between top edge of center support (4) and top of center support snapring groove (7) in transmission housing (5) with snapring gauge.

NOTE

Refer to table 5-14 to select. correct size snapring to be installed in step 12.

9. Release pressure from center support (4) by loosening compressor screw (8).

10. Remove compressor bar screws (11), compressor bar (10) and screw (8), and compressor base (9) from center support (4).

11. Install third clutch piston (2) in center support (4).

12. Install center support snapring (1) in transmission housing (5). Ensure snapring (1) is fully seated.

13. Tighten anchor bolt (12) 39-46 lb-ft (53-62 N•m)

14. Apply liberal amount of oil-soluble grease on two new step-joint seal rings (3), and install two new step-joint seal rings (3) on hub of center support (4).

Table 5-14. Center Support Snapring.

MEASURED CLEARANCE	SNAPRING THICKNESS	SNAPRING COLOR
0.150-0.154 in. (3.81-3.91 mm)	0.148-0.150 in. (3.76-3.81 mm)	White
0.154-0.157 in. (3.91-3.99 mm)	0.152-0.154 in. (3.86-3.91 mm)	Yellow
0.157-0.160 in. (3.99-4.06 mm)	0.155-0.157 in. (3.94-3.99 mm)	Green
0.160-0.164 in. (4.06-4.17 mm)	0.158-0.160 in. (4.01-4.06 mm)	Red

5-50. CENTER SUPPORT MAINTENANCE (Contd)

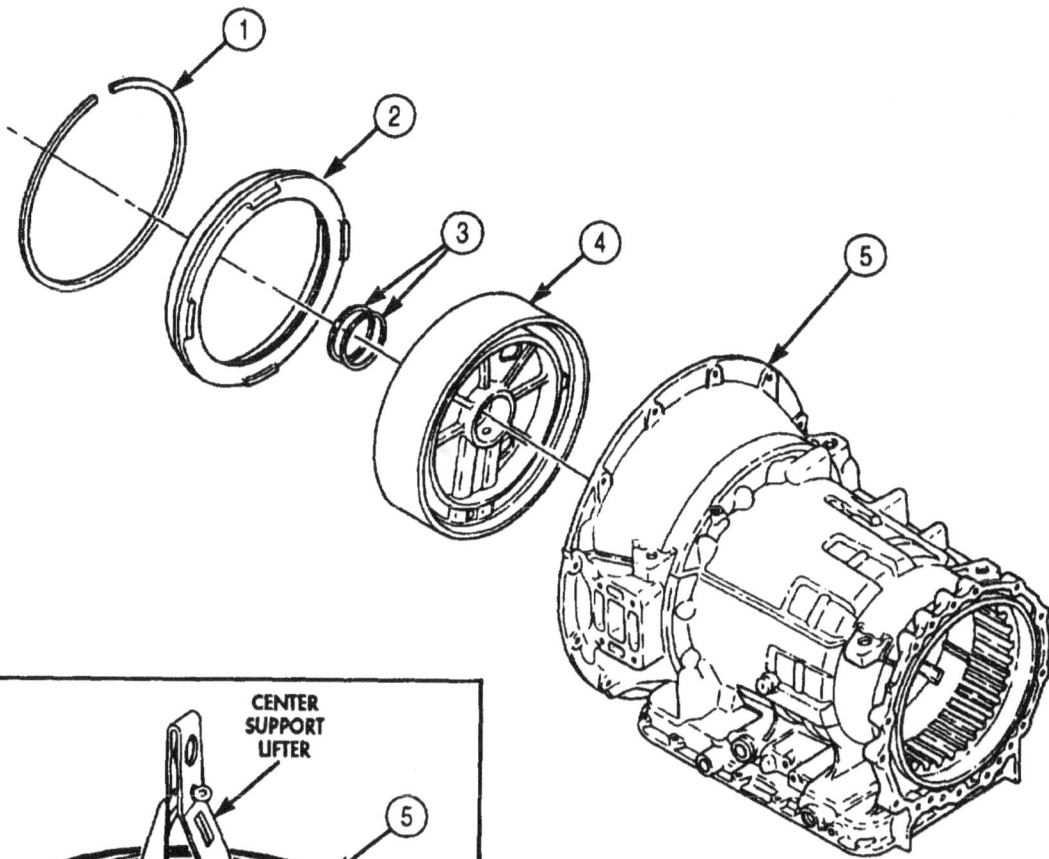

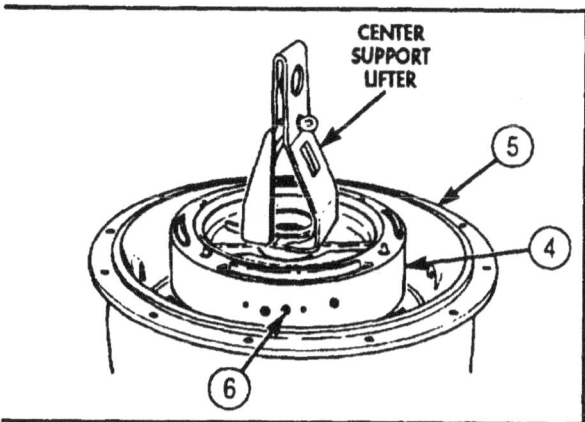

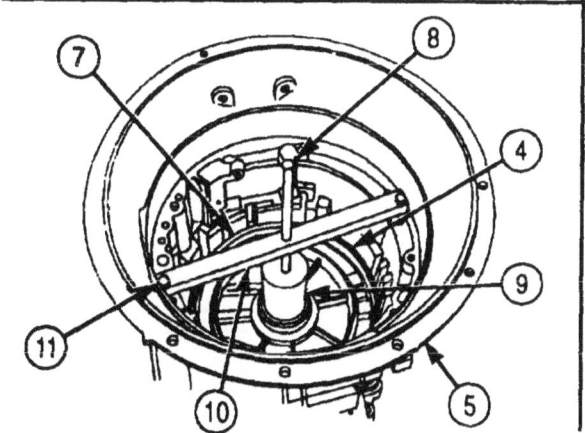

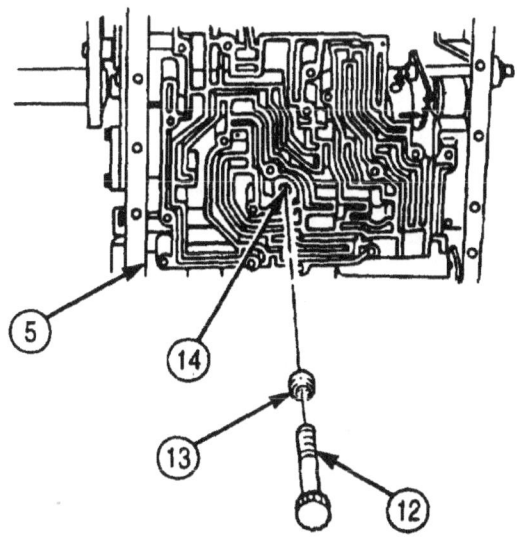

CENTER SUPPORT LIFTER

FOLLOW-ON TASK: Install third clutch (para. 5-49).

5-51. GEAR UNIT AND MAIN SHAFT MAINTENANCE

THIS TASK COVERS:

a. Removal
b. Disassembly
c. Cleaning and Inspection

d. Assembly
e. Installation

INITIAL SETUP:

APPLICABLE MODELS
All

SPECIAL TOOLS
Gear unit lifter (Appendix E, Item 58)

Tools
General mechanic's tool kit (Appendix E, Item 1)
Outside micrometer (Appendix E, Item 80)
Vernier caliper (Appendix E, Item 159)
Snapring pliers (Appendix E, Item 125)

MATERIALS/PARTS
Oil-soluble grease (Appendix C, Item 53)

REFERENCES (TM)
TM 9-2320-272-24P

EQUIPMENT CONDITION
Center support removed (para. 5-50).

SPECIAL ENVIRONMENTAL CONDITIONS
Work area clean and free from blowing dirt and dust.

---

a. Removal

Attach gear unit lifter to gear unit main shaft (1) and remove gear unit (2) from transmission housing (3).

b. Disassembly

1. Remove thrust washer (4), front sun gear (5), sun gear shaft (7), and thrust washer (6) from front planetary carrier (8).

2. Remove front planetary carrier (8) and thrust washer (9) from planetary carrier drum (10).

3. Remove snapring (12) from ring gear (13) and lift ring gear (13) out of planetary carrier drum (10).

4. Remove center planetary carrier (11) from planetary carrier drum (10).

5. Remove roller bearing (14) and bearing race (15) from main shaft (18).

6. Remove snapring (19) from main shaft (18) and slide lower planetary sun gear (20) off main shaft (18).

7. Remove main shaft (18) from rear planetary sun gear (23).

8. Remove center ring gear (17) and rear planetary sun gear (23) from rear planetary carrier (22).

9. Remove snapring (16) from center ring gear (17) and slide rear planetary sun gear (23) out of center ring gear (17).

10. Remove snapring (21) from rear planetary carrier (22) and lift rear planetary carrier (22) out of planetary carrier drum (10).

5-51. GEAR UNIT AND MAIN SHAFT MAINTENANCE (Contd)

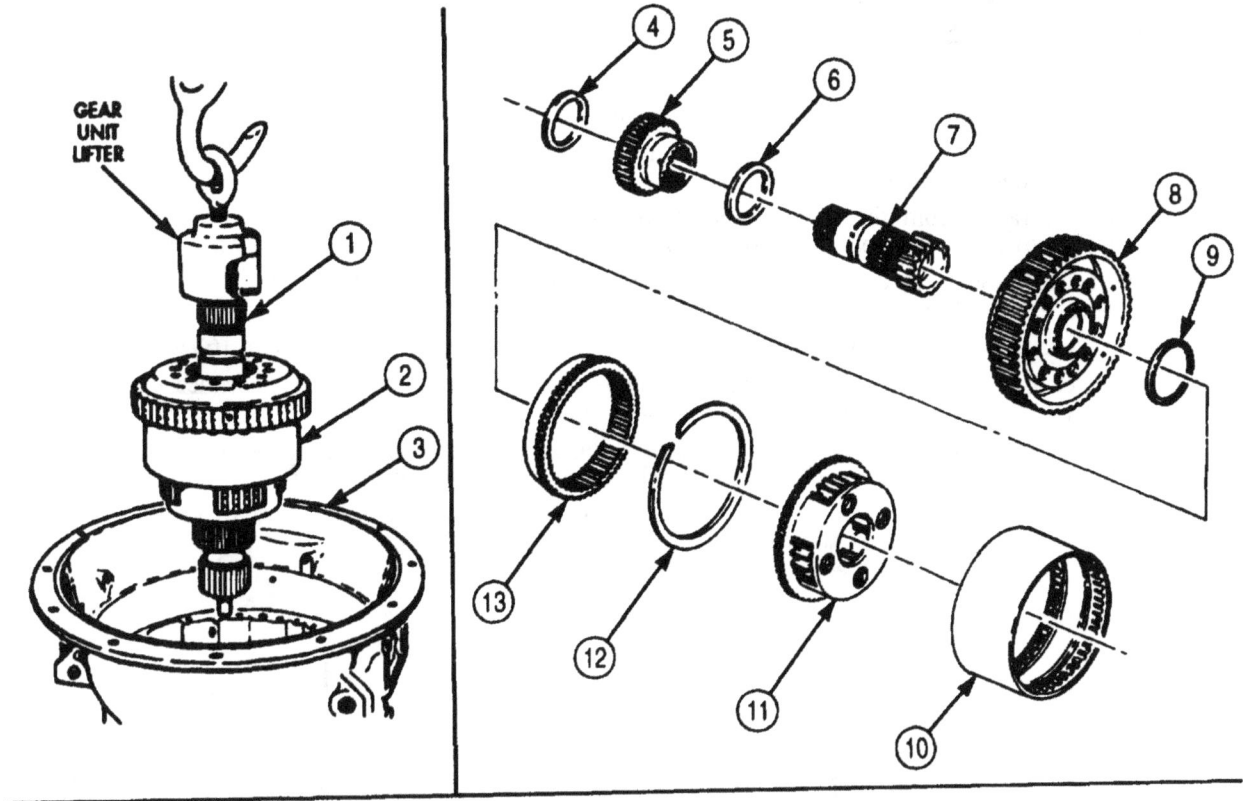

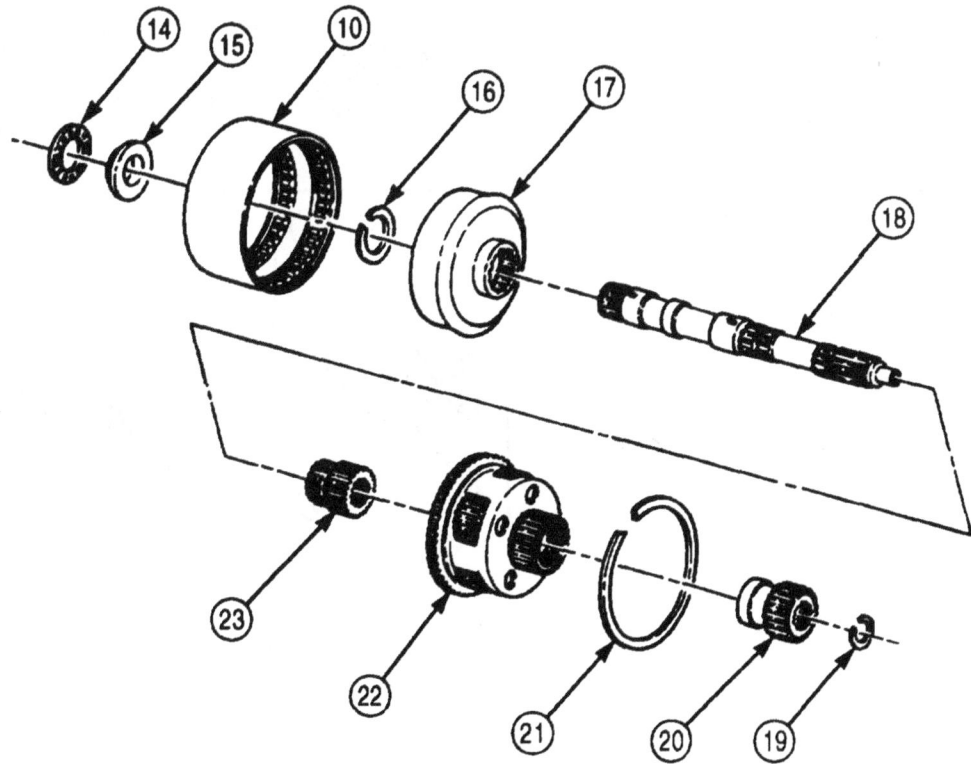

GEAR UNIT LIFTER

5-51. GEAR UNIT AND MAIN SHAFT MAINTENANCE (Contd)

c. Cleaning and Inspection

1. For general cleaning instructions, refer to para. 2-14.

2. For general inspection instructions, refer to para. 2-15.

3. Measure thickness of thrust washer (11). If thrust washer (11) thickness is less than 0.092 in. (2.354 mm), discard thrust washer (11).

4. Measure clearance from thrust washer (11) to front sun gear (10). If clearance from thrust washer (11) to front sun gear (10) exceeds 0.005 in. (0.127 mm), discard thrust washer (11).

5. Measure thickness of thrust washers (6) and (9). If thickness is less than 0.091 in. (2.31mm), discard thrust washers (6) and (9).

6. Inspect thrust washers (6) and (9) for scoring. If scored, discard thrust washers (6) and (9).

7. Measure clearance of two sun gear shaft bushings (4) on main shaft (5). If clearance exceeds 0.006 in. (0.152 mm), discard two sun gear shaft bushings (4).

8. Measure clearance of front planetary carrier (7) bushing (8) on front sun gear (10). If clearance exceeds 0.005 in. (0.127 mm), discard front planetary carrier (7) bushing (8).

9. Measure clearance of sun gear shaft (3) on center support bushing. (2). If clearance exceeds 0.006 in. (0.152 mm), discard sun gear shaft (3), or replace bushing (2) in center support (1) (para. 5-50).

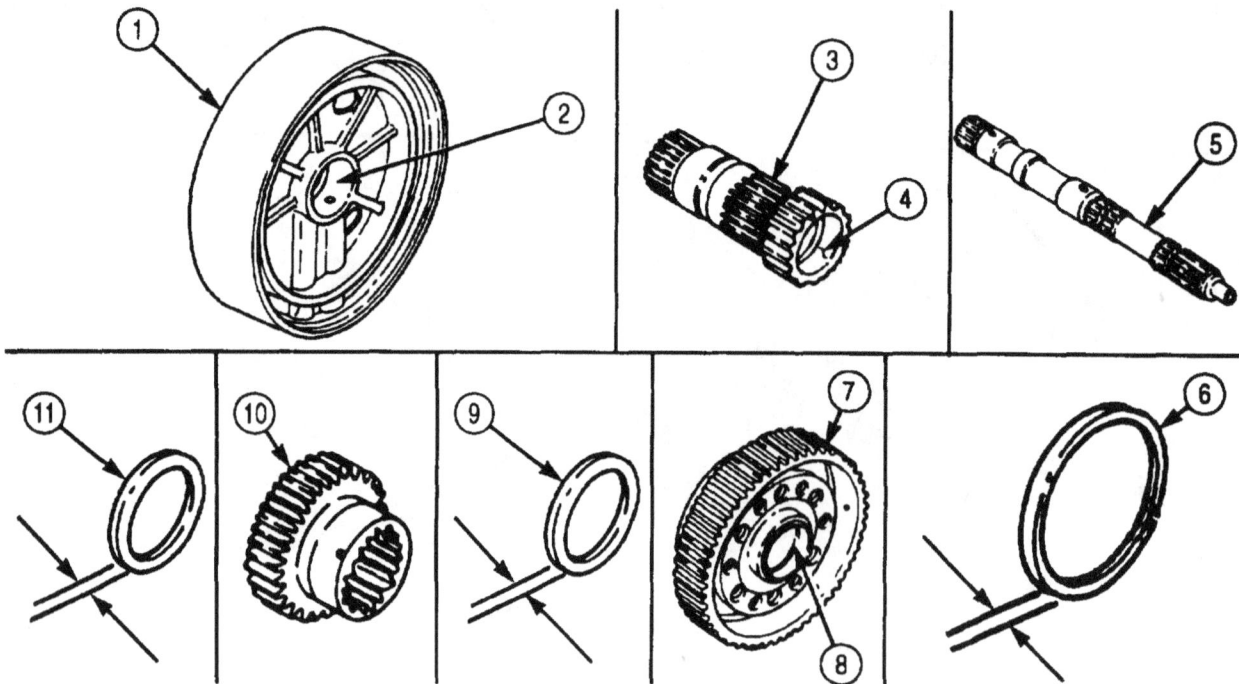

5-51. GEAR UNIT AND MAIN SHAFT MAINTENANCE (Contd)

d. Assembly

1. Position rear planetary carrier (18) in short splined end of planetary carrier drum (17) and install with snapring (21).
2. Position rear planetary sun gear (16) in center ring gear (15) until seated and install with snapring (14).
3. Install center ring gear (15) in planetary carrier drum (17), until seated against rear planetary carrier (18).
4. Install main shaft (22) into rear planetary sun gear (16), ensuring smaller end is installed first.
5. Install lower planetary sun gear (20) on rear of main shaft (22) with snapring (19).

NOTE

Use oil-soluble grease to hold bearing race and bearing in place.

6. Install bearing race (13) and bearing (12) on main shaft (22). Ensure lip of bearing race (13) faces away from center ring gear (15).

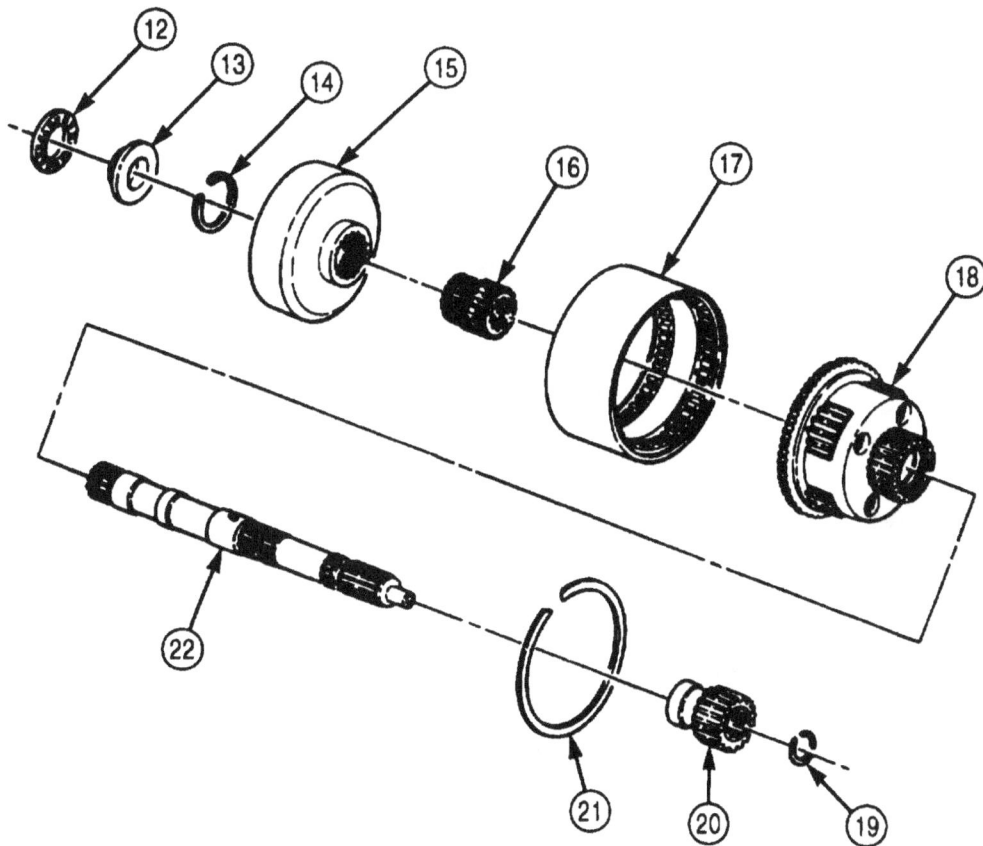

5-51. GEAR UNIT AND MAIN SHAFT MAINTENANCE (Contd)

7. Install center planetary carrier (11) in long splined end of planetary carrier drum (9).

8. Place front planetary ring gear (7) in planetary carrier drum (9) over center planetary carrier (11) and install with snapring (6).

NOTE
Use oil-soluble grease to hold thrust washers in place.

9. Install thrust washer (5) on front planetary carrier (4).

10. Install front planetary carrier (4) in planetary carrier drum (9).

11. Install thrust washer (3) on sun gear (2).

12. Install front sun gear (2) on sun gear shaft (10). Ensure spring pin on sun gear shaft (10) is aligned with wide spline on front sun gear (2).

NOTE
Ensure sun gear teeth and front planetary pinion teeth are flush and the front of the sun gear shaft is past groove between double splines on the main shaft.

13. Install sun gear shaft (10) over main shaft (8) in planetary carrier (9) until seated against bearing.

14. Install thrust washer (1) on sun gear (2) and over the gear shaft (10).

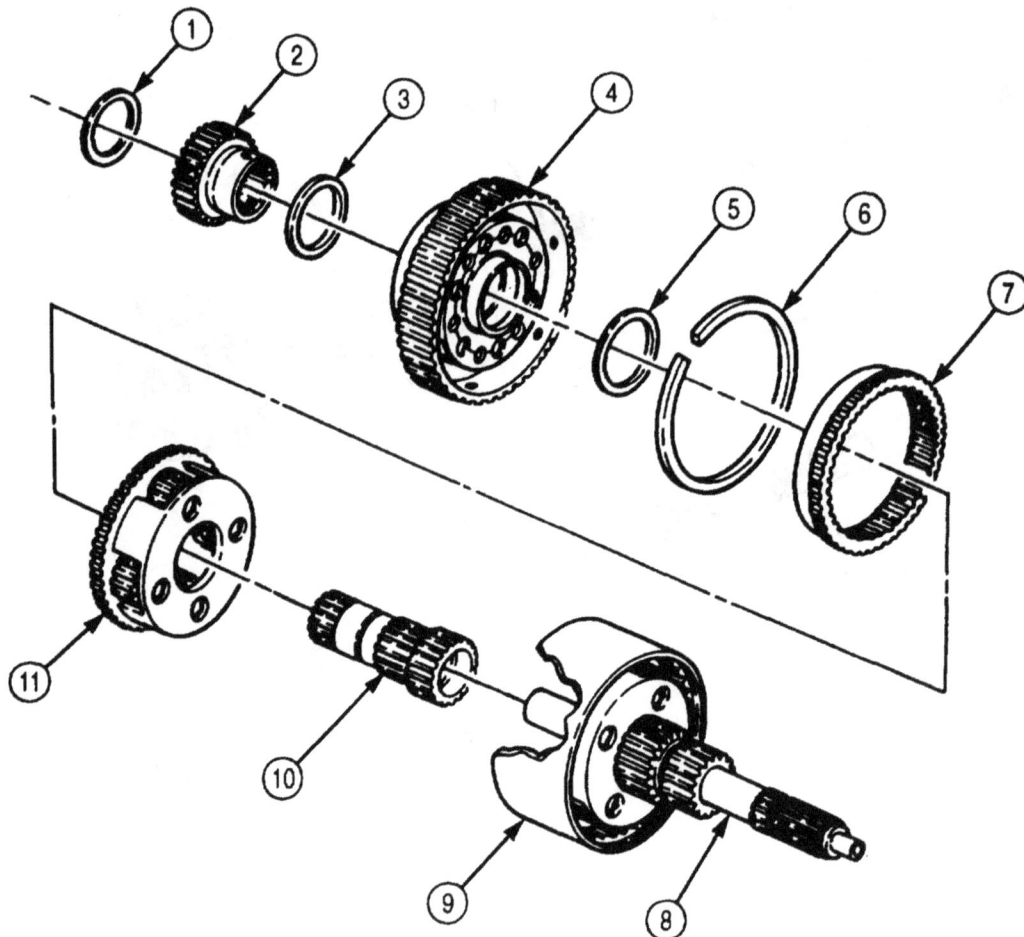

5-51. GEAR UNIT AND MAIN SHAFT MAINTENANCE (Contd)

e. Installation

1. Attach gear unit lifter to gear unit main shaft (12).

2. Carefully lower gear unit (13) into transmission housing (14) with hoist. Ensure all gear teeth (15) mesh.

3. Remove gear unit lifter from gear unit main shaft (12).

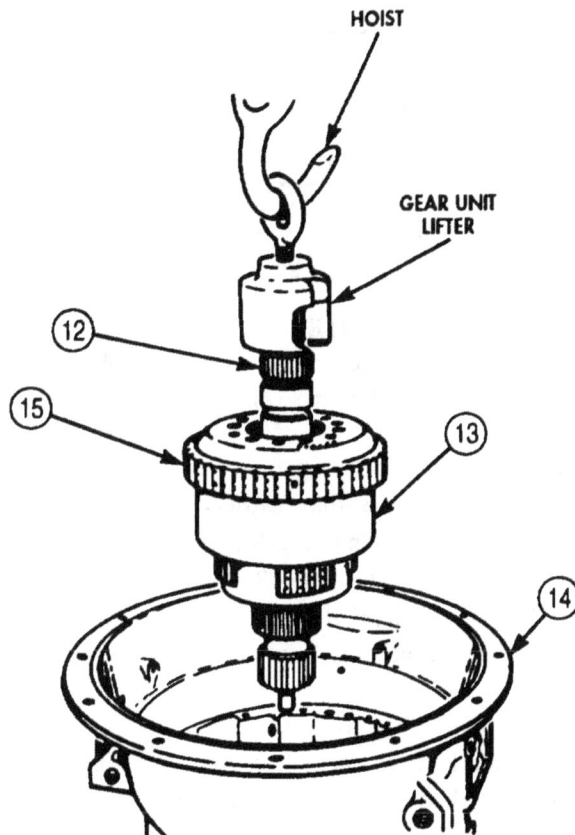

FOLLOW-ON TASK: Install center support (para. 5-50).

5-52. SECOND CLUTCH MAINTENANCE

THIS TASK COVERS:

a. Removal
b. Cleaning and Inspection

c. Installation
d. Establishing Clearance

INITIAL SETUP:

APPLICABLE MODELS
All

SPECIAL TOOLS
Second clutch clearance gauge (Appendix E, Item 118)

TOOLS
General mechanic's tool kit (Appendix E, Item 1)
Outside micrometer (Appendix E, Item 80)
Vernier caliper (Appendix E, Item 159)

EQUIPMENT CONDITION
Gear unit and main shaft removed (para. 5-51).

SPECIAL ENVIRONMENTAL CONDITIONS
Work area clean and free from blowing dirt and dust.

a. Removal

CAUTION

Second clutch consists of a backplate, four clutch plates, and four clutch discs. Keep all clutch parts together. Intermixing second clutch parts with any other clutch pack will cause transmission damage.

NOTE
Tag all second clutch parts for reassembly.

Remove snapring (1), four clutch discs (5), clutch plates (2), and backplate (4) from transmission housing (3).

b. Cleaning and Inspection

Flying rust and metal particles may cause injury to personnel.

1. For general cleaning instructions, refer to para. 2-14.
2. For general inspection instructions, refer to para. 2-15.
3. Inspect clutch plate (2) for burned surfaces. If burned, discard clutch plate (2).
4. Inspect clutch discs (5) for burned surfaces. If burned, discard clutch disc (5).
5. Measure backplate (4) thickness.
   a. If thickness of backplate (4) marked No. 10 is less than 0.234 in. (5.94 mm), discard backplate (4).
   b. If thickness of backplate (4) marked No. 11 is less than 0.208 in. (5.28 mm), discard backplate (4).
   b. If thickness of backplate (4) marked No. 12 is less than 0.221 in. (5.61 mm), discard backplate (4).

5-52. SECOND CLUTCH MAINTENANCE (Contd)

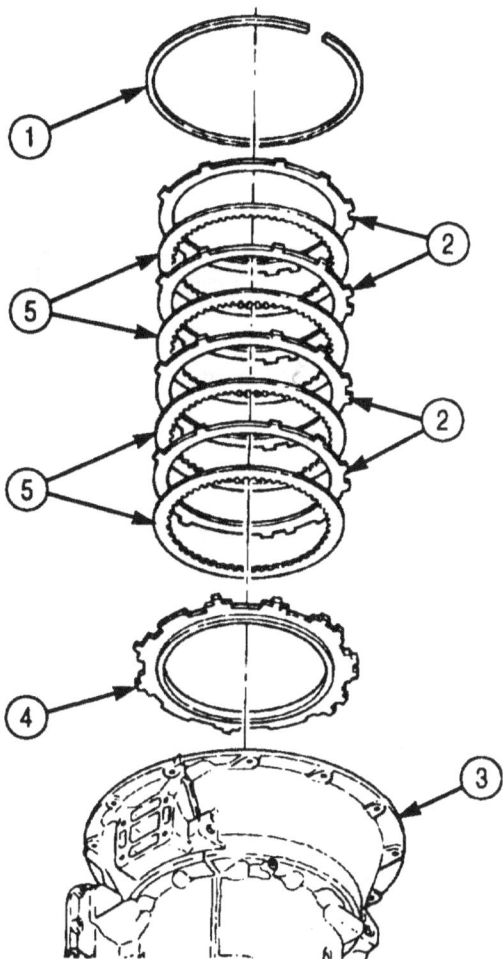

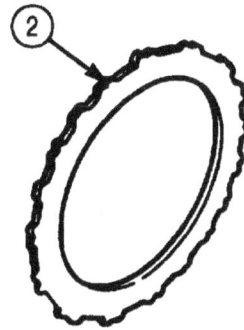

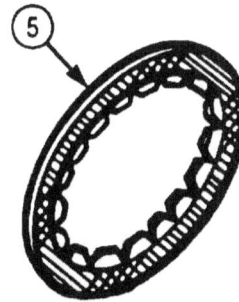

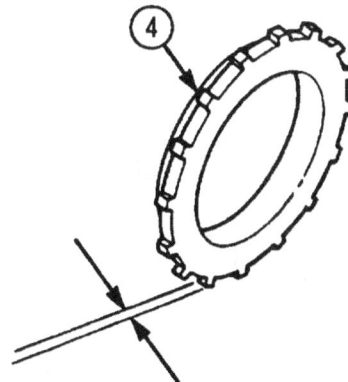

5-52. SECOND CLUTCH MAINTENANCE (Contd)

c. Installation

1. Install backplate (3) in transmission housing (2). Ensure that flat side of backplate (3) is up.
2. Starting with clutch disc (1), alternately install four clutch discs (1) and four clutch plates (5) in transmission housing (2).
3. Install snapring (4) on top of clutch plates (5).

d. Establishing Clearance

NOTE

- Adapter housing and first clutch must be removed to establish second clutch clearance. Refer to paras. 5-57 and 5-53.
- Tilt rear end of transmission housing upward in order to establish clearance.

1. Measure clearance between backplate (3) and transmission housing (2). Clearance should be 0.059-0.129 in. (1.49-3.28 mm).
2. If second clutch clearance exceeds 0.059-0.129 in. (1.49-3.28 mm), replace clutch discs (1) with new, thicker clutch discs (1).
3. If clutch clearance still exceeds 0.129 in. (3.28 mm) after completing step 1, replace backplate (3) with new, thicker backplate.
4. If clearance is less than 0.059 in. (1.49 mm), replace backplate (3) with new, thinner backplate (3).
5. If first clutch and adapter housing were removed to establish second clutch clearance, install them (paras. 5-57 and 5-53).

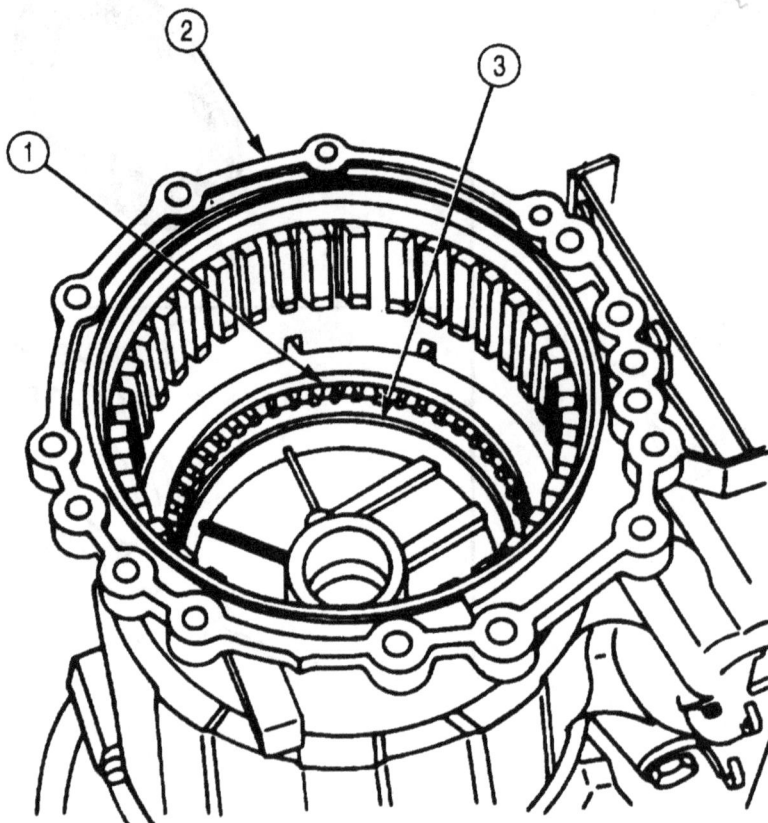

5-52. SECOND CLUTCH MAINTENANCE (Contd)

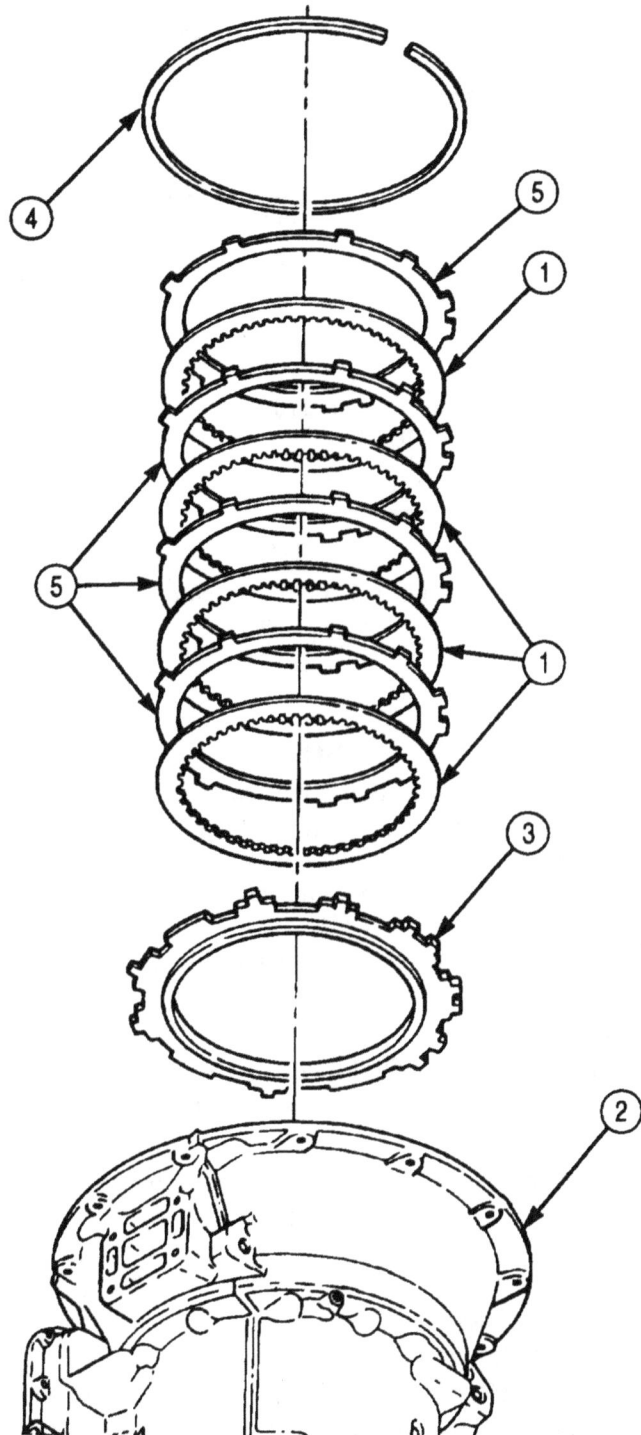

FOLLOW-ON TASK: Install gear unit and main shaft (para. 5-51).

5-53. FIRST CLUTCH MAINTENANCE

THIS TASK COVERS:

a. Removal
b. Cleaning and Inspection

c. Installation
d. Establishing Clearance

INITIAL SETUP:

APPLICABLE MODELS
All

SPECIAL TOOLS
First clutch clearance gauge (Appendix E, Item 45)

TOOLS
General mechanic's tool kit (Appendix E, Item 1)
Outside micrometer (Appendix E, Item 80)
Vernier caliper (Appendix E, Item 159)
Snapring pliers (Appendix C, Item 125)

REFERENCES (TM)
TM 9-2320-272-24P

EQUIPMENT CONDITION
Second clutch removed (para. 5-52).

SPECIAL ENVIRONMENTAL CONDITIONS
Work area clean and free from blowing dirt and dust.

CAUTION
First clutch consists of a backplate, six clutch plates, and six clutch discs. Keep all clutch parts together. Intermixing first clutch parts with any other clutch pack will cause transmission damage.

a. Removal

NOTE
- Tag all first clutch parts for reassembly.
- Backplate, five clutch discs, four clutch plates, and rear planetary ring gear are removed as one unit.

1. Remove snapring (5), backplate (6), five clutch discs (3), four clutch plates (2), and rear planetary ring gear (1) from transmission housing (4).
2. Remove remaining two clutch plates (2) and clutch disc (3) from transmission housing (4).

b. Cleaning and Inspection

1. For general cleaning instructions, refer to para. 2-14.
2. For general inspection instructions, refer to para. 2-15.
3. Inspect clutch discs (3) for burned surfaces. If surfaces are burned, discard clutch discs (3).
4. Inspect clutch plate (2) for burned surfaces. If surfaces are burned, discard clutch plate (2).

NOTE
Cast iron backplates marked number 1, 2, and 3 must have same thickness as corresponding new malleable iron backplates 4,5, and 6.

5. Measure backplate (6) thickness.
   a. If thickness of backplate (6) marked No. 4 is less than 0.702 in. (17.83 mm), discard backplate (6).
   b. If thickness of backplate (6) marked No. 5 is less than 0.671 in. (17.04 mm), discard backplate (6).
   c. If thickness of backplate (6) marked No. 6 is less than 0.640 in. (16.26 mm), discard backplate (6).

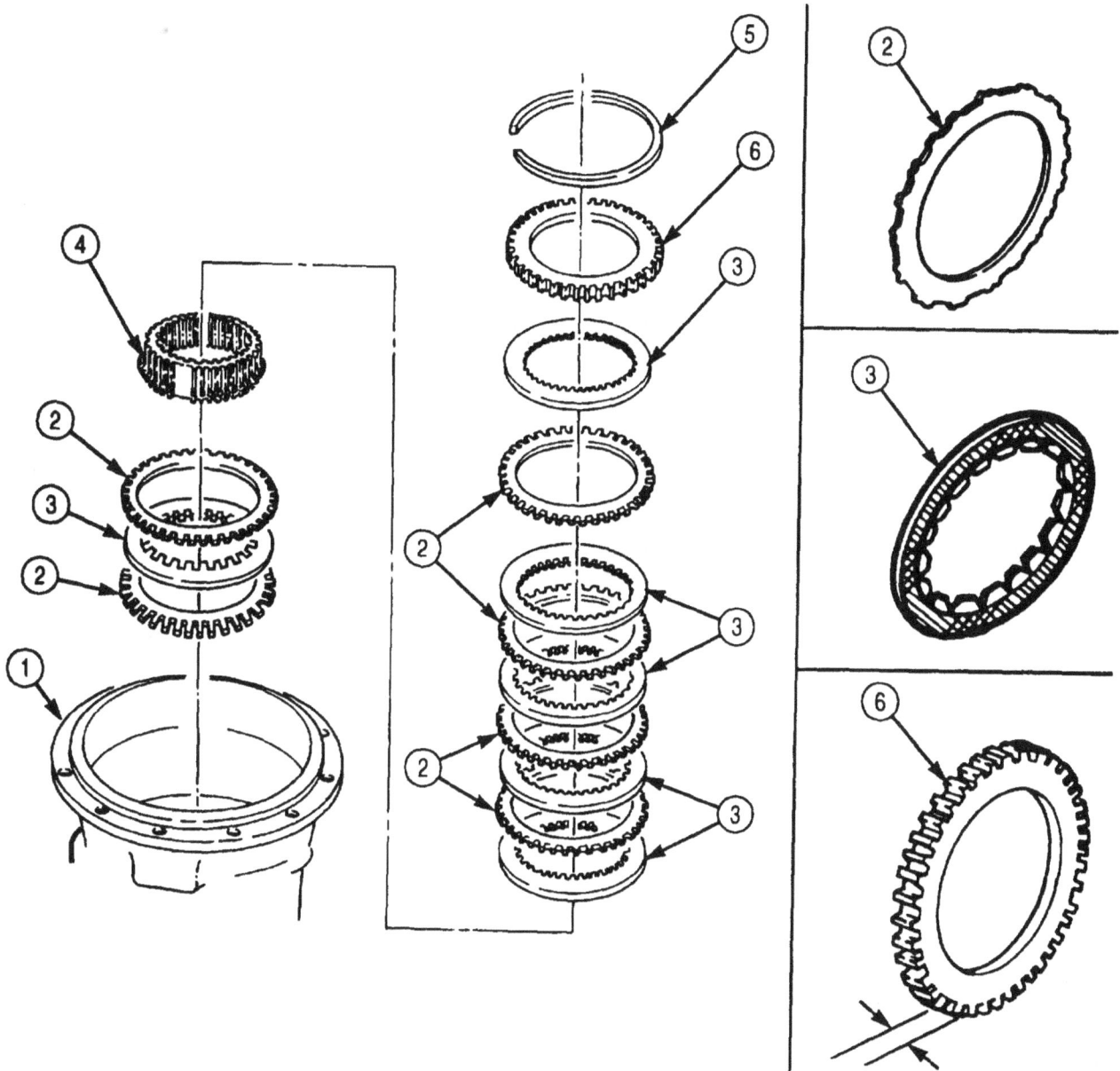

5-53. FIRST CLUTCH MAINTENANCE (Contd)

c. Installation

1. Starting with clutch plate (2), alternately install two clutch plates (2) and one clutch disc (3) in transmission housing (1).
2. Install rear planetary ring gear (4) in transmission housing (1). Ensure all clutch plate (2) tangs and clutch disc (3) splines mesh.
3. Starting with clutch disc (3), alternately install five clutch discs (3) and four clutch plates (2) on rear planetary ring gear (4).
4. Install backplate (6) and snapring (5) in transmission housing (1). Ensure flat side of backplate (6) is installed first.

d. Establishing Clearance

1. Measure clearance between snapring (5) and backplate (6) with first clutch clearance gauge. Clutch running clearance should be 0.074-0.147 in. (1.880-3.374 mm).
2. If clutch running clearance exceeds 0.147 in. (3.734 mm), replace clutch plates (2) and discs (3) with new, thicker plates (2) and discs (3).
3. If clutch running clearance still exceeds 0.147 in. (3.374 mm) after completing step 2, replace backplate (6) with new, thicker backplate (6).
4. If clutch running clearance is less than 0.074 in. (3.374 mm), replace backplate (6) with new, thinner backplate (6).

5-53. FIRST CLUTCH MAINTENANCE (Contd)

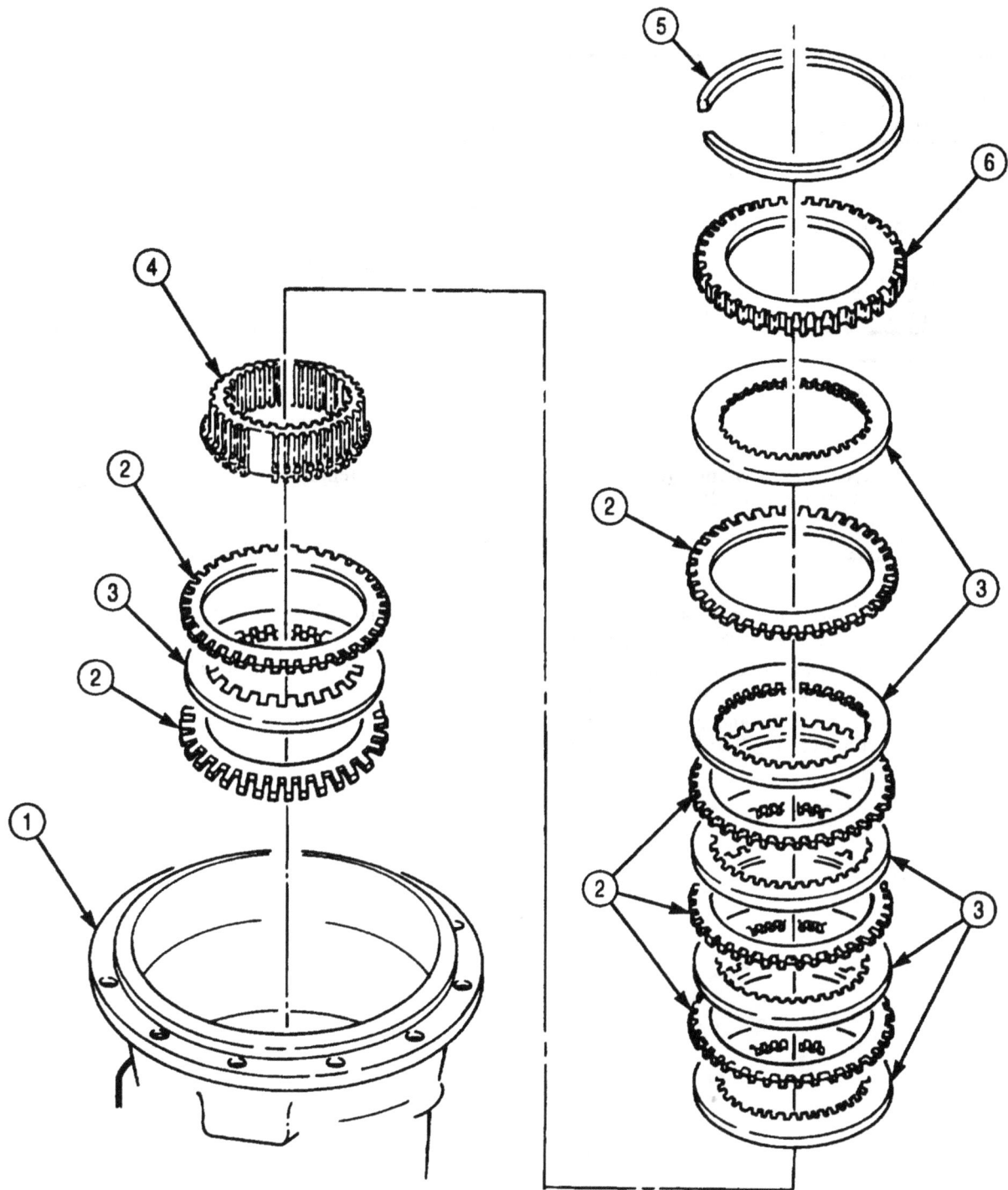

FOLLOW-ON TASK: Install second clutch (para. 5-52).

5-54. TRANSMISSION GOVERNOR REPLACEMENT

THIS TASK COVERS
a. Removal   b. Installation

INITIAL SETUP:

APPLICABLE MODELS
All

TOOLS
General mechanic's tool kit (Appendix E, Item 1)
Torque wrench (Appendix E, Item 146)

MATERIALS/PARTS
Governor cover gasket (Appendix D, Item 241)

REFERENCES (TM)
TM 9-2320-272-24P

EQUIPMENT CONDITION
First clutch removed (para. 5-53)

SPECIAL ENVIRONMENTAL CONDITIONS
Work area clean and free from blowing dirt and dust.

---

a. Removal

1. Remove four screws (3), governor cover (4), and gasket (5) from transmission rear cover (2). Discard gasket (5).
2. Rotate transmission governor (6) clockwise to disengage, and remove transmission governor (6) from transmission rear cover (2).

b. Installation

NOTE
Governor is properly seated only after counterclockwise rotation of governor gear is felt during installation.

1. Insert plastic gear (7) end of transmission governor (6) into transmission governor bore (1) and push inward. Ensure governor (6) is properly seated.
2. Install new gasket (5) and governor cover (4) on transmission rear cover (2) with four screws (3). Tighten screws (3) 15-20 lb-R (20-27 N•m).

5-54. TRANSMISSION GOVERNOR REPLACEMENT (Contd)

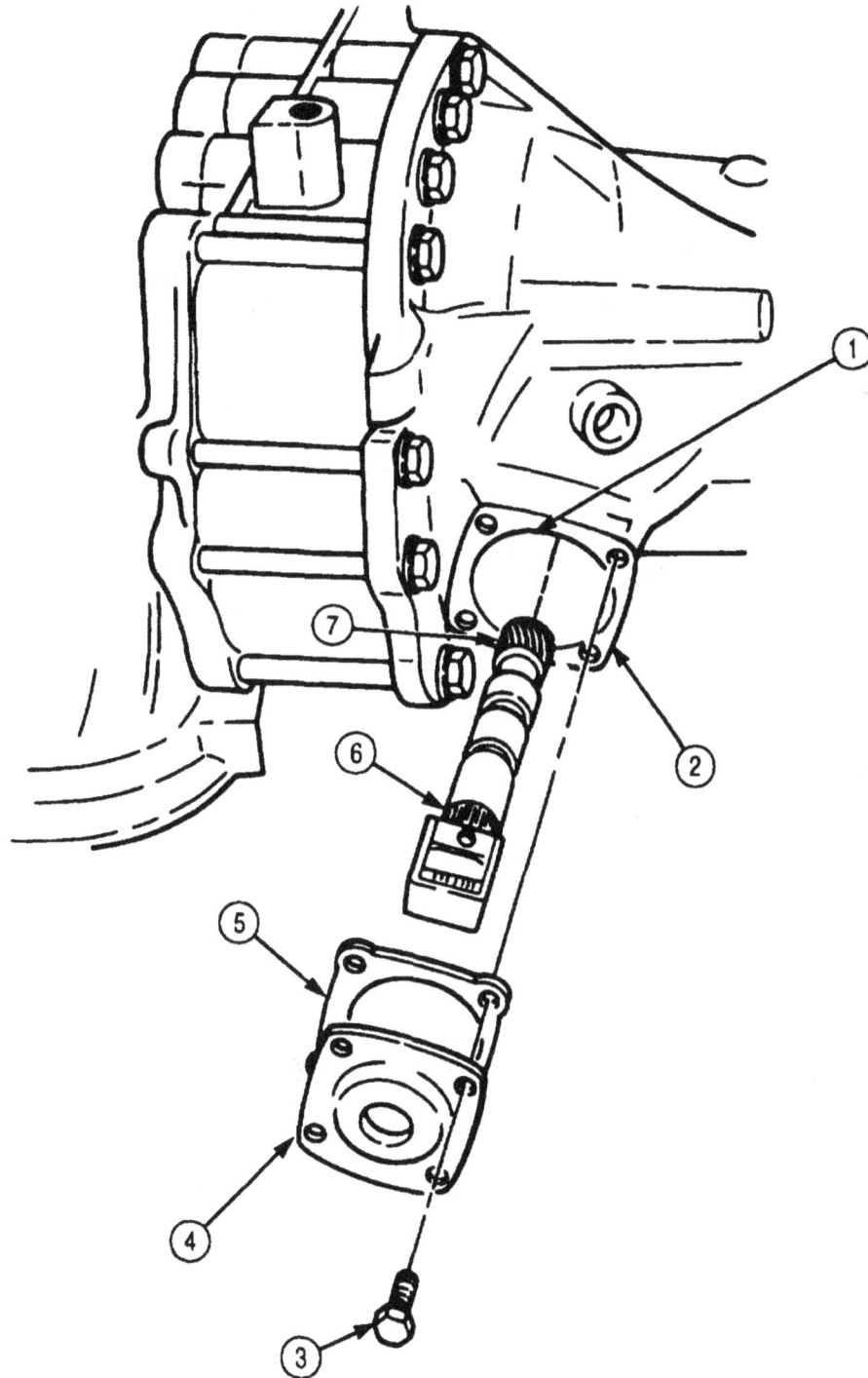

FOLLOW-ON TASK: Install first clutch (para. 5-53).

5-55. REAR COVER AND LOW PLANETARY CARRIER MAINTENANCE

THIS TASK COVERS:

a. Removal
b. Disassembly

c. Cleaning and Inspection
d. Assembly

INITIAL SETUP:

APPLICABLE MODELS
All

SPECIAL TOOLS
Driver handle (Appendix E, Item 38)
Spring compressor (Appendix E, Item 31)
Rear bearing installer (Appendix E, Item 115)
Driver handle (Appendix E, Item 37)
Seal remover (Appendix E, Item 92)
Converter turbine bearing puller
  (Appendix E, Item 105)
Pin remover (Appendix E, Item 94)
Bearing installer (Appendix E, Item 10)
'Iwo guide pliers (Appendix E, Item 59)

Tools
General mechanic's tool kit (Appendix E, Item 1)
Outside micrometer (Appendix-E, Item 80)
Vernier calipers (Appendix E, Item 159)
Spring tester (Appendix E, Item 131)
Torque wrench (Appendix E, Item 144)
Arbor press

MATERIALS/PARTS
Low clutch piston inner seal ring
  (Appendix D, Item 259)
Low clutch piston outer seal ring
  (Appendix D, Item 509)
Gasket (Appendix D, Item 228)
Crocus cloth (Appendix C, Item 20)
Oil-soluble grease (Appendix C, Item 53)

REFERENCES (TM)
TM 9-2320-272-24P

EQUIPMENT CONDITION
• Transmission governor removed (para. 5-54).
• Transmission output shaft oil seal removed
  (para. 4-78).

SPECIAL ENVIRONMENTAL CONDITIONS
Work area clean and free from blowing dirt and dust.

a. Removal

NOTE

Turn transmission housing with rear cover facing upward.

1. Remove fourteen screws (1) and washers (5) from rear cover (2).

NOTE

Clean gasket remains from mating surfaces.

2. Remove rear cover (2), low planetary carrier (9), gasket (4), and thrust washer (10) from transmission housing (3). Discard gasket (4).

3. Remove sleeve spacer (6), speedometer drive gear (7), and governor drive gear (8) from low planetary carrier (9).

5-55. REAR COVER AND LOW PLANETARY CARRIER MAINTENANCE (Contd)

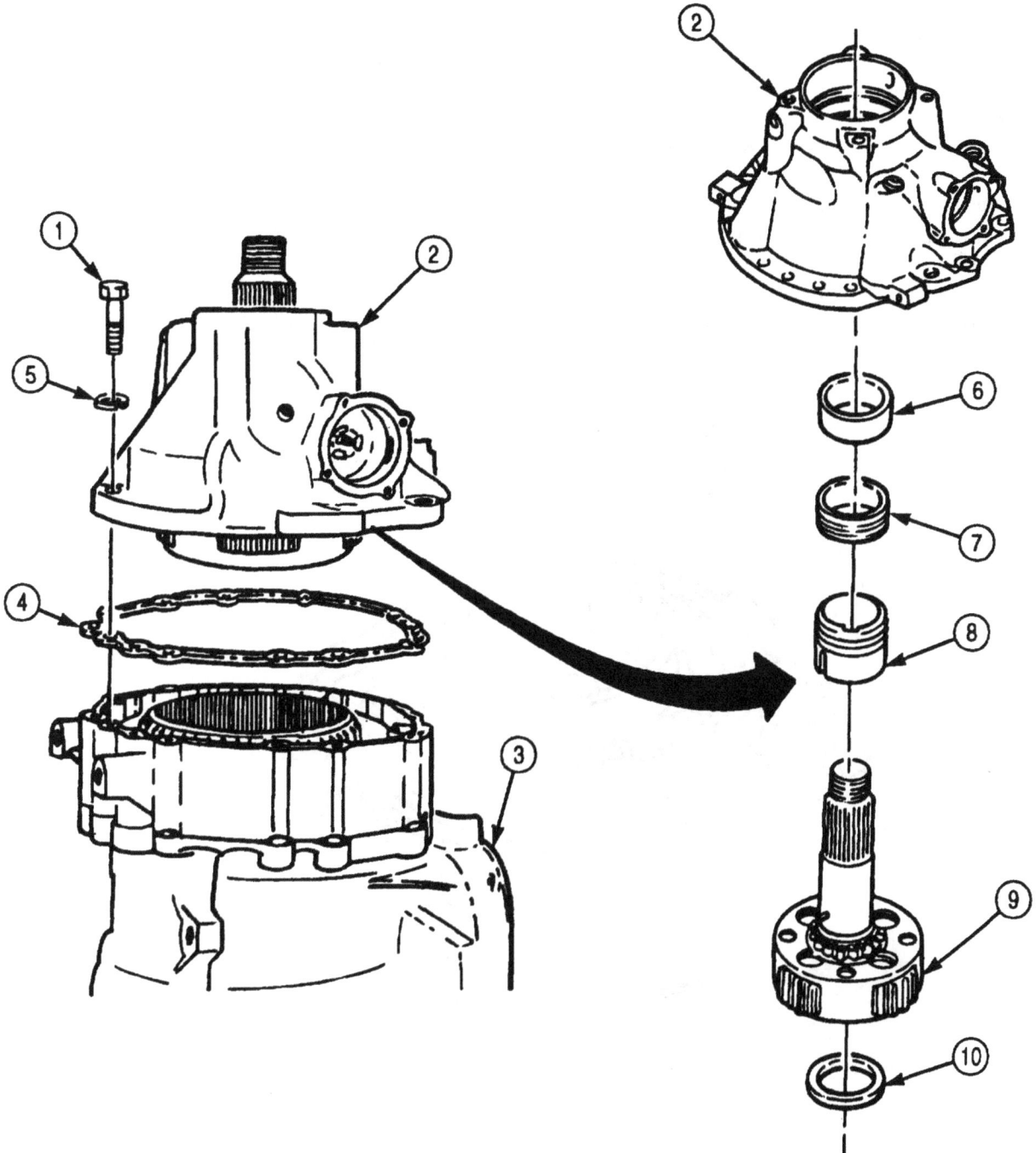

5-55. REAR COVER AND LOW PLANETARY CARRIER MAINTENANCE (Contd)

b. Disassembly

1. Remove dust cover (11) and oil seal (10) from rear cover (1) with seal puller and hook.
2. Remove beveled snapring (9) and rear output shaft bearing (8) from rear cover (1) with converter turbine bearing puller.
3. Position spring compressor and arbor press on low clutch spring retainer (3) and apply pressure to spring retainer (3).
4. Remove snapring (2) and spring compressor from low clutch spring retainer (3).
5. Remove low clutch spring retainer (3) and twenty-six piston return springs (4) from low clutch piston (5).
6. Remove low clutch piston (5) from rear cover (1).
7. Remove outer seal ring (6) and inner seal ring (7) from low clutch piston (5). Discard seal rings (6) and (7).
8. Remove plug (13), drain tube (12), and plug (17) from rear cover (1).
9. Remove filter plug (14), O-ring (15), and governor filter (16) from rear cover (1).

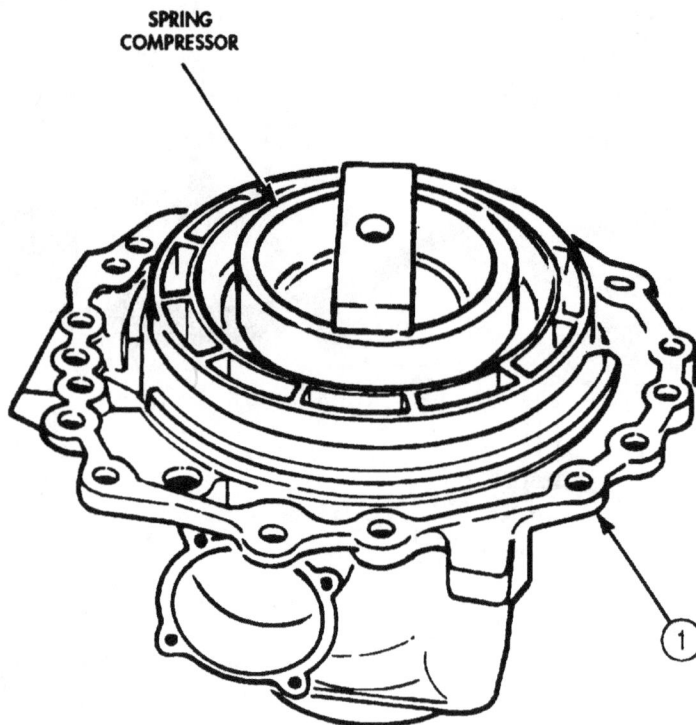

SPRING COMPRESSOR

5-55. REAR COVER AND LOW PLANETARY CARRIER MAINTENANCE (Contd)

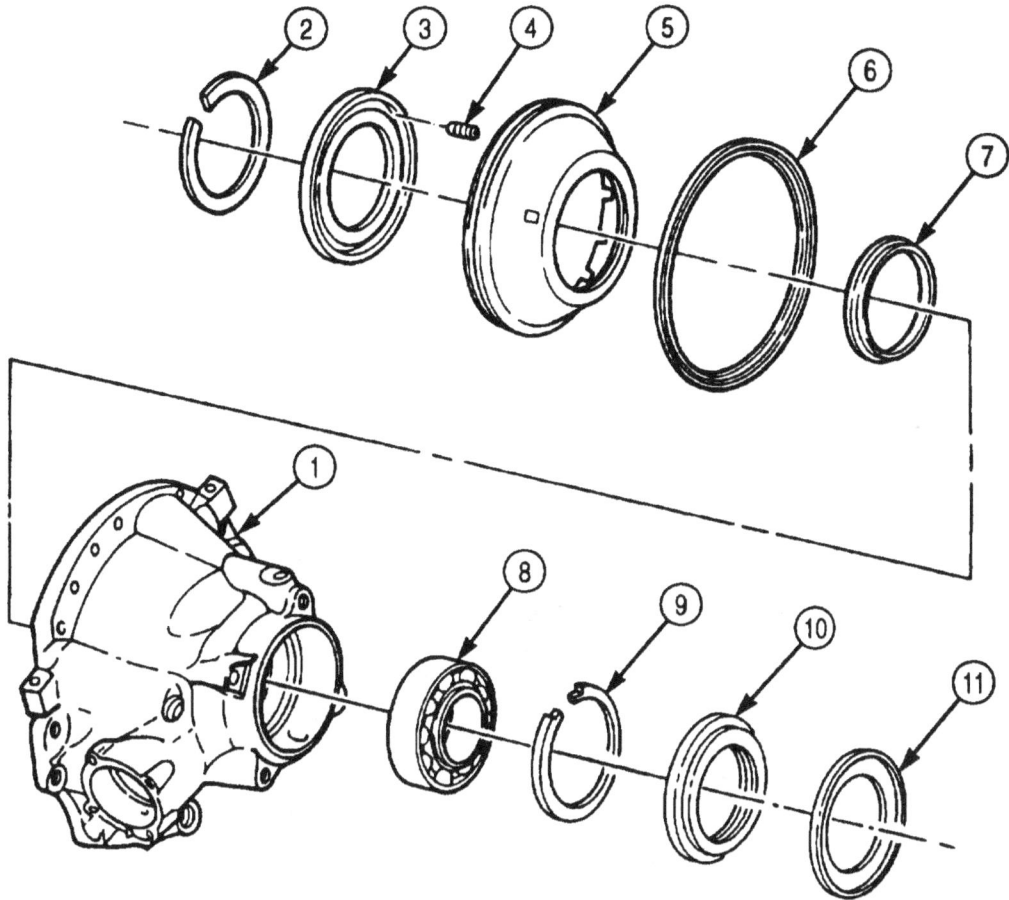

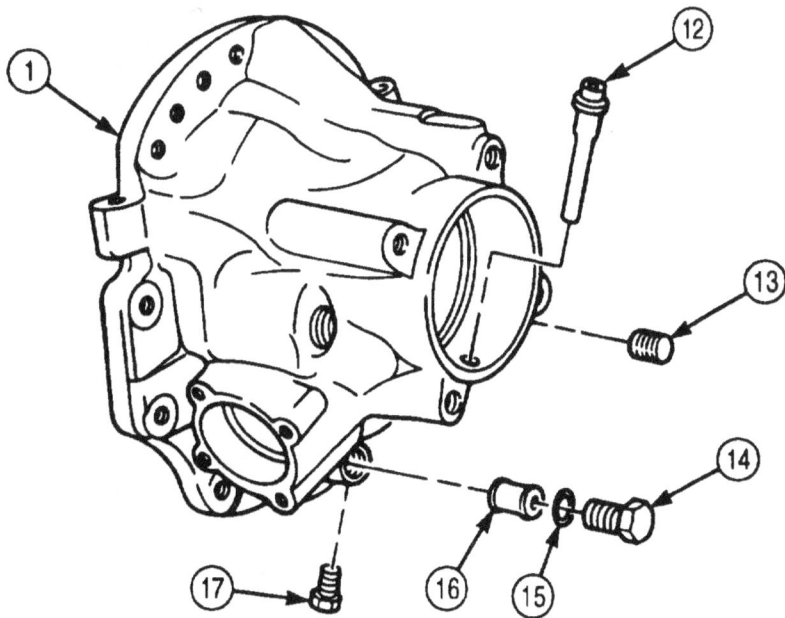

5-55. REAR COVER AND LOW PLANETARY CARRIER MAINTENANCE (Contd)

c. Cleaning and Inspection

1. For general cleaning instructions, refer to para. 2-14.
2. For general cleaning instructions, refer to para. 2-15.
3. Inspect rear cover (1) for breaks and cracks. Replace rear cover (1) if broken or cracked.
4. Measure governor clearance in bore (2) of rear cover (1). Replace rear cover (1) if clearance exceeds .0035 in. (.0089 mm).
5. Inspect governor support pin (9) for looseness, bends, and breaks. Replace governor support pin (9) with pin remover if loose, bent, or broken.
6. Inspect low planetary carrier (5) for missing or twisted splines (6). Replace low planetary carrier (5) if splines (6) are missing or twisted.
7. Inspect low planetary carrier (5) for burrs. If burrs are found, remove using crocus cloth.
8. Inspect roller bearing (7) on low planetary carrier (5) and race (8) in rear cover (1). Replace bearing (7) and race (8) as a set if defective.

NOTE
- Perform steps 9 and 10 only if bearing or race is defective.
- Pin must be removed to remove bearing.

9. Remove pin (4) and roller bearing (7) from low planetary carrier (5). Discard pin (4).
10. Remove race (8) from rear cover (1) using bearing puller.
11. Measure thrust washer (3) thickness. Discard thrust washer (3) if thickness is less than 0.091 in. (2.31mm).
12. Inspect thrust washer (3) for scoring. Discard if scored.
13. Inspect twenty-six piston return springs (10) for discoloration due to overheating. Discard piston return springs (10) if discolored.
14. Inspect twenty-six piston return springs (10) for broken coils or coils distorted due to wear. Discard piston return springs (10) if broken or distorted.
15. Inspect twenty-six piston return springs (10) for serviceability by checking load when compressed with spring tester. Discard spring (10) if spring (10) does not give load according to table 5-15, Spring Data.

Table 5-15. Spring Data.

SPRING	COLOR	FREE LENGTH	COMPRESSED LENGTH	UNDER LOAD
Piston return spring	Solid orange, yellow stripe	1.28 in. (32.5 mm)	0.95 in. (24.1 mm)	13.6-16.4 lb (60.5-72.9 N)

5-55. REAR COVER AND LOW PLANETARY CARRIER MAINTENANCE (Contd)

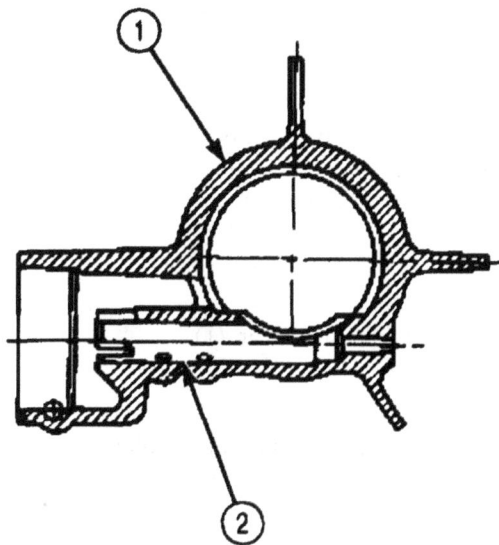

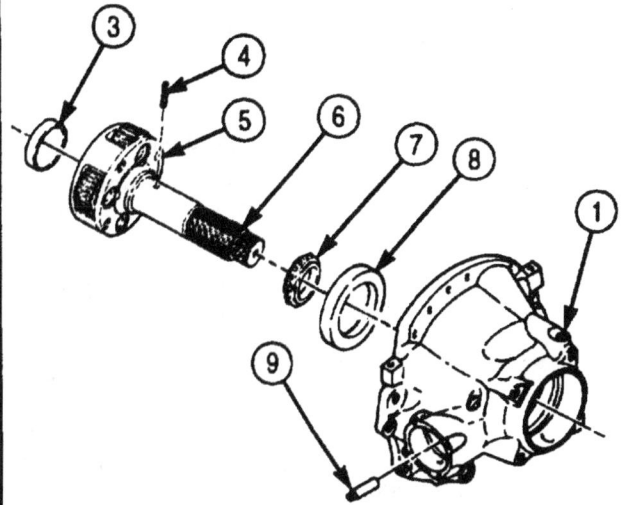

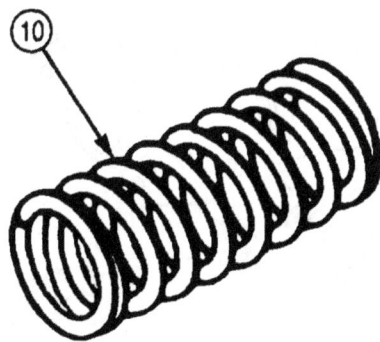

5-55. REAR COVER AND LOW PLANETARY CARRIER MAINTENANCE (Contd)

d. Assembly

1. Install governor filter (6), O-ring (5), and filter plug (4) in rear cover (1). Tighten filter plug (4) 50-70 lb-ft (68-95 N•m).

2. Install drain tube (2) and plug (3) in rear cover (1). Tighten plug (3) 12-16 lb-ft (16-22 N•m).

3. Install plug (7) in rear cover (1). Tighten plug (7) 4-8 lb-ft (5-11 N•m).

NOTE

Perform step 4 only if support pin was previously removed.

4. Using pin installer, press new governor support pin (11) into rear cover (1) until end of pin (11) is 5.886-5.896 in. (14.950-14.975 cm) from outside face (12) of rear cover (1).

NOTE

Perform steps 5 and 6 only if bearing and bearing race were previously removed.

5. Using bearing race installer, install bearing outer race (13) in rear cover (1).

NOTE

Install roller bearing before installing pin.

6. Install roller bearing (10) and new pin (8) on low planetary carrier (14). Seat roller bearing (10) to shoulder on shaft (9).

7. Install new thrust washer (15) if discarded during inspection.

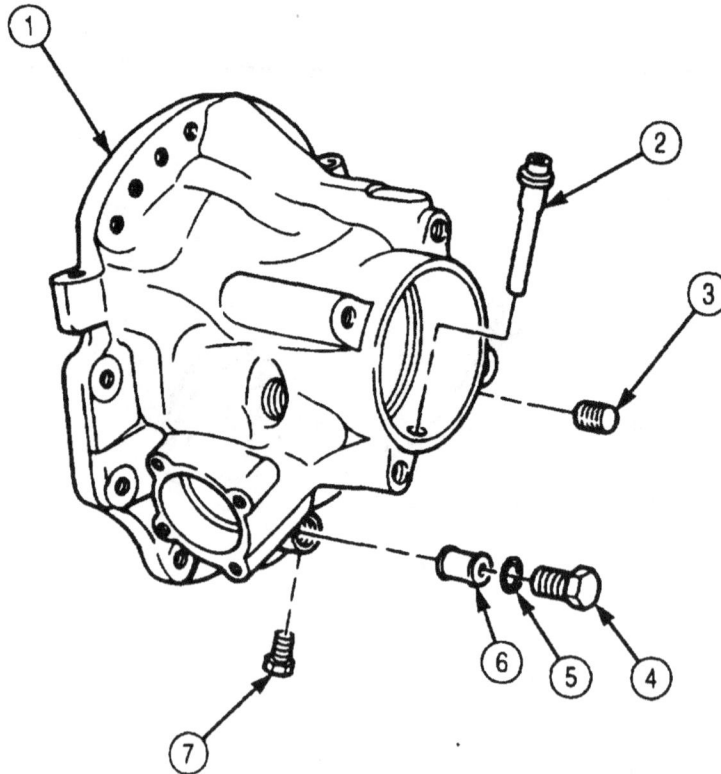

5-55. REAR COVER AND LOW PLANETARY CARRIER MAINTENANCE (Contd)

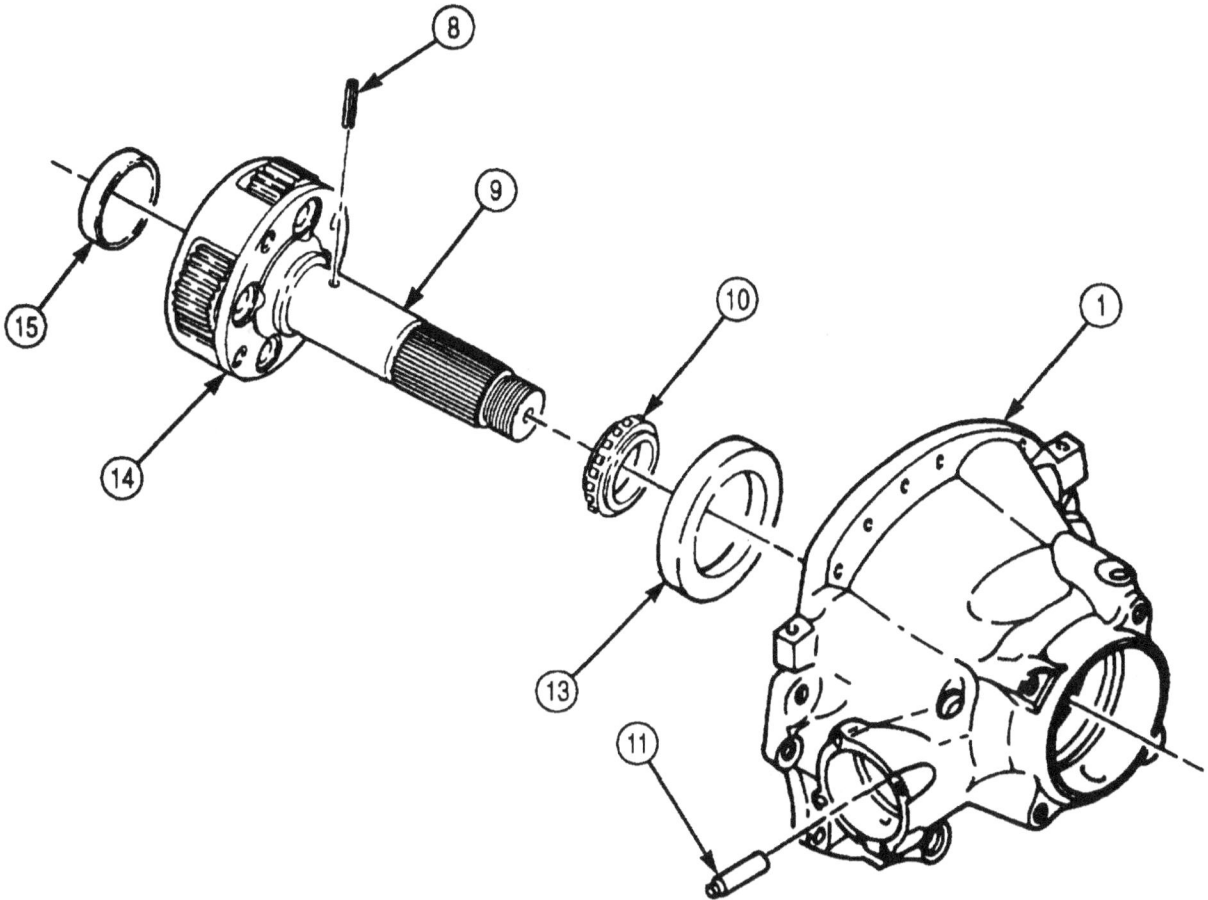

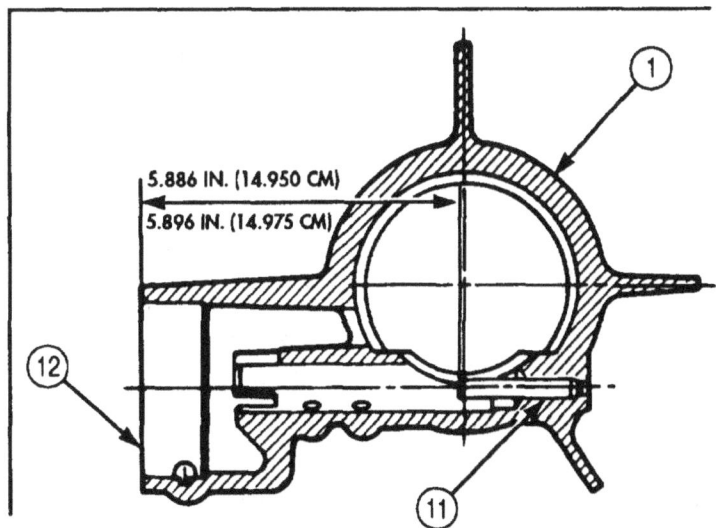

5.886 IN. (14.950 CM)

5.896 IN. (14.975 CM)

5-55. REAR COVER AND LOW PLANETARY CARRIER MAINTENANCE (Contd)

NOTE
- Ensure seal rings and lips face piston oil pressure side.
- Use oil-soluble grease to hold in place.

8. Install new outer seal ring (6) and new inner seal ring (7) on low clutch piston (5).

9. Place low clutch piston (5) in rear cover (1).

10. Place twenty-six piston return springs (4) in low clutch piston (5).

11. Place cupped side of clutch spring retainer (3) over piston return springs (4).

12. Place spring compressor and snapring (2) on clutch spring retainer (3) and compress twenty-six piston return springs (4).

13. Install clutch spring retainer (3) with snapring (2).

14. Remove spring compressor from clutch spring retainer (3).

15. Install rear output shaft bearing (8) in rear cover (1) with bearing installer until seated.

16. Install beveled snapring (9) in rear cover (1) with beveled side toward rear of transmission.

17. Install oil seal (10) and dust shield (11) in rear cover (1).

18. Install rear cover and low planetary carrier (para. 5-58).

5-55. REAR COVER AND LOW PLANETARY CARRIER MAINTENANCE (Contd)

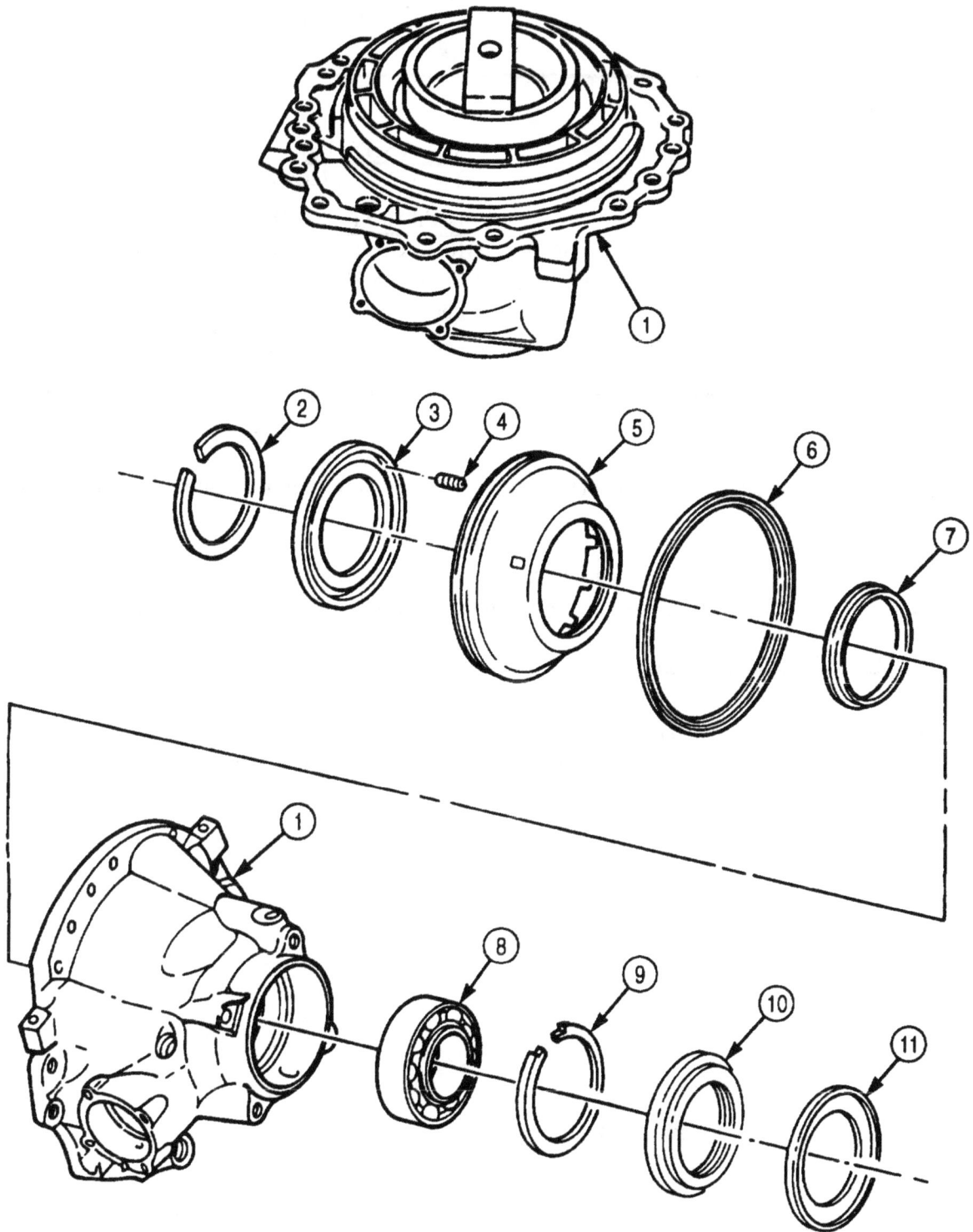

FOLLOW-ON TASKS: Install transmission governor (para. 5-54).
- Install transmission output shaft oil seal (para. 4-78).

5-56. LOW CLUTCH REMOVAL AND CLEARANCE

THIS TASK COVERS:

a. Removal   c. Establishing Clearance
b. Cleaning and Inspection

INITIAL SETUP:

APPLICABLE MODELS
All

REFERENCES (TM)
TM 9-2320-272-24P

TOOLS
General mechanic's tool kit (Appendix E, Item 1)
Depth micrometer (Appendix E, Item 81)

EQUIPMENT CONDITION
Rear cover and low planetary carrier removed
(para. 5-55).

MATERIALS/PARTS
Gasket (Appendix D, Item 228)

SPECIAL ENVIRONMENTAL CONDITIONS
Work area clean and free from blowing dirt and
dust.

---

a. Removal

1. Remove low planetary ring gear (1), snapring (3), and low gear ring hub (2) from adapter
    housing (6).

NOTE

Mark position of clutch plates in adapter housing for installation.

2. Remove eight clutch plates (4) and seven clutch discs (5) from adapter housing (6).

b. Cleaning and Inspection

1. For general cleaning instructions, refer to para. 2-14.
2. For general inspection instructions, refer to para. 2-15.
3. Inspect clutch discs (5) for burned surfaces. If burned, discard clutch discs (5).
4. Inspect clutch plates (4) for burned surfaces. If burned, discard clutch plates (4).

5-56. LOW CLUTCH REMOVAL AND CLEARANCE (Contd)

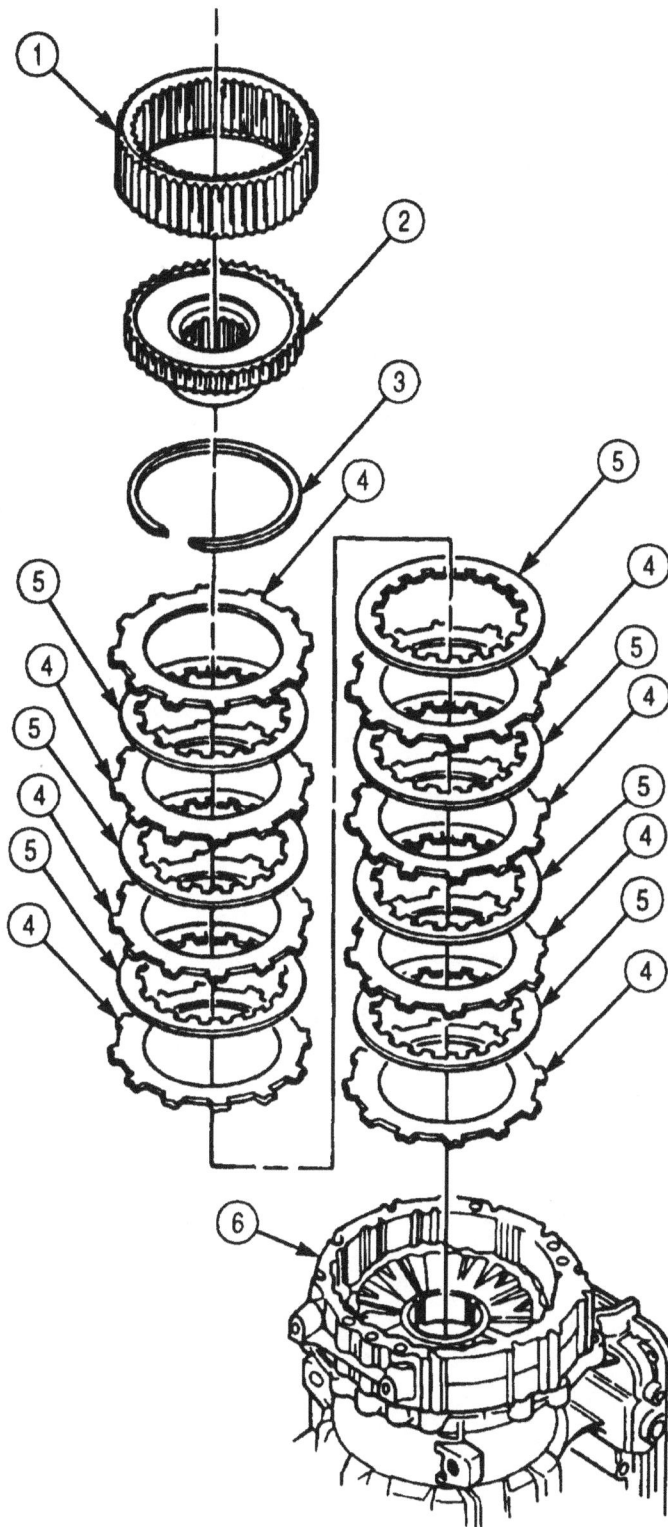

5-56. LOW CLUTCH REMOVAL AND CLEARANCE (Contd)

c. Establishing Clearance

1. Starting with clutch plate (2), alternately install eight clutch plates (2) and seven clutch discs (3) in adapter housing (1).

2. Measure distance from top edge of adapter housing (1) to top of clutch plate (2) with depth micrometer. Record measurement.

3. Position new rear cover gasket (4) on rear cover (5).

4. Measure distance from top of edge of piston (6) to gasket (4) with depth micrometer. Record measurement.

5. Subtract step 4 measurement from step 2 measurement. Record the difference. This measurement is low clutch clearance.

6. Low clutch clearance, as determined in step 5, should be 0.073-0.141 in. (1.85-3.58 mm). If clutch clearance exceeds 0.141 in. (3.58 mm), replace clutch plates (2) and discs (3) with new, thicker plates (2) and discs (3).

7. Remove eight clutch plates (2) and seven clutch discs (3) from adapter housing (1).

5-56. LOW CLUTCH REMOVAL AND CLEARANCE (Contd)

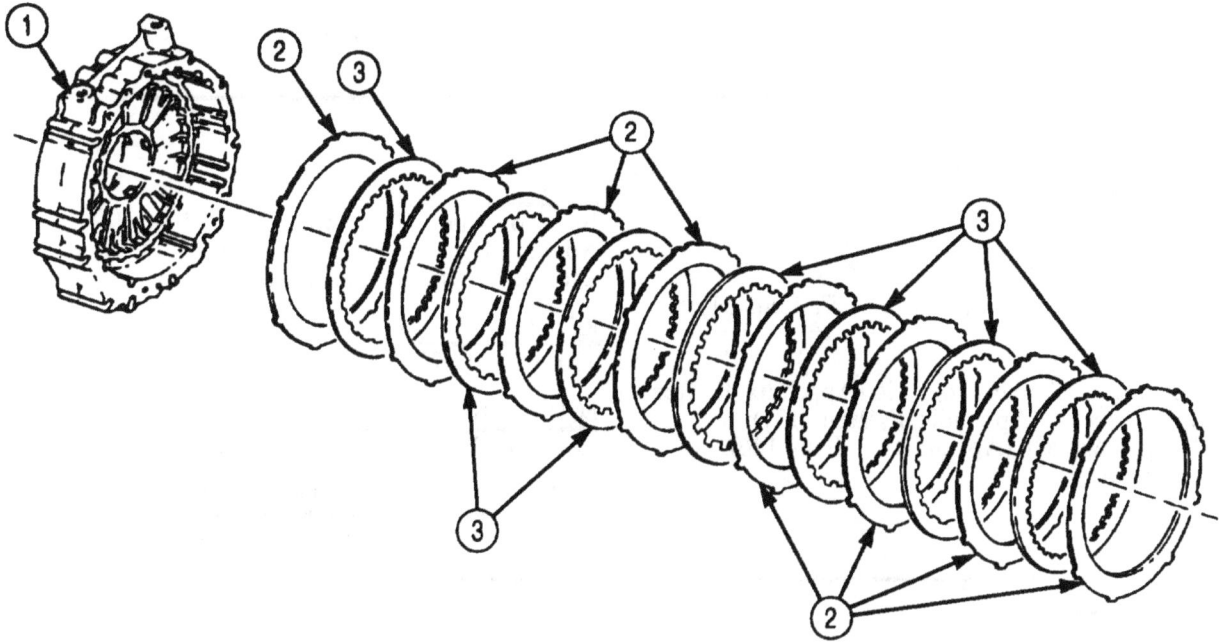

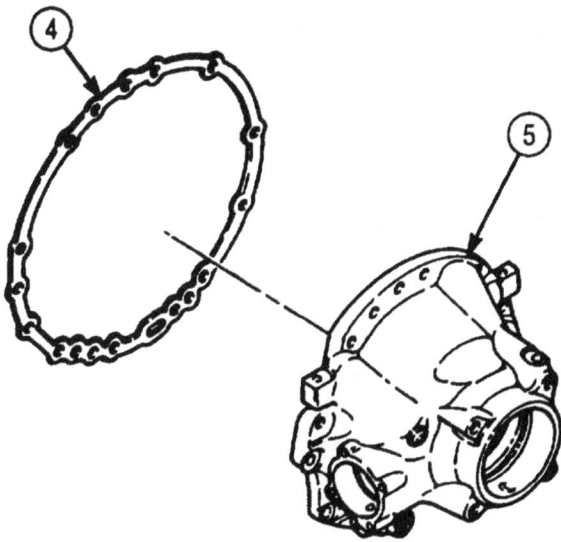

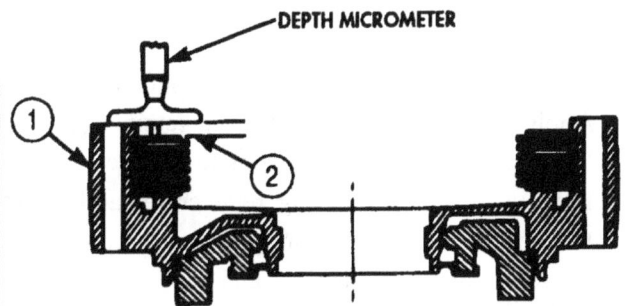

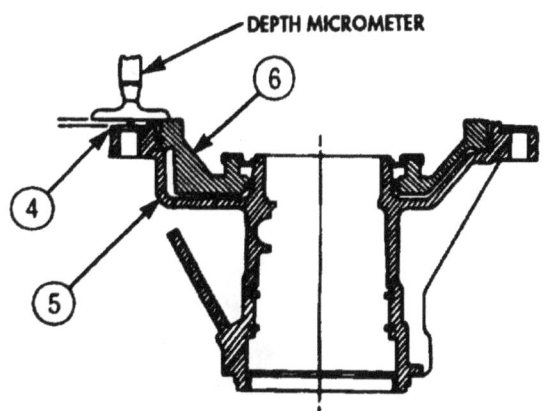

FOLLOW-ON TASK: Install rear cover and low planetary carrier (para. 5-58).

5-57. ADAPTER HOUSING AND FIRST CLUTCH PISTON MAINTENANCE

THIS TASK COVERS
a. Removal
b. Disassembly

c. Cleaning and Inspection
d. Assembly

INITIAL SETUP:

APPLICABLE MODELS
All

SPECIAL TOOLS
Spring compressor (Appendix E, Item 31)

TOOLS
General mechanic's tool kit (Appendix E, Item 1)
Spring tester (Appendix E, Item 131)
Torque wrench (Appendix E, Item 146)
Arbor press

MATERIALS/PARTS
Basic overhaul kit (Appendix D. Item 10)
Adapter housing gasket (Appendix D, Item 228)
Oil-soluble grease (Appendix C, Item 53)

REFERENCES (TM)
TM 9-2320-272-24P

EQUIPMENT CONDITION
Low clutch removed (para. 5-56)

SPECIAL ENVIRONMENTAL CONDITIONS
Work area clean and free from blowing dirt and dust.

a. Removal

Remove adapter housing (1) and gasket (2) from transmission housing (3). Discard gasket (2).

b. Disassembly

1. Position compressor tool on first clutch spring retainer (5) and arbor press.
2. Apply pressure to first clutch ring retainer (5), and remove snapring (4) from adapter housing (1).
3. Remove compressor tool from first clutch spring retainer (5).
4. Remove first clutch spring retainer (5) and twenty-six springs (6) from first clutch piston (7).
5. Remove first clutch piston (7) from adapter housing (1).
6. Remove outer seal ring (8) and inner seal ring (9) from first clutch piston (7). Discard inner seal ring (9) and outer seal ring (8).
7. Remove plug (10) from adapter housing (1).

5-57. ADAPTER HOUSING AND FIRST CLUTCH PISTON MAINTENANCE (Contd)

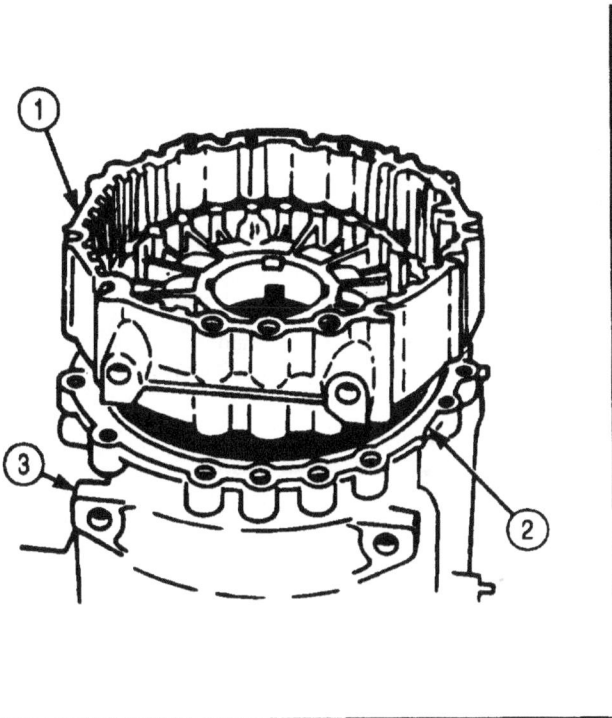

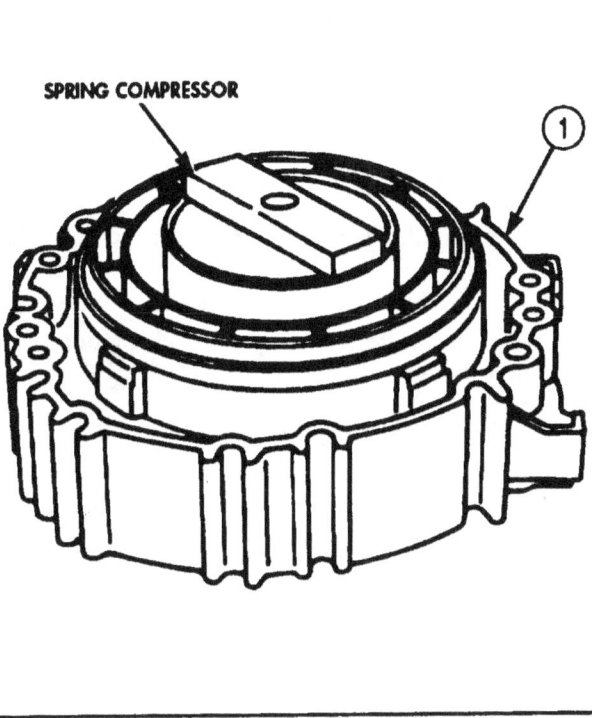

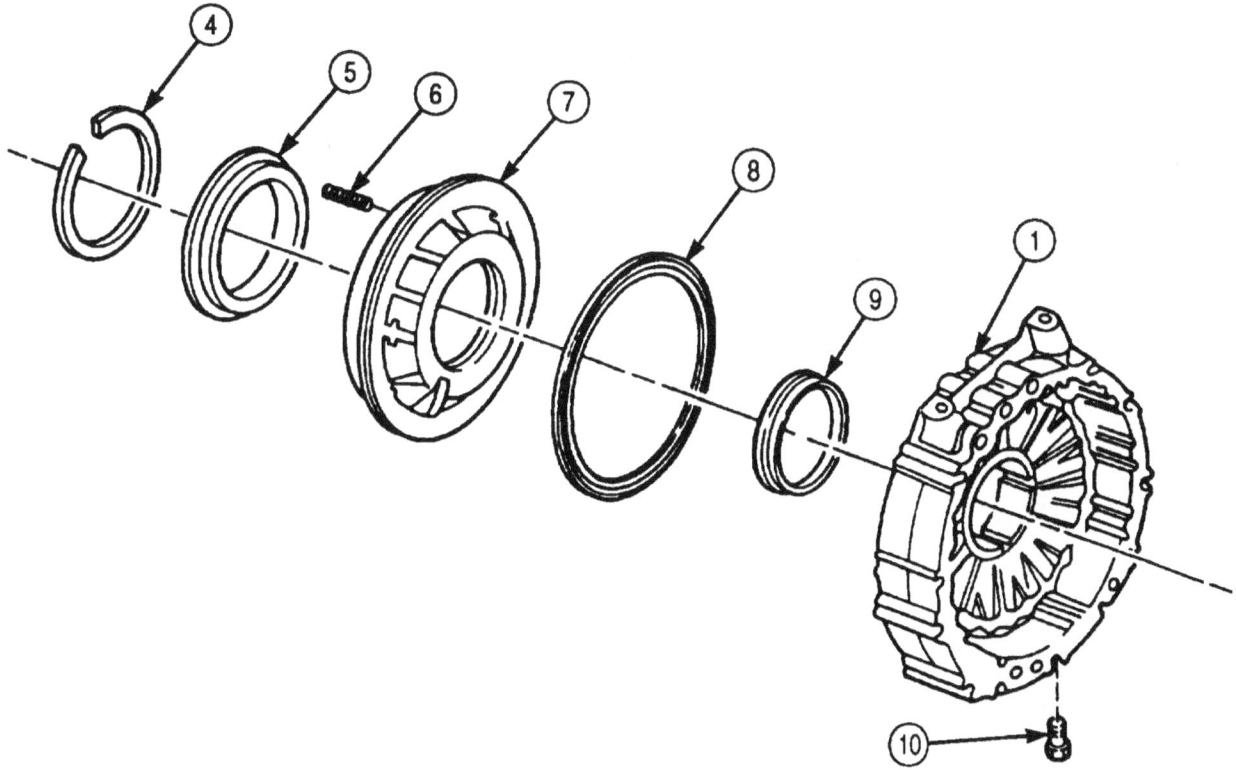

SPRING COMPRESSOR

5-57. ADAPTER HOUSING AND FIRST CLUTCH PISTON MAINTENANCE (Contd)

c. Cleaning and Inspection

1. For general cleaning instructions, refer to para. 2-14.

2. For general inspection instructions, refer to para. 2-15.

3. Inspect twenty-six piston return springs (4) for discoloration due to overheating. If discolored, discard piston return springs (4)

4. Inspect twenty-six piston return springs (4) for broken coils or coils distorted due to wear. If broken or distorted, discard piston return springs (4).

NOTE

Perform step 5 if springs pass visual inspection.

5. Using spring tester, compress each of twenty-six piston return springs (4) to 0.95 in. (24.1 mm) and check load. If piston return spring (4) does not give the correct load, discard piston return spring (4). Refer to table 5-16, Spring Data.

Table 5-16. Spring Data.

SPRING	COLOR	FREE LENGTH	COMPRESSED LENGTH	UNDER LOAD
Piston return spring	Solid orange, yellow stripe	1.28 in. (32.5 mm)	0.95 in. (24.1 mm)	13.6-16.4 lb. (60.5-72.9 N)

d. Assembly

NOTE

• Ensure seal rings and lips face piston oil pressure side.

• Use oil-soluble grease to hold seal rings in place.

1. Install new outer seal ring (6) and new inner seal ring (7) on first clutch piston (5).

2. Place first clutch piston (5) in adapter housing (1).

3. Place twenty-six piston return springs (4) in first clutch piston (5).

4. Place cupped side of spring retainer (3) over piston return springs (4).

5. Place compressor tool on spring retainer (3) and arbor press, and compress twenty-six piston return springs (4).

6. Install spring retainer (3) with snapring (2).

7. Remove compressor tool from spring retainer (3).

8. Install plug (8) in adapter housing (1). Tighten plug (8) 4-8 lb-ft (5-11 N•m).

5-57. ADAPTER HOUSING AND FIRST CLUTCH PISTON MAINTENANCE (Contd)

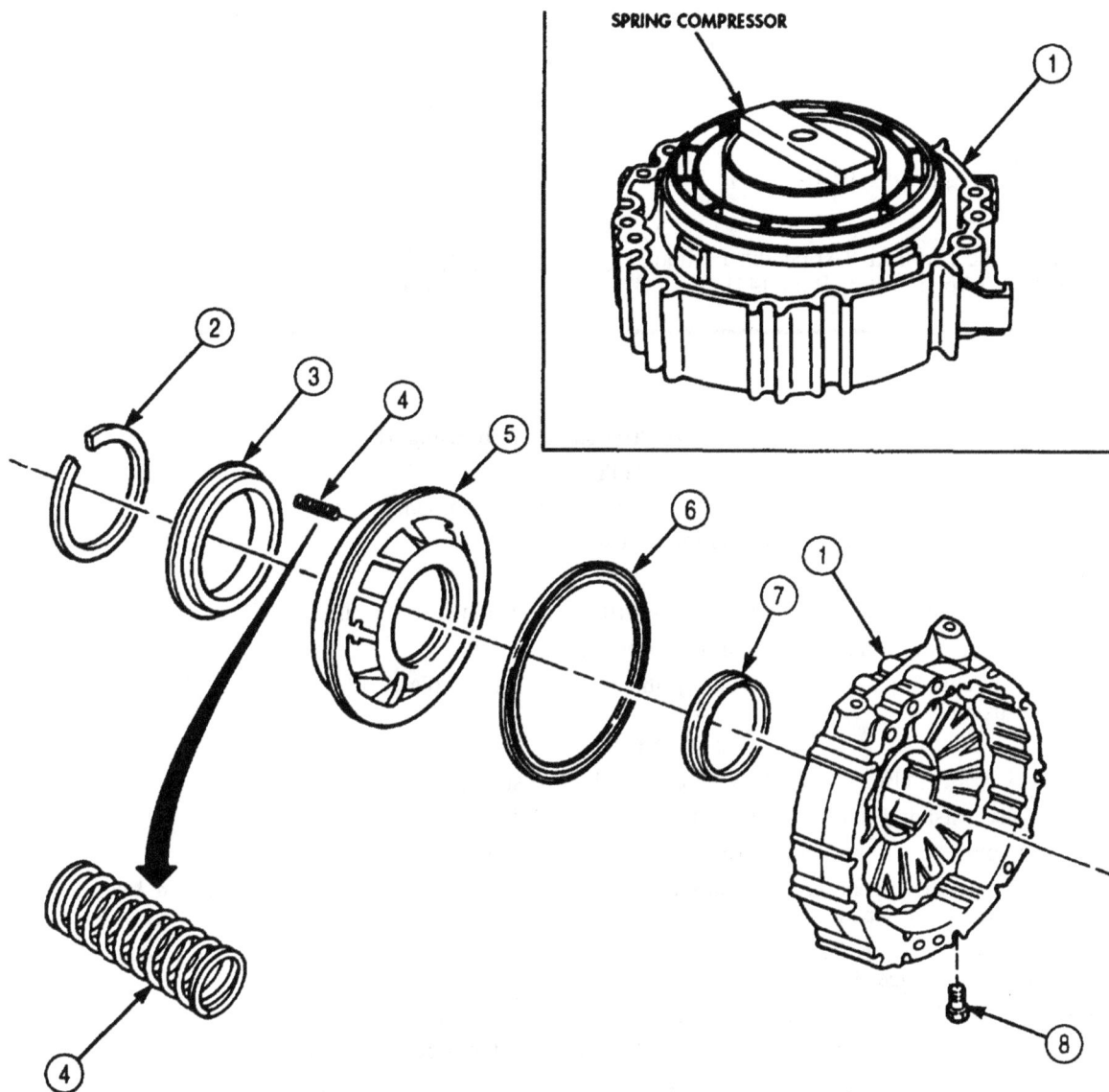

SPRING COMPRESSOR

FOLLOW-ON TASK: Install adapter housing, low clutch, low planetary carrier, and rear cover (para. 5-58).

5-58. ADAPTER HOUSING, LOW CLUTCH, LOW PLANETARY CARRIER, AND REAR COVER INSTALLATION

THIS TASK COVERS:
Installation

INITIAL SETUP:

APPLICABLE MODELS
All

SPECIAL TOOLS
Two guide pins (Appendix E, Item 59)

TOOLS
General mechanic's tool kit (Appendix E, Item 1)
Torque wrench (Appendix E, Item 144)

MATERIALS/PARTS
Two gaskets (Appendix D, Item 228)
Oil-soluble grease (Appendix C, Item 53)

REFERENCES (TM)
TM 9-2320-272-24P

SPECIAL ENVIRONMENTAL CONDITIONS
Work area clean and free from blowing dirt and dust.

Installation

1. Install new adapter housing gasket (8) on transmission housing (6).

NOTE
Guide pins maintain gasket alignment.

2. Install two guide pins (7) in transmission housing (6).
3. Install adapter housing (9) over guide pins (7).
4. Install low ring gear hub (11) and snapring (10) on low planetary ring gear (1).
5. Install low planetary ring gear (1) into adapter housing (9).
6. Starting with clutch plate (4), alternately install eight clutch plates (4) and seven clutch discs (5) in adapter housing (9).

NOTE
Use oil-soluble grease to hold thrust washer in place.

7. Install thrust washer (3) on low planetary carrier (2) hub.
8. Install low planetary carrier (2) into adapter housing (9).
9. Install governor drive gear (17), speedometer drive gear (16), and sleeve spacer (15) on low planetary carrier (2). Ensure governor drive gear (17) is seated over pin (19).
10. Install new rear cover gasket (18) and rear cover (14) over guide pins (7) with twelve washers (12) and screws (13). Finger-tighten screws (13).
11. Remove two guide pins (7) and install remaining two washers (12) and screws (13).
12. Alternately tighten fourteen screws (13) 180° apart. Tighten screws (13) 30 lb-ft (41 N•m).
13. Repeat tightening sequence on fourteen screws (13) 180° apart to achieve final torque. Tighten screws (13) 81-97 lb-ft (110-132 N•m).

5-58. ADAPTER HOUSING, LOW CLUTCH, LOW PLANETARY CARRIER, AND REAR COVER INSTALLATION (Contd)

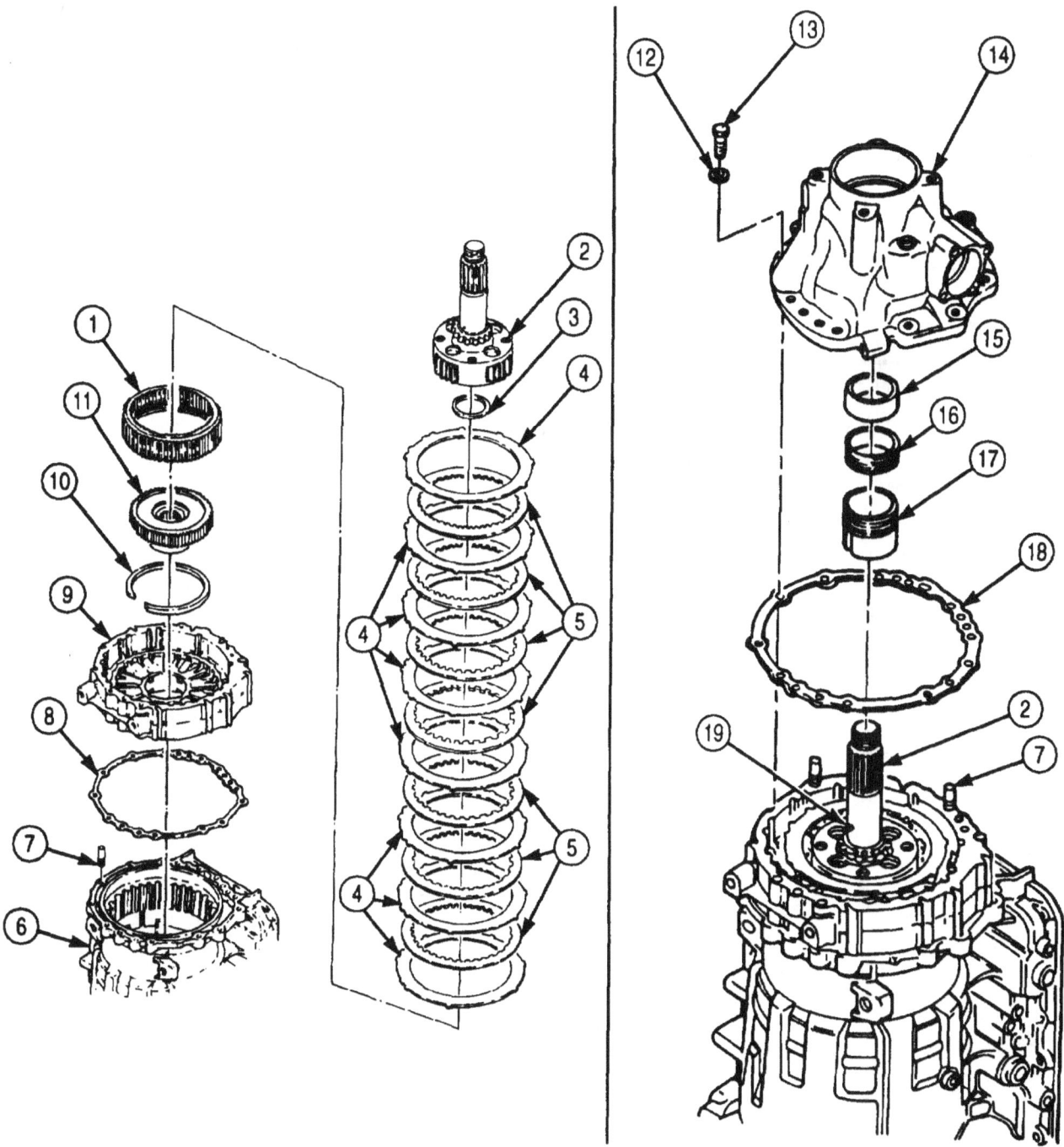

5-59. TRANSMISSION HOUSING MAINTENANCE

THIS TASK COVERS:

a. Disassembly   c. Assembly
b. Cleaning and Inspection

INITIAL SETUP:

APPLICABLE MODELS   **REFERENCES (TM)**
All   TM 9-2320-272-24P

TOOLS   **SPECIAL ENVIRONMENTAL CONDITIONS**
General mechanic's tool kit (Appendix E, Item 1) Work area clean and free from blowing dirt and
Torque wrench (Appendix E, Item 146)   dust.

---

a. Disassembly

Remove four plugs (3) from transmission housing (1).

b. Cleaning and Inspection

1. For general cleaning instructions, refer to para. 2-14.
2. For general inspection instructions, refer to para. 2-15.
3. If adapter (2) is defective, replace adapter (2).

c. Assembly

1. Install four plugs (3) in transmission housing (1). Tighten 4-5 lb-ft (5-7 N-m).

NOTE

Perform step 2 if early model transmission is to have wide oil
pump installed (para. 5-39).

2. Change data plate (4) part number to 23040127.

5-59. TRANSMISSION HOUSING MAINTENANCE (Contd)

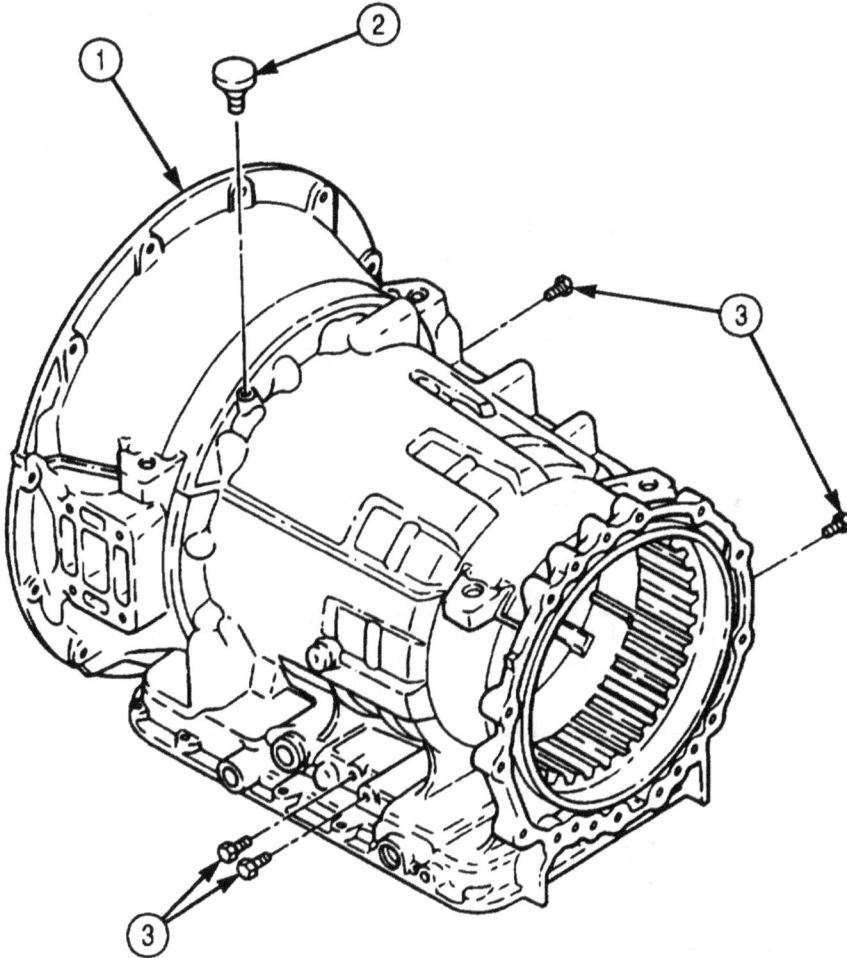

Allison Transmission

Manufactured By

Detroit Diesel Allison

Division of General Motors Corporation

Indianapolis, Indiana

SERIAL NO.	PART NO.
2420033993	~~6885292~~ 23040127 85C12

MODEL NO.	MT 654CR

5-60. TRANSMISSION OIL PRESSURE TESTING

THIS TASK COVERS:

a. Oil Cooler Pressure Test
b. Main Pressure and Governor
   Pressure Test

c. Automatic Shift Speed Test

INITIAL SETUP:

APPLICABLE MODELS
All

SPECIAL TOOLS
Pressure gauge set (300 psi) (Appendix E, Item 51)

TOOLS
General mechanic's tool kit (Appendix E, Item 1)
Drill press
21/64 drill bit
1/4-18 NPTF tap
Pipe plug
Two pipe plugs

MATERIALS/PARTS
O-ring (Appendix D, Item 430)
Cap and plug set (Appendix C, Item 14)

REFERENCES (TM)
TM 9-2320-272-10
TM 9-2320-272-24P

EQUIPMENT CONDITION
• Parking brake set (TM 9-2320-272-10).
• Transmission oil at proper level (TM 9-2320-272-10).
• Vehicle at curb weight (empty) (TM 9-2320-272-10).
• Wheels blocked (chocked) (TM 9-2320-272-10).

SPECIAL ENVIRONMENTAL CONDITIONS
Work area clean and free from blowing dirt and dust.

GENERAL SAFETY INSTRUCTIONS
Personnel must be clear from underside and front of vehicle when engine is running.

---

a. Oil Cooler Pressure Test

CAUTION
Before disconnecting transmission pressure lines, clean surrounding surfaces and plug all openings to prevent entry of dirt or debris into transmission. Damage will occur if dirt or debris enters transmission.

1. Disconnect oil cooler supply hose (1) from elbow (2) on top of oil cooler (7).

NOTE
Note elbow alignment for connection.

2. Loosen nut (3) and remove elbow (2) and O-ring (4) from oil cooler (7). Discard O-ring (4).
3. Using drill press and 21/64 drill bit, drill hole in side of elbow (2) opposite hose port.
4. Tap hole in elbow (2) with 1/4-18 NPTF tap.

CAUTION
After tapping, elbow must be thoroughly cleaned and all burrs and shavings removed. Any shavings entering system will damage transmission.

5. Remove elbow (2) from drill press and thoroughly clean.
6. Install new O-ring (4) and elbow (2) in oil cooler (7) until aligned as noted in step 2.
7. Tighten nut (3) until O-ring (4) is seated.
8. Install pipe plug (8) on elbow (2).
9. Connect oil cooler supply hose (1) to elbow (2).

5-60. TRANSMISSION OIL PRESSURE TESTING (Contd)

10. Remove oil sampling valve (6) from elbow (5).
11. Install pressure gauge and gauge hose (9) on elbow (5).

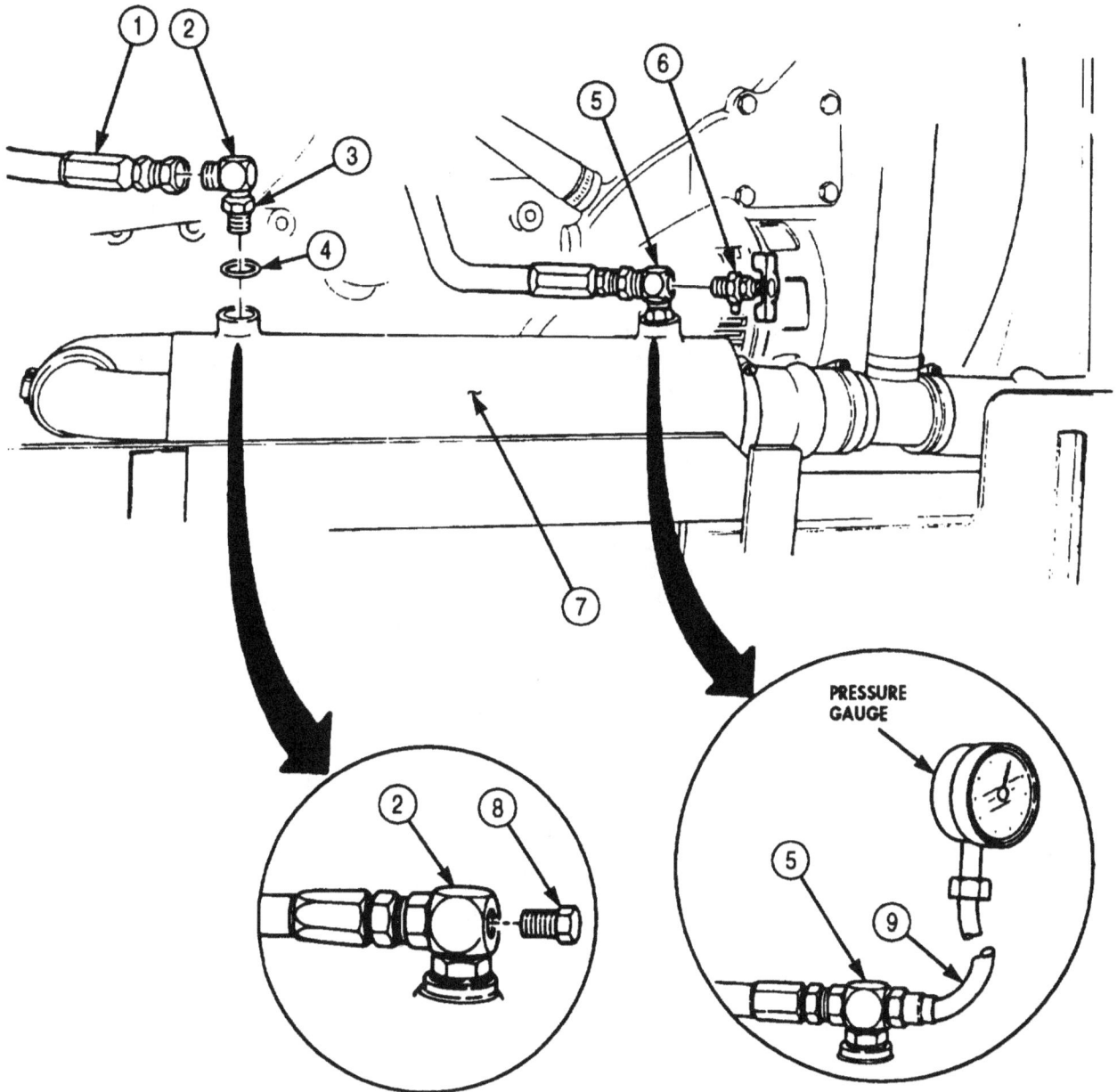

PRESSURE
GAUGE

5-60. TRANSMISSION OIL PRESSURE TESTING (Contd)

NOTE
Refer to TM 9-2320-272-10 when performing steps 12 through 20.

12. Set parking brake.

WARNING
Ensure all personnel are clear from underside and front of vehicle before starting engine. Transmission slipping into gear may cause injury to personnel.

13. Start engine and check oil cooler connections for leaks.

CAUTION
Do not maintain stalled condition for longer than 30-second intervals. Transmission oil may overheat and cause transmission damage.

14. Place transmission shift lever in 1-5 (drive) and operate engine at 1,200 rpm.

15. Place transmission shift lever in N (neutral) and operate engine at normal operating temperature.

16. Check transmission oil level.

17. Operate engine at 1,650 rpm with transmission shift lever in N (neutral) and parking brake set. Check pressure gauge and note reading.

18. If pressure is less than 26 psi (179.2 kPa), check for hose or internal oil cooler leakage.

19. If pressure exceeds 26 psi (179.2 kPa), check for cooler, cooler filter, or cooler hose restriction.

20. Stop engine and disconnect gauge and hose (2) from elbow (1).

21. Install oil sampling valve (3) in elbow (1).

22. Remove pipe plug (5) from supply hose elbow (4).

23. Install pressure gauge and hose (2) in supply hose elbow (4).

WARNING
Ensure all personnel are clear from underside and front of vehicle before starting engine. Transmission slipping into gear may cause injury to personnel.

NOTE
Refer to TM 9-2320-272-10 when performing steps 24 through 29.

24. Set parking brake.

25. Start engine and check oil cooler connections for leaks.

26. Operate engine at 1,650 rpm with transmission shift lever in N (neutral) and parking brake set. Check pressure gauge and note reading.

27. If pressure is below 30 psi (207 kPa), check for hose or internal oil cooler leakage.

28. If pressure exceeds 50 psi (345 kPa), check for cooler or cooler hose restriction.

29. Stop engine and disconnect gauge and hose (2) from supply hose elbow (4).

30. Install pipe plug (5) in elbow (4).

5-60. TRANSMISSION OIL PRESSURE TESTING (Contd)

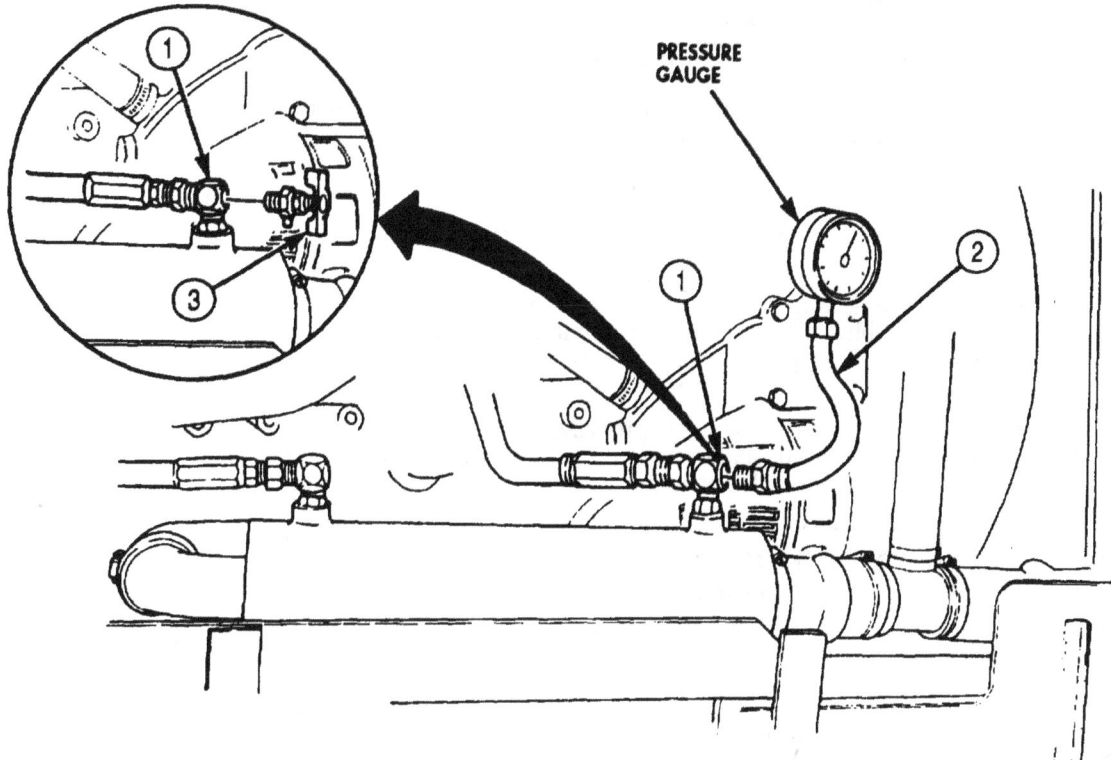

PRESSURE GAUGE

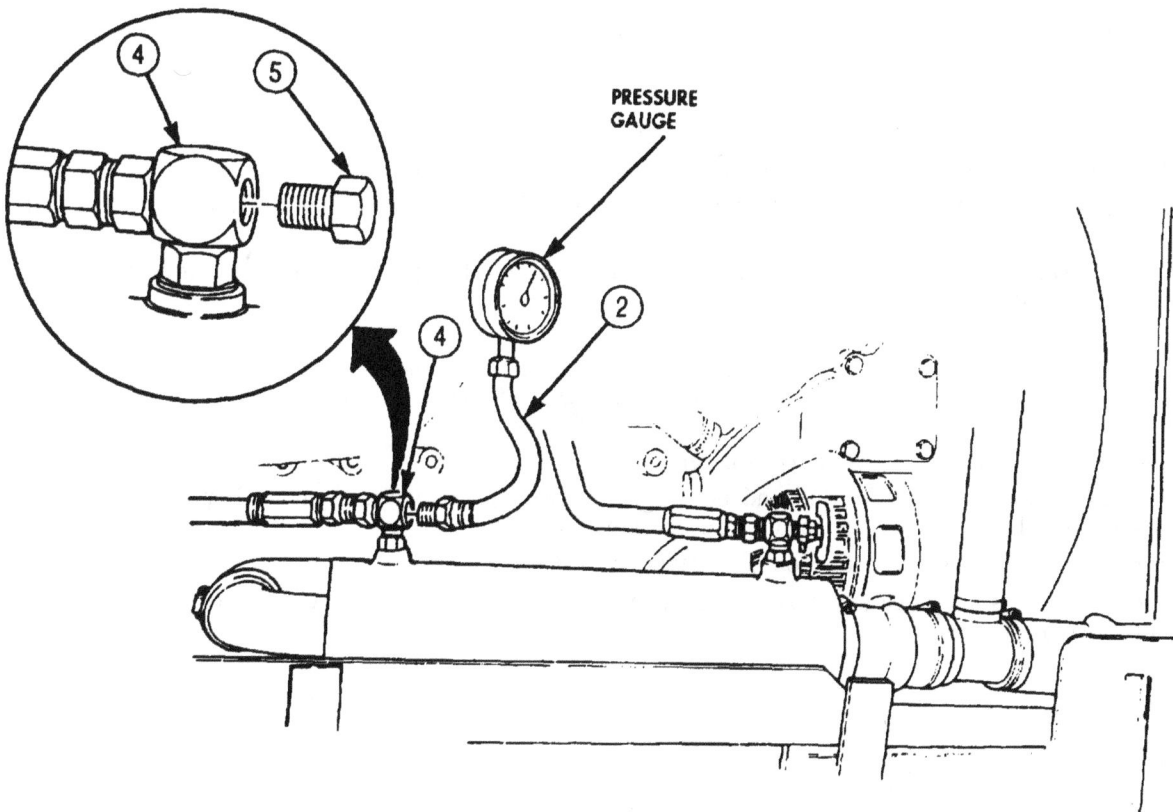

PRESSURE GAUGE

5-60. TRANSMISSION OIL PRESSURE TESTING (Contd)

b. Main Pressure and Governor Pressure Test

WARNING

Ensure transmission-to-transfer case propeller shaft has been removed before performing this test. If test is performed with propeller shaft installed, wheel blocks (chocks) will not prevent vehicle from rolling and causing injury to personnel.

1. Remove transmission-to-transfer case propeller shaft (para. 3-148).

CAUTION

Before disconnecting transmission pressure lines, clean surrounding surfaces and plug all openings to prevent entry of dirt or debris into transmission. Damage will occur if dirt or debris enters transmission.

NOTE

- Transmission-to-transfer case propeller shaft removal allows transmission output shaft to rotate to build up governor oil pressure. Upshift will not occur without governor pressure.

- Have drainage container ready to catch oil.

2. Disconnect main pressure line (1) from adapters (2) and (5) and remove main pressure line (1).

3. Remove adapter (5) from main pressure port (6).

4. Disconnect governor pressure line (4) from adapters (3) and (9) and remove governor pressure line (4).

NOTE

Only M936/A1/A2 vehicles are equipped with check valve.

5. Remove adapter (9) and check valve (8) from transmission auxiliary governor pressure port (7).

6. Install 1/8-27 NPTF Thd pipe plug (13) in auxiliary governor pressure port (7).

7. Connect pressure gauge and hose (12) to main pressure port (6) with 1/8-27 NPTF Thd hose fitting (11).

NOTE

Refer to TM 9-2320-272-10 for steps 8 through 28.

8. Start engine, check pressure port connections for leaks, and check oil.

9. Operate engine at 625± 25 rpm with transmission (10) shift lever in N (neutral) and parking brake set.

10. Note pressure reading on pressure gauge. If pressure is not 125 psi (802 kPa), repair transmission (10) as necessary.

NOTE

Do not maintain stalled condition for longer than 30-second intervals. Transmission oil may overheat and cause transmission damage.

11. Place transmission shift lever in 1-5 (drive) position.

12. Operate engine at 1,200± 25 rpm and note pressure reading on pressure gauge. If pressure is not 180-205 psi (1,241-1,413 kPa), repair transmission (10) as necessary.

13. Stop engine and place transmission shift lever in N (neutral).

14. Disconnect pressure gauge and hose (12) from main pressure port (6) on left side of transmission (10).

5-60. TRANSMISSION OIL PRESSURE TESTING (Contd)

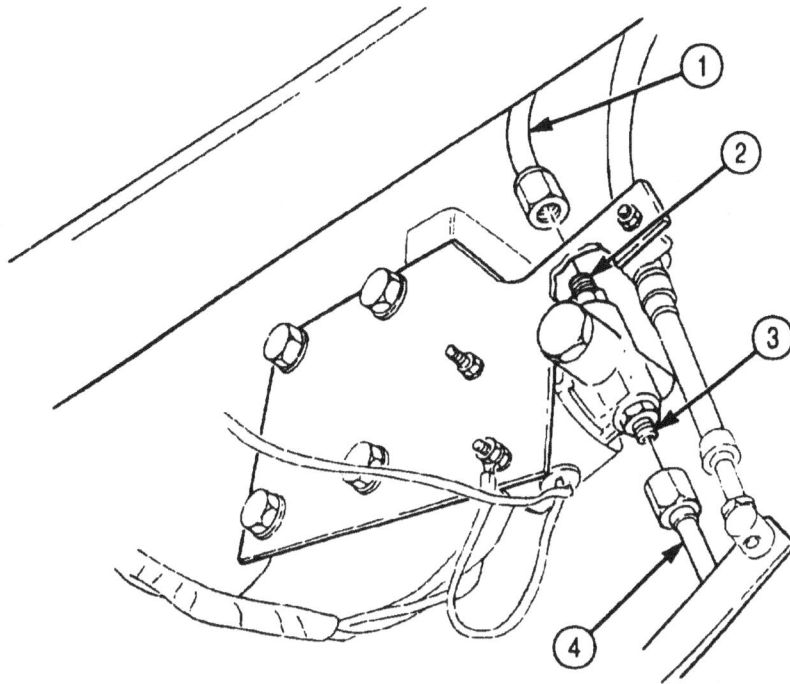

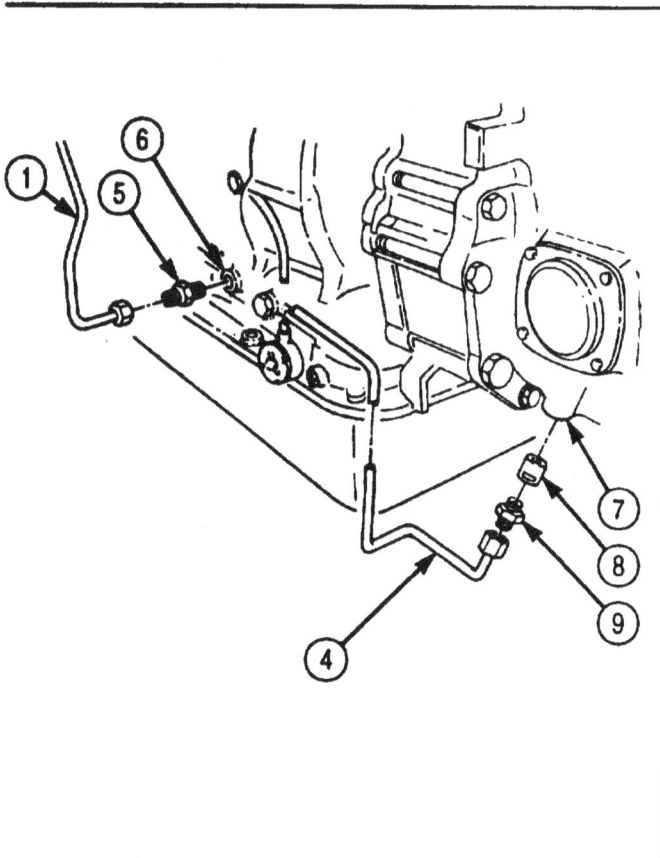

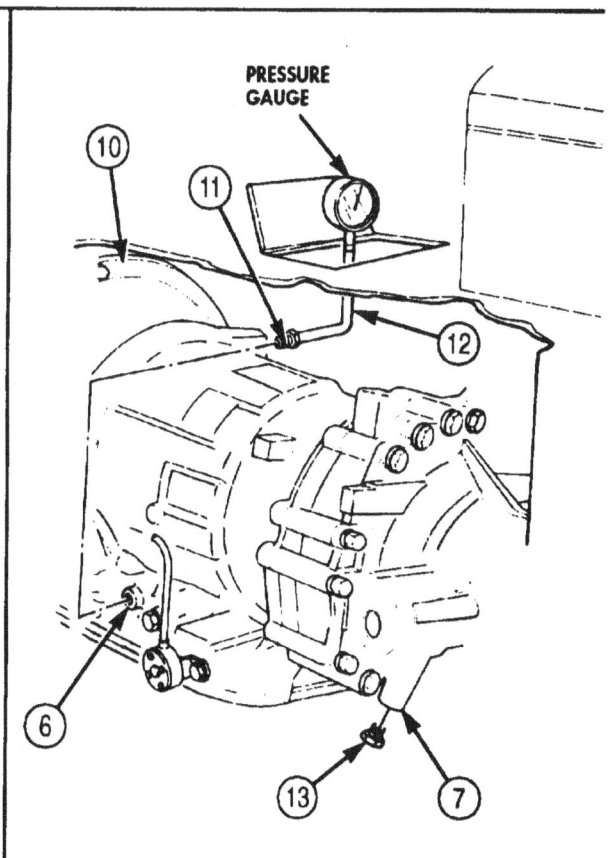

PRESSURE GAUGE

5-60. TRANSMISSION OIL PRESSURE TESTING (Contd)

15. Remove pipe plug (4) from auxiliary governor pressure port (3) and install in main pressure port (5).
16. Install pressure gauge and hose (2) in auxiliary governor pressure port (3).
17. Start engine and check pressure port connection for leaks.
18. Check transmission oil level.
19. Place transmission shift lever in 1-5 (drive) position and operate engine at 1,650 rpm. Note reading on pressure gauge.
20. If pressure is not 82-91 psi (565-627 kPa), repair transmission (1) as necessary
21. Stop engine and disconnect pressure gauge and hose (2) from auxiliary governor pressure port (3).
22. Remove pipe plug (4) from main pressure port (5).

NOTE

Only M936/A1/A2 vehicles are equipped with check valve.

23. Install check valve (8) and adapter (9) in auxiliary governor pressure port (3). Ensure bleed hole end of check valve (8) is inserted into pressure port (3) first.
24. Install governor pressure line (10) on adapters (9) and (12).
25. Install adapter (7) in main pressure port (5).
26. Install main pressure line (6) on adapters (7) and (11).
27. Start engine and check pressure port connections for leaks.
28. Check transmission oil level.
29. Install transmission-to-transfer case propeller shaft (para. 3-148).

c. Automatic Shift Speed Test

1. Road test vehicle and record engine rpm at shift points (TM 9-2320-272-10).
2. With vehicle in operation, shift transmission through the range sequence. See table 5-17 for lever range sequence.
3. Check recorded engine shift point rpm with table 5-17.
4. If shift points are incorrect, check modulator adjustment (para. 5-61). If modulator adjustment does not correct shift speed, repair transmission (1) as necessary.

Table 5-17. Transmission Shift Point Check.

SELECTOR LEVER RANGE	THROTTLE	AUTOMATIC SHIFTING	ENGINE (RPM)
1-2	Fully open	1-2	1,900-2,050
		2-2	1,900-2,050 (before converter lock-in)
		2-2	1,600-1,825 (after converter lock-in)
1-3		2-3	2,000-2,150
1-4		3-4	2,030-2,140
1-5 (drive)		4-5	2,015-2,130

5-60. TRANSMISSION OIL PRESSURE TESTING (Contd)

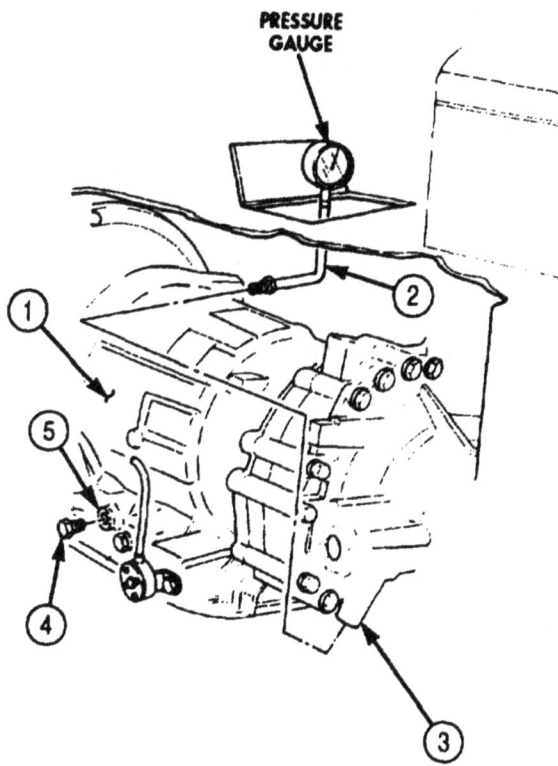

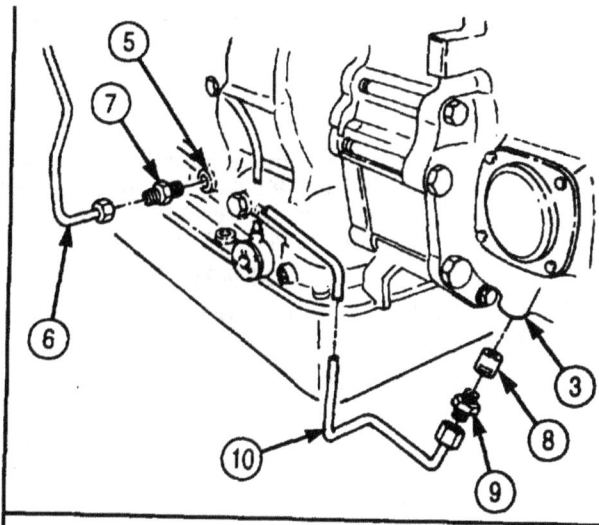

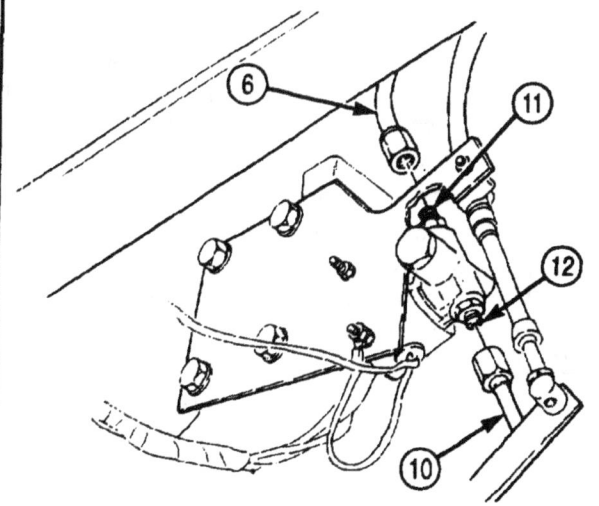

PRESSURE
GAUGE

5-61. TRANSMISSION MODULATOR MAINTENANCE

THIS TASK COVERS:

a. Testing   b. Adjustment

INITIAL SETUP:

APPLICABLE MODELS
All

TOOLS
General mechanic's tool kit (Appendix E, Item 1)

REFERENCES (TM)
LO 9-2320-272-12
TM 9-2320-272-10
TM 9-2320-272-24P

EQUIPMENT CONDITION
• Parking brake set (TM 9-2320-272-10).
• Vehicle at curb weight (empty) (TM 9-2320-272-10).

SPECIAL ENVIRONMENTAL CONDITIONS
Dry conditions, open roads, and easy grades.

---

a. Testing

NOTE

- Vehicle engine must be at normal operating temperature of 175-195°F (79-90°C) as indicated by temperature gauge. Transmission oil temperature must be at normal operating temperature of 120-220°F (49-104°C) as indicated by temperature gauge.
- Refer to TM 9-2320-272-10 for steps 1 through 7.

1. Allow vehicle engine and transmission to reach normal operating temperatures with transmission shift lever in N (neutral) and parking brake set.

2. After warmup, depress accelerator pedal until engine reaches 2,100 rpm. If engine does not reach 2,100 r-pm, see table 4-1, fuel system malfunction 15. Proceed with testing if engine reaches 2,100 rpm.

NOTE

Assistant will operate vehicle as directed by mechanic. Mechanic will observe and record engine rpm indicated by tachometer during shift changes.

3. Place transmission shift lever in 1-2 (second) and road test vehicle.

4. Accelerate at full throttle, 1,900-2,050 rpm, and note shift change. Record engine rpm at moment of shift change.

5. Stop vehicle and repeat steps 3 and 4 in 1-3 (third).

6. Stop vehicle and repeat steps 3 and 4 in 1-4 (fourth).

7. Stop vehicle and repeat steps 3 and 4 in 1-5 (drive).

8. Compare recorded engine rpm at shift points with correct shift point ranges given in table 5-17. If all shift points are too high or too low by approximately same amount, adjust modulator (4).

b. Adjustment

1. Inspect modulator (4) and cable (3) for looseness and improper installation.

5-61. TRANSMISSION MODULATOR MAINTENANCE (Contd)

NOTE

Modulator will be adjusted if properly installed. Refer to para. 3-145 for correct modulator and cable installation and adjustment instructions.

2. If looseness and/or improper installation are found, correct and retest modulator (4) and cable (3).

3. Remove modulator link (2) from throttle lever (1) and reinstall modulator link (2).

4. Retest modulator (4). If retest indicates defective modulation, replace modulator (4).

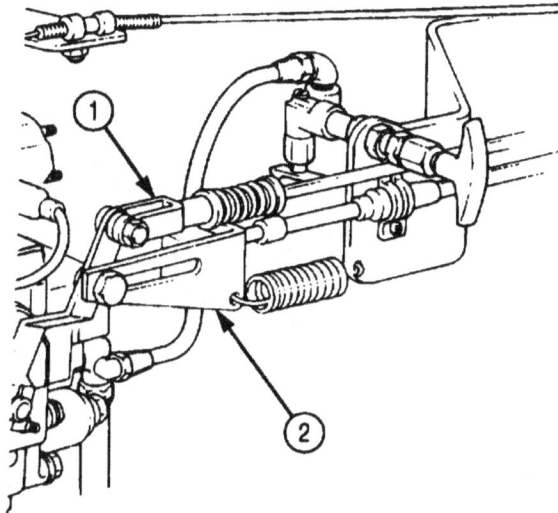

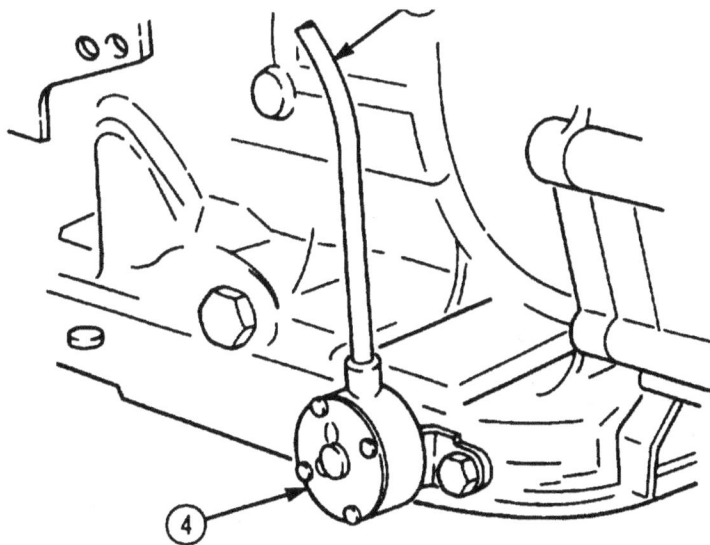

FOLLOW-ON TASKS:• Fill transmission to proper oil level (LO 9-2320-272-12).
   • Start engine (TM 9-2320-272-10) and road test vehicle.

5-62. TRANSMISSION CONVERTER STALL TEST

THIS TASK COVERS:
Forward Stall Test

INITIAL SETUP:

APPLICABLE MODELS
All

TOOLS
General mechanic's tool kit (Appendix E, Item 1)

REFERENCES (TM)
LO 9-2320-272-12
TM 9-2320-272-10
TM 9-2320-272-24P

EQUIPMENT CONDITION
• Parking brake set (TM 9-2320-272-10).
• Wheels blocked (chocked) (TM 9-2320-272-10).

SPECIAL ENVIRONMENTAL CONDITIONS
Work area clean and free from blowing dirt and dust.

GENERAL SAFETY INSTRUCTIONS
Do not allow anyone to stand in front of vehicle when conducting a stall test.

WARNING
Do not allow anyone to stand in front of vehicle when conducting stall test. Vehicle movement may cause injury to personnel.

CAUTION
• Do not maintain the stalled condition longer than 30-second intervals due to rapid heating of transmission oil. Observe transmission oil temperature gauge. Normal operating range is 120-220°F (49-104°C).
• Observe engine coolant temperature gauge. Operating temperature is 175-195°F (79-90°C).
• If oil temperature reaches 300°F (148°C), or if thirty seconds is insufficient time to complete needed tests, transmission oil temperature must be lowered or damage to transmission may result.
• Run engine at 1,200-1,500 rpm with transmission in neutral for two minutes to cool oil between tests.

NOTE
• The stall test is conducted when engine and/or transmission are not performing satisfactorily The purpose of the stall test is to determine if transmission or engine is the malfunctioning unit.
• The vehicle's transmission will stay in first speed during the stall tests regardless of transmission 5-4-3-2 and 1 quadrant position. Transmission does not and cannot upshift because the internal mechanism, output shaft, and governor are not turning. The stall test checks the engine performance, converter clutch operation or installation, and holding ability of the converter one-way clutch.
• Refer to TM 9-2320-272-10 for the following steps.

5-62. TRANSMISSION CONVERTER STALL TEST (Contd)

1. Apply service brakes and place transmission shift lever in any forward drive position 5-4-3-2-1.
2. Accelerate engine to full throttle (1,900-2,050 rpm).

NOTE

Perform step 3 only if engine speed exceeds 1,700 rpm.

3. If engine speed exceeds 1,700 rpm, check transmission oil level.

   a. If oil level is low, fill to proper level.

   b. If oil level is correct, repair transmission as necessary.

NOTE

- Perform steps 4 and 5 only if engine speed is less than 1,900 rpm.
- Refer to table 4-1, Mechanical Troubleshooting.

4. If engine speed is less than 1,400 rpm, troubleshoot engine for loss of power.
5. If engine is performing satisfactorily, repair transmission torque converter (para. 5-41).

FOLLOW-ON TASK: Start engine (TM 9-2320-272-10) and road test vehicle.

Section IV. TRANSFER CASE MAINTENANCE

5-63. TRANSFER CASE REPAIR

THIS TASK COVERS:

a. Disassembly
b. Cleaning and Inspection

c. Assembly and Adjustment

INITIAL SETUP:

APPLICABLE MODELS
All

SPECIAL TOOLS
Crowfoot wrench (Appendix E, Item 163)
Two flange puller standoffs
   (Appendix F, Item 1)

TOOLS
General mechanic's tool kit (Appendix E. Item 1)
Mechanical puller (Appendix E, Item 102)
Torque wrench (Appendix E, Item 144)
Outside micrometer (Appendix E, Item 80)
Dial indicator (Appendix E, Item 36)
Arbor press
Feeler gauge
Prybar
Soft-faced hammer
Torque multiplier

MATERIALS/PARTS
Cylinder assembly (Appendix D, Item 92)
Seal (Appendix D, Item 254)
Three seals (Appendix D, Item 500)
Three locknuts Appendix D, Item 320)
Snapring Appendix D, Item 663)
Woodruff key (Appendix D, Item 727)
Two seals (Appendix D, Item 627)
Adhesive sealant (Appendix C, Item 4)
GAA grease (Appendix C, Item 28)
Gasket sealant (Appendix C, Item 30)
Lubricating oil (Appendix C, Item 47)
Sealing compound (Appendix C, Item 62)

PERSONNEL REQUIRED
TWO

REFERENCES (TM)
TM 9-2320-272-24P
TM 9-214

EQUIPMENT CONDITION
Transfer case removed (para. 4-94 or para. 4-95).

GENERAL SAFETY INSTRUCTIONS
Support transfer case with wood blocks before performing disassembly.

---

a. Disassembly

WARNING
Transfer case is heavy. Use wood blocks to prevent transfer case from tipping over and causing injury to personnel.

NOTE
Clean exterior of transfer case thoroughly before performing procedure.

1. Using crowfoot wrench, remove interlock air cylinder (1) and push rod (2) from transfer case (3).

2. Remove eight screws (5) and inspection cover (4) from transfer case (3).

3. Remove locknut (6) and washer (7) from output shaft (14). Discard locknut (6).

4. Using mechanical puller, remove output flange (8) and parking brake drum (9) from output shaft (14).

5-63. TRANSFER CASE REPAIR (Contd)

5. Remove actuating plate (10) from brakeshoe backing plate (11).

6. Remove four screws (16), dustcover (12), and brakeshoe assembly (15) from companion flange (13).

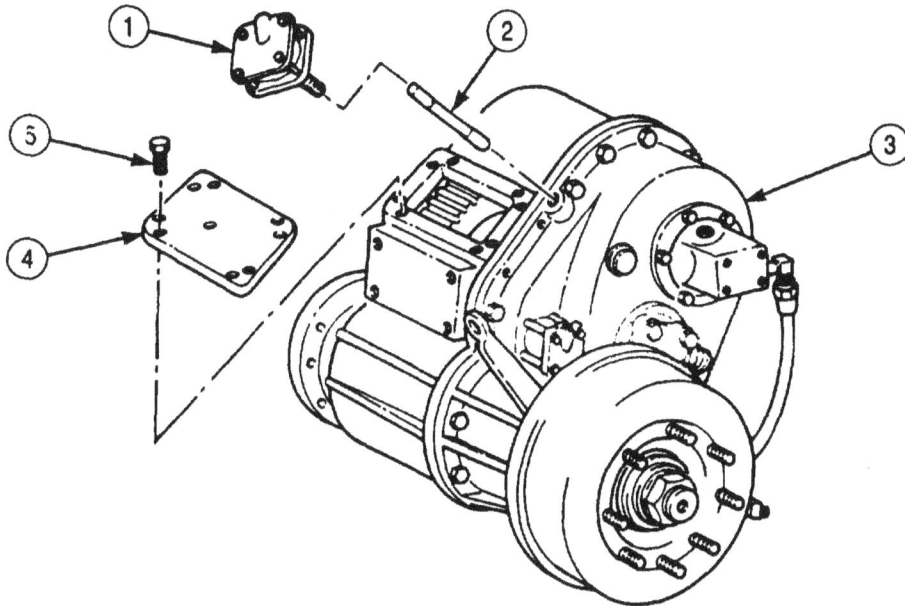

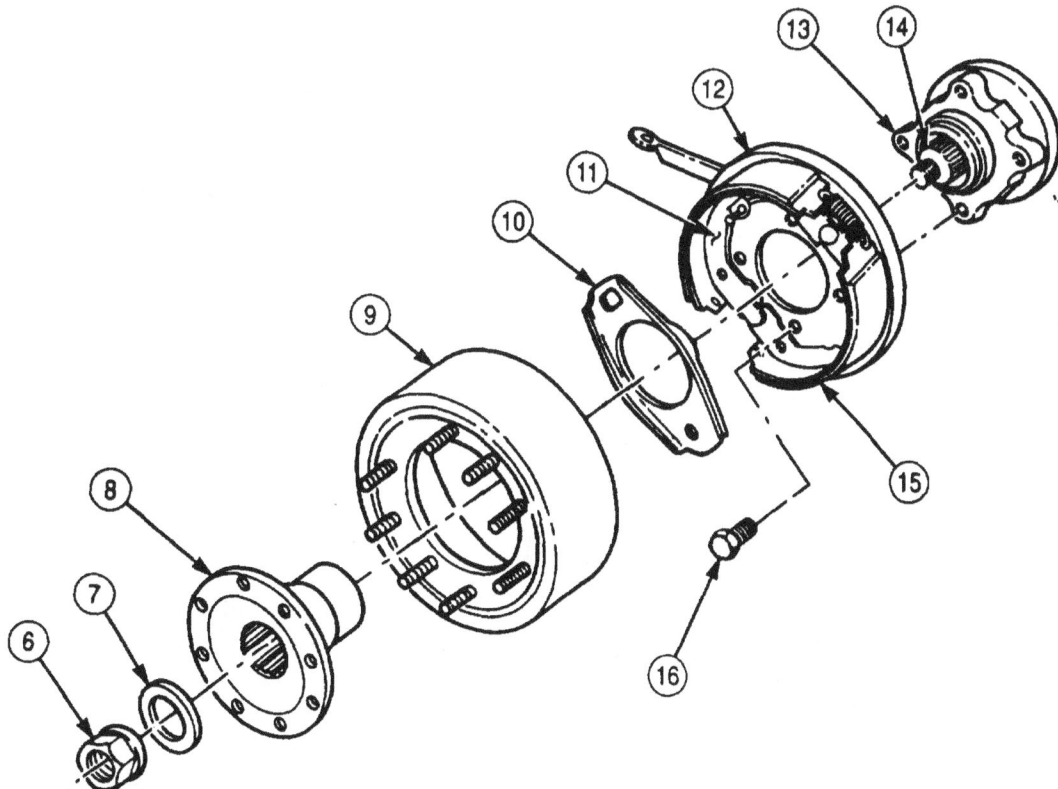

5-63. TRANSFER CASE REPAIR (Contd)

7. Disconnect oil line (4) from elbows (3) and (5).

8. Remove elbow (3) from oil pump (2).

NOTE

Perform steps 9 and 10 for vehicles equipped with transfer case power takeoff (PTO).

9. Disconnect oil line (9) from elbows (5) and (8).

10. Remove elbow (8) from PTO (10).

11. Remove elbow (5), adapter (6), and filter screen (7) from transfer case (1). Clean filter screen (7).

NOTE

Mark position of oil pump for installation.

12. Remove six screws (13), washers (12), and oil pump (11) from transfer case (1).

13. Bend tabs of locking plates (15) away from screws (16).

14. Remove four screws (16), locking plates (15), cover (14), cylinder (18), piston (19), and gasket (21) from transfer case (1) and declutch shaft (22). Discard gasket (21), piston (19), cylinder (18), and locking plates (15).

15. Remove seals (17) and (20) from cover (14) and piston (19). Discard seals (17) and (20).

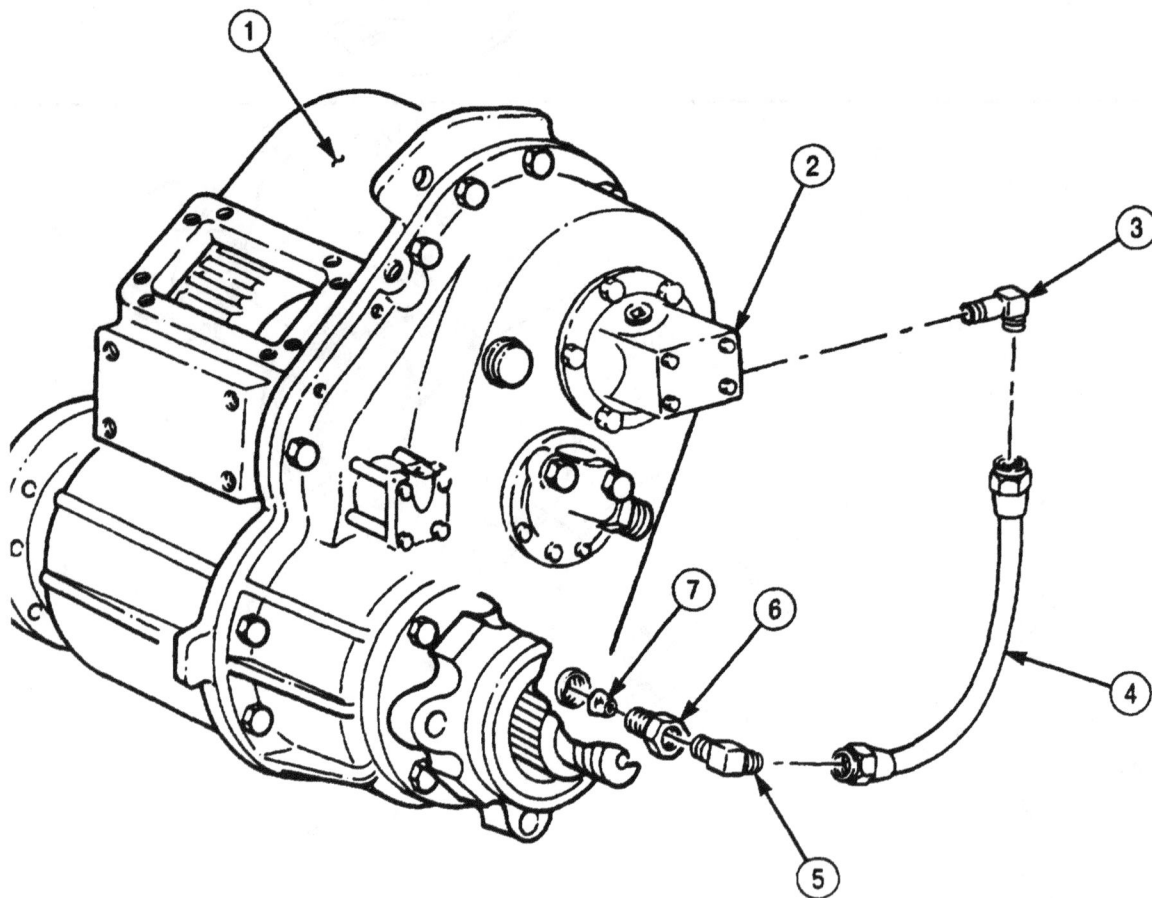

WITHOUT TRANSFER CASE PTO

5-63. TRANSFER CASE REPAIR (Contd)

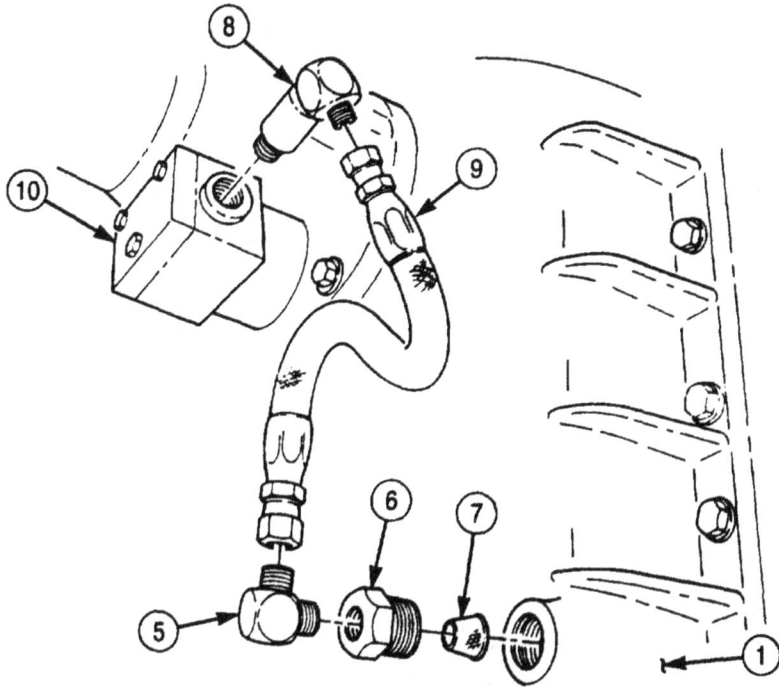

WITH TRANSFER CASE PTO

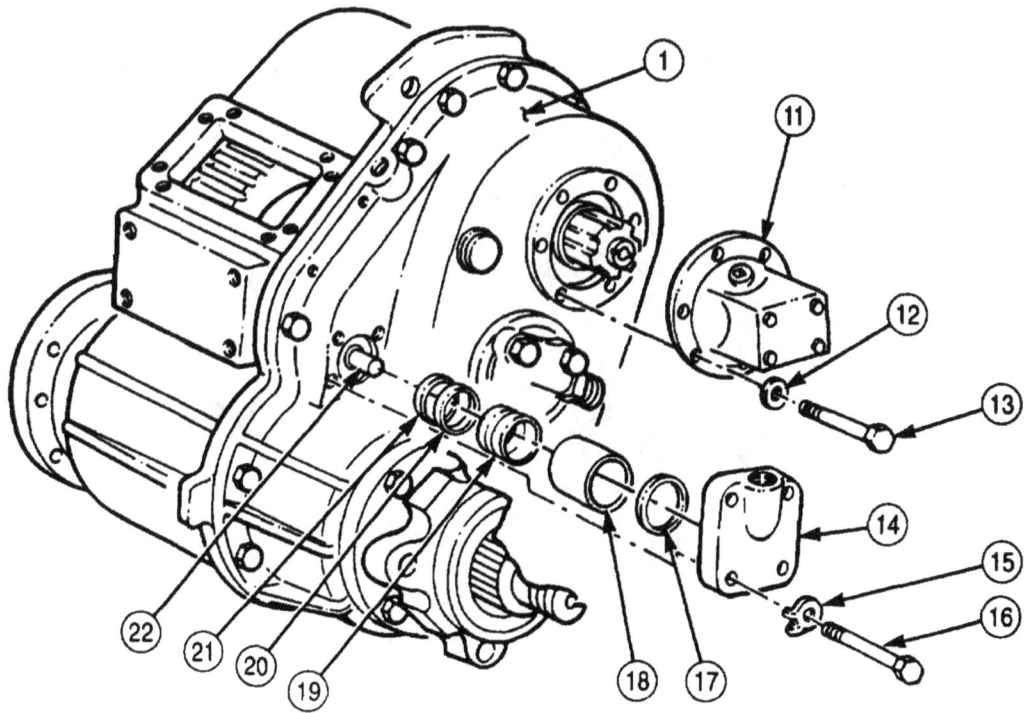

5-63. TRANSFER CASE REPAIR (Contd)

NOTE

- Chain is necessary to prevent main and output flanges from rotating when removing hardware.
- Tag main and front output flanges for installation.

16. Install chain on main flange (5) and output flange (11) with two washers (9), screw (10), two washers (3), screw (2), and nut (4).

17. Remove locknut (8) and washer (7) from main input shaft (6). Discard locknut (8).

18. Remove locknut (1) and washer (13) from front output shaft (12). Discard locknut (1).

19. Remove screw (10), two washers (9), nut (4), washer (3), screw (2), washer (3), and chain from main flange (5) and output flange (11).

20. Using puller, remove main flange (5) and output flange (11) from main input shaft (6) and front output shaft (12).

NOTE

Perform steps 21 and 22 for vehicles equipped with PTO.

21. Remove six screws (15), washers (16), and PTO (14) from transfer case (17).

22. Remove setscrew (19), PTO drive gear (18). and seal (20) from main input shaft (6). Discard seal (20).

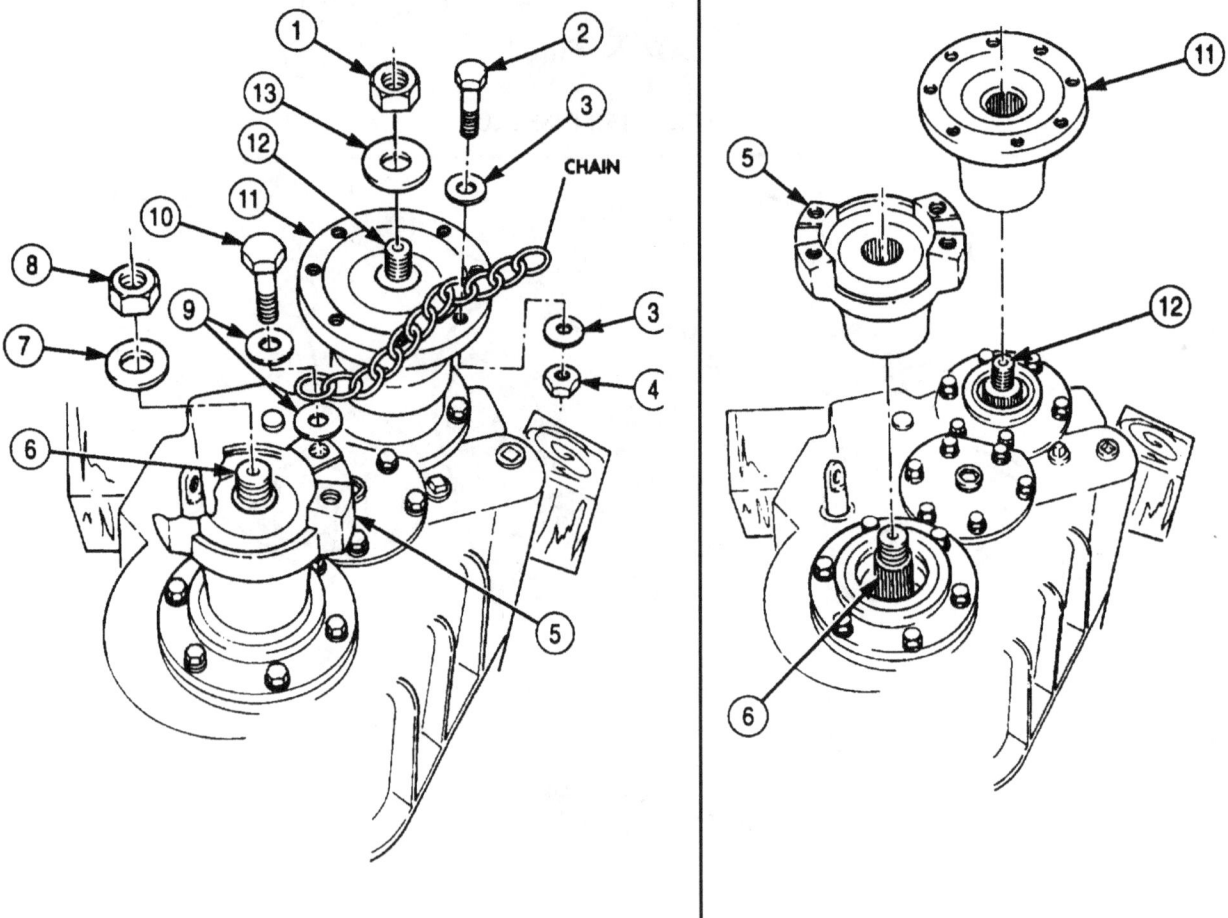

5-63. TRANSFER CASE REPAIR (Contd)

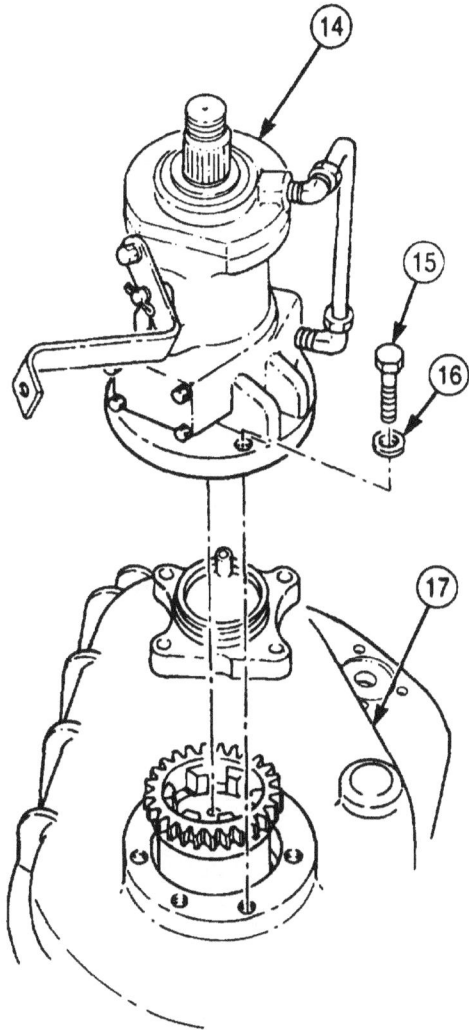

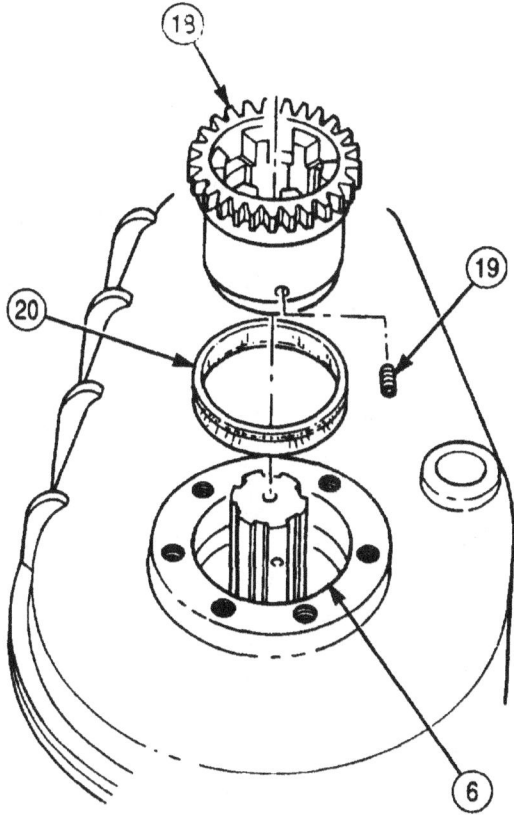

WITH TRANSFER CASE PTO

5-63. TRANSFER CASE REPAIR (Contd)

23. Loosen setscrew (13) and remove collar (12), seal (ll), and baffle (10) from main input shaft (7). Discard seal (11).

24. Remove seven screws (20), washers (19), flange (3), and shim pack (4) from transfer case (5).

25. Remove slinger (1) and seal (2) from flange (3). Discard seal (2).

26. Remove four screws (17), washers (18), two screws (16), washers (15), and gear cover (14) from transfer case (5).

27. Remove nineteen screws (9) and washers (8) from transfer case cover (21).

28. Position puller on transfer case cover (21) with drive screw of puller over intermediate shaft (7) and two long screws through puller and into transfer case cover (21).

29. Remove transfer case cover (21) from transfer case housing (22).

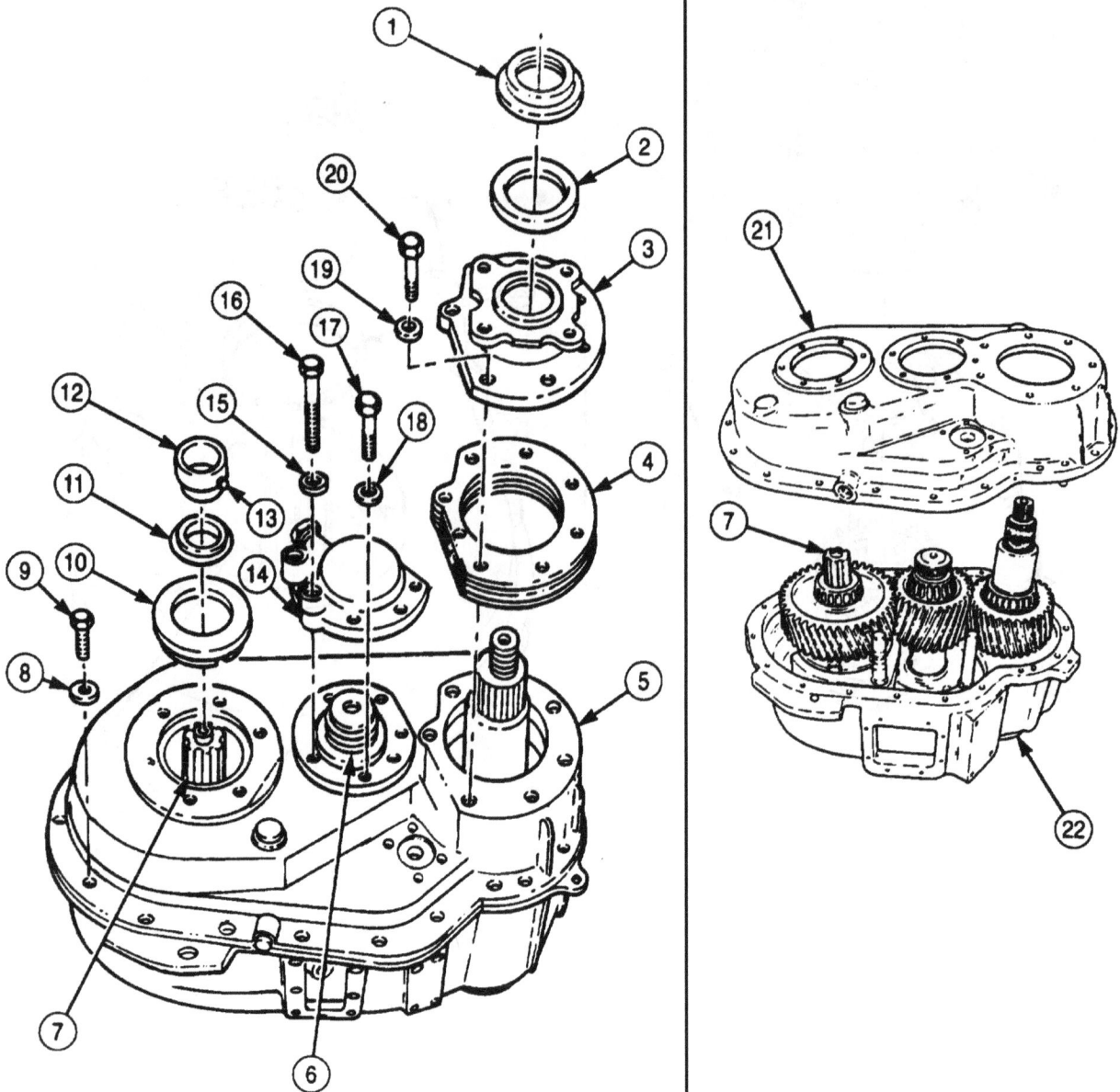

30. Remove setscrew (24), hi-lo shift fork (23), and hi-lo shift shaft (25) from synchronizer gear (26) and transfer case housing (22).

31. Remove main input shaft (7), rear output shaft (28), and intermediate shaft (27) from bearing races (29) and transfer case housing (22).

32. Remove setscrew (32), declutch shaft (34), declutch fork (31), sliding clutch (30), and spring (33) from transfer case housing (22). Discard spring (33) if broken or distorted.

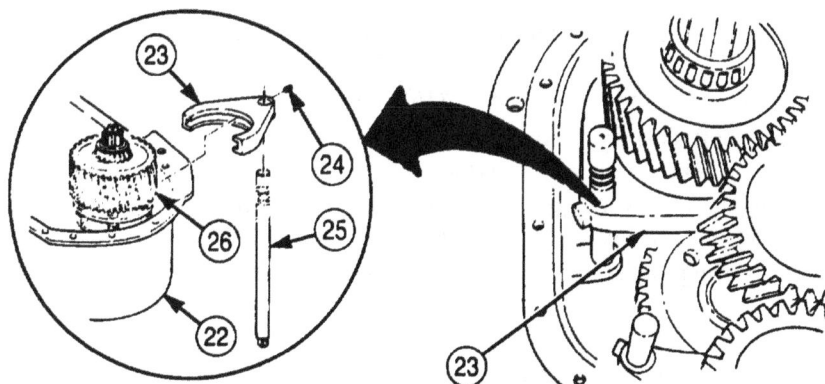

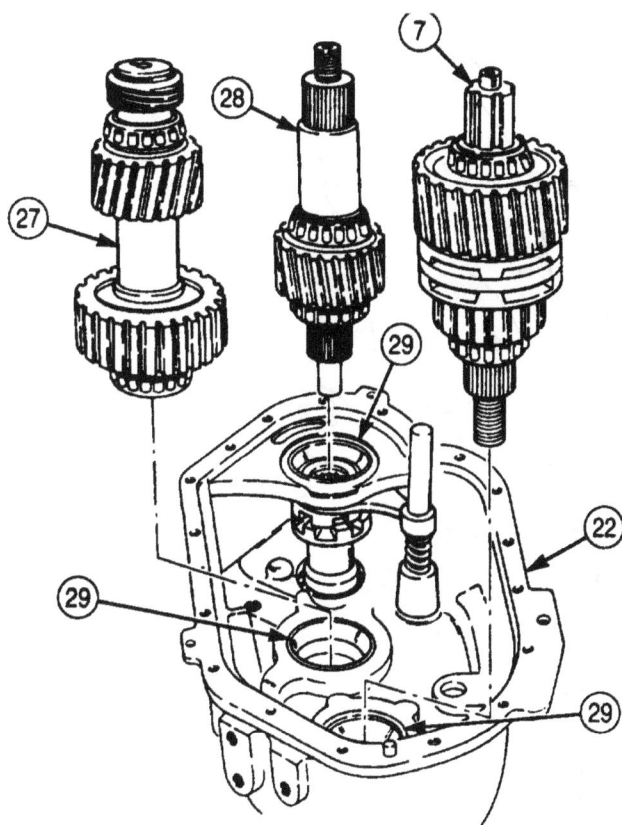

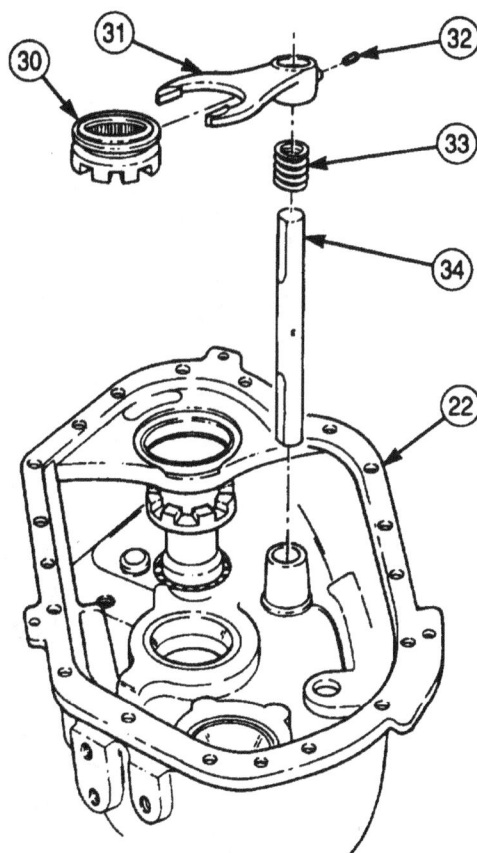

5-63. TRANSFER CASE REPAIR (Contd)

33. Remove hi-lo shaft seal (2) from transfer case housing (1). Discard seal (2).

34. Remove six screws (6), washers (5), main input shaft cover (4), and shim pack (3) from transfer case housing (1).

35. Remove seal (7) from main input shaft cover (4). Discard seal (7).

36. Remove plug (12) from intermediate shaft cover (9).

37. Remove six screws (11), washers (10), intermediate shaft cover (9), and shim pack (8) from transfer case housing (1).

38. Remove six screws (14), washers (15), and front output shaft cover (16) from transfer case housing (1).

39. Remove seal (13) from front output shaft cover (16). Discard seal (13).

40. Remove front output shaft (17) from transfer case housing (1).

NOTE

When disassembling front output, rear output, and intermediate shafts, tag all bearings, spacers, and gears for installation.

41. Using flat-axle type bearing puller, remove front output shaft bearing (18) from front output shaft (17).

42. Remove snapring (19) from front output shaft bearing (18).

43. Remove spacer (24) from rear output shaft (21).

44. Using arbor press, remove rear output shaft bearings (20) and (23), and gear (22) from rear output shaft (21).

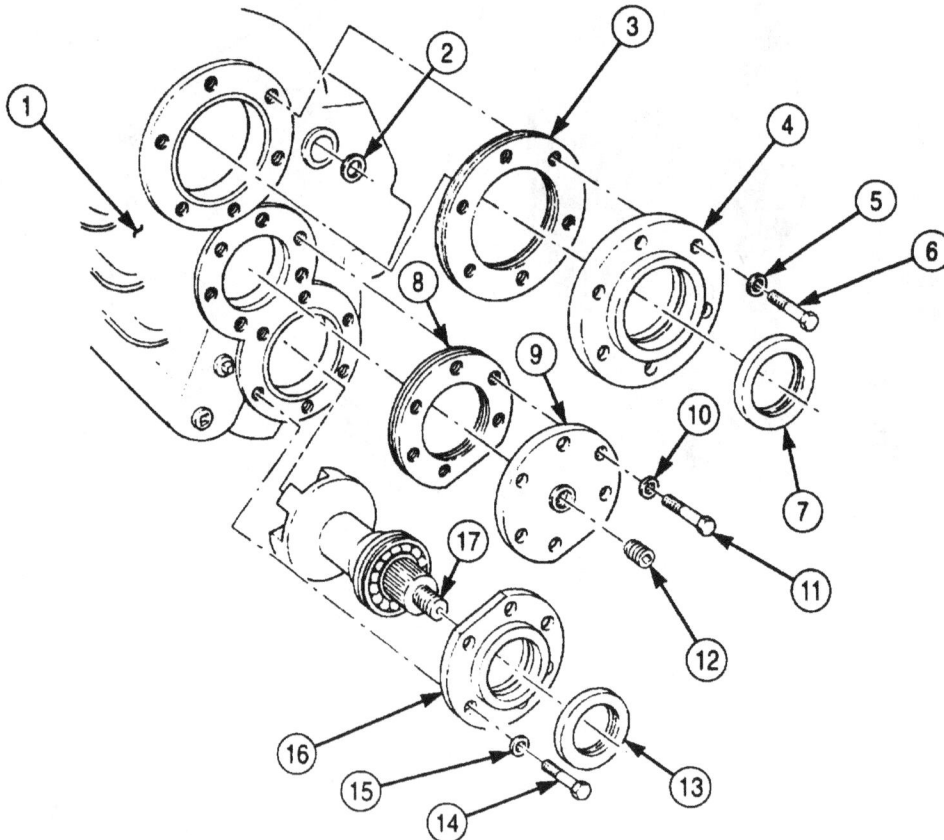

5-63. TRANSFER CASE REPAIR (Contd)

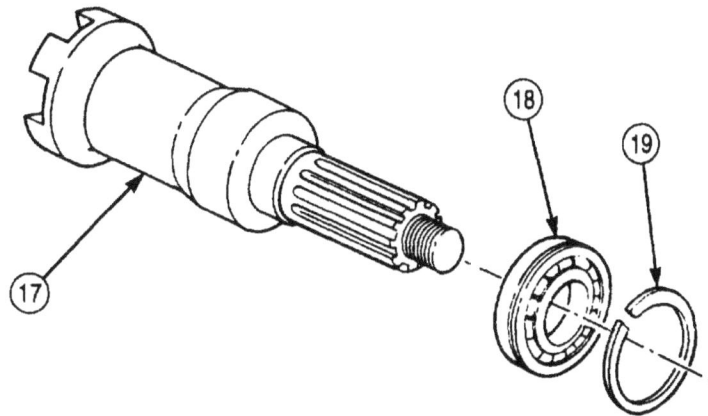

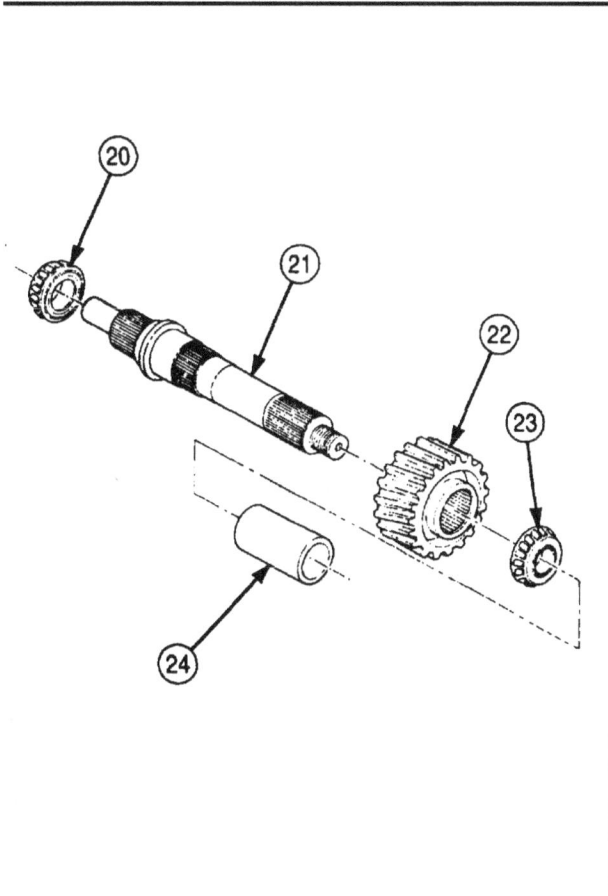

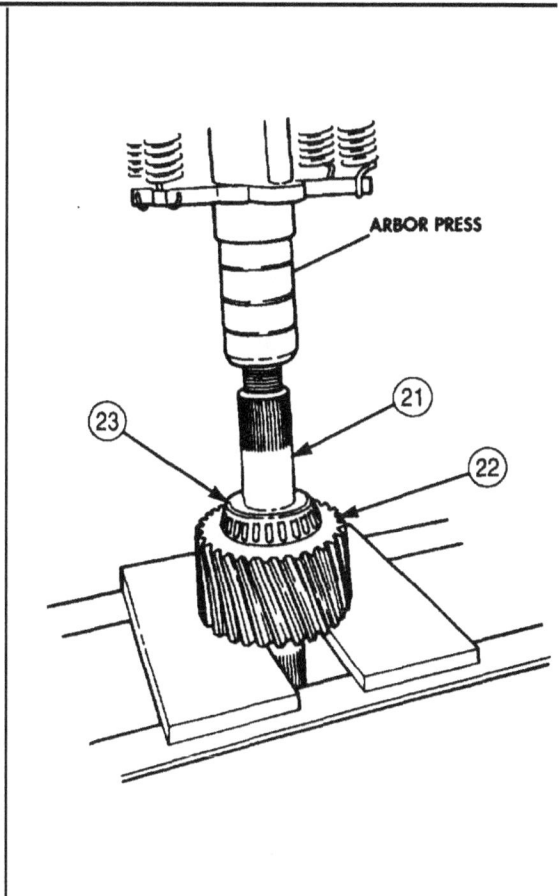

ARBOR PRESS

5-63. TRANSFER CASE REPAIR (Contd)

45. Remove snapring (8), speedometer drive gear (7), and woodruff key (4) from intermediate shaft (3). Discard woodruff key (4) and snapring (8).

46. Using arbor press, remove intermediate bearing (6), intermediate high-speed gear (5), intermediate bearing (1), and intermediate low-speed gear (2) from intermediate shaft (3).

47. Remove companion flange spacer (9) from main input shaft (14).

48. Using arbor press, remove main input shaft bearing (16), bearing spacer (17), and high-speed gear (18) from main input shaft (14).

49. Remove high-speed gear spacer (19) and synchronizer gear (20) from main input shaft (14).

50. Using arbor press, remove main input shaft bearing (10), bearing spacer (11), low-speed gear (12), and low-speed gear spacer (13) from main input shaft (14).

b. Cleaning and Inspection

1. For general cleaning instructions, refer to para. 2-14.
2. For general inspection instructions, refer to para. 2-15.
3. Inspect two dowel pins (15) of main input shaft (14) for breaks. If broken, replace dowel pins (15).

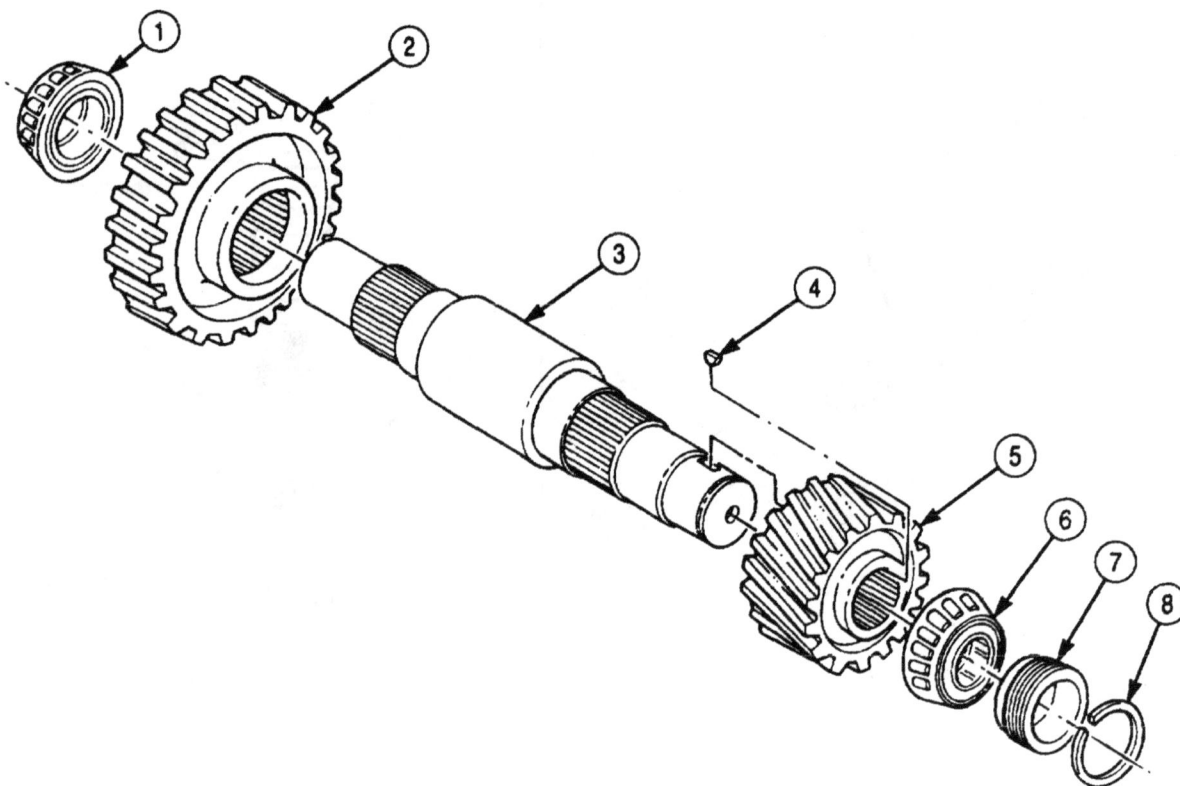

5-63. TRANSFER CASE REPAIR (Contd)

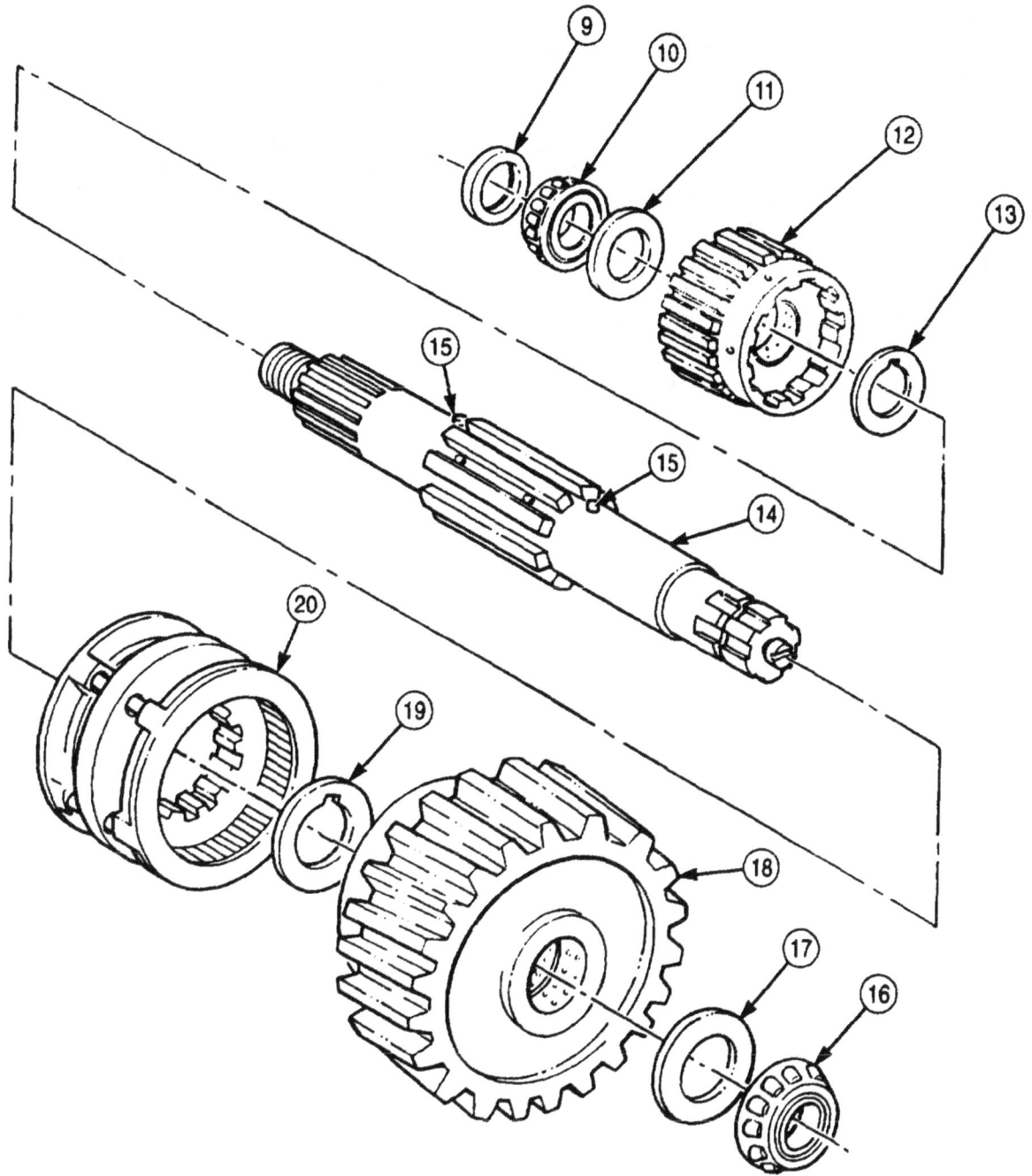

4. Inspect three dowel pins (2) for bends and breaks. Replace dowel pins (2) if bent or broken.

5. Remove filler plug (5) and drainplug (6) from transfer case housing (8) and inspect for stripped threads. Replace plugs (5) or (6) if threads are stripped.

6. Inspect freeze plugs (7) and (9) for looseness. Replace freeze plugs (7) or (9) if loose.

7. Inspect transfer case cover (3) and housing (8) for breaks, cracks, and stripped threads. Replace transfer case cover (3) or housing (8) if broken, cracked, or threads are not repairable (para. 2-5).

NOTE

- Transfer case housing and cover bearings and bearing races are matched parts and must be replaced as a set.
- Perform steps 9 and 10 if bearings or bearing races fail inspection.

8. Inspect bearings (4) and bearing races (1) in accordance with TM 9-214. If damaged, replace bearing (4) and matched bearing race (1).

9. Remove bearing race (1) from transfer case housing (8) or cover (3) by tapping alternately around outer edge of bearing race (1). Discard bearing race (1).

10. Using arbor press, install new bearing race (1) on transfer case housing (8) or cover (3).

5-63. TRANSFER CASE REPAIR (Contd)

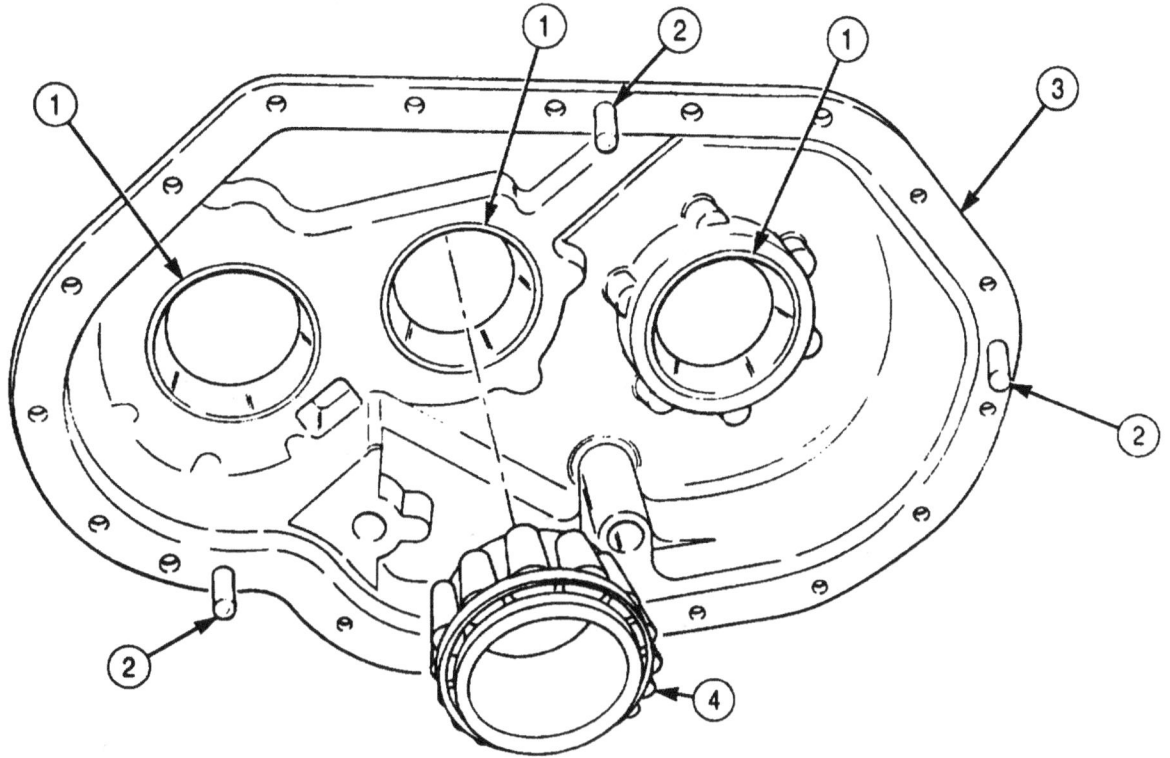

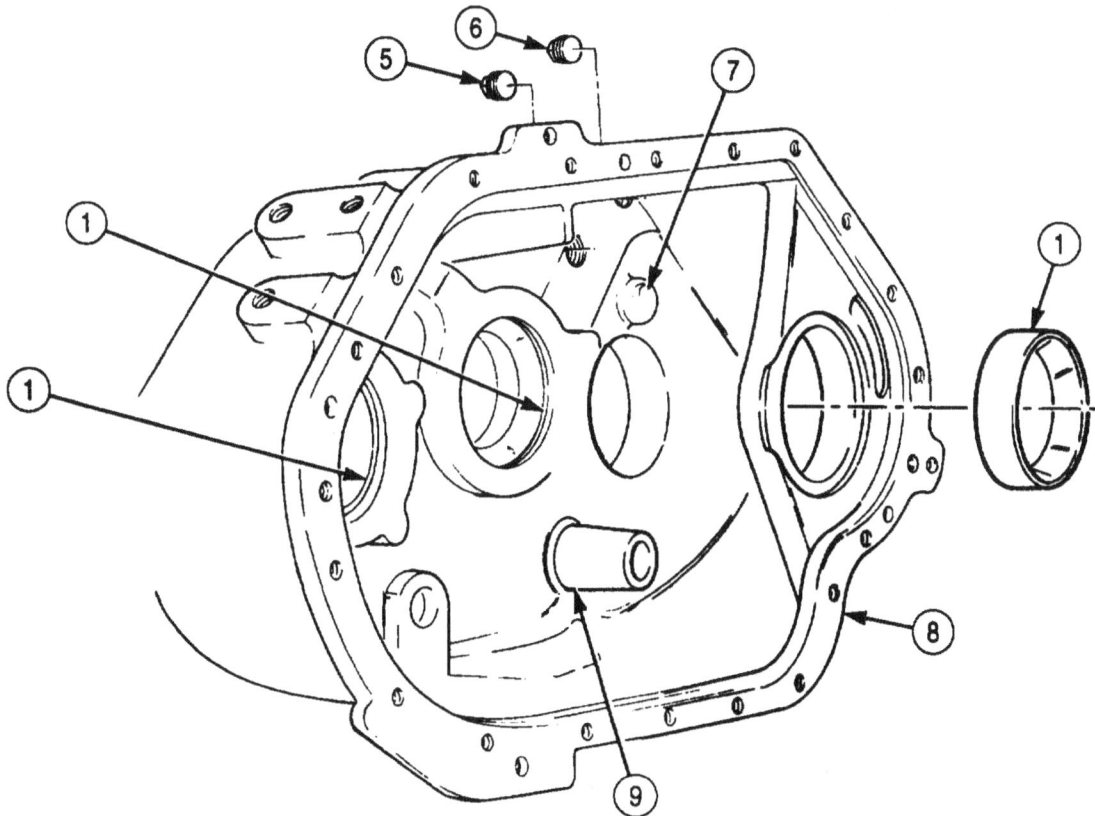

5-63. TRANSFER CASE REPAIR (Contd)

c. Assembly and Adjustment

NOTE

Prior to assembly, apply coat of oil to lubricate all gears, bearings, and shafts.

1. Measure thickness of bearing spacer (3). Replace bearing spacer (3) if thickness is less than 0.265 in. (6.73 mm).

2. Install low-speed gear spacer (5), low-speed gear (4), and bearing spacer (3) on main input shaft (6). Ensure low-speed gear spacer (5) is aligned on dowel pin (8).

3. Using arbor press, install main input shaft bearing (2) on main input shaft (6).

4. Measure thickness of bearing spacer (10). Replace bearing spacer (10) if thickness is less than 0.143 in. (3.63 mm).

5. Install synchronizer gear (13), high-speed gear spacer (12), high-speed gear (11), and bearing spacer (10) on main input shaft (6). Ensure gear spacer (12) is aligned with dowel pin (7).

6. Using arbor press, install main input shaft bearing (9) on main input shaft (6).

7. Install companion flange spacer (1) on main input shaft (6).

8. Using arbor press, install intermediate low-speed gear (15) and intermediate hearing (14) on intermediate shaft (16).

9. Install intermediate high-speed gear (18) and roller bearing (21) on intermediate shaft (16).

10. Install new woodruff key (17) and speedometer gear (20) on intermediate shaft (16) with new snapring (19).

11. Using arbor press, install rear output shaft driven gear (24) and rear output shaft bearings (25) and (22) on rear output shaft (23).

12. Install rear output shaft spacer (26) on rear output shaft (23).

13. Using arbor press, install front output shaft bearing (28) on front output shaft (27).

14. Install snapring (29) on front output shaft bearing (27).

5-63. TRANSFER CASE REPAIR (Contd)

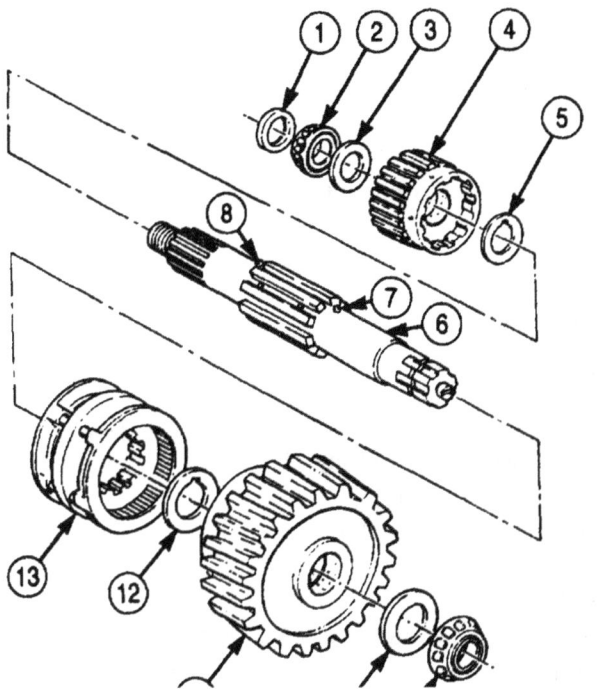

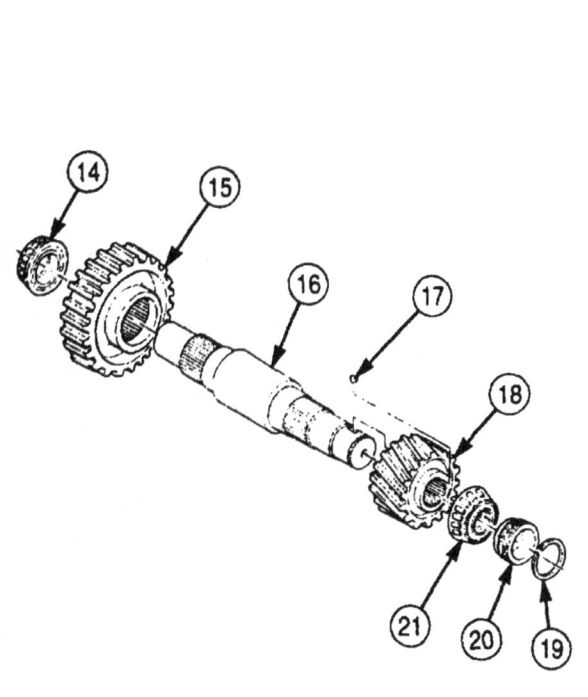

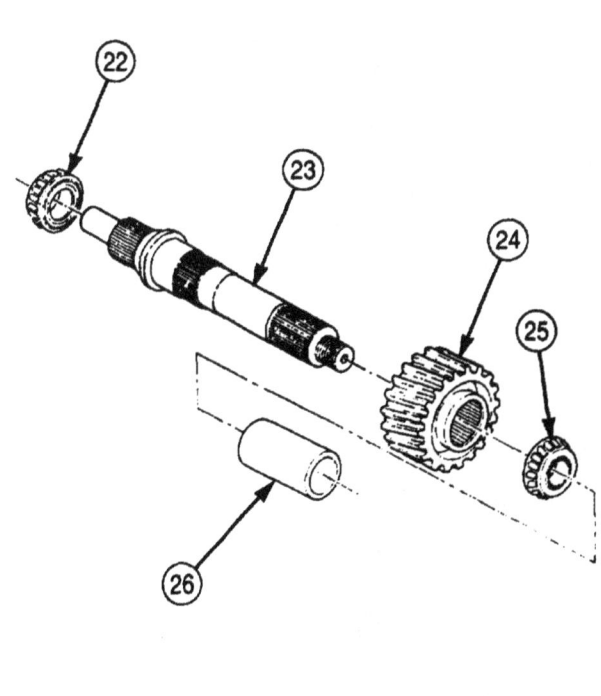

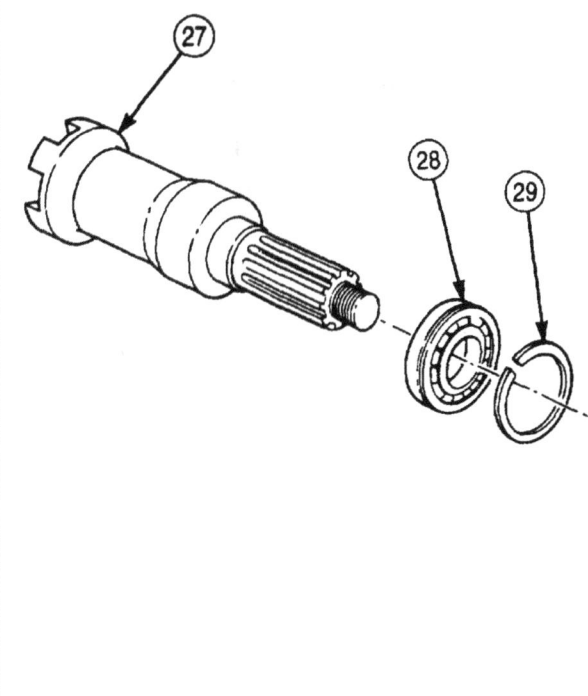

5-63. TRANSFER CASE REPAIR (Contd)

NOTE

Apply GAA grease to inside diameter of all seals.

15. Install front output shaft (8) in transfer case housing (1) until snapring (9) is flush with transfer case housing (1).

16. Install new front output shaft seal (4) in front output shaft cover (7).

17. Apply a thin coating of gasket sealer to mating surfaces of front output shaft cover (7) and transfer case housing (1).

18. Install front output shaft cover (7) on transfer case housing (1) with six washers (6) and screws (5). Tighten screws (5) 40-65 lb-R (54-88 N•m).

19. Install new input shaft seal (3) on main input shaft cover (2).

20. Install sliding clutch (10) on front output shaft (8).

21. Position declutch fork (11) on sliding clutch (10).

22. Position declutch spring (13) under declutch fork (11) and insert declutch shaft (14) through declutch fork (11), declutch spring (13), and into transfer case housing (1).

23. Install declutch fork (11) on declutch shaft (14) with setscrew (12).

24. Install intermediate shaft (15), rear output shaft (16), and main input shaft (18) in three bearing races (17) on transfer case housing (1).

25. Position hi-lo shift fork (19) on synchronizer gear (22).

26. Insert hi-lo shift shaft (21) through hi-lo shift fork (19) and into transfer case housing (1).

27. Install hi-lo shift fork (19) on hi-lo shift shaft (21) with setscrew (20).

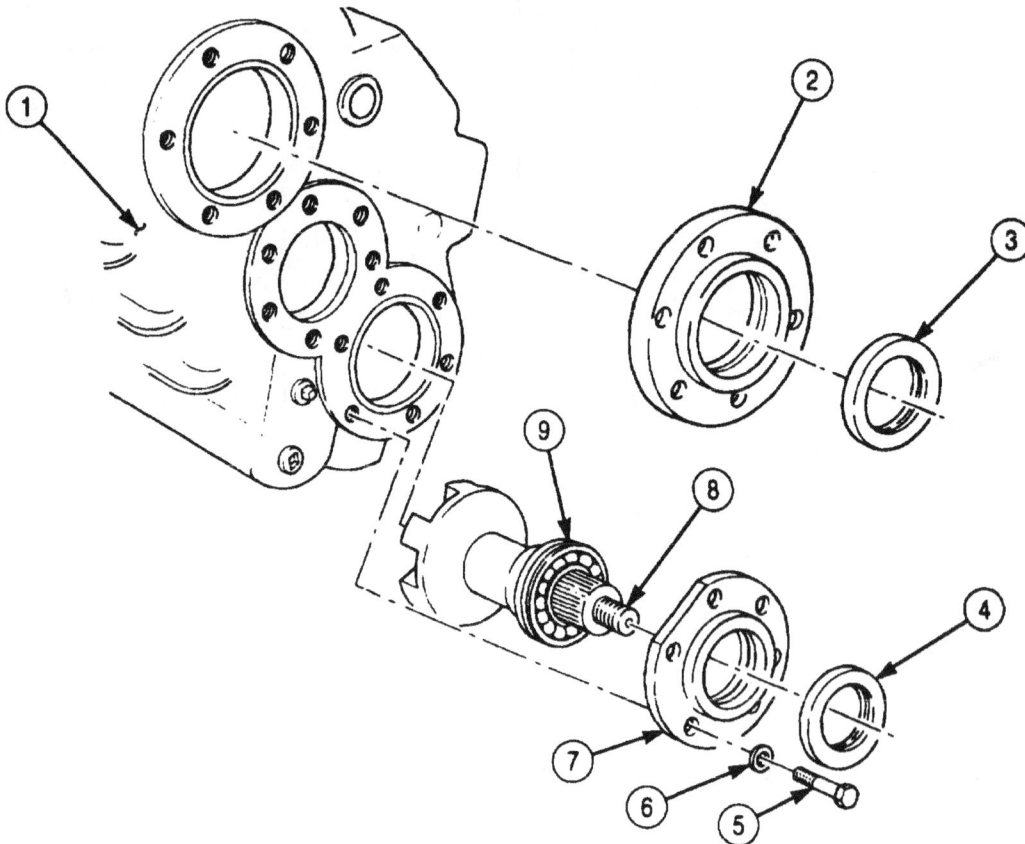

5-63. TRANSFER CASE REPAIR (Contd)

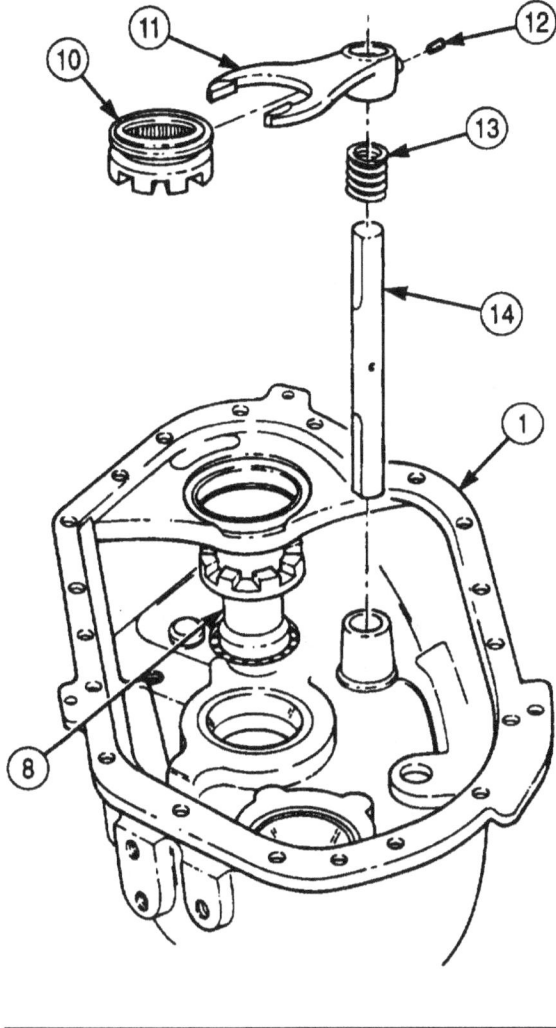

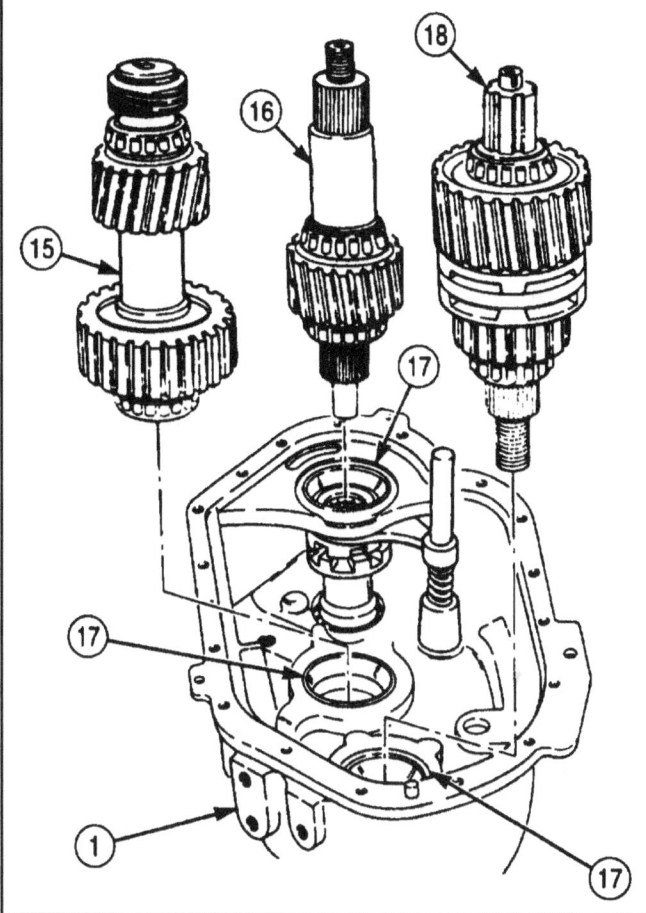

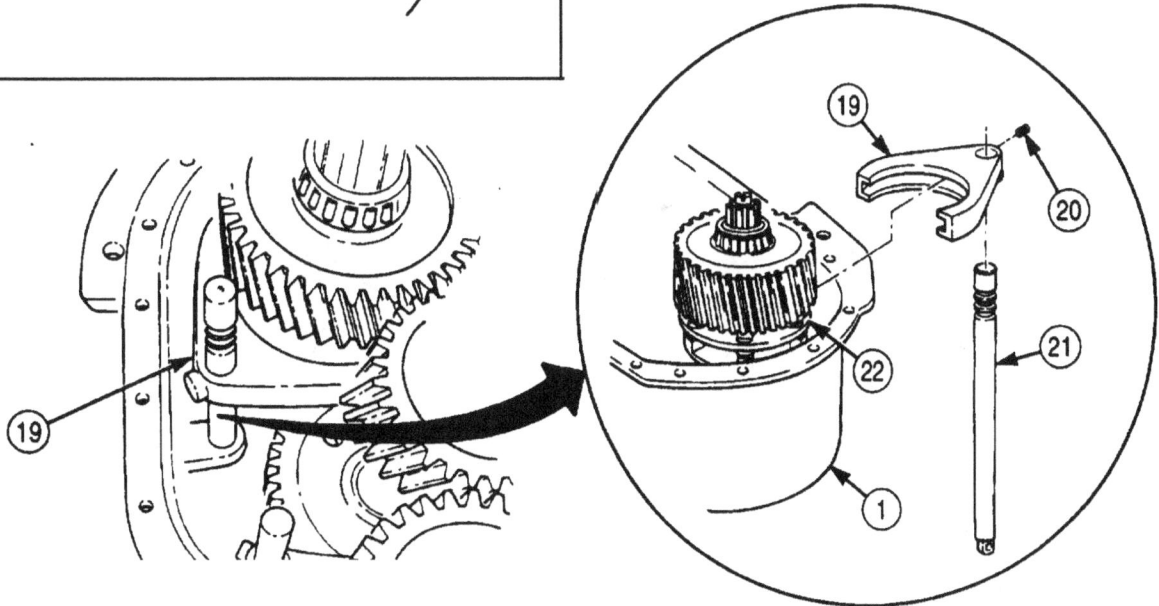

5-63. TRANSFER CASE REPAIR (Contd)

28. Apply gasket sealant to mating surfaces of transfer case cover (1) and housing (5).

29. Install three dowel pins (2) on transfer case cover (1).

30. Align dowel pins (2) on transfer case cover (1) with holes in transfer case housing (5) and install transfer case cover (1) on transfer case housing (5) with nineteen washers (4) and screws (3). Tighten screws (3) 40-65 lb-ft (54-88 N•m).

31. Tap around outer edge of bearing race (8) to seat against main input shaft bearing (9).

32. Install baffle (7) on main input shaft (6).

NOTE

Perform steps 32 through 34 for vehicles equipped with PTO.

33. Apply adhesive sealant to setscrew (11).

34. Install new seal (12) and PTO drive gear (10) on recess of main input shaft (9) with setscrew (11). Tighten setscrew (11) 12-16 lb-ft (16-22 N•m).

35. Apply sealing compound to PTO (16) and transfer case (15) and install PTO (16) on transfer case (15) with six washers (14) and screws (13). Tighten screws (13) 140-65 lb-ft (54-36 N•m).

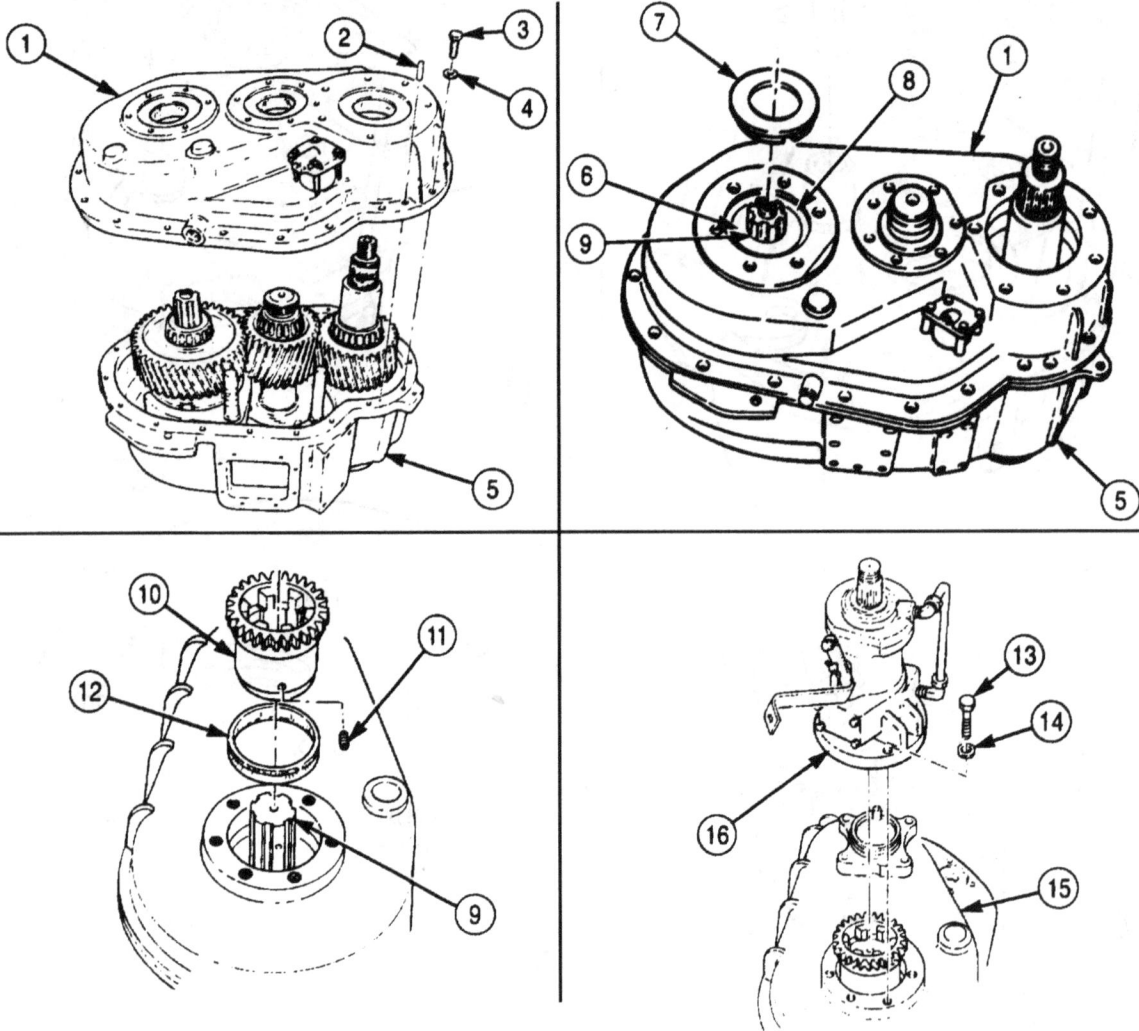

5-63. TRANSFER CASE REPAIR (Contd)

36. Apply adhesive sealant to setscrew (24).

37. Install new seal (23) on retaining collar (25) with large lip toward baffle (7).

38. Install retaining collar (25) on recess of main input shaft (6) with setscrew (24). Tighten setscrew (24) 12-16 lb-ft (16-22 N•m).

39. Apply a light coating of gasket sealant to mating surfaces of oil pump (26) and transfer case (15).

NOTE

Ensure to align pump drive with slot on main input shaft.

40. Install oil pump (26) on transfer case (15) with six washers (18) and screws (17). Tighten screws (17) 40-65 lb-ft (54-88 N-m).

41. Apply a light coating of gasket sealant to mating surfaces of speedometer drive cover (22) and transfer case (15).

42. Install speedometer drive cover (32) on transfer case (15) with six washers (21), four screws (20), and two screws (19). Tighten screws (19) and (20) 40-65 lb-ft (54-88) N•m).

43. Install new seal (28) and slinger (27) on flange (29).

44. Apply gasket sealant to new hi-lo shift shaft seal (30) and install on transfer case (15).

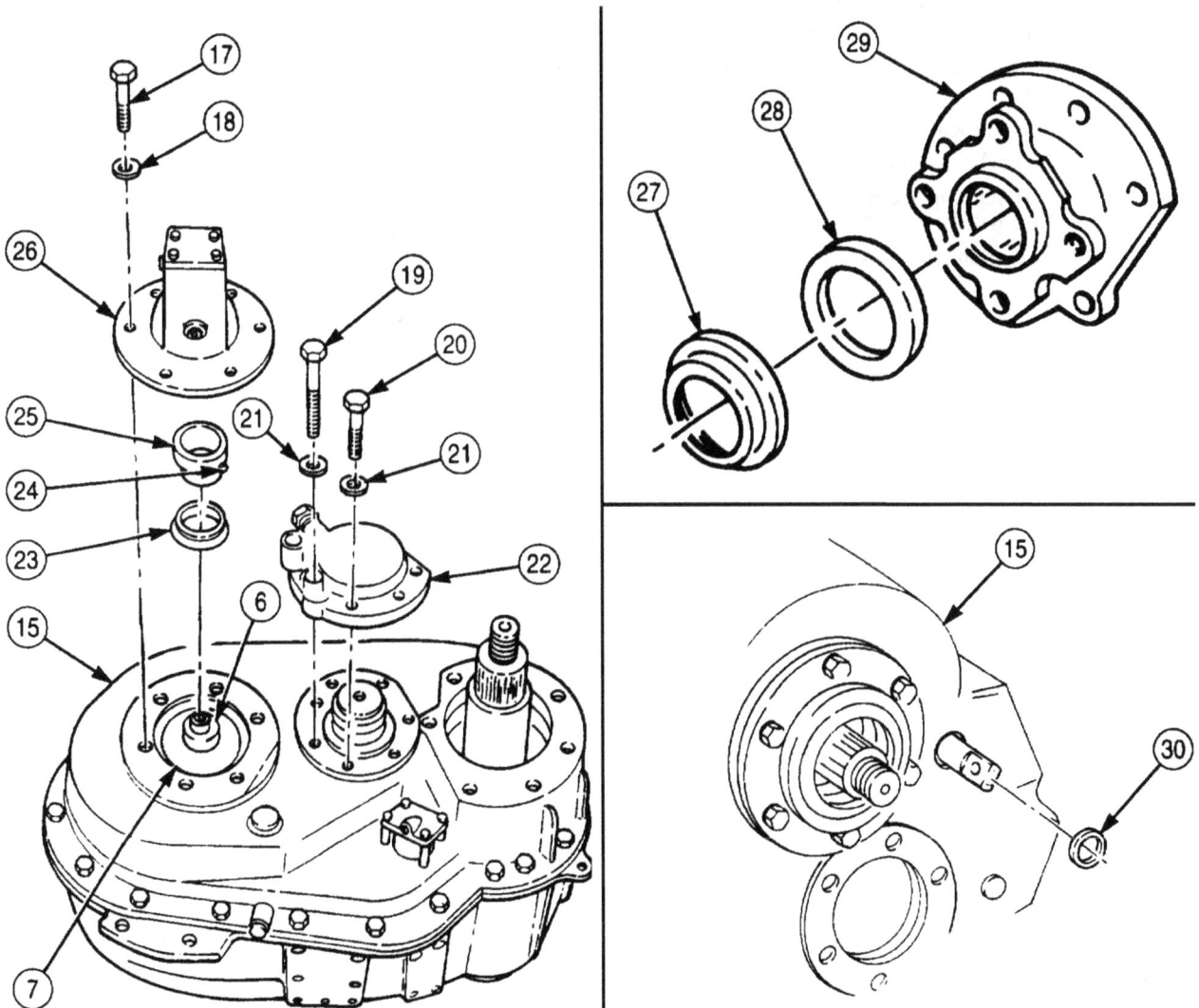

5-63. TRANSFER CASE REPAIR (Contd)

NOTE

Step 45 establishes initial shim pack thickness for intermediate shaft.

45. Install intermediate shaft cover (2) on transfer case (1) and tap alternately around outer edge with soft-faced hammer.

46. Using feeler gauge, measure and record clearance between intermediate shaft cover (2) and transfer case (1).

47. Remove intermediate shaft cover (2) from transfer case (1).

NOTE

Use measurement from step 46 plus a .003 in. (.7692 mm) shim for initial shim pack thickness.

48. Install shim pack (3) and intermediate shaft cover (2) on transfer case (1) with six washers (4) and screws (5). Tighten screws (5) 40-65 lb-ft (54-88 N•m).

49. Mount dial indicator on transfer case (1) with tip of plunger positioned through hole in intermediate shaft cover (2) on intermediate shaft (6).

NOTE

Perform step 50 to ensure intermediate shaft will rotate freely without end play.

50. Check intermediate shaft end play as follows:

a. Insert prybar through inspection hole (7) and force intermediate shaft (6) to rear of transfer case (1).

b. Set dial indicator to zero.

c. Using prybar, force intermediate shaft (6) to front of transfer case (1).

d. Record reading from dial indicator and add 0.003-0.006 in. (0.076-0.152 mm) to reading. This reading will be used for number of shims (3) to be removed.

51. Remove dial indicator from transfer case (1).

5-63. TRANSFER CASE REPAIR (Contd)

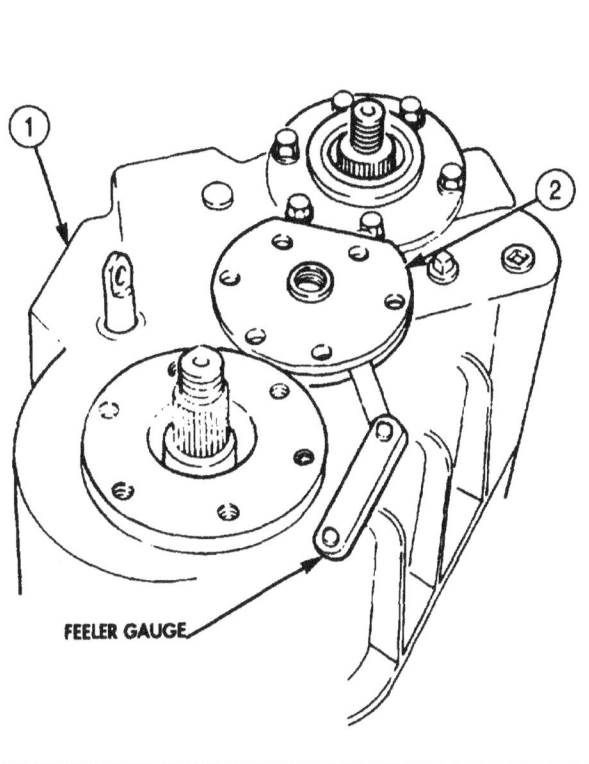

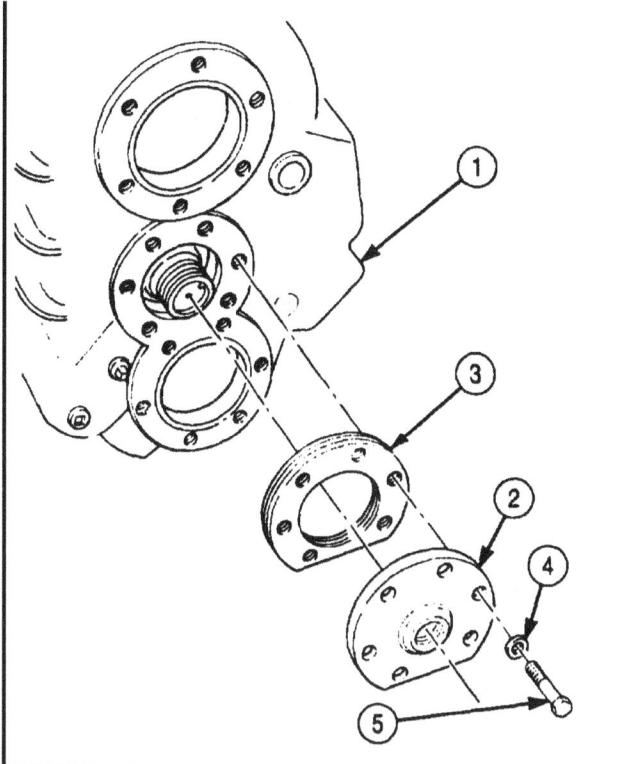

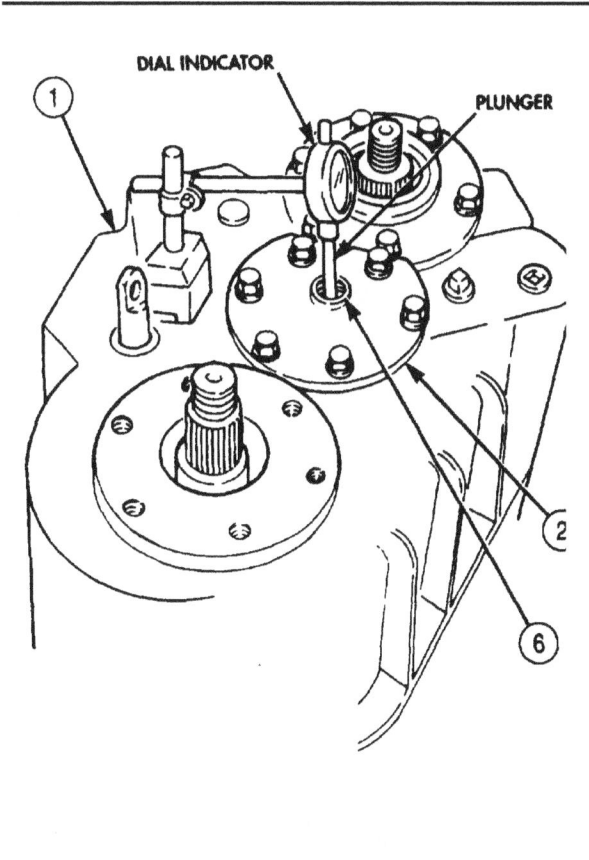

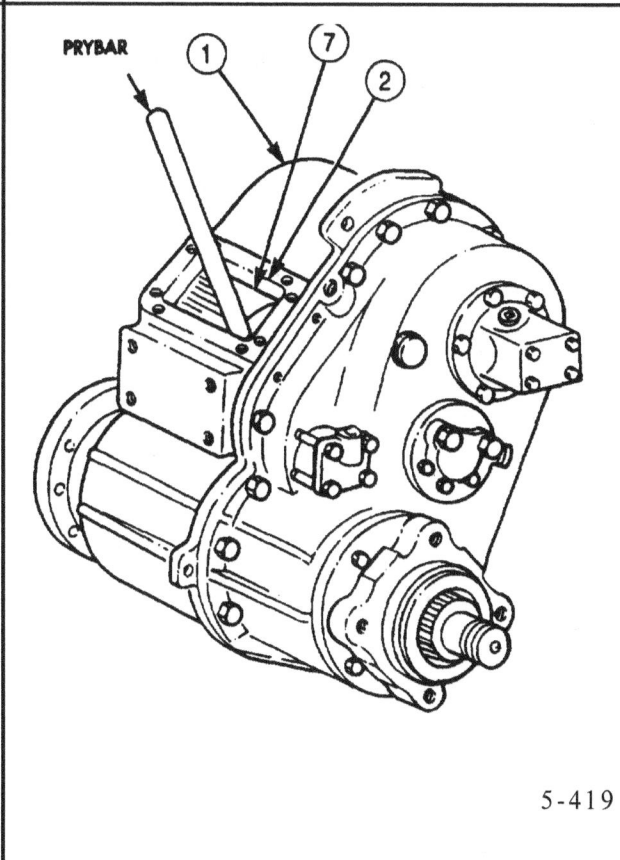

FEELER GAUGE

DIAL INDICATOR

PLUNGER

PRYBAR

5-63. TRANSFER CASE REPAIR (Contd)

52. Remove six screws (5), washers (4), intermediate shaft cover (3), and shim pack (2) from transfer case (1).

NOTE

Use reading obtained in step 50d for number or thickness of shims to be removed or added for intermediate shaft adjustment.

53. Remove or add amount of shims (2) as recorded.

CAUTION

Do not apply sealing compound to shims.

54. Apply a sealing compound to lightly coat mating surfaces of transfer case (1) and intermediate shaft cover (3).

55. Install shim pack (2) and intermediate shaft cover (3) on transfer case (1) with six washers (4) and screws (5). Tighten screws (5) 40-65 lb-ft (54-88 N•m).

56. Install plug (6) on intermediate shaft cover (3). Tighten plug (6) 35 lb-ft 48 N•m)

NOTE

Step 57 establishes initial shim pack thickness for main input shaft.

57. Install main input shaft cover (7) on transfer case (1) and tap alternately around outer edge with soft-faced hammer.

58. Using feeler gauge, measure and record clearance between main input shaft cover (7) and transfer case (1).

59. Remove main input shaft cover (7) from transfer case (1).

NOTE

Use measurement from step 58 plus a .003 in. .0762 mm) shim for initial shim pack thickness.

60. Install shim pack (8) and main input shaft cover (7) on transfer case (1) with six washers (10) and screws (9). Tighten screws (9) 40-65 lb-ft (54-88 N•m).

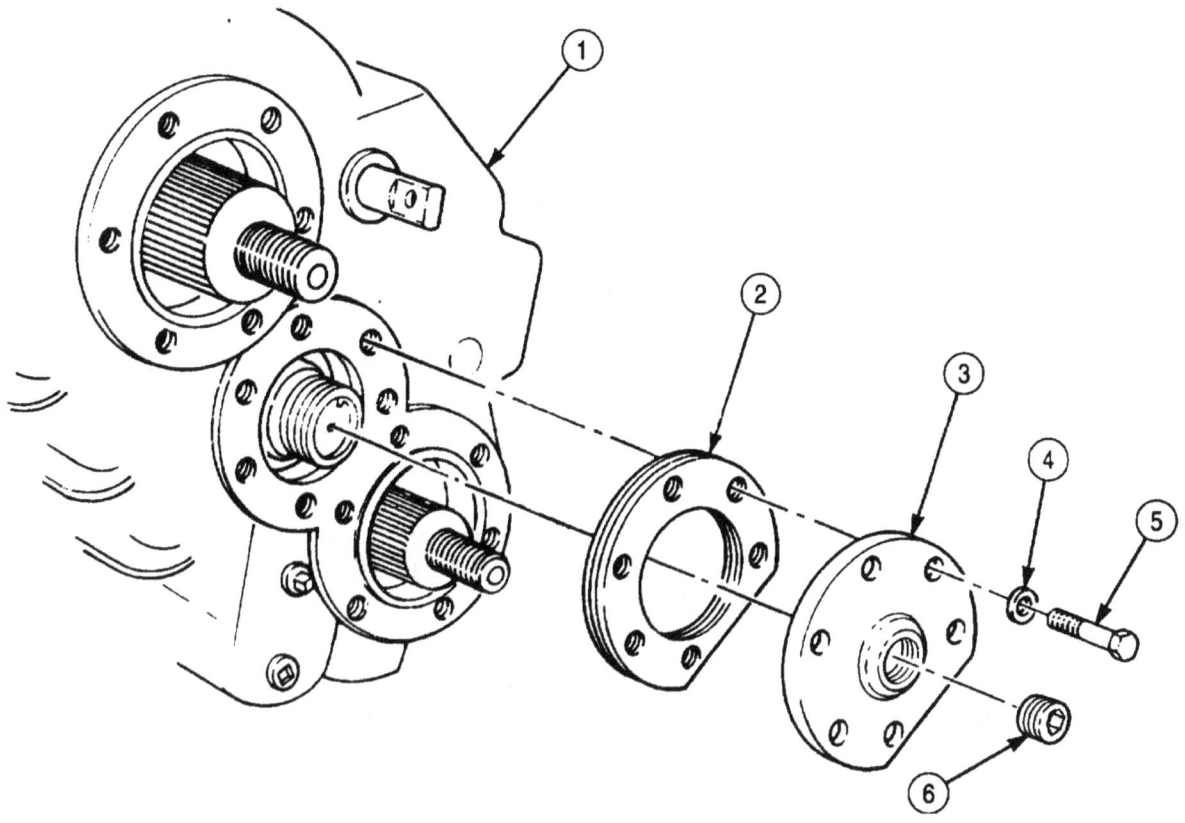

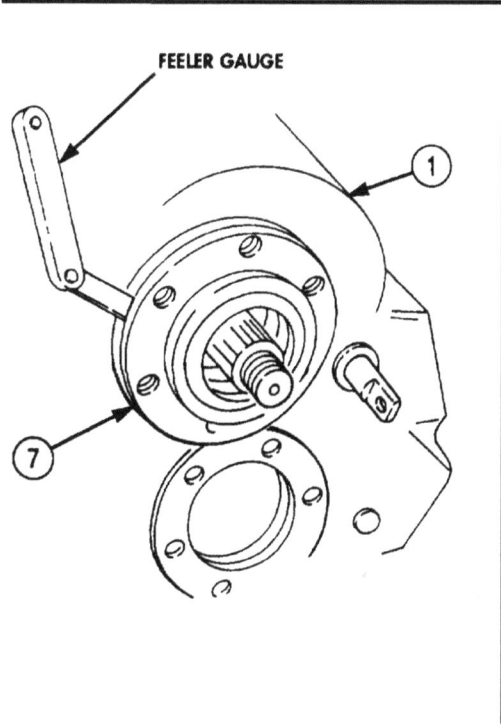

FEELER GAUGE

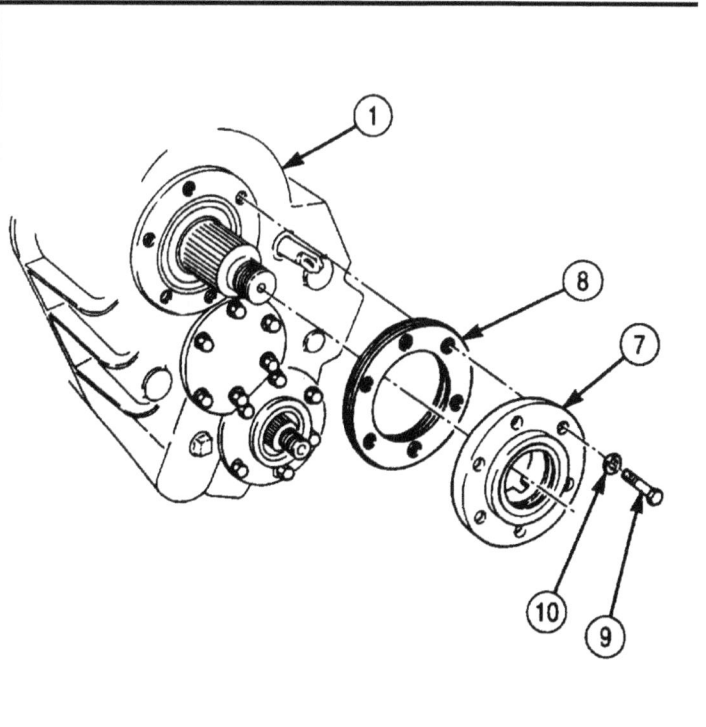

5-63. TRANSFER CASE REPAIR (Contd)

61. Mount dial indicator on transfer case (1) with tip of plunger on main input shaft (2).

NOTE

Perform steps 62 through 64 to ensure main shaft rotates freely without end play.

62. Check main input shaft end play as follows:

   a. Using prybar through inspection hole (3), force main input shaft (2) to rear of transfer case (1) and hold.

   b. Set dial indicator to zero.

   c. Using prybar through inspection hole (3), force main input shaft (2) to front of transfer case (1).

   d. Record reading on dial indicator. End play should be 0.0001-0.005 in. (0.00254-0.127 mm).

63. Remove dial indicator from transfer case (1).

64. Remove six screws (7), washers (6), main input shaft cover (5), and shim pack (4) from transfer case (1).

NOTE

Use reading obtained in step 62d for number or thickness of shims to be removed or added for main input shaft adjustment.

65. Remove or add amount of shims (4) as recorded.

CAUTION

Do not apply sealing compound to shims.

66. Apply a light coating of sealing compound to mating surfaces of transfer case (1) and main input shaft cover (5).

67. Install shim pack (4) and main input shaft cover (5) on transfer case (1) with six washers (6) and screws (7). Tighten screws (7) 40-65 lb-ft (54-88 N•m).

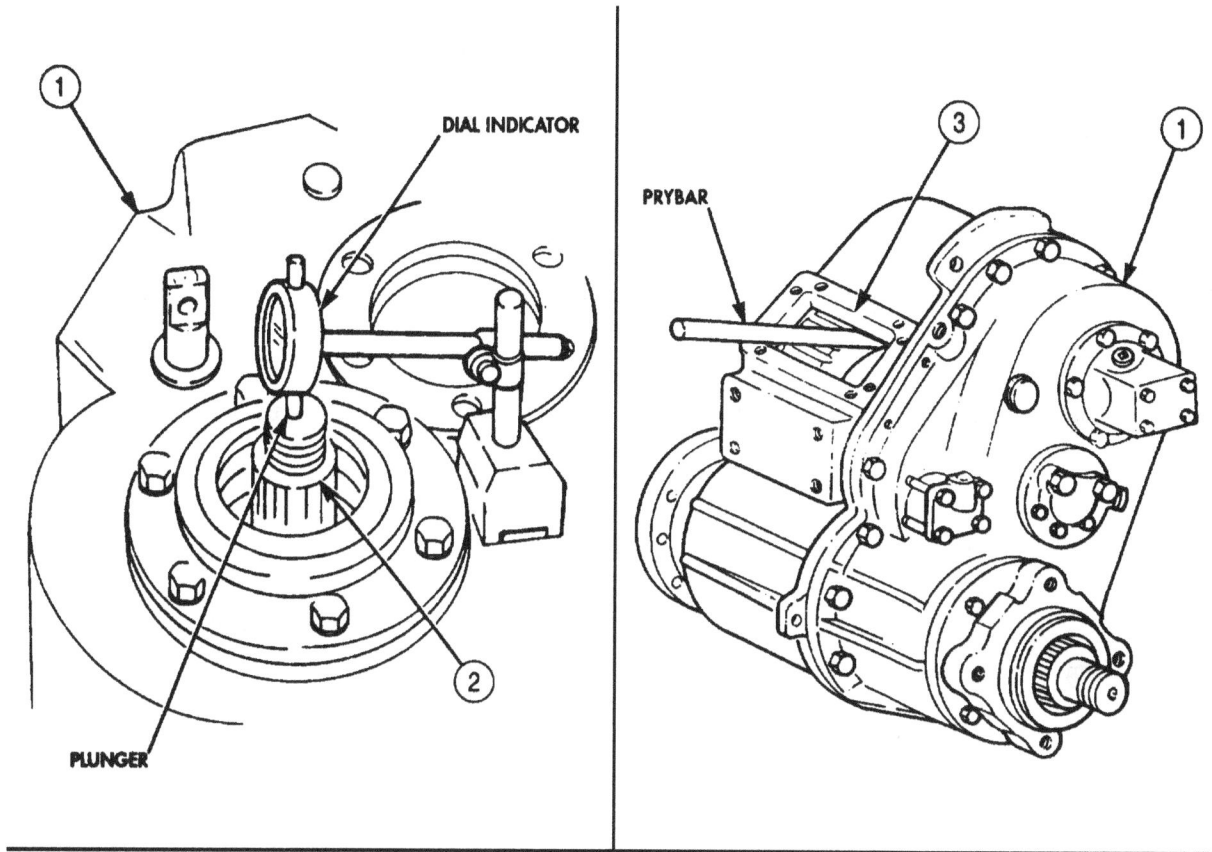

DIAL INDICATOR

PRYBAR

PLUNGER

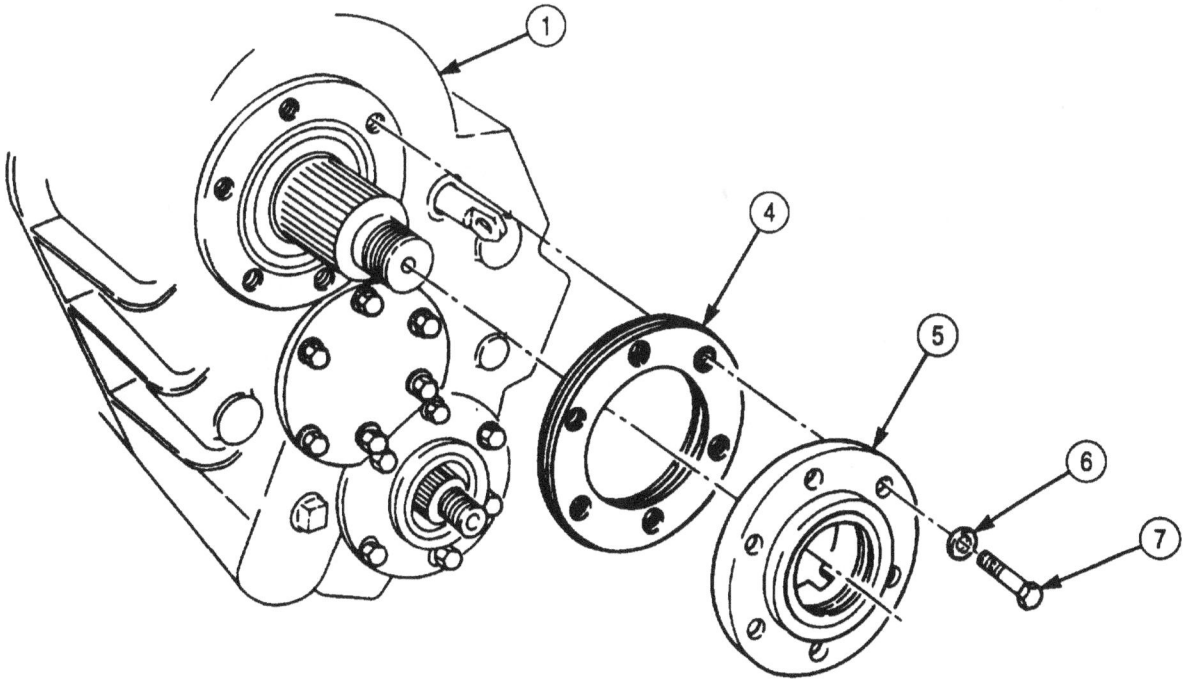

5-63. TRANSFER CASE REPAIR (Contd)

NOTE

Steps 68 and 69 establish initial shim pack thickness for rear
output shaft.

68. Install companion flange (2) on transfer case (1) and tap alternately around outer edge with soft-faced hammer.

69. Using feeler gauge, measure and record clearance between companion flange (2) and transfer case (1).

70. Remove companion flange (2) from transfer case (1).

NOTE

Use measurement from step 69 plus a 0.003 in. (0.0762 mm) shim
for initial shim pack thickness.

71. Install shim pack (3) and companion flange (2) on transfer case (1) with seven washers (4) and screws (5). Tighten screws (5) 60-100 lb-ft (81-136 N-m).

72. Mount dial indicator on transfer case (1) with tip of plunger positioned on rear output shaft (6).

73. Check rear output shaft (6) end play as follows:

   a. Using prybar through inspection hole (7), force rear output shaft (6) to front of transfer case (1) and hold.

   b. Set dial indicator to zero.

   c. Using prybar through inspection hole (7), force rear output shaft (6) to rear of transfer case (1).

   d. Record reading on dial indicator. End play should be 0.0001-0.005 in. (0.00254-0.127 mm).

74. Remove dial indicator from transfer case (1).

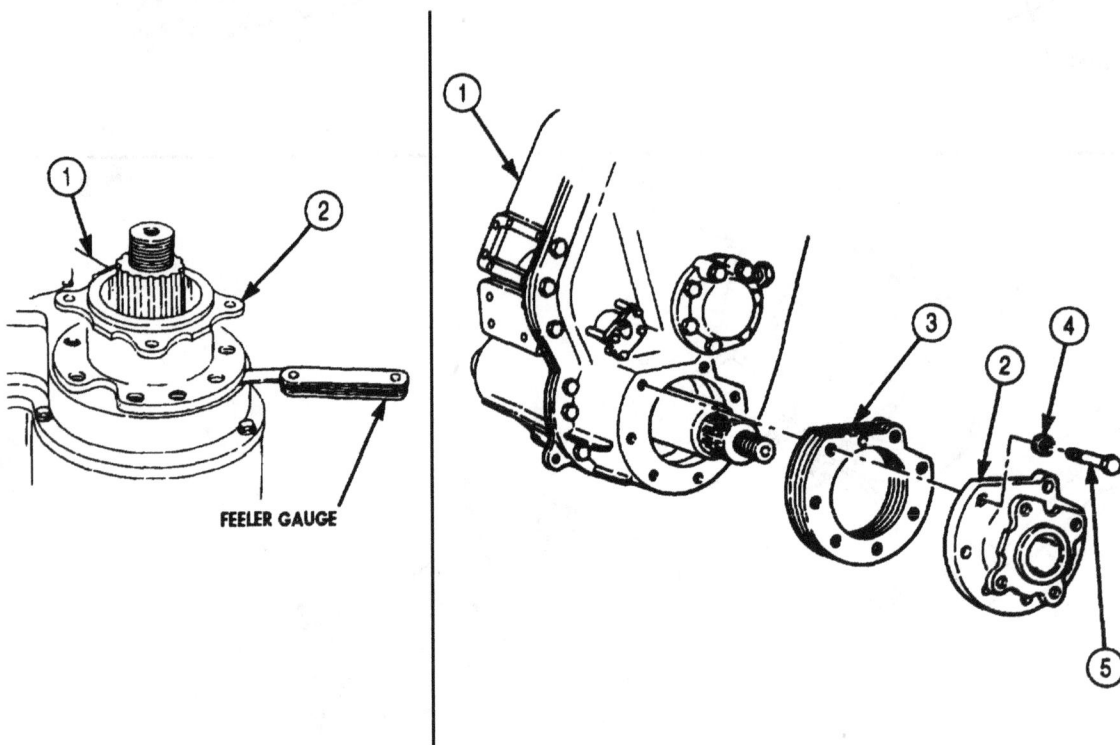

FEELER GAUGE

5-63. TRANSFER CASE REPAIR (Contd)

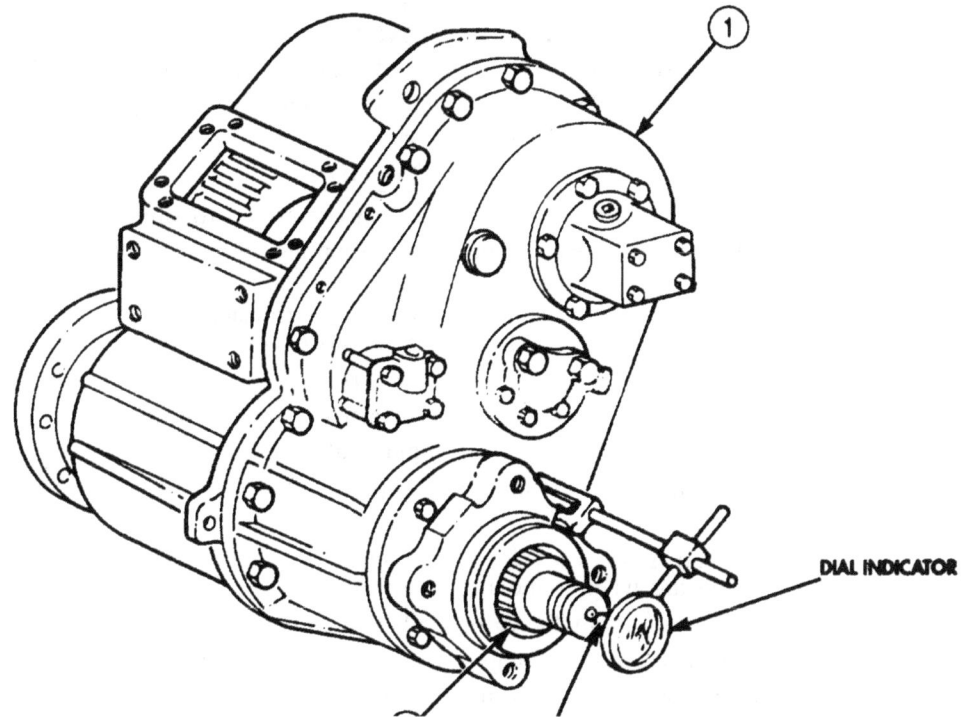

DIAL INDICATOR

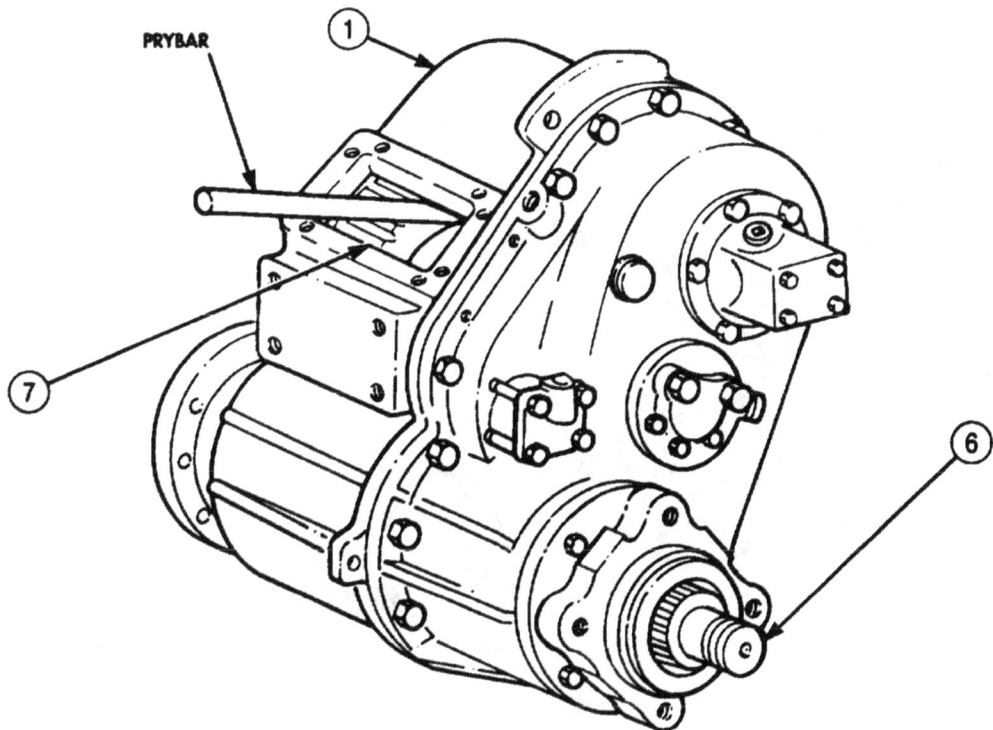

PRYBAR

5-63. TRANSFER CASE REPAIR (Contd)

75. Remove seven screws (4), washers (3), and companion flange (5) from transfer case (1).

NOTE

Use reading obtained in step 73d for number or thickness of shims
to be removed or added for rear output shaft adjustment.

76. Remove or add amount of shims (2) as recorded.

CAUTION

Do not apply sealing compound to shims.

77. Apply a light coating of sealing compound to mating surfaces of transfer case (1) and companion flange (5).

78. Install shim pack (2) and companion flange (5) on transfer case (1) with seven washers (3) and screws (4). Tighten screws (4) 110-145 lb-ft (149-197 N•m).

NOTE

Apply GAA grease to inside diameter of all seals.

79. Install main input flange (6) on main input shaft (9).

80. Install front output flange (7) on front output shaft (8).

NOTE

Chain is necessary to prevent main and output flanges from
rotating when removing hardware.

81. Install chain on main input flange (6) with two washers (16) and screw (17).

82. Install chain on front output flange (7) with washer (12), screw (11), washer (12), and nut (13).

83. Apply a thin coating of gasket sealant to mating surfaces of washers (18) and (14), and install washers (18) and (14) and new locknuts (10) and (15) on front output shaft (8) and main input shaft (9). Tighten locknuts (10) and (15) 450-600 lb-ft (610-814 N•m).

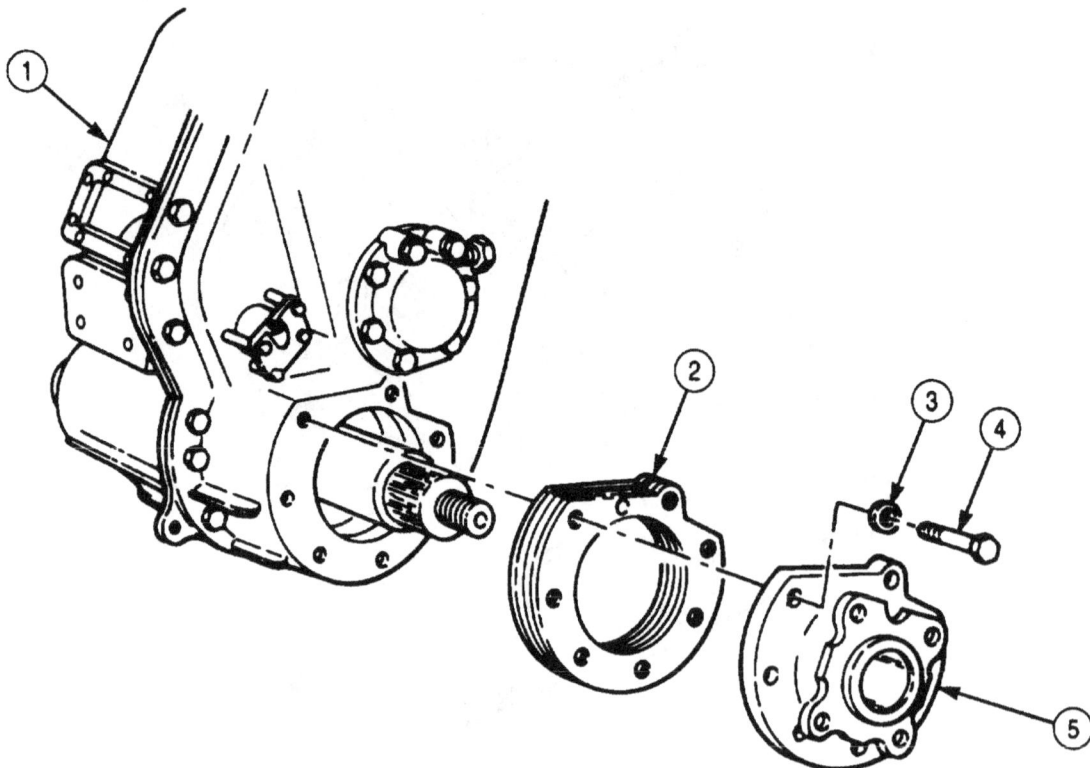

5-63. TRANSFER CASE REPAIR (Contd)

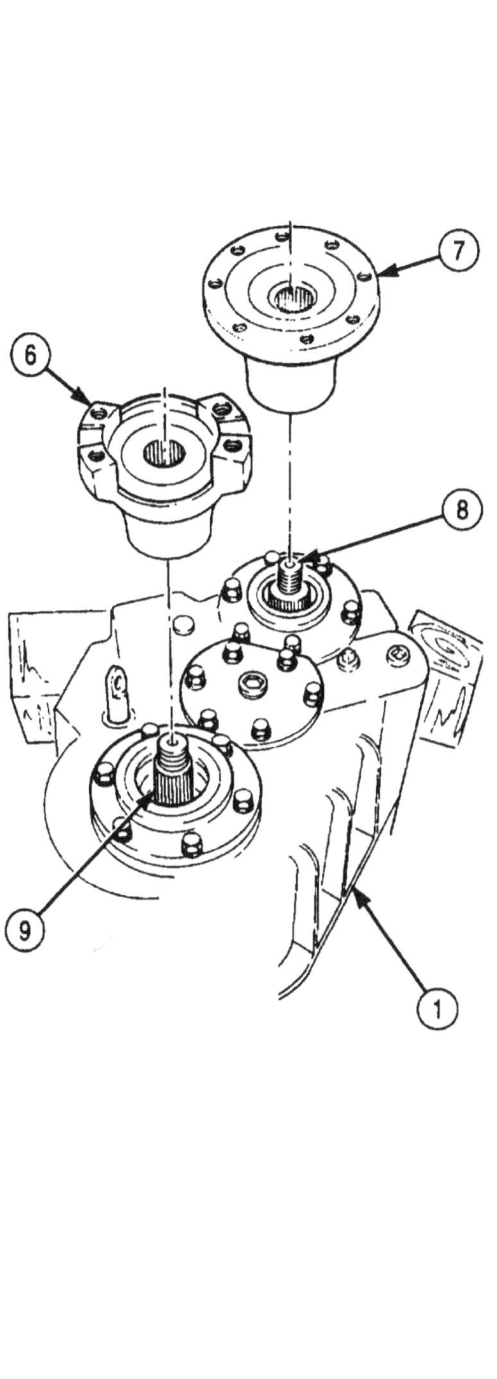

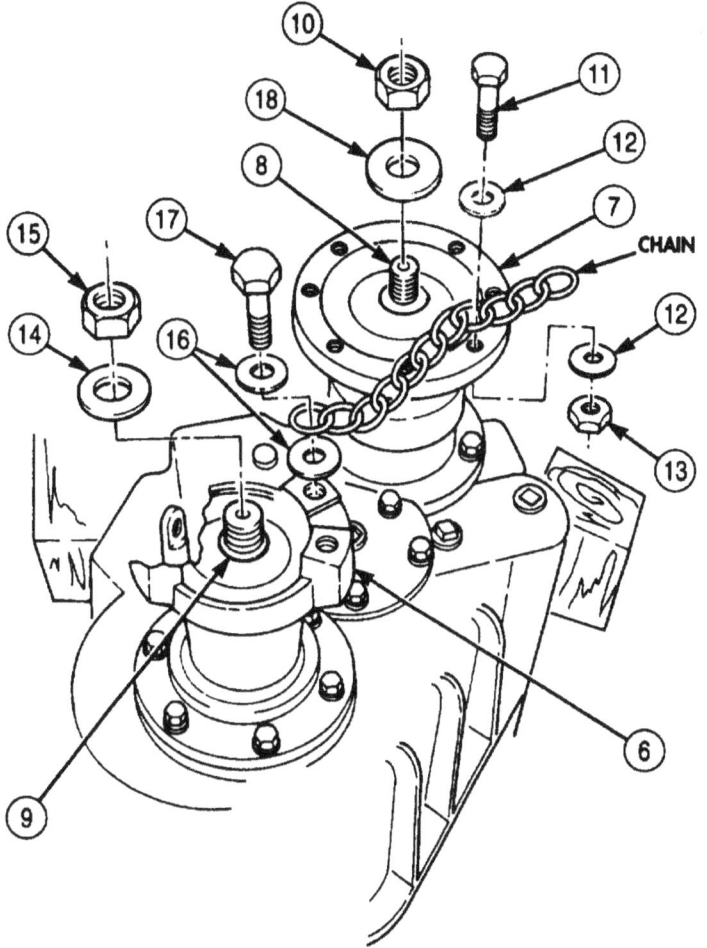

5-63. TRANSFER CASE REPAIR (Contd)

84. Install filter screen (3), adapter fitting (4), and elbow (5) on transfer case (2). Tighten elbow (5) 15 lb-ft (20 N•m).

NOTE
Perform steps 85 and 86 for vehicles equipped with PTO.

85. Install elbow (1) on PTO (7). Tighten elbow (1) 26 lb-ft (35 N•m).

86. Connect oil line (6) to elbows (5) and (1).

87. Install nipple end of elbow (13) on oil pump (8). Tighten elbow (13) 20 lb-ft (27 N•m).

88. Connect oil line (14) to elbows (5) and (13).

89. Install new seal (9) on cover (12).

90. Install new seal (17) on piston (16).

91. Install new gasket (18). new piston (16), new cylinder (15), and cover (12) on transfer case (2) with four new locking plates (11) and screws (10). Tighten screws (10) 5-10 lb-ft (8-13 N•m).

92. Bend tabs of locking plates (11) over heads of screws (10).

5-63. TRANSFER CASE REPAIR (Contd)

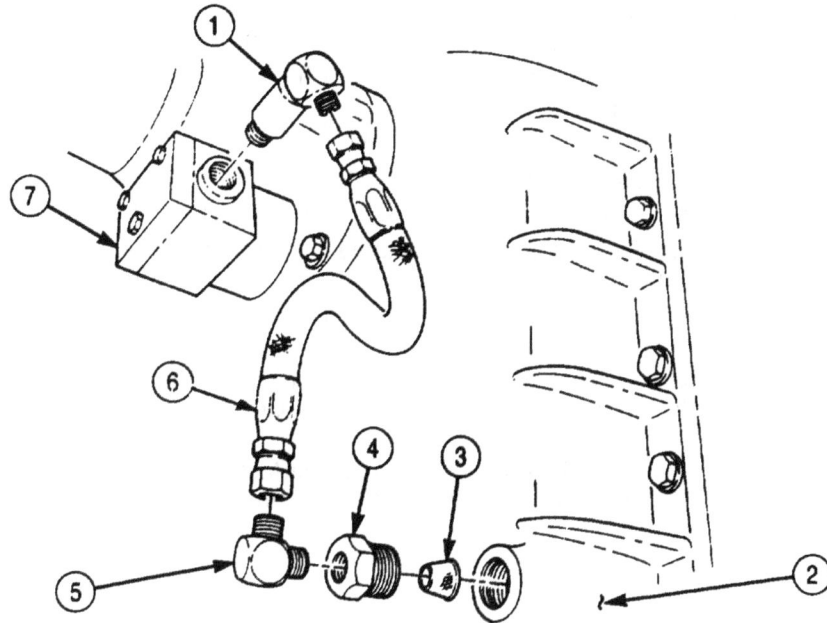

WITH TRANSFER CASE PTO

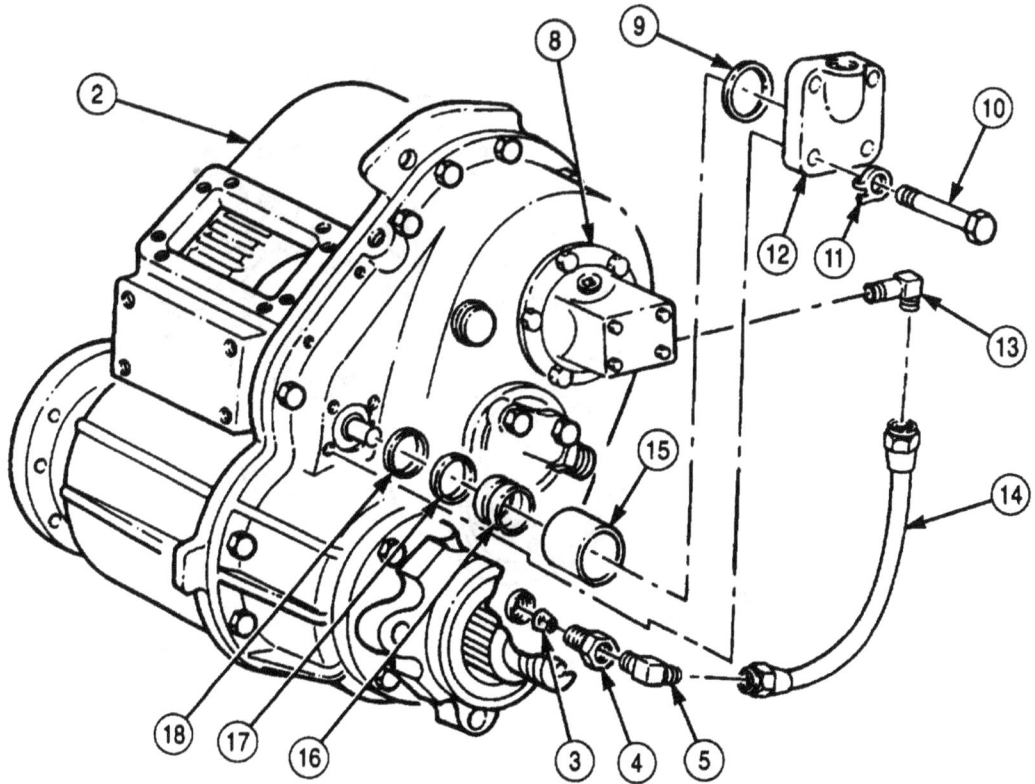

WITHOUT TRANSFER CASE PTO

5-63. TRANSFER CASE REPAIR (Contd)

93. Install brakeshoe assembly (12) and dust cover (9) on companion flange (10) with four screws (13). Tighten screws (13) 180-230 lb-ft (244-312 N•m).

94. Position actuating plate (6) against backing plate (7) so retainer opening (5) fits over brake lever stud (8) of backing plate (7).

NOTE

Apply GAA grease to inside diameter of all seals.

95. Install parking brake drum (4) and output flange (3) on output shaft (11).

96. Apply a thin coating of gasket sealant to mating surfaces of washer (2) and install washer (2) and new locknut (1) on output shaft (11). Tighten locknut (1) 450-600 lb-ft (610-814 N•m).

97. Remove nut (18), washer (17), screw (16), washer (17), screw (14), washer (19), and chain from main input flange (20) and front output flange (15).

98. Apply a thin coating of gasket sealant to mating surfaces of inspection plate (24) and transfer case (23) and install inspection plate (24) on transfer case (23) with eight screws (25). Do not tighten screws (25).

99. Using crowfoot wrench, install pushrod (22) and interlock air cylinder (21) on transfer case (23). Tighten interlock air cylinder (21) 30-40 lb-ft (41-54 N•m).

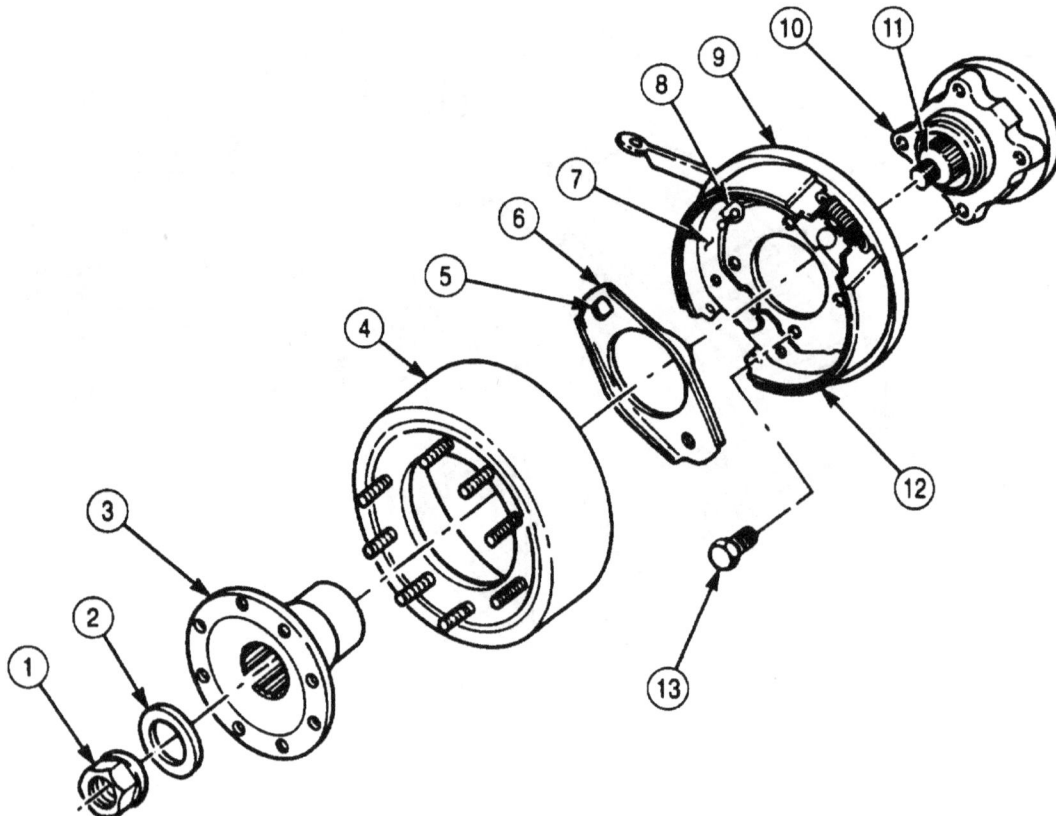

5-63. TRANSFER CASE REPAIR (Contd)

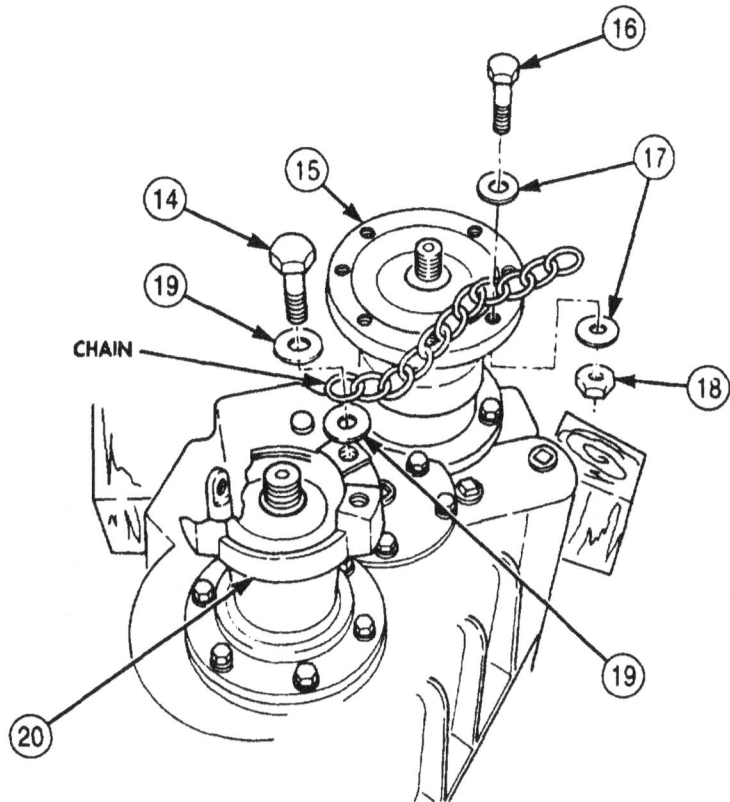

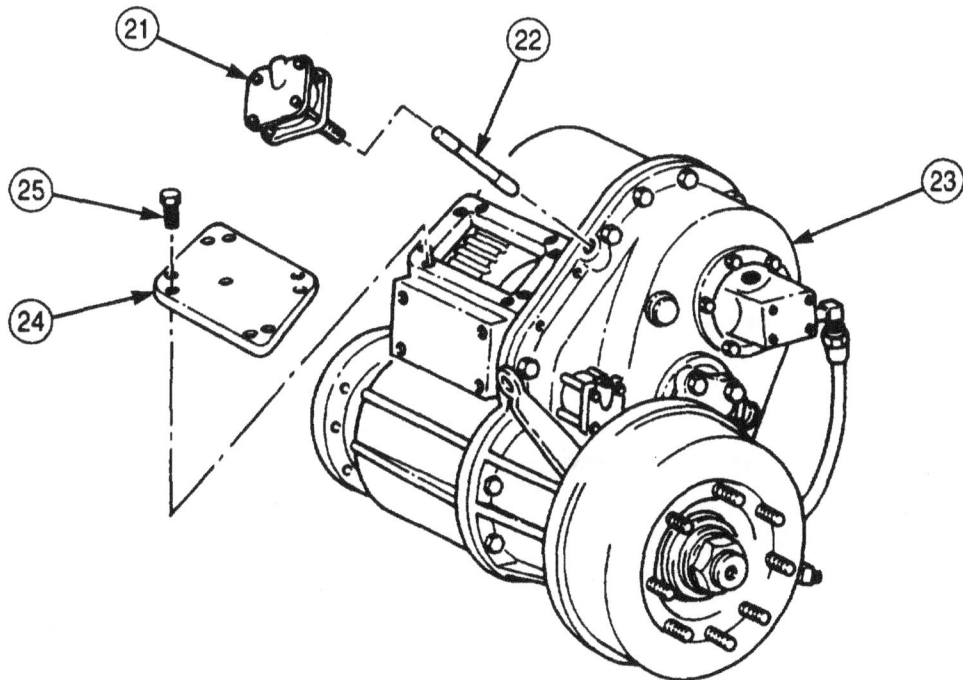

CHAIN

FOLLOW-ON TASK: Install transfer case (para. 4-94 or para. 4-95).

Section V. FRONT AND REAR AXLE MAINTENANCE

5-64. DIFFERENTIAL AND DIFFERENTIAL CARRIER REPAIR

THIS TASK COVERS:

a. Disassembly
b. Cleaning, Inspection, and Repair

c. Assembly and Adjustment

INITIAL SETUP:

APPLICABLE MODELS
All

SPECIAL TOOLS
Bearing replacer (Appendix E, Item 14)
Two puller screws (Appendix E, Item 104)

TOOLS
General mechanic's tool kit
  (Appendix E, Item 1)
Mechanical puller kit (Appendix E, Item 102)
Inside micrometer
  (Appendix E, Item 83)
Torque wrench (Appendix E, Item 146)
Torque wrench (Appendix E, Item 145)
Dial indicator (Appendix E, Item 36)
Outside micrometer (Appendix E, Item 80)
Arbor press
Torque multiplier
Vise
Soft-head hammer
Chain
Lifting device
Bearing preload tester

MATERIALS/PARTS
Gasket (Appendix D, Item 206)
Gasket (Appendix D, Item 207)
Gasket (Appendix D, Item 208)
Three safety wires, 12 in. (Appendix D, Item 568)
Wear sleeve kit (Appendix D, Item 717)
Gasket and shim set (Appendix D, Item 248)
Six lockwashers (Appendix D, Item 350)
Locktab washer (Appendix D, Item 344)
Two cotter pins (Appendix D, Item 61)
Two oil seals (Appendix D, Item 500)
Gasket (Appendix D, Item 653)
Six lockwashers (Appendix D, Item 354)
Safety wire, 36 in. (Appendix D, Item 568)
Crocus cloth (Appendix C, Item 20)
GAA grease (Appendix C, Item 28)
Lubricating oil (Appendix C, Item 47)
Blue oil-base pigment (Appendix C, Item 54)
Sealing compound (Appendix C, Item 61)
Sealing compound (Appendix C, Item 62)
Twine (Appendix C, Item 77)
White carbonate pigment (Appendix C, Item 78)

REFERENCES (TM)
LO 9-2320-272-12
TM 9-214
TM 9-2320-272-24P

EQUIPMENT CONDITION
Differential and carrier assembly removed
(para. 4-100).

a. Disassembly

NOTE
Front and rear differentials are repaired the same way.

1. Remove safety wires (1) from two screws (2) and four screws (4). Discard safety wires (1).

5-64. DIFFERENTIAL AND DIFFERENTIAL CARRIER REPAIR (Contd)

2. Remove two screws (2) and adjusting nut locks (3) from bearing caps (6).

CAUTION

Scribe bearing caps and saddles for installation. These items are machine-matched and damage will result if they are intermixed.

3. Remove four screws (4), washers (5), and two bearing caps (6) from differential carrier housing (11).

NOTE

Tag bearings and races as matched parts for inspection and installation.

4. Remove two adjusting nuts (9) and bearing races (8) from bearing saddles (10) and caps (6).
5. Remove differential gear assembly (7) from differential carrier housing (11) and bearing saddles (10).
6. Using puller, remove two tapered roller bearings (12) from differential housings (13).
7. Remove safety wire (18) from eight slotted nuts (19). Discard safety wire (18).

CAUTION

Two gear housings and helical drive gear must be scribed for installation. These items are machine-matched and damage may result if they are intermixed.

8. Remove eight slotted nuts (19) and screws (22) from two differential housings (13). Mark two differential housings (13) and helical ring gear (16) for installation.
9. Remove four bevel spider gears (20), thrust washers (21), and spider (17) from differential housing (13).
10. Remove two bevel side gears (15) and thrust washers (14) from differential housings (13).

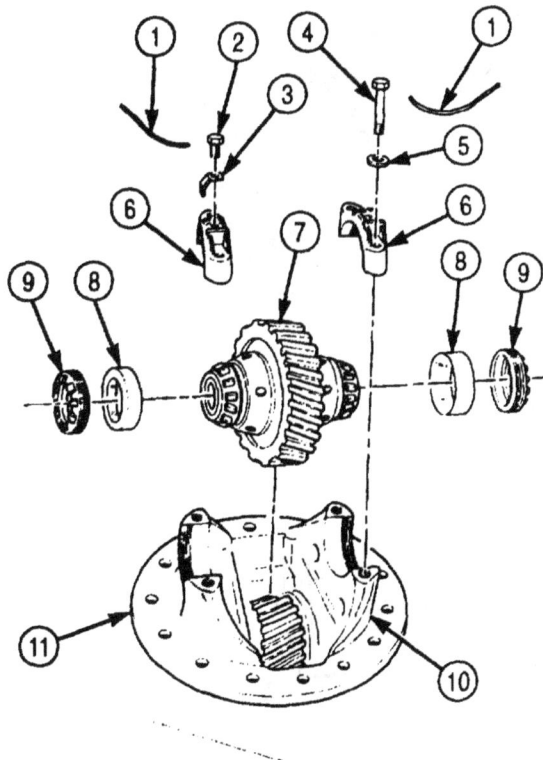

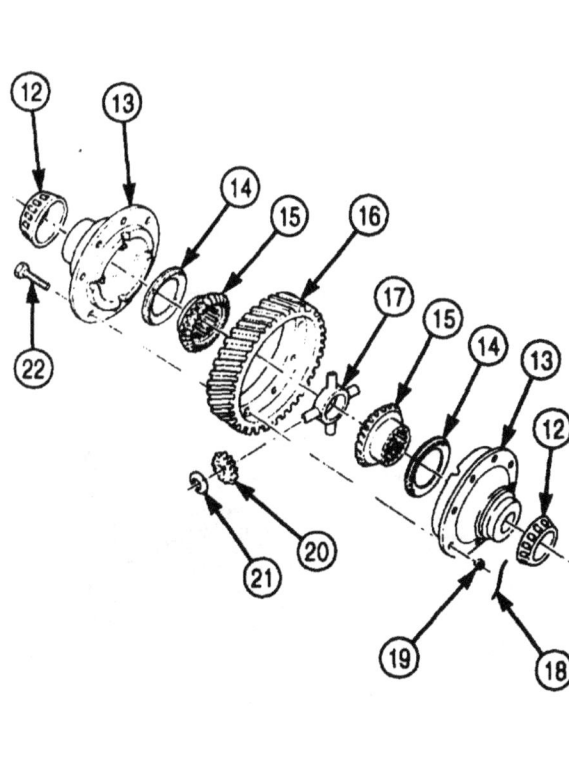

5-64. DIFFERENTIAL AND DIFFERENTIAL CARRIER REPAIR (Contd)

11. Remove two cotter pins (1), nuts (2), and companion flanges (3) from mainshaft (4). Discard cotter pins (1).

12. Remove six screws (9) and lockwashers (10) from rear bearing cover (8). Discard lockwashers (10).

13. Remove rear bearing cover (8), gasket (6), and thrust washer (5) from differential carrier housing (11). Discard gasket (6).

14. Remove oil seal (7) from rear bearing cover (8). Discard oil seal (7).

15. Remove eight screws (16) and washers (17) from front bearing cover (13).

16. Remove front bearing cover (15) and gasket (12) from differential carrier housing (11). Discard gasket (12).

17. Using puller, remove oil seal (13) and gasket (14) from front bearing cover (15). Discard oil seal (13) and gasket (14).

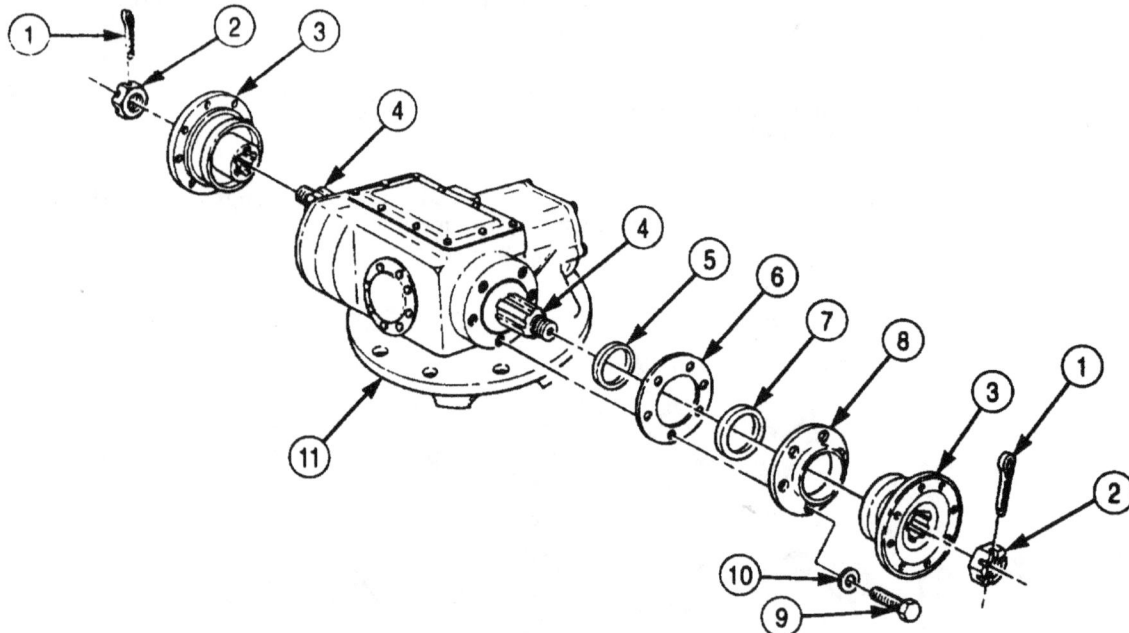

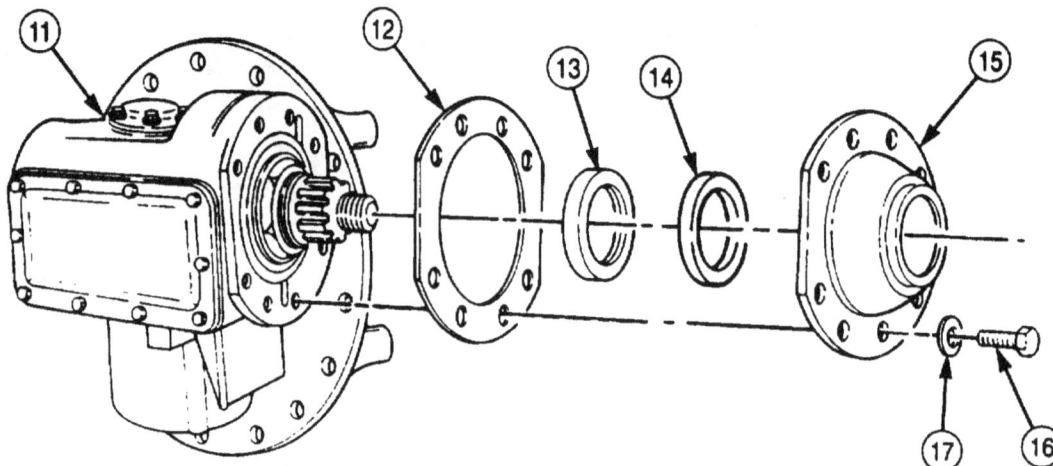

5-64. DIFFERENTIAL AND DIFFERENTIAL CARRIER REPAIR (Contd)

18. Remove mainshaft and pinion gear assembly (4) from differential carrier housing (11).

NOTE

Tag and retain shims for measurement and installation.

19. Using bearing remover, remove shims (19) from front of differential carrier housing (11).
20. Remove bearing (18) from rear of differential carrier housing (11).
21. Place mainshaft (4) in soft-jawed vise.
22. Using puller, remove rear bearing race (24) from rear of mainshaft (4).
23. Open tab(s) on locktab washer (21) and remove outer nut (20), locktab washer (21), keywasher (22), and inner nut (23) from mainshaft (4). Discard locktab washer (21).

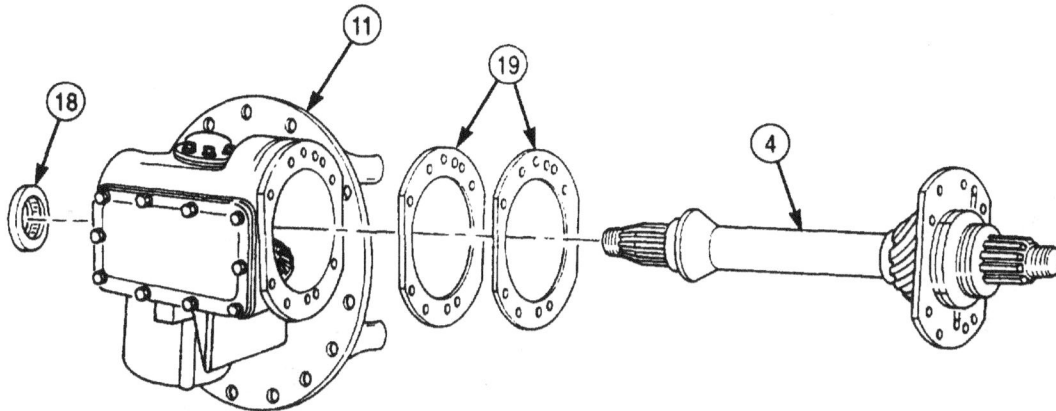

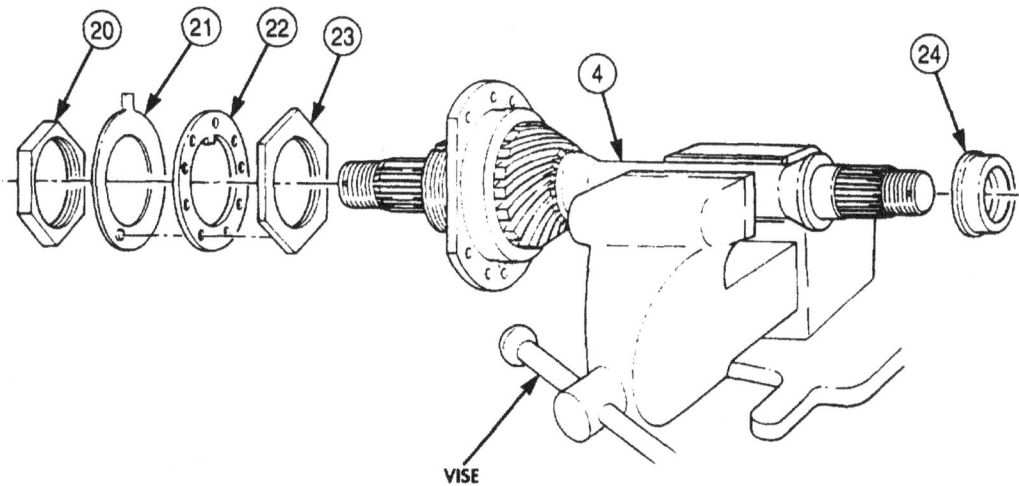

5-64. DIFFERENTIAL AND DIFFERENTIAL CARRIER REPAIR (Contd)

NOTE

Bearings and races are matched sets and must be kept together for inspection and installation.

24. Using arbor press, remove outer bearing (1), retainer (3), and spacer (5) from pinion gear (6) and mainshaft (7).

NOTE

Races may require light tapping to remove from retainer.

25. Using bearing replacer, remove races (2) and (4) from retainer (3).
26. Remove mainshaft (7) from vise.
27. Using arbor press, remove pinion gear (6) with bearing (8) from mainshaft (7).
28. Using puller, remove bearing (8) from pinion gear (6).

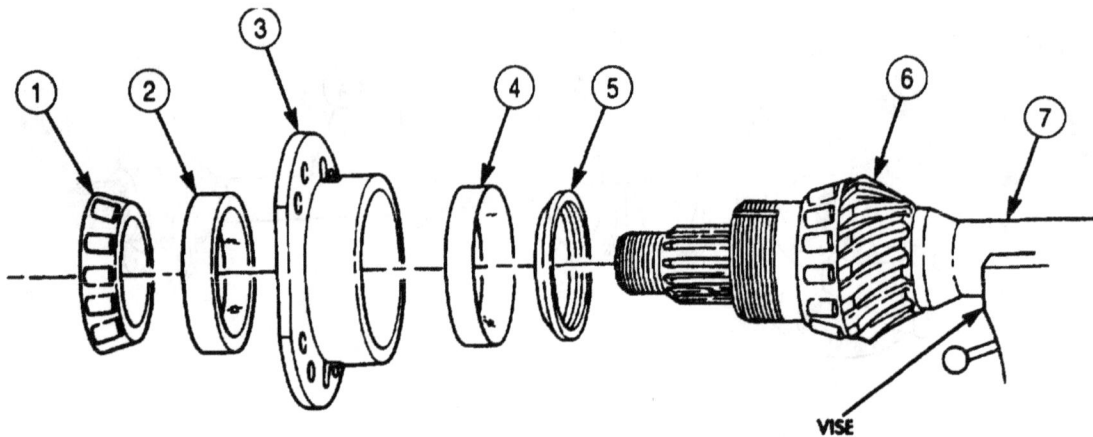

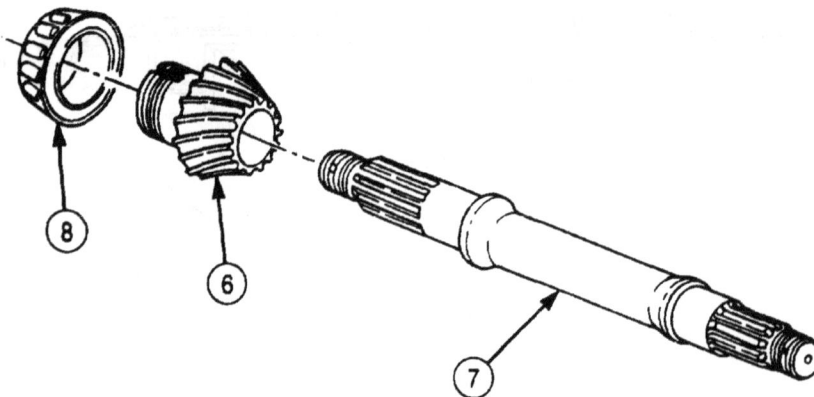

5-64. DIFFERENTIAL AND DIFFERENTIAL CARRIER REPAIR (Contd)

29. Remove eight screws (9) and washers (10) from side cover (11).

30. Remove side cover (11) and gasket (12) from differential carrier housing (13). Discard gasket (12).

31. Remove six screws (17) and lockwashers (16) from bearing retaining plate (15). Discard lockwashers (16).

NOTE
Record number and thickness of shims for installation.

32. Remove bearing retaining plate (15) and shims (14) from differential carrier housing (13). Discard shims (14).

33. Remove safety wire (22) from three screws (23) on retaining plate (21). Discard safety wire (22).

34. Remove three screws (23) and retaining plate (21) from shaft (18).

35. Install two puller screws (24) in tapped jacking holes (25) in bearing cap (20) and remove bearing cap (20). Tighten screws (24) evenly while removing bearing cap (20). Remove puller screws (24) from bearing cap (20).

NOTE
Tag and retain shims for measurement and installation.

36. Remove shims (19) from differential carrier housing (13).

37. Using arbor press, remove two bearings (27) and races (26) from bearing cap (20).

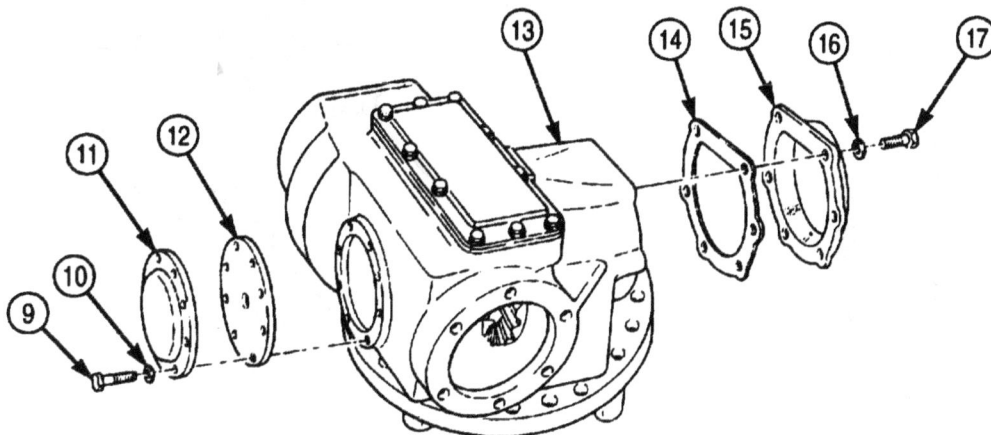

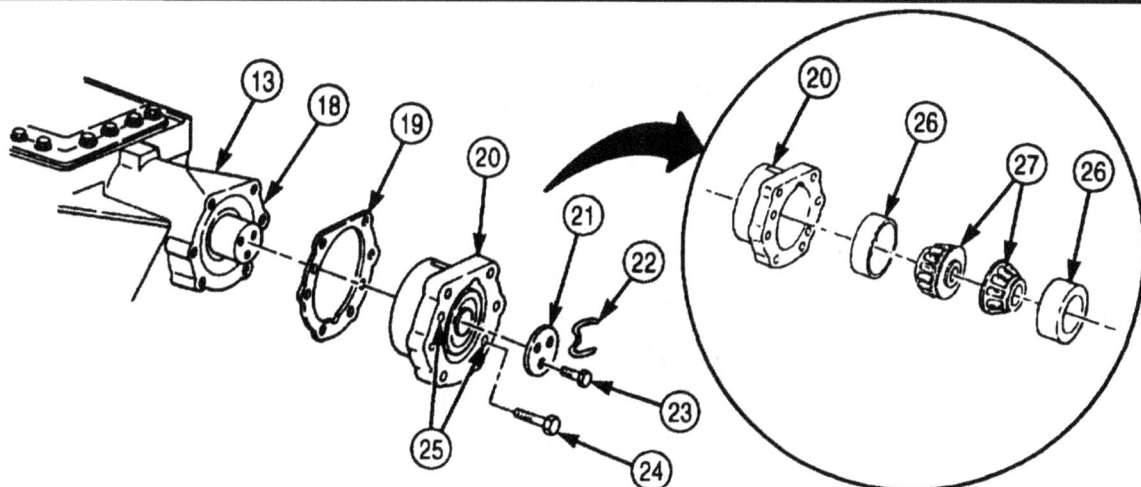

5-64. DIFFERENTIAL AND DIFFERENTIAL CARRIER REPAIR (Contd)

38. Remove ten screws (1), washers (5), access cover (2), and gasket (3) from differential carrier housing (4). Discard gasket (3).

39. Place two soft-iron spacers (10) between bevel ring gear (6) and differential carrier housing (4).

40. Using arbor press, remove helical pinion gear (7) from bevel ring gear (6).

41. Remove spacer (9) and machine key (8) from shaft of helical pinion gear (7).

42. Remove bevel ring gear (6) with bearing (11) from differential carrier housing (4).

43. Remove setscrew (13) from differential carrier housing (4).

CAUTION

Drive bearing sleeve out by tapping evenly. Scoring will occur if sleeve tilts in housing.

44. Using puller, remove bearing sleeve (12) from differential carrier housing (4).

45. Using puller, remove bearing (11) from bevel ring gear (6).

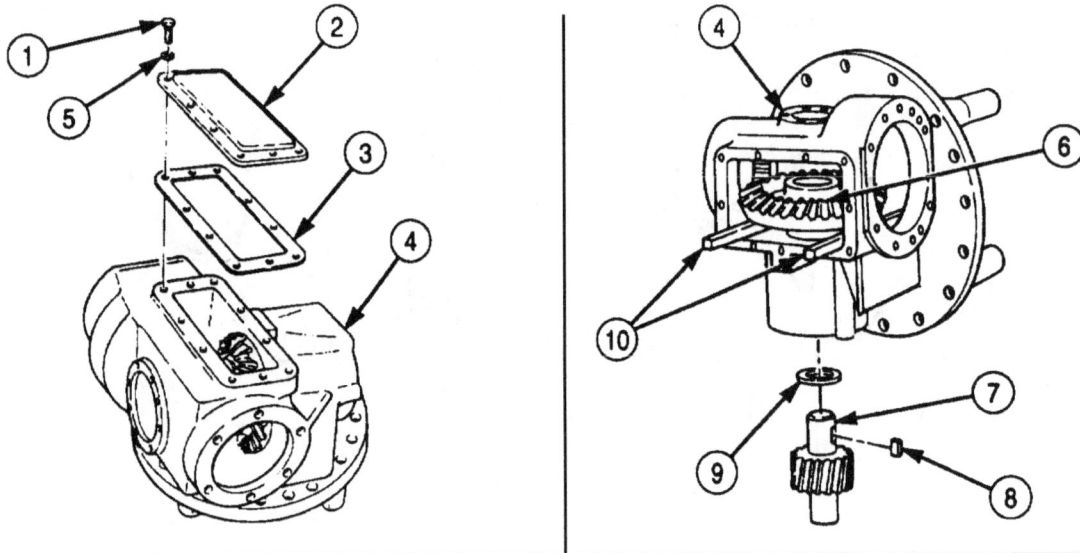

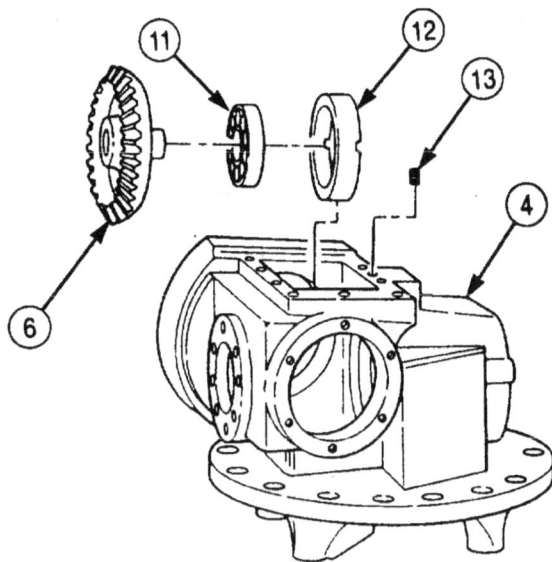

5-64. DIFFERENTIAL AND DIFFERENTIAL CARRIER REPAIR (Contd)

b. Cleaning, Inspection, and Repair

1. For general cleaning instructions, refer to para. 2-14.
2. For general inspection instructions, refer to para. 2-15.

CAUTION

Differential carrier housing and cap, bearings, and bearing races are machined or matched parts. Damage may result if they are intermixed. Replace all matched parts as sets if damaged.

3. Inspect differential carrier housing (14) and bearing caps (16) for scores, cracks, distortion, breaks, burrs, and wear. For repair of minor damage, refer to para. 2-16. Replace housing (14) and caps (16) if either part is damaged.
4. Inspect two adjusting nuts (15) for cracks, breaks, missing locking lugs, and damaged threads. Replace adjusting nut(s) (15) if cracked, broken, locking lugs are missing, or threads are damaged.

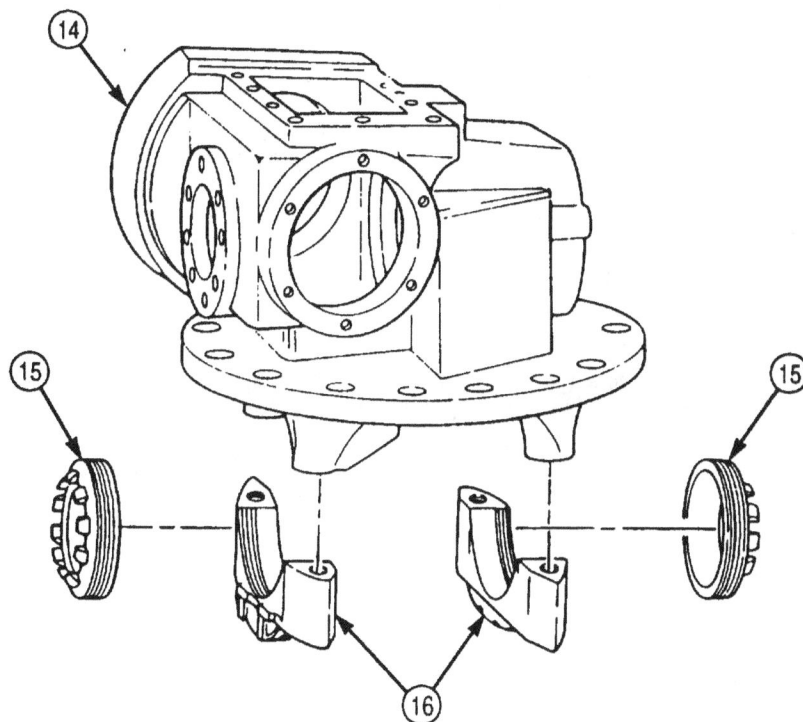

NOTE
Replace entire differential gear assembly if helical ring gear,
spider, spider gears, bevel gears, or thrust washer fails inspection.

5. Inspect two bearings (2) and races (1) (TM 9-214). Replace bearing (2) and race (1) as a set if either part fails inspection.

6. Inspect two differential gear housings (4) for chips, burrs, cracks, pitting, scoring, and breaks. For repair of minor damage, refer to para. 2-16. Replace entire differential gear assembly if differential gear housings (4) are damaged.

7. Measure inside diameter of bearing races (5) and outside diameter of differential gear housing hubs (3). The difference between measurements (press fit) must be within 0.0015-0.0035 in. (0.0381-0.0889 mm). Replace entire differential gear assembly if press fit is not within limits.

8. Inspect four thrust washers (12), bevel spider gears (ll), two thrust washers (6), bevel side gears (7), spider (9), and helical ring gear (8) for cracks, breaks, chips, burrs, scoring, and pitting. For repair of minor damage, refer to para. 2-16. Replace entire differential gear assembly if any part fails inspection.

9. Measure outside diameter of four spider arms (10). Measurement should be 1.122-1.123 in. (28.50-28.52 mm). Replace entire differential gear assembly if spider arm(s) (10) are not within limits.

10. Measure inside diameter of four bevel spider gears (11). Measurement should be 1.128-1.130 in. (28.65-28.70 mm). Replace entire differential gear assembly if bevel spider gear(s) (11) are not within limits.

11. Inspect two bearings (19) and races (18) (TM 9-214). Replace bearing (19) and race (18) as a set if either part fails inspection.

12. Inspect bevel ring gear (13), sleeve (15), spacer (16), and helical pinion gear (17) for chips, breaks, cracks, burrs, nicks, and scores. For repair of minor damage, refer to para. 2-16. Replace bevel ring gear (13), sleeve (15), spacer (16), and helical pinion gear (17) if damaged.

13. Measure inside diameter of bearing races (20) and outside diameter of pinion gear shaft (21). The difference between measurements (press fit) must be within 0.0000-0.0015 in. (0.0000-0.0381 mm). Replace helical pinion gear (17) and bearings(s) (19) if press fit is not within limits.

14. Inspect bearing (14) (TM 9-214). Replace bearing (14) if damaged.

15. Measure outside diameter of bevel ring gear (13) and inside diameter of race on bearing (14). The difference between measurements (press fit) must be within 0.0006-0.0011 in. (0.015-0.028 mm). Replace bevel ring gear (13) and bearing (14) if press fit is not within limits.

16. Measure outside diameter of bearing (14) and inside diameter of spacer (15). The difference between measurements (press fit) must be within 0.0030-0.0058 in. (0.076-0.147 mm). Replace bearing (14) and sleeve (15) if press fit is not within limits.

5-64. DIFFERENTIAL AND DIFFERENTIAL CARRIER REPAIR (Contd)

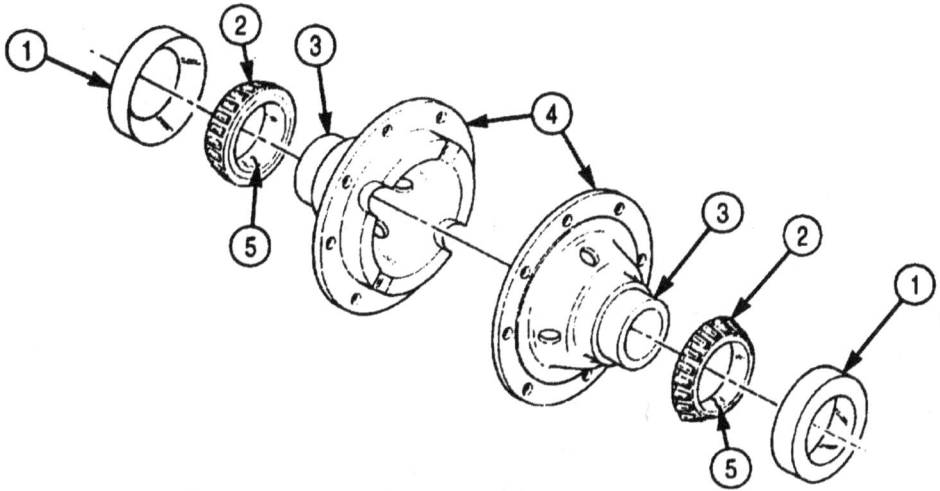

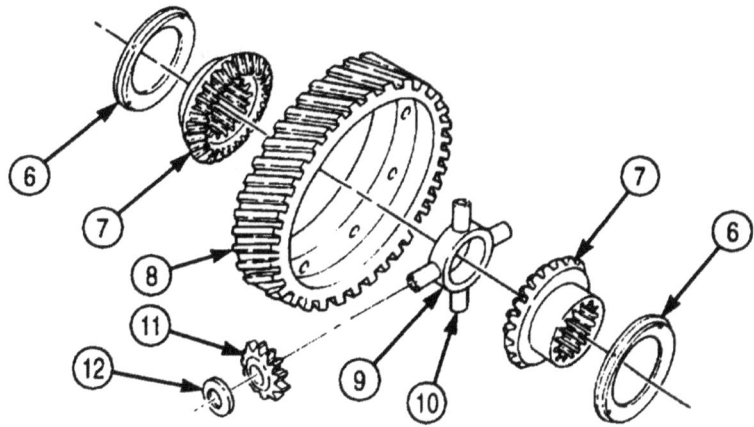

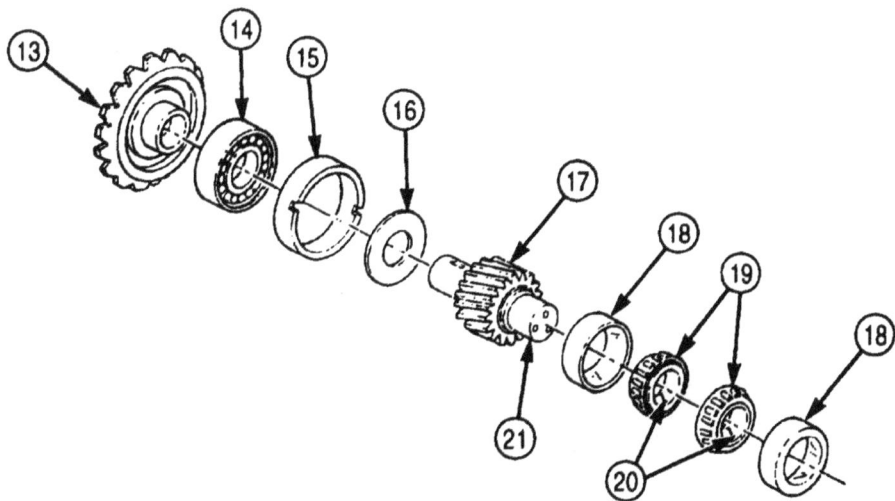

5-64. DIFFERENTIAL AND DIFFERENTIAL CARRIER REPAIR (Contd)

17. Inspect two bearings (1) and races (2), bearing (7), and race (6) (TM 9-214). Replace bearings (1) or (7) and appropriate races (2) or (6) if damaged.

18. Inspect bevel pinion gear (3) and mainshaft (4) for chips, burrs, breaks, bends, and scoring. For repair of minor damage, refer to para. 2-16. Replace bevel pinion gear (3) or mainshaft (4) if damaged.

19. Measure inside diameter of bearing race (2) and outside diameter of middle step (8) of bevel pinion gear (3). The difference between measurements (press fit) must be within 0.0000-0.0020 in. (0.000-0.051 mm). Replace bearing race (2) and bevel pinion gear (3) if press fit is not within limits.

20. Measure inside diameter of bearing race (10) and outside diameter of front step (9) of bevel pinion gear (3). The difference between measurements (press fit) must be within 0.0010-0.0025 in. (0.025-0.064 mm). Replace bearing (1) and bevel pinion gear (3) if press fit is not within limits.

21. Measure inside diameter of bearing race (6) and outside diameter of mainshaft shoulder (5). The difference between measurements (press fit) must be within 0.0006-0.0019 in. (0.015-0.048 mm). Replace bearing race (6) and mainshaft (40) if press fit is nut within limits.

22. Measure outside diameter of bearing (6) and inside diameter of rear bore (12) of differential carrier housing (11). The difference between measurements (press fit) must be +0.0009 to -0.0007 in. (+0.023 to -0.018 mm). Replace bearing (6) and differential carrier housing (11) if press tit is not within limits.

23. Inspect companion flange (13), deflector (14), and wear sleeve (15) for cracks, breaks, bends, scoring, and wear. Replace companion flange (13), deflector (14), or wear sleeve (15) if cracked, broken, bent, scored, or worn. Follow instructions in wear sleeve kit, P/N 12375353, for replacement of wear sleeve (15).

5-64. DIFFERENTIAL AND DIFFERENTIAL CARRIER REPAIR (Contd)

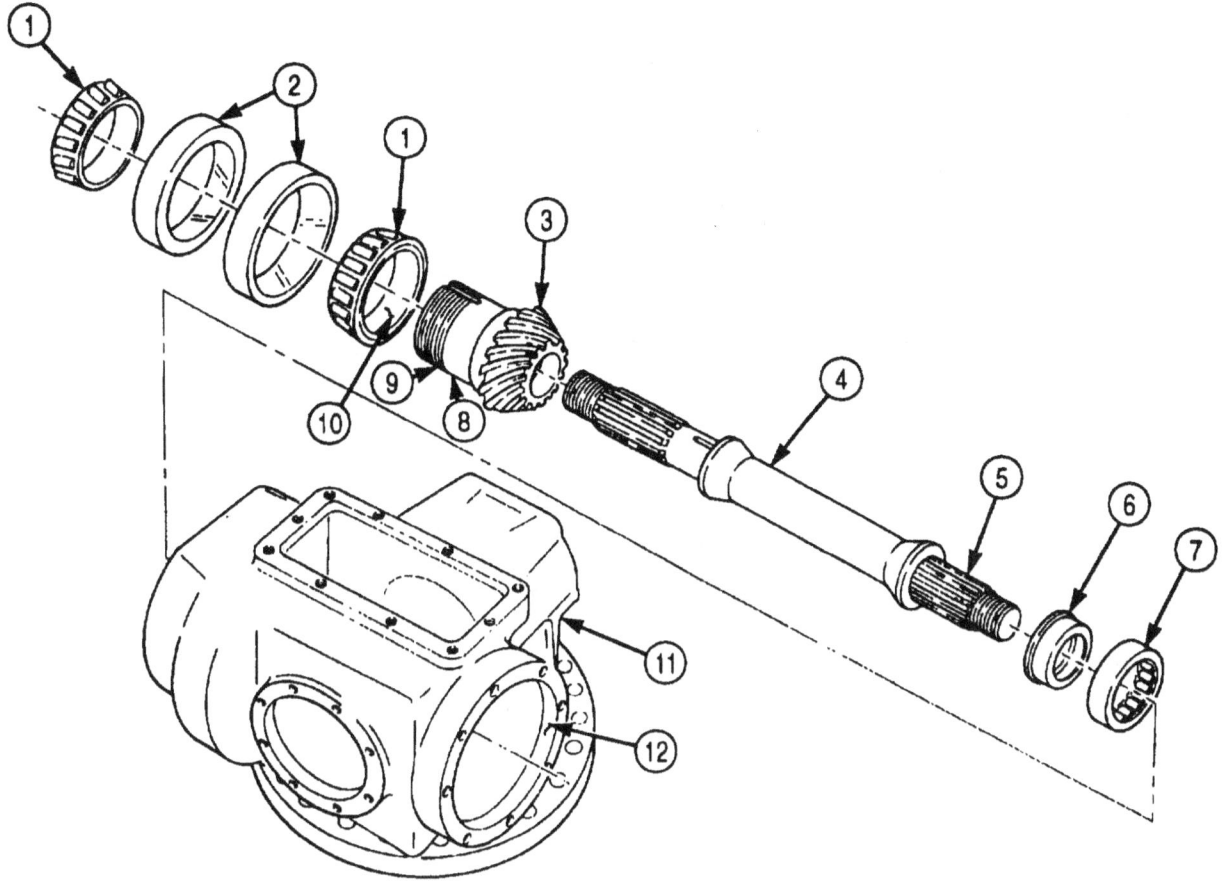

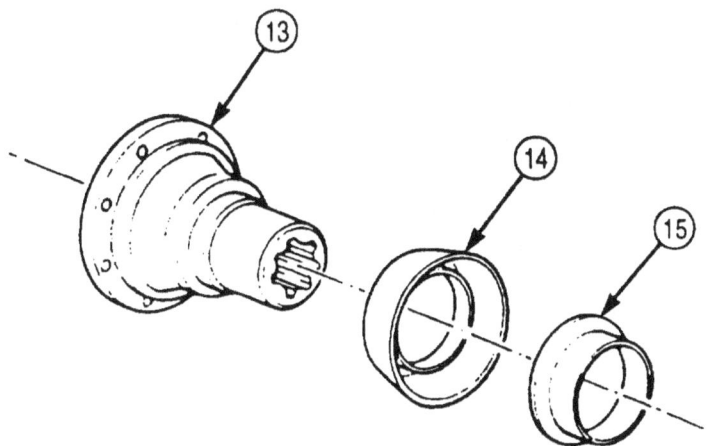

5-64. DIFFERENTIAL AND DIFFERENTIAL CARRIER REPAIR (Contd)

c. Assembly and Adjustment

NOTE

Coat all bearings and gears with lubricating oil during reassembly.

1. Using bearing replacer, install inner bearing race (3) and outer bearing race (1) in retainer (2). Ensure thick edges of races (1) and (3) are toward center of retainer (2).

2. Install bearing (4) on hub of bevel pinion gear (6). Thick inner race (10) must be seated against back edge of teeth (5) on bevel pinion gear (6).

NOTE

White carbonate pigment is used as a high-pressure lubricant.

3. Coat long splined end (7) of mainshaft (9) with white carbonate pigment.

4. Position bevel pinion gear (6) over mainshaft (9), ensuring internal splines in bevel pinion gear (6) align with splined end (7), and press bevel pinion gear (6) and bearing (4) onto mainshaft (9). Nose of bevel pinion gear (6) must seat against shoulder (8).

5. Place mainshaft (7) in soft-jawed vise.

NOTE

Collar is used to set preload on bevel pinion gear bearings. Ensure collar removed in disassembly is installed for preload test.

6. Install spacer (14) on hub (15) of bevel pinion gear (6) next to bearing (16).

7. Position small end (13) of retainer (12) over hub (15) and bearing (161 and place bearing (11) in retainer (12) over hub (15).

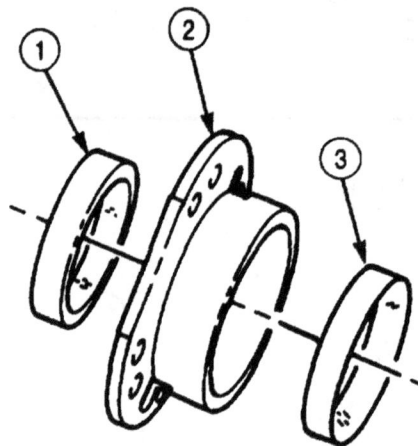

5-64. DIFFERENTIAL AND DIFFERENTIAL CARRIER REPAIR (Contd)

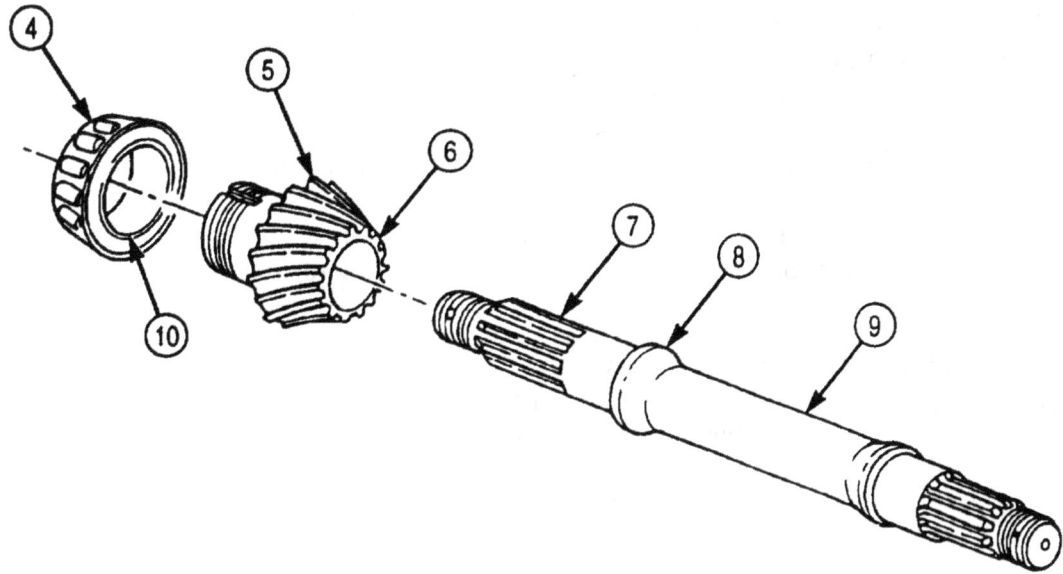

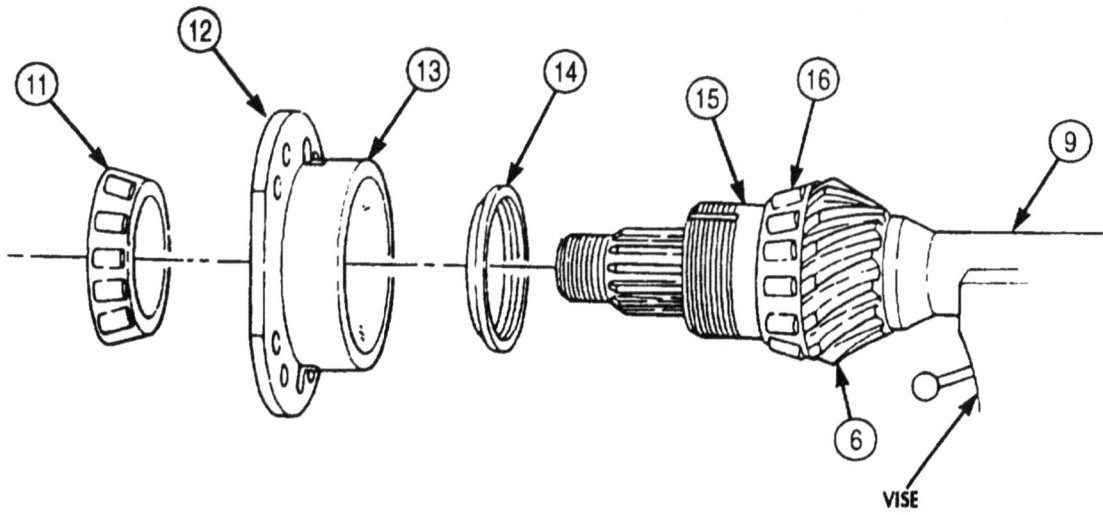

5-64. DIFFERENTIAL AND DIFFERENTIAL CARRIER REPAIR (Contd)

8. Install inner nut (4) on threaded portion of hub (5) of bevel pinion gear (7). Ensure stud (9) on inner nut (4) is facing away from retainer (6) and bevel pinion gear (7). Tighten inner nut (4) 800-1,000 lb-ft (1,085-1,356 N-m).

9. Install keywasher (3) on hub (5). Ensure hole in keywasher (3) aligns over stud (9) of inner nut (4).

10. Install new locktab washer (2) on hub (5). Ensure stud (9) sets into a hole in keywasher (3).

11. Install outer nut (1) on hub (5). Tighten nut (1) 1,OOO-1,200 lb-ft (1,356-1,627 N•m).

NOTE

Pinion bearings must be preloaded for proper operation of differential assembly.

12. Wrap twine around retainer (6) and attach to bearing preload tester to determine bearing preload.

13. Observe pointer and pull on bearing preload tester until retainer (6) starts to rotate. Record reading.

14. If original bearings are installed, bearing preload pull must be 1.2-2.4 lb (0.54-1.09 kg). If torque method is used to measure bearing preload, torque must be 4-8 lb-in. (0.5-0.9 N•m).

15. If new bearings are installed, bearing preload pull must be 3.6-5.5 lb (1.63-2.50 kg). If torque method is used to measure bearing preload, torque must be 5-25 lb-in. (0.41-0.62 N•m).

NOTE

- If preload is not within given limits, perform steps 16 through 19.
- If preload is within given limits, go to step 21.

16. Remove outer nut (1), locktab washer (2), keywasher (3), and inner nut (4) from hub (5) of bevel pinion gear (7).

17. Using arbor press, remove outer bearing (10), retainer (6), and spacer (11) from hub (5).

18. If preload valve (steps 13, 14, and 15) exceeds limits, use a thicker spacer (11) from bevel pinion gear spacer kit. If preload value is less than limits, use a thinner spacer (11).

19. Install parts on hub (5) following steps 5 through 11. Measure bearing preload using either twine and bearing preload tester or torque method. Repeat steps 16 through 19 until bearing preload is within limits of either step 14 or 15.

20. Remove mainshaft (8) from soft-jawed vise.

5-64. DIFFERENTIAL AND DIFFERENTIAL CARRIER REPAIR (Contd)

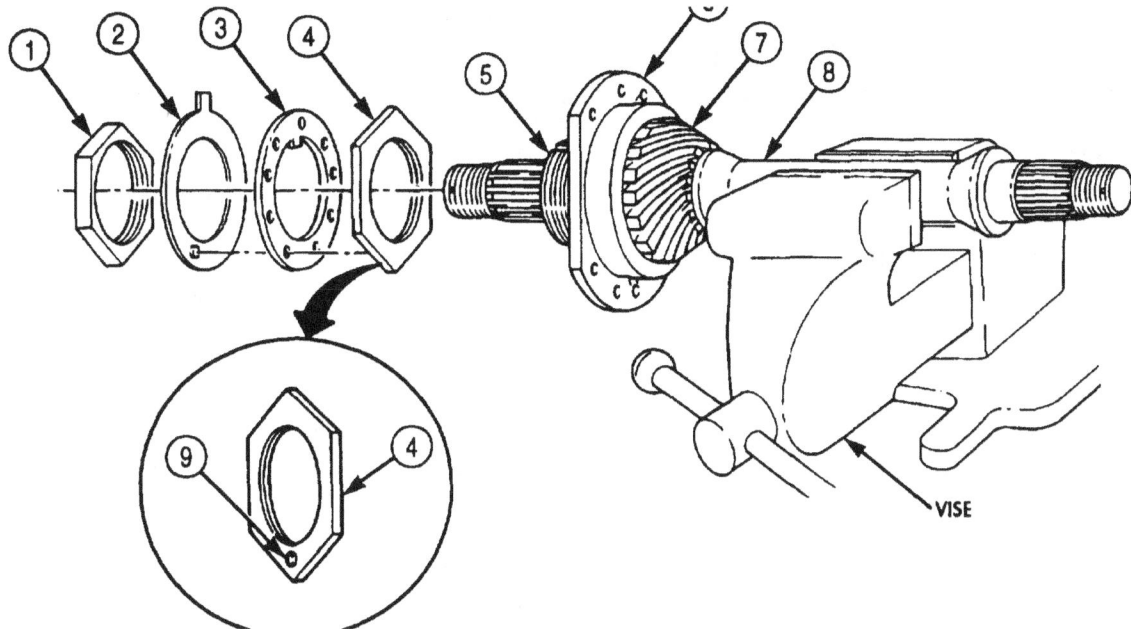

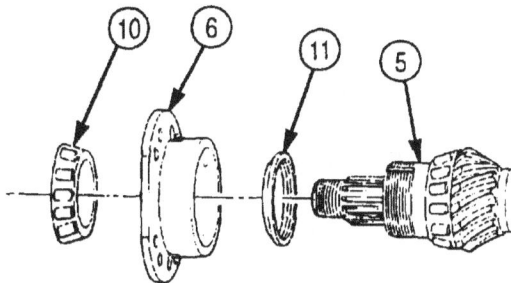

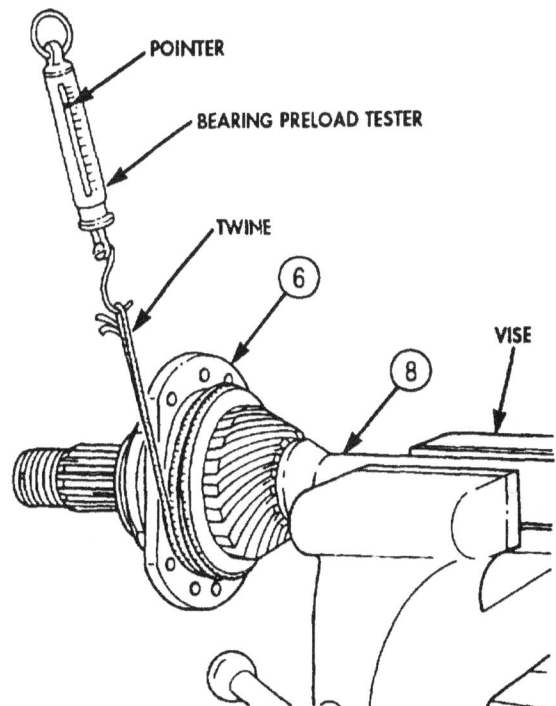

VISE

POINTER

BEARING PRELOAD TESTER

TWINE

VISE

5-64. DIFFERENTIAL AND DIFFERENTIAL CARRIER REPAIR (Contd)

21. Bend tab(s) (4) on locktab washer over flats on outer nut (1).
22. Using arbor press, install rear bearing inner race (3) on short splined end of mainshaft (2).
23. Using arbor press, install bearing (7) in rear bore (5) of differential carrier housing (6).
24. Install sleeve (8) in bore (13) with notches (10) toward shoulder of bore (13). Ensure hole (9) in sleeve (8) aligns with hole (12) in differential carrier housing (6).
25. Install setscrew (11) through hole (12) until seated in sleeve (8).
26. Press bearing (15) on hub (20) of bevel ring gear (21). Bearing (15) must rest against shoulder (14) on hub (20).
27. Install spacer washer (16) on keyed end (19) of helical pinion gear (18).
28. Install machine key (17) in keyed end (19) of helical pinion gear (18). Ensure exposed edges of machine key (17) are not burred after installation. Remove any burrs with fine mill file.

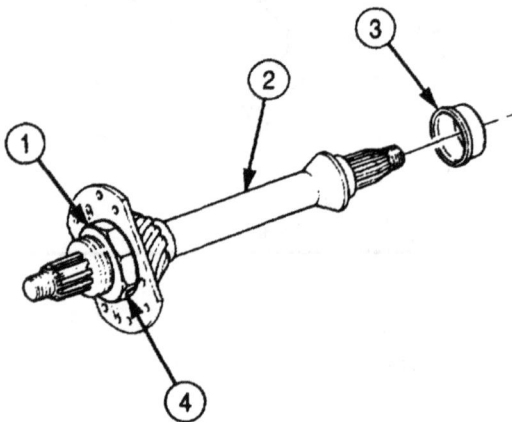

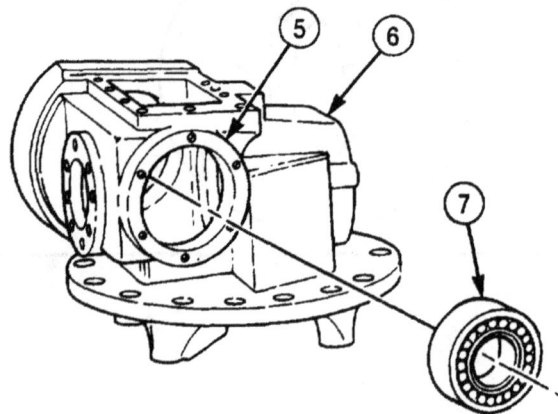

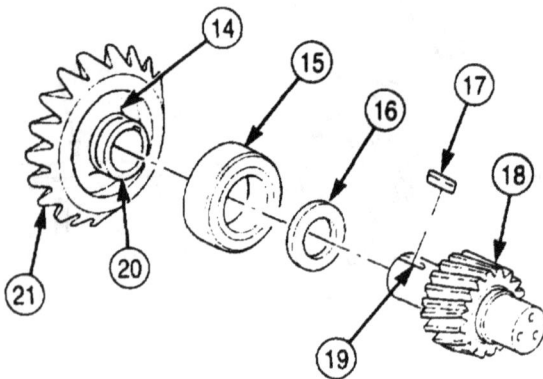

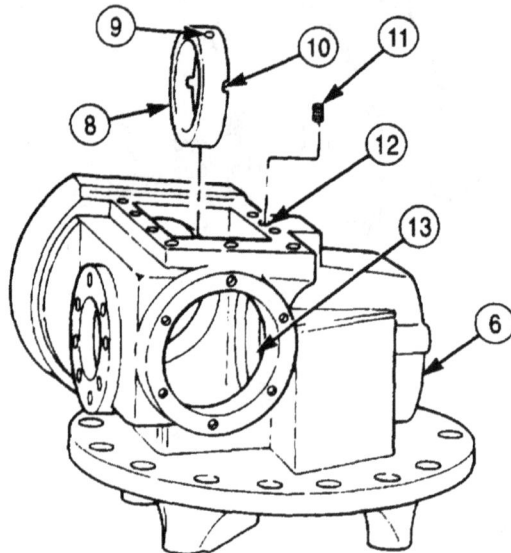

5-64. DIFFERENTIAL AND DIFFERENTIAL CARRIER REPAIR (Contd)

29. Place bevel ring gear (21) in differential carrier housing (6) with hub and bearing (15) pointing toward bore (22).

30. Support bevel ring gear (21) with two soft-iron semicircular blocks. Use support adapter (23) as an alternate method.

NOTE

White carbonate pigment is used as high-pressure lubricant.

31. Coat keyed end (19) of helical pinion gear (18) with pigment lubricant and align over bore in bevel ring gear (21). Ensure machine key (17) is aligned with keyway in bevel ring gear (21).

32. Using arbor press, install keyed end (19) of helical pinion gear (18) in bore of bevel ring gear (21).

NOTE

Light tapping with soft-head hammer or use of arbor press may be needed to install bearing.

33. Remove blocks, turn differential carrier housing (6) over, and push bearing (15) into sleeve (16).

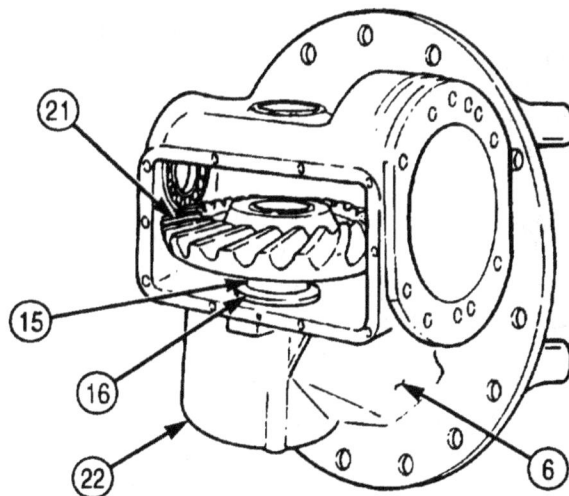

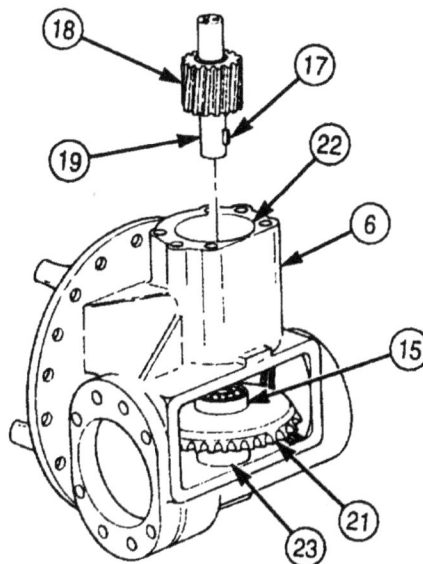

5-64. DIFFERENTIAL AND DIFFERENTIAL CARRIER REPAIR (Contd)

NOTE

Coat bearings and races with clean gear oil before installation.

34. Using bearing replacer, install race (2) in bearing cap (1). Thick edge of race (2) must be toward small end (6) of bearing cap (1).

35. Install bearings (3) and (4) and race (5) in bearing cap (1). Use replacer to seat race (5) in bearing cap (1). Wide edge of race (5) must face away from face of cap (1).

36. Starting with same thickness of shims (9) as removed, place shims (9) over shaft (8) and on differential carrier housing (7). Carefully align holes in shims (9) to holes in carrier housing (7).

37. Place bearing cap (1) over shaft (8) and press bearing cap (1) into position on shaft (8) to differential carrier housing (7).

38. Install retaining plate (10) on shaft (8) with three screws (12). Tighten screws (12) 42-54 lb-ft (57-73 N-m) and install new safety wire (11) in screws (12).

39. Install new shims (13) and bearing cover (14) on differential carrier housing (7) with six new lockwashers (15) and screws (16). Select shims (13) of same thickness as removed. Tighten screws (16) 60.100 lb-ft (81-136 N-m).

40. To measure helical pinion gear (17) bearing preload, wrap twine around helical pinion gear (17) and attach to bearing preload tester. Observe pointer and pull on bearing preload tester until helical pinion gear (17) starts to rotate. Record value.

41. If original bearings are installed, bearing preload must be 4-8 lb-in. (.5-.9 N-m) If new bearings are installed, bearing preload must be 12-18 lb-in. (1.36-2.03 N-m).

NOTE

- If preload is not within given limits, perform steps 42 through 45.
- If preload is within given limits, go to step 47.

42. Remove six screws (16), lockwashers (15), bearing cover (14), and shims (13) from differential carrier housing (7). Remove shims (13) carefully to avoid creasing or damaging shims (13).

43. If preload scale (or torque) reading exceeds limits, add shims (13).

44. If preload scale (or torque) reading is less than lower limit, remove shims (13).

45. Install shims (13) and bearing cover (14) on differential carrier housing (7) with six lockwashers (15) and screws (16). Carefully align shims (13) with holes in differential carrier housing (7). Tighten screws (16) 60-100 lb-ft (81-136 N-m).

46. Repeat steps 40 through 45 as necessary.

5-64. DIFFERENTIAL AND DIFFERENTIAL CARRIER REPAIR (Contd)

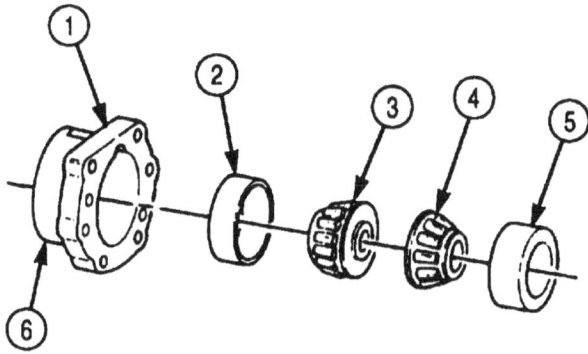

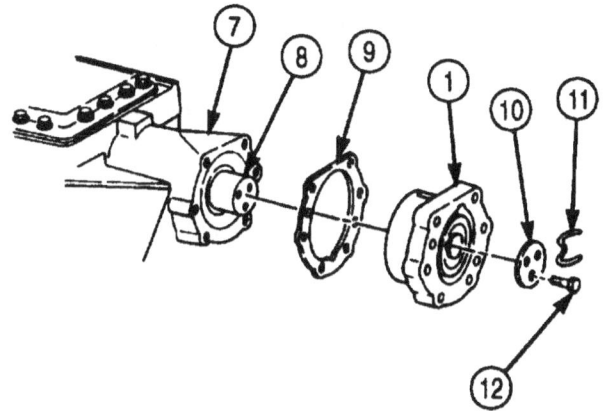

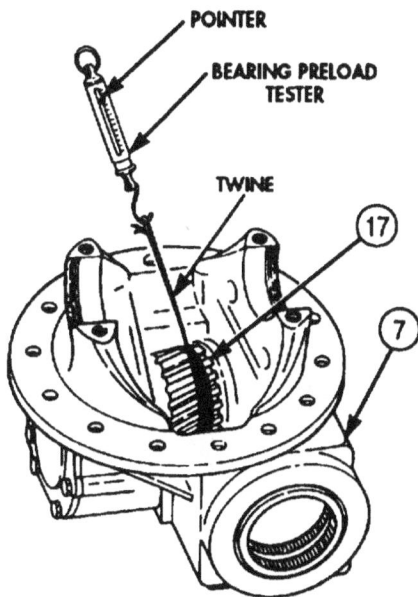

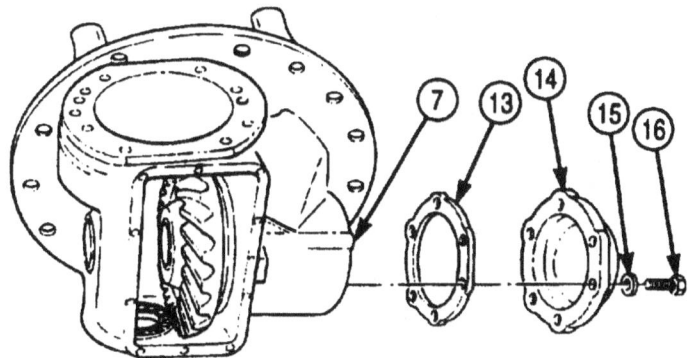

POINTER

BEARING PRELOAD TESTER

TWINE

5-64. DIFFERENTIAL AND DIFFERENTIAL CARRIER REPAIR (Contd)

47. Position shims (2), same thickness pack as removed, on retainer (7) and slide mainshaft (4) into differential carrier housing (8). Align inner race (3) with rear roller bearing (9).

48. Align holes in shims (2), differential carrier housing (8), bore (1), and retainer (7) and install with eight washers (6) and screws (5). Tighten screws (5) 93-120 lb-ft (126-163 N•m).

CAUTION

Correct bevel ring gear and pinion gear contact pattern is critical to correct operation of differential and differential carrier assembly. Incorrect contact pattern can result in noisy operation or premature failure of these two gears.

49. Check bevel ring gear (10) and pinion gear (11) tooth contact pattern by applying blue oil-base pigment to at least three teeth of clean bevel ring gear (10). Turn mainshaft (4) for three full rotations of bevel ring gear (10).

50. Examine teeth contact pattern (13) of bevel ring gear (10) with pinion gear (11). Pattern should be centered both ways on teeth (12) and cover over two-thirds of tooth (12) contact surface.

NOTE

• If pattern is not correct, perform steps 51 through 69.

• If pattern is correct, go to step 71.

51. If contact pattern resembles pattern (14), perform steps 55-59 and 63-69.

52. If contact pattern resembles pattern (15), perform steps 55-58, 60, and 63-69.

53. If contact pattern resembles pattern (16), perform steps 55-58, 61, and 63-69.

54. If contact pattern resembles pattern (17), perform steps 55-58, 62, and 63-69.

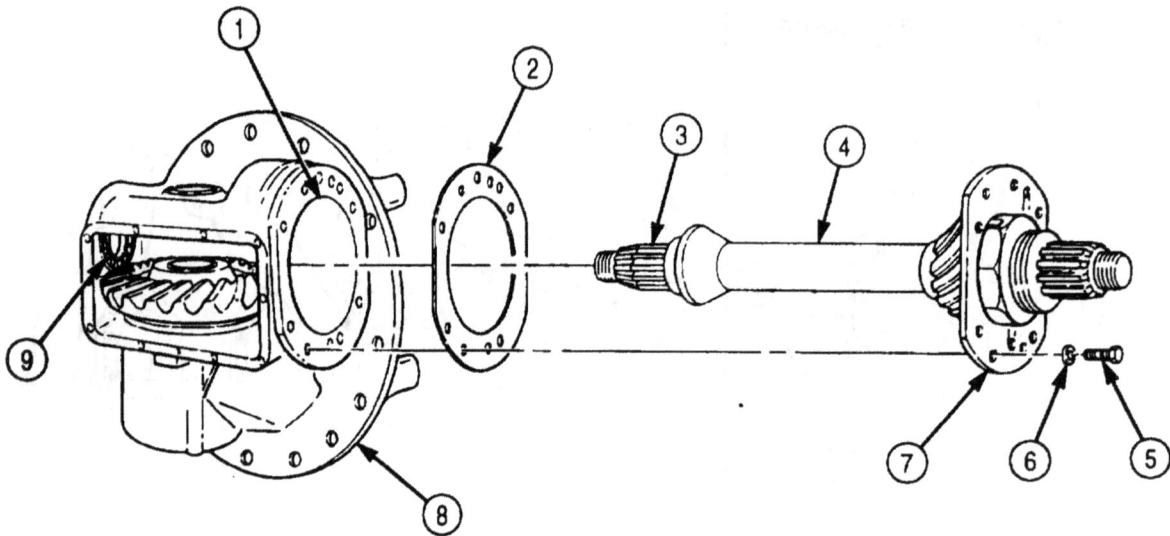

5-64. DIFFERENTIAL AND DIFFERENTIAL CARRIER REPAIR (Contd)

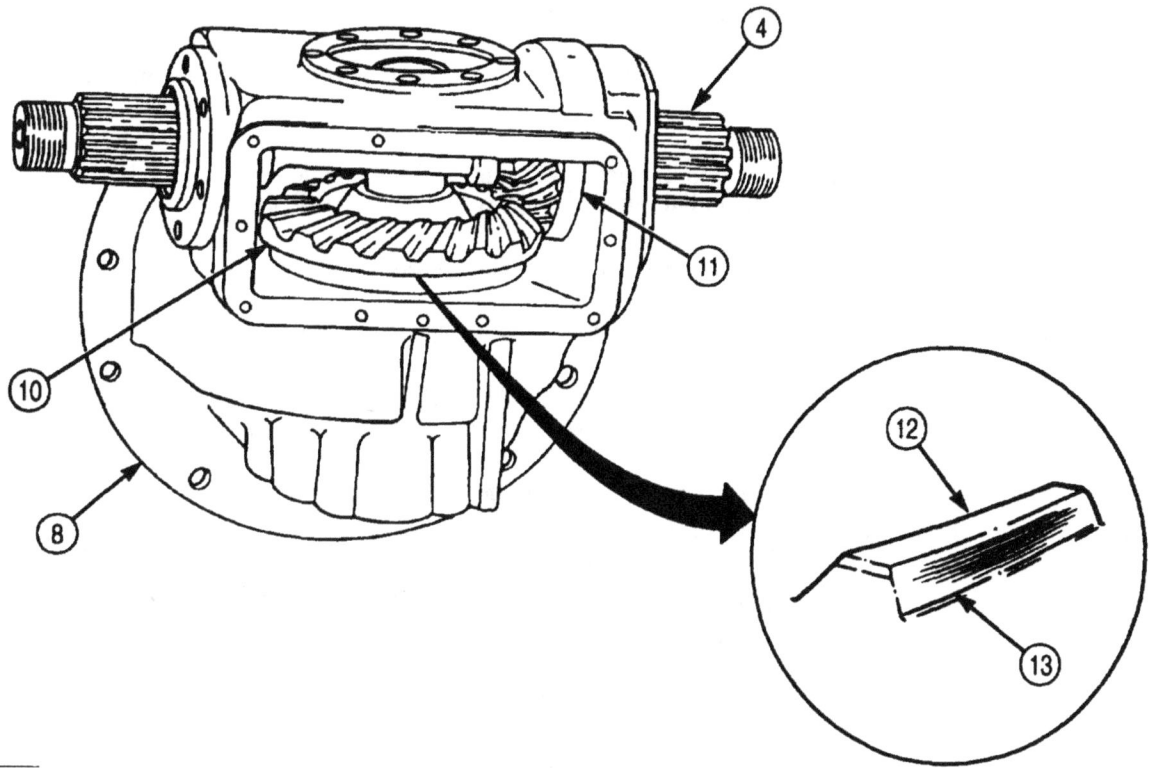

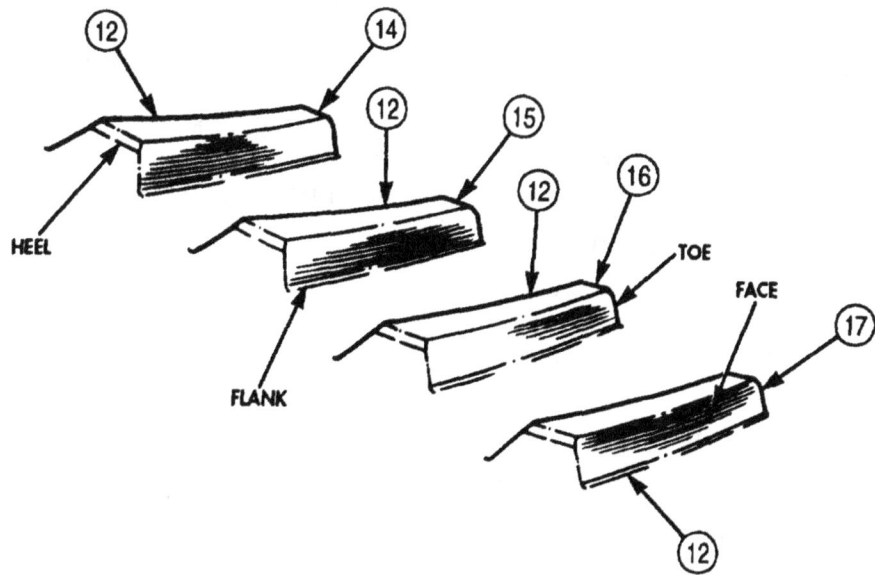

HEEL

FLANK

TOE

FACE

TM 9-2320-272-24-4

5-64. DIFFERENTIAL AND DIFFERENTIAL CARRIER REPAIR (Contd)

55. Remove six screws (6), lockwashers (7), cover (8), and shims (9) from differential carrier housing (1).
56. Remove safety wire (15), three screws (16), and retaining plate (14). Discard safety wire (15).
57. Using two jacking screws (17), remove bearing retainer (13) and shim pack (12) from shaft (11) and shaft housing shoulder (10).
58. Remove eight screws (4), washers (5), mainshaft (3), and shim pack (2) from differential carrier housing (1).

NOTE

If adjustment is needed to correct bevel ring gear and pinion gear contact pattern, make initial shim changes of at least 0.010 in. (0.25 mm).

59. For contact pattern of type in step 51, add shims to shim pack (2) and remove shims from shim pack (12).
60. For contact pattern of type in step 52, add shims to shim pack (12) and remove shims from shim pack (2).
61. For contact pattern of type in step 53, remove shims from pack (2) and add shims to shim pack (12).
62. For contact pattern of type in step 54, remove shims from shim pack (12) and add shims to shim pack (2).
63. Using dial indicator, check backlash (20) of bevel ring gear (18) and pinion gear (19).
64. Backlash (20) should be 0.005-0.015 in. (0.127-0.381 mm).

NOTE

• If backlash needs correction, perform steps 55, 56, 57, and 58.
• If bevel ring gear and pinion gear backlash are within limits, go to step 71.
• If bevel ring gear and pinion gear backlash are not within limits, perform steps 65 and 66.

65. If backlash (20) is less than limits, add equal amounts of shims to shim packs (2) and (12).
66. If backlash (20) is greater than limits, remove equal amounts of shims from shim packs (2) and (12).
67. Carefully place mainshaft (3) and shim pack (2) in differential carrier housing (1) and install with eight washers (5) and screws (4). Tighten screws (4) 93-120 lb-ft (126-163 N•m).
68. Carefully align shim pack (12) to shaft housing shoulder (10) and press on bearing retainer (13) over shaft (11).
69. Install retaining plate (14) on sheet (11) with three screws (16). Tighten screws (16) 42-54 lb-ft (57-73 N•m). Fasten new safety wire (15) to all three screws (16).
70. Carefully align shims (9) to bearing retainer (13) and install cover (8), shims (9), retainer (13), and shims (12) on shoulder (10) of carrier housing (1) with six new lockwashers (7) and screws (6). Tighten screws (6) 60-100 lb-ft (81-136 N•m).

5-64. DIFFERENTIAL AND DIFFERENTIAL CARRIER REPAIR (Contd)

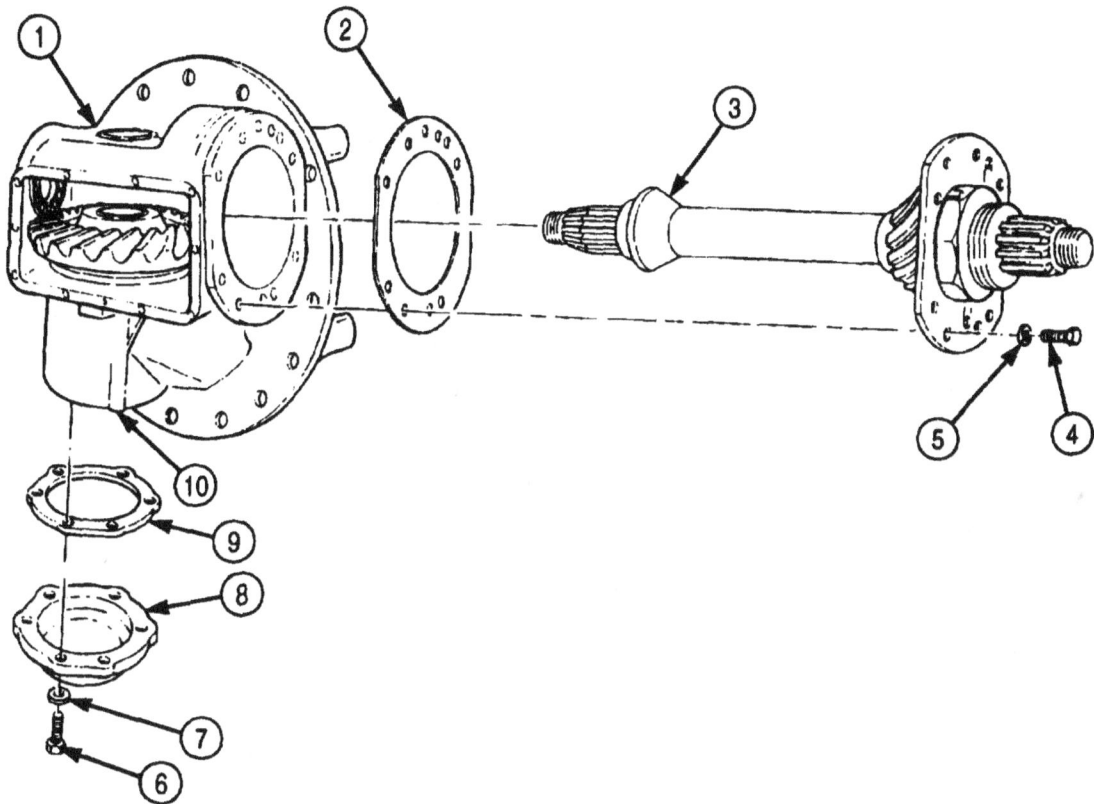

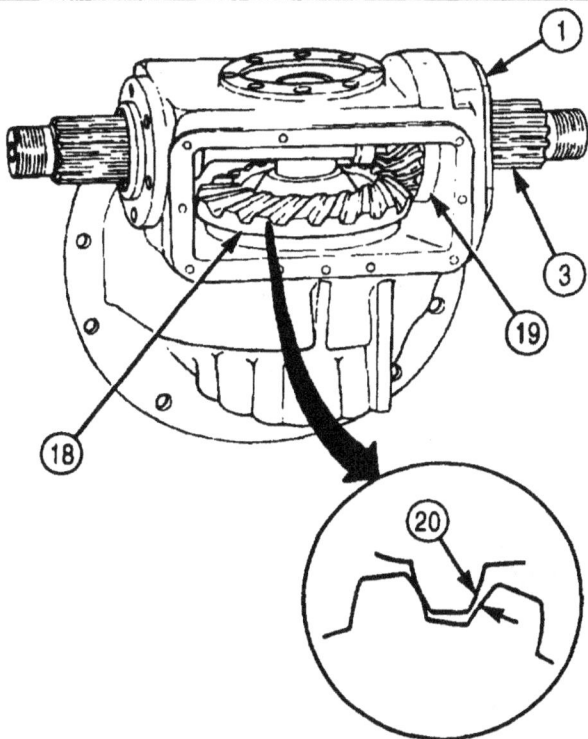

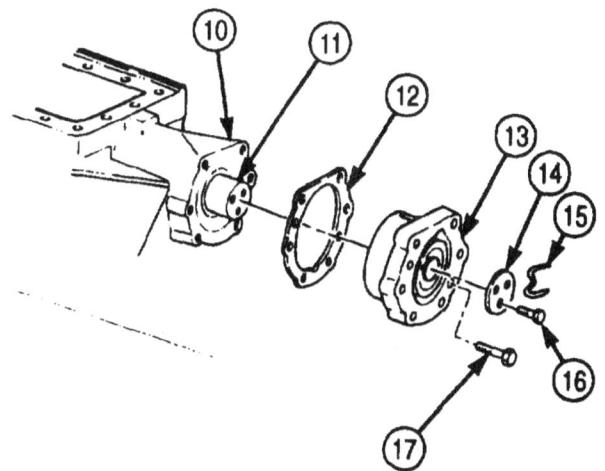

5-64. DIFFERENTIAL AND DIFFERENTIAL CARRIER REPAIR (Contd)

71. Coat top of new gasket (4) and gasket surface (11) with light coat of sealing compound and position gasket (4) on differential carrier housing (5).

72. Install access cover (3) on gasket (4) and differential carrier housing (5) with ten washers (2) and screws (1). Tighten screws (1) 27-40 lb-ft (37-54 N•m).

73. Coat gasket surface (6) with light coat of sealing compound.

74. Install new gasket (7) and side cover (8) on differential carrier housing (5) with eight washers (10) and screws (9). Tighten screws (9) 27-40 lb-ft (37-54 N•m).

75. Lubricate sealing surface of new oil seal (13) with GM grease, coat metal contact outer diameter with sealing compound, and install new gasket (14) and seal (13) in front bearing cover (15) with seal installer tool and arbor press.

NOTE
Ensure oil slots on retainer are not blocked with sealing compound or covered by new gasket.

76. Coat both sides of new gasket (12) with light coat of sealing compound and position on retainer (18).

77. Remove eight screws (16) and washers (17) from retainer (18).

78. Install front bearing cover (15), seal (13), gasket (12), and retainer (18) on carrier housing (5) with eight washers (17) and screws (16). Tighten screws (16) 93-120 lb-ft (126-163 N•m).

79. Lubricate sealing surface of new seal (25) with GAA grease, coat outside metal contact outer diameter with sealing compound, and install in bearing cover (26).

80. Coat new gasket (24) with light coat of sealing compound and position on differential carrier housing (5).

81. Place thrust washer (23) over mainshaft (22) and install rear bearing cover (26) with six new lockwashers (28) and screws (27). Tighten screws (27) 24-40 lb-ft (33-54 N•m).

CAUTION
Do not tap or hammer companion flanges on driveshaft assembly. Damage to bearings will result.

82. Install two companion flanges (21) on mainshaft (22) with two nuts (20). Tighten nuts (20) 300-400 lb-ft (407-542 N•m) and install two new cotter pins (19).

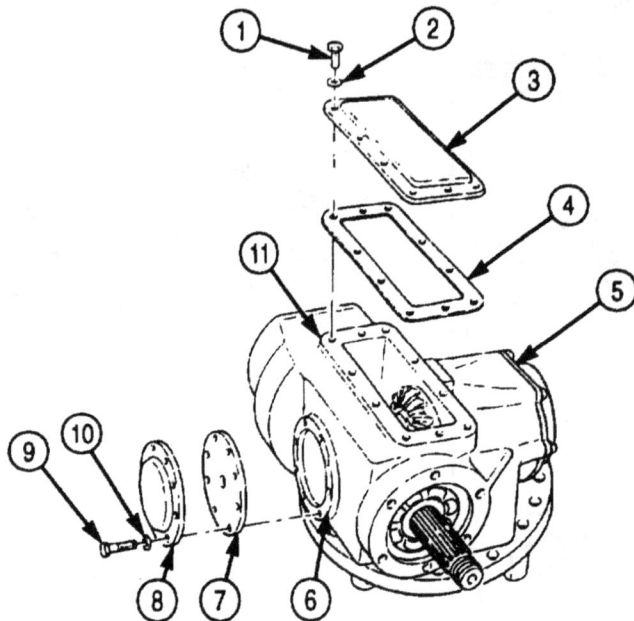

5-64. DIFFERENTIAL AND DIFFERENTIAL CARRIER REPAIR (Contd)

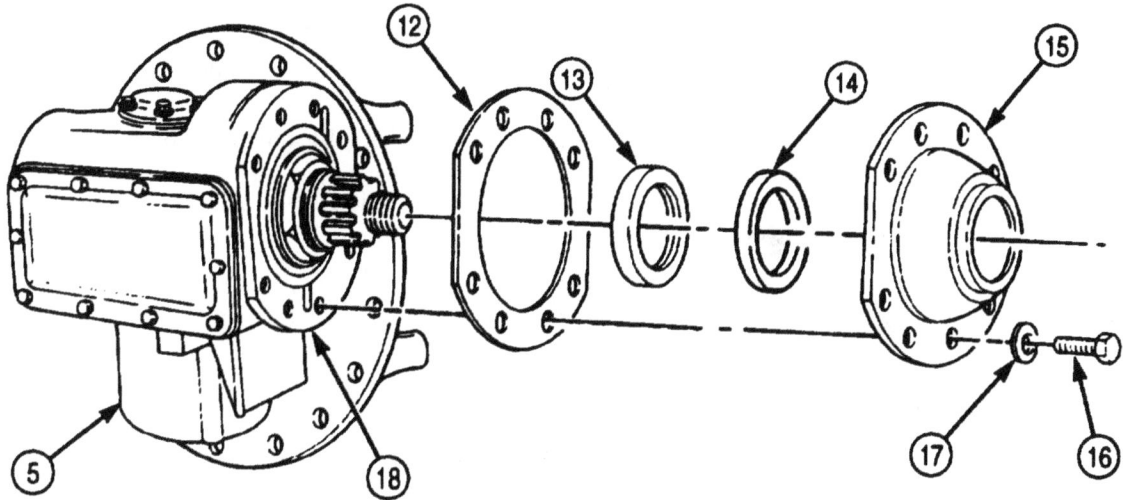

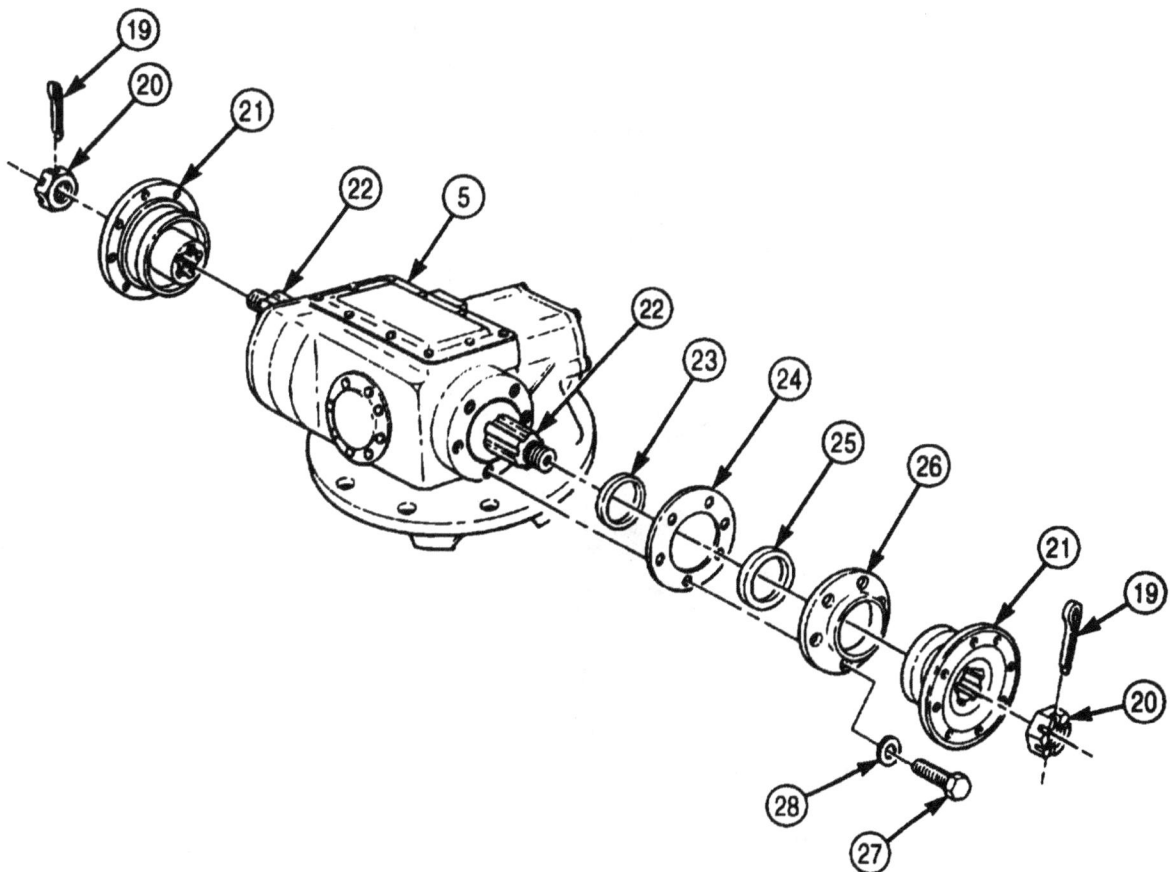

5-64. DIFFERENTIAL AND DIFFERENTIAL CARRIER REPAIR (Contd)

NOTE

Coat all bearings, races, gears, and thrust washers with clean gear
oil before assembly.

83. Install two bearings (1) and (4) on hubs of differential gear housings (2) and (3).

84. Lay one differential gear housing (3), bearing side down, on work surface and install one thrust washer (8) and side gear (9) in differential gear housing (3).

85. Place four bevel spider gears (6) and thrust washers (7) on spider (10) and place over bevel side gear (9) in housing (3).

86. Place thrust washer (5) over hub of bevel side gear (11) and position bevel side gear (11) over spider gear assembly in differential gear housing (3).

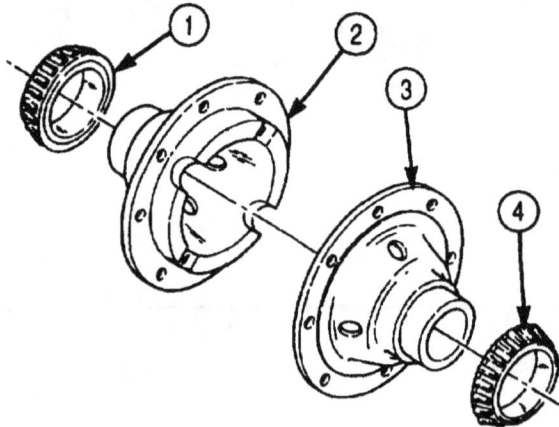

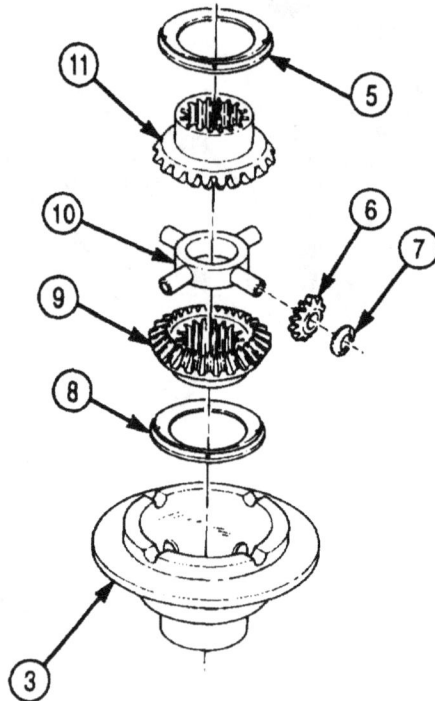

5-64. DIFFERENTIAL AND DIFFERENTIAL CARRIER REPAIR (Contd)

87. Carefully align index marks made during disassembly and position two differential gear housings (2) and (3) on helical ring gear (14) and install with eight screws (15) and slotted nuts (13). Tighten nuts (13) 130-170 lb-ft (176-231 N•m). Secure with new safety wire (12). Twist ends of safety wire (12) together until wire is tight. Remove and discard excess safety wire (12).

88. Place two races (16) over bearings (1) and (4) on differential gear (17) and set assembly in saddles (18) of differential carrier housing (20). Ensure races (16) seat inside threads of saddles (18).

89. If races (16) do not fully seat in saddles (18), check that helical ring gear (14) is in mesh with helical pinion gear (19).

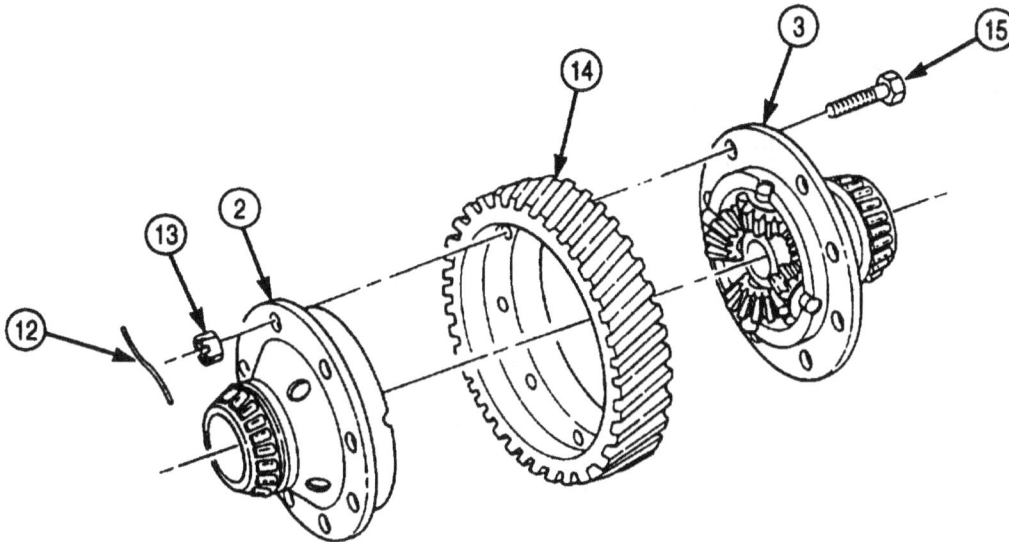

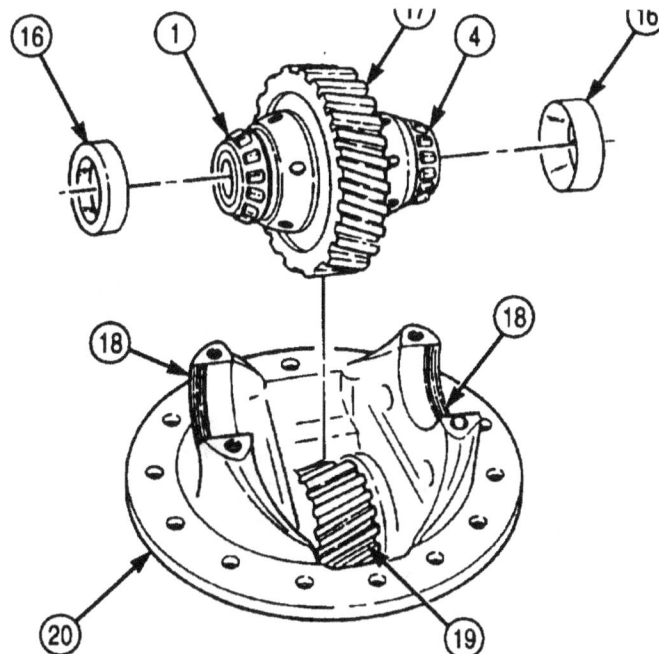

5-64. DIFFERENTIAL AND DIFFERENTIAL CARRIER REPAIR (Contd)

90. Position two caps (3) on saddles (5) and install with four washers (2) and screws (1). Finger-tighten screws (1).

91. Start two adjusting nuts (6) in threads of saddles (5) and caps (3). Move caps (3) as necessary to align threads to adjusting nuts (6).

92. Turn adjusting nuts (6) by hand until contact is made with races (4).

93. Ensure helical ring gear (7) aligns with mating helical pinion gear (9). If necessary, loosen and tighten adjusting nuts (6) to align gears.

94. Tighten screws (1) 7-12 lb-ft (9-16 N•m).

NOTE

Ensure a notch of each adjusting nut aligns with top center of cap.

95. To establish helical bearing preload, perform step a. or b.

    a. Alternately tighten adjusting nuts (6) a total 1.5 to 2.75 notch widths.

    b. Tighten each adjusting nut (6) 15-35 lb-in. (1.7-4.0 N•m).

96. Install dial indicator on differential carrier housing (8) with indicator plunger perpendicular to edge of helical ring gear (7).

97. Turn helical ring gear (7) one revolution in each direction. Runout reading on dial indicator must not exceed 0.008 in. (0.20 mm). If runout exceeds this value, loosen adjusting nuts (6) and screws (1) and repeat steps 90 through 95.

98. Reposition dial indicator so plunger contacts tooth (10) and detects motion in plane of rotation of helical ring gear (7) as the gear rotates.

99. Hold helical pinion gear (9) still and rock helical ring gear (7) back and forth. Observe backlash reading on dial indicator.

106. Backlash should be 0.005-0.015 in. (0.127-0.381 mm). If reading is outside of these limits, replace helical ring gear (7) and helical pinion gear (9).

101. Tighten four screws (1) 290-370 lb-ft (393-502 N•m).

102. Install two locks (12) of adjusting nut (6) on caps (3) with screws (13). Tighten screws (13) 66-85 lb-ft (90-115 N•m).

103. Install new safety wire (11) through four screws (1) and two screws (13). Twist safety wire (11) ends together, cut off and discard excess safety wire (11), and bend twisted ends of safety wire (11) out of way.

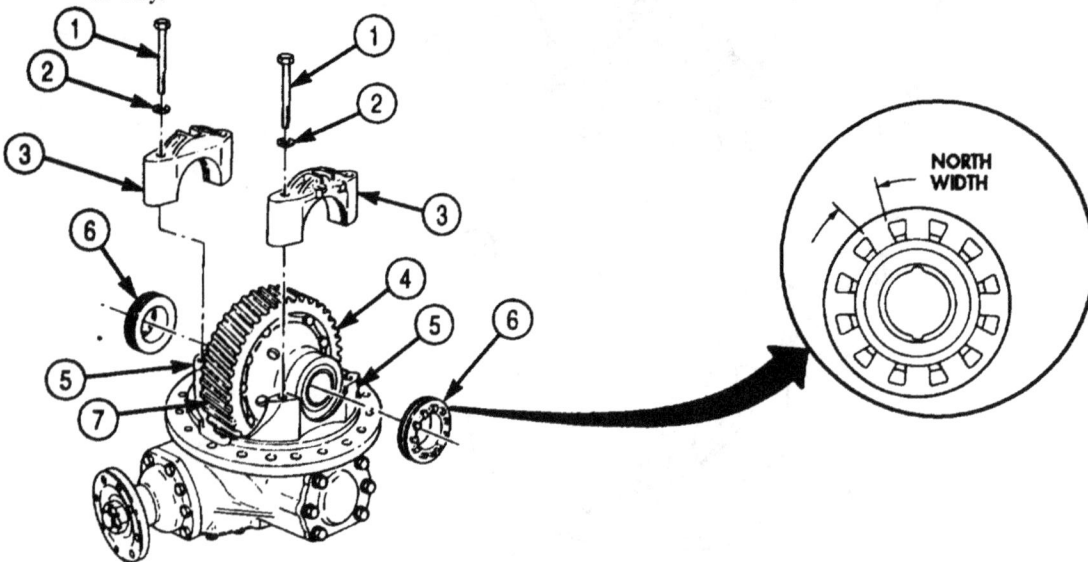

5-64. DIFFERENTIAL AND DIFFERENTIAL CARRIER REPAIR (Contd)

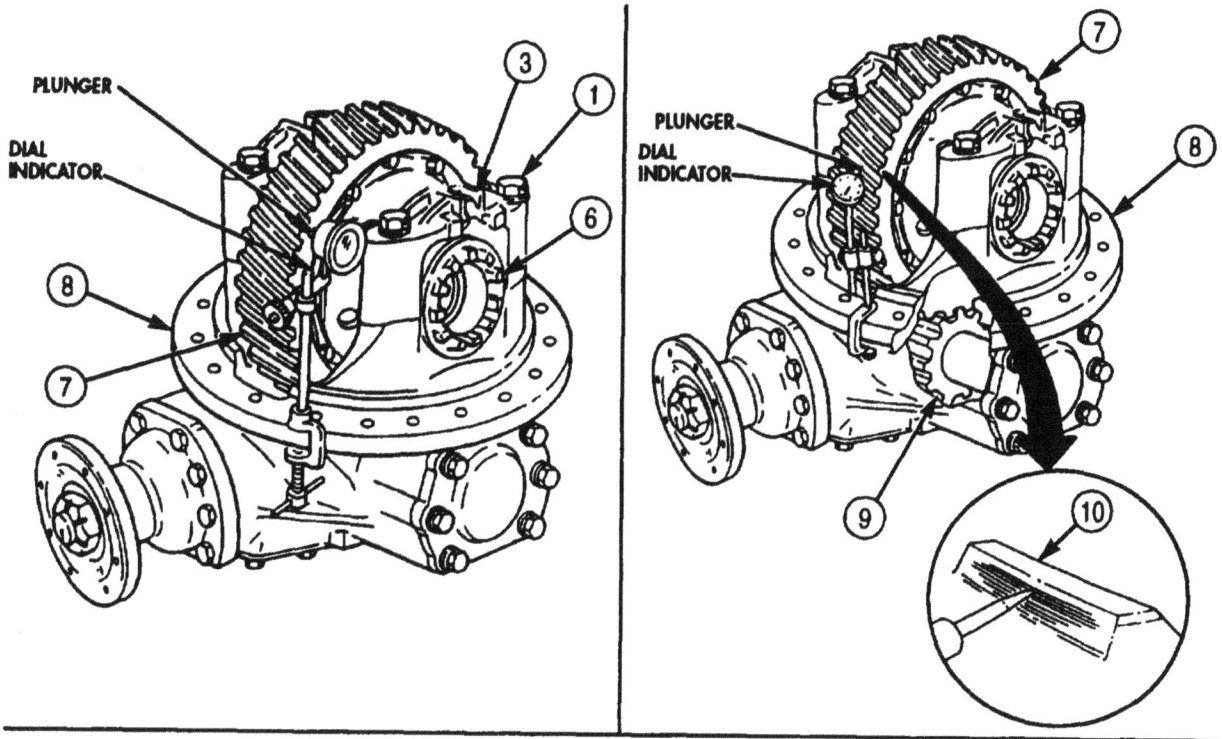

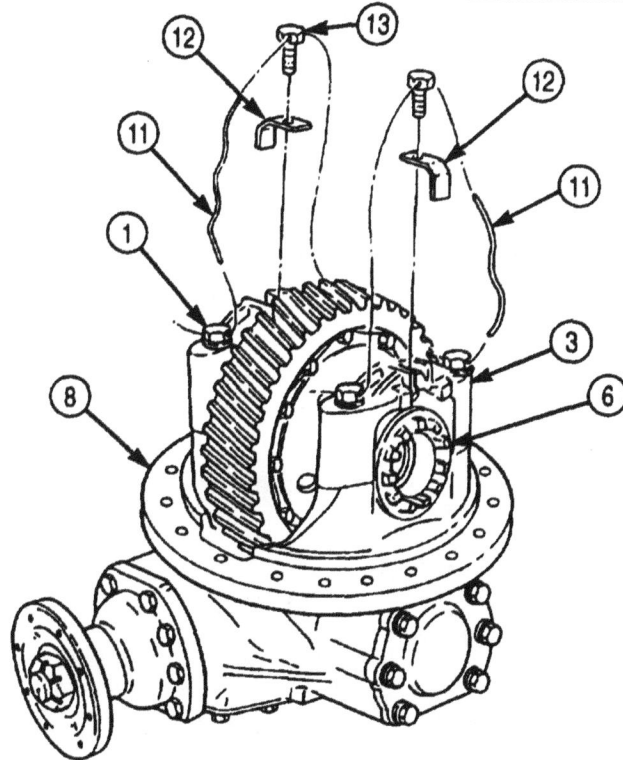

FOLLOW-ON TASK: Install differential and carrier assembly (para. 4-100).

Section VI. SPECIAL PURPOSE BODIES MAINTENANCE

5-65. SPECIAL PURPOSE BODIES MAINTENANCE INDEX

5-66. VAN BODY REPLACEMENT

THIS TASK COVERS:

a. Removal   b. Installation

INITIAL SETUP:

APPLICABLE MODELS
M934/A1/A2

TOOLS
General mechanic's tool kit (Appendix E, Item 1)

MATERIALS/PARTS
Ten locknuts (Appendix D, Item 291)
Twelve locknuts (Appendix D, Item 321)

PERSONNEL REQUIRED
Two

REFERENCES (TM)
TM 9-2320-272-10
TM 9-2320-272-24P

EQUIPMENT CONDITION
1 Parking brake set (TM 9-2320-272-10).
1 Ladders removed (TM 9-2320-272-10).

GENERAL SAFETY INSTRUCTIONS
All personnel must stand clear during lifting
operations.

a. Removal

1. Remove four locknuts (18), screws (13), washers (14), springs (15), and springs (16) from van body (1) and two frame brackets (8). Discard locknuts (18).
2. Remove eight locknuts (12) and screws (9) from van body (1) and four frame brackets (8). Discard locknuts (12).
3. Attach chain to four lifting brackets (2) and lifting device.

WARNING

All personnel must stand clear during lifting operations. A
snapped cable, or swinging or shifting load, may result in injury to
personnel.

NOTE

Second assistant will help with steps 4 and 5.

4. Raise van body (1) clear of frame (7), release parking brake, and remove vehicle from under van body (1).
5. Lower van body (1) onto jack stands positioned evenly under van body (1).
6. Remove four blocks (5) from frame brackets (8).

NOTE

Sills and hardware will be in position for removal after van body
has been removed.

7. Remove ten locknuts (3), bevel washers (4), screws (10, washers (11), and two sills (6) and (17) from van body (1). Discard locknuts (3).
8. Remove chain from lifting device and four lifting brackets (2).

5-66. VAN BODY REPLACEMENT (Contd)

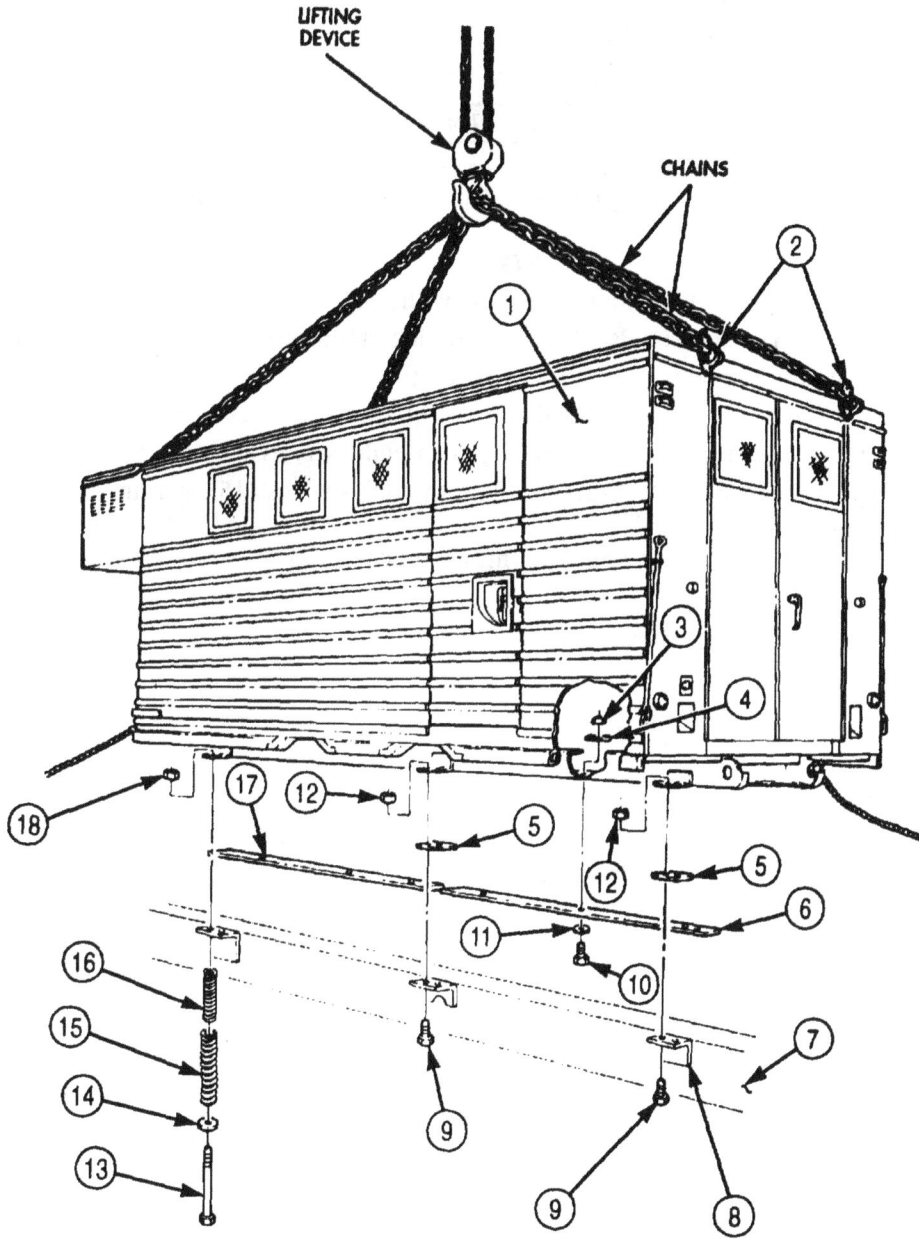

5-66. VAN BODY REPLACEMENT (Contd)

b. Installation

1. Install two sills (6) and (17) on van body (1) with ten washers (11), screws (10), bevel washers (4), and new locknuts (3).
2. Position four blocks (5) on frame brackets (8).
3. Attach chain to four lifting brackets (2) and lifting device.

WARNING

All personnel must stand clear during lifting operations. A snapped cable, or swinging or shifting load, may result in injury to personnel.

NOTE

Second assistant will help with steps 4 and 5.

4. Raise van body (1) from jack stands, and remove jack stands from under van body (1).
5. Position vehicle under van body (1) and set parking brake.
6. Lower van body (1) onto frame (7) and align holes of frame brackets (8) and van body (1).
7. Install van body (1) on four frame brackets (8) with eight screws (9) and new locknuts (12).
8. Install van body (1) on two frame brackets (8) with four springs (15), springs (16), washers (14), screws (13), and new locknuts (18).
9. Remove chains from lifting device and four lifting brackets (2).

5-66. VAN BODY REPLACEMENT (Contd)

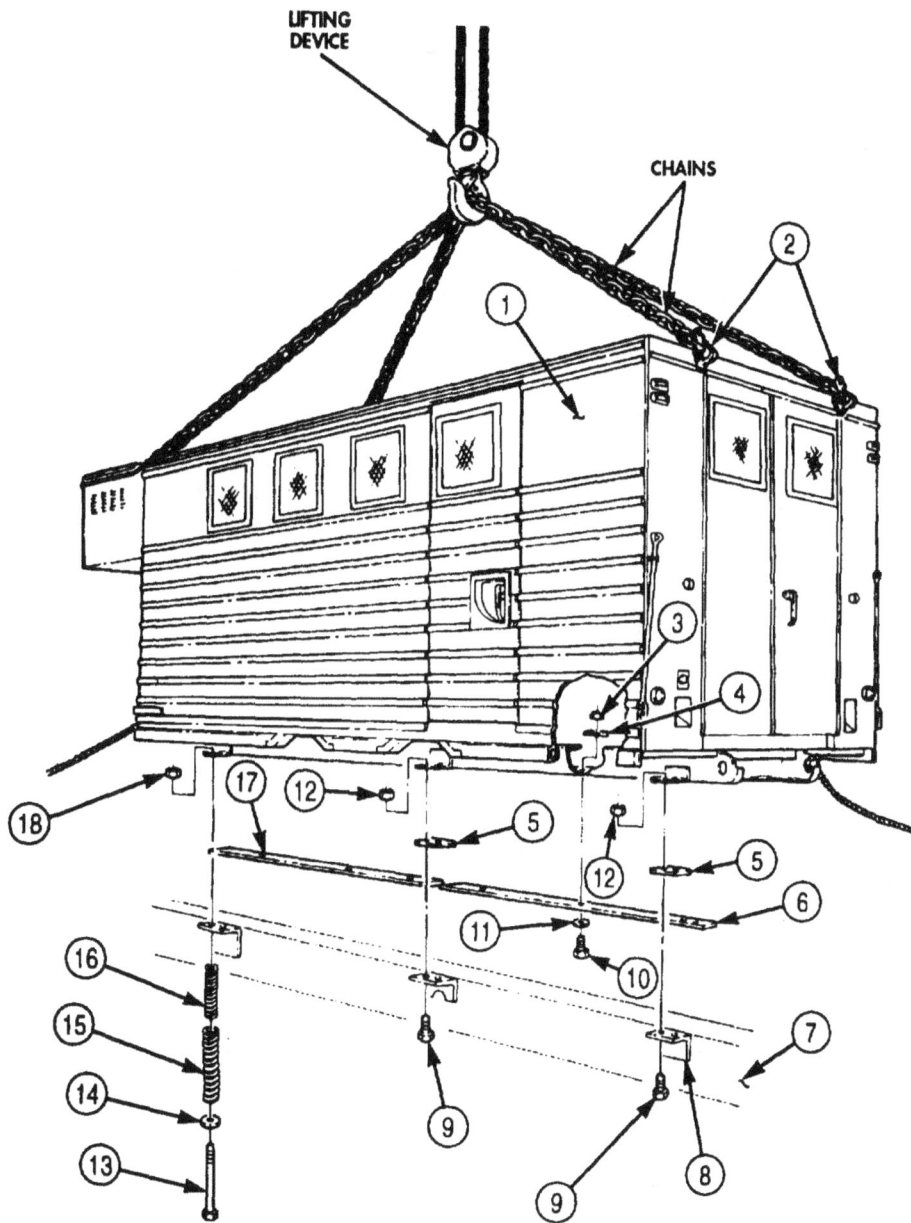

FOLLOW-ON TASK: Install ladders (TM 9-2320-272-10).

5-67. RETRACTABLE BEAM REPLACEMENT

THIS TASK COVERS:

a. Removal

b. Installation

INITIAL SETUP:

APPLICABLE MODELS
M9341/A1/A2

TOOLS
General mechanic's tool kit (Appendix E, Item 1)

MATERIALS/PARTS
Two locknuts (Appendix D, Item 274)
Seal (Appendix D, Item 614)
Seal (Appendix D, Item 615)
Seal (Appendix D, Item 616)

REFERENCES (TM)
TM 9-237
TM 9-2320-272-10
TM 9-2320-272-24P

EQUIPMENT CONDITION
• Parking brake set (TM 9-2320-272-10).
• Van body sides fully expanded and secured (TM 9-2320-272- 10).
• Retractable beam rollers removed (para. 5-69).
• Retractable beam drive shaft and lock removed (para. 5-68).

SPECIAL ENVIRONMENTAL CONDITIONS
Vehicle must be on level surface.

NOTE

All ten retractable beams are replaced the same way. This procedure is for the left-rear retractable beam.

a. Removal

1. Break welds between side panel (1) and side panel support (13) and remove retractable beam (12) (TM 9-237).

CAUTION

Remove retractable beam slowly from underframe. Failure to do so may result in damage to equipment.

2. Slide retractable beam (12) under side panel (1) and remove from underframe (4) and channel (5).

3. Remove two locknuts (7), washers (8), screws (11), seven screws (10), retainer (9), and seals (2), (3), and (6) from underframe (4). Discard locknuts (7) and seals (2), (3), and (6).

b. Installation

1. Install new seals (6), (3), and (2), and retainer (9) on underframe (4) with seven screws (10), two screws (11), washers (8), and new locknuts (7).

CAUTION

Install retractable beam slowly into underframe. Failure to do so may result in damage to equipment .

2. Slide retractable beam (12) under side panel (1) and install in under-frame (4) at channel (5).

3. Install retractable beam (12) by welding retractable beam (12) to side panel (1) and side panel support (13) (TM 9-237).

5-67. RETRACTABLE BEAM REPLACEMENT (Contd)

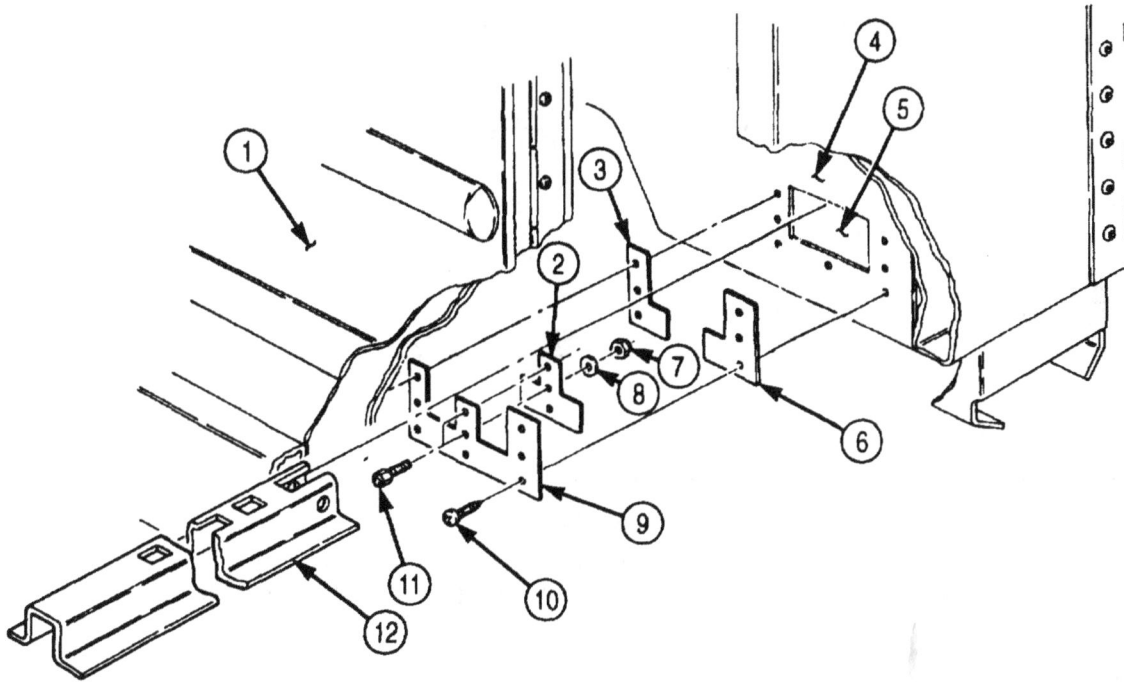

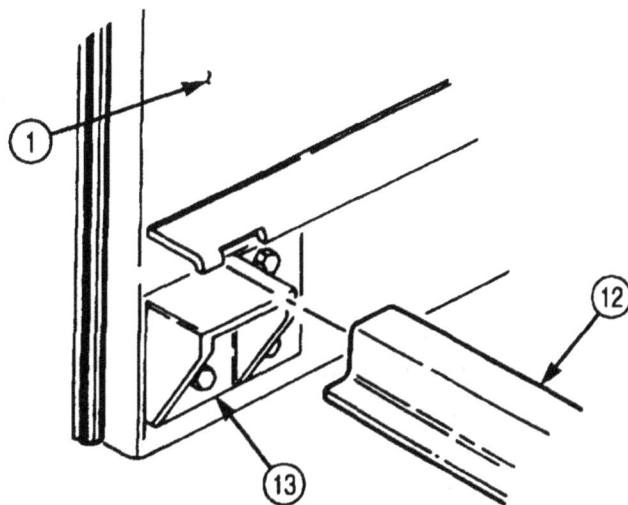

FOLLOW-ON TASKS:• Install retractable beam drive shaft and lock (para. 5-68).
• Install retractable beam rollers (para. 5-69).
• Retract van body sides (TM 9-2320-272-10).

5-68. RETRACTABLE BEAM DRIVE SHAFT AND LOCK MAINTENANCE

THIS TASK COVERS:

a. Removal

b. Cleaning, Inspection, and Repair

c. Installation

INITIAL SETUP:

APPLICABLE MODELS
M934/A1/A2

TOOLS
General mechanic's tool kit (Appendix E, Item 1)

MATERIALS/PARTS
Five keys (Appendix D, Item 266)
Locknut (Appendix D, Item 297)
GAA grease (Appendix C, Item 28)

PERSONNEL REQUIRED
Two

REFERENCES (TM)
LO 9-2320-272-12
TM 9-2320-272-10
TM 9-2320-272-24P

EQUIPMENT CONDITION
Parking brake set (TM 9-2320-272-10).

SPECIAL ENVIRONMENTAL CONDITIONS
Vehicle must be on level surface.

NOTE

Left and right retractable beam drive shafts and locks are replaced the same. This procedure is for the left retractable beam drive shaft and lock.

a. Removal

1. Remove locknut (1) and lock (2) from under-frame stud (8). Discard locknut (1).
2. Remove screw (4), pawl (3), and lock (2) from underframe (5).
3. Remove thirty screws (7) and five covers (6) from underframe (5).
4. Remove four setscrews (15) from bushings (14).

CAUTION

Do not bend or strain ratchet shaft during removal. Doing so may result in damage to equipment.

NOTE

• Remove all nicks, burrs, and corrosion from ratchet shaft and lubricate before removal.

• Direct assistant to catch ratchet shaft components under van body during removal.

5. Slowly pull ratchet shaft (9) and remove four bushings (14), nine bushings (13), spacers (12), sprockets (11), and five spacers (10) from under-frame (5).

5-68. RETRACTABLE BEAM DRIVE SHAFT AND LOCK MAINTENANCE (Contd)

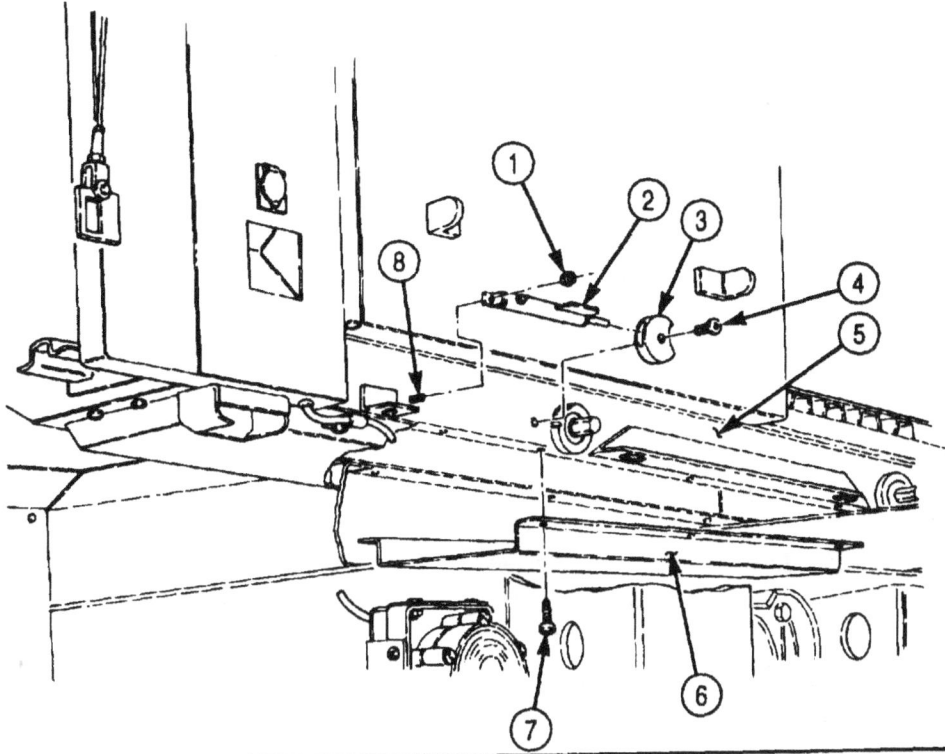

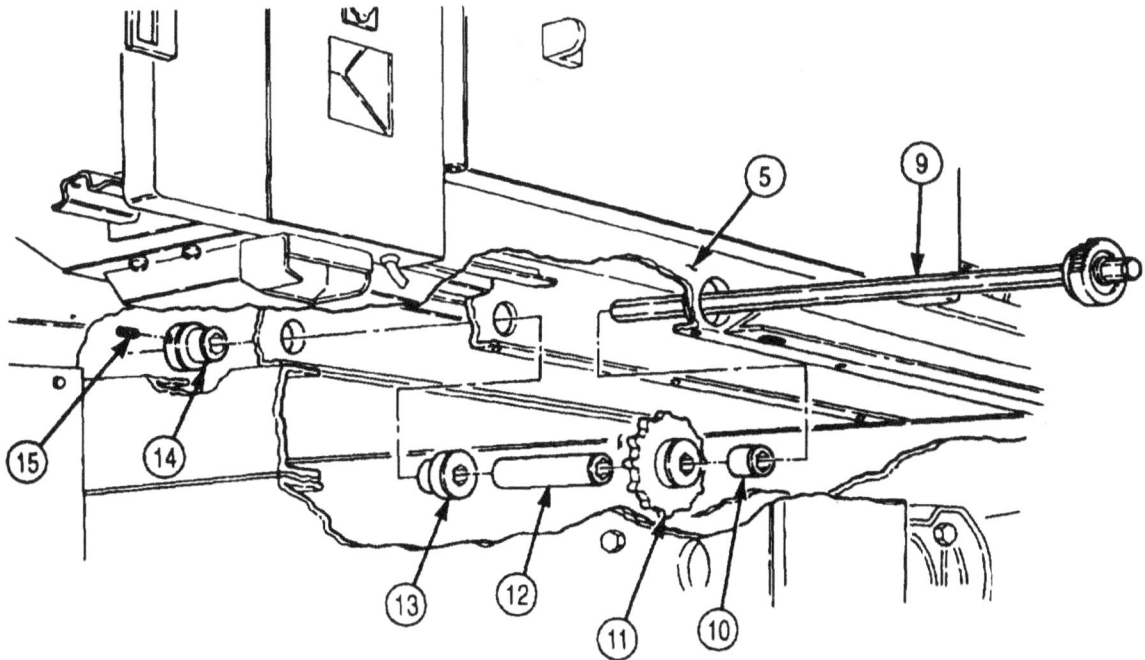

5-68. RETRACTABLE BEAM DRIVE SHAFT AND LOCK MAINTENANCE (Contd)

c. Cleaning, Inspection, and Repair

1. For general cleaning instructions, refer to para. 2-14.
2. For general inspection instructions, refer to para. 2-15.
3. Remove five keys (11) and bushings (10) from sprockets (4). Discard keys (11).
4. Inspect bushings (10), (6), and (8) for wear and damage. If worn or damaged replace bushings (10), (6), or (8).
5. Inspect ratchet shaft (2) for burrs, nicks, and corrosion. If damaged, repair with file or emery cloth. If damage is excessive, replace.
6. Test spring action of lock plunger (14). If spring action is weak, replace lock (14).
7. Install five new keys (11) and bushings (10) on sprockets (4).

c. Installation

CAUTION

Do not bend or strain ratchet shaft during installation. Doing so may result in damage to equipment.

NOTE

• Bevel forward end of ratchet shaft for installation.

• Apply GAA grease on two feet of forward end of ratchet shaft for installation.

• Direct assistant to install ratchet shaft components under van body when installing ratchet shaft.

1. Slowly insert ratchet shaft (2) into underframe (1) and install five spacers (3), sprockets (4), spacers (5), nine bushings (6), and four bushings (8).
2. Install four setscrews (9) in bushings (8) in underframe channels (7).
3. Lubricate ratchet shaft (2) with GAA grease (LO 9-2320-272-12).
4. Install five covers (17) on underframe (1) with thirty screws (18).
5. Install lock (14) and pawl (15) on underframe stud (12) and underframe (1) with new locknut (13) and screw (16).

5-68. RETRACTABLE BEAM DRIVE SHAFT AND LOCK MAINTENANCE (Contd)

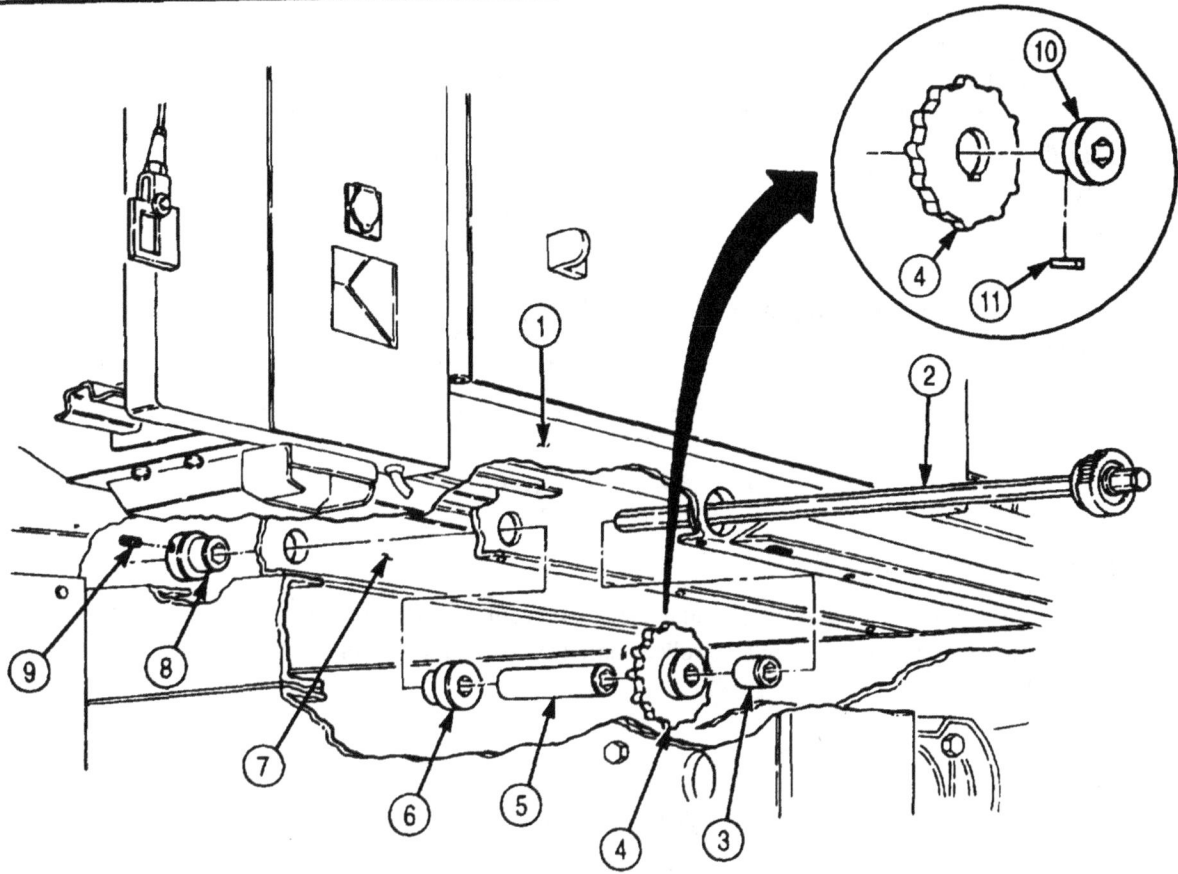

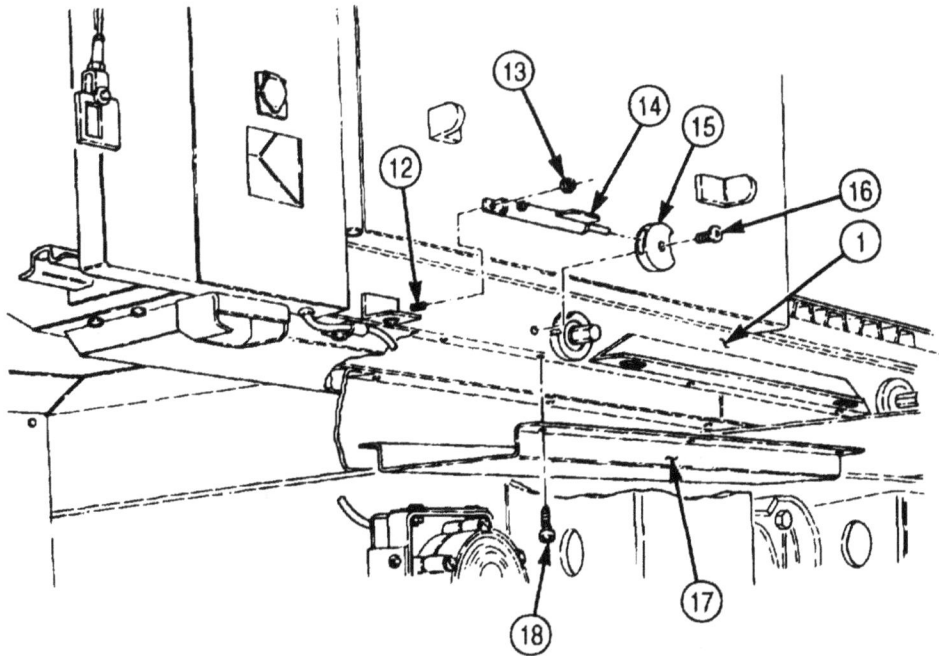

5-69. RETRACTABLE BEAM ROLLERS REPLACEMENT

THIS TASK COVERS:

a. Support Roller Removal
b. Support Roller Installation

c. End Roller Removal
d. End Roller Installation

INITIAL SETUP:

APPLICABLE MODELS
M934/A1/A2

TOOLS
General mechanic's tool kit (Appendix E, Item 1)

MATERIALS/PARTS
Four cotter pins (Appendix D, Item 72)

REFERENCES (TM)
TM 9-2320-272-10
TM 9-2320-272-24P

EQUIPMENT CONDITION
• Parking brake (TM 9-2320-272-10).
• Van body sides fully expanded and secured (TM 9-2320-272-10).

NOTE

All support rollers are removed basically the same way, This procedure is for left-rear support rollers.

a. Support Roller Removal

1. Remove seven screws (2), clamps (3), and cover (4) from underframe (1)
2. Remove two cotter pins (5), support roller shaft (7), and support roller (6) from underframe (1). Discard cotter pins (5).

b. Support Roller Installation

1. Install support roller (6) on underframe (1) with support roller shaft (7) and two new cotter pins (5).
2. Install cover (4) on underframe (1) with seven clamps (3) and screws (2).

5-69. RETRACTABLE BEAM ROLLERS REPLACEMENT (Contd)

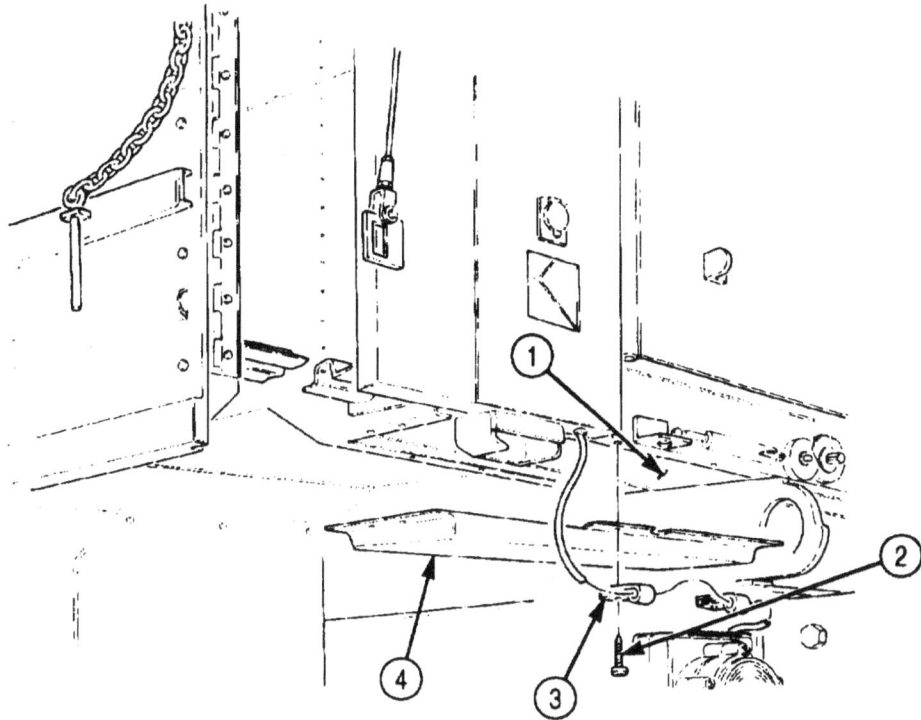

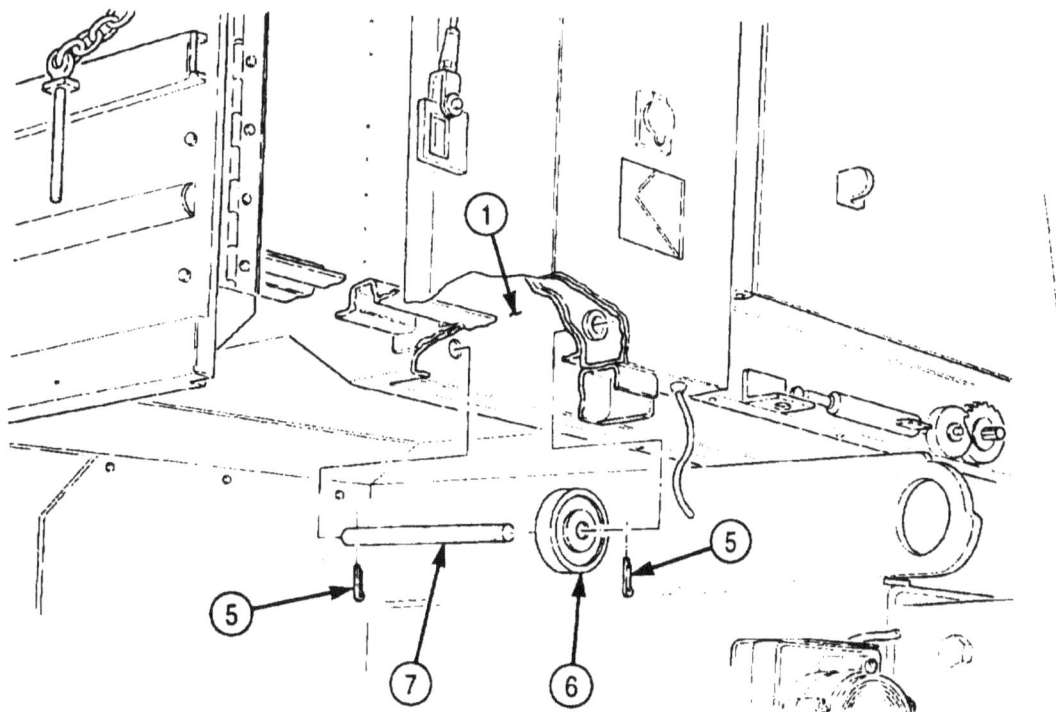

5-69. RETRACTABLE BEAM ROLLERS REPLACEMENT (Contd)

NOTE

All end rollers are replaced basically the same way. This procedure is for left-rear end rollers.

c. End Roller Removal

1. Remove support roller (task a.).
2. Remove two cotter pins (2), end roller shaft (3), and end roller (4) from underframe (1). Discard cotter pins (2).

d. End Roller Installation

1. Install end roller (4) on underframe (1) with end roller shaft (3) and two new cotter pins (2).
2. Install support roller (task b.).

5-69. RETRACTABLE BEAM ROLLERS REPLACEMENT (Contd)

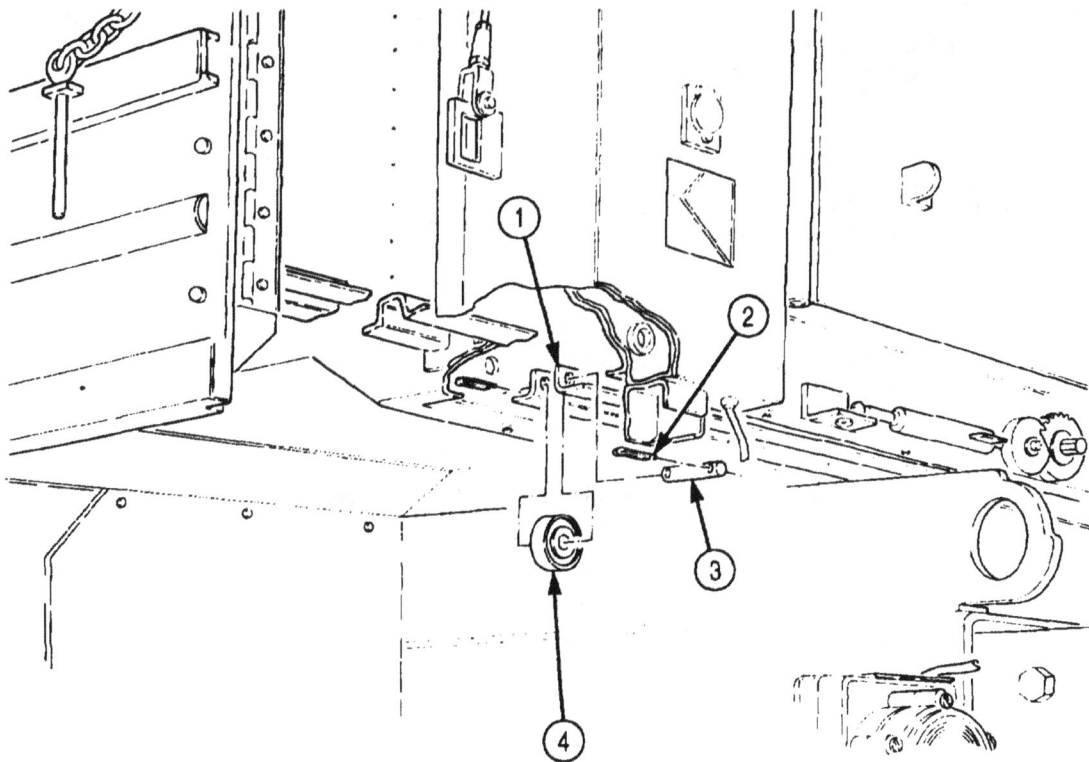

FOLLOW-ON TASK: Retract van body sides (TM 9-2320-272-10).

5-70. SIDE DOORS MAINTENANCE

THIS TASK COVERS:

a. Removal
b. Disassembly
c. Cleaning and Inspection

d. Assembly
e. Installation

INITIAL SETUP:

APPLICABLE MODELS
M934/A1/A2

TOOLS
General mechanic's tool kit (Appendix E, Item 1)
Rivet gun

MATERIALS/PARTS
Two cotter pins (Appendix D, Item 51)
Two lockwashers (Appendix D, Item 364)
Three rivets (Appendix D, Item 548)
Sixteen rivets (Appendix D, Item 547)
Gasket (Appendix D, Item 229)
Seal (Appendix D, Item 605)
Seal (Appendix D, Item 633)
Adhesive sealant (Appendix C, Item 7)
Sealing compound (Appendix C, Item 67)

REFERENCES (TM)
TM 43-0213
TM 9-2320-272-10
TM 9-2320-272-24P

EQUIPMENT CONDITION
Parking brake set (TM 9-2320-272-10).
Van body sides fully expanded and secured
(TM 9-2320-272-10).
Side door window removed (para. 3-350).

NOTE
Left and right side doors are replaced the same way. This
procedure is for the left side door.

a. Removal

1. Remove two screws (1), lockwashers (2), bracket (3), and door check lever (4) from door (5). Discard lockwashers (2).

NOTE
Assistant will help with step 2.

2. Remove fifteen screws (6) and door (5) from side panel of van body (7).

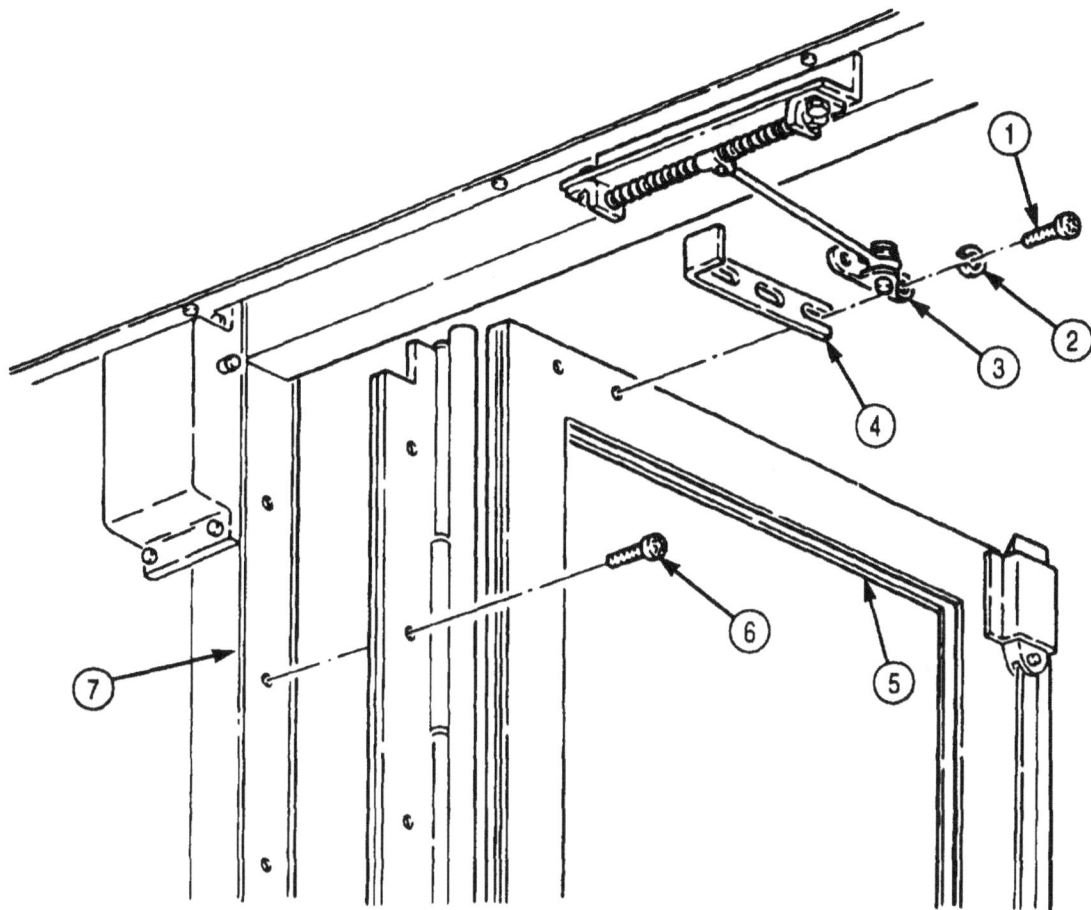

5-70. SIDE DOORS MAINTENANCE (Contd)

b. Disassembly

1. Remove three rivets (13), handle (12), and gasket (11) from pocket (10). Discard rivets (13) and gasket (11).
2. Remove sixteen rivets (14) and pocket (10) from outer panel (9). Discard rivets (14).
3. Remove two cotter pins (23), screws (19), washers (20), and rods (21) from upper case (2), lower case (18), and center case (22). Discard cotter pins (23).
4. Remove pin (26) and handle (25) from center case (22).
5. Remove four screws (27) and center case (22) from inner panel (24).
6. Remove eight screws (1), upper case (2), and lower case (18) from inner panel (24).
7. Remove fourteen screws (8), molding (7), seal (6), hinge (5), and seal (4) from door frame (17). Discard seals (4) and (6).
8. Break adhesive seal and remove two moldings (3) from inner panel (24).
9. Remove two screws (16) and moldings (15) from inner panel (24).

c. Cleaning and Inspection

1. For general cleaning instructions, refer to para. 2-14.
2. For general inspection instructions, refer to para. 2-15.
3. Inspect all movable components of center case (22) and lower and upper cases (2) and (18) for proper operation. Replace center case (22), lower case (18), or upper case (2) if damaged or defective.
4. Inspect rods (21) for breaks and bends. Replace rods (21) if broken or bent.
5. Inspect moldings (3) and (15) and pocket (10) for cracks and bends. Replace moldings (3), (15), or pocket (10) if cracked or bent.
6. Inspect outer panel (9), door frame (17), and inner panel (24) for damage. Replace entire side door if either is damaged.

d. Assembly

NOTE
- Apply rustproofing compound to all inside surfaces and boxed-in areas (TM 43-0213).
- Apply sealing compound to all exterior joints for installation.

1. Install two moldings (15) on inner panel (24) with two screws (16).
2. Apply adhesive to two moldings (3) and install on inner panel (24).
3. Install new seal (4), hinge (5), new seal (6), and molding (7) on door frame (17) with fourteen screws (18).
4. Install upper case (2) and lower case (18) on inner panel (24) with eight screws (1).
5. Install center case (22) on inner panel (24) with four screws (27).
6. Install handle (25) on center case (22) with pin (26).
7. Install two rods (21) on center (22), upper case (2), and lower case (18) with two new cotter pins (23), washers (20), and screws (19).
8. Install pocket (10) on outer panel (9) with sixteen new rivets (14).
9. Install new gasket (11) and handle (12) on pocket (10) with three new rivets (13).

5-70. SIDE DOORS MAINTENANCE (Contd)

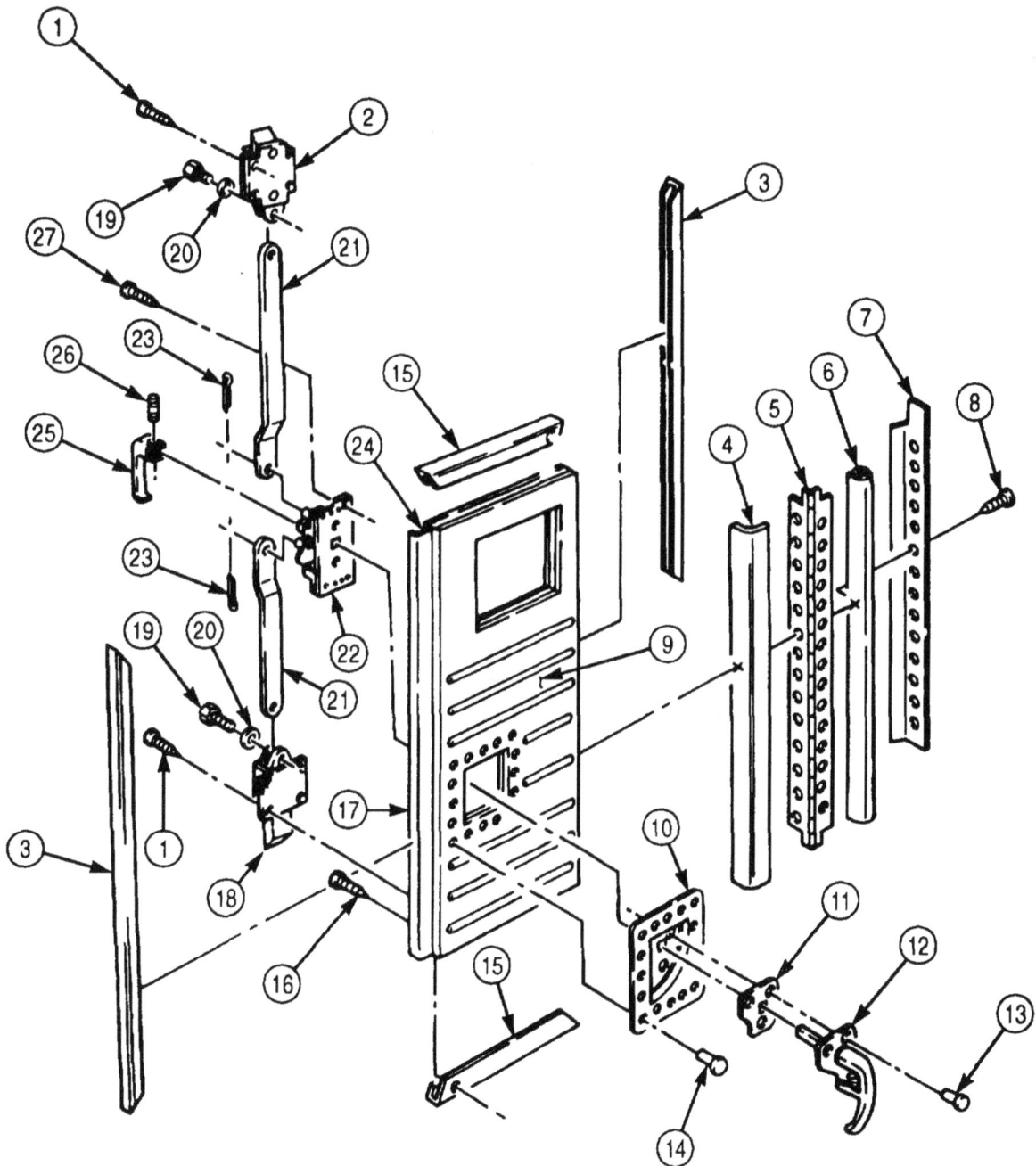

5-70. SIDE DOORS MAINTENANCE (Contd)

e. Installation

NOTE

Assistant will help with step 1.

1. Install door (5) on side panel of van body (7) with fifteen screws (6).
2. Install door check lever (4) and bracket (3) on door (5) with two new lockwashers (2) and screws (1).

5-70. SIDE DOORS MAINTENANCE (Contd)

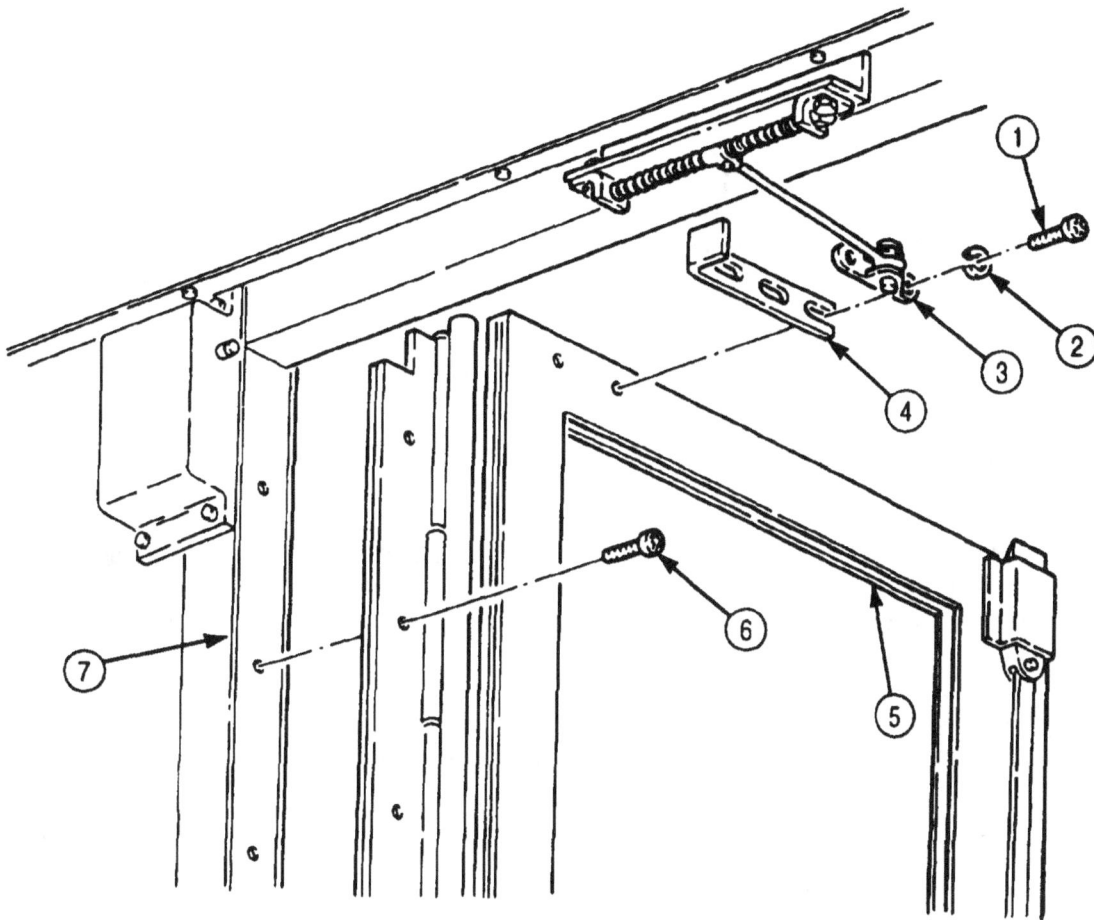

FOLLOW-ON TASKS: Install side door window (para. 3-350).
- Retract van body sides (TM 9-2320-272-10).

5-71. UNDERFRAME PARTS REPLACEMENT

THIS TASK COVERS:

a. Removal   b. Installation

INITIAL SETUP:

APPLICABLE MODELS	**REFERENCES (TM)**
M934/A1/A2	TM 9-2320-272-10
	TM 9-2320-272-24P
TOOLS	
General mechanic's tool kit (Appendix E, Item 1)	**EQUIPMENT CONDITION**
Rivet gun	Parking brake set (TM 9-2320-272-10).
MATERIALS/PARTS	**SPECIAL ENVIRONMENTAL CONDITIONS**
Eight lockwashers (Appendix D, Item 400)	Vehicle must be on level surface.
Two seals, 206 in. (Appendix D, Item 617)	
Thirty-two rivets (Appendix D, Item 558)	
Sealing compound (Appendix C, Item 67)	

a. Removal

NOTE
Assistant will help with step 1.

1. Remove one hundred thirty-eight screws (14), two retainers (15), channels (12), and seals (13) from underframe (1). Discard seals (13).

2. Remove thirty-two screws (4) and four covers (5) from under-frame (1).

3. Remove eight nuts (3), lockwashers (2), screws (7), and four plates (6) from under-frame (1). Discard lockwashers (2).

4. Remove thirty-two rivets (8) and eight tiedowns (9) from underframe (1). Discard rivets (8).

5. Remove four screws (11) and angle bracket (10) from underframe (1).

b. Installation

1. Install angle bracket (10) on underframe (1) with four screws (11).

2. Install eight tiedowns (9) on underframe (1) with thirty-two new rivets (8).

3. Install four plates (6) on underframe (1) with eight screws (7), new lockwashers (2), and nuts (3).

4. Install four covers (5) on underframe (1) with thirty-two screws (4).

NOTE
Apply sealing compound to exterior joints for installation.

5. Install two new seals (13), channels (12), and retainers (15) on underframe (1) with one hundred thirty-eight screws (14).

5-71. UNDERFRAME PARTS REPLACEMENT (Contd)

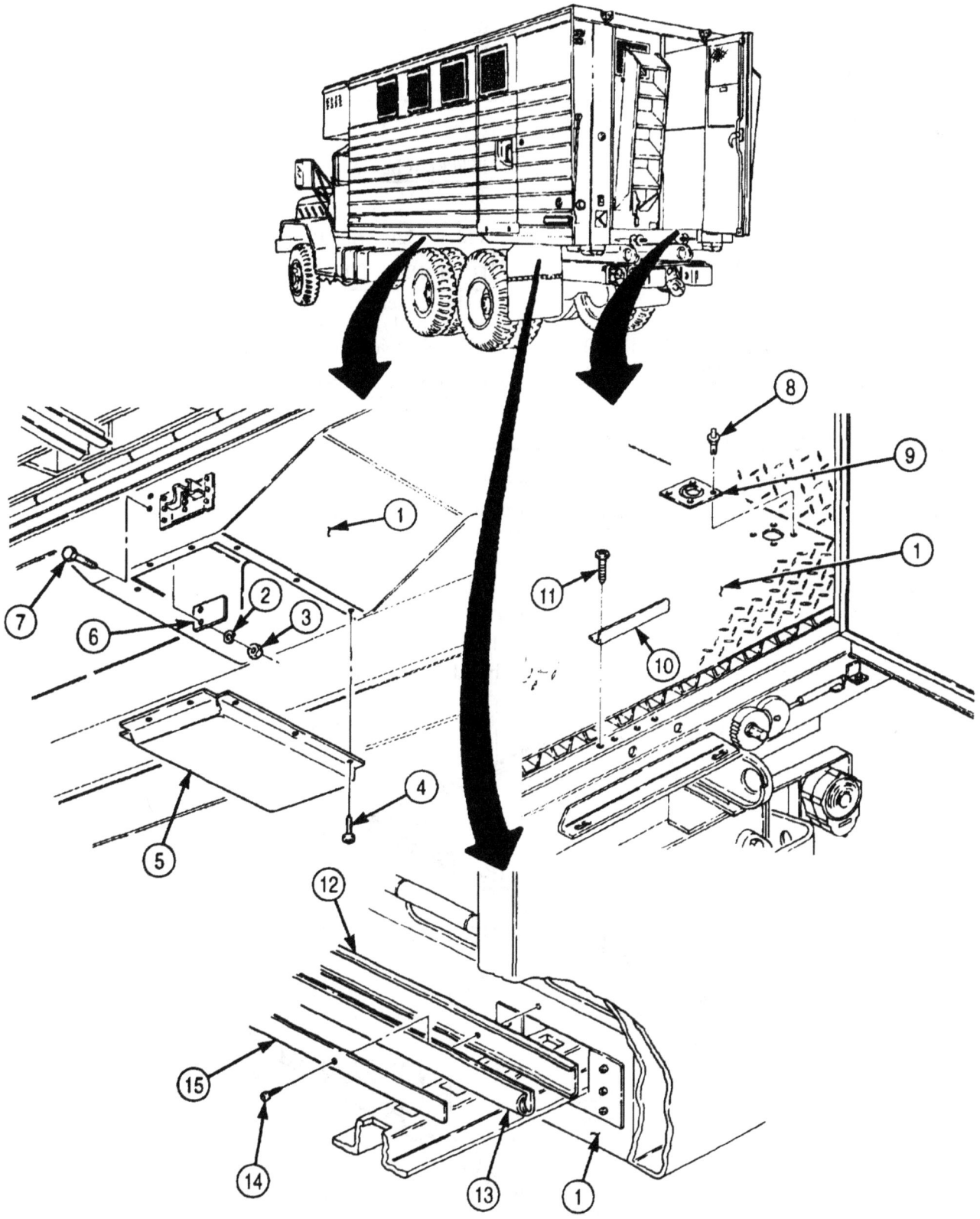

5-72. HINGED FLOOR MAINTENANCE

THIS TASK COVERS:

a. Removal   d. Assembly
b. Disassembly   e. Installation
c. Cleaning and Inspection

INITIAL SETUP:

APPLICABLE MODELS
M934/A1/A2

TOOLS
General mechanic's tool kit (Appendix E, Item 1)
Rivet gun

MATERIALS/PARTS
Fifty-six rivets (Appendix D Item 549)
Four seals, 39 in. (Appendix D, Item 617)
Seal, 203 in. (Appendix D, Item 631)
Adhesive (Appendix C, Item 3)
Sealing compound (Appendix C, Item 67)

REFERENCES (TM)
TM 9-237
TM 9-2320-272-10
TM 9-2320-272-24P

EQUIPMENT CONDITION
• Hinged end panel open (TM 9-2320-272-10)
• Van body sides extended (TM 9-2320-272-10).
• Counterbalance removed (para. 3-356).

SPECIAL ENVIRONMENTAL CONDITIONS
Vehicle must be on level surface.

NOTE
Left and right hinged floors are maintained the same way. This procedure is for left hinged floor.

a. Removal

NOTE
Assistant will help with step.

Remove fifty-five screws (3), four screws (2), and hinged floor (1) from hinge (4).

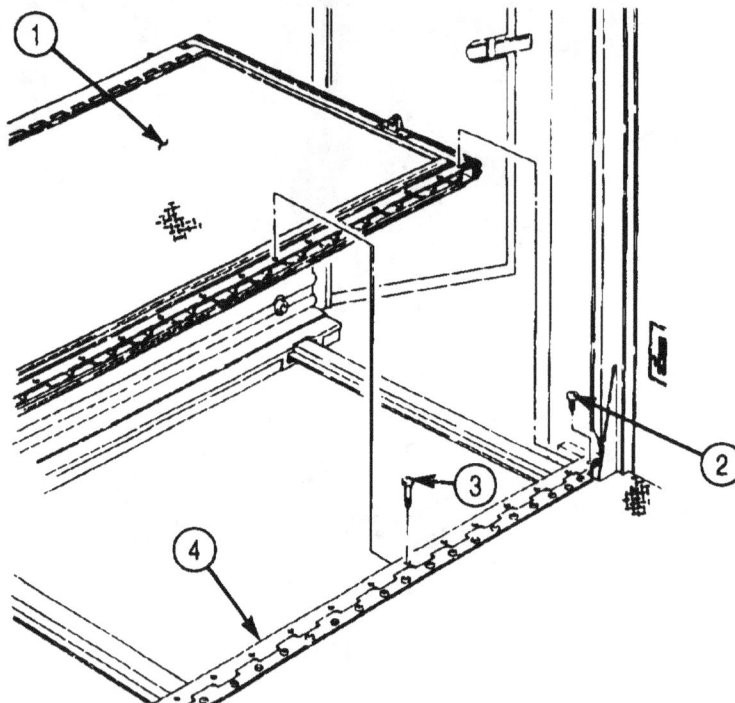

5-72. HINGED FLOOR MAINTENANCE (Contd)

b. Disassembly

1. Remove four screws (5) and two pivots (6) from hinged floor (1).
2. Remove thirty screws (7), two retainers (8), four seals (9), and two channels (15) from hinged floor (1). Discard seals (9).
3. Remove ten screws (13) and five pads (14) from hinged floor (1).
4. Remove four screws (11) and handle (12) from hinged floor (1).
5. Remove seal (10) from hinged floor (1). Discard seal (10).

NOTE

Perform step 6 only if inspection requires replacement of hinge.

6. Remove fifty-six rivets (16) and hinge (4) from under-frame (17). Discard rivets (16).

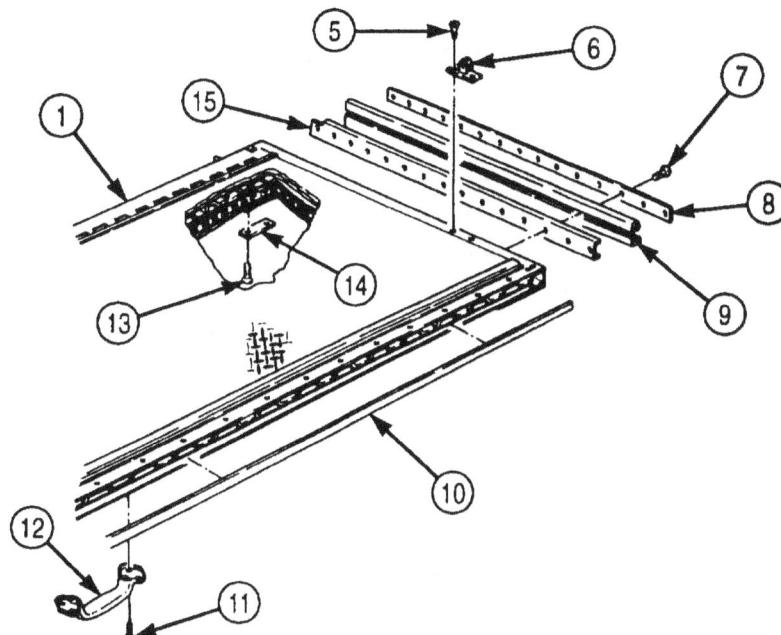

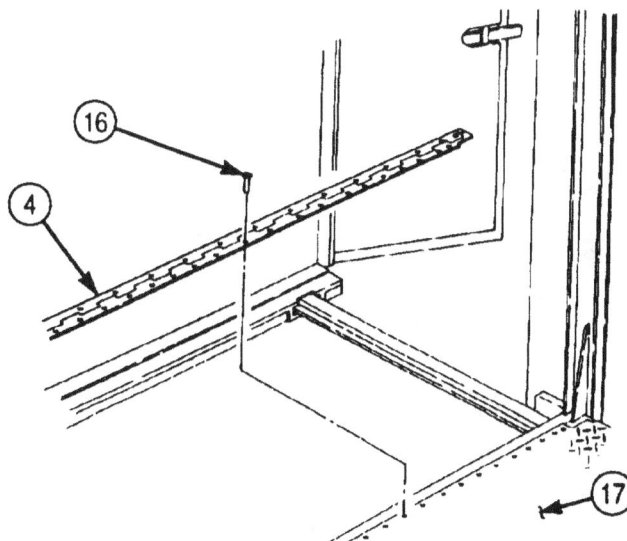

5-72. HINGED FLOOR MAINTENANCE (Contd)

c. Cleaning and Inspection

1. For general cleaning instructions, refer to para. 2-14.
2. For general inspection instructions, refer to para. 2-15.
3. Inspect hinged floor (1) for cracked, broken, and torn floor plates. If cracked, broken, or torn, replace hinged floor (1) (TM 9-237).
4. Inspect retainers (8) and channels (15) for bends and breaks. If bent or broken, replace retainers (8) or channels (15).
5. Inspect hinge (4) for breaks, cracks, and smooth operation. If cracked, broken, or not operating properly, replace hinge (4).

d. Assembly

NOTE

Apply sealing compound to exterior joints prior to installation.

Insulate exterior joints and areas of metal-to metal contact with adhesive.

Perform step 1 only if hinge is to be replaced .

1. Install hinge (4) on under-frame (17) with fifty-six new rivets (16).
2. Install new seal (10) on hinged floor (1).
3. Install handle (12) on hinged floor (1) with four screws (11).
4. Install five pads (14) on hinged floor (1) with ten screws (13).
5. Install two channels (15), four new seals (9), and two retainers (8) on hinged floor (1) with thirty screws (7).
6. Install two pivots (6) on hinged floor (1) with four screws (5).

e. Installation

NOTE

Assistant will help with step.

Install hinged floor (1) on hinge (4) with fifty-five screws (3) and four screws (2).

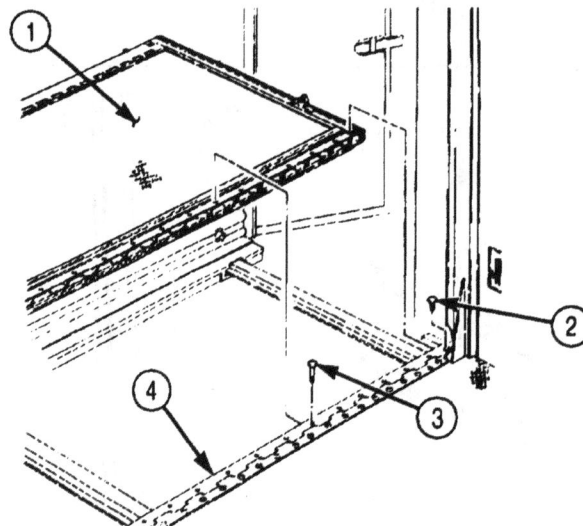

5-72. HINGED FLOOR MAINTENANCE (Contd)

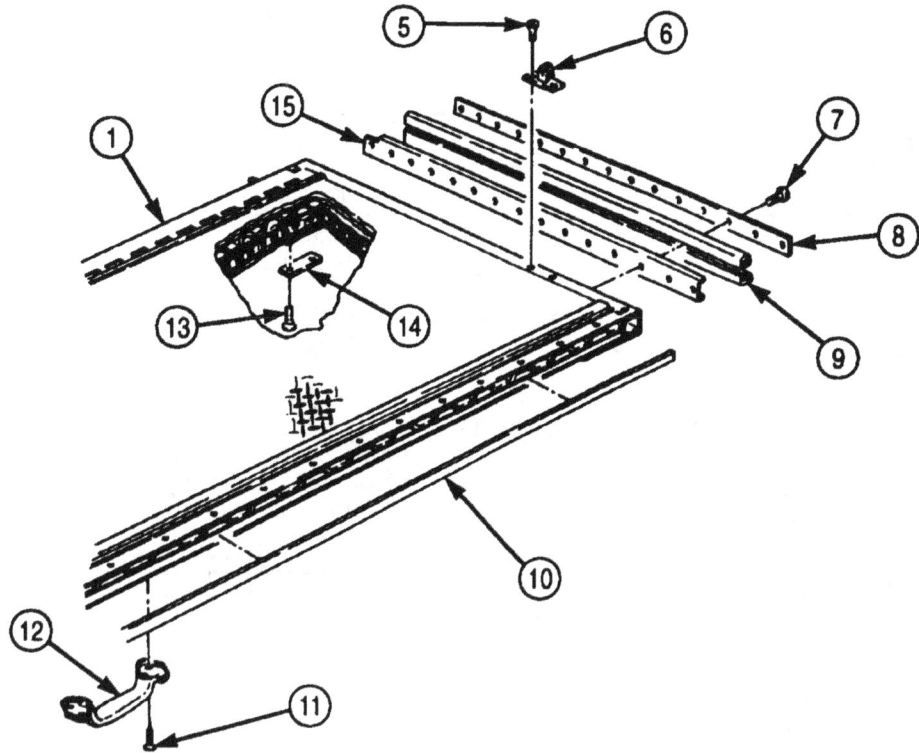

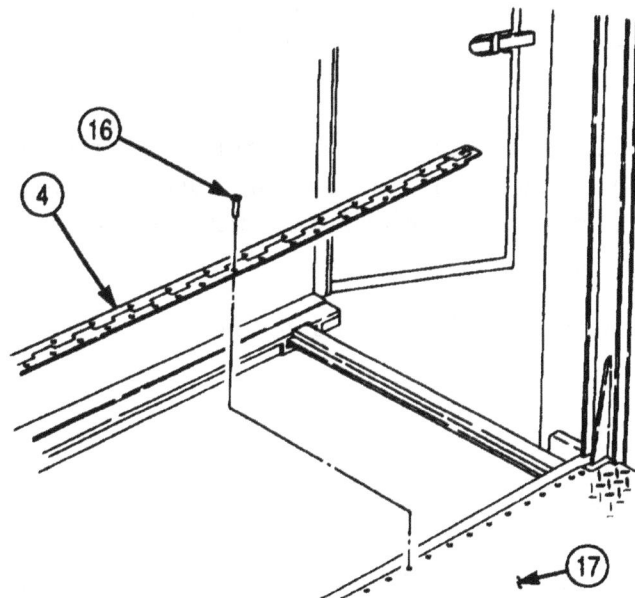

FOLLOW-ON TASKS: Install counterbalance (para. 3-356).
- Close hinged end panel (TM 9-2320-272-10).
- Retract van body sides (TM 9-2320-272-10).

5-73. EXTERIOR SIDE PANEL MAINTENANCE

THIS TASK COVERS:

a. Disassembly   c. Assembly
b. Cleaning and inspection

INITIAL SETUP:

APPLICABLE MODELS
M934/A1/A2

TOOLS
General mechanic's tool kit (Appendix E, Item 1)
Rivet gun

MATERIALS/PARTS
Two hundred forty-eight rivets (Appendix D, Item 547)
Twenty-four rivets (Appendix D, Item 551)
One hundred eighty-nine rivets (Appendix D, Item 550)
Two seals, 86 in. (Appendix D, Item 617)
Two seals, 206 in. (Appendix D, item 617)
Two seals, 6 in. (Appendix D, Item 617)
Seal, 242 in. (Appendix D, Item 617)
Seal, 204 in. (Appendix D, Item 618)
Channel seal (Appendix D, Item 36)
Adhesive (Appendix C, Item 7)
Sealing compound (Appendix C, Item 67)

REFERENCES (TM)
TM 9-2320-272-10
TM 9-2320-272-24P

EQUIPMENT CONDITION
• Parking brake set (TM 9-2320-272-10).
• Van body fully expanded and secured (TM 9-2320-272-10).
• Side panel front lock removed (para. 3-369).
• Side panel rear lock removed (para. 3-368).
• Side door removed (para. 5-70).
• Retractable window removed (para. 3-351).

SPECIAL ENVIRONMENTAL CONDITIONS
Vehicle must be on level surface.

NOTE
Left and right exterior side panels are disassembled the same.
This procedure covers the right exterior side panel.

a. Disassembly

1. Remove twenty-five screws (5) and retainers (3) and (15) from side panel frame (14).

2. Remove seals (2) and (4) and channel (1) from side panel frame (14). Discard seals (2) and (4).

3. Remove thirty screws (12) and five retainers (13) from seal (10) at side panel frame (14).

4. Remove sixty-two screws (11), six retainers (9), and seal (10) from side panel frame (14). Discard seal (10).

5. Remove six screws (8), two retainers (7), and seals (6) from side panel frame (14). Discard seals (6).

5-73. EXTERIOR SIDE PANEL MAINTENANCE (Contd)

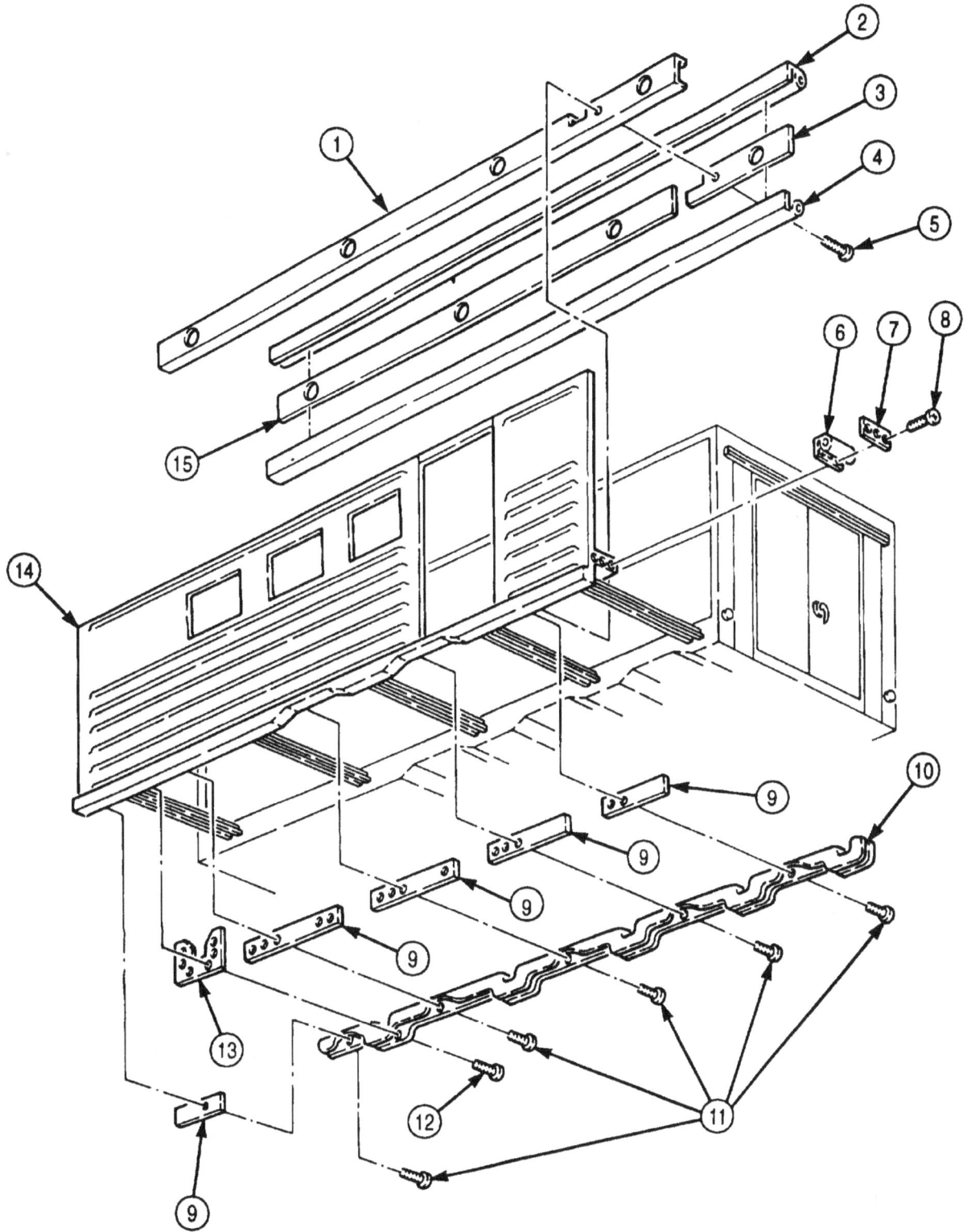

5-73. EXTERIOR SIDE PANEL MAINTENANCE (Contd)

6. Remove one hundred eighty-nine rivets (19), bracket (17), and seal (18) from skins (2) and (11) and side panel frame (8). Discard rivets (19) and seal (18).

7. Remove eighty-two nuts (5), screws (1), four screws (10), two retainers (4), and seals (3) from side panel frame (8). Discard seals (3).

8. Remove channel seal (9) from doorway of side panel frame (8). Discard channel seal (9).

9. Remove nine screws (6) and molding (7) from doorway of side panel frame (8).

NOTE

- Perform steps 10 through 12 if skin is to be replaced.
- Assistant will help with steps 10 through 12.

10. Remove two hundred forty-eight rivets (12) and skins (2) and (11) from side panel frame (8). Discard rivets (12).

11. Remove eight screws (15) and two ladder hangers (14) from side panel frame (8).

12. Remove twenty-four rivets (16) and lower skin (13) from side panel frame (8). Discard rivets (16).

5-73. EXTERIOR SIDE PANEL MAINTENANCE (Contd)

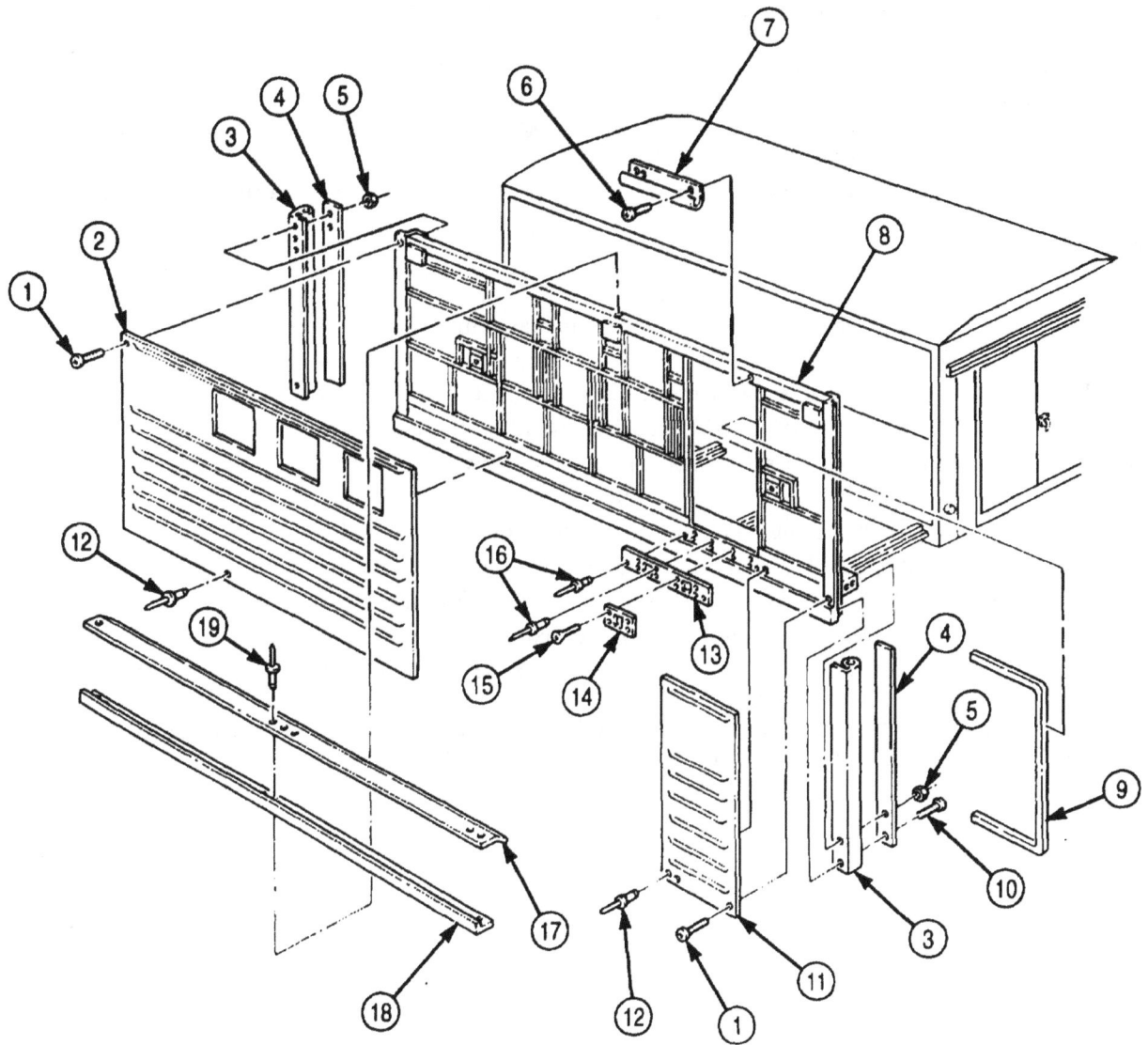

5-73. EXTERIOR SIDE PANEL MAINTENANCE (Contd)

b. Cleaning and Inspection

1. For general cleaning instructions, refer to para. 2-14.
2. For general inspection instructions, refer to para. 2-15.
3. Inspect five retainers (12), retainers (13), (5), and (6) for cracks and bends. If cracked or broken, replace.
4. Inspect bracket (3), retainers (9) and (10), and channel (11) for rust corrosion and breaks. If rusted corroded, or broken, replace.
5. Inspect molding (4), two ladder hangers (8), and side panel frame (2) for cracks, breaks, and warpage. If cracked, broken, or warped, replace.
6. Inspect skins (1) and (14) and lower skin (7) for tears and punctures. If torn or punctured, replace.

c. Assembly

NOTE
- Perform steps 1 through 3 if skin was removed.
- Assistant will help with steps 1 through 3.

1. Install lower skin (7) on side panel frame (2) with twenty-four new rivets (23).
2. Install two ladder hangers (8) on side panel frame (2) with eight screws (22).
3. Install skins (1) and (14) on side panel frame (2) with two hundred forty-eight new rivets (21).

NOTE
- Apply sealing compound to exterior joints for installation.
- Apply adhesive to rubber and metal surfaces for installation.

4. Install molding (4) on doorway of side panel frame (2) with nine screws (18).
5. Install new channel seal (19) on doorway of side panel frame (2).
6. Install two new seals (16) and retainers (5) on side panel frame (2) with eighty-two screws (15), nuts (17), and four screws (20).
7. Install new seal (25) and bracket (24) on skins (1) and (14) and side panel frame (2) with one hundred eighty-nine new rivets (26).

5-73. EXTERIOR SIDE PANEL MAINTENANCE (Contd)

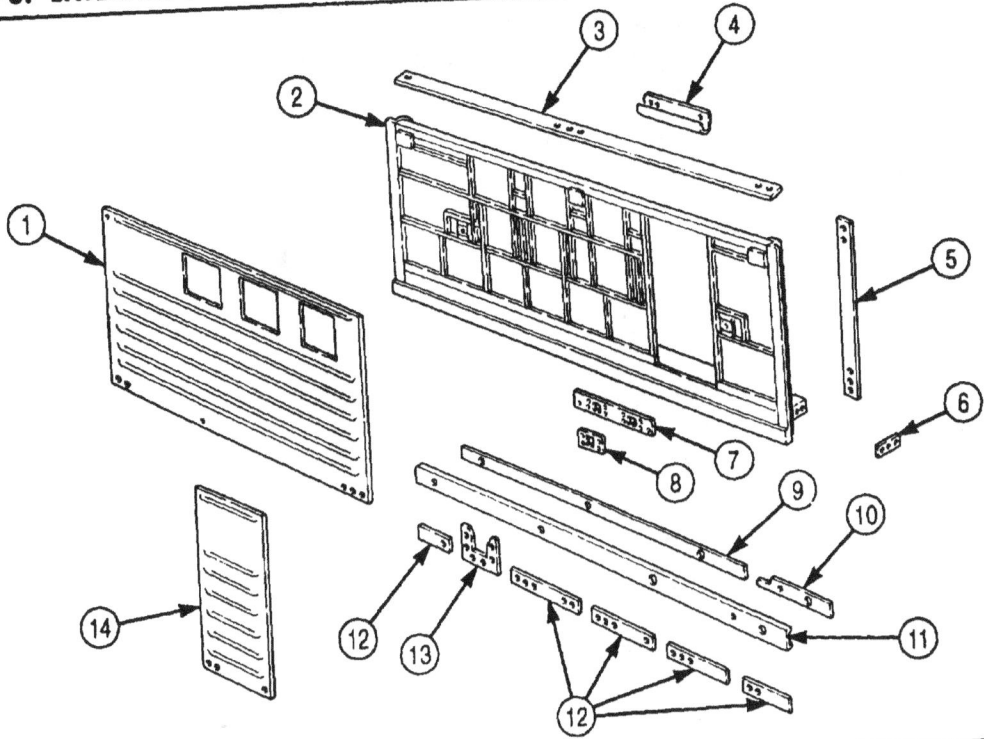

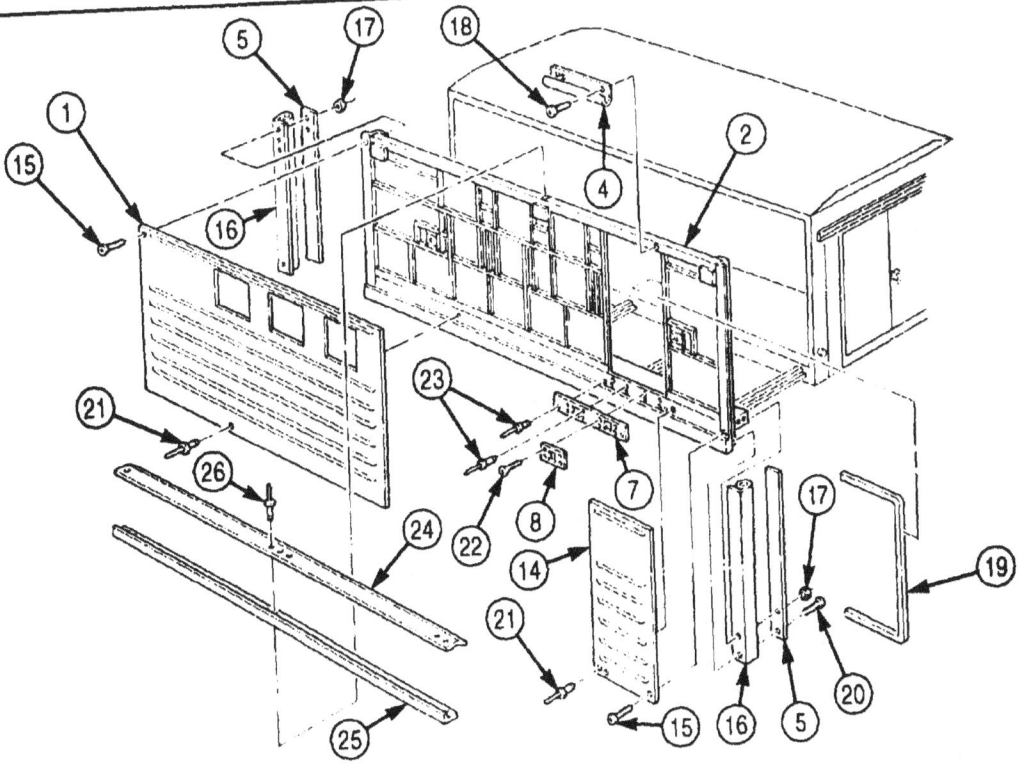

5-73. EXTERIOR SIDE PANEL MAINTENANCE (Contd)

8. Install two new seals (6) and retainers (7) on side panel frame (14) with six screws (8).

9. Install new seal (10) and six retainers (9) on side panel frame (14) with sixty-two screws (11).

10. Install five retainers (13) on side panel frame (14) and seal (10) with thirty screws (12).

NOTE
New seals may need to be formed to fit on side panel frame.

11. Install channel (1), new seals (2) and (4), retainer (3), and retainer (15) on side panel frame (14) with twenty-five screws (5).

5-73. EXTERIOR SIDE PANEL MAINTENANCE (Contd)

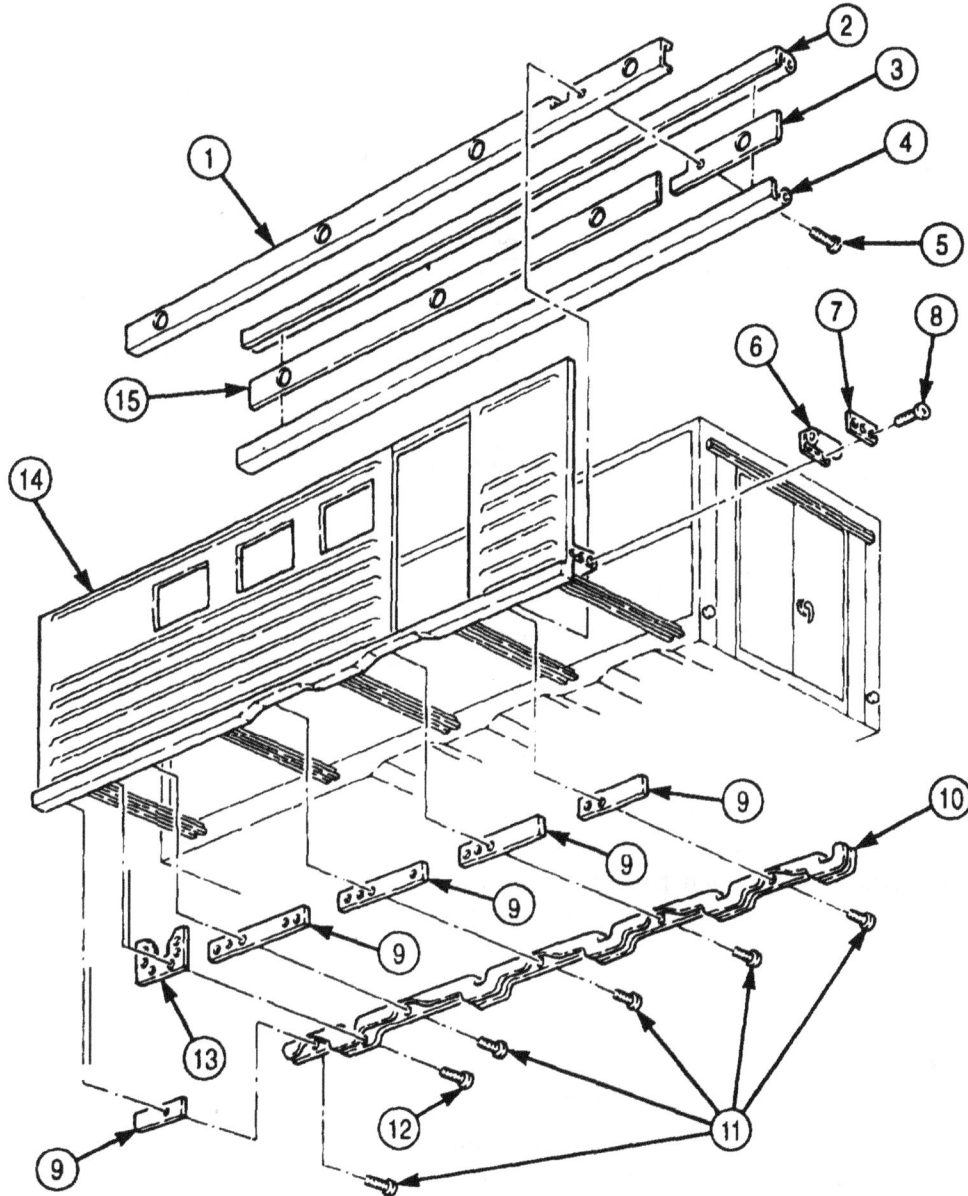

FOLLOW-ON TASKS:
- Install side panel rear lock (para. 3-368).
- Install side panel front lock (para. 3-369).
- Install retractable window (para. 3-351).
- Install side door (para. 5-70).
- Retract van body sides (TM 9-2320-272-10).

5-74. REAR WALL INTERIOR PANELS REPLACEMENT

THIS TASK COVERS:

a. Removal   b. Installation

INITIAL SETUP:

APPLICABLE MODELS
M934/A1/A2

TOOLS
General mechanic's tool kit (Appendix E, Item 1)

MATERIALS/PARTS
Sealing compound (Appendix C, Item 67)

REFERENCES (TM)
TM 9-2320-272-10
TM 9-2320-272-24P

EQUIPMENT CONDITION
• Parking brake set (TM 9-2320-272-10).
• Rear doors opened (TM 9-2320-272-10).
• Van body sides fully expanded and secured (TM 9-2320-272-10).
• Inside telephone jack post removed (para. 3-374).
• Blackout light switch removed (para. 3-373).
• Fire extinguisher bracket removed (para. 3-416).
  Electrical load center removed (para. 4-161).
• Electrical junction box removed (para. 4-163).
• Converter outlet box removed (para. 4-169).
• Heater thermostat removed (para. 4-165).

SPECIAL ENVIRONMENTAL CONDITIONS
Vehicle must be on a level surface.

NOTE

- Rear wall exterior panels are not serviceable and if damaged are replaced as a unit with rear wall assembly.

- Left and right rear wall interior panels are replaced the same way. This procedure is for the left rear wall interior panel.

a. Removal

NOTE

Assistant will help with step 2.

1. Remove thirteen screws (4) and molding (1) from panel (2) and header (3).
2. Remove twenty-two screws (5) and panel (2) from header (3).
3. Remove eight screws (9) and two gussets (8) from skin (7) and underframe (6).

b. Installation

NOTE

- Apply sealing compound to exterior joints prior to installation.
- Assistant will help with step 2.

1. Install two gussets (8) on skin (7) and underframe (6) with eight screws (9).
2. Install panel (2) on header (3) with twenty-two screws (5).
3. Install molding (1) on panel (2) and header (3) with thirteen screws (4).

5-74. REAR WALL INTERIOR PANELS REPLACEMENT (Contd)

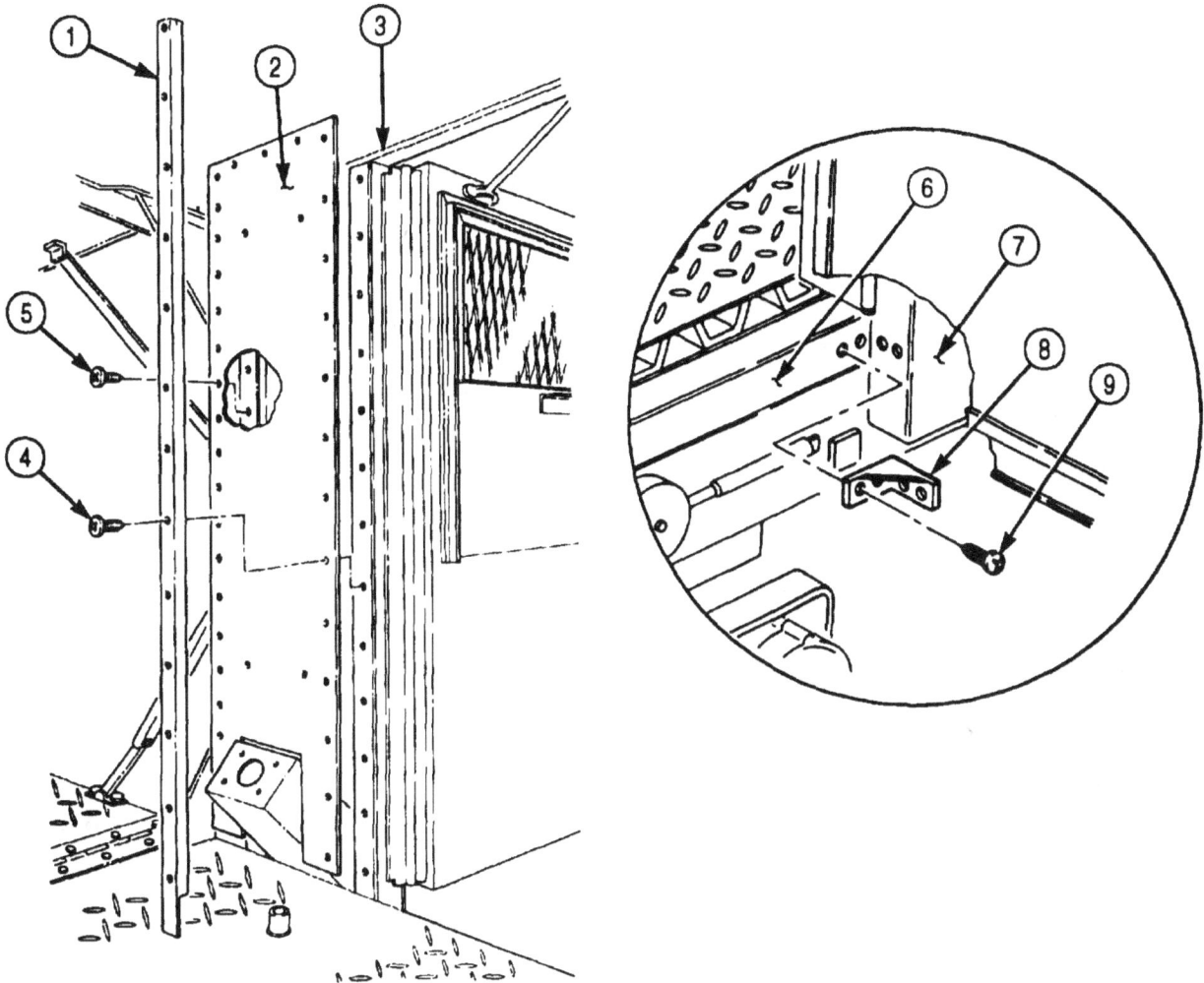

FOLLOW-ON TASKS: Install heater thermostat (para. 4-165).
- Install converter outlet box (para. 4-169).
- Install electrical junction box (para. 4-163).
- Install electrical load center (para. 4-161).
- Install fire extinguisher bracket (para. 3-416).
- Install blackout light switch (para. 3-373).
- Install inside telephone jack post (para. 3-374).
- Retract van body sides (TM 9-2320-272-10).
- Close rear doors (TM 9-2320-272-10).

5-75. INTERIOR SIDE PANELS AND LATCHES REPLACEMENT

THIS TASK COVERS:

a. Removal   b. Installation

INITIAL SETUP:

APPLICABLE MODELS

M934/A1/A2

TOOLS

General mechanic's tool kit (Appendix E, Item 1)

MATERIALS/PARTS

Two cotter pins (Appendix D, Item 51)
Gasket (Appendix D, Item 231)
Three grommets (Appendix D, Item 251)
Six lockwashers (Appendix D, Item 364)
O-ring (Appendix D, Item 479)
Four rubber bumpers (Appendix D, Item 563)
Primer (Appendix C, Item 57)
Sealing compound (Appendix C, Item 67)
Adhesive (Appendix C, Item 7)

REFERENCES (TM)

TM 9-2320-272-10
TM 9-2320-272-24P

EQUIPMENT CONDITION

- Parking brake set (TM 9-2320-272-10).
- Van body sides fully expanded and secured (TM 9-2320-272-10).
- Hinged end panels removed (para. 4-155).
- Hinged floor removed (para. 5-72).
- Hinged roof removed (para. 5-77).
- Side doors removed (para. 5-70).

SPECIAL ENVIRONMENTAL CONDITIONS

Vehicle must be on a level surface.

NOTE

Left and right interior side panels are replaced the same way. This procedure is for right interior side panel.

a. Removal

1. Remove six screws (1) and three hooks (2) from side panel frame (3).
2. Remove six screws (5) and three clips (6) from side panel frame (3).
3. Remove two screws (8) and striker (7) from side panel frame (3).
4. Remove two nuts (4), screws (10), lockwashers (11), and door check assembly (9) from door angle (15). Discard lockwashers (11).
5. Remove two screws (12), striker (13), and two spacer plates (14) from side panel frame (3).

5-75. INTERIOR SIDE PANELS AND LATCHES REPLACEMENT (Contd)

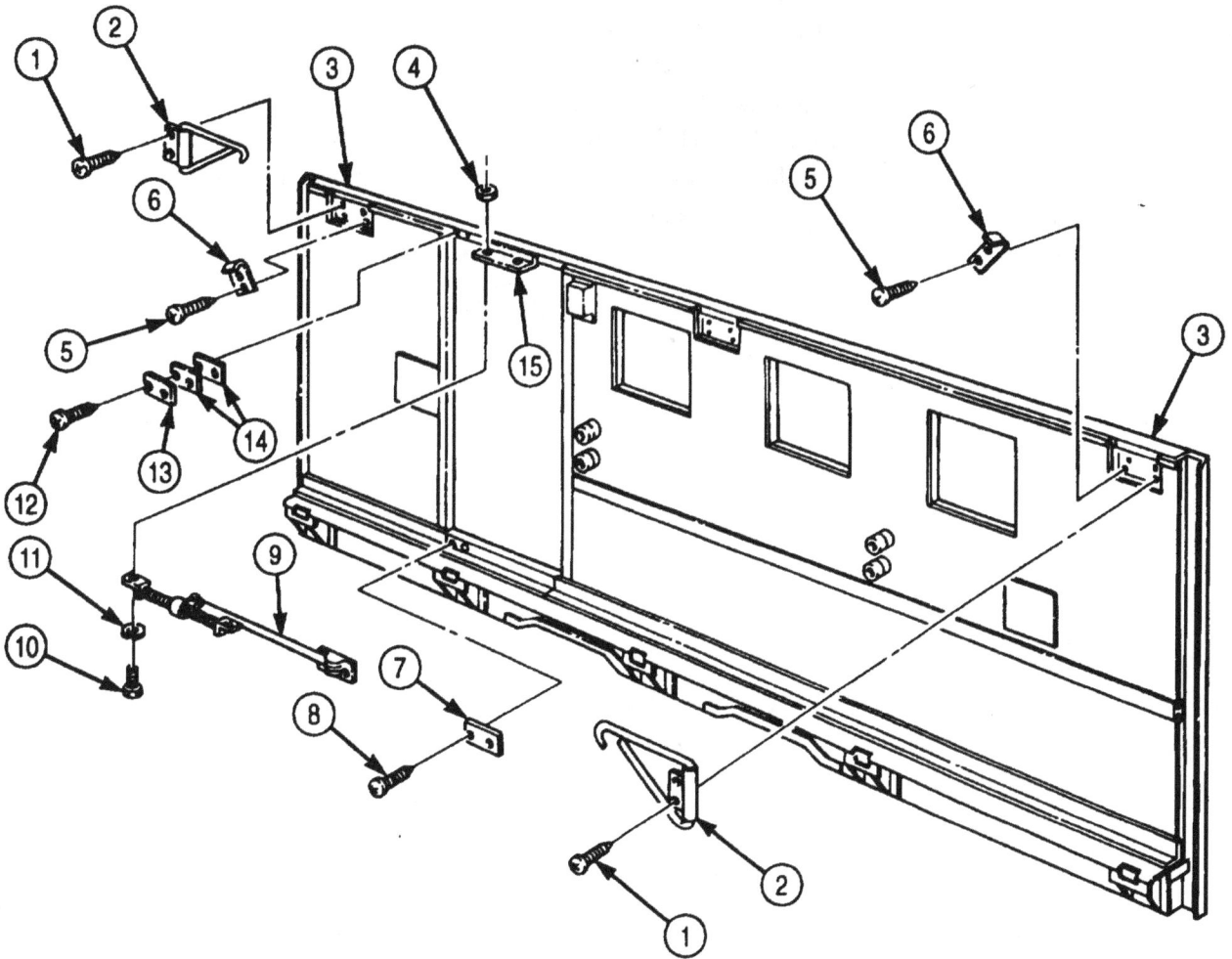

5-75. INTERIOR SIDE PANELS AND LATCHES REPLACEMENT(Contd)

6. Remove two screws (2) and grommets (3) from side panel frame (1). Discard grommets (3).

7. Remove four screws (11) and two clips (12) from panel (13) and side panel frame (1).

8. Remove four screws (16), washers (17), and rubber bumpers (18) from panel (5). Discard rubber bumpers (18).

9. Remove eight screws (15) and four wood spacers (19) from panel (5).

10. Remove eight screws (21) and four hangers (20) from panel (5).

11. Remove four screws (23) and junction box (24) from panel (5).

12. Remove eight screws (7) and plate (8) from panel (5).

13. Remove eleven screws (22) and molding (25) from panel (5).

NOTE

Assistants will help with steps 14 and 15.

14. Remove thirty-three screws (10) and panel (13) from side panel frame (1).

15. Remove fifty-eight screws (9), molding (14), and panel (5) from side panel frame (1).

16. Remove moldings (26), (4), and (6) from panel (5).

17. Remove two moldings (27), moldings (28), and moldings (29) from panel (5).

5-75. INTERIOR SIDE PANELS AND LATCHES REPLACEMENT (Contd)

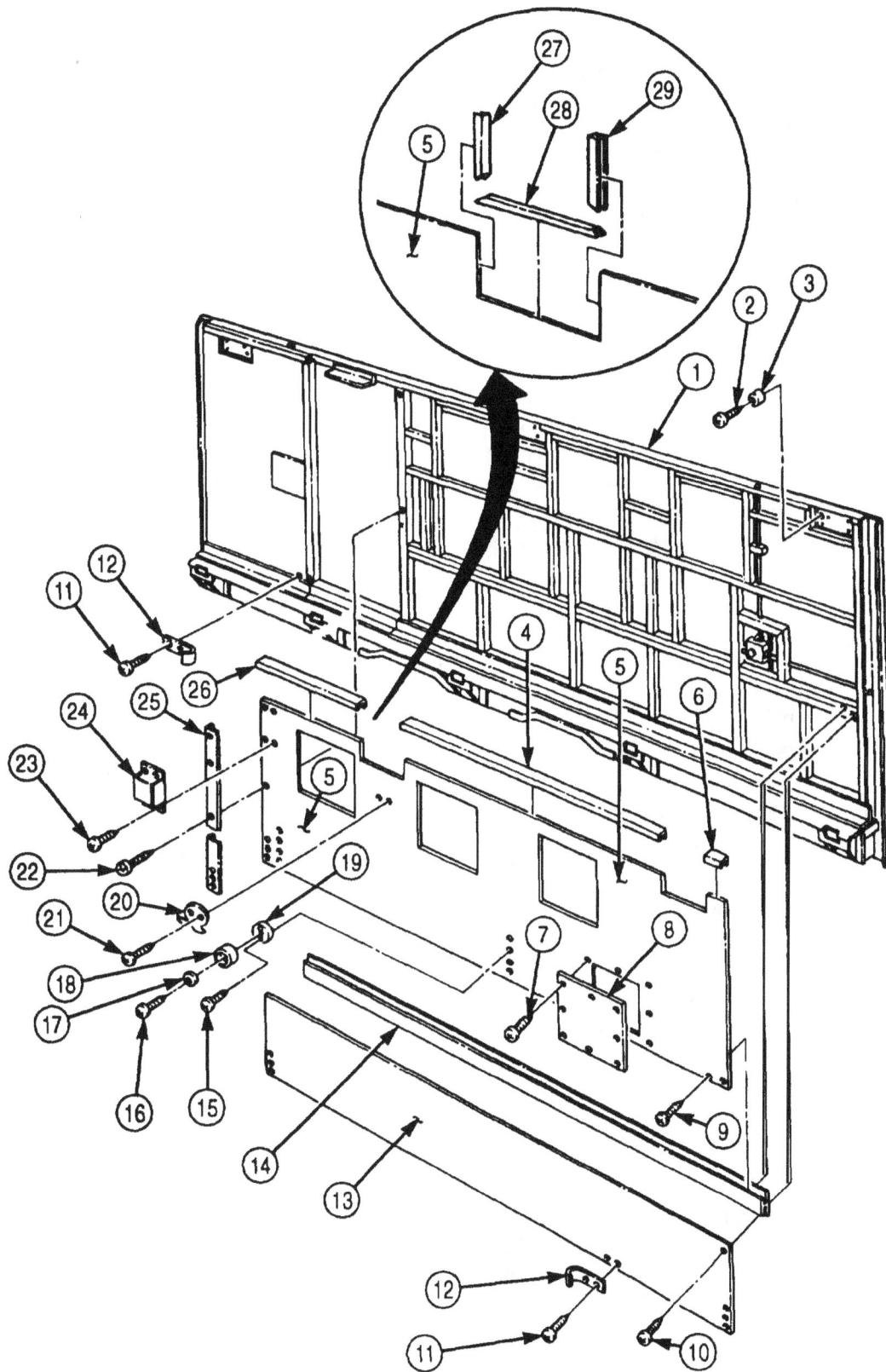

5-75. INTERIOR SIDE PANELS AND LATCHES REPLACEMENT (Contd)

18. Remove eight screws (15) and plate (14) from panel (16).

NOTE
Assistant will help with step 19.

19. Remove thirty-three screws (13), moldings (27), (28), (29), (12), (17), and (18), and panel (16) from side panel frame (4).
20. Remove nut (25) and washer (24) from latch (23).
21. Remove two screws (7), plate (8), gasket (9), shank (5), and O-ring (6) from latch (23). Discard O-ring (6) and gasket (9).
22. Remove two cotter pins (22), pins (20), clevis (26), and clevis (21) from latch (23). Discard cotter pins (22).
23. Remove four nuts (11), lockwashers (10), and latch (23) from side panel frame (4). Discard lockwashers (10).
24. Remove two screws (31) and retainer (30) from side panel frame (4).
25. Remove clevis (21) from flush bolt (3).
26. Remove clevis (26) from horizontal bar (19).
27. Remove horizontal bar (19) and flush bolt (3) from side panel frame (4).

NOTE
Repeat steps 18 through 27 for removal of each front latch assembly.

28. Remove screw (1) and grommet (2) from side panel frame (4). Discard grommet (2).

b. Installation

NOTE
- Insulate areas of dissimilar metal-to-metal contact with zinc chromate primer.
- Apply sealing compound to exterior joints prior to installation.
- Apply adhesive to rubber and metal surfaces.

1. Install new grommet (2) on side panel frame (4) with screw (1).
2. Position horizontal bar (19) and flush bolt (3) in side panel frame (4).
3. Install clevis (26) on horizontal bar (19).
4. Install clevis (21) on flush bolt (3).
5. Install retainer (30) on side panel frame (4) with two screws (31).
6. Install latch (23) on side panel frame (4) with four new lockwashers (10) and nuts (11).
7. Install clevises (26) and (21) on latch (23) with two pins (20) and new cotter pins (22).
8. Install new O-ring (6), shank (5), new gasket (9), and plate (8) on latch (23) with two screws (7).
9. Install washer (24) and nut (25) on shank (5) and latch (23).

NOTE
- Insulate entire structure with fibrous glass felt insulation.
- Assistant will help with step 10.

10. Install panel (16) and moldings (27), (28), (29), (12), (17), and (18) on side panel frame (4) with thirty-three screws (13).
11. Install plate (14) on panel (16) with eight screws (15).

NOTE
Repeat steps 2 through 11 for installation of each front latch assembly.

5-75. INTERIOR SIDE PANELS AND LATCHES REPLACEMENT (Contd)

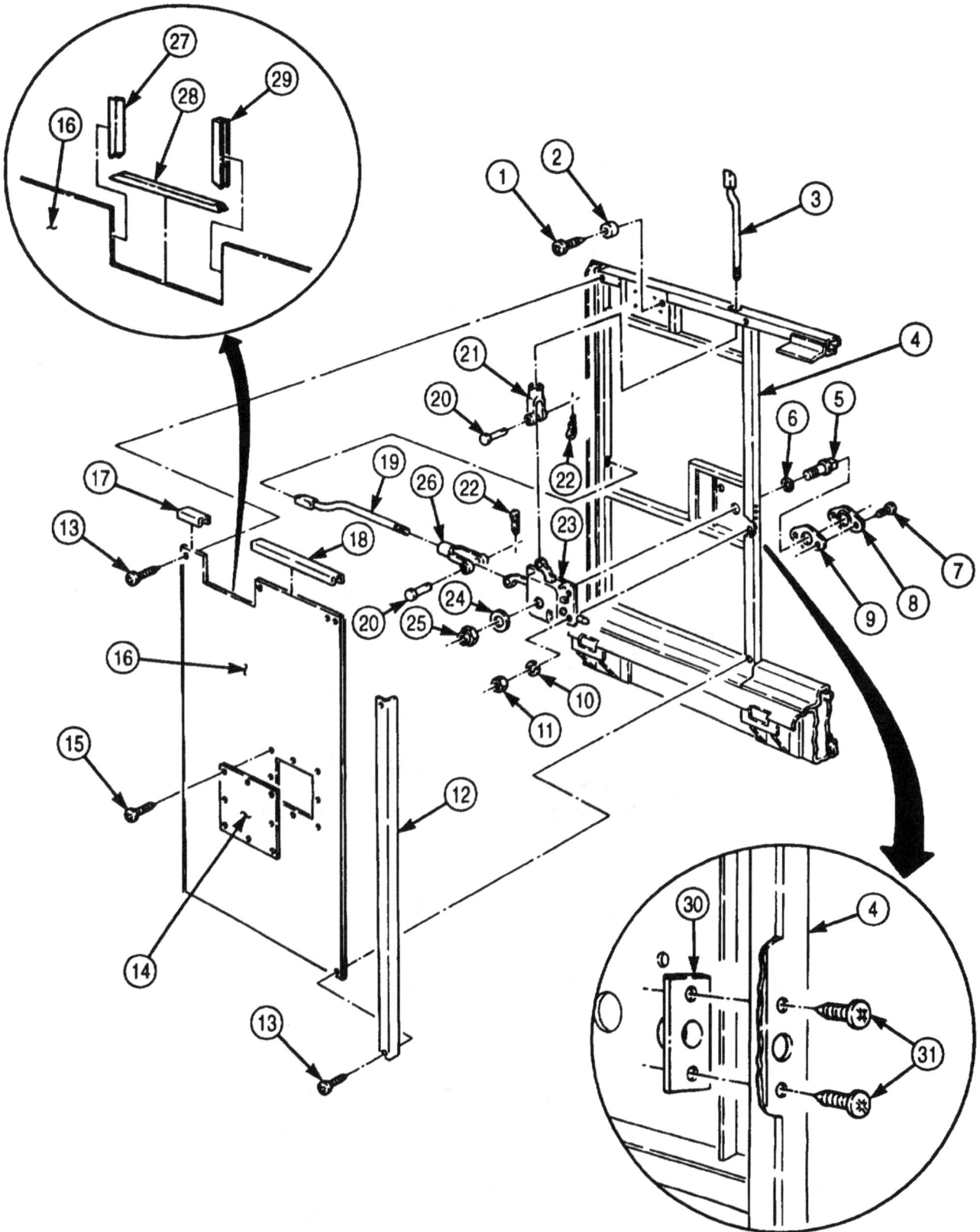

5-75. INTERIOR SIDE PANELS AND LATCHES REPLACEMENT (Contd)

12. Install moldings (6), (4), and (25) on panel (5).

NOTE
- Assistants will help with steps 13 and 14.
- Insulate entire structure with fibrous glass felt insulation.

13. Install panel (5) and molding (14) on side panel frame (1) with fifty-eight screws (9).
14. Install panel (13) on side panel frame (1) with thirty-three screws (10).
15. Install molding (24) on panels (5) and (13) with eleven screws (21).
16. Install plate (8) on panel (5) with eight screws (7).
17. Install junction box (23) on panel (5) with four screws (22).
18. Install four hangers (20) on panel (5) with eight screws (19).
19. Install four wood spacers (26) on panel (5) with eight screws (15).
20. Install four new rubber bumpers (18) on panel (5) with four washers (17) and screws (16).
21. Install two clips (12) on panel (13) with four screws (11).
22. Install two new grommets (3) on side panel frame (1) with two screws (2).

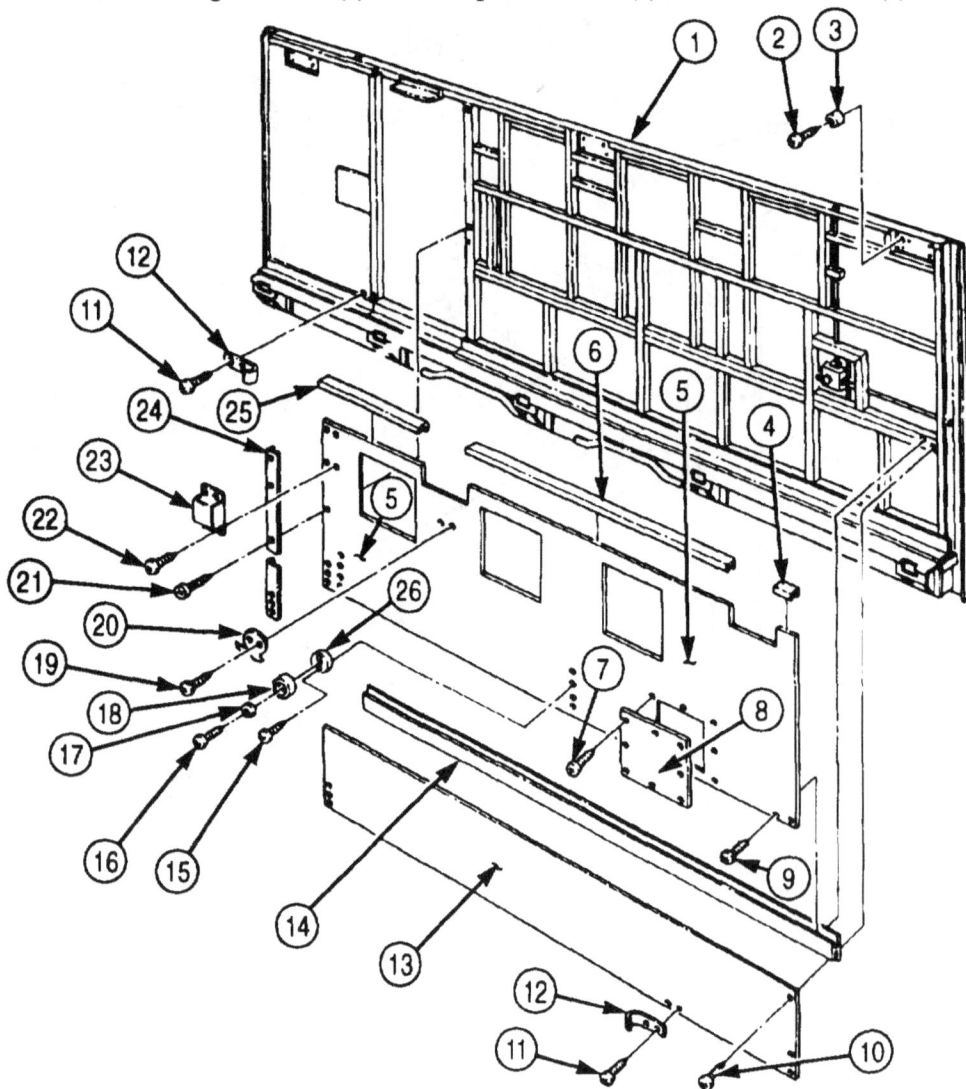

5-75. INTERIOR SIDE PANELS AND LATCHES REPLACEMENT (Contd)

23.   Install two spacer plates (39) and striker (38) on side panel frame (1) with two screws (37).

24.   Install door check assembly (34) on door angle (40) with two new lockwashers (36), screws (35), and nuts (30).

25.   Install striker (32) on side panel frame (1) with two screws (33).

26.   Install three clips (29) on side panel frame (1) with six screws (31).

27.   Install three hooks (28) on side panel frame (1) with six screws (27).

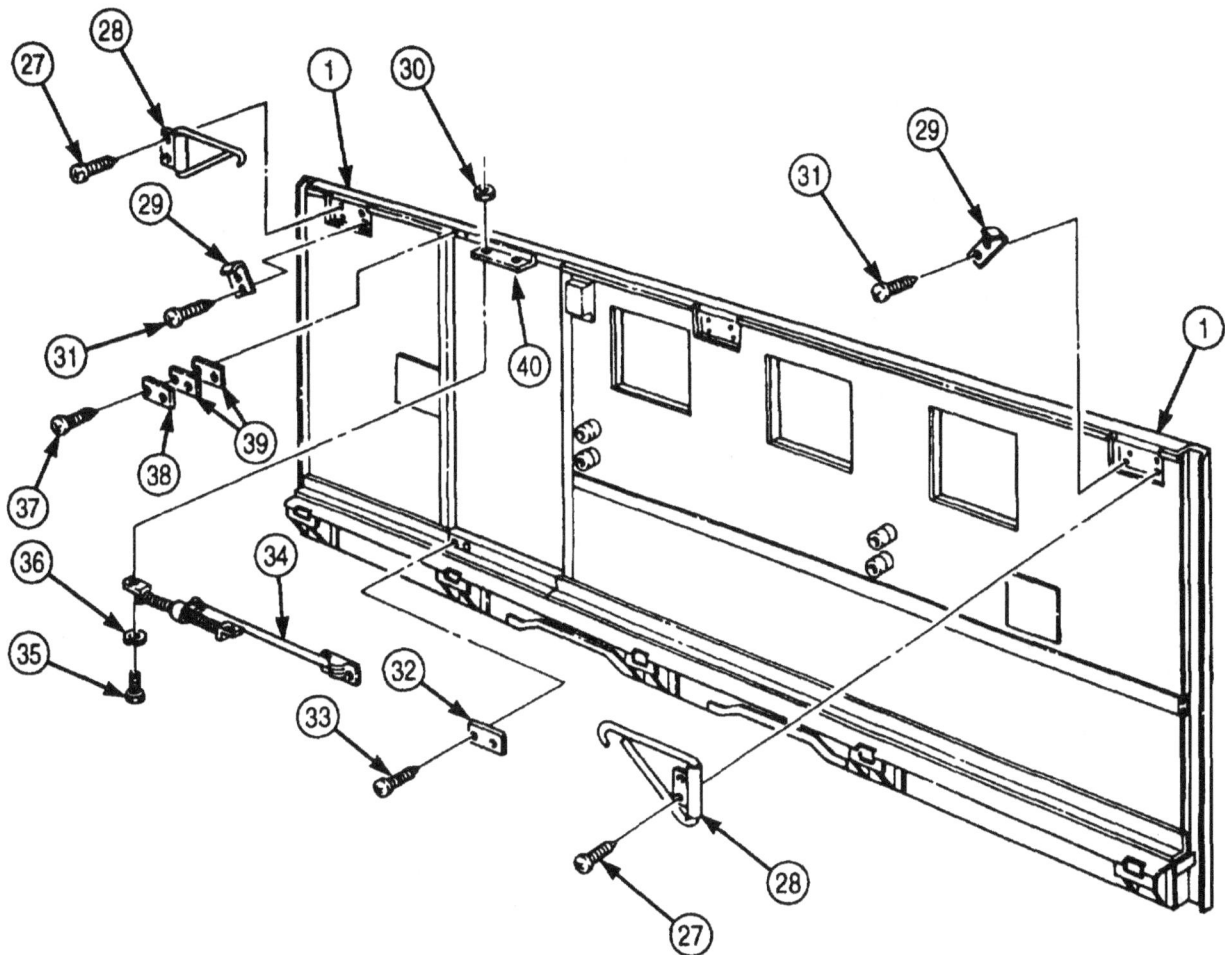

FOLLOW-ON TASKS:• Install side doors (para. 5-70).
  • Install hinged roof (para. 5-77).
  • Install hinged floor (para. 5-72).
  • Install hinged end panels (para. 4-155).
  • Retract van body sides (TM 9-2320-272-10).

5-76. CEILING AND FRAME MAINTENANCE

THIS TASK COVERS:

a. Removal
b. Cleaning and Inspection

c. Installation

INITIAL SETUP:

APPLICABLE MODELS
M934/A1/A2

TOOLS
General mechanic's tool kit (Appendix E, Item 1)
Rivet gun

MATERIALS/PARTS
Six lockwashers (Appendix D, Item 392)
One-hundred sixteen rivets (Appendix D, Item 552)
One-hundred forty-four rivets (Appendix D, Item 553)
Eighty-eight rivets (Appendix D, Item 551)
Two-hundred eight rivets (Appendix D, Item 542)
Seal (Appendix D, Item 619)
Seal (Appendix D, Item 620)
Adhesive (Appendix C, Item 3)
Primer (Appendix C, Item 57)

REFERENCES (TM)
TM 9-2320-272-10
TM 9-2320-272-24P

EQUIPMENT CONDITION
- Parking brake set (TM 9-2320-272-10).
- Van body sides fully expanded and secured (TM 9-2320-272-10).
- Ceiling air ducts removed (para. 5-81).
- Ceiling filler and side panels removed (para. 5-79).
- Ceiling rear cover removed (para. 5-80).
- Rear doors removed (para. 4-154).
- Ceiling transition removed (para. 5-82).

a. Removal

NOTE
Assistant will help with steps 1 through 4.

1. Remove one-hundred two rivets (6), ceiling panel (5), and two liner strips (1) from roof frame (2). Discard rivets (6).

2. Remove ninety-three rivets (4), ceiling panel (3), and two liner strips (1) from roof frame (2). Discard rivets (4).

5-76. CEILING AND FRAME MAINTENANCE (Contd)

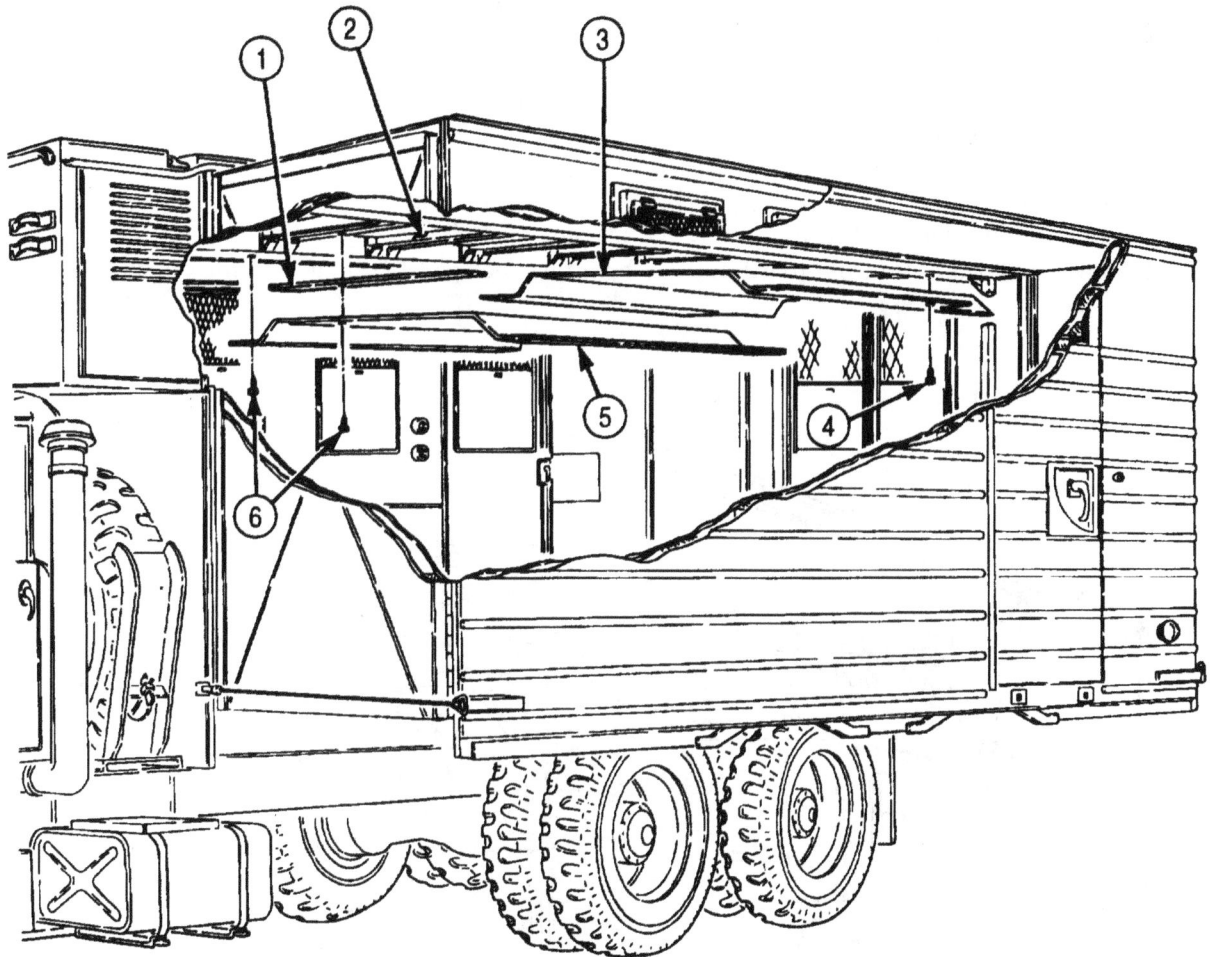

5-76. CEILING AND FRAME MAINTENANCE (Contd)

3. Remove ninety-two rivets (3), ceiling panel (4), and two liner strips (2) from roof frame (1). Discard rivets (3).

4. Remove seventy-six rivets (5), ceiling panel (6), and three liner strips (2) from roof frame (1). Discard rivets (5).

NOTE
Left and right side moldings are removed the same way. This procedure covers the right side.

5. Remove fifty-nine screws (11), molding (10), molding (12), and seal (9) from roof frame (1). Discard seal (9).

6. Remove six nuts (13), screws (7), lockwashers (18), and two lifting brackets (17) from roof frame (1). Discard lockwashers (18).

NOTE
- Perform steps 7 and 8 only if roof is damaged and is to be removed (task b).
- Assistant will help with steps 7 and 8.

7. Remove one-hundred twenty-six rivets (15), eighteen rivets (14), and one-hundred sixteen rivets (8) connecting roof panel (16) to roof frame (1). Discard rivets (8), (14), and (15).

8. Remove roof panel (16) from roof frame (1) by lifting roof panel (16) and placing on supports.

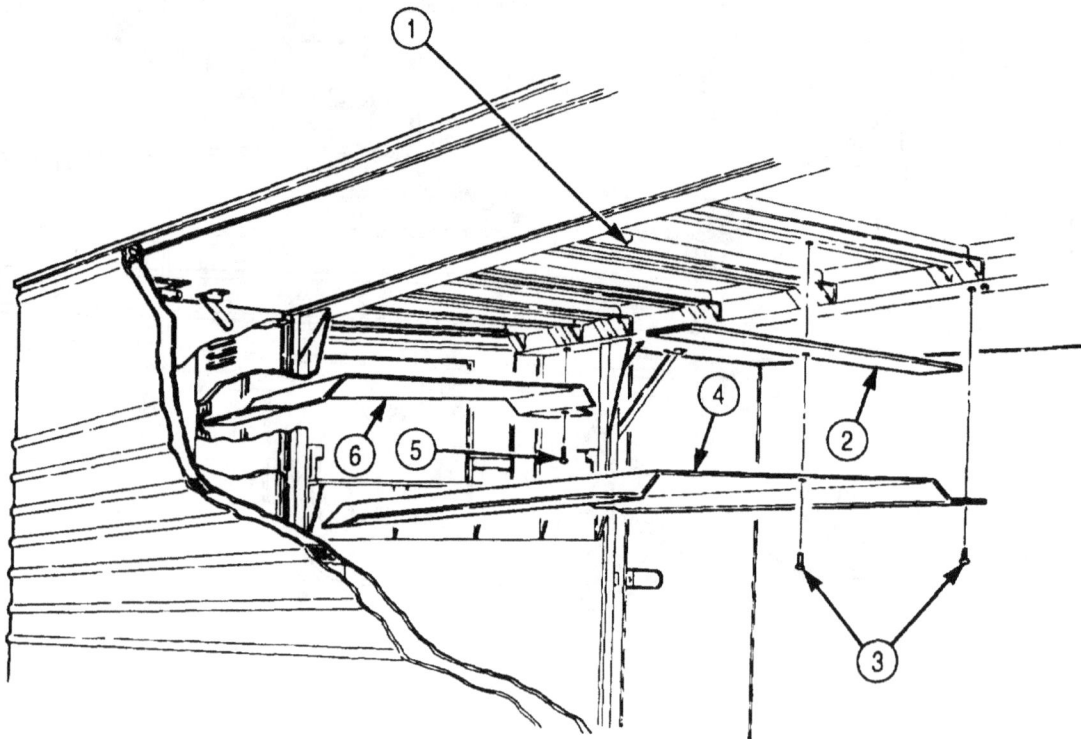

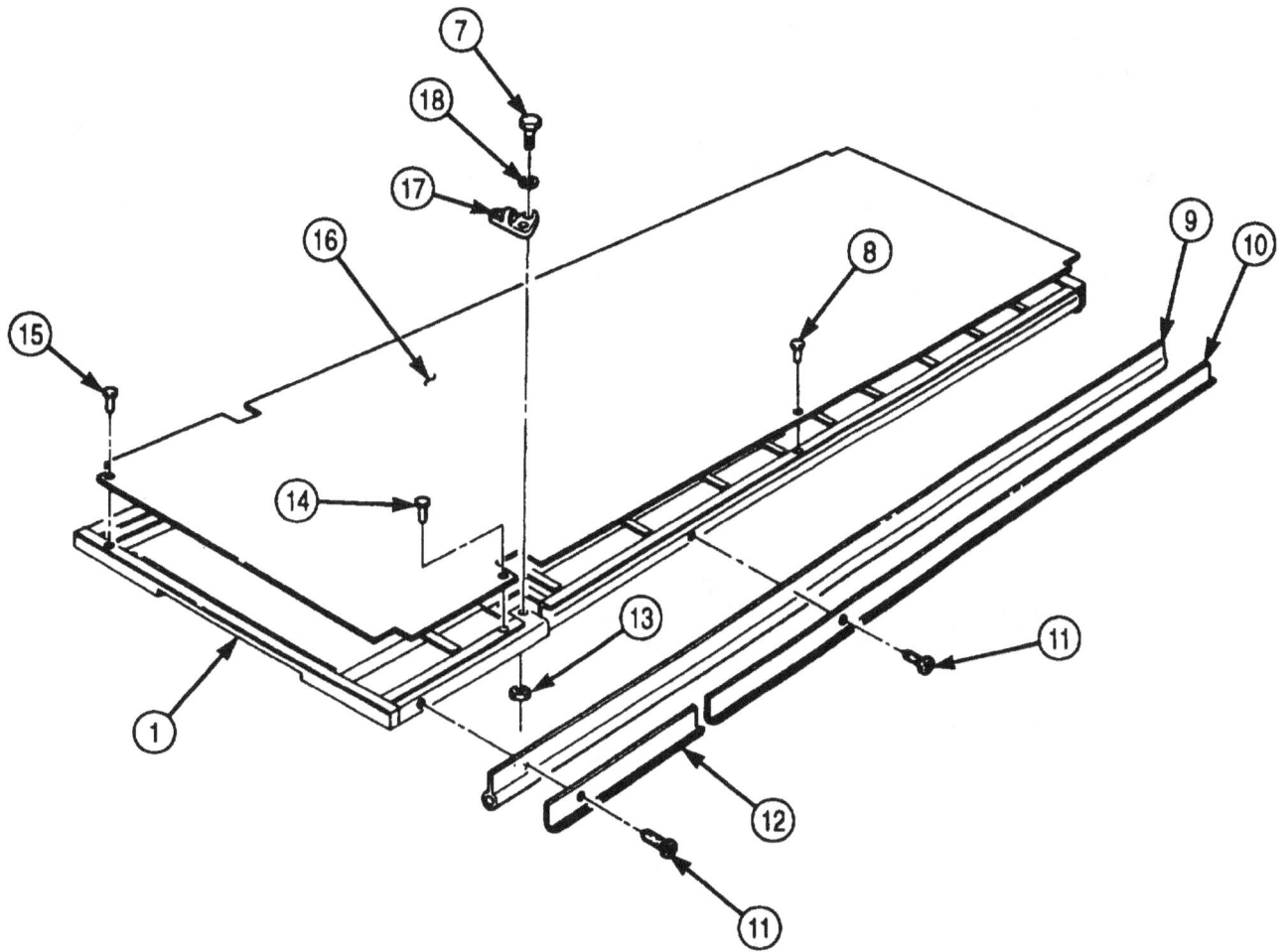

5-76. CEILING AND FRAME MAINTENANCE (Contd)

9. Remove twenty-three screws (2) and moldings (1), (3), and (4) from van body (11).
10. Remove three screws (6) and angle bracket (7) from retainer (8)
11. Remove fifteen screws (5), seal (9), retainer (8), and woodblock (10) from van body (11). Discard seal (9).

b. Cleaning and Inspection

1. For general cleaning instructions, refer to para. 2-14.
2. Inspect moldings (1), (3), and (4) for bends and breaks. If bent or broken, replace molding (l), (3), or (4).
3. Inspect angle bracket (7) and retainer (8) for cracks and breaks. If cracked or broken, replace.
4. Inspect woodblock (10) for breaks. If broken, replace woodblock (10).
5. Inspect moldings (13) and (14) for bends, breaks, and warpage. If bent, broken, or warped, replace molding (13) or (14).
6. Inspect lifting bracket (12) for breaks and cracks. If broken or cracked, replace lifting bracket (12).
7. Inspect roof panel (16) for tears and punctures. If torn or punctured, replace roof panel (16).
8. Inspect roof frame (15) for breaks, warpage, and broken bows. If roof frame (15) is damaged, replace entire roof assembly.
9. Inspect ceiling panels (17), (18), (19), and (20) for cracks and punctures. If cracked or punctured, replace.

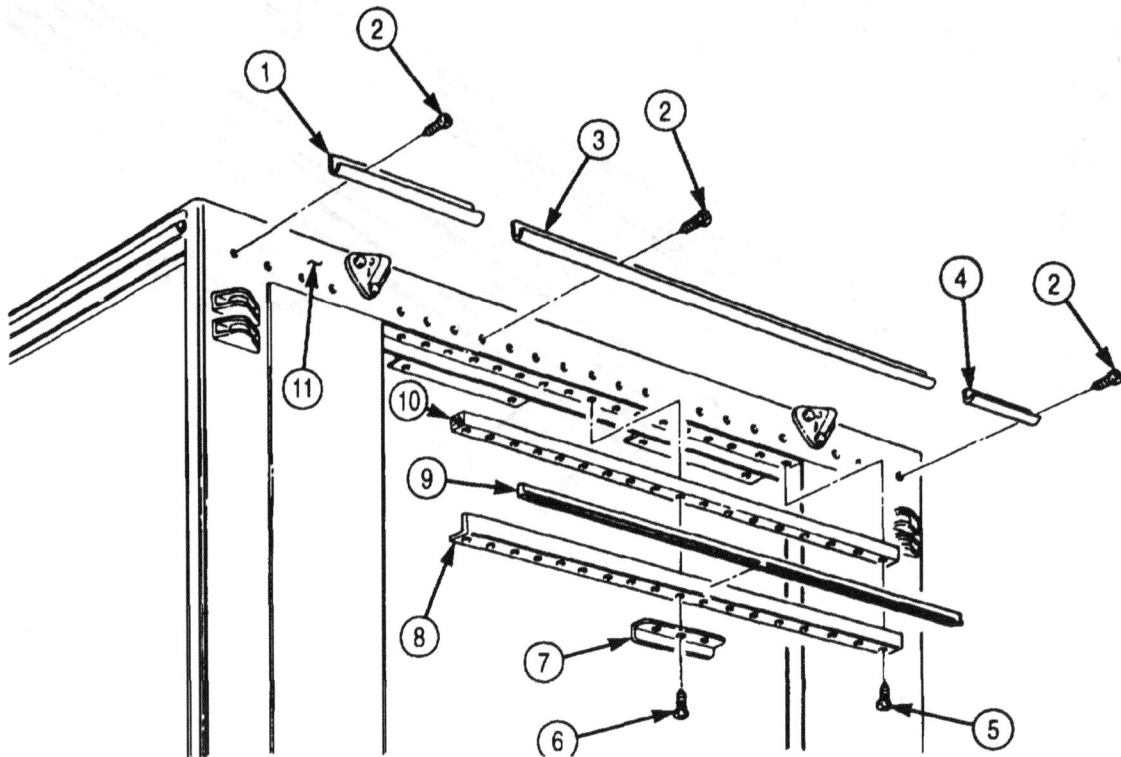

5-76. CEILING AND FRAME MAINTENANCE (Contd)

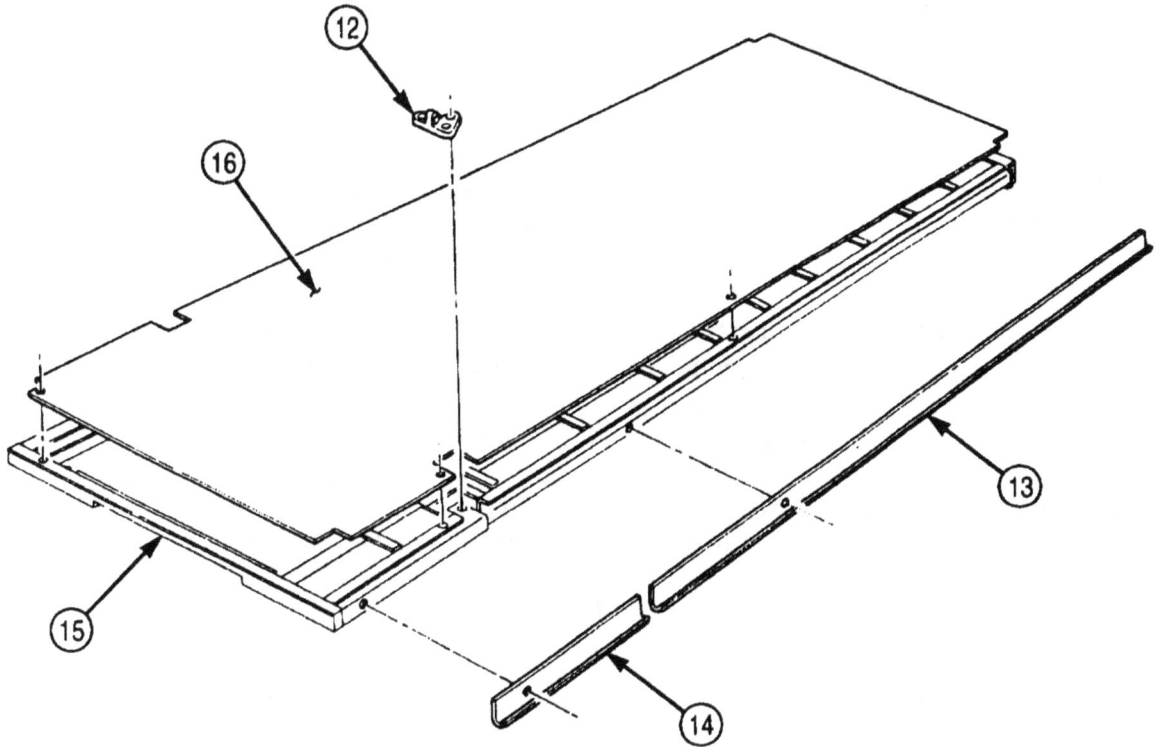

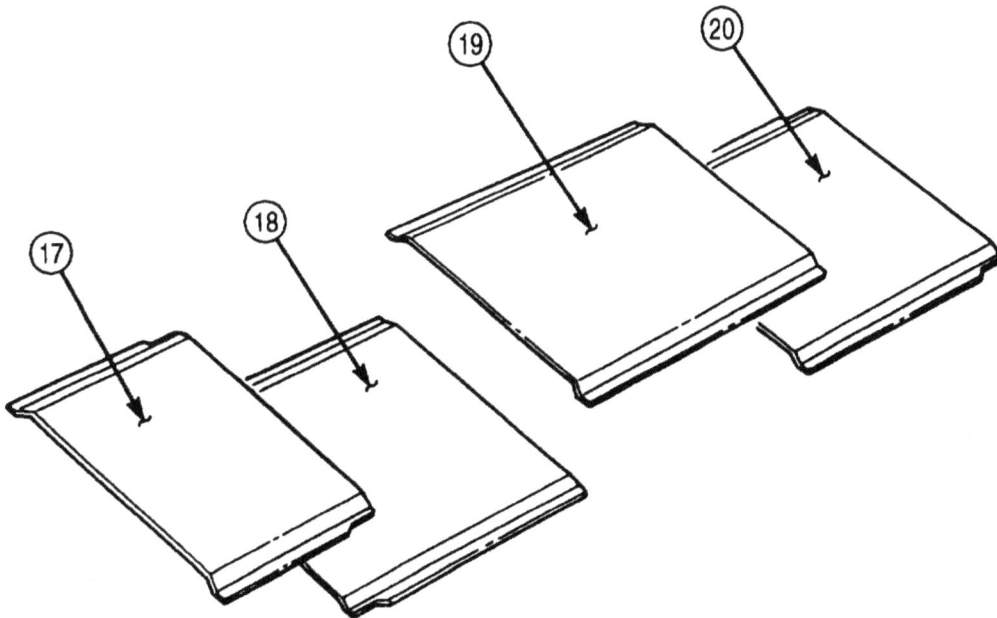

5-76. CEILING AND FRAME MAINTENANCE (Contd)

c. Installation

NOTE
- Apply adhesive prior to installation.
- Insulate areas of dissimilar metal-to-metal contact with zinc chromate primer.

1. Install woodblock (10), new seal (9), and retainer (8) on van body (11) with fifteen screws (5). Leave three center holes of retainer open for angle bracket (7) installation.

2. Install angle bracket (7) on retainer (8) with three screws (6).

3. Install moldings (3), (4), and (1) on van body (11) with twenty-three screws (2).

NOTE
- Perform steps 4 and 5 only if roof panel was removed.
- Insulate entire structure with fibrous glass felt insulation.
- Assistant will help with steps 4 and 5.

4. Lift roof panel (22) from supports and position on roof frame (20).

5. Install roof panel (22) on roof frame (20) with one-hundred sixteen new rivets (13), eighteen new rivets (19), and one-hundred twenty-six new rivets (21).

6. Install two lifting brackets (23) on roof frame (20) with six new lockwashers (24), screws (12), and nuts (18).

NOTE
Left and right side moldings are installed the same way. This procedure covers the right side.

7. Install new seal (14), molding (15), and molding (17) on roof frame (20) with fifty-nine screws (16).

5-76. CEILING AND FRAME MAINTENANCE (Contd)

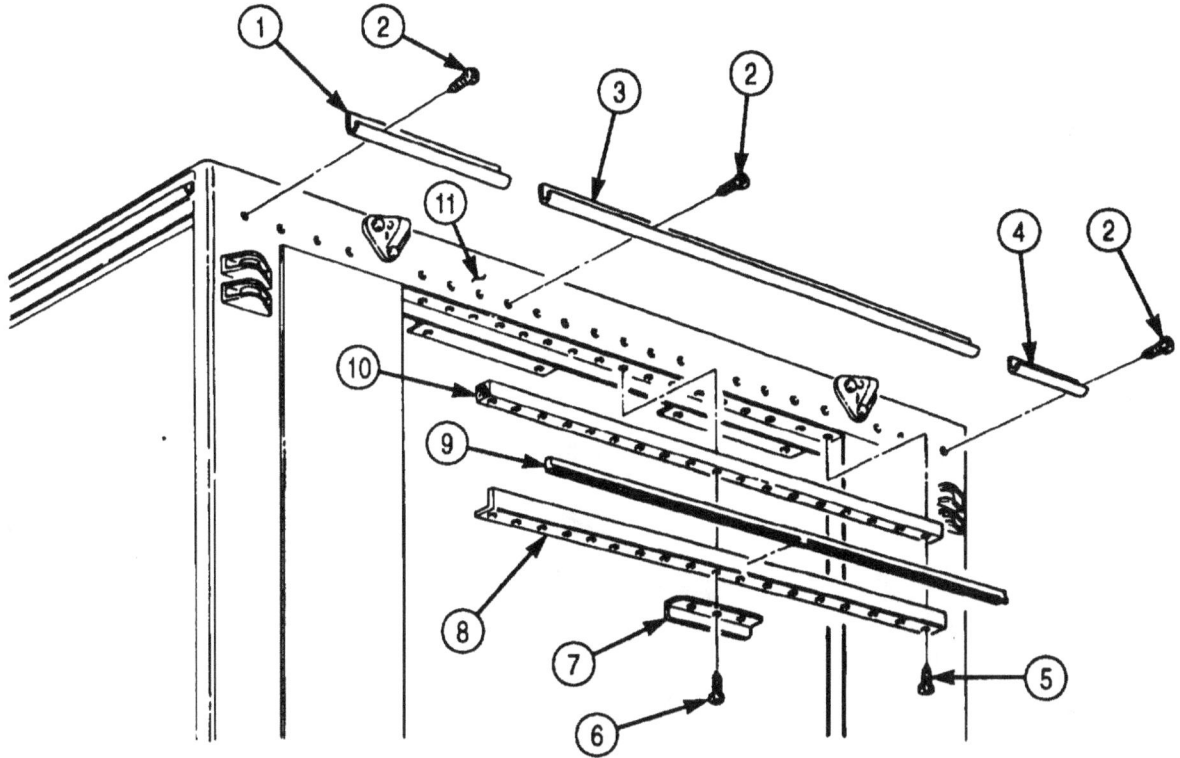

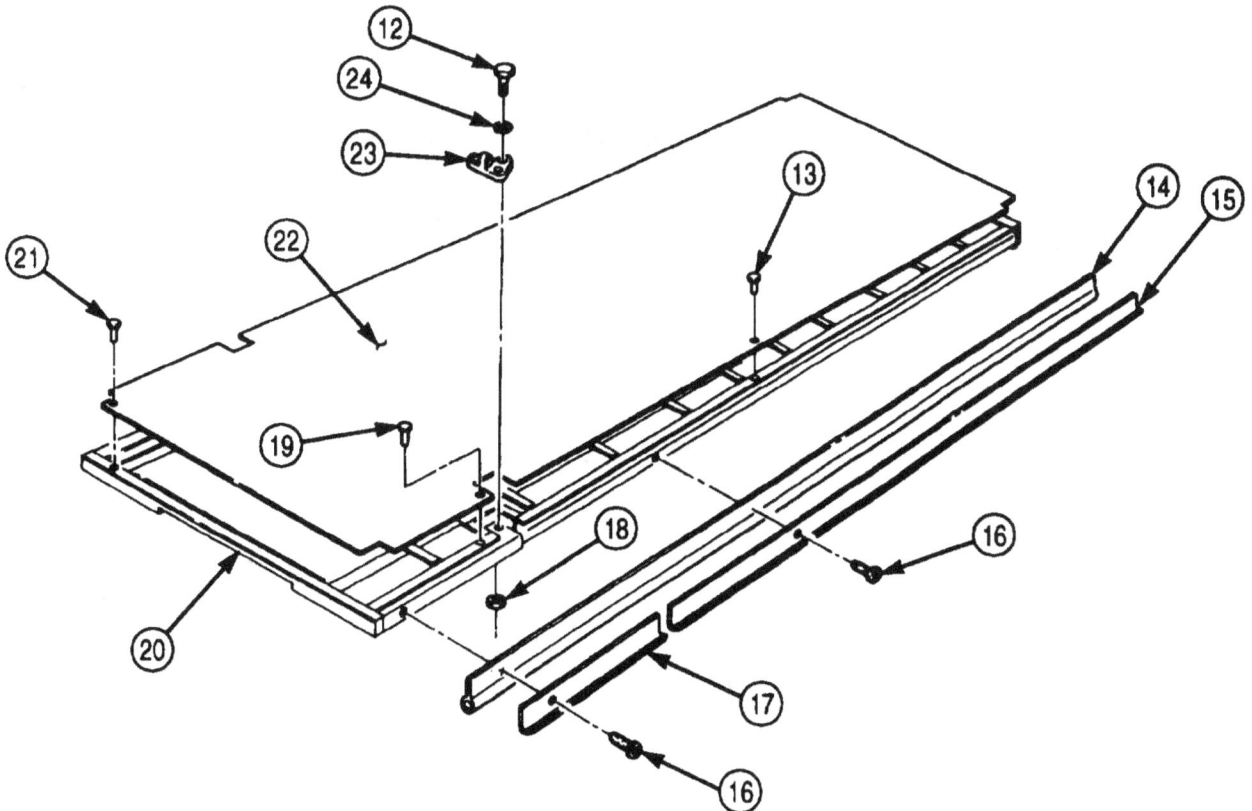

5-76. CEILING AND FRAME MAINTENANCE (Contd)

NOTE
Assistant will help with steps 8 through 11.

8. Install ceiling panel (6) and three liner strips (2) on roof frame (1) with seventy-six new rivets (5).

9. Install ceiling panel (4) and two liner strips (2) on roof frame (1) with ninety-two new rivets (3).

10. Install ceiling panel (7) and two liner strips (2) on roof frame (1) with ninety-three new rivets (8).

11. Install ceiling panel (9) and two liner strips (2) on roof frame (1) with one-hundred two new rivets (10).

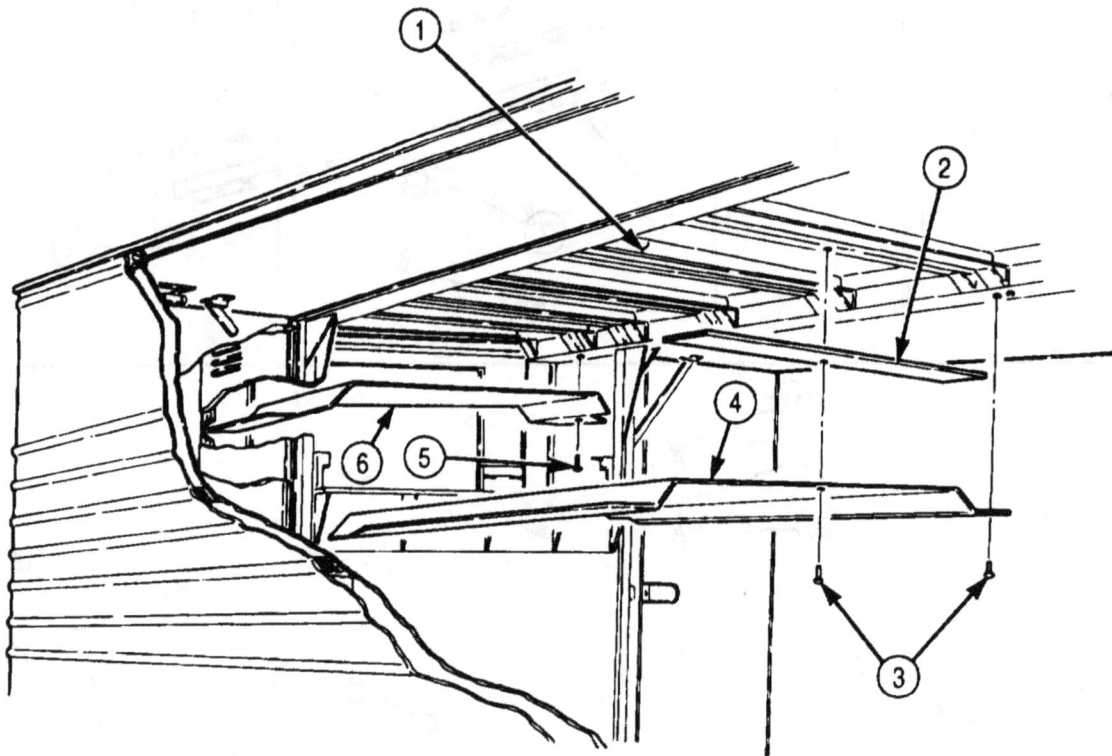

5-76. CEILING AND FRAME MAINTENANCE (Contd)

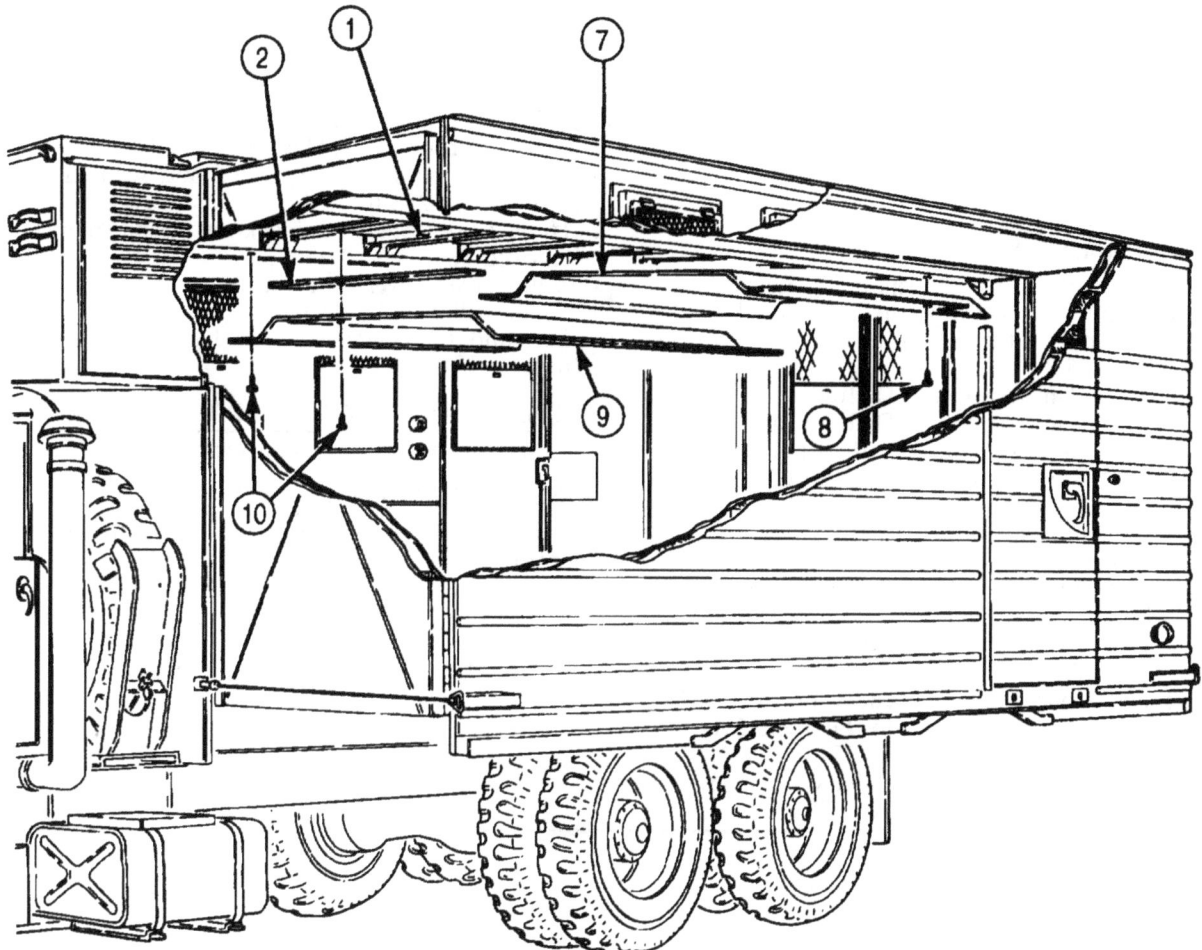

FOLLOW-ON TASKS: Install rear doors (para. 4-154).
- Install ceiling transition (para. 5-82).
- Install ceiling rear cover (para. 5-80).
- Install ceiling filler and side panels (para. 5-79).
- Install ceiling air ducts (para. 5-81).
- Retract van body sides (TM 9-2320-272-10).

5-77. HINGED ROOF MAINTENANCE

THIS TASK COVERS:

a. Removal
b. Disassembly
c. Cleaning and Inspection

d. Assembly
e. Installation

INITIAL SETUP:

APPLICABLE MODELS
M93/A1/A2

TOOLS
General mechanic's tool kit (Appendix E, Item 1)
Lifting device
Rivet gun
Four chains

MATERIALS/PARTS
Three cotter pins (Appendix D, Item 56)
Fifty-two rivets (Appendix D, Item 555)
One-hundred fifty-seven rivets (Appendix D, Item 553)
Seal, 206 in. (Appendix D, Item 633)
Seal, 206 in. (Appendix D, Item 618)
Seal, 35 in. (Appendix D, Item 621)
Seal, 206 in. (Appendix D, Item 622)
Seal, 35 in. (Appendix D, Item 606)
Two cushion pads (left side) (Appendix D, Item 90)
Two cushion pads (Right side) (Appendix D, Item 91)
Gasket (Appendix D, Item 232)
O-ring (Appendix D, Item 464)
Twenty-four rivets (Appendix D, Item 554)
Four grommets (Appendix D, Item 251)
Primer (Appendix C, Item 57)
Sealing compound (Appendix C, Item 67)
Adhesive (Appendix C, Item 7)

REFERENCES (TM)
TM 9-2320-272-10
TM 9-2320-272-24P

EQUIPMENT CONDITION
- Parking brake set (TM 9-2320-272-10).
- Van body sides fully expanded and secured (TM 9-2320-272-10).
- Hinged roof-operated blackout circuit plungers removed (para. 3-377).
- Ceiling rear cover removed (para. 5-80).
- Ceiling filler and side panels removed (para. 5-79).
- Left and right side blackout harness removed (para. 5-98).

SPECIAL ENVIRONMENTAL CONDITIONS
Vehicle must be on a level surface.

GENERAL SAFETY INSTRUCTIONS
All personnel must stand clear during lifting operations.

NOTE

Left and right hinged roofs are maintained the same way. This procedure covers right side.

a. Removal

1. Attach four chains to lifting device and hinged roof (1). Remove slack from chains.
2. Remove four screws (3), two angle brackets (2), and holding rods (4) from hinged roof (1).

5-77. HINGED ROOF MAINTENANCE (Contd)

WARNING
All personnel must stand clear during lifting operations. A
swinging or shifting load may cause injury to personnel.

NOTE
Assistant will help with step 3.

3. Remove sixty-nine screws (5) from van body (6) and hinged roof (1). Lift hinged roof (1) and position on supports.

4. Remove chains from lifting device and hinged roof (1).

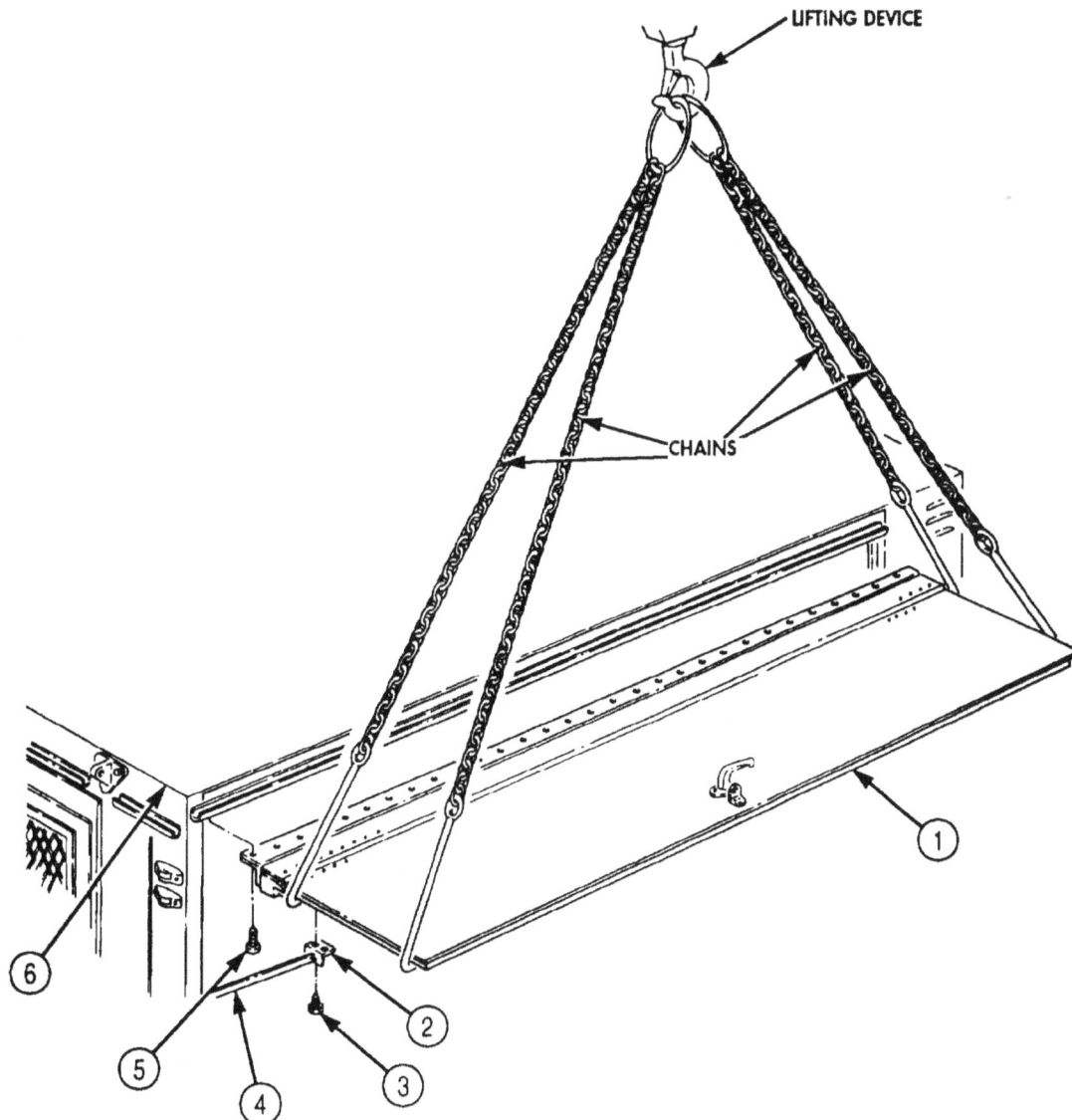

5-77. HINGED ROOF MAINTENANCE (Contd)

b. Disassembly

1. Remove twelve screws (27) and three clamps (28) from hinged roof frame (1).
2. Remove six screws (25) and three holder assemblies (26) from hinged roof frame (1).
3. Remove thirty screws (5) and moldings (4), (6), and (7) from panels (2) and (18).
4. Remove four screws (11) and handle (12) from panel (2).
5. Remove ten screws (13) and cover (14) from panel (2).
6. Remove six screws (9) and plate (8) from panel (2).
7. Remove six screws (19) and three bars (20) from fillers (22).
8. Remove six screws (21) and three fillers (22) from panels (2) and (8).
9. Remove two screws (16) and molding (15) from panels (2) and (18).
10. Remove moldings (3), (23), and (24) from panels (2) and (18).

NOTE

Assistant will help with steps 11 and 12.

11. Remove fifty-five screws (10) and panel (2) from hinged roof frame (1).
12. Remove forty-one screws (17) and panel (18) from hinged roof frame (1).

5-77. HINGED ROOF MAINTENANCE (Contd)

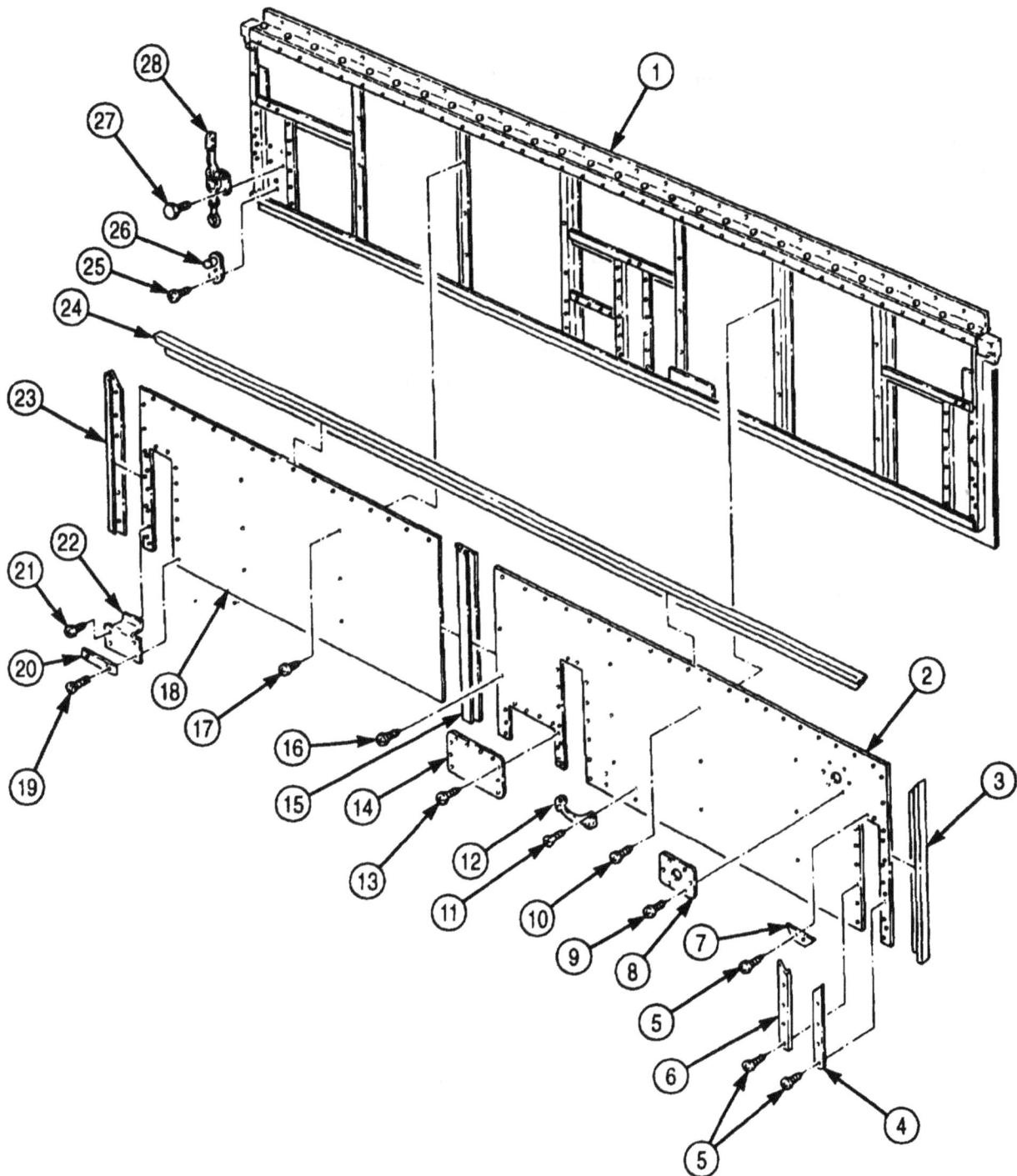

5-77. HINGED ROOF MAINTENANCE (Contd)

13. Remove sixty-eight screws (26), hinge (14), and seal (13) from hinged roof frame (11). Discard seal (13).

14. Remove twenty-four rivets (29), retainer (30), and seal (28) from hinged roof frame (11). Discard seal (28) and rivets (29).

15. Remove cushion pads (12) and (27) from hinged roof frame (11). Discard cushion pads (12) and (27).

16. Remove fifty-two rivets (32), retainer (34), and seal (35) from hinged roof frame (11). Discard rivets (32) and seal (35).

17. Remove seals (31) and (33) from hinged roof frame (11). Discard seals (31) and (33).

18. Remove padlock and chain (2) and chain hook (1) from angle bracket (37).

19. Remove three screws (38) and angle bracket (37) from skin (7).

20. Remove cotter pin (24), nut (23), and washer (22) from handle (3). Discard cotter pin (24).

21. Remove two screws (4), handle (3), plate (5), gasket (6), and O-ring (36) from skin (7). Discard O-ring (36) and gasket (6).

22. Remove two cotter pins (19), pins (16), and clevises (18) from latch (17). Discard cotter pins (19).

23. Remove four nuts (21), washers (20), and latch (17) from hinged roof frame (11).

24. Remove clevises (18) from bars (15) and (25).

25. Remove bars (15) and (25) from hinged roof frame (11).

26. Remove four screws (9) and grommets (10) from hinged roof frame (11). Discard grommets (10).

NOTE

- Perform step 27 only if skin is to be removed (task c).
- Assistant will help with step 27.

27. Remove one-hundred fifty-seven rivets (8) and skin (7) from hinged roof frame (11). Discard rivets (8).

5-77. HINGED ROOF MAINTENANCE (Contd)

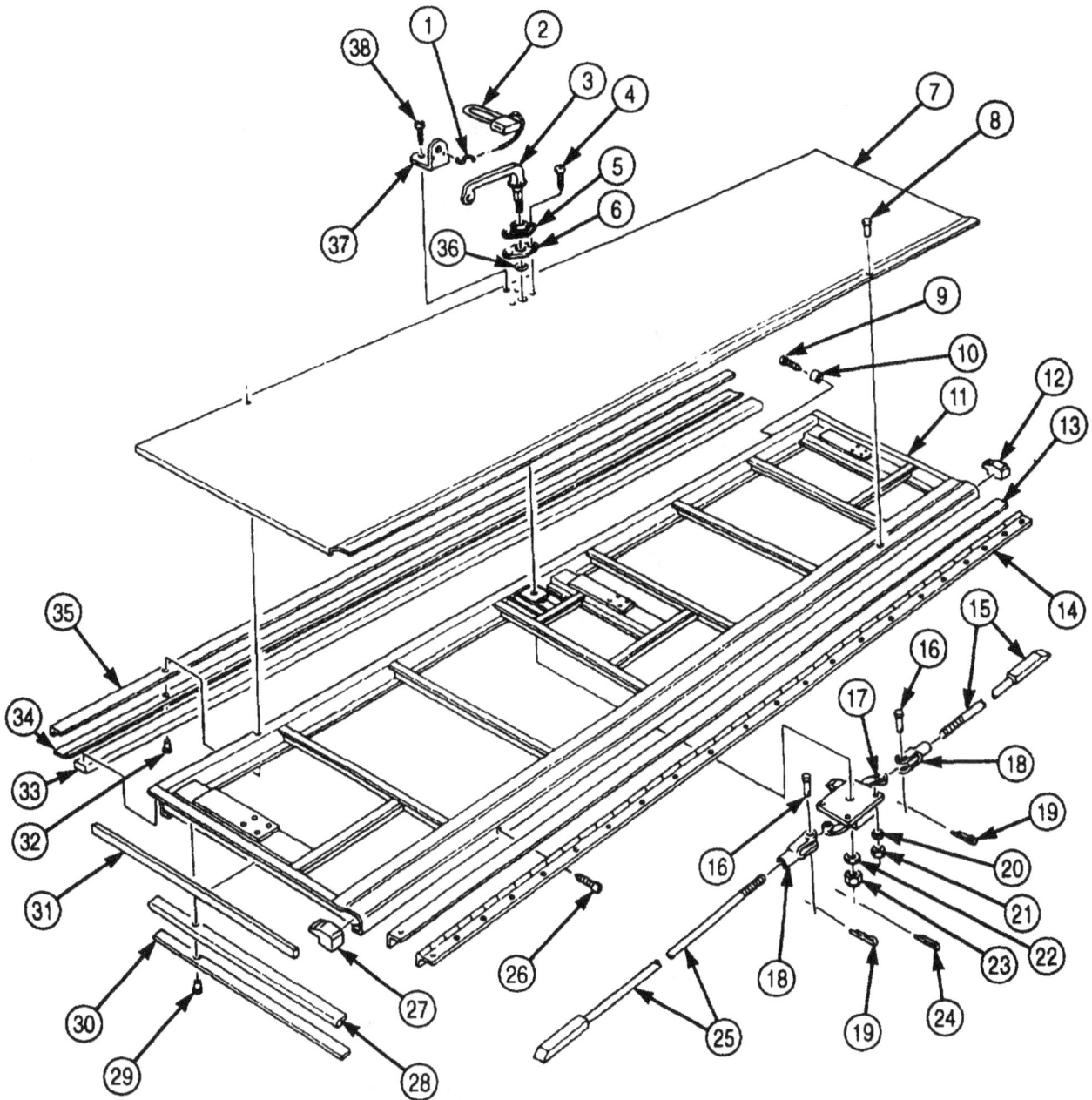

5-77. HINGED ROOF MAINTENANCE (Contd)

c. Cleaning and Inspection

1. For general cleaning instructions, refer to para. 2-14.

2. Inspect panels (2) and (11) for cracks and breaks. Replace panels (2) and (11) if cracked or broken.

3. Inspect moldings (3), (4), (5), (6), (10), (14), and (15) for warpage and breaks. Replace moldings (3), (4), (5), (6), (10), (14), and (15) if moldings are warped or broken.

4. Inspect plate (7), handle (8), cover (9), three bars (12), and fillers (13) for bends and breaks. Replace plate (7), handle (8), cover (9), bars (12), and fillers (13) if bent or broken.

5. Inspect three clamps (17) and holder assemblies (16) for breaks and proper operation. Replace clamps (17) and holder assemblies (16) if damaged.

6. Inspect hinge (23) for breaks and proper operation. Replace hinge (23) if damaged.

7. Inspect retainers (28) and (29) for cracks and breaks. Replace retainer (28) or (29) if cracked or broken.

8. Inspect angle bracket (18), padlock and chain (19), chain hook (20), and handle (21) for breaks. Replace angle bracket (18), padlock and chain (19), chain hook (20), and handle (21) if broken.

9. Inspect bars (24) and (27), and two clevises (25) for bends and breaks. Replace bars (24), (27) and clevises (25) if bent or broken.

10. Inspect latch (26) for proper operation. Replace latch (26) if damaged.

11. Inspect skin (22) for tears and punctures. Replace skin (22) if torn or punctured.

12. Inspect frame (1) for breaks. Replace entire hinged roof if damaged.

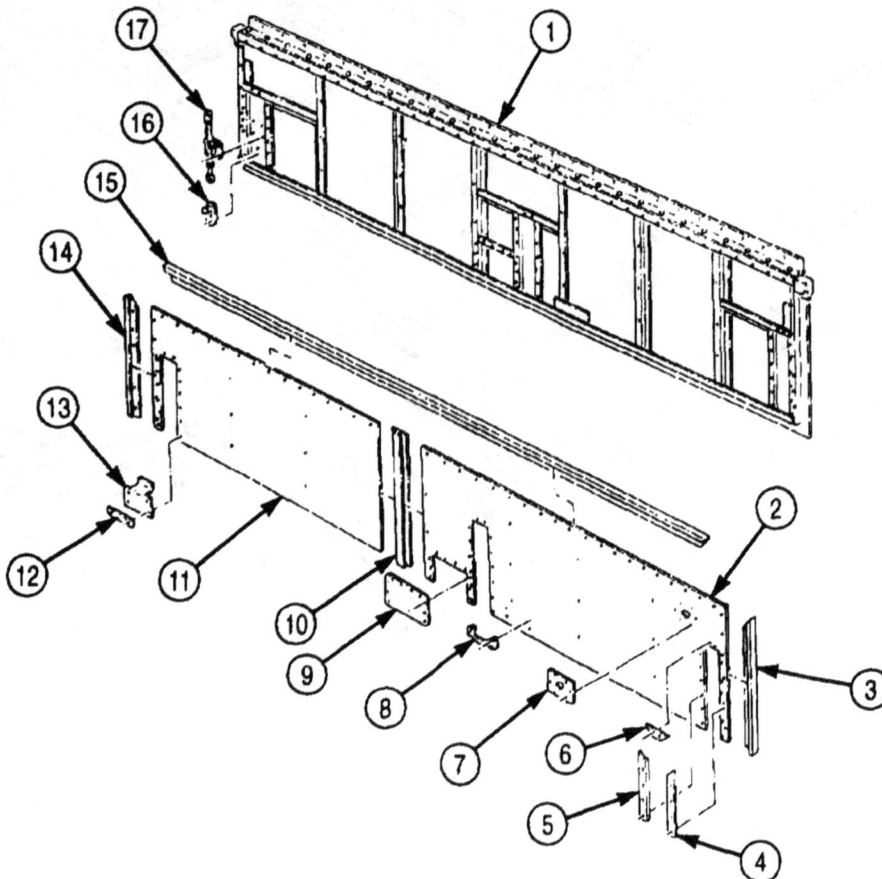

5-77. HINGED ROOF MAINTENANCE (Contd)

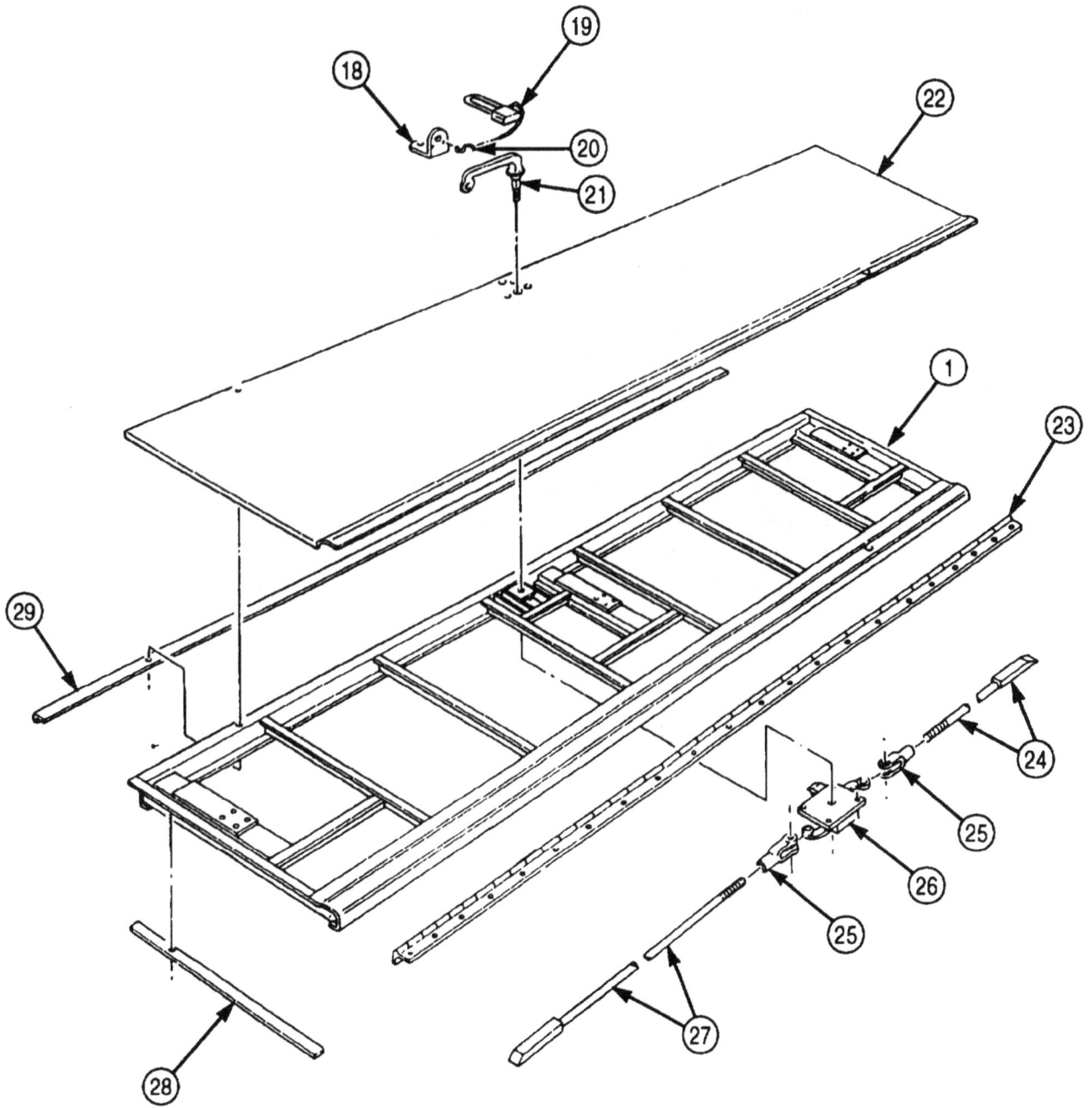

5-77. HINGED ROOF MAINTENANCE (Contd)

d. Assembly

NOTE

- Insulate areas of dissimilar metal-to-metal contact with zinc chromate primer.
- Apply sealing compound to exterior joints prior to installation.
- Apply adhesive to both rubber and metal surfaces.
- Perform step 1 only if skin was removed.
- Assistant will help with step 1.

1. Install skin (7) on hinged roof frame (11) with one-hundred fifty-seven new rivets (8).
2. Install four new grommets (10) on hinged roof frame (11) with four screws (9).
3. Install clevises (18) on bars (15) and (25) and position in slots of hinged roof frame (11).
4. Install latch (17) on hinged roof frame (11) with four washers (20) and nuts (21).
5. Install two clevises (18) on latch (17) with two pins (16) and new cotter pins (19).
6. Install new O-ring (36), new gasket (6), plate (5), and handle (3) on skin (7) with two screws (4).
7. Install handle (3) on latch (17) with washer (22), nut (23), and new cotter pin (24).
8. Install angle bracket (37) on skin (7) with three screws (38).
9. Install chain hook (1) and padlock and chain (2) on angle bracket (37).
10. Install new seals (31) and (33) on hinged roof frame (11).
11. Install new seal (35) and retainer (34) on hinged roof frame (11) with fifty-two new rivets (32).
12. Install new cushion pads (12) and (27) on hinged roof frame (11).
13. Install new seal (28) and retainer (30) on hinged roof frame (11) with twenty-four new rivets (29).
14. Install new seal (13) and hinge (14) on hinged roof frame (11) with sixty-eight screws (26).

5-77. HINGED ROOF MAINTENANCE (Contd)

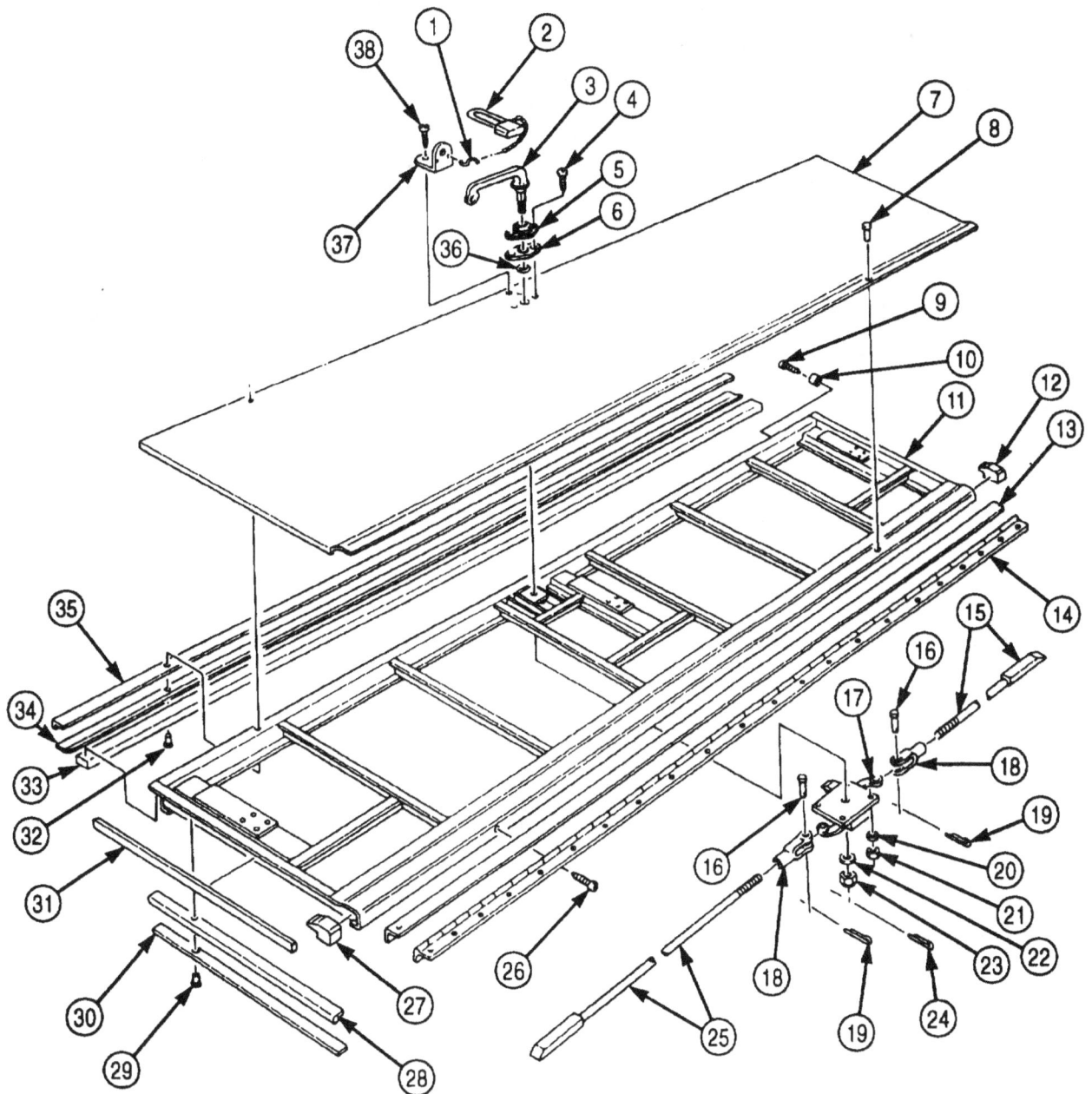

5-77. HINGED ROOF MAINTENANCE (Contd)

NOTE
- Insulate entire structure with fibrous glass felt insulation.
- Assistant will help with steps 15 and 16.

15. Install panel (18) on hinged roof frame (1) with forty-one screws (17).
16. Install panel (2) on hinged roof frame (1) with fifty-five screws (10).
17. Install molding (15) on panels (2) and (18) with two screws (16).
18. Install moldings (3), (23), and (24) on panels (2) and (18).
19. Install three fillers (22) on panels (2) and (18) with six screws (21).
20. Install three bars (20) on fillers (22) with six screws (19).
21. Install plate (8) on panel (2) with six screws (9).
22. Install cover (14) on panel (2) with ten screws (13).
23. Install handle (12) on panel (2) with four screws (11).
24. Install moldings (4), (6), and (7) on panels (2) and (18) with thirty screws (5).
25. Install three holder assemblies (26) on hinged roof frame (1) with six screws (25).
26. Install three clamps (28) on hinged roof frame (1) with twelve screws (27).

5-77. HINGED ROOF MAINTENANCE (Contd)

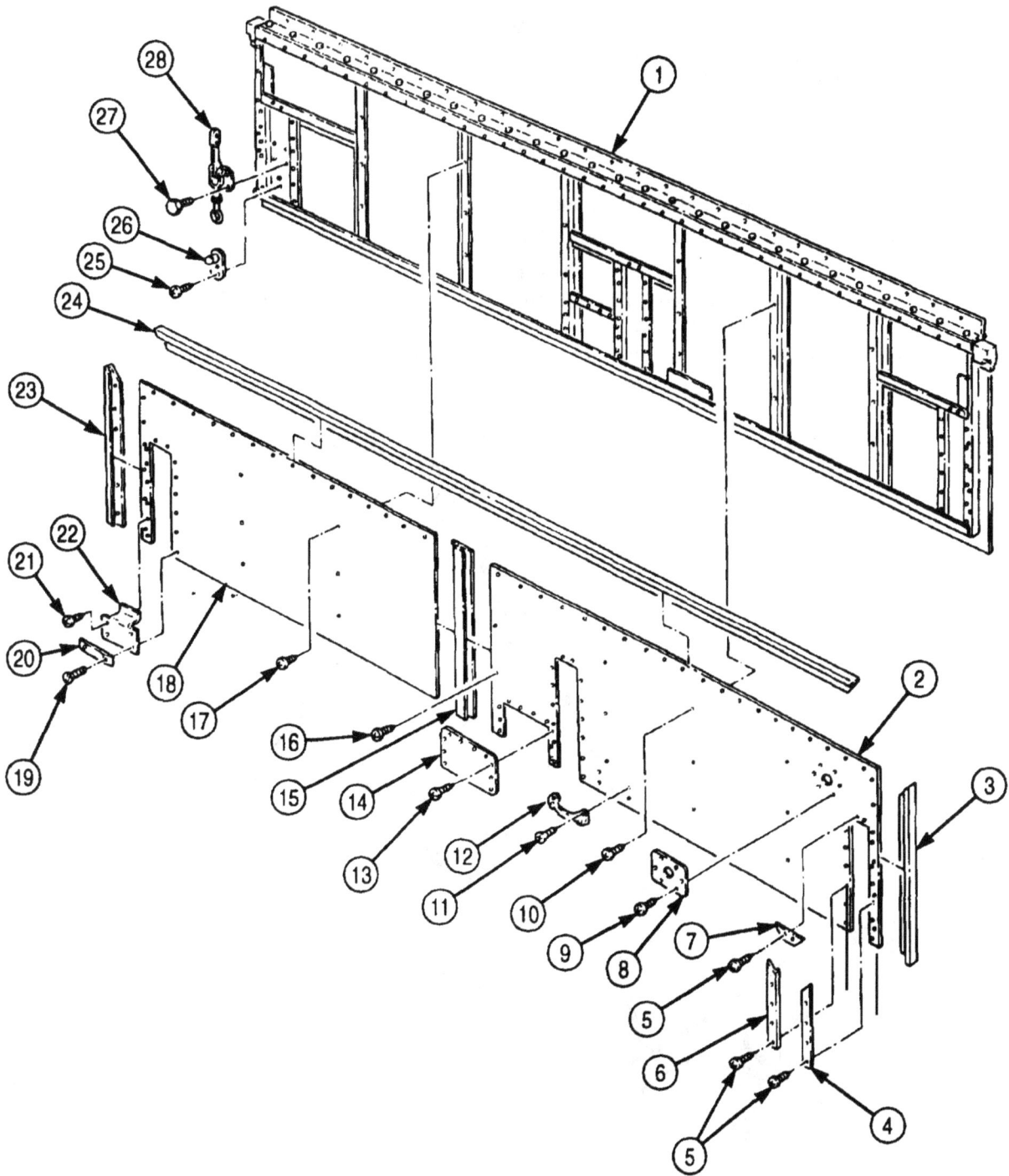

5-77. HINGED ROOF MAINTENANCE (Contd)

e. Installation

1. Attach four chains to lifting device and hinged roof (1).

WARNING

All personnel must stand clear during lifting operations. A swinging or shifting load may cause injury to personnel.

NOTE

Assistant will help with step 2.

2. Lift hinged roof (1) from supports and install on van body (6) with sixty-nine screws (5).
3. Install two holding rods (4) and angle brackets (2) on hinged roof (1) with four screws (3).
4. Remove chains from lifting device and hinged roof (1).

NOTE

Hinged roof-operated blackout circuit plungers must be installed before side blackout harnesses are installed in load center.

5. Install hinged roof-operated blackout circuit plungers (para. 3-377).

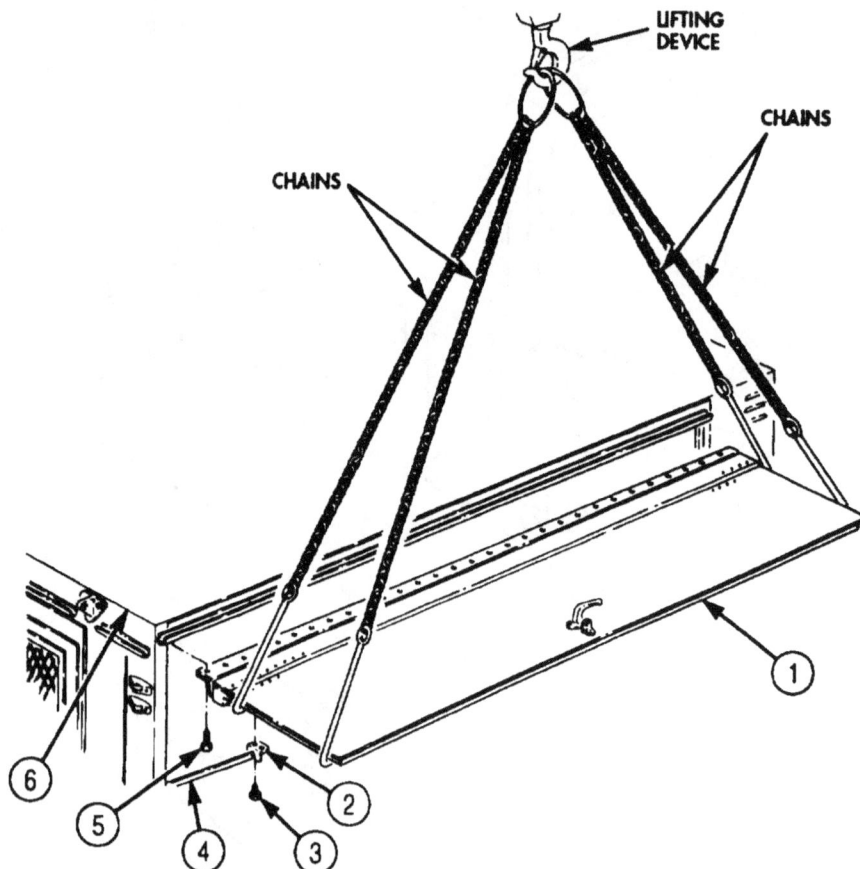

FOLLOW-ON TASKS: Install ceiling filler and side panels (para. 5-79).
- Install ceiling rear cover (para. 5-80).
- Install left and right side blackout harness (para. 5-98).
- Retract van body sides (TM 9-2320-272-10).

5-78. FRONT WALL REGISTERS REPLACEMENT

THIS TASK COVERS:

a. Removal

b. Installation

INITIAL SETUP:

APPLICABLE MODELS
M934/A1/A2

TOOLS
General mechanic's tool kit (Appendix E, Item 1)

REFERENCES (TM)
TM 9-2320-272-10
TM 9-2320-272-24P

EQUIPMENT CONDITION
Parking brake set (TM 9-2320-272-10).

NOTE
All four front wall registers are replaced the same way. This procedure covers one register.

a. Removal

Remove four screws (2) and register (3) from front wall (1).

b. Installation

Install register (3) on front wall (1) with four screws (2).

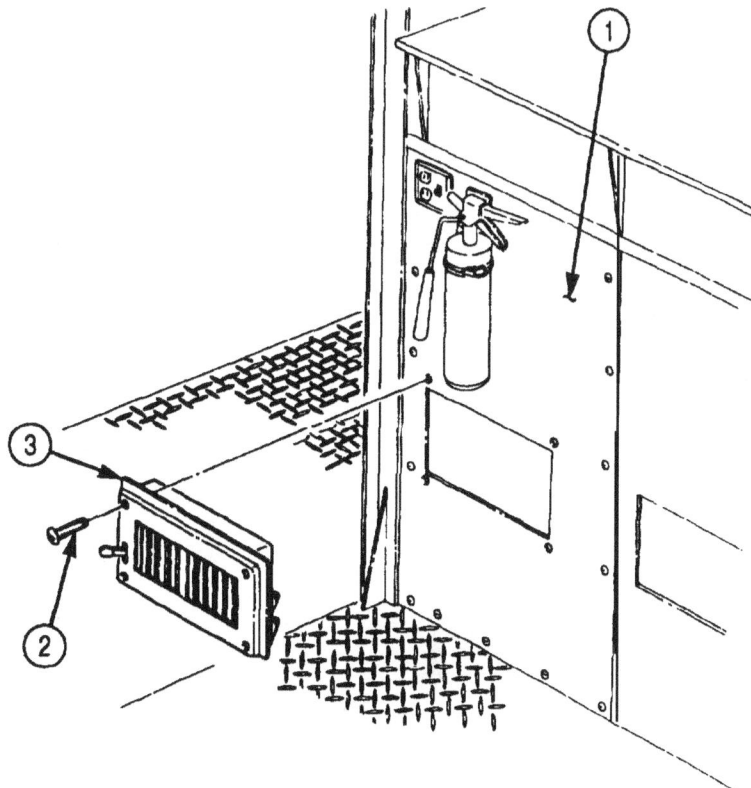

5-79. CEILING FILLER AND SIDE PANELS REPLACEMENT

THIS TASK COVERS:

a. Removal   b. Installation

INITIAL SETUP:

APPLICABLE MODELS
M934/A1/A2

TOOLS
General mechanic's tool kit (Appendix E, Item 1)

MATERIALS/PARTS
Adhesive (Appendix C, Item 7)
Primer (Appendix C, Item 57)

PERSONNEL REQUIRED
TWO

REFERENCES (TM)
TM 9-2320-272-10
TM 9-2320-272-24P

EQUIPMENT CONDITION
- Parking brake set (TM 9-2320-272-10).
- Inside telephone jacks (ceiling) removed (para. 3-374).
- Blackout light switches and receptacles removed (para. 3-373).
- Ceiling deflectors and registers removed (para. 5-83).
- Blackout and emergency light fixtures removed (para. 3-372).
- Fluorescent light fixtures removed (para. 4-168).
- Rear splice plate removed (para. 5-84).

SPECIAL ENVIRONMENTAL CONDITIONS
Vehicle must be on a level surface.

NOTE

Left and right ceiling filler and side panels are replaced the same way. This procedure covers the left side.

a. Removal

1. Remove thirty screws (9) and side panel (8) from lintel (6), air ducts (5) and (7), and filler panel (1).

2. Remove seven screws (10) and filler panel (1) from filler angles (2) and (3) and transition (4).

3. Remove forty-four screws (14) and side panel (15) from lintel (6), rear header (13), and air ducts (11) and (12).

b. Installation

NOTE

- Insulate areas of dissimilar metal-to-metal contact with zinc chromate primer.
- Apply adhesive to both rubber and metal surfaces.
- Insulate entire structure with fibrous glass felt insulation.

1. Install side panel (15) on lintel (6), rear header (13), and air ducts (12) and (11) with forty-four screws (14).

2. Install tiller panel (1) on filler angles (3) and (2) and transition (4) with seven screws (10).

3. Install side panel (8) on lintel (6), air ducts (7) and (5), and filler panel (1) with thirty screws (9).

5-79. CEILING FILLER AND SIDE PANELS REPLACEMENT (Contd)

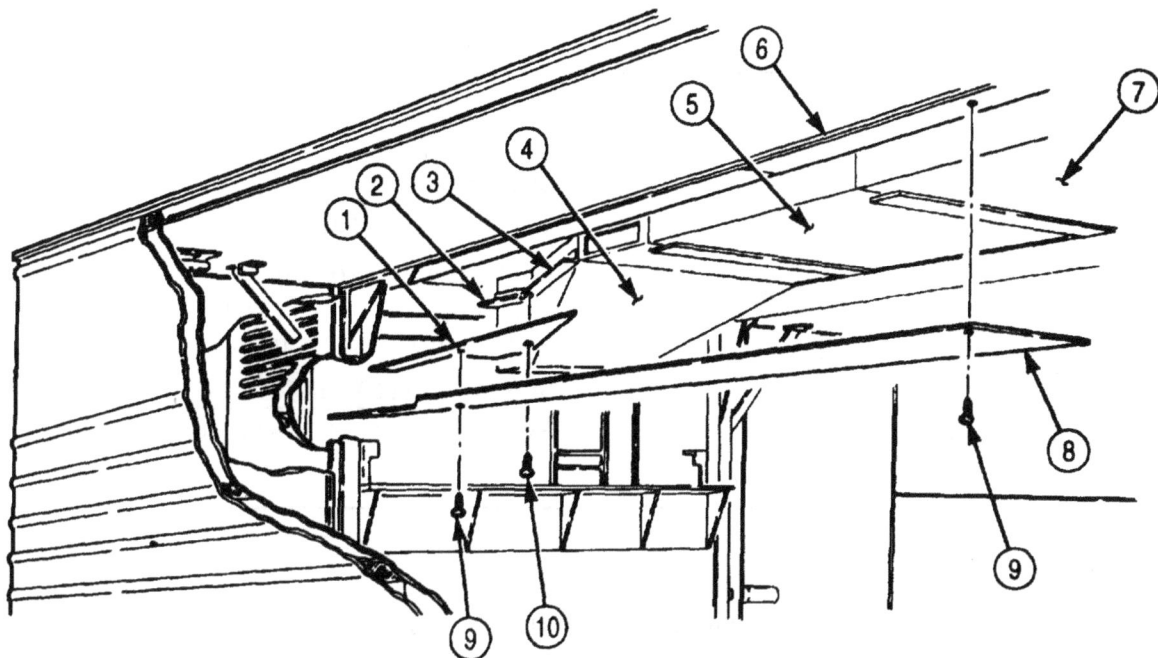

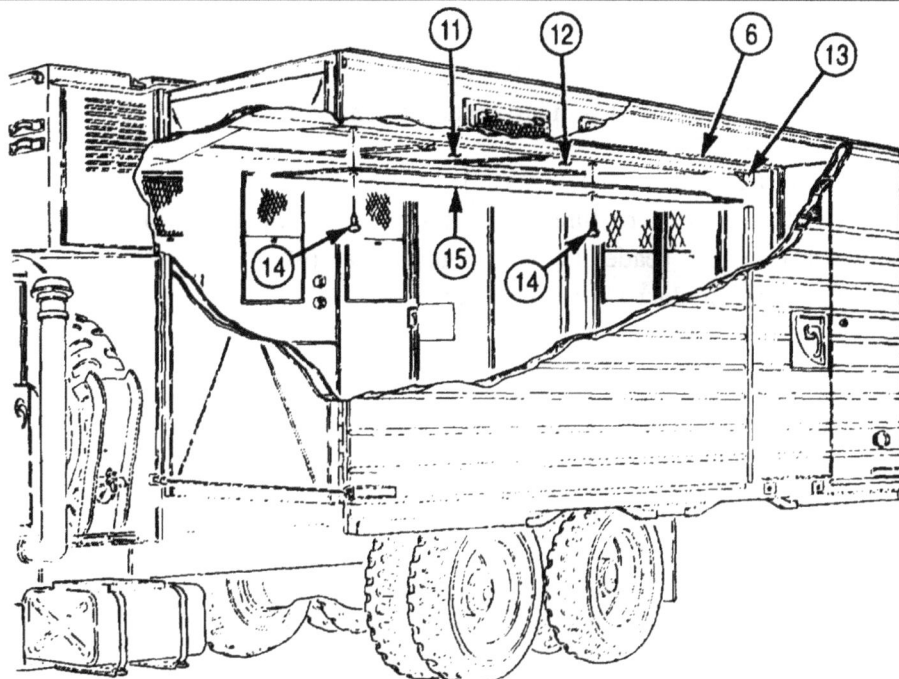

FOLLOW-ON TASKS: Install rear splice plate (para. 5-84).
- Install fluorescent light fixtures (para. 4-168).
- Install blackout and emergency light fixtures (para. 3-372).
- Install blackout light switches and receptacles (para. 3-373).
- Install ceiling deflectors and registers (para. 5-83).
- Install inside telephone jacks (ceiling) (para. 3-374).

5-80. CEILING REAR COVER REPLACEMENT

THIS TASK COVERS:

a. Removal   b. Installation

INITIAL SETUP:

APPLICABLE MODELS
M934/A1/A2

TOOLS
General mechanic's tool kit (Appendix E, Item 1)

MATERIALS/PARTS
Adhesive (Appendix C, Item 7)
Primer (Appendix C, Item 57)

PERSONNEL REQUIRED
TWO

REFERENCES (TM)
TM 9-2320-272-10
TM 9-2320-272-24P

EQUIPMENT CONDITION
- Parking brake set (TM 9-2320-272-10).
- Rear splice plate removed (par. 5-84).

SPECIAL ENVIRONMENTAL CONDITIONS
Vehicle must be on a level surface.

a. Removal

Remove eight screws (4), thirteen screws (5), and rear cover (3) from rear head (2) and air duct (1).

b. Installation

NOTE
- Insulate areas of dissimilar metal-to-metal contact with zinc chromate primer.
- Apply adhesive to metal surfaces.
- Insulate entire structure with fibrous glass felt insulation.

1. Install rear cover (3) on rear header (2) with eight screws (4).
2. Install rear cover (3) on air duct (1) with thirteen screws (5).

5-80. CEILING REAR COVER REPLACEMENT (Contd)

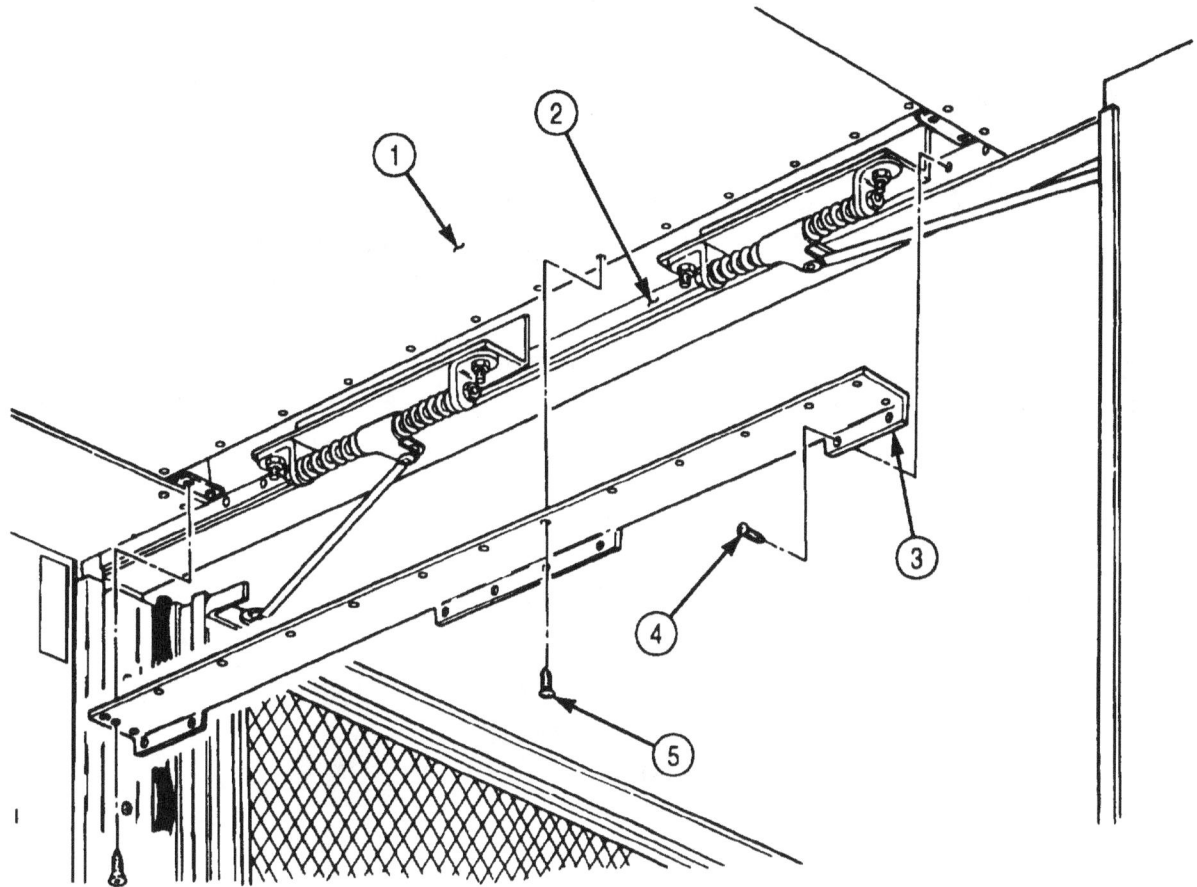

FOLLOW-ON TASK: Install rear splice plate (para. 5-84).

5-81. CEILING AIR DUCTS REPLACEMENT

THIS TASK COVERS:

a. Removal   b. Installation

INITIAL SETUP:

APPLICABLE MODELS
M934/A1/A2

TOOLS
General mechanic's tool kit (Appendix E, Item 1)

MATERIALS/PARTS
Primer (Appendix C, Item 57)

PERSONNEL REQUIRED
TWO

REFERENCES (TM)
TM 9-2320-272-10
TM 9-2320-272-24P

EQUIPMENT CONDITION
- Parking brake set (TM 9-2320-272-10).
- Van sides fully expanded and secured (TM 9-2320-272-10).
- Ceiling rear cover removed (para. 5-80).
- Ceiling filler and side panels removed (para. 5-79).

SPECIAL ENVIRONMENTAL CONDITIONS
Vehicle must be on a level surface.

a. Removal

1. Remove fifty-two screws (6) and two enclosures (7) from air ducts (3) and (5).
2. Remove fourteen screws (4) from air ducts (3) and (5) and support (1).
3. Remove forty-two screws (2) and air duct (3) from ceiling (9).
4. Remove eighteen screws (8) and air duct (5) from ceiling (9).
5. Remove nine screws (10) and support (1) from ceiling (9).
6. Remove fifty-two screws (20) and two enclosures (19) from air ducts (17) and (21).
7. Remove four screws (16) from air duct (17) and support (13).
8. Remove six screws (18) from air duct (21) and support (13).
9. Remove twenty-six screws (15) and air ducts (17) and (21) from ceiling (9).
10. Remove ten screws (14) and support (13) from ceiling (9).
11. Remove seven screws (12) and support (11) from ceiling (9).

b. Installation

NOTE
Insulate areas of dissimilar metal-to-metal contact with zinc chromate primer.

1. Install support (11) on ceiling (9) with seven screws (12).
2. Install support (13) on ceiling (9) with ten screws (14).
3. Install air ducts (17) and (21) on ceiling (9) with twenty-six screws (15).
4. Install air duct (21) on support (13) with six screws (18).
5. Install air duct (17) on support (13) with four screws (16).
6. Install two enclosures (19) on air ducts (21) and (17) with fifty-two screws (20).
7. Position support (1) over bow of roof frame and install on ceiling (9) with nine screws (10).
8. Install air duct (5) on ceiling (9) with eighteen screws (8).
9. Install air duct (3) on ceiling (9) with forty-two screws (2).
10. Install air ducts (3) and (5) on support (1) with fourteen screws (4).
11. Install two enclosures (7) on air ducts (5) and (3) with fifty-two screws (6).

5-81. CEILING AIR DUCTS REPLACEMENT (Contd)

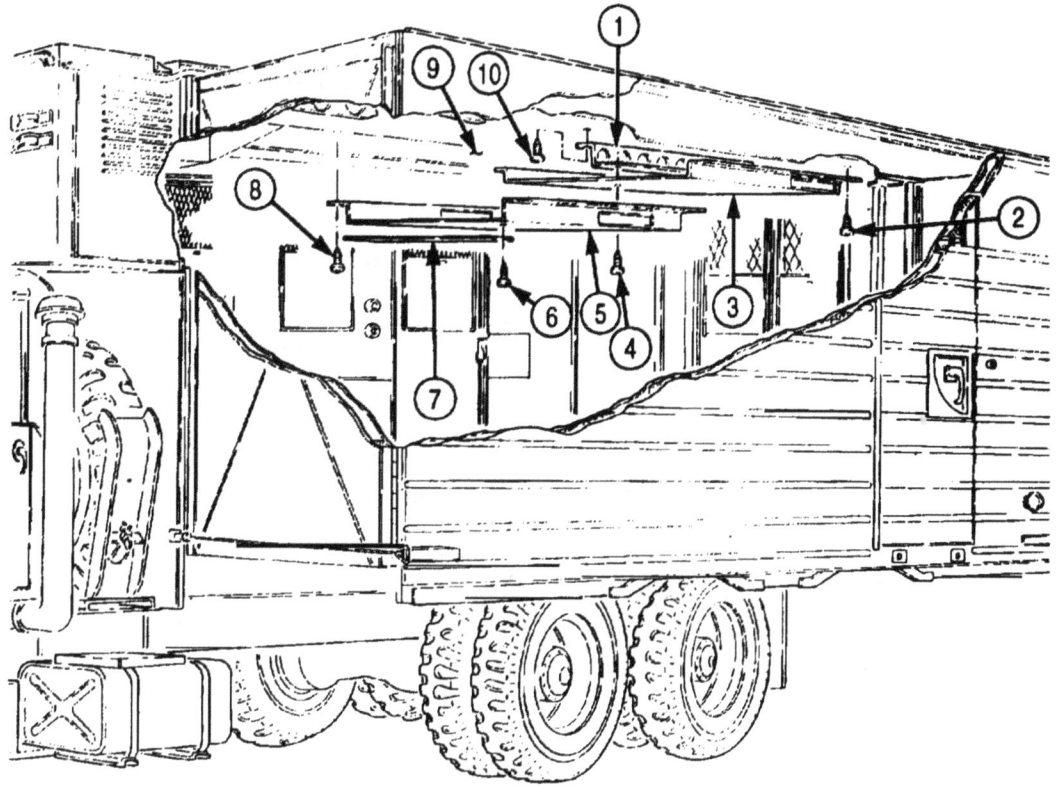

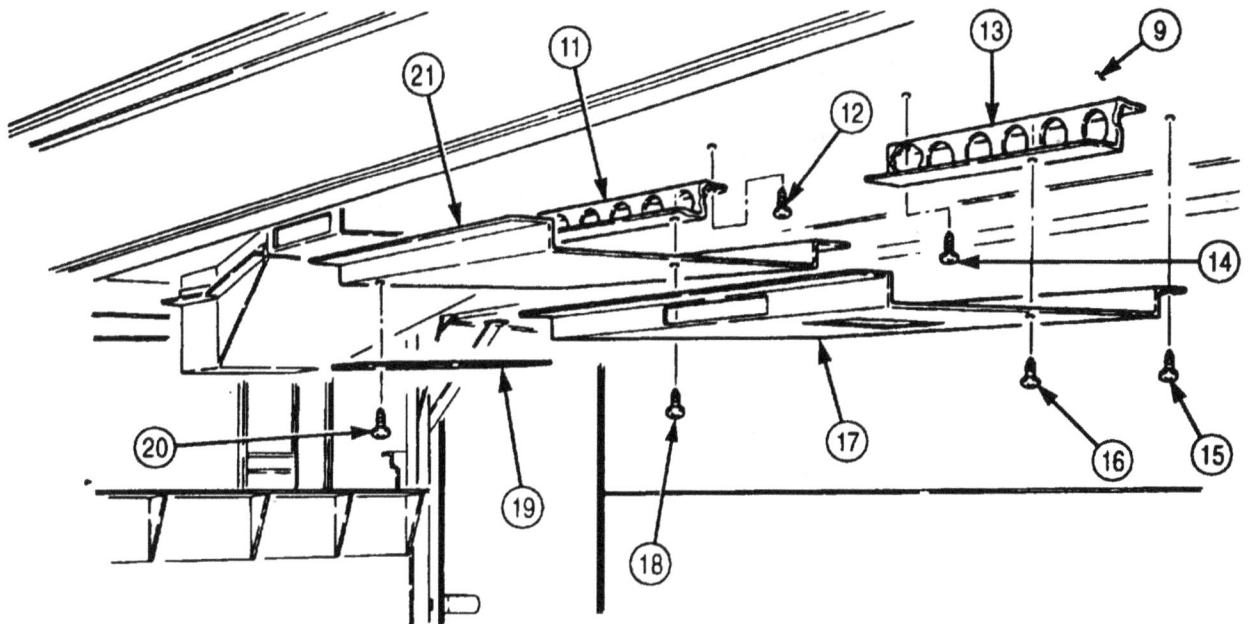

FOLLOW-ON TASKS: • Install ceiling filler and side panels (para. 5-79).
• Install ceiling rear cover (para. 5-80).
• Retract van body sides (TM 9-2320-272-10).

5-82. CEILING TRANSITION MAINTENANCE

THIS TASK COVERS:

a. Removal
b. Disassembly

c. Assembly
d. Installation

INITIAL SETUP:

APPLICABLE MODELS
M934/A1/A2

TOOLS
General mechanic's tool kit (Appendix E, Item 1)
Rivet gun

MATERIALS/PARTS
Two seals, 15 in. (Appendix D, Item 617)
Twenty-four rivets (Appendix D, Item 542)
Two seals, 28 in. (Appendix D, Item 617)
Adhesive (Appendix C, Item 7)

REFERENCES (TM)
TM 9-2320-272-10
TM 9-2320-272-24P

EQUIPMENT CONDITION
- Parking brake set (TM 9-2320-272-10).
- Van sides fully expanded and secured (TM 9-2320-272-10).
- Ceiling filler and side panels removed (para. 5-79).
- Ceiling air ducts removed (para. 5-81).

SPECIAL ENVIRONMENTAL CONDITIONS
Vehicle must be on a level surface.

a. Removal

1. Remove seven screws (4) from support (8).

NOTE
Assistants will help with step 2.

2. Remove thirty screws (2) and transition (3) from ceiling (1).

b. Disassembly

1. Remove seven rivets (6), retainer (7), seal (5), and support (8) from transition (3). Discard rivets (6) and seal (5).

2. Remove ten rivets (18), two retainers (9), and seals (10) from transition (3). Discard seals (10) and rivets (18).

3. Remove seven rivets (16), retainer (17), and seal (15) from transition (3). Discard seal (15) and rivets (16).

4. Remove four screws (11) and filler angle (12) from transition (3).

5. Remove three screws (13) and filler angle (14) from transition (3).

5-82. CEILING TRANSITION MAINTENANCE (Contd)

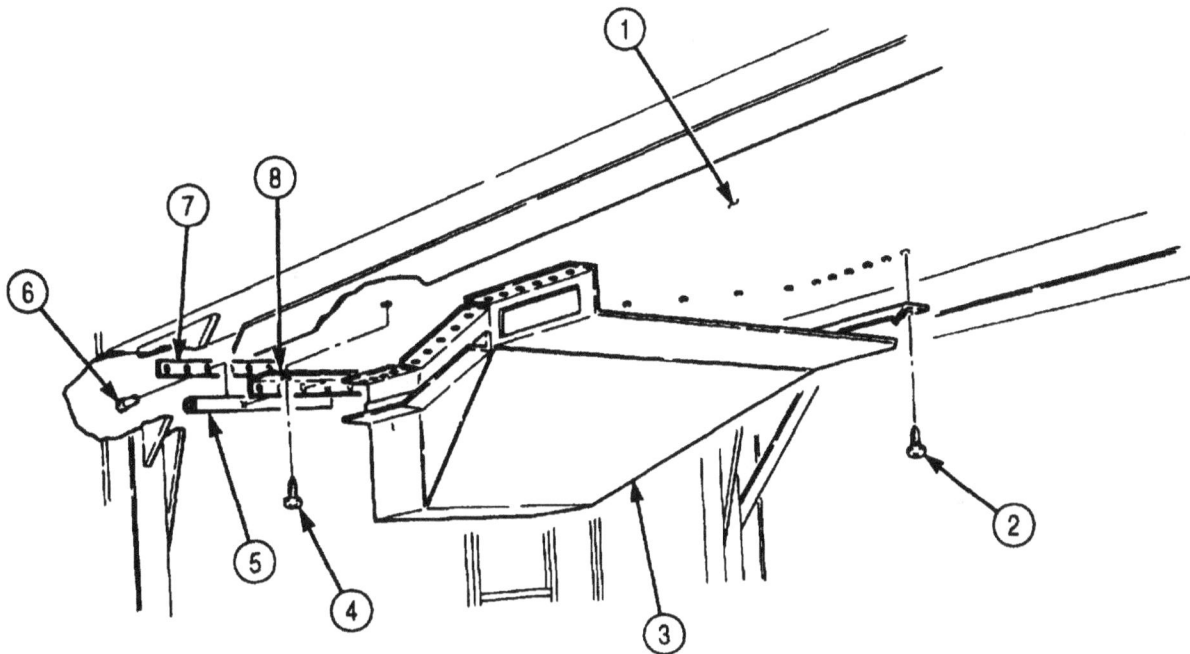

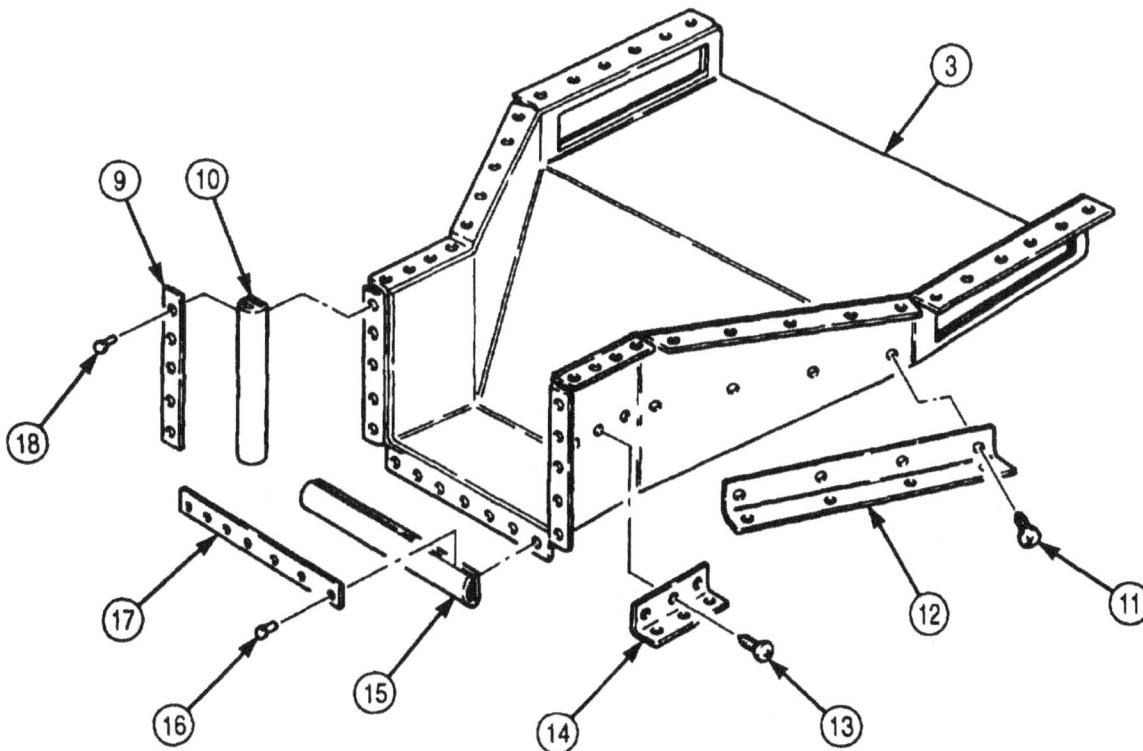

5-82. CEILING TRANSITION MAINTENANCE (Contd)

c. Assembly

1. Install filler angle (14) on transition (3) with three screws (13).
2. Install tiller angle (12) on transition (3) with four screws (11).

NOTE

Apply adhesive to both rubber and metal surfaces.

3. Install new seal (15) and retainer (17) on transition (3) with seven new rivets (16).
4. Install two new seals (10) and retainers (9) on transition (3) with ten new rivets (18).
5. Align support (8) with flanges on transition (3) and install support (8), new seal (5), and retainer (7) on transition (3) with seven new rivets (6).

d. Installation

NOTE

- Insulate inside of transition with duct lining insulation.
- Assistants will help with steps 1 and 2.

1. Install transition (3) on ceiling (1) with thirty screws (2).
2. Install support (8) on ceiling (1) with seven screws (4).

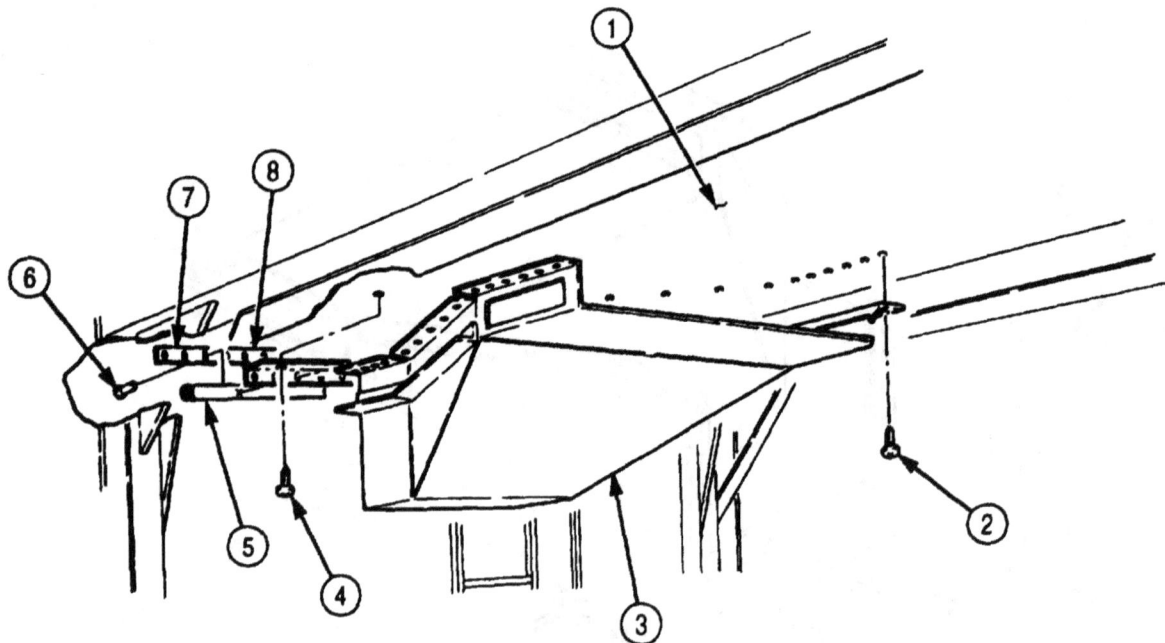

5-82. CEILING TRANSITION MAINTENANCE (Contd)

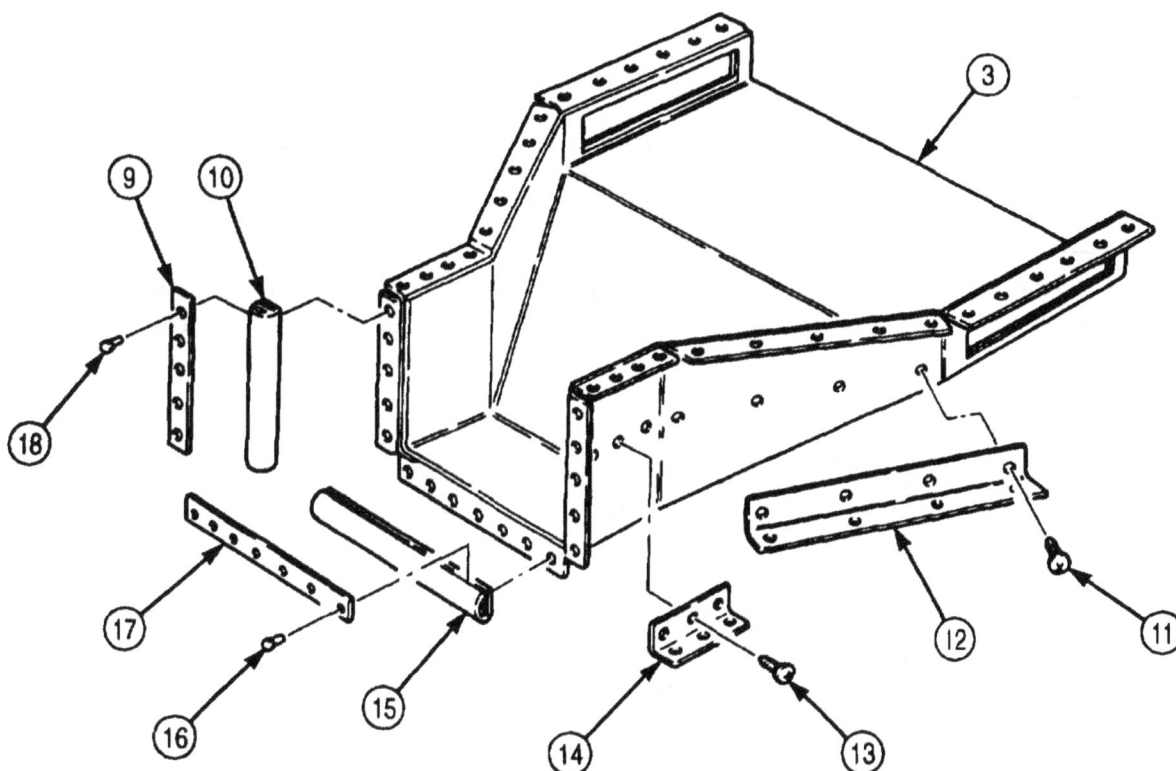

FOLLOW-ON TASKS: Install ceiling air ducts (par. 5-81).
- Install ceiling filler and side panels (para. 5-79).
- Retract van body sides (TM 9-2320-272-10).

5-83. CEILING DEFLECTOR AND REGISTERS REPLACEMENT

THIS TASK COVERS:

a. Removal

b. Installation

INITIAL SETUP:

APPLICABLE MODELS
M934/A1/A2

TOOLS
General mechanic's tool kit (Appendix E, Item 1)

REFERENCES (TM)
TM 9-2320-272-10
TM 9-2320-272-24P

EQUIPMENT CONDITION
Parking brake set (TM 9-2320-272-10).

a. Removal

1. Remove twenty screws (3) and ten registers (4) from ceiling side panels (2) and deflectors (6).
2. Remove one hundred screws (5) and ten deflectors (6) from air ducts (1) and ceiling side panels (2).

b. Installation

1. Install ten deflectors (6) on air ducts (1) and ceiling side panels (2) with one hundred screws (5).
2. Install ten registers (4) on ceiling side panels (2) and ten deflectors (6) with twenty screws (3).

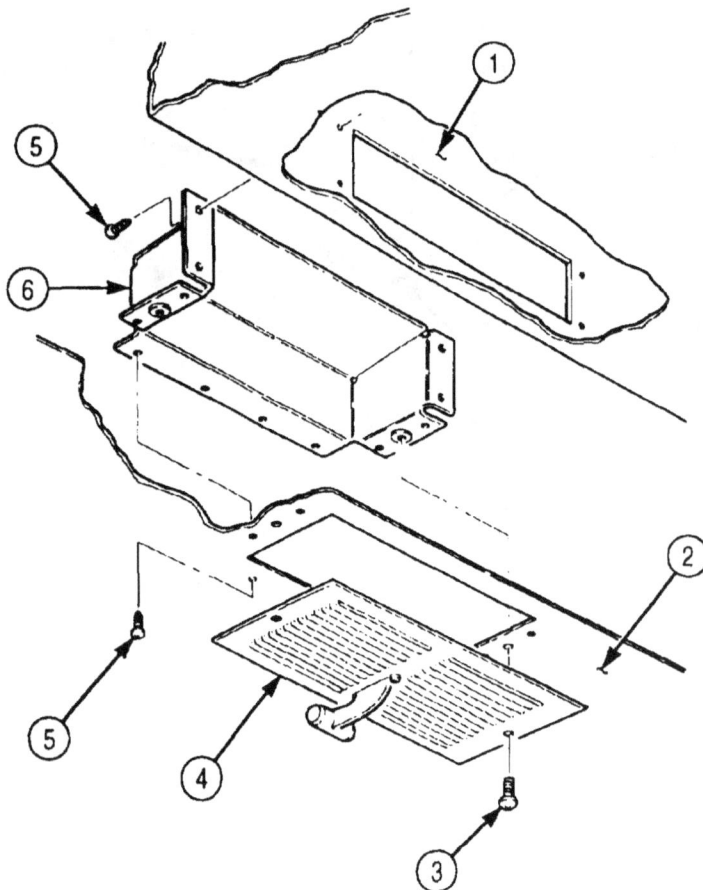

5-84. REAR SPLICE PLATE REPLACEMENT

THIS TASK COVERS:
a. Removal   b. Installation

INITIAL SETUP:

APPLICABLE MODELS
M934/A1/A2

TOOLS
General mechanic's tool kit (Appendix E, Item 1)

REFERENCES (TM)
TM 9-2320-272-10
TM 9-2320-272-24P

EQUIPMENT CONDITION
Parking brake set (TM 9-2320-272-10).

NOTE
Left and rear splice plates are replaced the same way. This procedure covers the right side.

a. Removal

Remove four screws (4) and splice plate (3) from side panel (1) and rear cover (2).

b. Installation

Install splice plate (3) on side panel (1) and rear cover (2) with four screws (4).

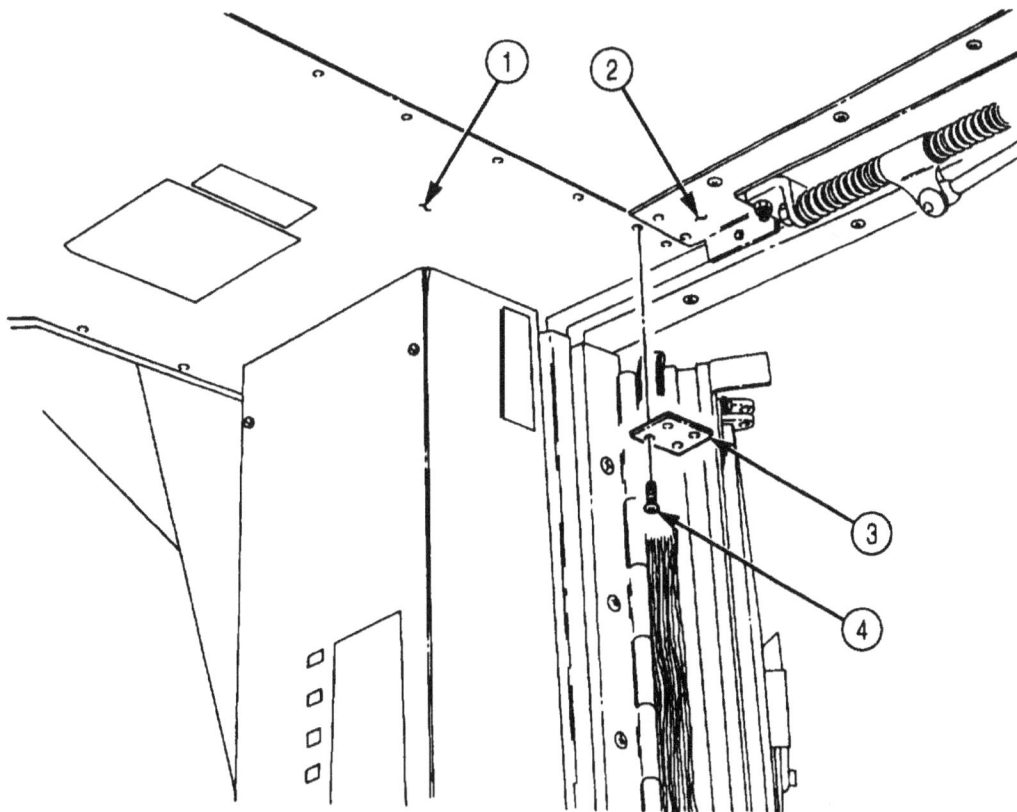

5-85. BONNET FRAME PARTS REPLACEMENT

THIS TASK COVERS:

a. Removal   b. Installation

INITIAL SETUP:

APPLICABLE MODELS
M934/A1/A2

TOOLS
General mechanic's tool kit (Appendix E, Item 1)
Rivet gun

MATERIALS/PARTS
Insulation (Appendix D, Item 260)
Two speed nuts (Appendix D, Item 667)
Eight rivets (Appendix D, Item 557)
Thirty-five rivets (Appendix D, Item 556)
One-hundred sixteen rivets (Appendix D,
  Item 547)
One-hundred seventy-six rivets (Appendix D,
  Item 542)
Sixteen rivets (Appendix D, Item 559)
Adhesive (Appendix C, Item 7)
Primer (Appendix C, Item 57)
Sealing compound (Appendix C, Item 67)

REFERENCES (TM)
TM 9-2320-272-10
TM 9-2320-272-24P

EQUIPMENT CONDITION
- Parking brake set (TM 9-2320-272-10).
- Bonnet control rod removed (para. 3-361).
- Van heater removed (para. 3-379).
- Air conditioner (if equipped) removed (para. 5-88).
- Bonnet access door removed (para. 5-86).
- Bonnet door removed (para. 5-87).

---

a. Removal

1. Remove twenty-seven screws (5) and drip molding (4) from bonnet frame (14).

2. Remove seventeen screws (13) and door panel (12) from bonnet frame (14).

3. Remove four screws (15) and plate (1) from outer panel (2) and bonnet frame (14).

4. Remove thirty-eight rivets (3), twenty-four rivets (7), and outer panels (6) and (2) from bonnet frame (14) and bonnet lower panel (11). Discard rivets (3) and (7).

NOTE
- Left and right fillers are removed the same way. This procedure is for the left filler.
- Assistant will help with step 5.

5. Remove sixty-four rivets (9), thirty-one rivets (10), bonnet lower panel (11), and two tillers (8) from bonnet frame (14). Discard rivets (9) and (10).

5-85. BONNET FRAME PARTS REPLACEMENT (Contd)

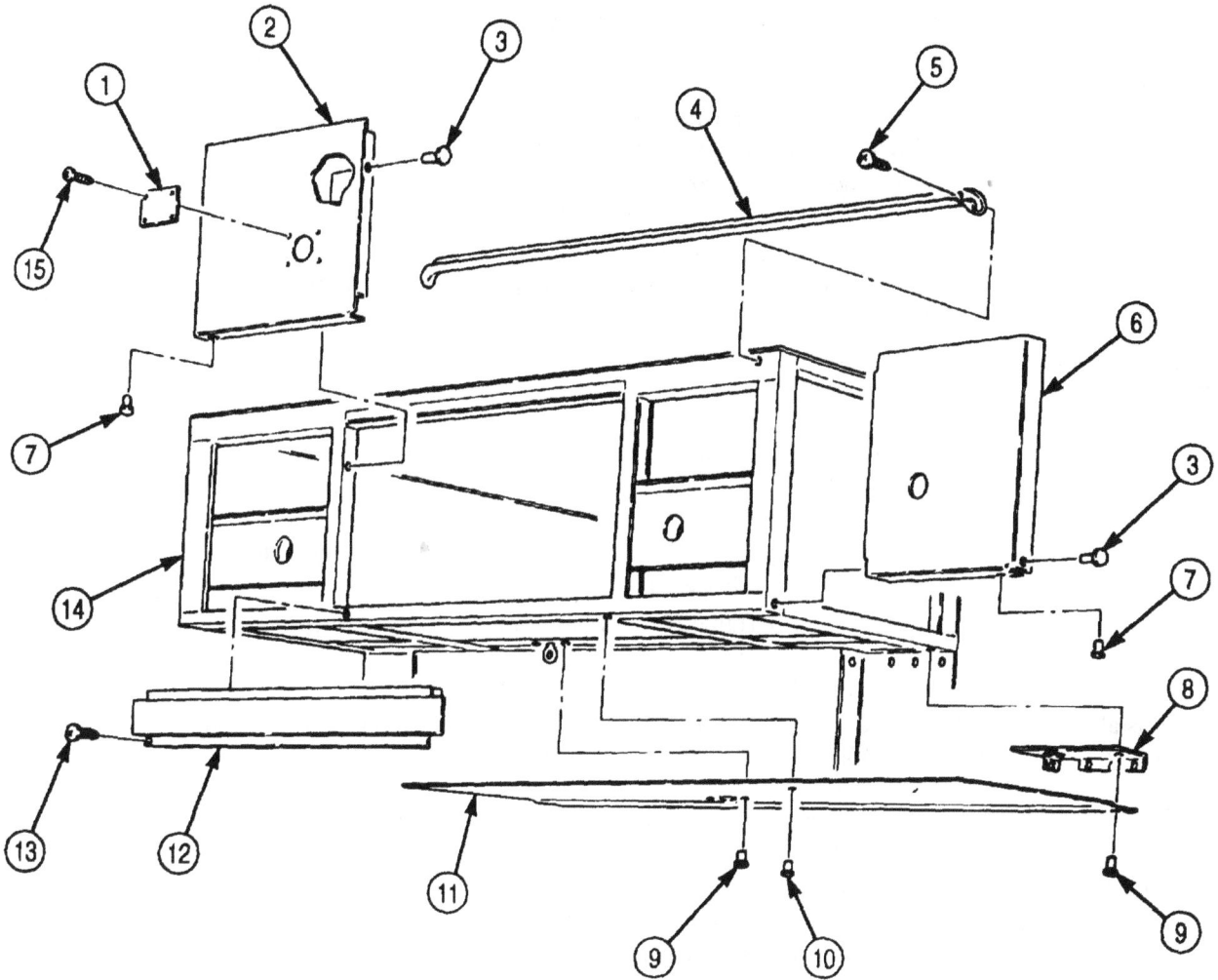

5-85. BONNET FRAME PARTS REPLACEMENT (Contd)

NOTE

Left and right heater duct assemblies are removed the same way.
This procedure covers the right side.

6. Remove two fasteners (5) from support (14) and heater duct (2).

7. Remove four rivets (12) and two speed nuts (13) from support (14). Discard rivets (12) and speed nuts (13).

8. Remove three screws (11) and support (14) from front wall (15).

9. Remove thirty-six rivets (3), heater duct (2), and insulation (4) from bonnet floor (19). Discard rivets (3) and insulation (4).

10. Remove sixteen rivets (7), twenty-four rivets (8), thirty-six rivets (10), and inner panels (9) and (16) from bonnet frame (6) and bonnet floor (19). Discard rivets (7), (8), and (10).

11. Remove three screws (20) and post (1) from bonnet floor (19).

NOTE

Assistant will help with step 12.

12. Remove forty-three rivets (17), thirty-five rivets (18), and bonnet floor (19) from bonnet frame (6) and front wall (15). Discard rivets (17) and (18).

b. Installation

NOTE

- Insulate areas of dissimilar metal-to-metal contact with zinc chromate primer.
- Apply sealing compound to exterior joints prior to installation.
- Insulate all enclosed structure with fibrous glass felt insulation.
- Assistant will help with step 1.

1. Install bonnet floor (19) on bonnet frame (6) and front wall (15) with forty-three new rivets (17) and thirty-five new rivets (18).

2. Install post (1) on bonnet floor (19) with three screws (20).

3. Install inner panels (9) and (16) on bonnet frame (6) and bonnet floor (19) with thirty-six new rivets (10), twenty-four new rivets (8), and sixteen new rivets (7).

NOTE

Left and right heater duct assemblies are installed the same way.
This procedure covers the right side.

4. Install new insulation (4) and heater duct (2) on bonnet floor (19) with thirty-six new rivets (3).

5. Install support (14) on front wall (15) with three screws (11).

6. Install two new speed nuts (13) on support (14) with four new rivets (12).

7. Install heater duct (2) on support (14) with two fasteners (5).

5-85. BONNET FRAME PARTS REPLACEMENT (Contd)

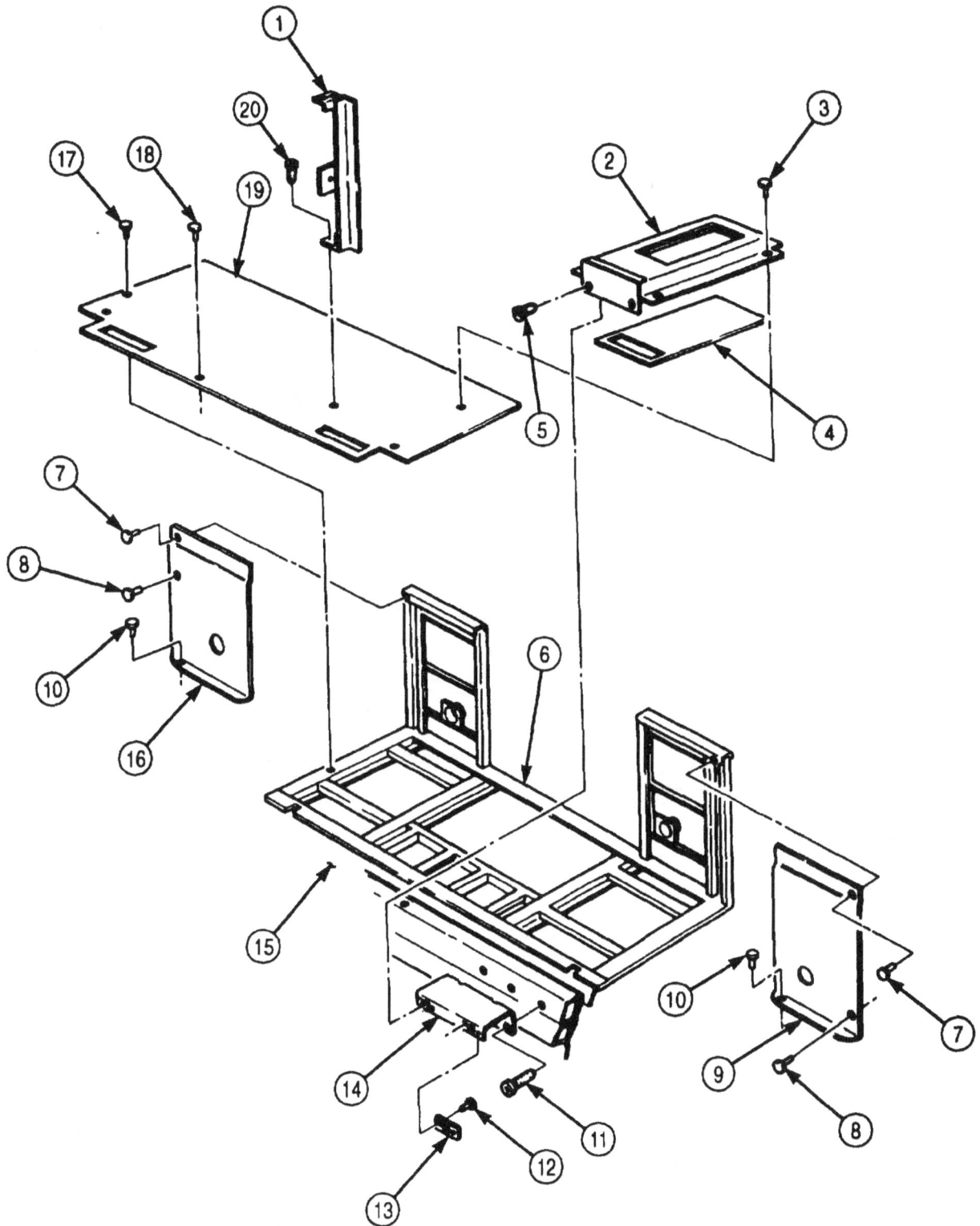

5-85. BONNET FRAME PARTS REPLACEMENT (Contd)

NOTE
- Left and right fillers are installed the same way. The left filler is shown.
- Assistant will help with step 8.

8. Install two fillers (8) and bonnet lower panel (11) on bonnet frame (14) with sixty-four new rivets (9) and thirty-one new rivets (10).

9. Install outer panels (6) and (2) on bonnet frame (14) and bonnet lower panel (11) with twenty-four new rivets (7) and thirty-eight new rivets (3).

10. Install plate (1) on outer panel (2) and bonnet frame (14) with four screws (15).

11. Install door panel (12) on bonnet frame (14) with seventeen screws (13).

12. Install drip molding (4) on bonnet frame (14) with twenty-seven screws (5).

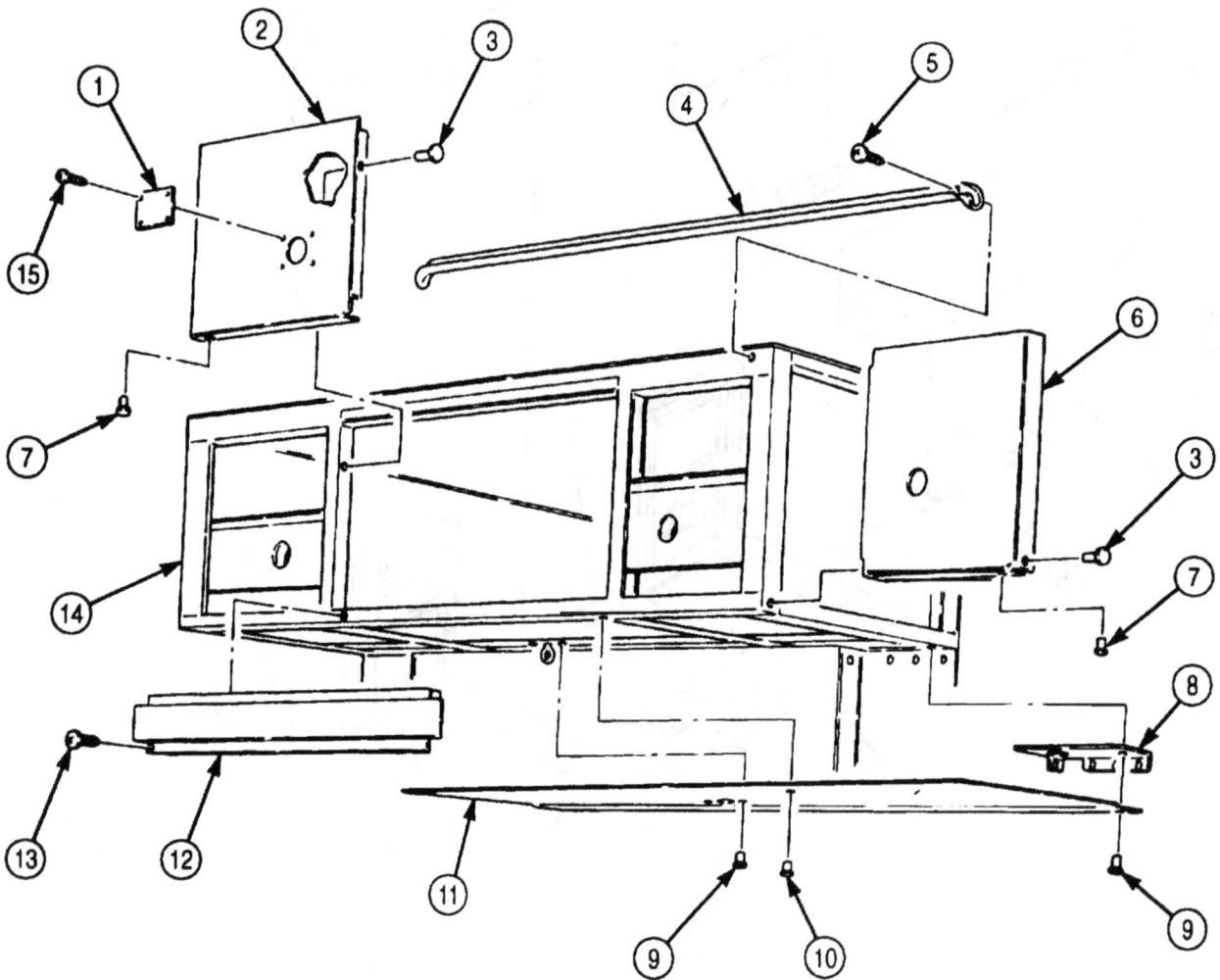

FOLLOW-ON TASKS:
- Install bonnet door (para. 5-87).
- Install bonnet access door (para. 5-86).
- Install van heater (para. 3-379).
- Install air conditioner (if equipped) (para. 5-88).
- Install bonnet control rod (para. 3-361).

5-86. BONNET ACCESS DOOR MAINTENANCE

THIS TASK COVERS:

a. Removal
b. Disassembly
c. Cleaning and Inspection

d. Assembly
e. Installation

INITIAL SETUP:

APPLICABLE MODELS
M934/A1/A2

TOOLS
General mechanic's tool kit (Appendix E, Item 1)
Rivet gun

MATERIALS/PARTS
One hundred forty-eight rivets (Appendix D,
  Item 551)
Seal, 51 in. (Appendix D, Item 633)
Seal (Appendix D, Item 623)
Primer (Appendix C, Item 57)
Sealing compound (Appendix C, Item 67)

REFERENCES (TM)
TM 9-2320-272-10
TM 9-2320-272-24P

EQUIPMENT CONDITION
• Parking brake set (TM 9-2320-272-10).
• Bonnet control rod removed (para. 3-361).

a. Removal

NOTE
Assistant will help with step.

Remove twelve screws (1) and access door assembly (2) from bonnet frame (3).

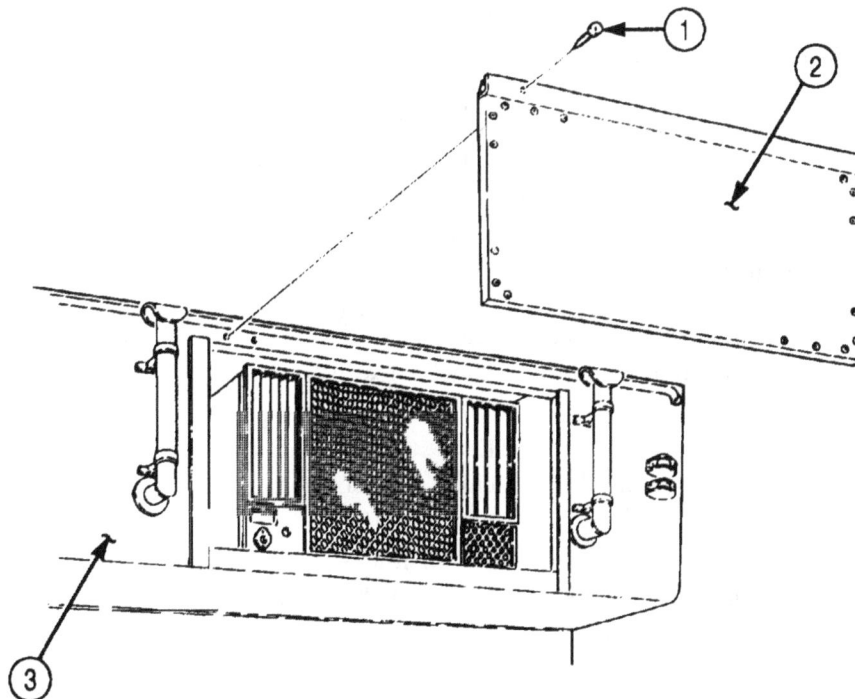

5-86. BONNET ACCESS DOOR MAINTENANCE (Contd)

b. Disassembly

1. Remove twelve screws (1), hinge (2), and seal (21) from door frame (17). Discard seal (21).
2. Remove four screws (5) and two angle brackets (4) from inner panel (3).
3. Remove four screws (7), rod (8), and holder bracket (6) from inner panel (3).
4. Remove screw (11), two nuts (9), screws (12), and bracket (10) from inner panel (3).

NOTE

Assistant will help with step 5.

5. Remove thirty-three nuts (13), screws (18), seventy-four rivets (19), outer panel (20), three retainers (14), and seal (16) from door frame (17). Discard seal (16) and rivets (19).

NOTE

Perform step 6 if inner panel is to be replaced (subtask c.).

6. Remove seventy-four rivets (15) and inner panel (3) from door frame (17). Discard rivets (15).

c. Cleaning and Inspection

1. For general cleaning instructions, refer to para. 2-14.
2. Inspect hinge (2) for breaks and proper operation. Replace hinge (2) if damaged.
3. Inspect three retainers (14) for bends and breaks. Replace if bent or broken.
4. Inspect angle brackets (4), holder bracket (6), and rod (8) for cracks and breaks. Replace if cracked or broken.
5. Inspect outer panel (20) for tears and punctures. Replace outer panel (20) if torn or punctured.
6. Inspect inner panel (3) for tears and punctures. Replace inner panel (3) if torn or punctured.
7. Inspect door frame (17) for bend, cracks, and breaks. Replace door frame (17) if bent, cracked, or broken.

d. Assembly

NOTE

- Insulate areas of dissimilar metal-to-metal contact with zinc chromate primer.
- Apply sealing compound to exterior joints prior to installation.
- Insulate all enclosed structure with fibrous glass felt insulation.
- Perform step 1 only if inner panel was removed.
- Assistant will help with steps 1 and 2.

1. Install inner panel (3) on door frame (17) with seventy-four new rivets (15).
2. Install new seal (16), three retainers (14), and outer panel (20) on door frame (17) with thirty-three screws (18), nuts (13), and seventy-four new rivets (19).
3. Install holder bracket (6) and rod (8) on inner panel (3) with four screws (7).
4. Install screw (11) and two nuts (9) on bracket (10).

NOTE

Angle bracket can be installed on either side of door.

5. Install bracket (10) on inner panel (3) with two screws (12).
6. Install two angle brackets (4) on inner panel (3) with four screws (5).
7. Install new seal (21) and hinge (2) on door frame (17) with twelve screws (1).

5-86. BONNET ACCESS DOOR MAINTENANCE (Contd)

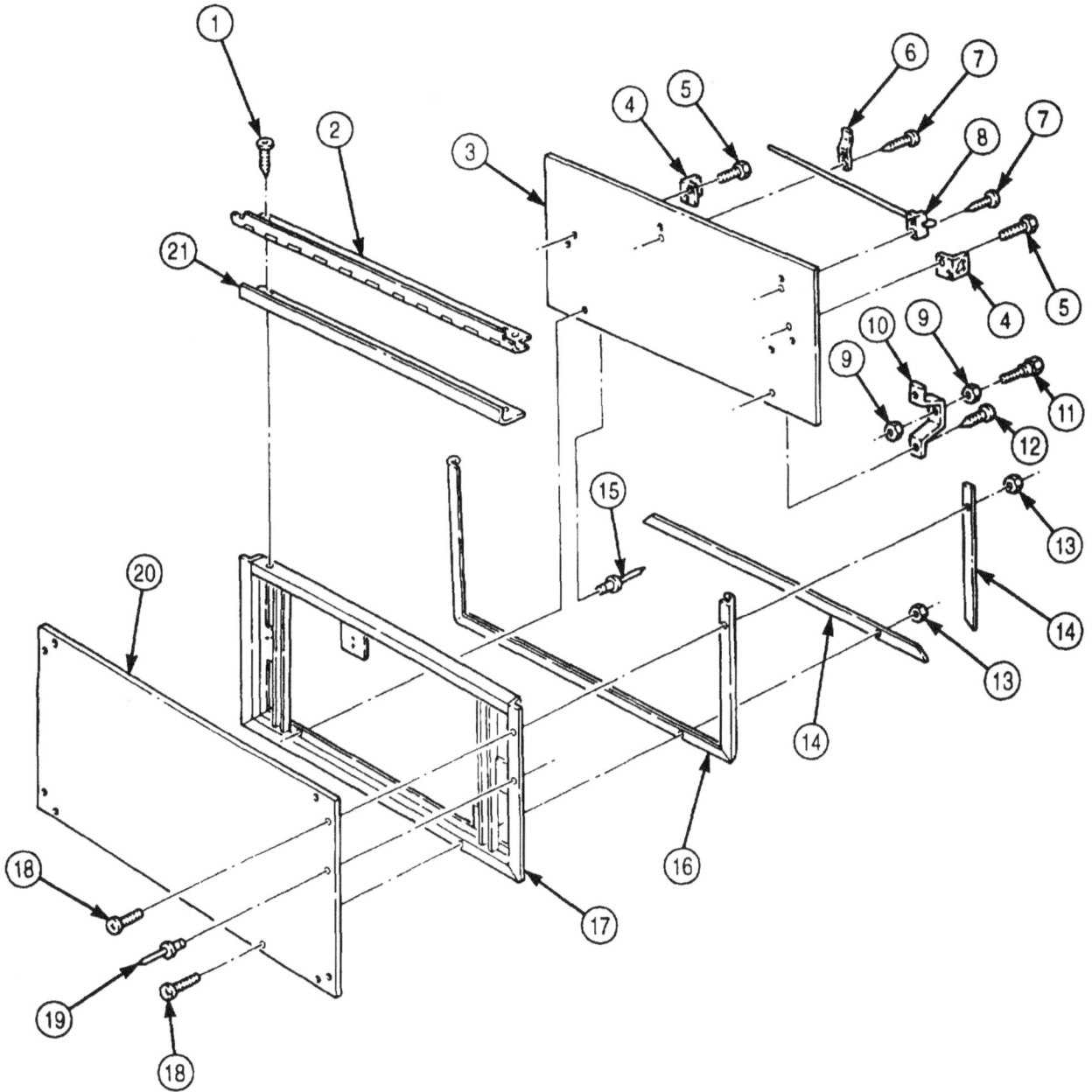

5-86. BONNET ACCESS DOOR MAINTENANCE (Contd)

e. Installation

NOTE

Assistant will help with step.

Install access door assembly (2) on bonnet frame (3) with twelve screws (1).

5-86. BONNET ACCESS DOOR MAINTENANCE (Contd)

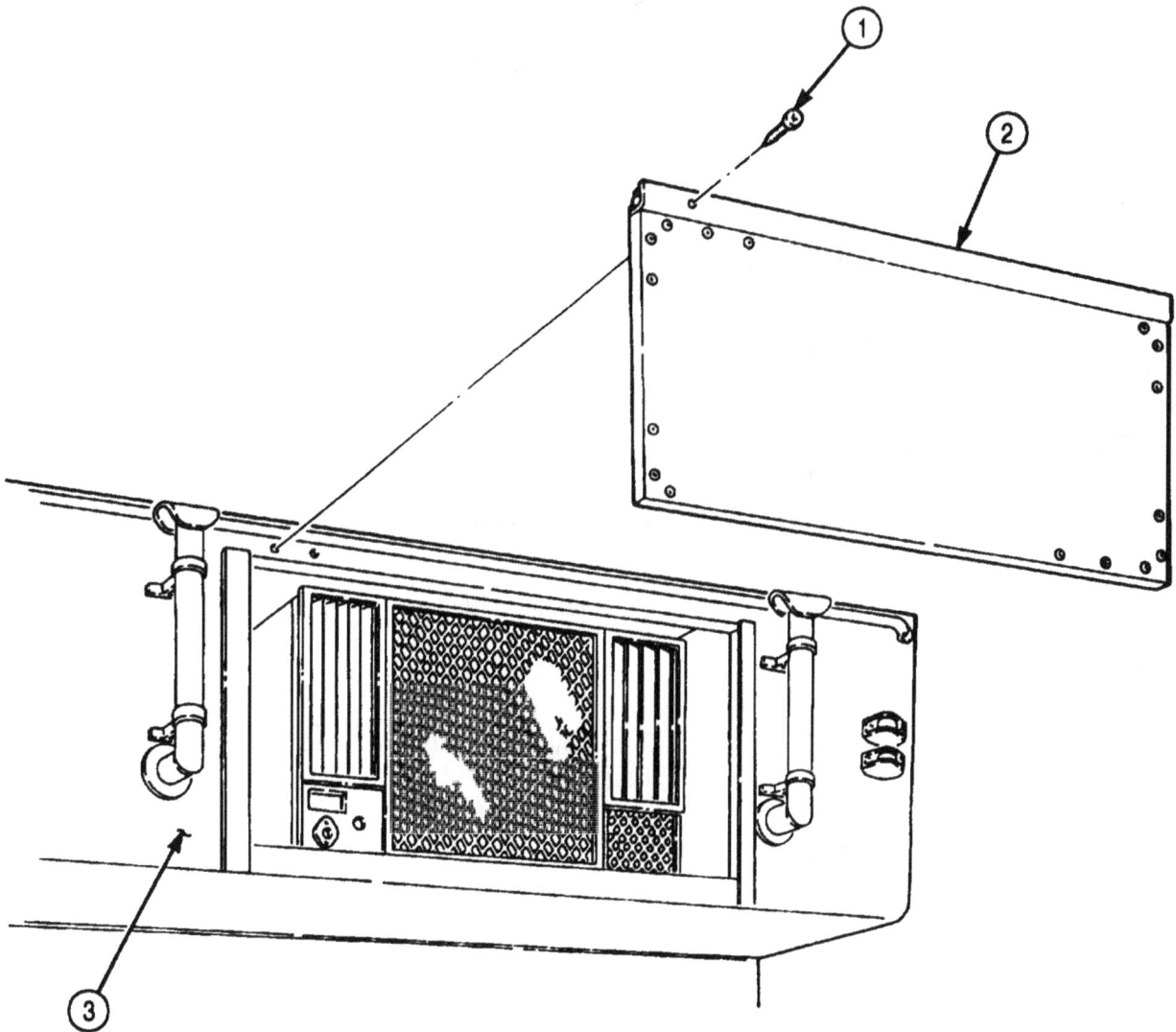

FOLLOW-ON TASK: Install bonnet control rod (para. 3-361).

TM 9-2320-272-24-4

5-87. BONNET DOOR MAINTENANCE

THIS TASK COVERS:

a. Removal
b. Disassembly
c. Cleaning and Inspection

d. Assembly
e. Installation

INITIAL SETUP:

APPLICABLE MODELS

M934/A1/A2

TOOLS

General mechanic's tool kit (Appendix E, Item 1)
Rivet gun

MATERIALS/PARTS

Channel seals (as required) (Appendix D, Item 623)
Forty-eight rivets (Appendix D, Item 547)
Fifty-two rivets (Appendix D, Item 551)
Seal, 78 in. (Appendix D, Item 605)
Seal, 24 in. (Appendix D, Item 605)
Seal, 27 in. (Appendix D, Item 633)
Cork (Appendix C, Item 24)
Primer (Appendix C, Item 57)
Sealing compound (Appendix C, Item 67)

REFERENCES (TM)

TM 9-2320-272-10
TM 9-2320-273-24P

EQUIPMENT CONDITION

Parking brake set (TM 9-2320-272-10).

NOTE
Left and right bonnet doors are replaced the same way. This procedure covers the left side.

| a. Removal |

1. Remove two screws (4) and washers (5) from bonnet door (3) and bonnet (1).
NOTE
Assistant will help with step 2.
2. Remove seven screws (2) and bonnet door (3) from bonnet (1).
3. Position bonnet door (3) squarely on four jack stands with outer panel facing upwards.

5-87. BONNET DOOR MAINTENANCE (Contd)

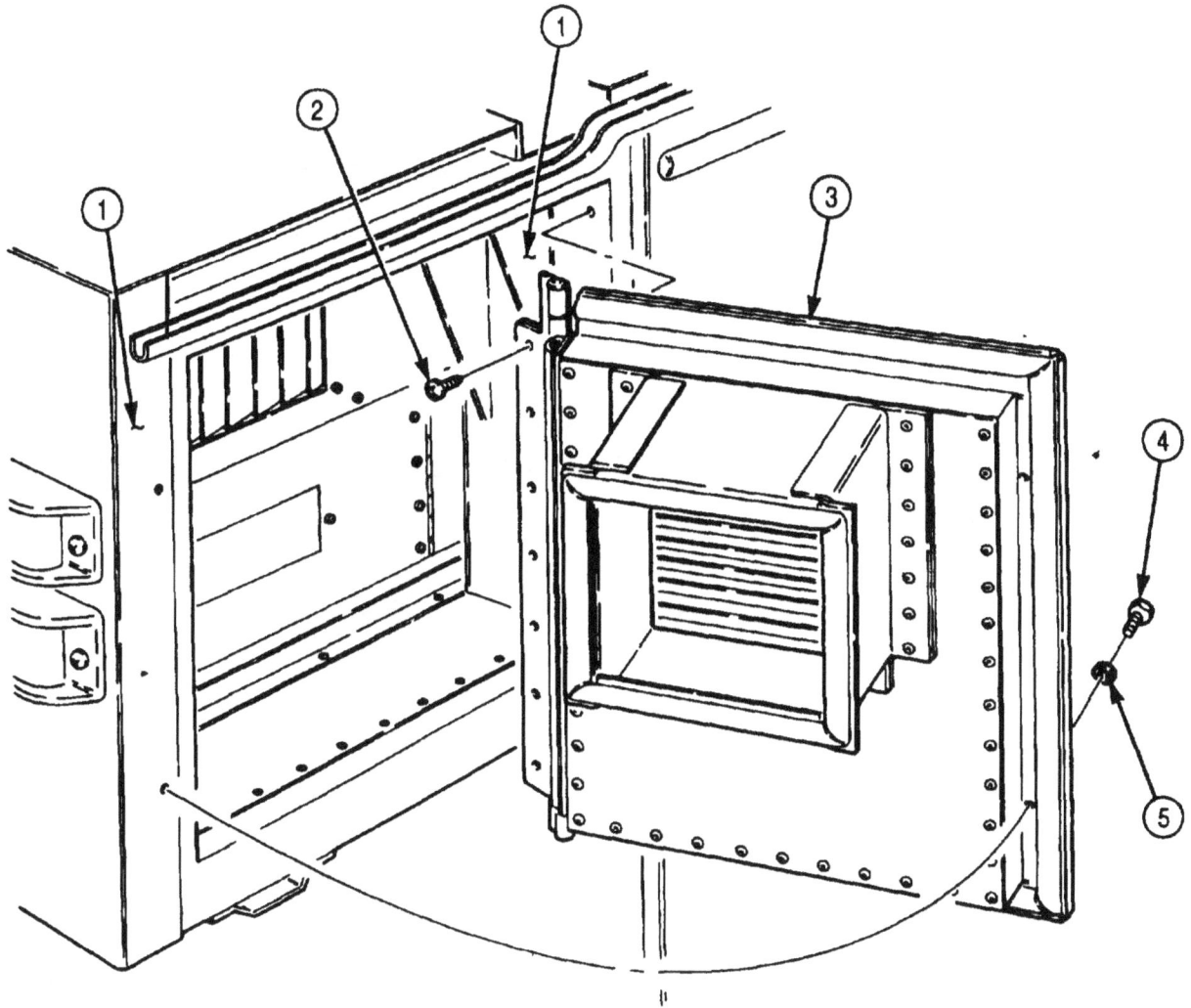

5-87. BONNET DOOR MAINTENANCE (Contd)

1. Remove six screws (16), spacer plate (15), seal (14), hinge (13), and seal (12) from frame (7). Discard seals (12) and (14).
2. Remove eighteen nuts (28), screws (11), and two retainers (4) from outer panel (10), frame (7). and seal (3).

NOTE
Assistant will help with step 3.

3. Remove seventeen rivets (9) and outer panel (10) from frame (7). Discard rivets (9).
4. Remove twenty-four rivets (27), intake assembly (17), intake panel (26) (if present), and preformed cork (1) from inner panel (2). Discard rivets (27) and cork (1).
5. Remove fourteen rivets (21), two channels (20), channel seals (19), and channels (18) from intake assembly (17). Discard rivets (21) and channel seals (19).
6. Remove ten rivets (22), two channels (23), channel seals (24), and channels (25) from intake assembly (17). Discard rivets (22) and channel seals (24).
7. Remove seven nuts (5), screws (8), retainer (6), and seal (3) from frame (7). Discard seal (3).

NOTE
Assistant will help with step 9.

8. Remove thirty-three rivets (29), two rivets (30), and inner panel (2) from frame (7). Discard rivets (29) and (30).

5-87. BONNET DOOR MAINTENANCE (Contd)

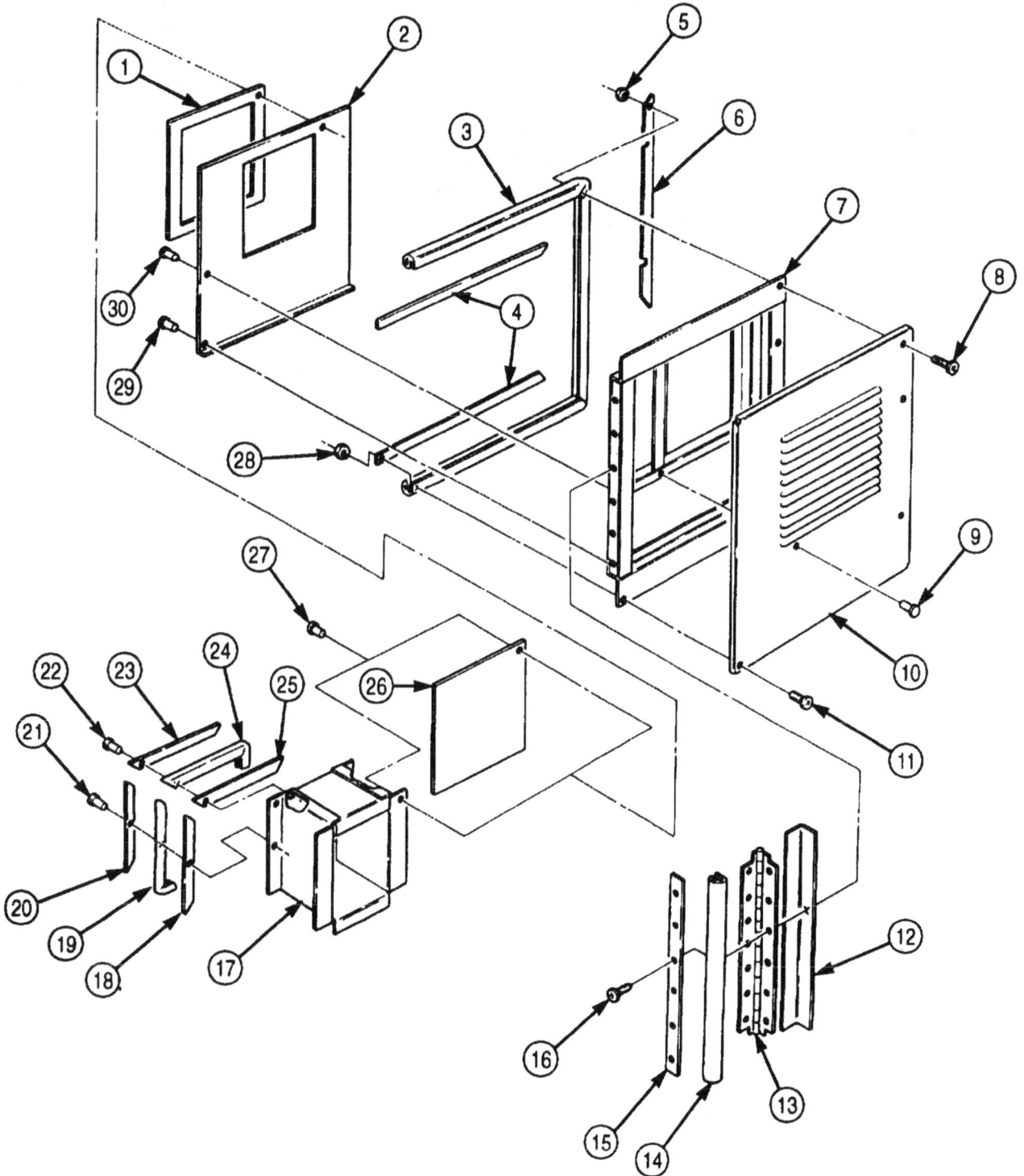

5-87. BONNET DOOR MAINTENANCE (Contd)

c. Cleaning and Inspection

1. For general cleaning instructions, refer to para. 2-14.
2. Inspect hinge (9) for breaks, corrosion, and proper operation. Replace hinge (9) if damaged.
3. Inspect retainers (8) and (5) and spacer plate (10) for bends and breaks. Replace retainers (8) and (5) and space plate (10) if bent or broken.
4. Inspect outer panel (4) and inner panel (7) for tears, cracks, and punctures. Replace outer panel (4) and inner panel (7) if torn, cracked, or punctured.
5. Inspect intake assembly (11) for cracks and bends. Replace intake assembly (11) if cracked or bent.
6. Inspect intake panel (3) for bends and breaks. Replace intake panel (3) if bent or broken.
7. Inspect channels (12), (13), (1), and (2) for bends and breaks. Replace if bent or broken.
8. Inspect frame (6) for bends, breaks, cracks, and warpage. Replace entire bonnet door assembly if frame (6) is damaged.

5-87. BONNET DOOR MAINTENANCE (Contd)

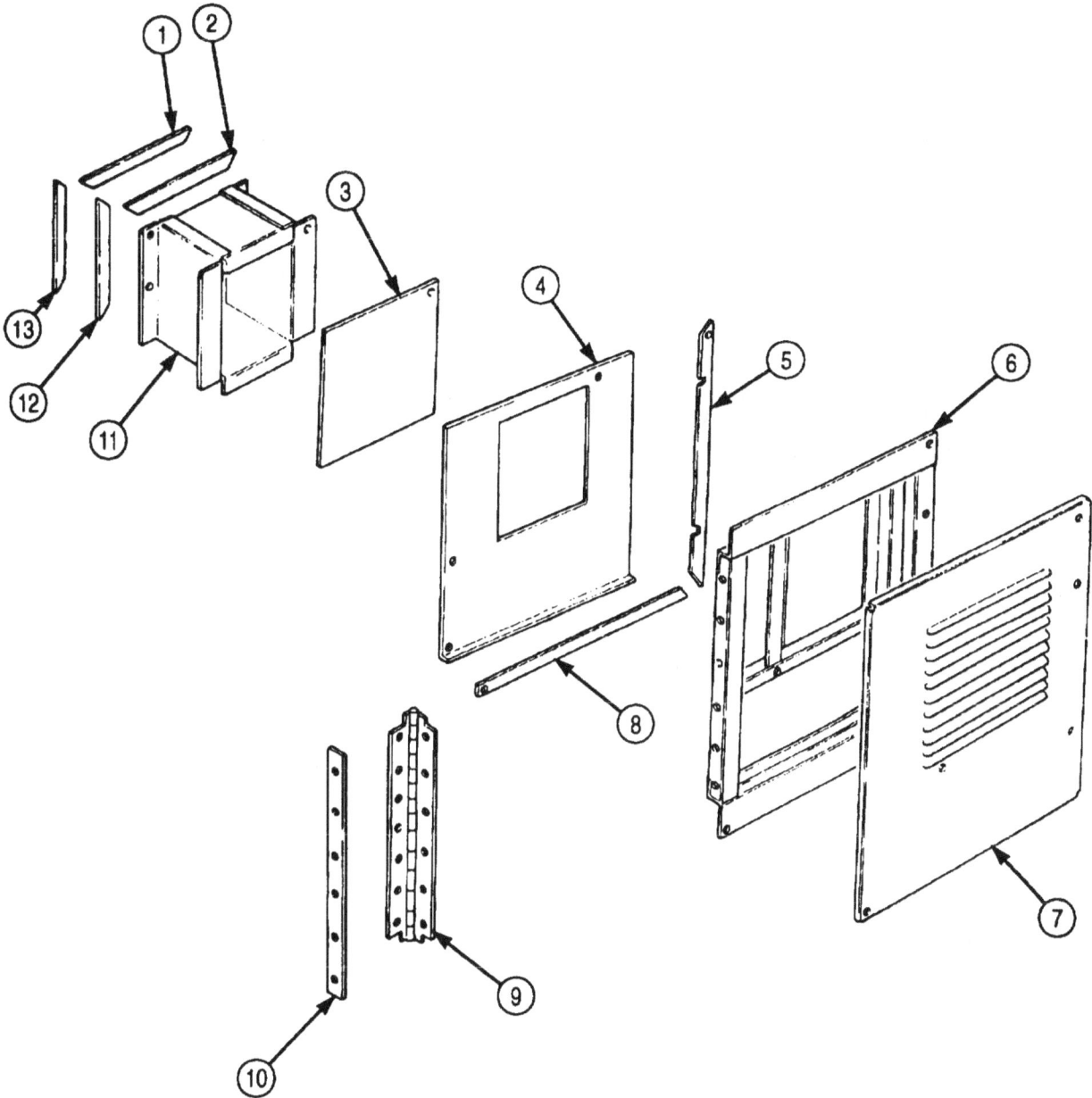

5-87. BONNET DOOR MAINTENANCE (Contd)

d. Assembly

NOTE

- Insulate areas of dissimilar metal-to-metal contact with zinc chromate primer.
- Apply sealing compound to exterior joints prior to installation.
- Insulate all enclosed structure with fibrous glass felt insulation.
- Assistant will help with step 1.

1. Install inner panel (2) on frame (7) with thirty-three new rivets (29) and two new rivets (30).
2. Install new seal (3) and retainer (6) on frame (7) with seven screws (8) and nuts (5).
3. Install two new channel seals (24), channels (23), and channels (25) on intake assembly (17) with ten new rivets (22).
4. Install two new channel seals (19), channels (18), and channels (20) on intake assembly (17) with fourteen new rivets (21).

NOTE

Assistant will help with steps 6 and 7.

5. Install new preformed cork (1), intake panel (26) (if present), and intake assembly (17) on inner panel (2) with twenty-four new rivets (27).
6. Install outer panel (10) on frame (7) with seventeen new rivets (9).
7. Install two retainers (4) in seal (3) with eighteen screws (11) and nuts (28).
8. Install new seal (12), hinge (13), spacer plate (15), and new seal (14) on frame (7) with six screws (16). Ensure hinge is flush with top of door.

5-87. BONNET DOOR MAINTENANCE (Contd)

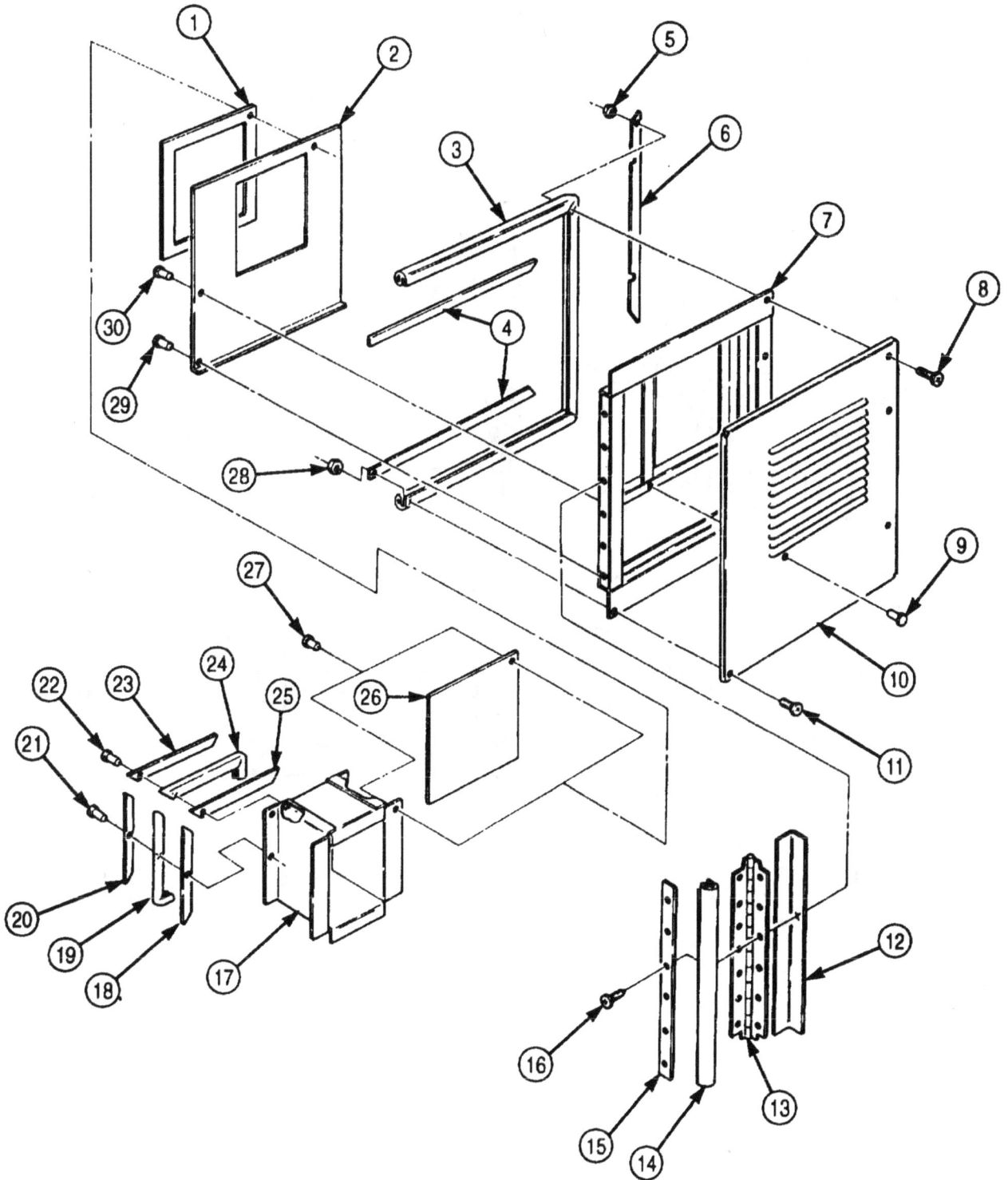

5-87. BONNET DOOR MAINTENANCE (Contd)

e. Installation

NOTE
Assistant will help with step.

Install bonnet door (3) on bonnet (1) with seven screws (2), two screws (4), and washers (5).

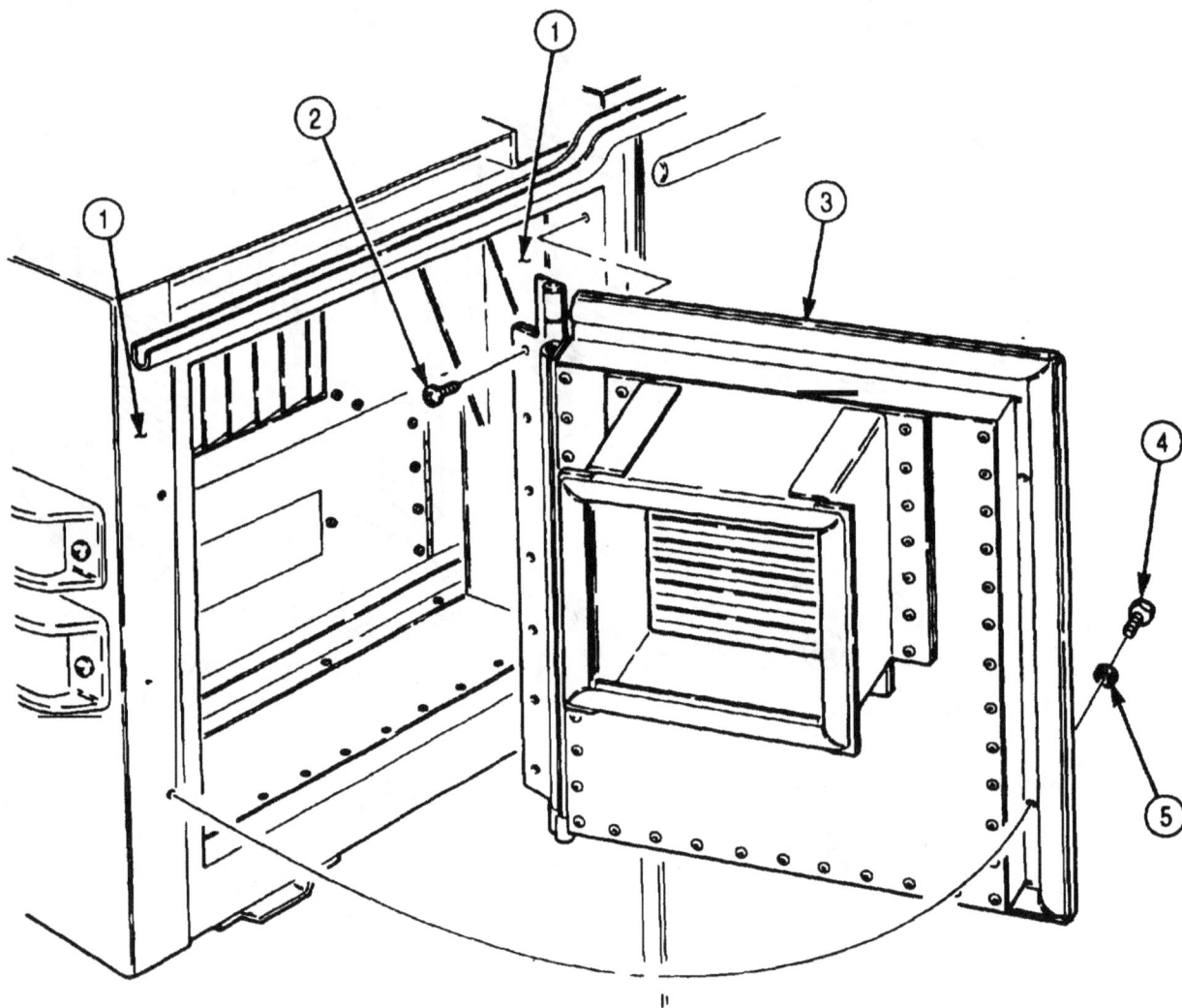

5-88. AIR CONDITIONER REPLACEMENT

THIS TASK COVERS:

a. Removal   b. Installation

APPLICATION MODELS
M934/A1/A2

TOOLS
General mechanic's tool kit (Appendix E, Item 1)
Chain
Lifting device

MATERIALS/PARTS
Cotter pin (Appendix D, Item 57)
Two lockwashers (Appendix D, Item 418)

REFERENCES (TM)
TM 9-2320-272-10
TM 9-2320-272-24P

EQUIPMENT CONDITION
- Parking brake set (TM 9-2320-272-10).
- Cab tarpaulin and bows removed (TM 9-2320-272-10).
- Windshield lowered (TM 9-2320-272-10).
- Companion seat lowered (TM 9-2320-272-10).
- External power source disconnected (TM 9-2320-272-10).
- Battery ground cables disconnected (para. 3-126).
- Air conditioner drain tube removed (para. 3-382).
- Bonnet door removed (para. 5-87).

GENERAL SAFETY INSTRUCTIONS
All personnel must stand clear during lifting operations.

a. Removal

NOTE

Screw quantity for steps 1, 2, and 4 may differ from vehicle to vehicle. Record quantity removed for installation.

1. Remove sixteen screws (6) and lower angle bracket (5) from lower panel mounting angle (7).
2. Remove twelve screws (9) and lower panel mounting angle (7) from bonnet (11).
3. Remove four screws (8) and two mounting angles (4) from bonnet (11).
4. Remove fourteen screws (3), snap (2), and condenser guard (1) from air conditioner (10).

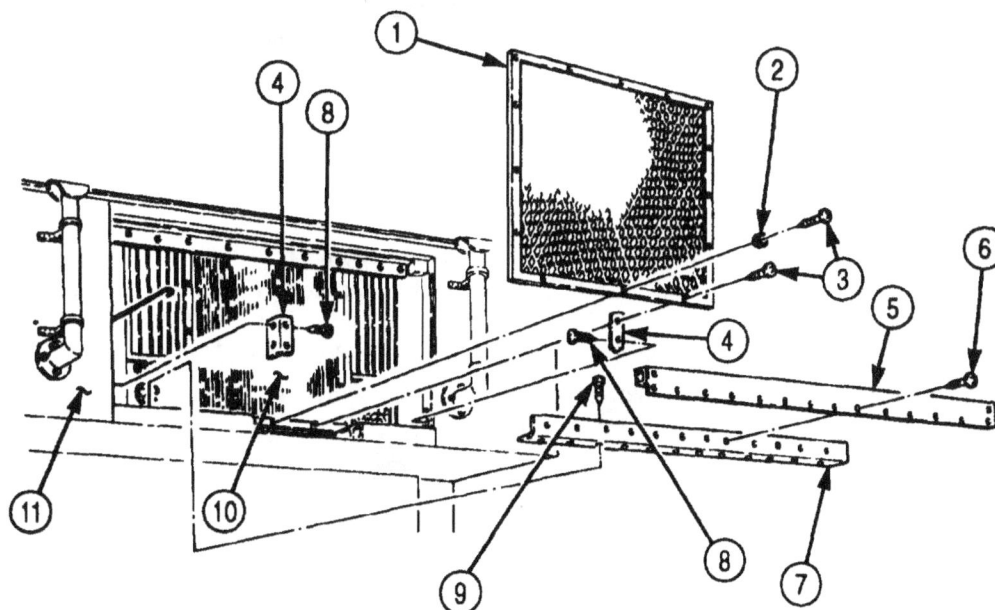

5-88. AIR CONDITIONER REPLACEMENT (Contd)

5. Remove sixteen screws (1) and top enclosure (2) from bonnet (8) and air conditioner (7).

6. Remove twelve screws (4) and right enclosure (3) from bonnet (8) and air conditioner (7).

7. Remove four screws (14), plate (13), and bellows (12) from left enclosure (11).

8. Remove ten screws (9) and left enclosure (11) from bonnet (8) and air conditioner (7).

9. Remove cotter pin (16) and door rod (10) from swing arm (15). Discard cotter pin (16).

10. Remove seven screws (6) and bottom enclosure (5) from bonnet (8) and air conditioner (7).

11. Disconnect power cable (24) from air conditioner (7).

12. Remove screw (26), ground cable (27), and lockwasher (25) from air conditioner (7). Discard lockwasher (25).

13. Remove screw (23), ground cable (27), and lockwasher (28) from mounting plate (18) and bonnet floor (29). Discard lockwasher (28).

14. Remove five screws (22) from mounting plate (18) and bonnet floor (29).

15. Remove six screws (21) from mounting plate (17) and bonnet floor (29).

WARNING

All personnel must stand clear during lifting operations.
A snapped cable, or swinging or shifting load may result
in injury or death to personnel.

NOTE

Two assistants will help with step 16.

16. Attach chains to lifting device and four lifting brackets (20) on air conditioner (7) and remove air conditioner (7) from bonnet (8). Position air conditioner (7) for access to bottom of unit.

17. Remove six screws (19) and mounting plates (17) and (18) from air conditioner (7).

b. Installation

1. Install mounting plates (17) and (18) on air conditioner (7) with six screws (19).

WARNING

All personnel must stand clear during lifting operations.
A snapped cable, or swinging or shifting load may result
in injury or death to personnel.

NOTE

Two assistants will help with step 2.

2. Attach chains to lifting device and four lifting brackets (20).

3. Position air conditioner (7) in bonnet (8).

4. Install mounting plate (17) on bonnet floor (29) with six screws (21).

5. Install ground cable (27) on mounting plate (18) and bonnet floor (29) with new lockwasher (28), and screw (23).

6. Install ground cable (27) on air conditioner (7) with new lockwasher (25), and screw (26).

7. Install mounting plate (18) on bonnet floor (29) with five screws (22).

8. Connect power cable (24) to air conditioner (7).

9. Remove lifting device and chains from air conditioner (7).

5-88. AIR CONDITIONER REPLACEMENT (Contd)

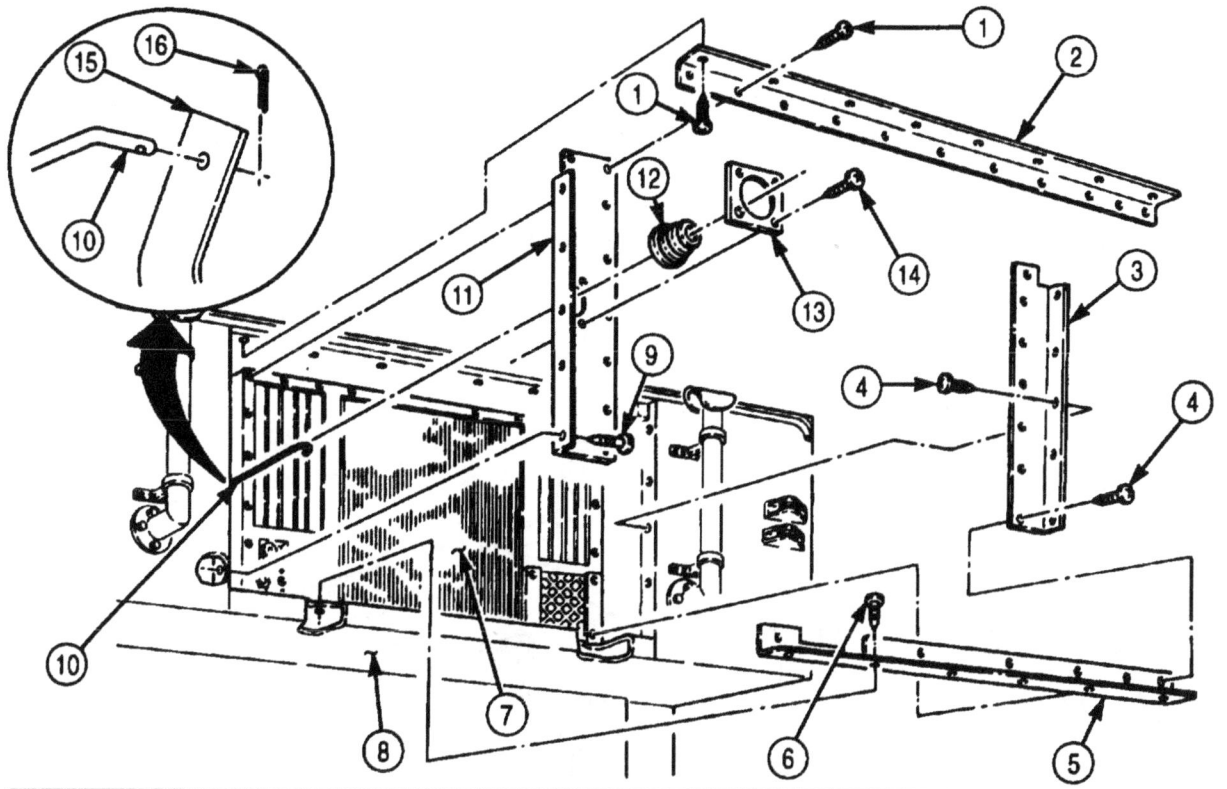

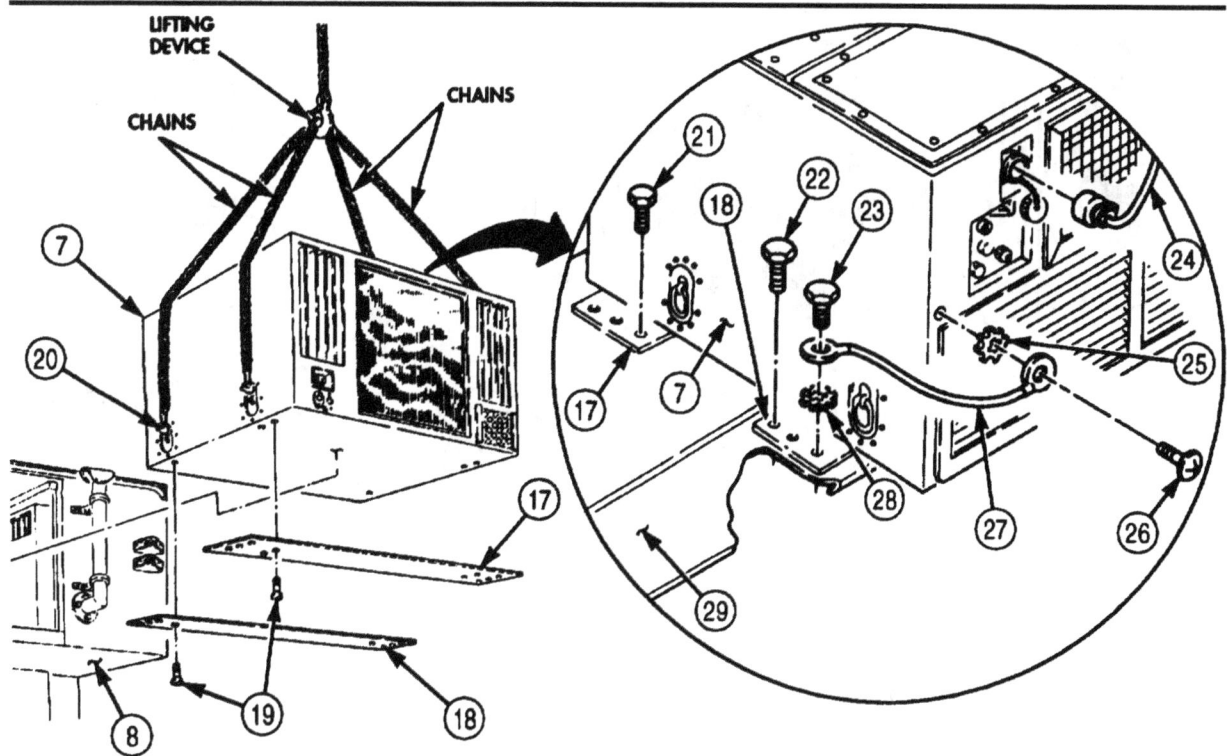

LIFTING DEVICE

CHAINS

CHAINS

5-88. AIR CONDITIONER REPLACEMENT (Contd)

10. Install bottom enclosure (5) on bonnet (7) and air conditioner (8) with seven screws (6).

11. Install door rod (9) on swing arm (15) with new cotter pin (16).

12. Install left enclosure (10) on bonnet (7) and air conditioner (8) with ten screws (11). Ensure door rod (9) is positioned through bellows (12).

13. Install bellows (12) on left enclosure (10) with plate (13) and four screws (14).

14. Install right enclosure (3) on bonnet (7) and air conditioner (8) with twelve screws (4).

15. Install top enclosure (2) on bonnet (7) and air conditioner (8) with sixteen screws (1).

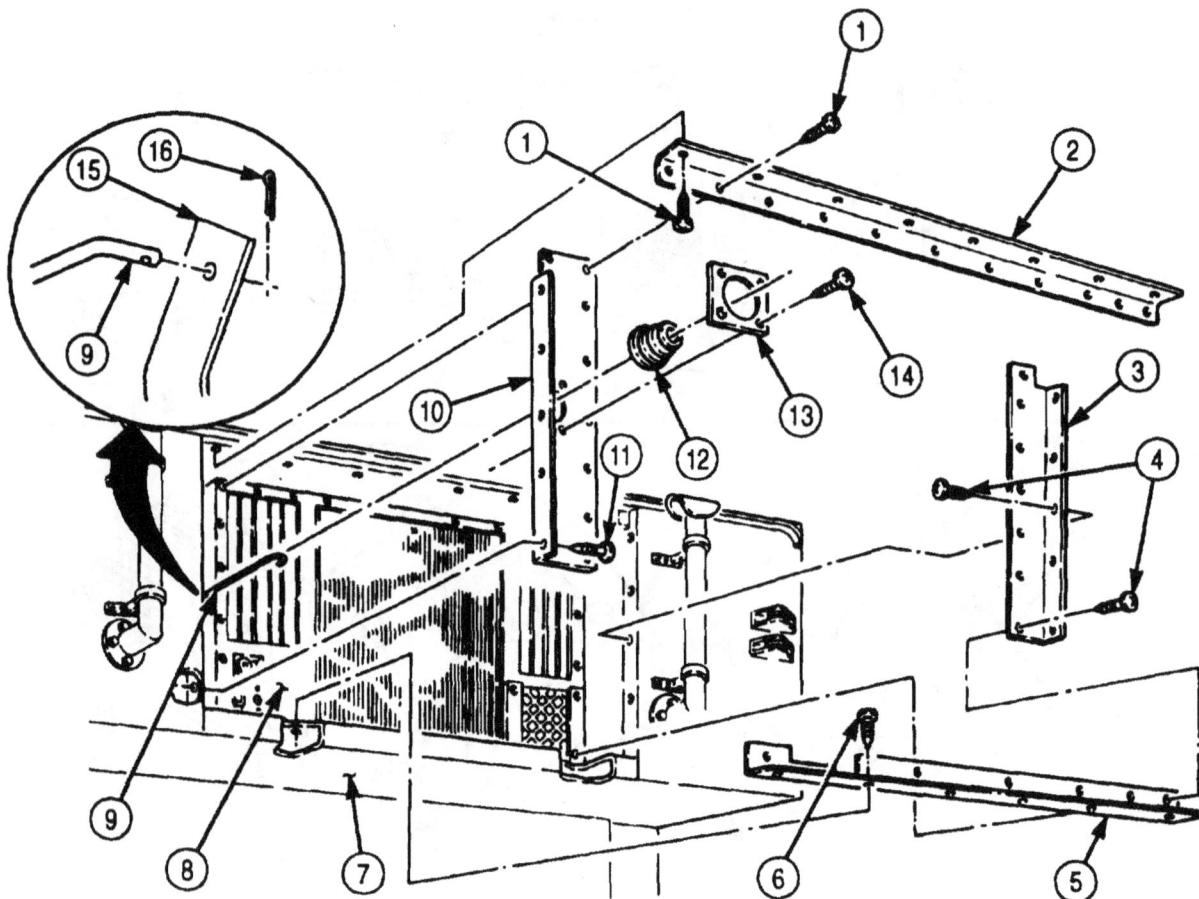

5-88. AIR CONDITIONER REPLACEMENT (Contd)

NOTE

Screw quantity for steps 16, 18, and 19 may differ from vehicle to vehicle. Install screws as recorded.

16. Install condenser guard (17) on air conditioner (8) with snap (18) and fourteen screws (19).

17. Install two mounting angles (22) on bonnet (7) with four screws (24).

18. Install lower panel mounting angle (23) on bonnet (7) with twelve screws (25).

19. Install lower angle bracket (21) on bonnet (7) and two mounting angles (22) with sixteen screws (20).

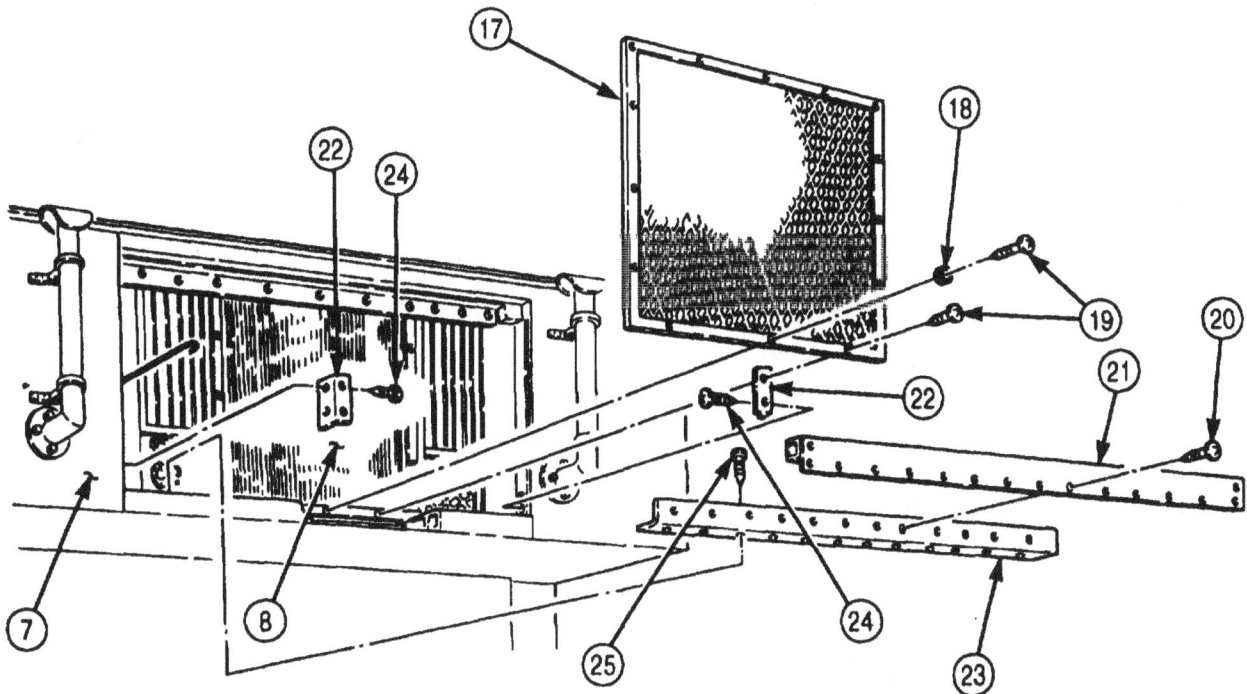

FOLLOW-ON TASKS:
- Install bonnet door (para. 5-87).
- Install air conditioner drain tube (para. 3-382).
- Connect battery ground cables (para. 3-126).
- Connect external power source (TM 9-2320-272-10).
- Raise companion seat (TM 9-2320-272-10).
- Raise windshield (TM 9-2320-272-10).
- Install cab tarpaulin and bows (TM 9-2320-272-10).

5-89. MAIN WIRING HARNESS REPLACEMENT

THIS TASK COVERS:

a. Removal   b. Installation

INITIAL SETUP:

APPLICABLE MODELS
M934/A1/A2

TOOLS
General mechanic's tool kit (Appendix E, Item 1)

REFERENCES (TM)
TM 9-2320-272-10
TM 9-2320-272-24P

EQUIPMENT CONDITION
- Parking brake set (TM 9-2320-272-10).
- Battery ground cables disconnected (para 3-126).
- Fluorescent light tubes removed (para 3-371).
- Emergency/blackout lamp and light removed (para 3-372).
- Blackout light switch and 110-volt receptacle removed (para 3-373).

NOTE
- The left and right main wiring harnesses are replaced the same way. This procedure covers the right main wiring harness.
- Tag wires for installation.

a. Removal

1. Remove six screws (1) and cover (2) from load center (8).
2. Remove three setscrews (9) and wires (3) from relay (10).
3. Remove setscrew (15) and wire (16) from neutral bus (11).
4. Remove three screws (13) and five wires (14) from 20 amp circuit breakers (12).
5. Remove screw (4) and clamp (5) from plate (7) and wiring harnesses (6).
6. Remove twelve screws (22) and clamps (21) from van ceiling (18) and main wiring harness (20).
7. Remove grommet (23) from van side panel (17).
8. Remove three screws (25) and wire clip (24) from van side panel (17).
9. Remove main wiring harness (20) from van body (19).

b. Installation

1. Position main wiring harness (20) on van body (19).
2. Install wire clip (24) on side panel (17) of van body (19) with three screws (25).
3. Install grommet (23) on side panel (17).
4. Install twelve clamps (21) on main wiring harness (20) and van ceiling (18) with twelve screws (22).
5. Install clamp (5) on harnesses (6) and plate (7) with screw (4).
6. Install five wires (14) on 20 amp circuit breakers (12) with three screws (13).
7. Install wire (16) on neutral bus (11) with setscrew (15).
8. Install three wires (3) on relay (10) with three setscrews (9).
9. Install cover (2) on load center (8) with six screws (1).

5-89. MAIN WIRING HARNESS REPLACEMENT (Contd)

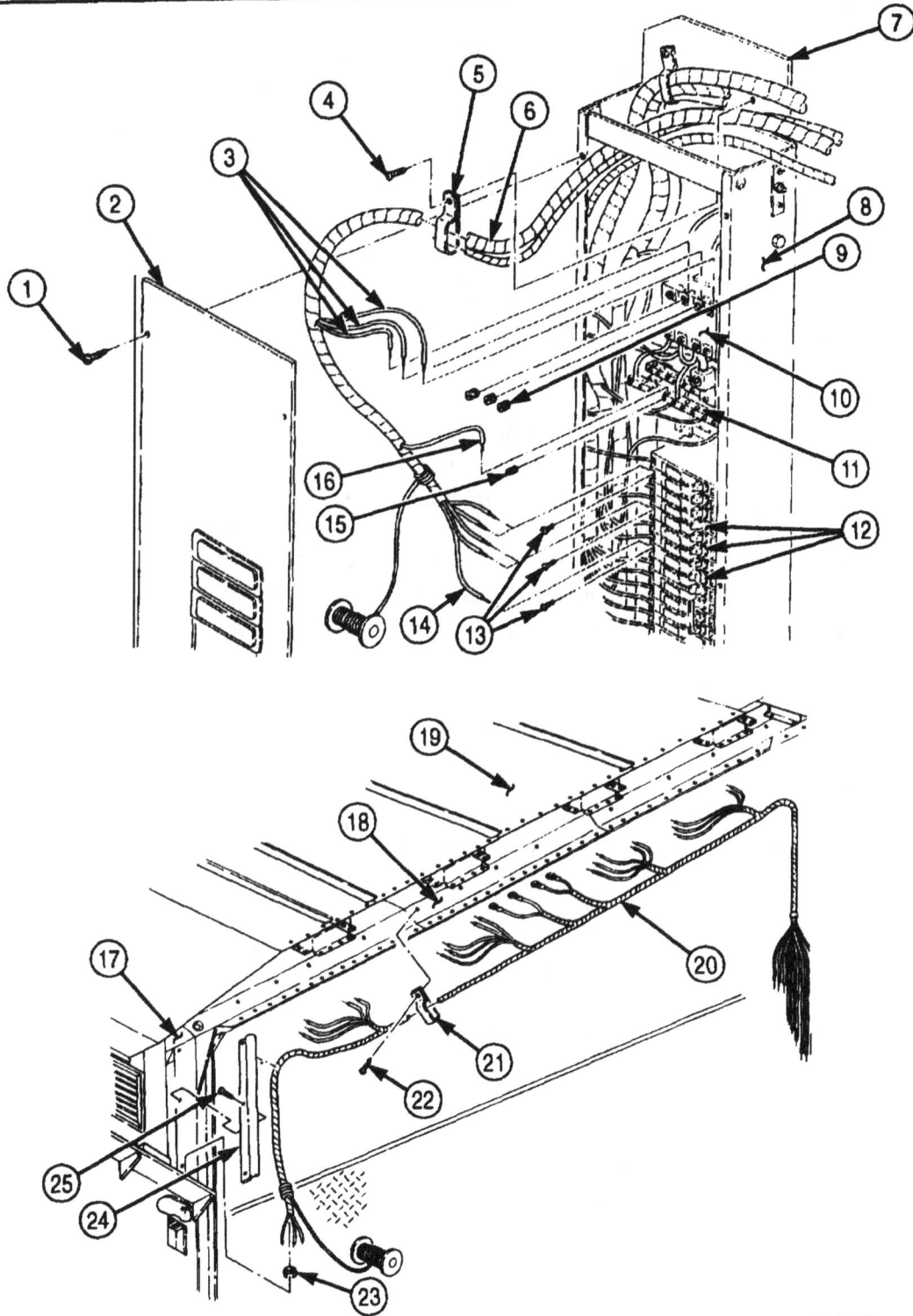

FOLLOW-ON TASKS• Install blackout light switch and 110-volt receptacle (para. 3-373
  • Install emergency/blackout lamp and light (para. 3-372).
  • Install fluorescent light tubes (para. 3-371).
  • Connect battery ground cables (para. 3-126).

5-90. VAN AIR CONDITIONER WIRING HARNESS REPLACEMENT

THIS TASK COVERS:
a. Removal   b. Installation

<u>INITIAL SETUP:</u>

<u>APPLICABLE MODEL</u>
M934/A1/A2

<u>TOOLS</u>
General mechanic's tool kit (Appendix E, Item 1)

<u>REFERENCES</u> (TM)
TM 9-2320-272-24P
TM 9-2320-272-24P

<u>EQUIPMENT CONDITION</u>
- Parking brake set (TM 9-2320-272-10)
- Battery ground cables disconnected (para. 3-126).
- Ceiling filler and side panels removed (para. 5-79).

a. Removal

NOTE
Tag cables for installation.

1. Remove twelve screws (4) and clamps (5) from wiring harness (6) and van ceiling (1).
2. Disconnect plug (2) from air conditioner (3).
3. Remove six screws (22) and cover (21) from load center (16).
4. Remove screw (7) and clamp (8) from wiring harnesses (20) and (23) and plate (12).
5. Remove nut (9), screw (15), two washers (10), cable assembly (14), ground cable (13), and wiring harness cable (11) from load center (16).
6. Remove three screws (19) and wiring harness cables (18) from 30 amp circuit breaker (17).
7. Remove wiring harness (20) from load center (16).

b. Installation

1. Position wiring harness (20) on load center (16).
2. Install three wiring harness cables (18) on 30 amp circuit breaker (17) with three screws (19).
3. Install wiring harness cable (11), ground cable (13), and cable assembly (14) on load center (16) with screw (15), two washers (10), and nut (9).
4. Install clamp (8) on wiring harnesses (23) and (20) and plate (12) with screw (7).
5. Install cover (21) on load center (16) with six screws (22).
6. Connect plug (2) to air conditioner (3).
7. Install twelve clamps (5) on wiring harness (6) and van ceiling (1) with twelve screws (4).

5-90. VAN AIR CONDITIONER WIRING HARNESS REPLACEMENT (Contd)

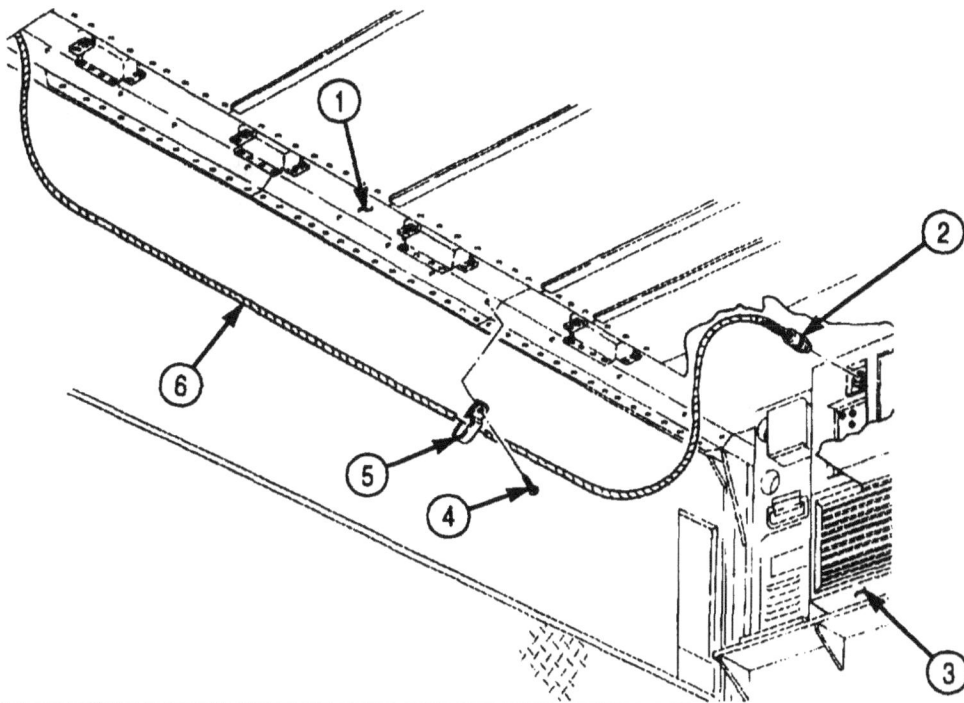

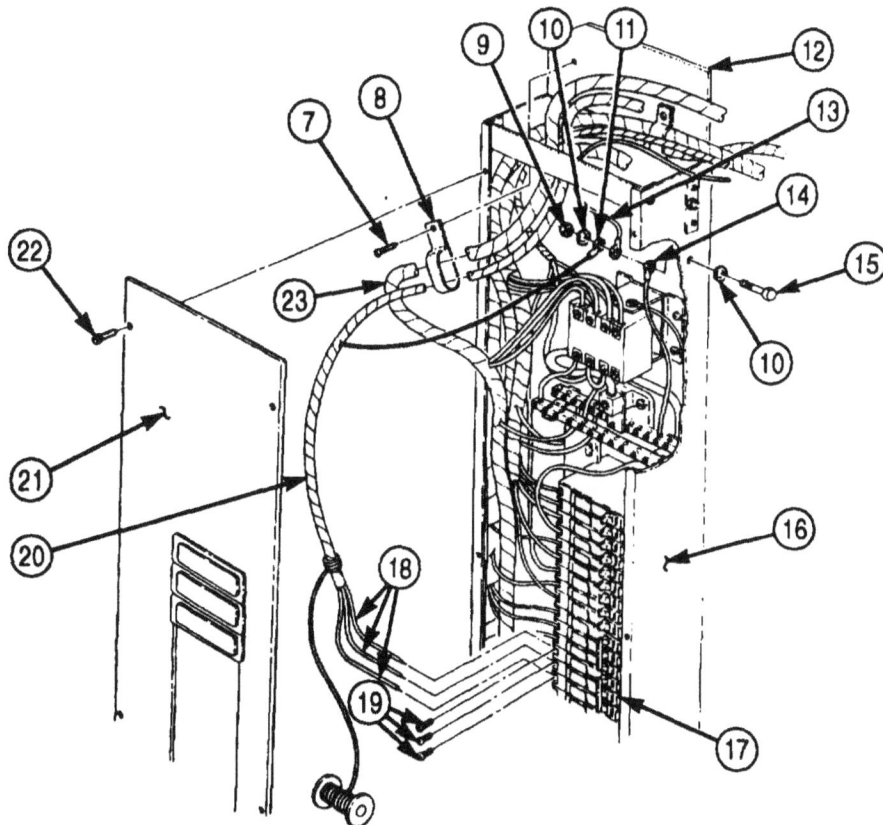

FOLLOW-ON TASKS: Install ceiling filler and side panels (para. 5-79).
• Connect battery ground cables (para. 3-126).

5-91. VAN HEATER WIRING HARNESS REPLACEMENT

THIS TASK COVERS:

a. Removal   b. Installation

INITIAL SETUP:

APPLICABLE MODELS
M934/A1/A2

TOOLS
General mechanic's tool kit (Appendix E, Item 1)

REFERENCES (TM)
TM 9-2320-272-10
TM 9-2320-272-24P

EQUIPMENT CONDITION
- Parking brake set (TM 9-2320-272-10)
- Battery ground cables disconnected (para. 3-126).
- Ceiling filler and side panels removed (para. 5-79).

---

a. Removal

NOTE
Tag wires for installation.

1. Remove twelve screws (6) and clamps (7) from van ceiling (1) and wiring harness (5).

2. Remove two grommets (3) from van ceiling (1) and wiring harness (5).

3. Disconnect two plugs (2) from heater (4).

5-91. VAN HEATER WIRING HARNESS REPLACEMENT (Contd)

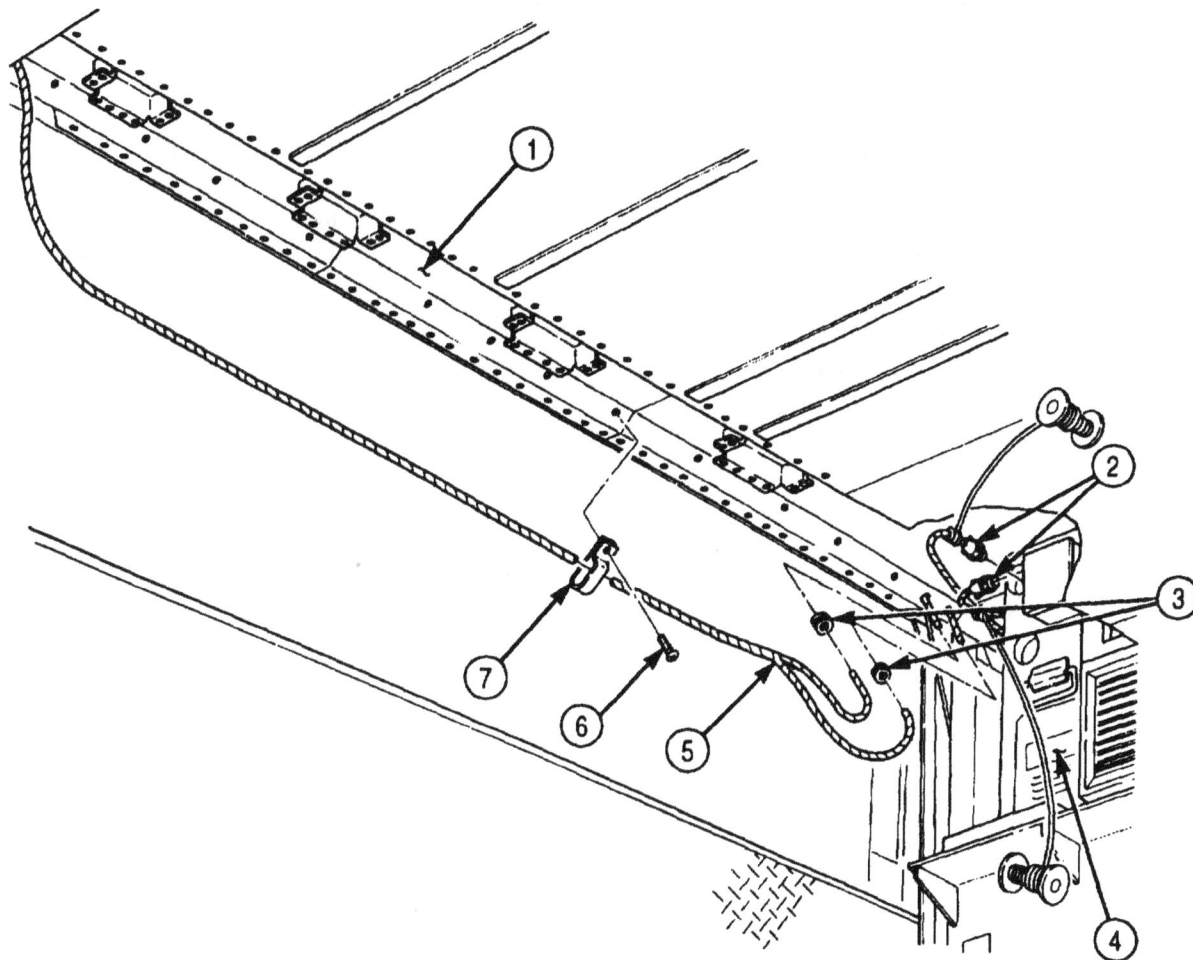

5-91. VAN HEATER WIRING HARNESS REPLACEMENT (Contd)

4. Remove thermostat cover (5) from thermostat (6).
5. Remove four screws (4) and thermostat (6) from control center box (9).
6. Remove two screws (2) and wires (3) from control center box (9).
7. Remove four screws (7) and upper cover (1) from control center box (9).
8. Remove six screws (17) and cover (16) from load center (12).
9. Remove screw (10) and clamp (11) from wiring harnesses (18).
10. Remove setscrew (14), screw (13), and two wires (15) from load center (12).
11. Remove wiring harnesses (18) from van body (8).

b. Installation

1. Position wiring harness (18) on van body (8) and install two wires (15) on load center (12) with screw (13) and setscrew (14).
2. Place clamp (11) on wiring harnesses (18) and install on load center (12) with screw (10).
3. Install cover (16) on load center (12) with six screws (17).
4. Install upper cover (1) on center control box (9) with four screws (7).
5. Install two wires (3) on thermostat (6) with two screws (2).
6. Install thermostat (6) on upper cover (1) with two screws (4).
7. Install thermostat cover (5) on thermostat (6).

5-91. VAN HEATER WIRING HARNESS REPLACEMENT (Contd)

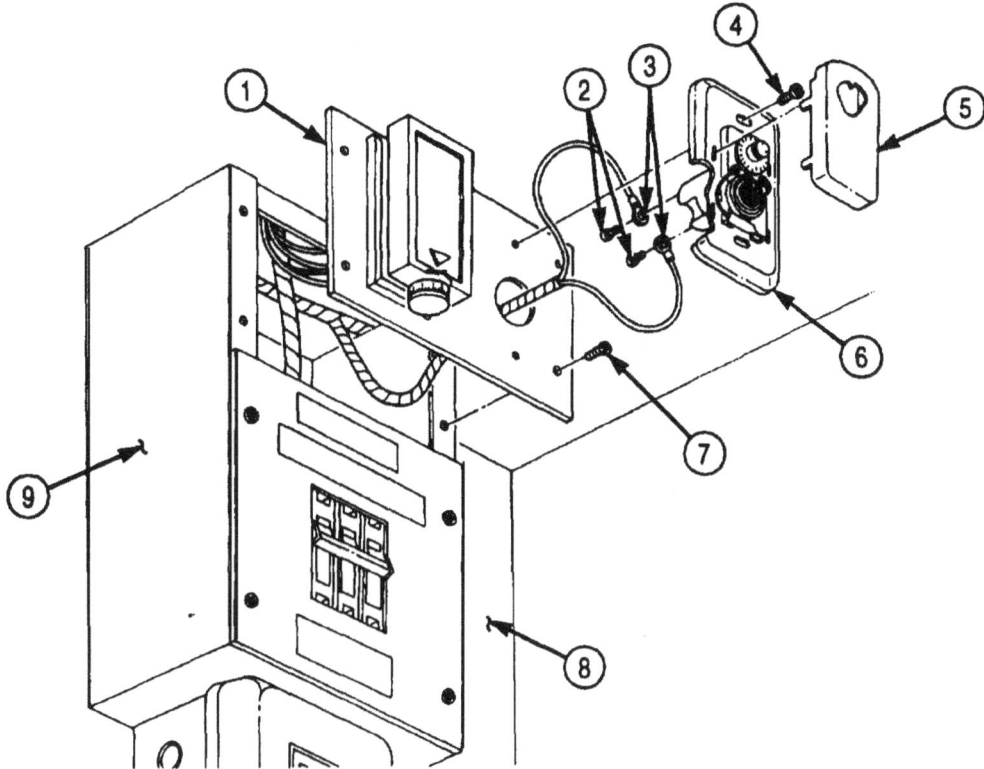

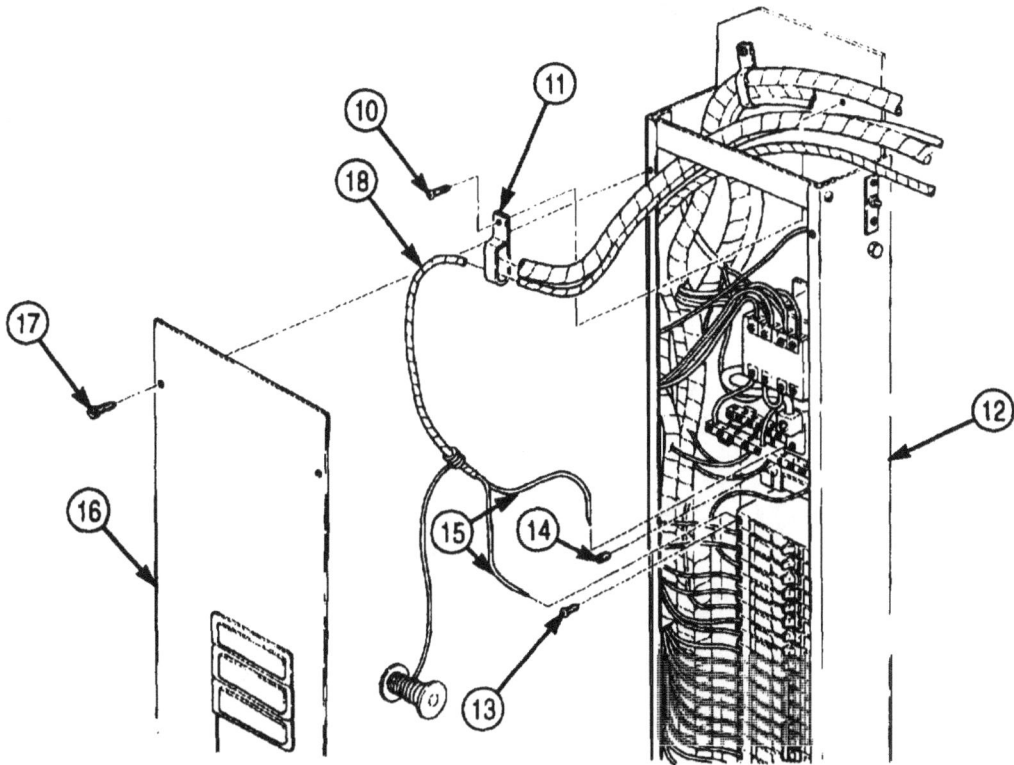

5-91. VAN HEATER WIRING HARNESS REPLACEMENT (Contd)

8. Connect two plugs (2) to heater (4).

9. Install two grommets (3) on wiring harness (5) and van ceiling (1).

10. Install twelve clamps (7) on wiring harness (5) and van ceiling (1) with twelve screws (6).

5-91. VAN HEATER WIRING HARNESS REPLACEMENT(Contd)

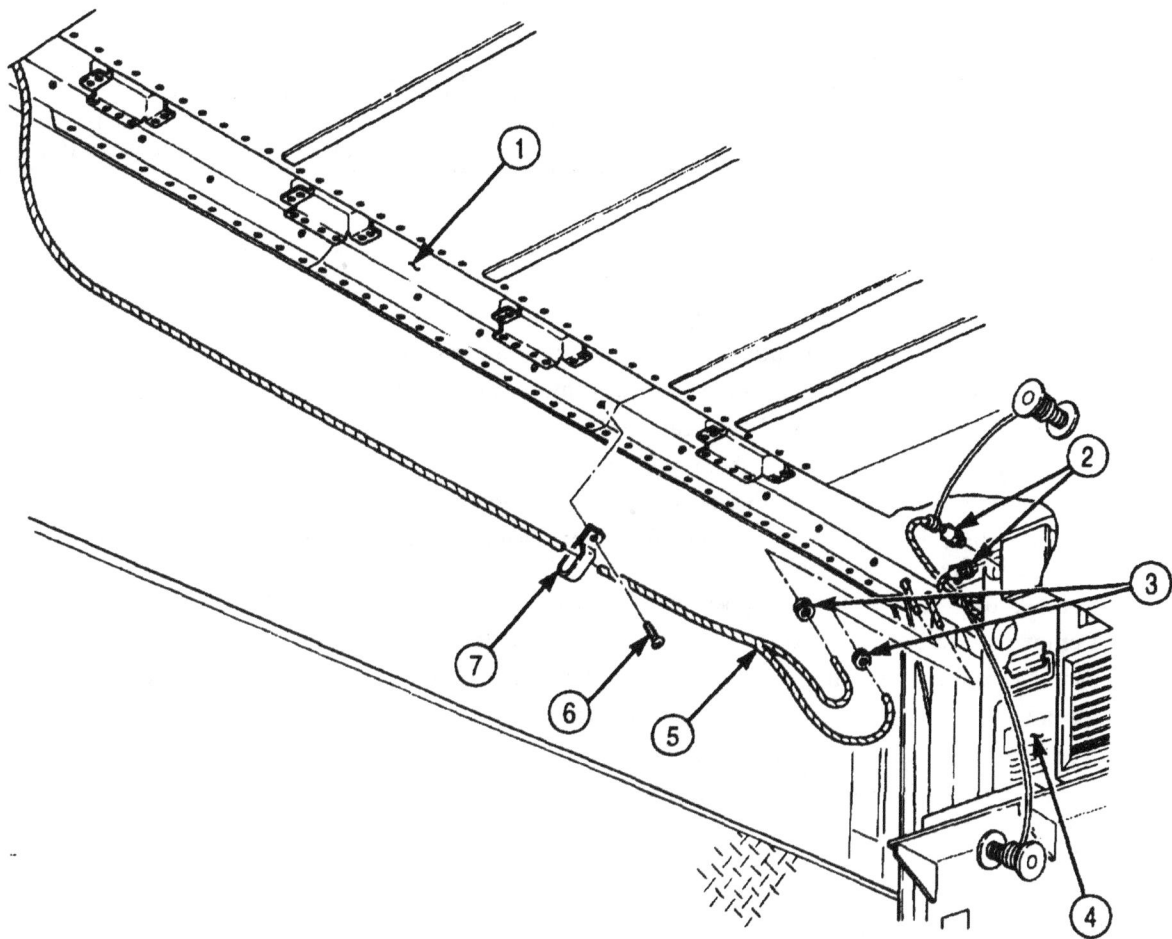

FOLLOW-ON TASKS:• Install ceiling filler and side panels (para. 5-79).
• Connect battery ground cables (para. 3-126).

5-92. ELECTRIC HEATER (10 KW) WIRING HARNESS REPLACEMENT

THIS TASK COVERS:

a. Removal   b. Installation

INITIAL SETUP:

APPLICABLE MODELS   REFERENCES (TM)
M934/A1/A2   TM 9-2320-272-10
                                           TM 9-2320-272-24P
TOOLS
General mechanic's tool kit (Appendix E, Item 1) EQUIPMENT CONDITION
Electrical tool kit (Appendix E, Item 40)   • Parking brake set (TM 9-2320-272-10).
                                           • Battery ground cables disconnected (para. 3-126).
                                           • Ceiling filter and side panels removed (para. 5-79).

---

a. Removal

NOTE
Tag wires for installation.

1. Remove twelve screws (28) and clamps (27) from van ceiling (25) and wiring harness (26).
2. Remove grommet (24) from van ceiling (25) and wiring harness (26).
3. Remove wires (29) and (23) from heater (30).
4. Remove thermostat cover (4) from thermostat (5).
5. Remove two screws (3) and thermostat (5) from upper cover (7).
6. Remove three insulated wire splicers (6) from wiring harness (8) and thermostat leads (2). Discard insulated wire splicers (6).
7. Remove four screws (1) and upper cover (7) from control center box (10).
8. Remove six screws (22) and cover (11) from load center (15).
9. Remove screw (12), clamp (13), and wiring harnesses (21) and (8) from load center (15).
10. Remove screw (20) and wire (19) from neutral bus (14).
11. Remove three screws (17) and wires (18) from 20 amp circuit breakers (16).
12. Remove wiring harness (8) from van body (9).

b. Installation

1. Position wiring harness (8) on van body (9).
2. Install three wires (18) on 20 amp circuit breakers (16) with three screws (17).
3. Install wire (19) on neutral bus (14) with screw (20).
4. Install clamp (13) on wiring harnesses (8) and (21) and load center (15) with screw (12).
5. Install cover (11) on load center (15) with six screws (22).
6. Install upper cover (7) on control center box (10) with four screws (1).
7. Connect thermostat leads (2) to wiring harness (8) with three new insulated wire splicers (6).
8. Install thermostat (5) on upper cover (7) with two screws (3).
9. Install thermostat cover (4) on thermostat (5).
10. Install wires (23) and (29) on heater (30).

5-92. ELECTRIC HEATER (10 KW) WIRING HARNESS REPLACEMENT (Contd)

11. Install grommet (24) on wiring harness (26) and van ceiling (25).

12. Install twelve clamps (27) on wiring harness (26) and van ceiling (25) with twelve screws (28).

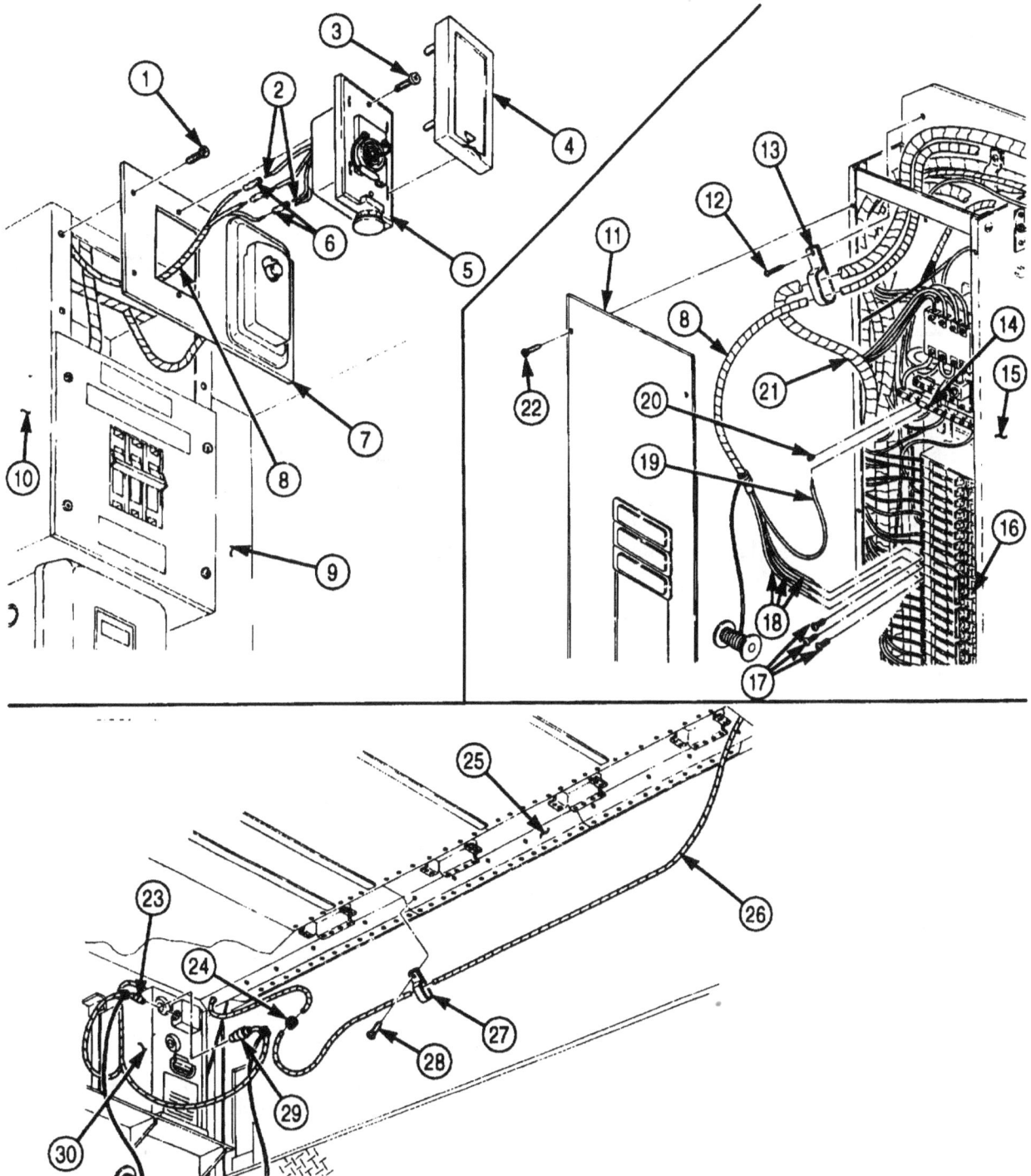

FOLLOW-ON TASKS: Install ceiling filler and side panels (para. 5-79).
  • Connect battery ground cables (para. 3-126).

5-93. BLACKOUT BYPASS WIRING HARNESS MAINTENANCE

THIS TASK COVERS:

a. Removal   c. Installation
b. Repair

INITIAL SETUP:

APPLICABLE MODELS   REFERENCES (TM)
M934/A1/A2   TM 9-2320-272-10
                                     TM 9-2320-272-24P
TOOLS
General mechanic's tool kit (Appendix E, Item 1) EQUIPMENT CONDITION
                                     • Parking brake set (TM 9-2320-272-10).
                                     • Battery ground cabled disconnected (para. 3-126)

---

a. Removal

NOTE
Tag wires for installation

1. Remove six screws (1) and cover (2) from load center (6).
2. Remove two setscrews (10) and wires (3) from 3-pole circuit breaker (5).
3. Remove screw (8) and wire (9) from 30 amp circuit breaker (7).
4. Remove two screws (13) and cover (14) from switch (11).
5. Remove two screws (12) and switch (11) from van body (16).
6. Loosen three screws (15) and remove wires (3) from switch (11).
7. Remove wiring harness (4) from van body (16).

b. Repair

For wiring harness repair, refer to para. 3-131.

c. Installation

1. Position wiring harness (4) on van body (16).
2. Install three wires (3) on switch (11) and tighten three screws (15).
3. Install switch (11) on van body (16) with two screws (12).
4. Install cover (14) on switch (11) with two screws (13).
5. Install wire (9) on 30 amp circuit breaker (7) with screw (8).
6. Install two wires (3) on 3-pole circuit breaker (5) with two setscrews (10).
7. Install cover (2) on load center (6) with six screws (1).

5-93. BLACKOUT BYPASS WIRING HARNESS MAINTENANCE (Contd)

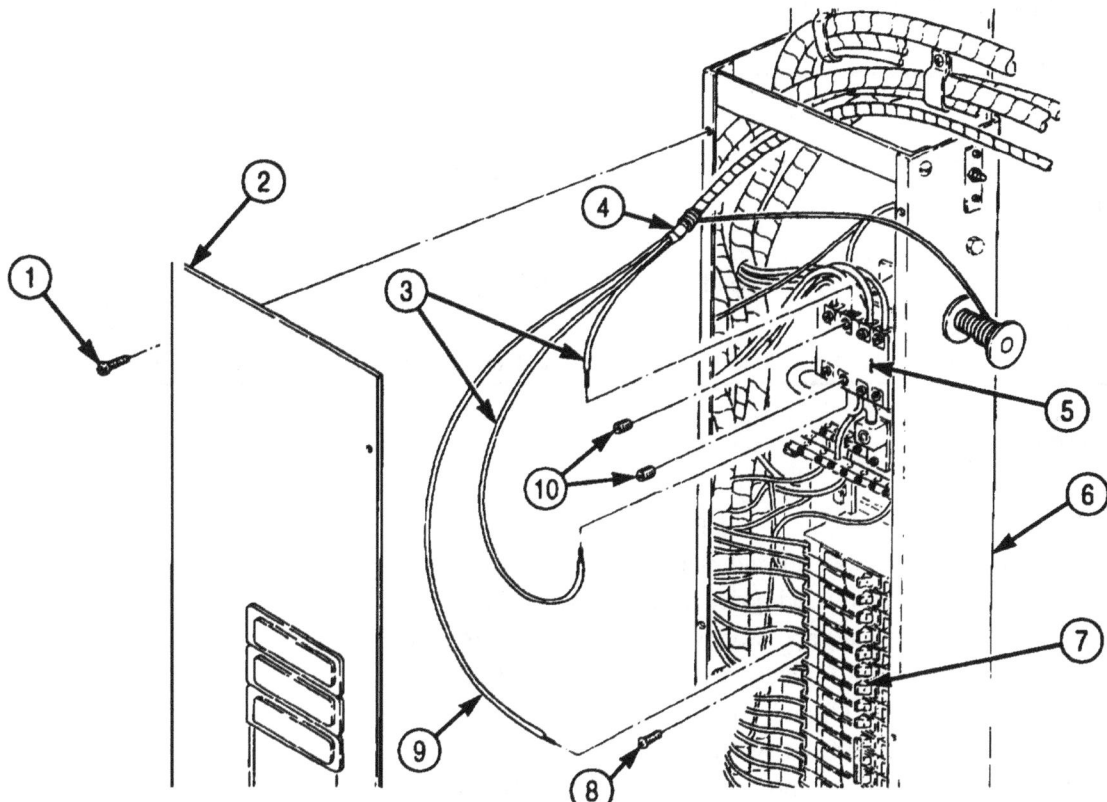

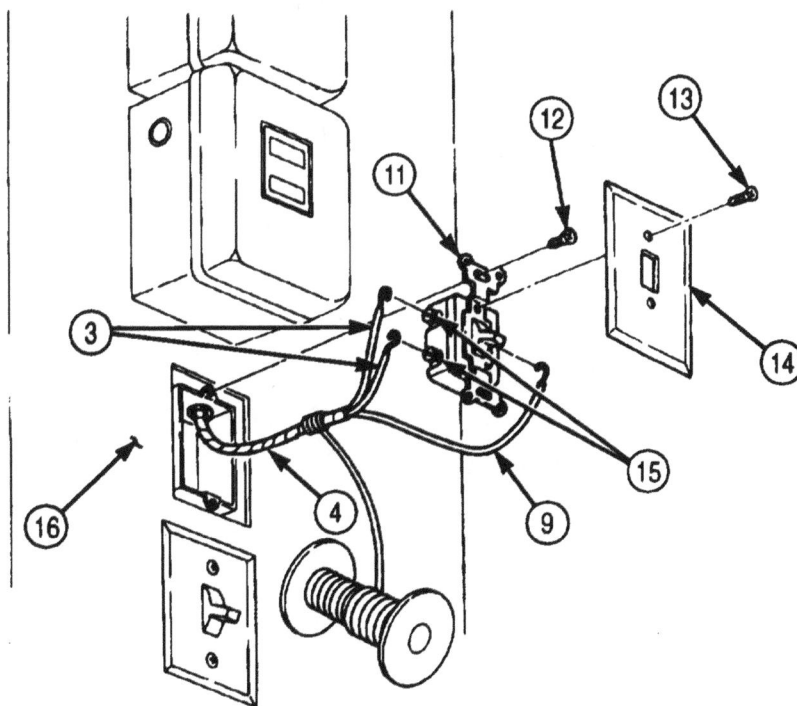

FOLLOW-ON TASK: Connect battery ground cables (para. 3-126).

5-94. EMERGENCY LAMP WIRING HARNESS REPLACEMENT

THIS TASK COVERS:

a. Removal   b. Installation

INITIAL SETUP:

APPLICABLE MODELS
M934/A1/A2

TOOLS
General mechanic's tool kit (Appendix E, Item 1)

MATERIALS/PARTS
Locknut (Appendix D, Item 294)

REFERENCES (TM)
TM 9-2320-272-10
TM 9-2320-272-24P

EQUIPMENT CONDITION
• Parking brake set (TM 9-2320-272-10).
• Battery ground cables disconnected (para. 3-126).
• Ceiling filter and side panels removed (para. 5-79).

---

a. Removal

NOTE
Tag wires for installation.

1. Remove twelve screws (16) and clamps (15) from van ceiling (13) and wiring harness (14).
2. Remove two screws (6) and cover (5) from switch (8).
3. Remove two screws (7) and switch (8) from switch box (10).
4. Disconnect two wires (4) and wire (9) from switch (8).
5. Remove four screws (3) and box (10) from van wall (2).
6. Remove grommet (1) from switch box (10).
7. Disconnect emergency lamp wiring harness lead (11) from emergency lamp wiring harness lead (12).
8. Remove twelve screws (16) and clamps (15) from wiring harness (17) in van ceiling (13).
9. Remove grommet (18) and wiring harness (17) from van ceiling (13).

5-94. EMERGENCY LAMP WIRING HARNESS REPLACEMENT (Contd)

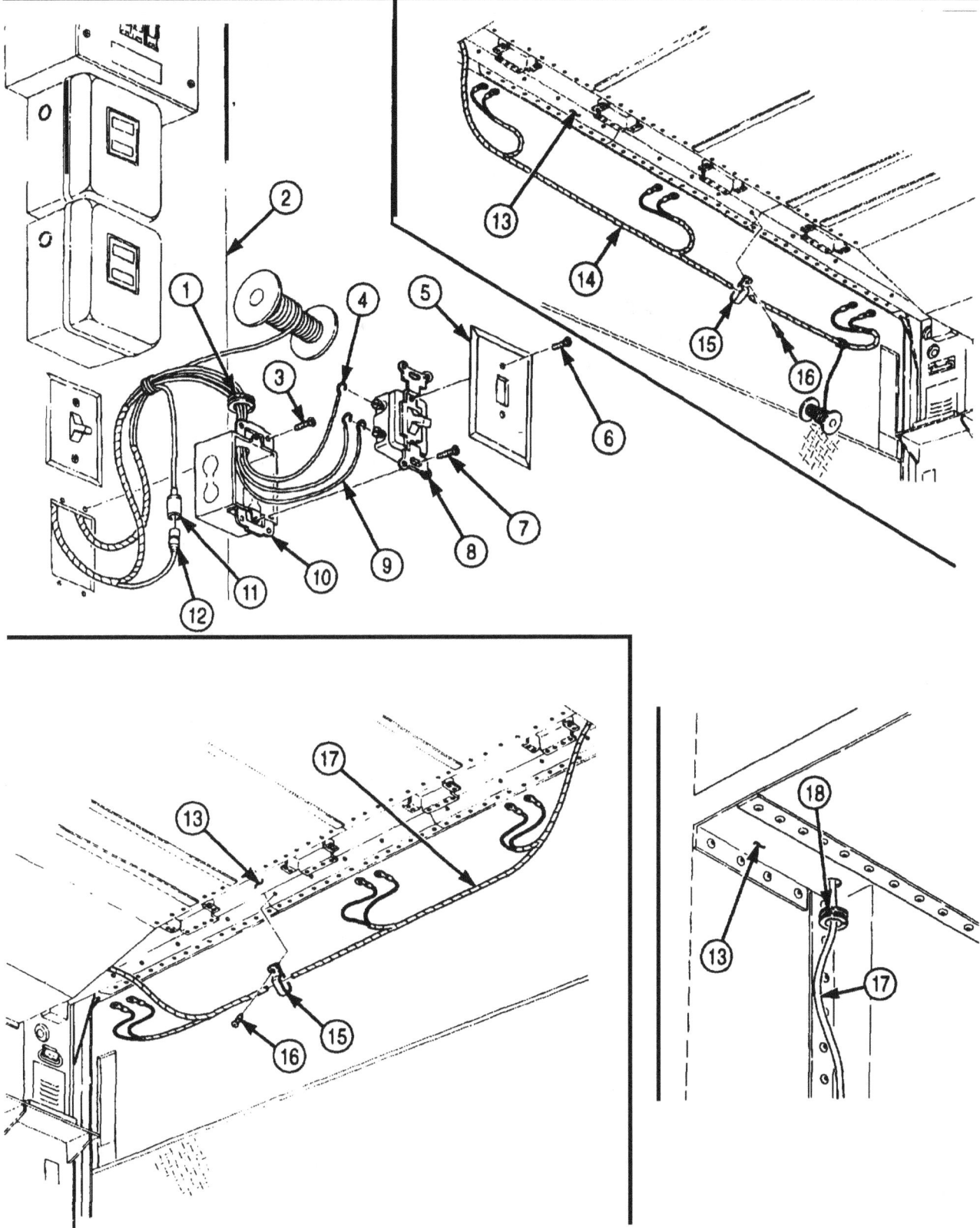

5-94. EMERGENCY LAMP WIRING HARNESS REPLACEMENT (Contd)

10. Disconnect lead (2) from circuit breaker (1).
11. Remove two screws (5) and clamps (4) from wiring harness (3) and van body (12).
12. Remove two screws (11) and clamps (6) from wiring harness (3) and van body (12).
13. Remove locknut (8), screw (10), clamp (9), and wiring harness (3) from spare tire carrier bracket (7). Discard locknut (8).
14. Remove nut (13) and wire (15) from positive battery terminal (14).

b. Installation

1. Install wire (15) on positive battery terminal (14) with nut (13).
2. Install wiring harness (3) on spare tire carrier bracket (7) with clamp (9), screw (10), and new locknut (8).
3. Install two clamps (6) on wiring harness (3) and van body (12) with two screws (11).
4. Install two clamps (4) on wiring harness (3) and van body (12) with two screws (5).
5. Connect lead (2) to circuit breaker (1).

5-94. EMERGENCY LAMP WIRING HARNESS REPLACEMENT (Contd)

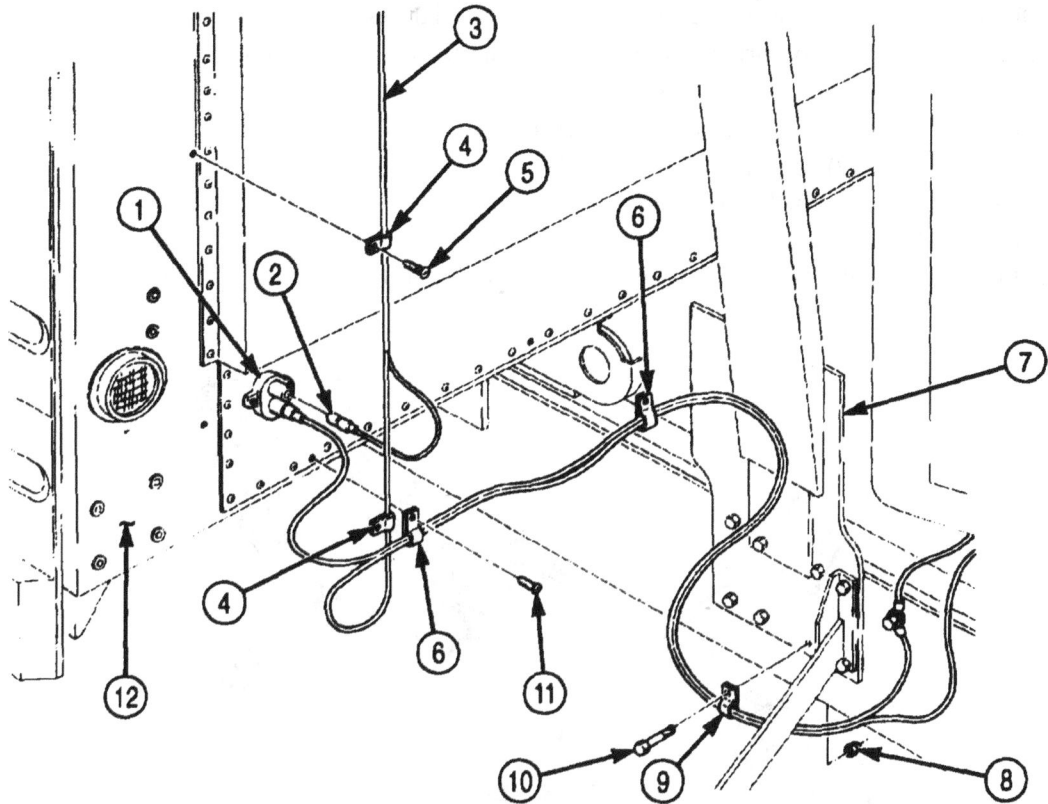

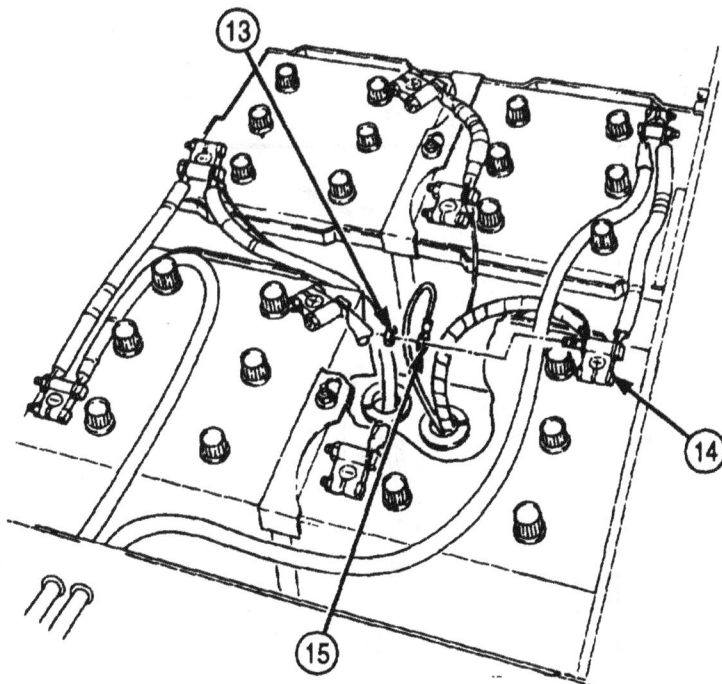

5-94. EMERGENCY LAMP WIRING HARNESS REPLACEMENT (Contd)

6. Install grommet (5) on van ceiling (1) and wiring harness (2).
7. Install twelve clamps (3) on van ceiling (1) and wiring harness (2) with twelve screws (4).
8. Connect emergency lamp wiring harness (14) to emergency lamp wiring harness lead (15).
9. Install grommet (17) on switch box (13).
10. Install two cables (7) and cable (12) on switch (11).
11. Install switch box (13) on van wall (6) with four screws (16).
12. Install switch (11) on switch box (13) with two screws (10).
13. Install cover (9) on switch (11) with two screws (8).
14. Install emergency lamp wiring harness (18) on van ceiling (1) with twelve clamps (19) and screws (20).

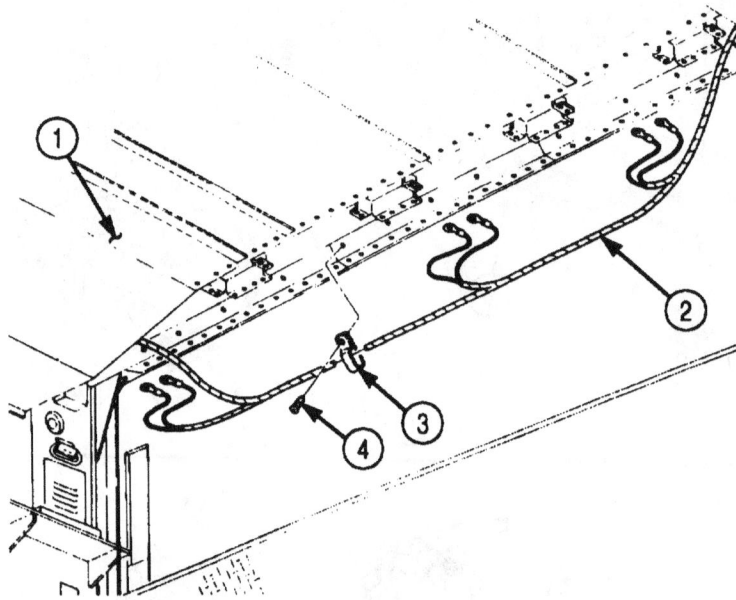

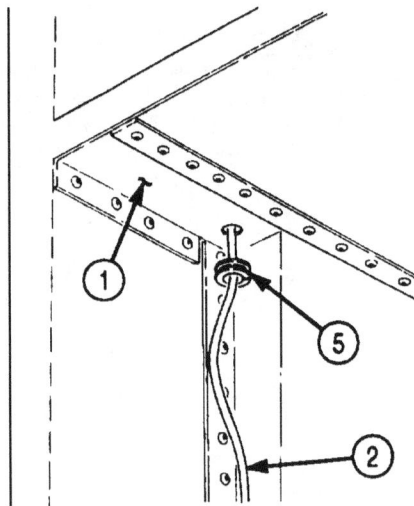

5-94. EMERGENCY LAMP WIRING HARNESS REPLACEMENT (Contd)

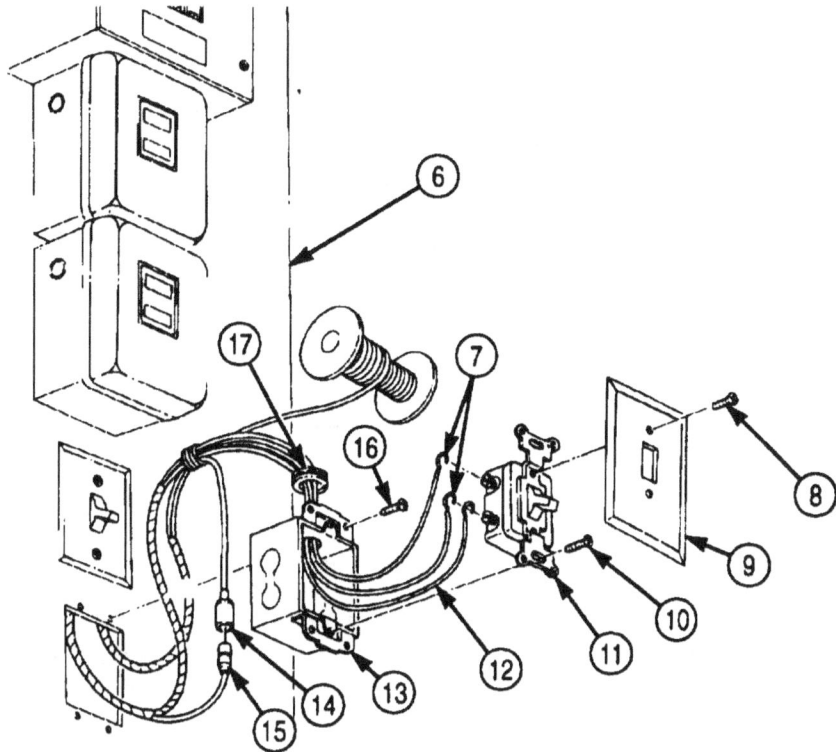

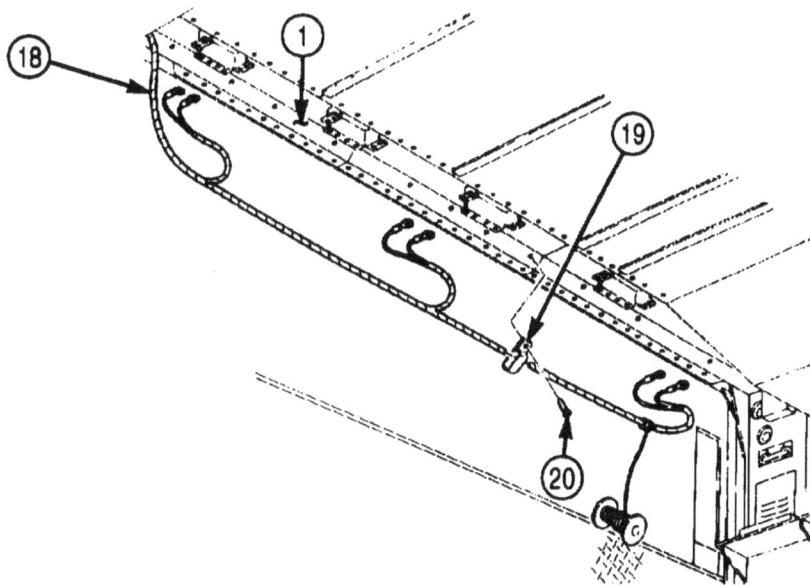

FOLLOW-ON TASKS: • Install ceiling filler and side panels (para. 5-79).
• Connect battery ground cables (para. 3-126).

5-95. BLACKOUT AND CLEARANCE LIGHTS WIRING HARNESS REPLACEMENT

THIS TASK COVERS:

a. Removal   b. Installation

INITIAL SETUP:

APPLICABLE MODELS
M934/A1/A2

TOOLS
General mechanic's tool kit (Appendix E, Item 1)

REFERENCES (TM)
TM 9-2320-272- 10
TM 9-2320-272-24P

EQUIPMENT CONDITION
- Parking brake set (TM 9-2320-272-10X
- Main power switch off (TM 9-2320-272-10).
- Clearance and blackout marker lights removed (para. 4-166).
- Rear wall interior panels removed (para. 5-74).
- Ceiling filler and side panels removed (para. 5-79).
- Ceiling air ducts removed (para. 5-81).
- Battery ground cables disconnected (para. 3-126).

---

a. Removal

NOTE
Tag wires for installation.

1. Remove grommet (2) from van body (3) and wiring harness (1).
2. Remove two screws (6) and clamps (7) from van body (3) and wiring harness (1).
3. Disconnect two wiring harness connectors (4) from rear wiring harness connectors (5).
4. Remove twelve screws (11) and clamps (12) from van ceiling (13) and wiring harness (1).
5. Remove five screws (10) and clamps (9) from van ceiling (13) and wiring harness (1).
6. Remove four grommets (8) from van ceiling (13) and wiring harness (1).
7. Remove wiring harness (1) from van body (14).

b. Installation

1. Position wiring harness (1) on van body (14) and install four grommets (8) on van ceiling (13) and wiring harness (1).
2. Install five clamps (9) on van ceiling (13) and wiring harness (1) with five screws (10).
3. Install twelve clamps (12) on van ceiling (13) and wiring harness (1) with twelve screws (11).
4. Connect two wiring harness connectors (4) to rear wiring harness connectors (5).
5. Install two clamps (7) on wiring harness (1) and van body (3) with two screws (6).
6. Install grommet (2) on wiring harness (1) and van body (3).

5-95. BLACKOUT AND CLEARANCE LIGHTS WIRING HARNESS REPLACEMENT (Contd)

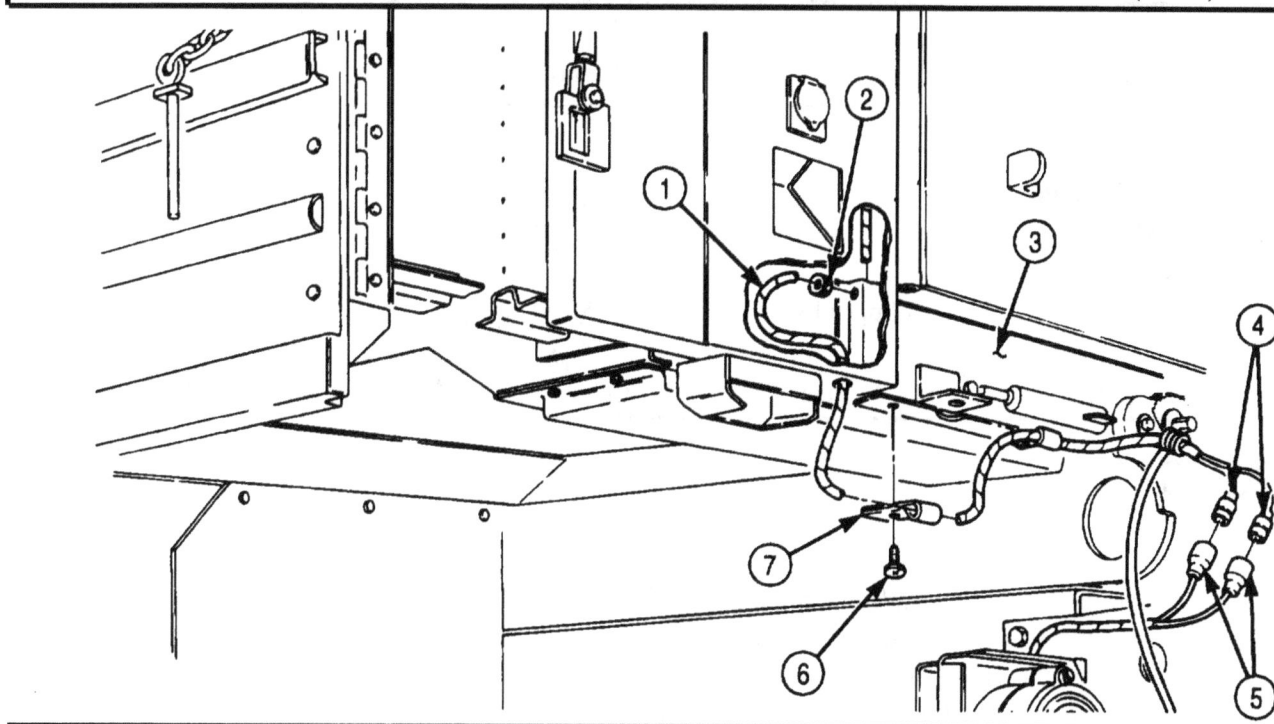

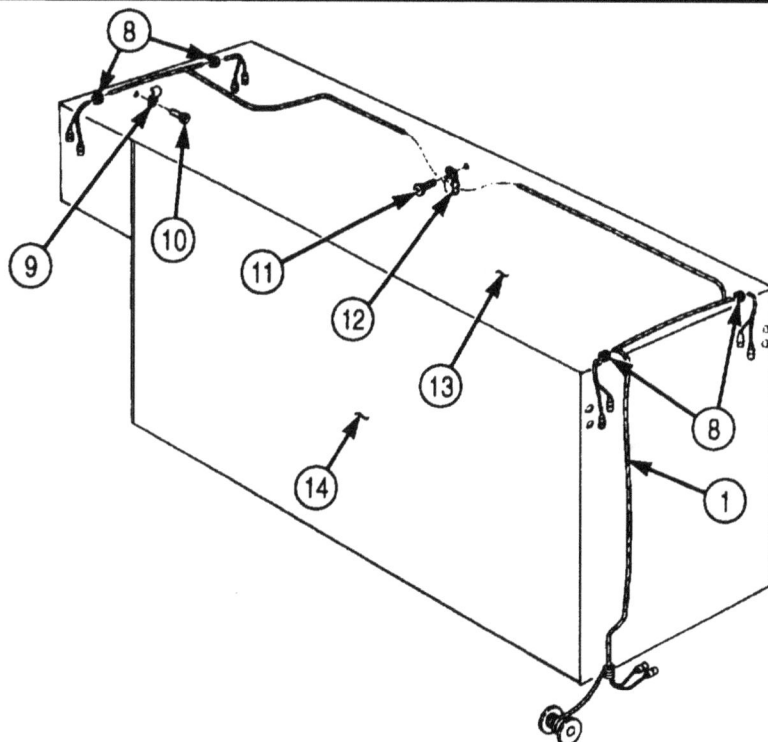

FOLLOW-ON TASKS: • Install ceiling air ducts (para. 5-81).
• Install ceiling filler and side panels (para. 5-79).
• Install clearance and blackout marker lights (para. 4-166).
• Install rear wall interior panels (para. 5-74).
• Connect battery ground cables (para. 3-126).
• Main power switch on (TM 9-2320-272-10).

5-96. 400 HZ SUPPLY WIRING HARNESS REPLACEMENT

THIS TASK COVERS:

a. Removal   b. Installation

INITIAL SETUP:

APPLICABLE MODELS
M934/A1/A2

TOOLS
General mechanic's tool kit (Appendix E, Item 1)

MATERIALS/PARTS
Sealing compound (Appendix C, Item 62)

REFERENCES (TM)
TM 9-2320-272-10
TM 9-2320-272-24P

EQUIPMENT CONDITION
• Parking brake set (TM 9-2320-272-10).
• Battery ground cables disconnected (para. 3-126).
• Ceiling filter and side panels removed (para. 5-79).

---

a. Removal

NOTE
Tag wires for installation.

1.   Remove four screws (1) and upper cover (2) from control center box (10).

2.   Remove four screws (5) and lower cover (4) from control center box (10).

3.   Remove screw (11) and wire (8) from terminal (9).

4.   Remove three screws (3) and wires (7) from circuit breaker (6).

5-96. 400 HZ SUPPLY WIRING HARNESS REPLACEMENT (Contd)

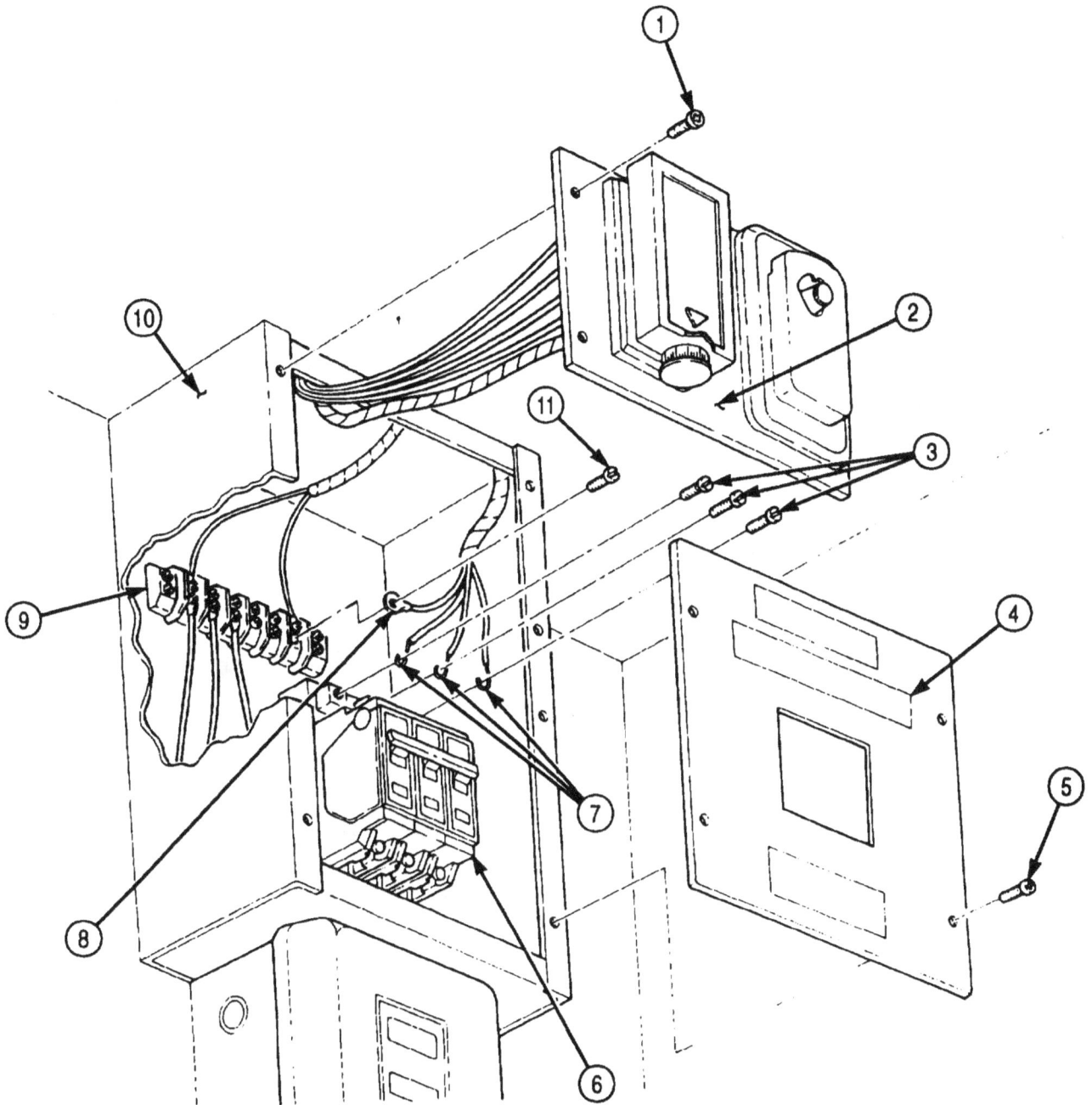

5-96. 400 HZ SUPPLY WIRING HARNESS REPLACEMENT (Contd)

5. Disconnect cables (13) and (12) from connectors (14) and (10).

6. Remove eight screws (11) and electrical connector box cover (15) from electrical connector box (16).

7. Disconnect four wires (7) from connector (14).

8. Remove six screws (9) and cover (1) from load center (5).

9. Remove screw (2) and clamp (3) from harnesses (4) and (8) and load center (5).

10. Pull wiring harness (8) out of coupling (6) and remove from load center (5) and electrical connector box (16).

b. Installation

1. Push wiring harness (8) through coupling (6), electrical connector box (16), and load center (5).

2. Connect four wires (7) to connector (14).

3. Install clamp (3) on harnesses (4) and (8) and load center (5) with screw (2).

4. Install cover (1) on load center (5) with six screws (9).

5. Apply sealing compound to screws (11) and install electrical connector box cover (15) on electrical connector box (16) with eight screws (11).

6. Connect cables (13) and (12) to connectors (10) and (14).

5-96. 400 HZ SUPPLY WIRING HARNESS REPLACEMENT (Contd)

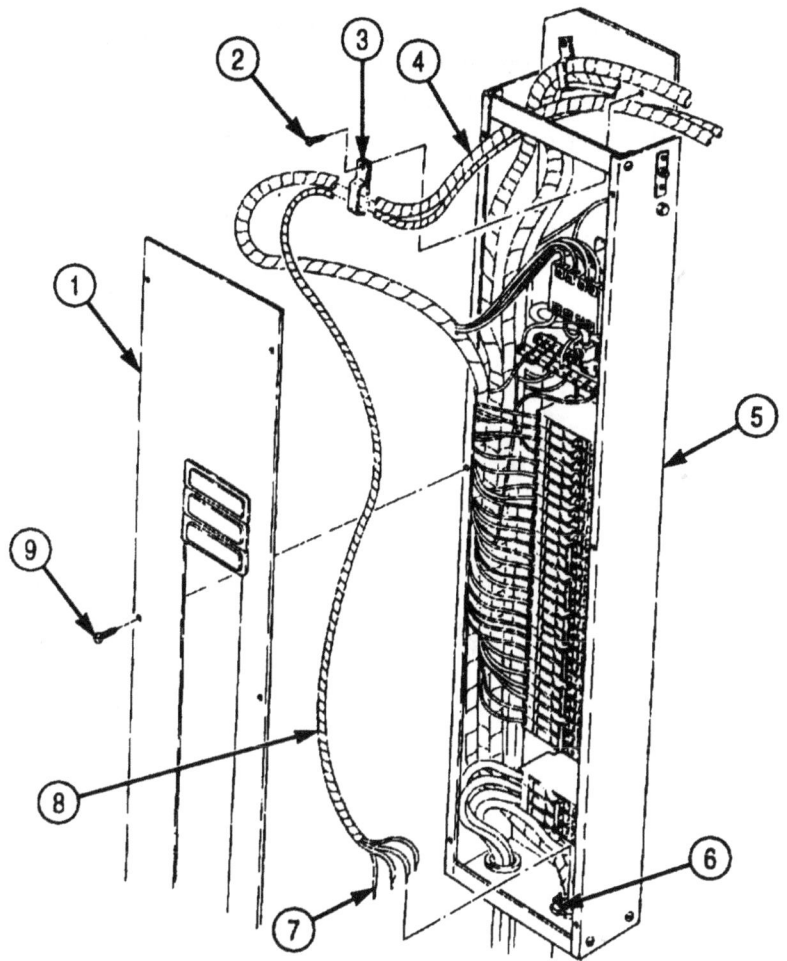

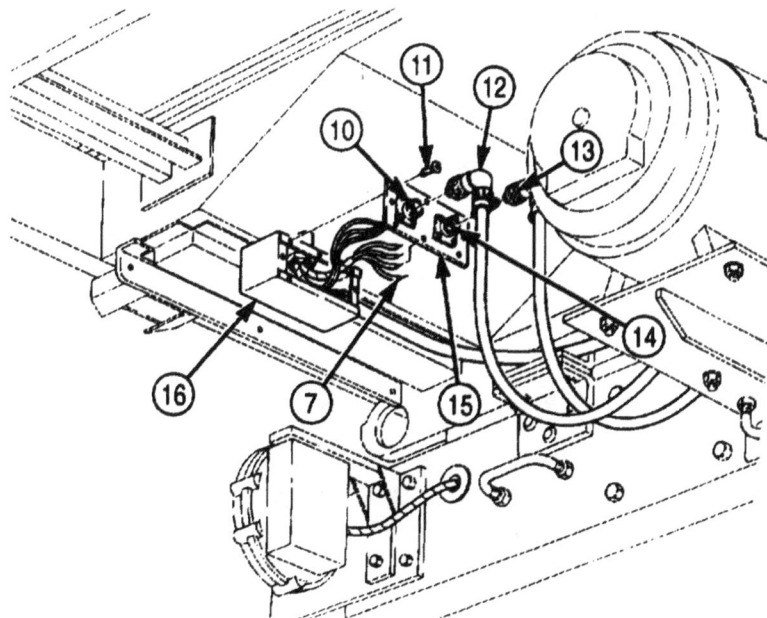

5-96. 400 HZ SUPPLY WIRING HARNESS REPLACEMENT (Contd)

7. Install three wires (7) on circuit breaker (6) with three screws (3).
8. Install wire (8) on terminal (9) with screw (11).
9. Install lower cover (4) on control center box (10) with four screws (5).
10. Install upper cover (2) on control center box (10) with four screws (1).

5-96. 400 HZ SUPPLY WIRING HARNESS REPLACEMENT (Contd)

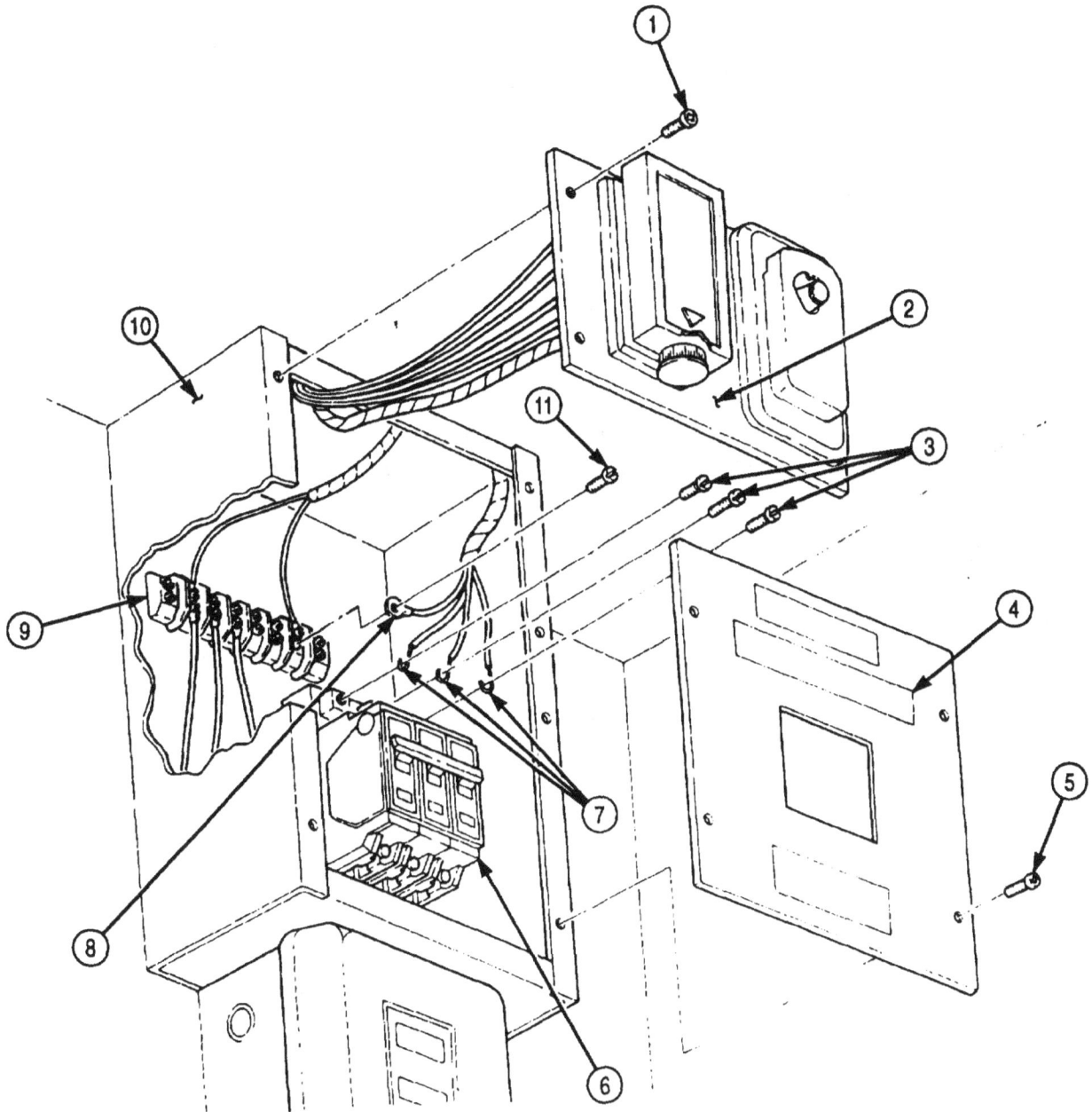

FOLLOW-ON TASKS:• Install ceiling filler and side panels (para. 5-79).
• Connect battery ground cables (para. 3-126).

5-97. BRANCHED 400 HZ RECEPTACLE WIRING HARNESS REPLACEMENT

THIS TASK COVERS:

a. Removal

b. Installation

<u>INITIAL SETUP:</u>

<u>APPLICABLE MODELS</u>
M934/A1/A2

<u>TOOLS</u>
General mechanic's tool kit (Appendix E, Item 1)

<u>REFERENCES (TM)</u>
TM 9-2320-272-10
TM 9-2320-272-24P

<u>EQUIPMENT CONDITION</u>
• Parking brake set (TM 9-2320-272-10).
• Battery ground cables disconnected (para. 3-126).
• Ceiling filter and side panels removed (para. 5-79).

---

a. Removal

NOTE
Tag wires for installation.

1. Remove four screws (1) and upper cover (2) from control center box (6).
2. Remove two screws (3) and wires (4) from terminal (5).
3. Remove six screws (10) and clamps (9) from branched 400 Hz receptacle harness (8) and van ceiling (7).

b. Installation

1. Install 400 Hz branched receptacle harness (8) on van ceiling (7) with six clamps (9) and screws (10).
2. Install two wires (4) on terminal (5) with two screws (3).
3. Install upper cover (2) on control center box (6) with four screws (1).

5-97. BRANCHED 400 HZ RECEPTACLE WIRING HARNESS REPLACEMENT (Contd)

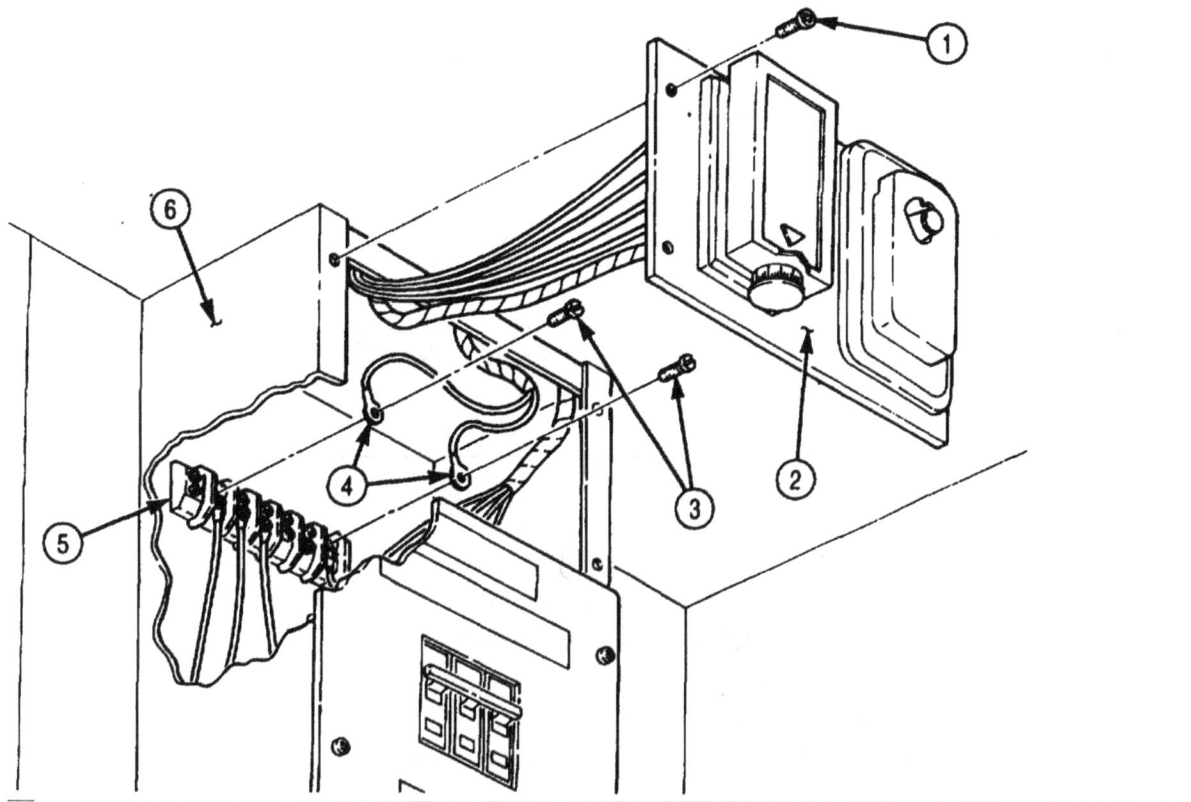

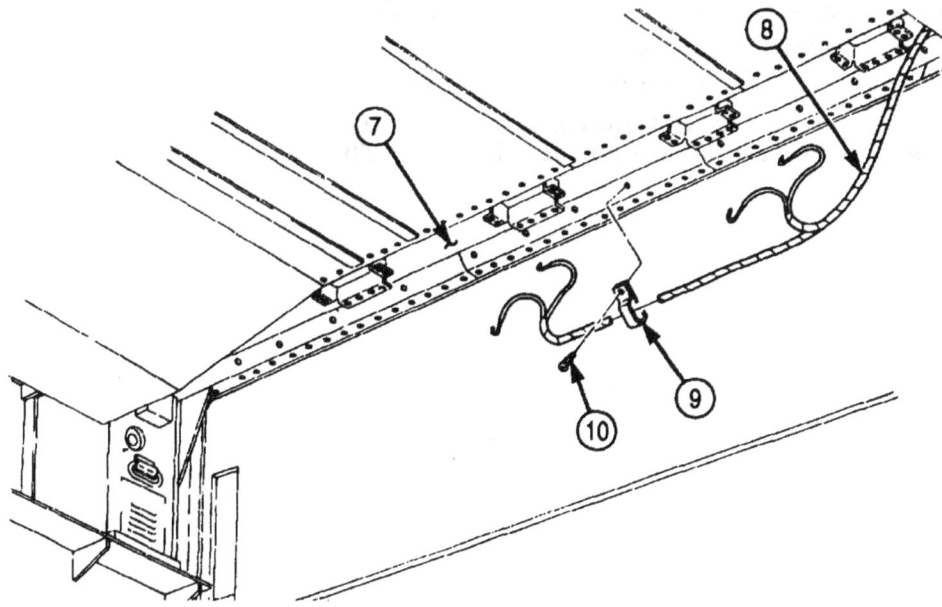

FOLLOW-ON TASKS:• Install ceiling filler and side panels (para. 5-79).
• Connect battery ground cables (para. 3-126).

5-98. RIGHT AND LEFT SIDE BLACKOUT HARNESS MAINTENANCE

THIS TASK COVERS:

a. Removal
b. Repair

c. Installation

INITIAL SETUP:

APPLICABLE MODELS
M934/A1/A2

TOOLS
General mechanic's tool kit (Appendix E, Item 1)

REFERENCES (TM)
TM 9-2320-272-10
TM 9-2320-272-24P

EQUIPMENT CONDITION
• Parking brake set (TM 9-2320-272-10).
• Van body sides fully expanded and secured (TM 9-2320-272-10).
• Hinged roof-operated blackout circuit plungers removed (para. 3-377).

a. Removal

NOTE
Tag wires for installation.

1. Remove six screws (20) and cover (19) from load center (16).
2. Remove screw (14) and wire (13) from neutral bus (15).
3. Remove screw (10) and wire (11) from relay (12).
4. Remove screws (5) and (9) and wires (7) and (8) from rear door blackout switch (6).

NOTE
Perform steps 5 through 7 for left and right side of van body.

5. Remove six screws (24), cover (22), and grommet (4) from van body (26).
6. Remove nut (21), two screws (17), and two connector halves (18) from harness (23) and cover (22).
7. Remove nut (3), two screws (25), and two connector halves (1) from harness (23) and ceiling truss (2) and pull harness (23) through hole in ceiling truss (2).

5-98. RIGHT AND LEFT SIDE BLACKOUT HARNESS MAINTENANCE (Contd)

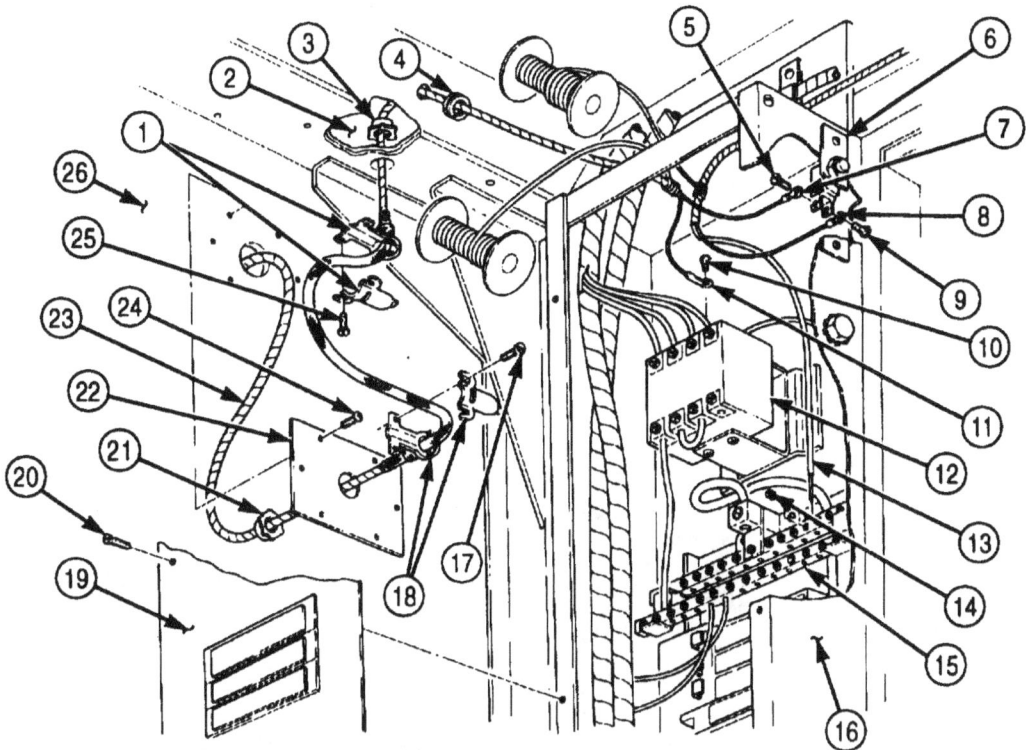

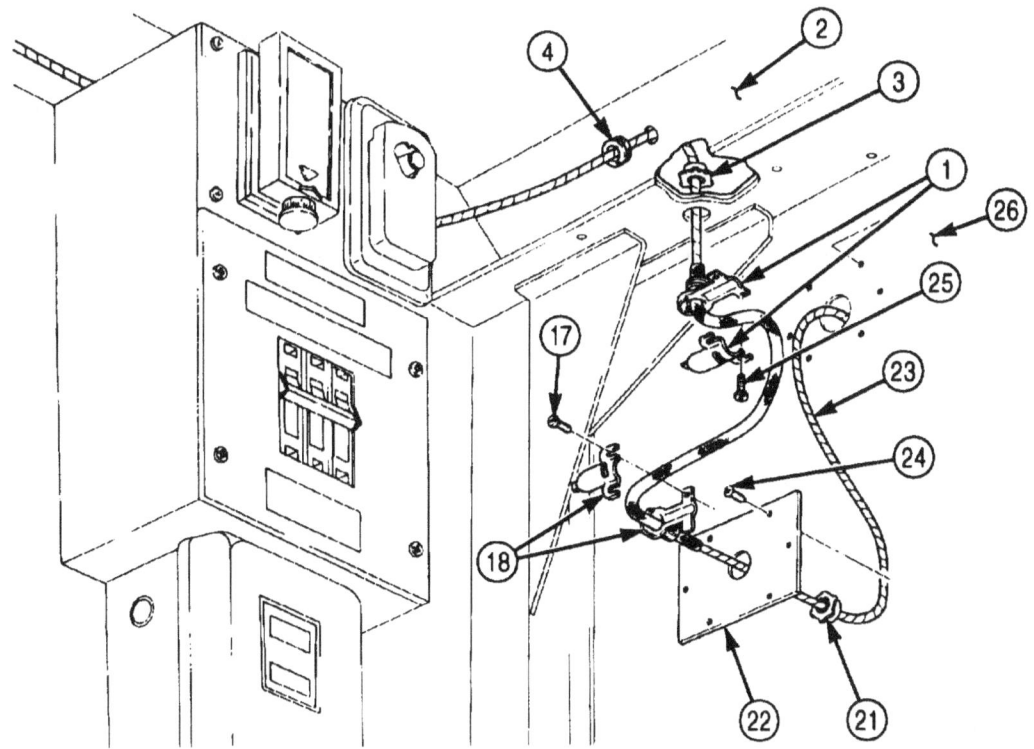

5-98. RIGHT AND LEFT SIDE BLACKOUT HARNESS MAINTENANCE (Contd)

b. Repair

For wiring harness repair, refer to para. 3-131.

c. Installation

1. Push harness (23) in through hole in ceiling truss (2) and position on load center (15).

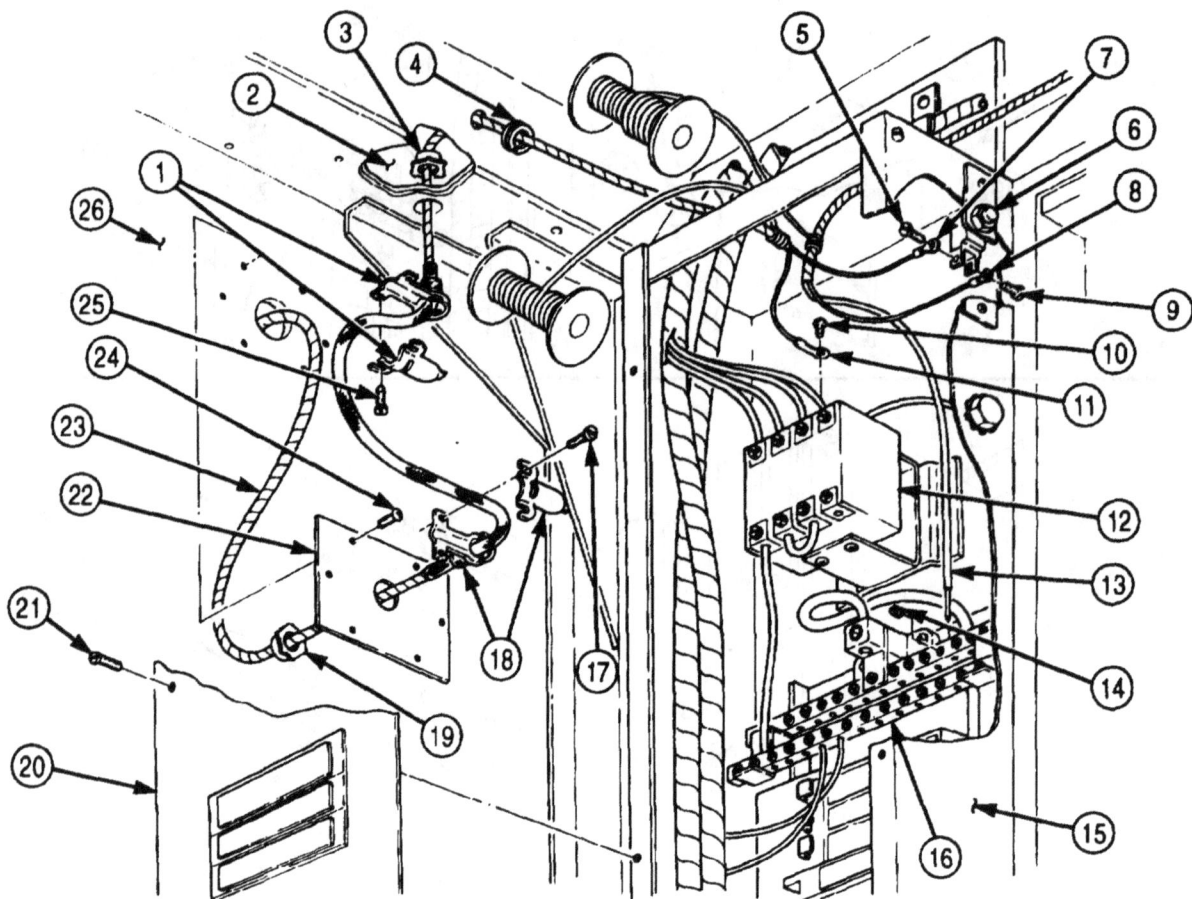

5-98. RIGHT AND LEFT SIDE BLACKOUT HARNESS MAINTENANCE (Contd)

NOTE

Perform steps 2 through 5 for left and right side of van body.

2. Install two connector halves (1) on harness (23) and ceiling truss (2) with nut (3) and two screws (25).
3. Install cover (22) on van body (26) with six screws (241.
4. Install two connector halves (18) on harness (23) and cover (22) with nut (19) and two screws (17).
5. Install grommet (4) on van body (26).
6. Install wires (7) and (8) on rear door blackout switch (6) with screws (5) and (9).
7. Install wire (11) on relay (12) with screw (10).
8. Install wire (13) on neutral bus (16) with screw (14).
9. Install cover (20) on load center (15) with six screws (21).

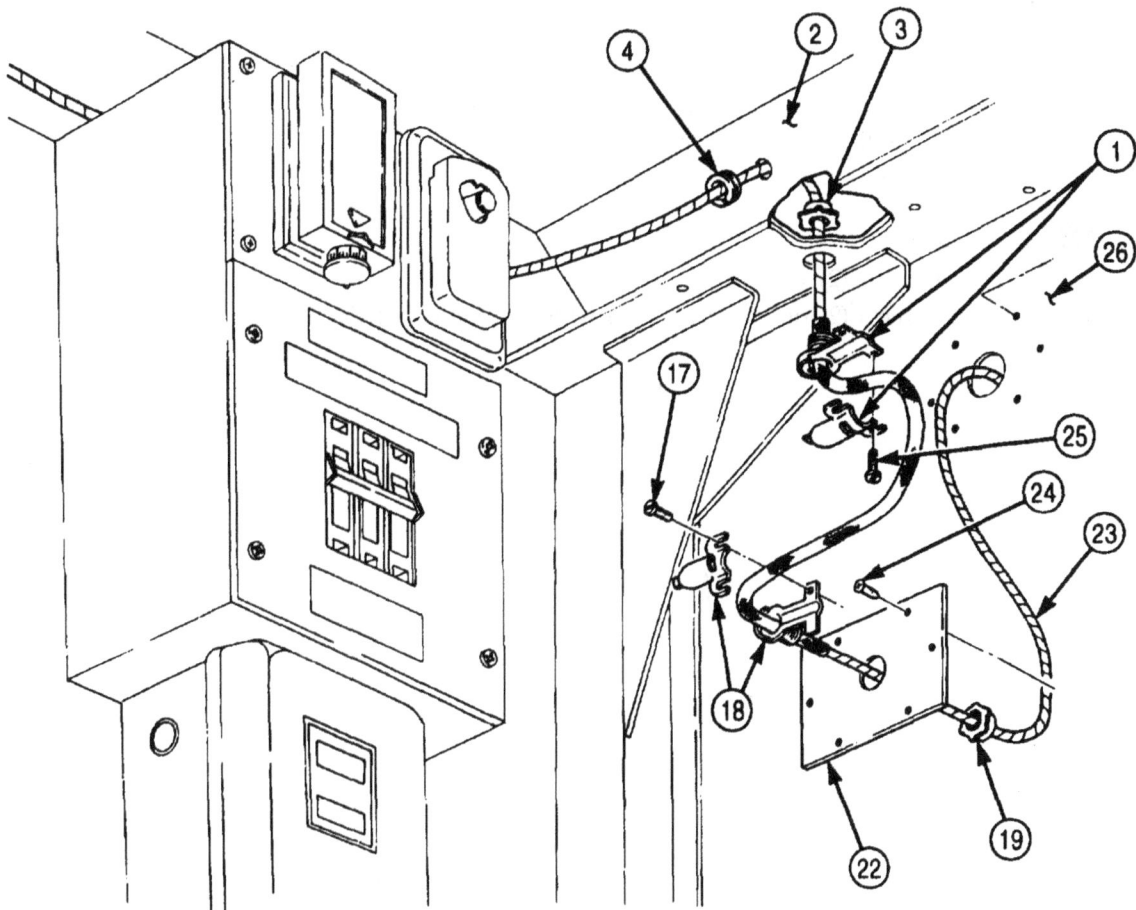

FOLLOW-ON TASKS:• Install hinged roof-operated blackout circuit plungers (para. 3-377).
• Retract van body sides (TM 9-2320-272-10).

5-99. TELEPHONE POST WIRING HARNESS REPLACEMENT

THIS TASK COVERS:

a. Removal   b. Installation

INITIAL SETUP:

APPLICABLE MODELS
M934/A1/A2

TOOLS
Gal mechanic's tool kit (Appendix E, Item 1)

REFERENCES (TM)
TM 9-2320-272-10
TM 9-2320-272-24P

EQUIPMENT CONDITIONS
• Parking brake set (TM 9-2320-272-10).
• Inside telephone jacks removed (para. 3-374).
• Ceiling filler and side panels removed (para. 5-79).

a. Removal

NOTE
Tag wires for installation.

1. Remove six screws (4) and clamps (3) from wiring harness (2) and van ceiling (1).
2. Remove wiring harness (2) from van body (5).

b. Installation

1. Install wiring harness (2) on van body (5).
2. Install wiring harness (2) on van ceiling (1) with six clamps (3) and screws (4).

5-99. TELEPHONE POST WIRING HARNESS REPLACEMENT (Contd)

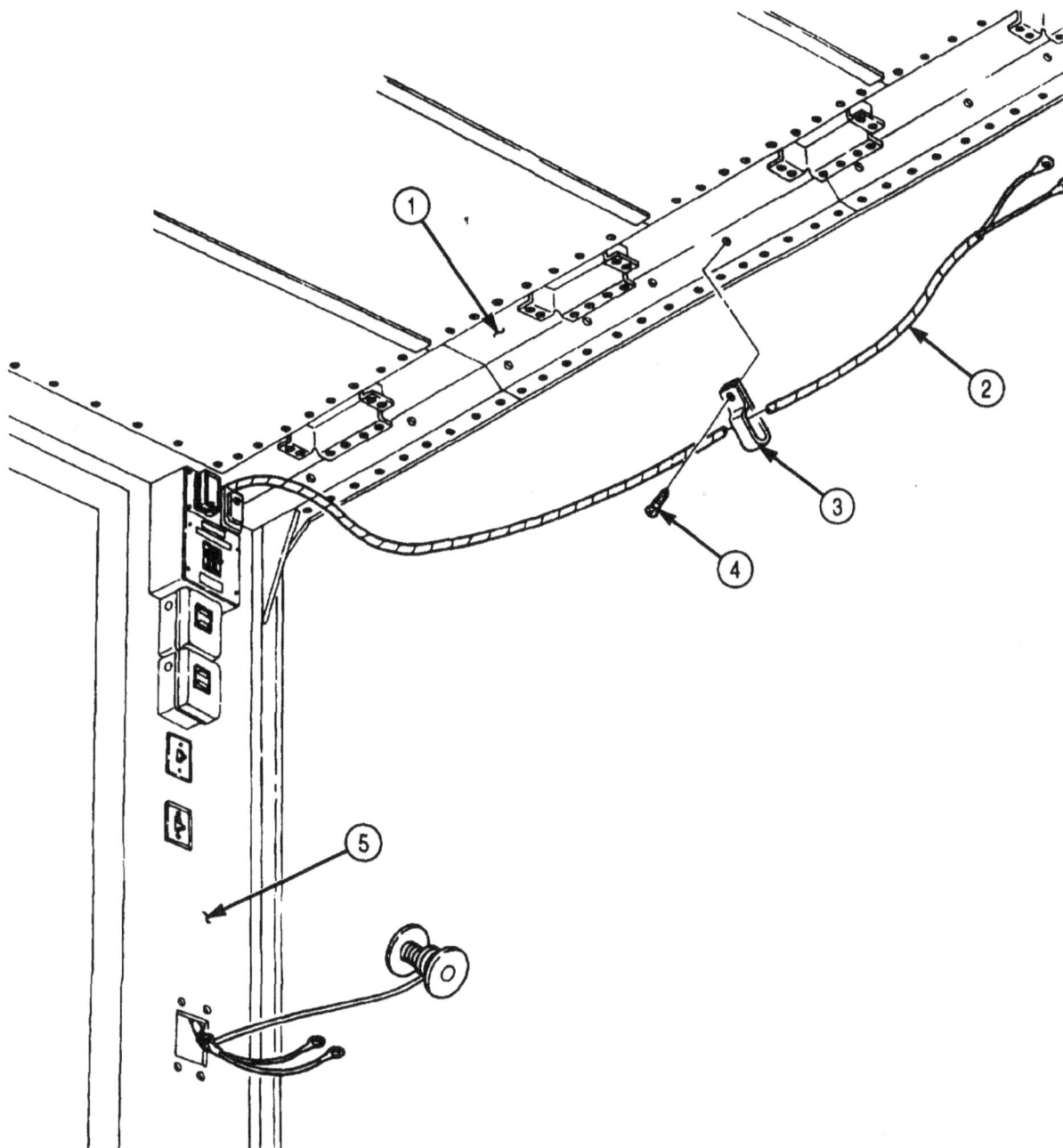

FOLLOW-ON TASKS: Install ceiling filler and side panels (para. 5-79).
• Install inside telephone jacks (para. 3-374).

5-100. 3 PHASE RECEPTACLE WIRING HARNESS REPLACEMENT

THIS TASK COVERS:

a. Removal   b. Installation

INITIAL SETUP:

APPLICABLE MODELS
M934/A1/A2

TOOLS
General mechanic's tool kit (Appendix E, Item 1)

REFERENCES (TM)
TM 9-2320-272-10
TM 9-2320272-24P

EQUIPMENT CONDITION
• Parking brake set (TM 9-2320-272-10).
• Battery ground cables disconnected (para. 3-126).
• Ceiling filter and side panels removed (para. 5-79).

a. Removal

NOTE
Tag wires for installation.

1. Remove six screws (1) and cover (12) from load center (5).
2. Remove screw (3) and clamp (13) from wiring harness (2) and van body (4).
3. Remove screw (8) and wire (6) from neutral bus (7).
4. Remove three screws (10) and wires (11) from circuit breakers (9).
5. Remove twelve screws (16) and clamps (15) from van ceiling (14) and wiring harness (2).
6. Remove wiring harness (2) from van body (4).

b. Installation

1. Position wiring harness (2) on van body (4).
2. Install wiring harness (2) on van ceiling (14) with twelve clamps (15) and screws (16).
3. Install three wires (11) on circuit breakers (9) with three screws (10).
4. Install wire (6) on neutral bus (7) with screw (8).
5. Install clamp (13) on wiring harness (2) and van body (4) with screw (3).
6. Install cover (12) on load center (5) with six screws (1).

5-100. 3 PHASE RECEPTACLE WIRING HARNESS REPLACEMENT (Contd)

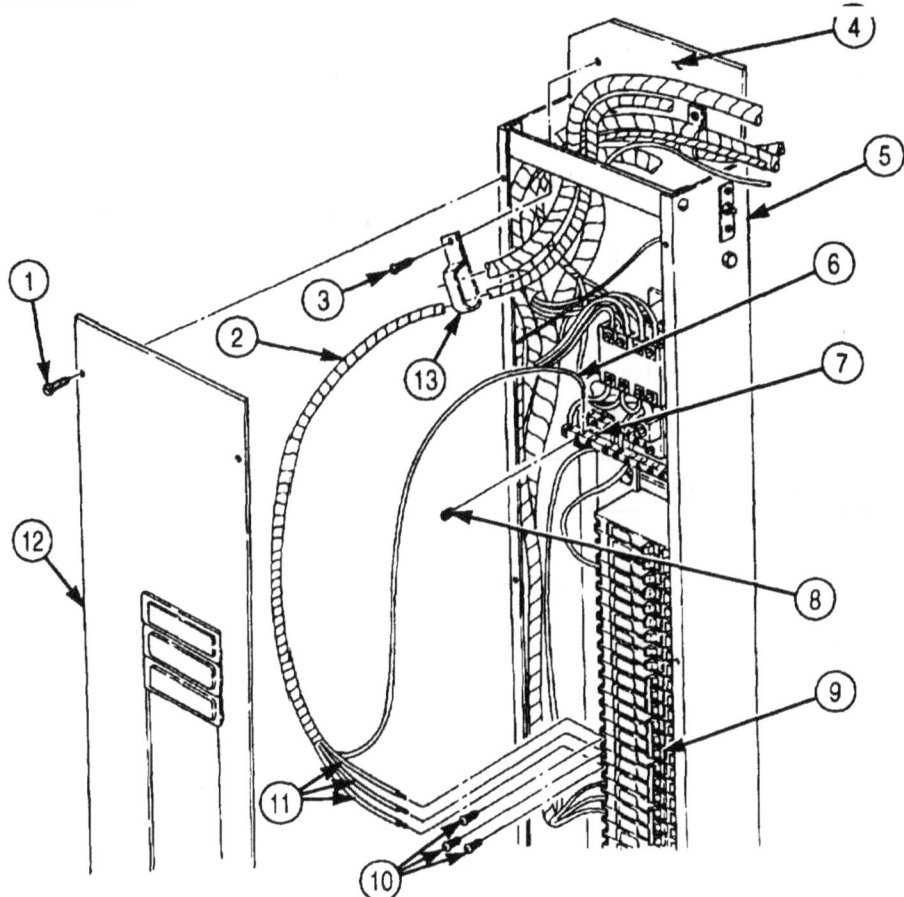

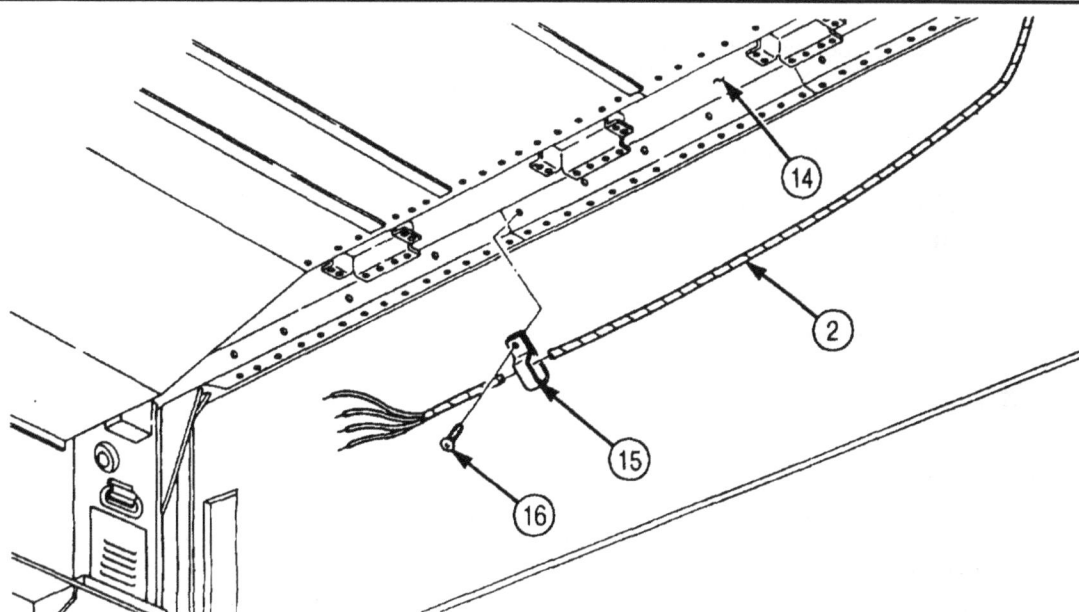

FOLLOW-&TASKS: • Install ceiling filler and side panels (para. 5-79).
• Connect battery ground cables (para. 3-126).

Section VII. WINCH, HOIST, AND POWER TAKEOFF MAINTENANCE

5-101. WINCH, HOIST AND POWER TAKEOFF MAINTENANCE INDEX

5-102. BOOM ELEVATING CYLINDER REPAIR

THIS TASK COVERS:

a. Disassembly

b. Cleaning, Inspection, and Repair

c. Assembly

INITIAL SETUP:

APPLICABLE MODELS
M936/A1/A2

SPECIAL TOOLS
Bushing installer (Appendix E, Item 23)
Handle (Appendix E, Item 61)

TOOLS
General mechanic's tool kit (Appendix E, Item 1)
Spanner wrench (Appendix E, Item 166)

MATERIALS/PARTS
Piston seal (Appendix D, Item 519)
Rod seal (Appendix D, Item 560)
O-ring (Appendix D, Item 465)
Lockwire (Appendix D, Item 420)
Wiper strip (Appendix D, Item 720)
Packing (Appendix D, Item 510)

MATERIALS/PARTS (CONTD)
Crocus cloth (Appendix C, Item 20)
Lint-free cloth (Appendix C, Item 21)
Drycleaning solvent (Appendix C, Item 71)

REFERENCES (TM)
LO 9-2320-272-12
TM 9-2320-272-10
TM 9-2320-272-24P

EQUIPMENT CONDITION
• Parking brake set (TM 9-2320-272-10).
• Boom elevating cylinder removed (para. 4-193)

GENERAL SAFETY INSTRUCTIONS
• Keep fire extinguisher nearby when using drycleaning solvent.
• Drycleaning solvent is flammable. Do not use near open flame.

---

a. Disassembly

1. Using spanner wrench, remove cylinder head (2) from elevating cylinder body (3).

2. Remove piston rod (1) from elevating cylinder body (3).

3. Remove lockwire (17), nut (18), washer (16), retainer (15), seal (14), piston (13), rod seal (5), cylinder head (2), and wiper strip (8) from piston rod (1). Discard lockwire (17), piston seal (14), rod seal (5), and wiper strip (8).

5-102. BOOM ELEVATING CYLINDER REPAIR (Contd)

4. Remove packing nut (9), packing (10), bushing (11), and O-ring (12) from cylinder head (2). Discard O-ring (12) and packing (10).

5. Remove two half rings (4) from piston (13).

6. Remove sleeve (7) from piston rod (1).

7. Remove two lubrication fittings (6) from cylinder body (3) and piston rod (1).

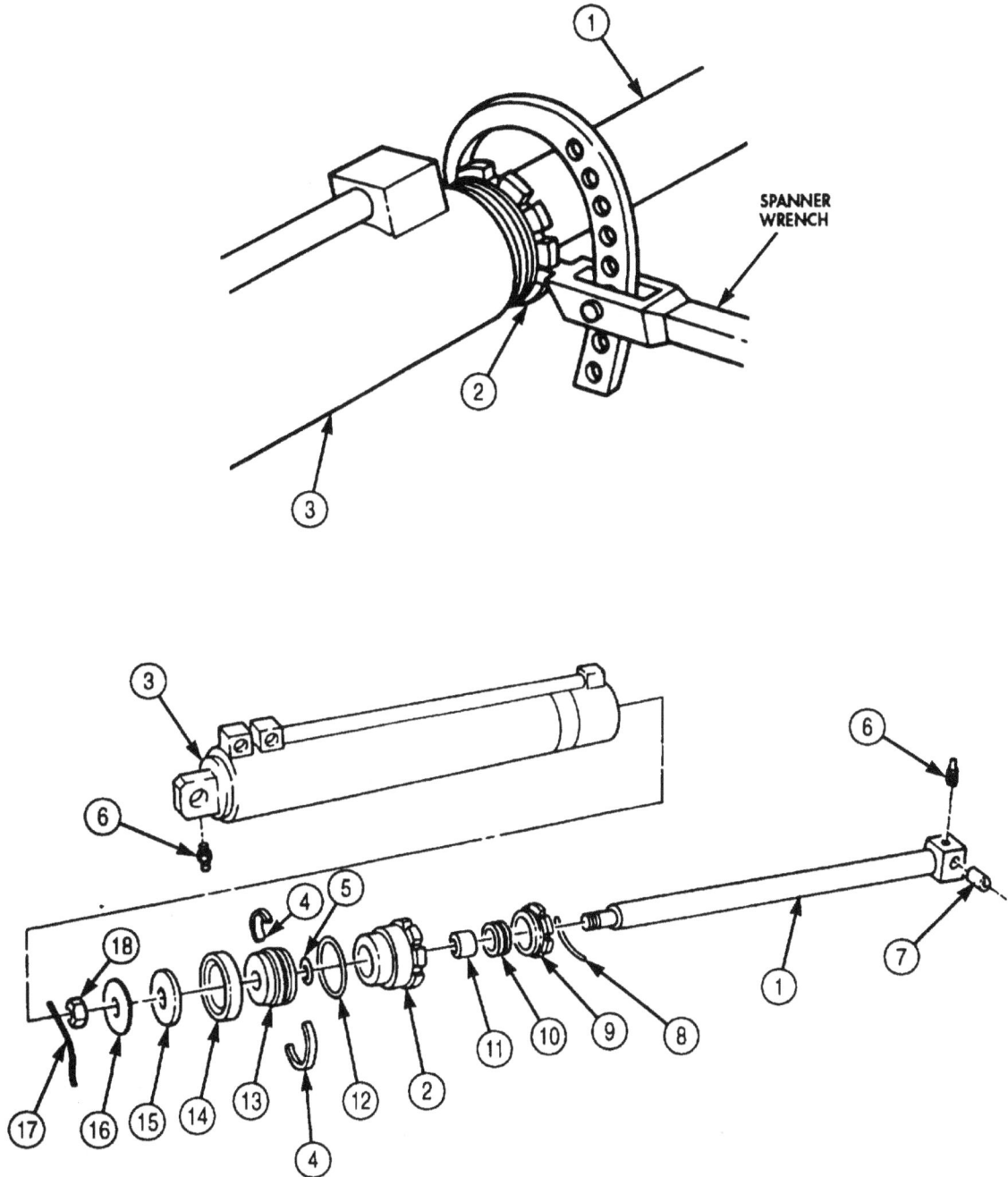

5-102. BOOM ELEVATING CYLINDER REPAIR (Contd)

b. Cleaning, Inspection, and Repair

WARNING

Drycleaning solvent is flammable and will not be used near open flame. Use only in well-ventilated places. Failure to do this may result in injury to personnel.

1. Clean all parts of elevating cylinder with drycleaning solvent and dry with lint-free cloth.
2. Inspect piston rod (9) for burrs, scoring, scratches, and stripped threads. If minor scratches and burrs are present, remove with crocus cloth.
3. If piston rod (9) is scored, scratched, or if stripped threads are evident, replace piston rod (9).
4. Inspect piston (13) and cylinder body (1) for scoring. If piston (13) is scored, replace. If cylinder body (1) is scored, replace elevating cylinder.

c. Assembly

1. Install two lubrication fittings (2) in cylinder body (1) and piston rod (9).
2. Install sleeve (10) in piston rod (9).
3. Install two half rings (3) on piston (13).
4. Install piston rod bushing (6) in cylinder head (5) with bushing installer and handle.
5. Install new O-ring (4), new packing (7), packing nut (8), and new wiper strip (11) on cylinder head (5).

CAUTION

Be careful not to damage wiper strip when passing over threads on piston rod.

6. Install cylinder head (5) on piston rod (9).
7. Install new rod seal (12), piston (13), new seal (14), retainer (15), and washer (16) on piston rod (9).
8. Install nut (18) on piston rod (9) and secure nut (18) with lockwire (17).
9. Slide piston rod (9) into cylinder body (1) and install cylinder head (5) on cylinder body (1).
10. Using spanner wrench, tighten cylinder head (5).

5-102. BOOM ELEVATING CYLINDER REPAIR (Contd)

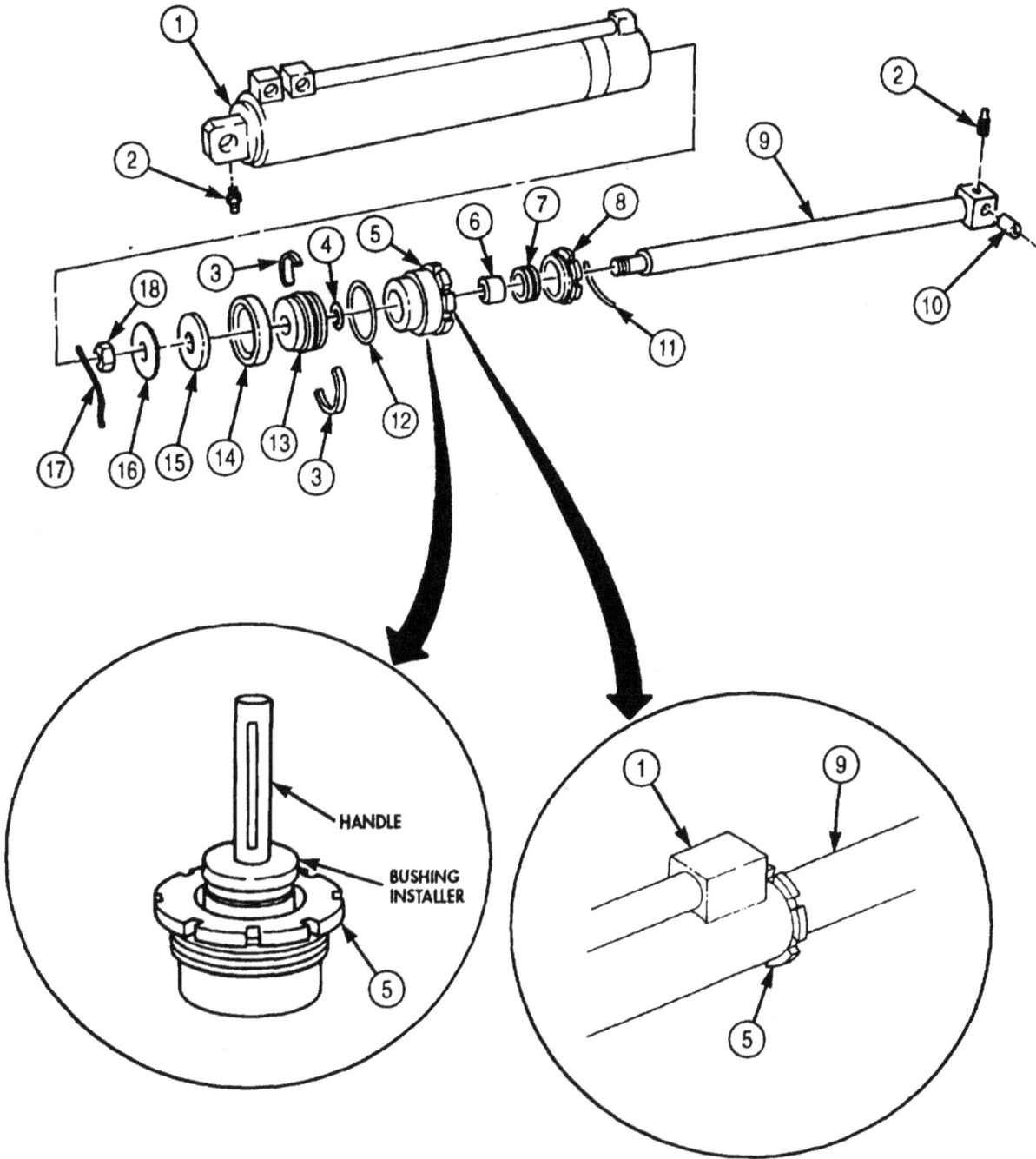

HANDLE

BUSHING INSTALLER

FOLLOW-ON TASKS:• Install boom elevating cylinder (para. 4-193).
• Lubricate elevating cylinder (LO 9-2320-272-12).

5-103. BOOM ELEVATING CYLINDER PACKING REPLACEMENT

THIS TASK COVERS:

a. Removal

b. Installation

INITIAL SETUP:

APPLICABLE MODELS
M936/A1/A2

SPECIAL TOOLS
Spanner wrench (Appendix E, Item 167)

TOOLS
General mechanic's tool kit (Appendix E, Item 1)
Boom lifting device

MATERIALS/PARTS
Packing (Appendix D, Item 510)
GAA grease (Appendix C, Item 28)

REFERENCES (TM)
LO 9-2320-272-12
TM 9-2320-272-10
TM 9-2320-272-24P

EQUIPMENT CONDITION
• Parking brake set (TM 9-2320-272-10).
• Boom lowered (TM 9-2320-272-10).

GENERAL SAFETY INSTRUCTIONS
All personnel must stand clear during lifting operations.

WARNING

All personnel must stand clear during lifting operations. A snapped cable, or shifting or swinging load ,may cause injury to personnel.

a. Removal

1. Place wrecker boom (2) cylinder control lever (7) in UP position.
2. Raise wrecker boom (2) with lifting device and install shipper braces (TM 9-2320-272-10).
3. Place boom cylinder control lever (7) in LOWER position to release pressure from elevating cylinder (6).
4. Using spanner wrench, loosen packing nut (3) from cylinder head (5).
5. Slide nut (3) up on piston rod (1) to expose packing (4).

CAUTION

Use care not to scratch piston rod surface when removing packing.

6. Slide packing (4) up piston rod (1) and remove. Discard packing (4).

b. Installation

1. Coat packing (4) with GAA grease and position around piston rod (1).
2. Install packing (4) into cylinder head (5) and slide nut (3) onto cylinder head (5).
3. Using spanner wrench, tighten nut (3).
4. Remove lifting device from wrecker boom (2).

5-103. BOOM ELEVATING CYLINDER PACKING REPLACEMENT (Contd)

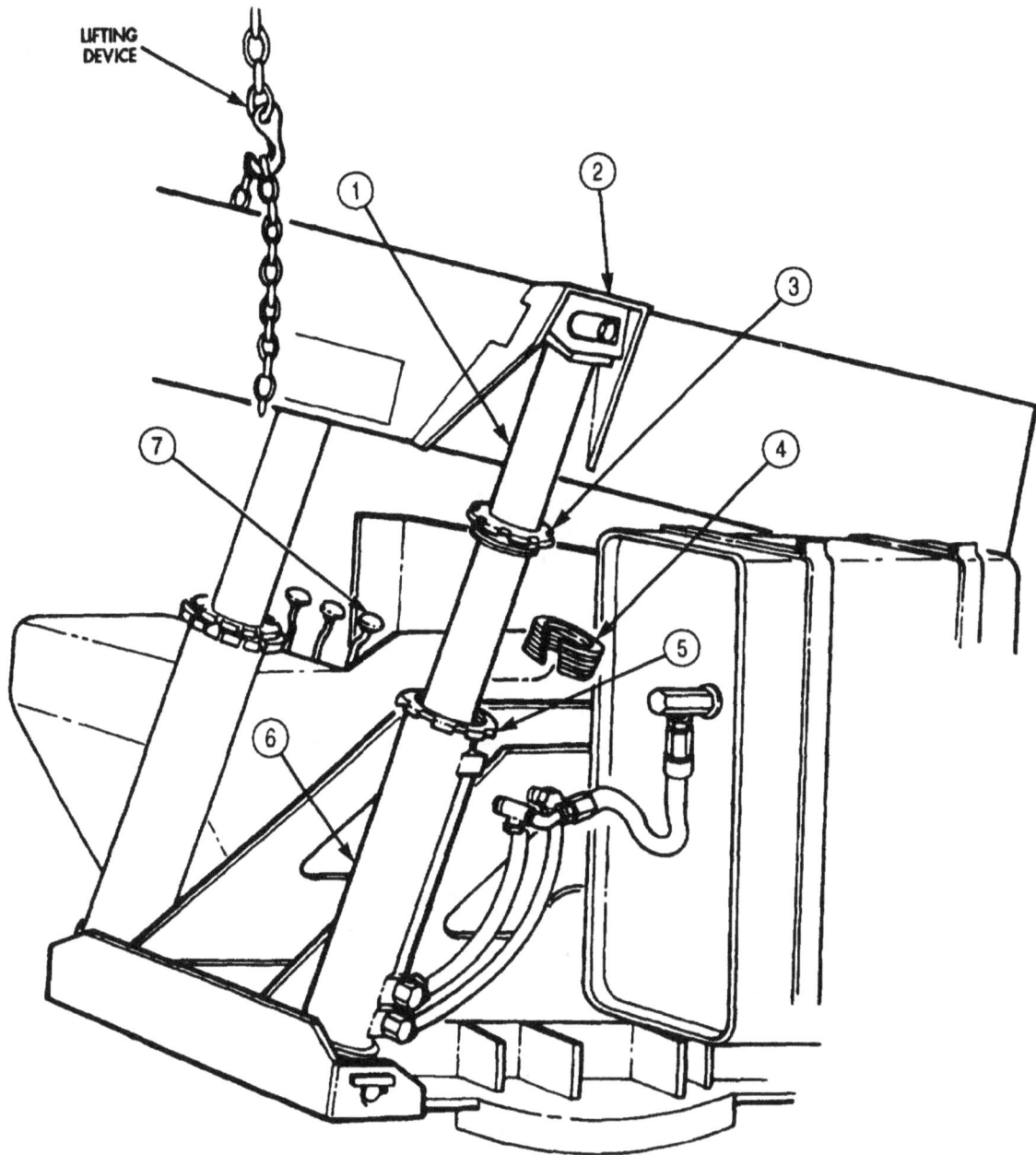

FOLLOW-ON TASK: Operate crane through full range (TM 9-2320-272-10) and check for leaks at cylinder.

5-104. DUMP ROLLER ARM MAINTENANCE

THIS TASK COVERS:

a. Removal
b. Disassembly
c. Inspection and Repair

d. Assembly
e. Installation

INITIAL SETUP:

APPLICABLE MODELS
M929/A1/A2, M930/A1/A2

TOOLS
General mechanic's tool kit (Appendix E, Item 1)
Brass drift
Soft-faced hammer

MATERIALS/PARTS
GAA grease (Appendix C, Item 28)

REFERENCES (TM)
LO 9-2320-272-12
TM 9-2320-272-10
TM 9-2320-272-24P

EQUIPMENT CONDITION
• Parking brake set (TM 9-2320-272-10).
• Dump body removed (para. 4-144).

GENERAL SAFETY INSTRUCTIONS
Do not operate dump controls when dump body is removed.

WARNING

Ensure dump control lever is in neutral and not moved. Injury to personnel may result if lift cylinder is operated when not secured.

NOTE

Left and right roller arms are replaced basically the same way. This procedure covers right roller arm.

a. Removal

NOTE

Use soft-faced hammer to tap roller arm if necessary to complete step 1.

Remove roller arm (3) from hoist cylinder crosshead shaft (1).

b. Disassembly

1. Using punch and hammer, remove two pins (2) from roller arm (3).
2. Using brass drift and hammer, drive roller pin (5) out of roller arm (3) and roller (6).
3. Remove roller (6) from roller arm (3).

c. Inspection and Repair

1. Inspect roller (6) and roller pin (5). If cracked or deeply grooved, replace roller (6) and roller pin (5).
2. Inspect roller arm (3). If broken or cracked, replace.
3. Inspect two bushings (4) in roller arm (3). If broken, cracked, or out-of-round, replace bushings (4).

NOTE

Perform steps 4 and 5 only if bushings need to be replaced.

4. Using brass drift and hammer, drive two bushings (4) out of roller arm (3).
5. Using wood block and hammer, drive one new bushing (4) in from each side of roller arm (3) until bushings (4) are flush with roller arm (3).

5-104. DUMP ROLLER ARM MAINTENANCE (Contd)

d. Assembly

1. Position roller (6) in roller arm (3).
2. Coat roller pin (5) with light film of GAA grease and insert through roller arm (3) and roller (6). Tap pins (5) into place with soft-faced hammer.
3. Install roller (6) on roller arm (3) with two pins (2)

e. Installation

Install roller arm (3) on hoist cylinder crosshead shaft.

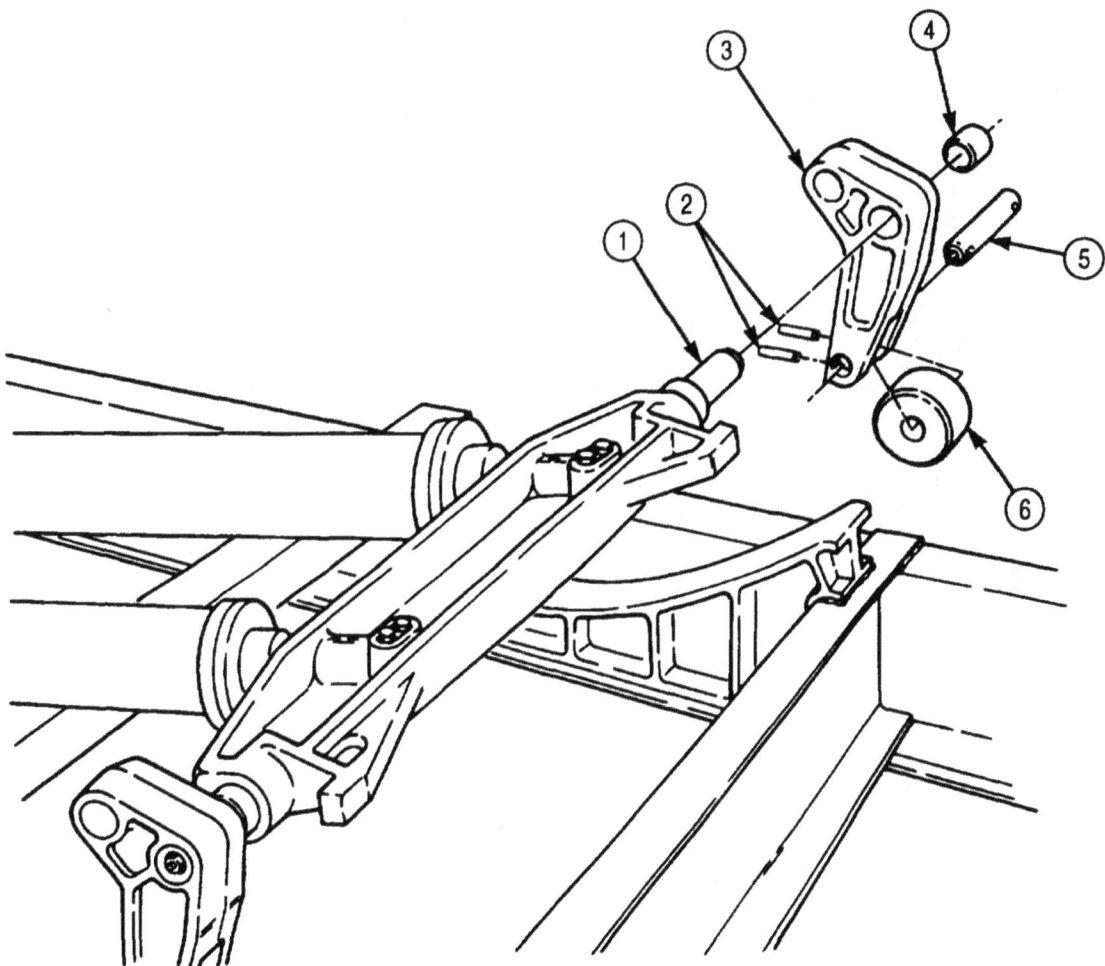

FOLLOW-ON TASKS:• Install dump body (para. 4-144).
　　　　　　　　• Lubricate roller arms (LO 9-2320-272-12).

5-105. DUMP HOIST CYLINDER MAINTENANCE

THIS TASK COVERS:

a. Removal
b. Disassembly
c. Cleaning and Inspection

d. Assembly
e. Installation

INITIAL SETUP:

APPLICABLE MODELS
M929/A1/A2, M930/A1/A2

TOOLS
General mechanic's tool kit (Appendix E, Item 1)
Spring tester (Appendix E, Item 131)
Soft-faced hammer
Outside micrometer (Appendix E, Item 80)
Inside micrometer (Appendix E, Item 83)
Torque wrench (Appendix E, Item 146)

MATERIALS/PARTS
Eight lockwashers (Appendix D, Item 393)
Preformed packing (Appendix D, Item 511)
O-ring (Appendix D, Item 466)
Cotter pin (Appendix D, Item 75)
Cap and plug set (Appendix C, Item 14)
Lint-free cloth (Appendix C, Item 21)
Clean hydraulic oil (Appendix C, Item 35)

REFERENCES (TM)
LO 9-2320-272-12
TM 9-2320-272-10
TM 9-2320-272-24P

EQUIPMENT CONDITION
• Dump body removed (para. 4-144).
• Dump roller arms removed (para. 5-104).

GENERAL SAFETY INSTRUCTIONS
Do not operate dump controls when dump body is removed.

WARNING

Ensure dump control is in neutral and not moved. Injury to personnel may result if lift cylinder is operated when not secured.

a. Removal

NOTE
A 4x4 block of wood is recommended for use as support.

1. Raise hoist cylinders (3) and (9) from subframe (4) and place wood support between subframe (4) and cylinders (3) and (9).

2. Remove four screws (2), lockwashers (1), two upper crosshead retainers (5), lower crosshead retainers (7), and crosshead (8) from two piston rods (6). Discard lockwashers (1).

CAUTION
Plug all hydraulic lines or openings to prevent dirt from entering and damaging components.

NOTE
• Cross fitting to hoist cylinder hoses must be disconnected from cross fittings first. Then hoses can be removed from hoist cylinders.

• Have drainage containers ready to catch oil.

• Tag all parts for installation.

3. Disconnect hydraulic hoses (18) and (14) from right cross fitting (17).

5-105. DUMP HOIST CYLINDER MAINTENANCE (Contd)

4. Disconnect hydraulic hoses (19) and (15) from left cross fitting (16).
5. Disconnect hydraulic hoses (18) and (14) from right and left cylinder ports A (10) and (12).
6. Disconnect hydraulic hoses (19) and (15) from right and left cylinder ports B (11) and (13).

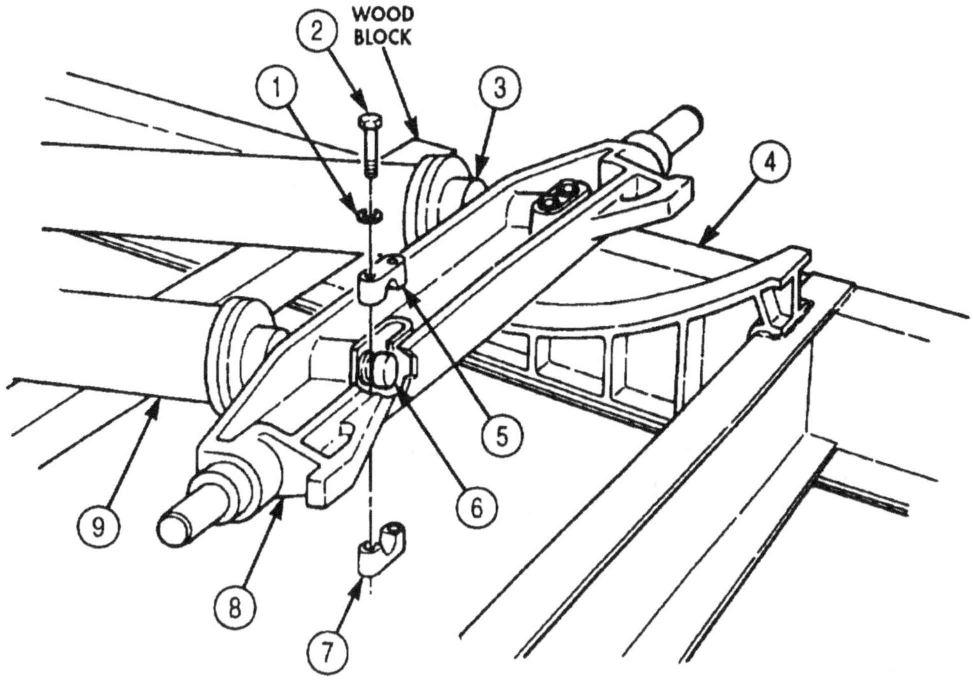

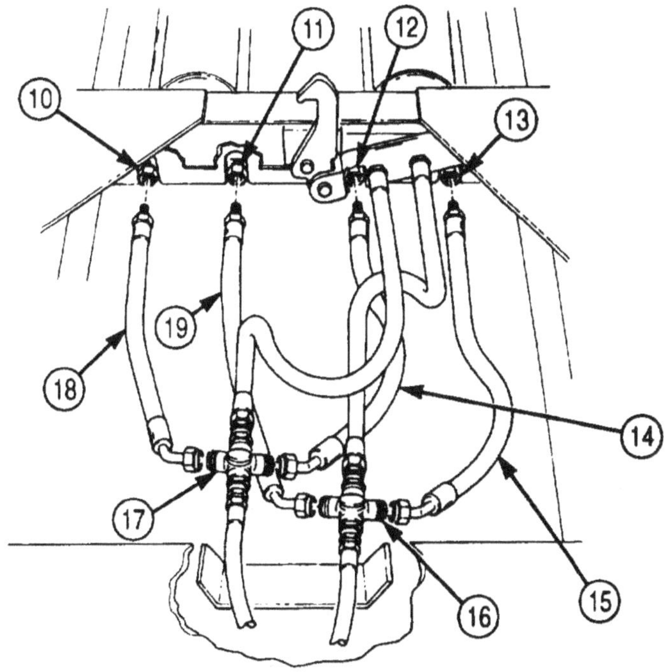

5-105. DUMP HOIST CYLINDER MAINTENANCE (Contd)

7. Remove eight screws (3), lockwashers (4), four bearing caps (2), and hoist cylinders (5) and (1) from subframe (6). Discard lockwashers (4).

b. Disassembly

NOTE

Both hoist cylinders are disassembled the same way. Steps 1 through 12 cover the left hoist cylinder.

1. Remove two square-head screws (18) from cylinder base (7) and slide hinge pin (19) out of cylinder base (7).
2. Remove three screws (11) from gland (12) and cylinder head (14).
3. Tap gland (12) free of cylinder head (14) and slide off piston rod (16).
4. Remove ten screws (10) and lockwashers (9) from cylinder head (14) and hoist cylinder housing (8). Discard lockwashers (9).
5. Tap cylinder head (14) free of hoist cylinder housing (8) and slide cylinder head (14) off piston rod (16).
6. Remove packing (13) and O-ring (15) from cylinder head (14). Discard packing (13) and O-ring (15).

CAUTION

Use care when performing step 7 to prevent damage to cylinder bore or piston.

7. Pull piston rod (16) and piston (17) straight out of hoist cylinder housing (8).

5-105. DUMP HOIST CYLINDER MAINTENANCE (Contd)

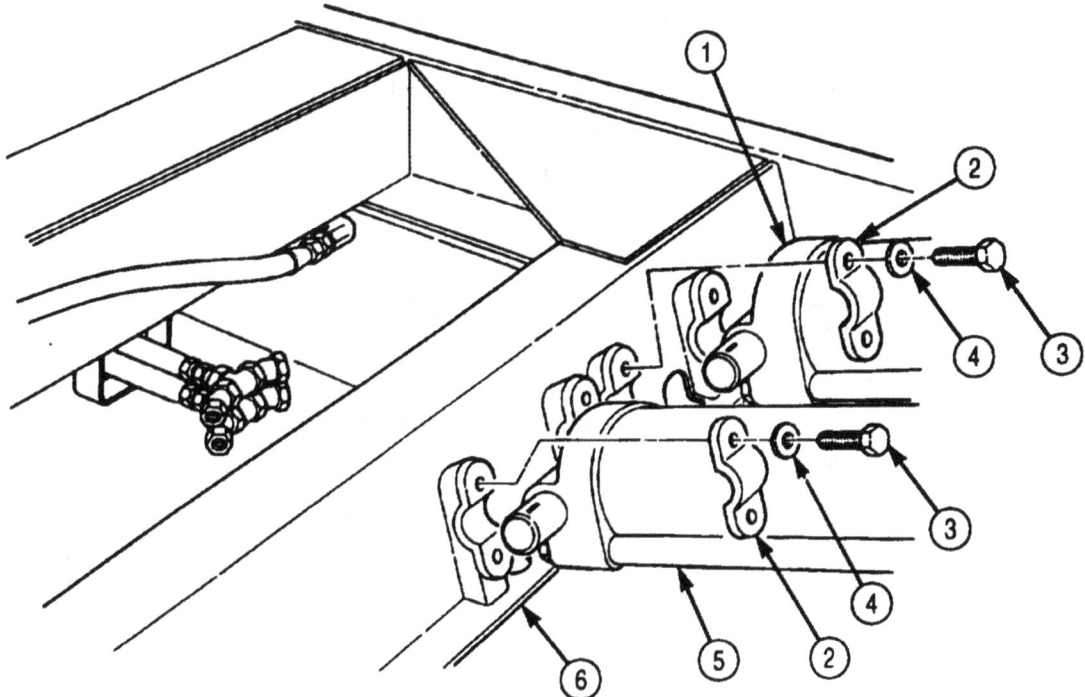

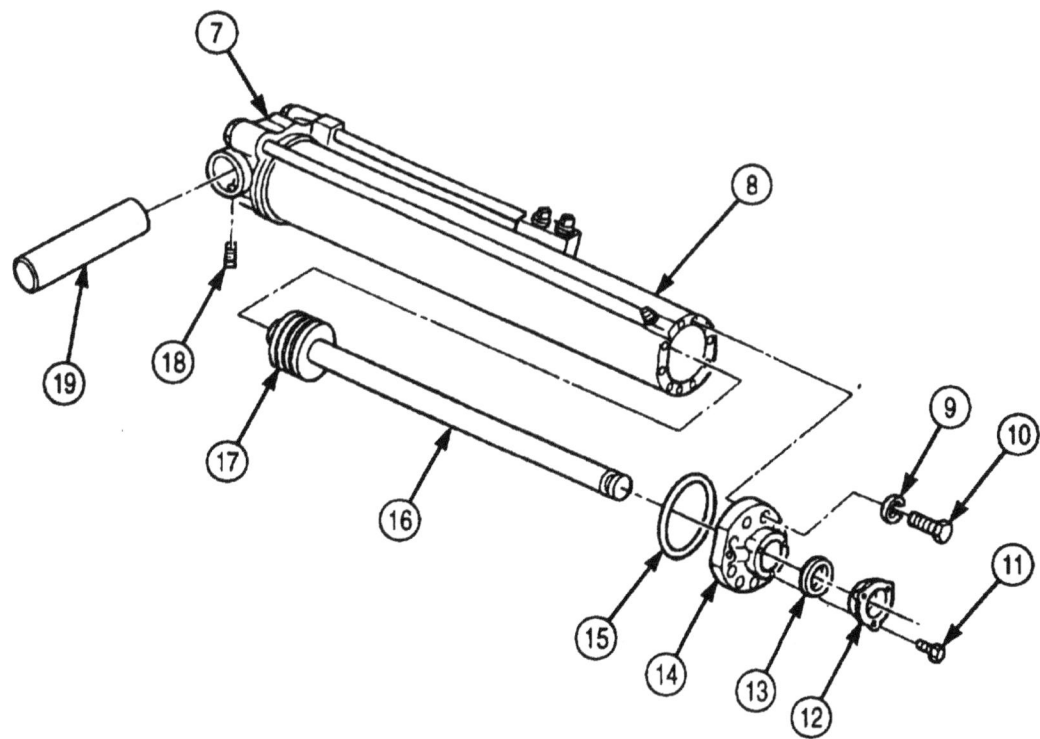

5-105. DUMP HOIST CYLINDER MAINTENANCE (Contd)

8. Remove cotter pin (12) and slotted nut (13) from piston rod (8) and piston (9). Discard cotter pin (12).
9. Slide piston (9) off piston rod (8).
10. Remove three piston rings (11) from piston grooves (10).

NOTE

Tag bypass plugs, springs, and check balls for installation.

11. Remove three bypass plugs (6), springs (5), and check balls (4) from hoist cylinder housing (7).
12. Remove orifice plug cover (1) and orifice plug (2) from cylinder base (3).

NOTE

Use clean lint-free cloth to wipe parts clean. After parts are cleaned and inspected, coat them with a light film of clean hydraulic oil.

1. Wipe check balls (4) clean and inspect. Replace if scratched, chipped, or marked.
2. Wipe piston (9) clean and inspect. Replace if cracked or broken.
3. Measure outside diameter of piston (9). If outside diameter is less than 5.236 in. (132.994 mm), replace piston (9).
4. Wipe piston rod (8) clean and inspect. Replace if scratched, chipped, or scored.
5. Measure outside diameter of piston rod (8). If outside diameter is less than 1.994 in. (50.648 mm), replace piston rod (8).

5-105. DUMP HOIST CYLINDER MAINTENANCE (Contd)

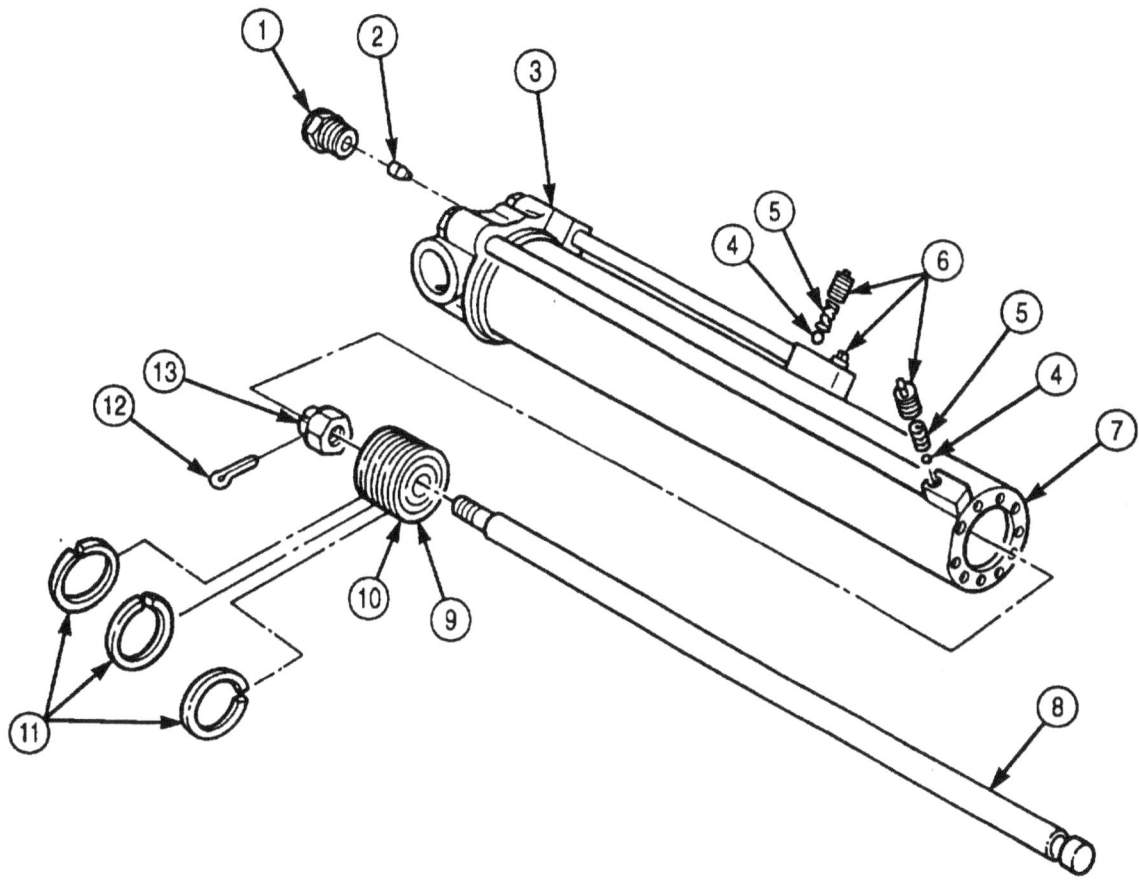

5-105. DUMP HOIST CYLINDER MAINTENANCE (Contd)

6. Wipe head (5) and gland (6) clean and inspect. If cracked or broken, replace head (5) or gland (6).

7. Measure inside diameter of head (5) and gland (6). If inside diameter is not 2.010 in. (51.05 mm), replace head (5) or gland (6).

8. Measure free length of three bypass springs (1). If length is not 1.25 in. (31.75 mm), replace bypass spring (1).

9. Compress each bypass spring (1) to 0.938 in. (23.8 mm) length. If torque reading is not 10 lb& (14 N•m), replace bypass spring (1).

10. Inspect three piston rings (11) for breaks. If any ring (11) is broken, replace all three rings (11).

11. Wipe orifice plug (2) and hoist cylinder walls (4) clean and inspect. If scratched or chipped, replace orifice plug (2) or hoist cylinder walls (4).

NOTE

Both hoist cylinders are assembled in the same way. Steps 1 through 12 cover the left hoist cylinder.

1. Install three piston rings (11) on piston grooves (10).

2. Slide piston (9) on piston rod (8) until seated and install with slotted nut (13) and new cotter pin (12).

CAUTION

Use care when performing step 3 to prevent damage to cylinder bore or piston.

3. Slide piston (9) and piston rod (8) in cylinder housing (3).

4. Install new O-ring (7) on cylinder head (5).

5-105. DUMP HOIST CYLINDER MAINTENANCE (Contd)

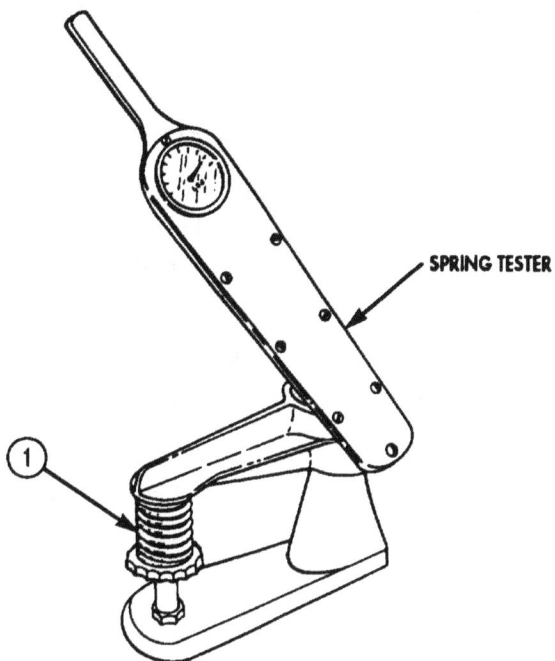

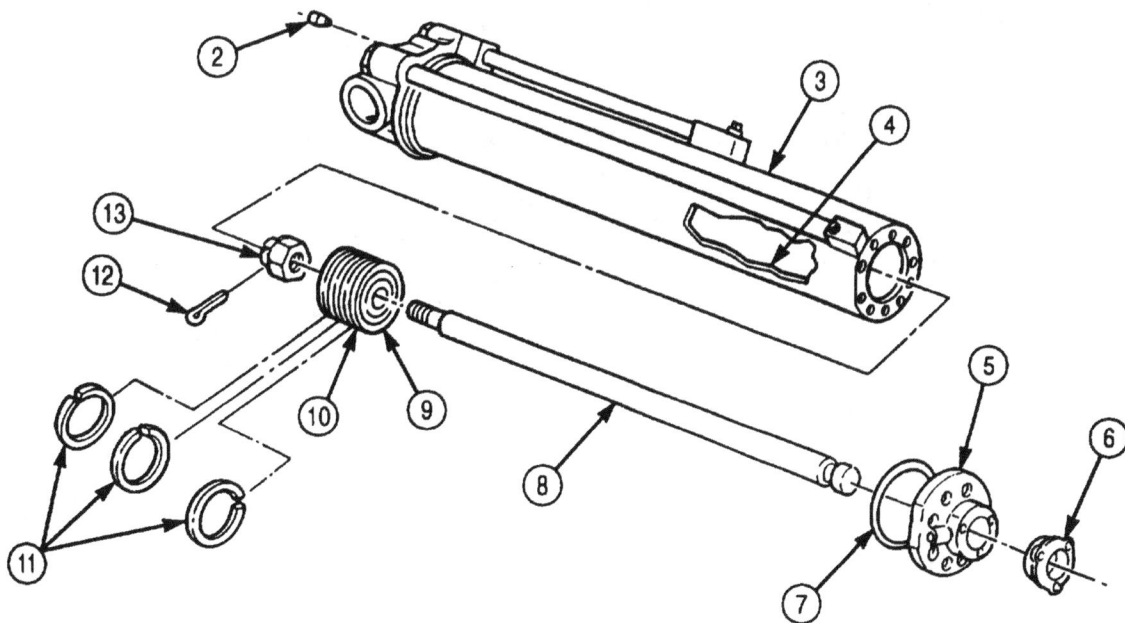

SPRING TESTER

5-105. DUMP HOIST CYLINDER MAINTENANCE (Contd)

5. Slide cylinder head (17) over piston rod (18) and position on cylinder housing (19).

6. Install cylinder head (17) on cylinder housing (19) with ten new lockwashers (12) and screws (13). Tighten screws (13) 8-10 lb-ft (11-14 N-m).

7. Install packing (16) around piston rod (18) in cylinder head (17). Use fingers to position new packing (16).

8. Slide gland (15) over piston rod (18) until seated against packing (16) and install on cylinder head (17) with three screws (14).

9. Install orifice plug (2) and orifice plug cover (1) in cylinder base (3).

NOTE

Short bypass tube connector block is located in middle of cylinder.

10. Install two bypass check balls (7), springs (6), and plugs (5) in short bypass tube connector block (4).

NOTE

Long bypass tube connector block is located at head end of cylinder.

11. Install one bypass check ball (10), spring (9), and plug (8) in long bypass tube connector block (11).

12. Position hinge pin (21) in cylinder base (3) until aligned with holes for screws (20) and install two square-head screws (20).

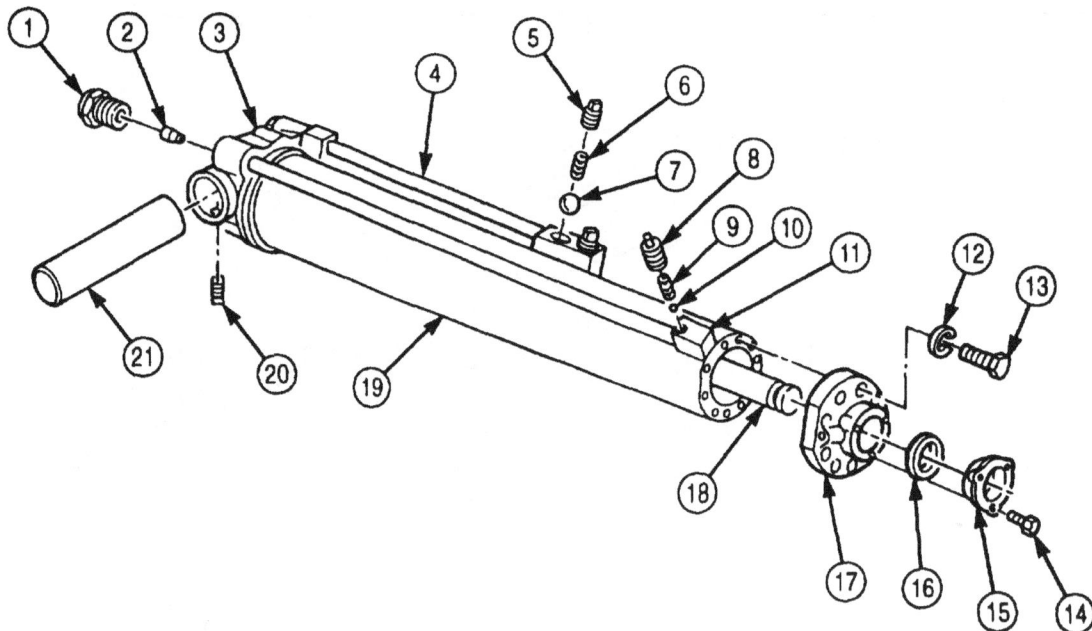

5-105. DUMP HOIST CYLINDER MAINTENANCE (Contd)

e. Installation

1. Position hoist cylinders (27) and (22) on subframe (28) with bypass tubes (26) facing down.
2. Install each hoist cylinder (27) and (22) with two bearing caps (23), four new lockwashers (25), and screws (24).

NOTE

To properly identify hose connection points, the cylinder port with bypass tube extending to the middle of the cylinder will be identified as port B. The cylinder port with bypass tube extending full length of the cylinder will be identified as port A.

3. Connect hydraulic hoses (31) and (35) to left cylinder port A (38), right cylinder port A (29), and right cross fitting (34).
4. Connect hydraulic hoses (32) and (36) to left cylinder port B (30), right cylinder port B (37), and left cross fitting (33).

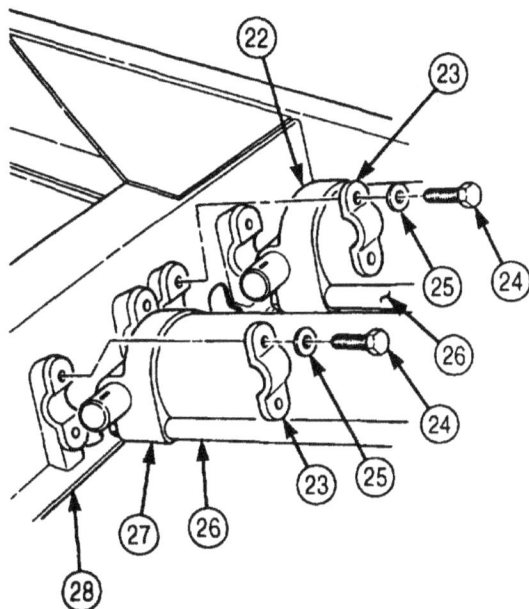

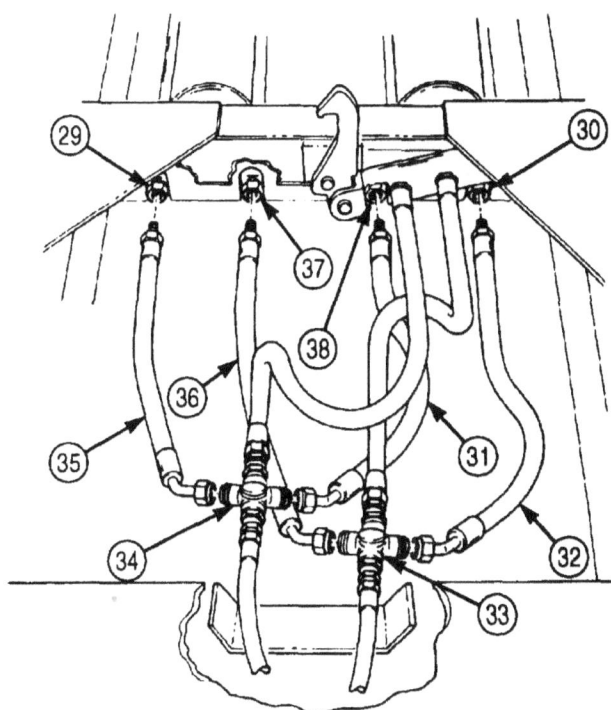

5-105. DUMP HOIST CYLINDER MAINTENANCE (Contd)

5. Position crosshead (2) on piston rods (1) and (3) and install with two upper crosshead retainers (7), lower crosshead retainers (8), four new lockwashers (4), and screws (5).

6. Raise hoist cylinders (6) and (9) and remove wood block.

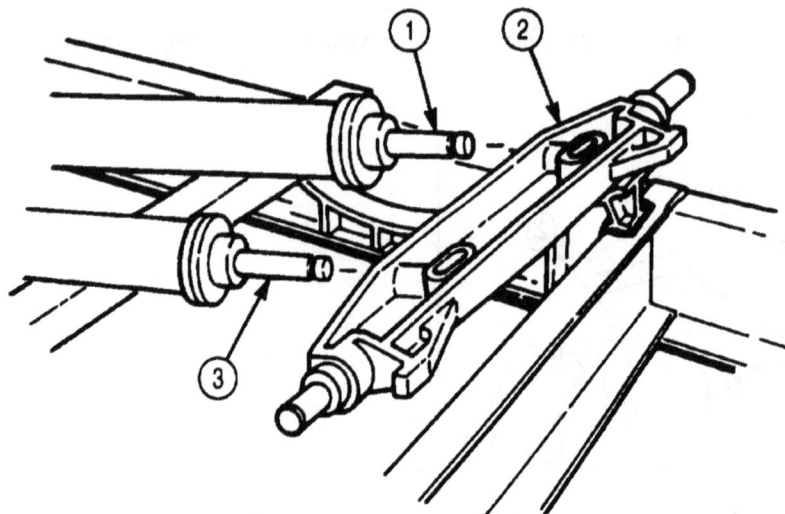

5-105. DUMP HOIST CYLINDER MAINTENANCE (Contd)

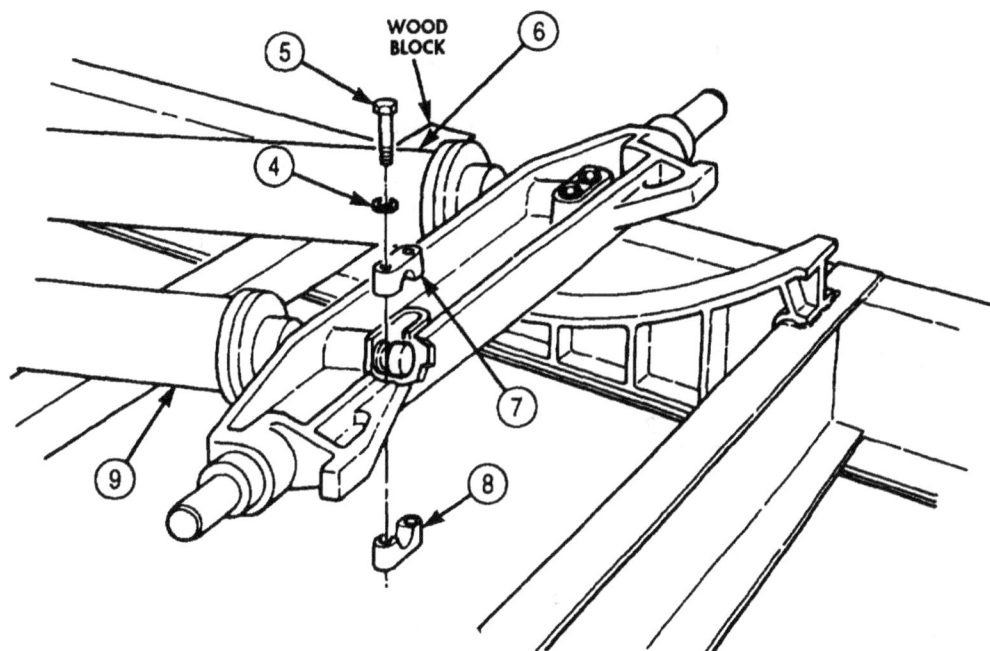

FOLLOW-ON TASKS:
- Install dump roller arms (para. 5-104).
- Install dump body (para. 4-144).
- Fill hydraulic reservoir to proper oil level (LO 9-2320-272-12).
- Start engine (TM 9-2320-272-10) and operate dump through full range. Check for leaks and proper operation.

5-106. BOOM EXTENSION CYLINDER REPAIR

THIS TASK COVERS:

a. Disassembly
b. Cleaning, Inspection, and Repair

c. Assembly

INITIAL SETUP:

APPLICABLE MODELS
M936/A1/A2

SPECIAL TOOLS
Bushing installer (Appendix E, Item 23)
Handle (Appendix E, Item 61)

TOOLS
General mechanic's tool kit (Appendix E, Item 1)
Spanner wrench (Appendix E, Item 166)

MATERIALS/PARTS
Wiper strip (Appendix D, Item 720)
Lockwire (Appendix D, Item 421)
O-ring (Appendix D, Item 467)
Packing (Appendix D, Item 510)
Crocus cloth (Appendix C, Item 20)
Drycleaning solvent (Appendix C, Item 71)
Lint-free cloth (Appendix C, Item 21)

REFERENCES (TM)
LO 9-2320-272-12
TM 9-2320-272-24P

EQUIPMENT CONDITION
Extension cylinder removed (para. 4-196)

GENERAL SAFETY INSTRUCTIONS
- Keep fire extinguisher nearby when using drycleaning solvent.
- Drycleaning solvent is flammable and toxic. Do not use near an open flame.

a. Disassembly

1. Using spanner wrench, remove cylinder head (2) from cylinder body (3).
2. Remove piston rod (1) from cylinder body (3).
3. Remove lockwire (14) and piston rod nut (15) from piston rod (1). Discard lockwire (14).
4. Remove washer (13), two U-cup retainers (9), U-cups (10), piston (11), cylinder head (2), and wiper strip (5) from piston rod (1). Discard wiper strip (5).
5. Remove packing nut (4), packing (6), O-ring (8), and piston rod bushing (7) from cylinder head (2). Discard O-ring (8) and packing (6).
6. Remove two half rings (12) from piston (11).

5-106. BOOM EXTENSION CYLINDER REPAIR (Contd)

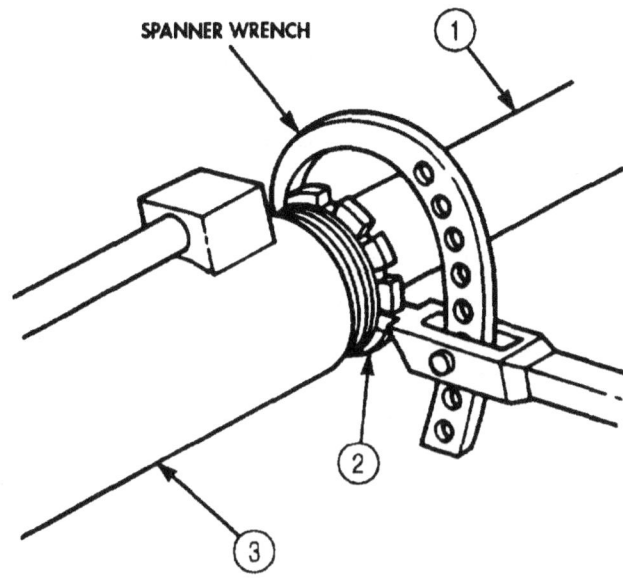

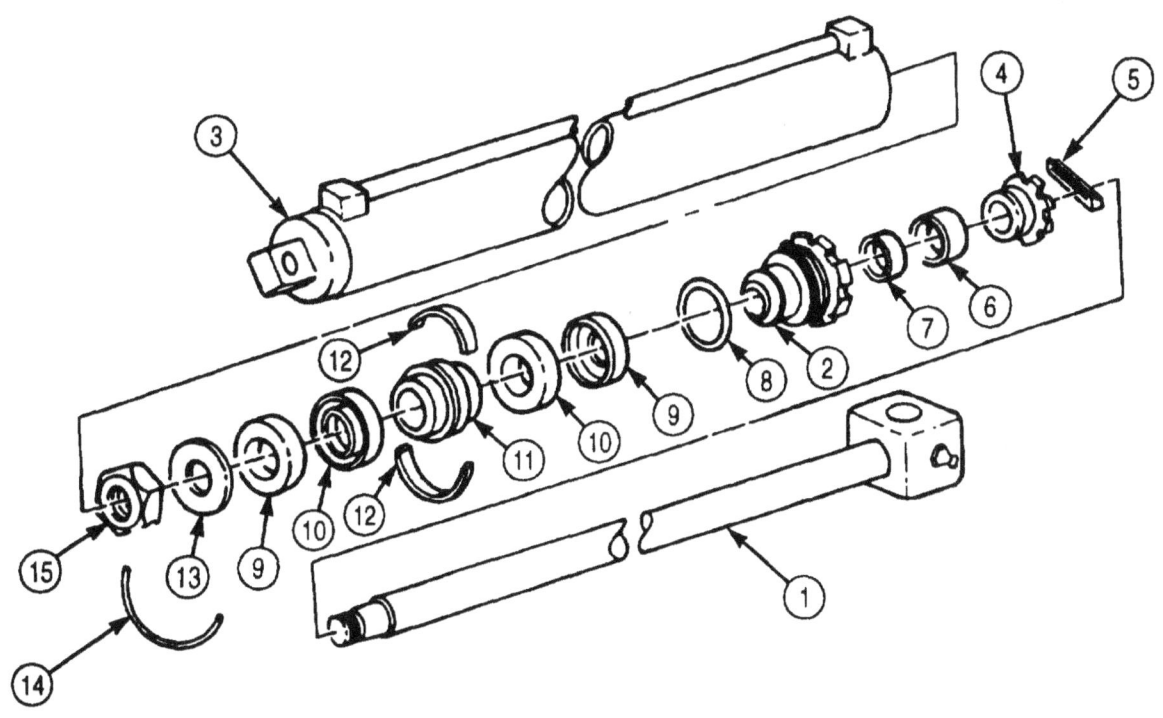

SPANNER WRENCH

5-106. BOOM EXTENSION CYLINDER REPAIR (Contd)

WARNING

Drycleaning solvent is flammable and toxic. Do not use near open
flame and always have a fire extinguisher nearby when solvents
are used. Use only in well-ventilated places, wear protective
clothing, and dispose of cleaning rags in approved container.
Failure to do this may result in injury or death to personnel and/or
damage to equipment.

1. Clean all boom extension cylinder components with drycleaning solvent and dry with lint-free cloth (para. 2-14).
2. Inspect piston rod (6). Remove scratches and burrs with crocus cloth. Replace if scored or threads are stripped.
3. Inspect piston (11) and cylinder body (1). Replace piston (11) or cylinder body (1) if scored.

c. Assembly

1. Install two half rings (12) on piston (11).
2. Using bushing installer and handle, install piston rod bushing (5) in cylinder head (7).
3. Install new O-ring (8), new packing (4), packing nut (2), and new wiper strip (3) on cylinder head (7).

NOTE

Use care not to damage wiper strip when passing over threads on
piston rod.

4. Install cylinder head (7) on piston rod (6).
5. Install two U-cup retainers (9), U-cups (10), piston (11), and washer (13) on piston rod (6) with piston rod nut (15) and new lockwire (14).
6. Slide piston rod (6) into cylinder body (1) and install cylinder head (7) on cylinder body (1). Using spanner wrench, tighten cylinder head (7).

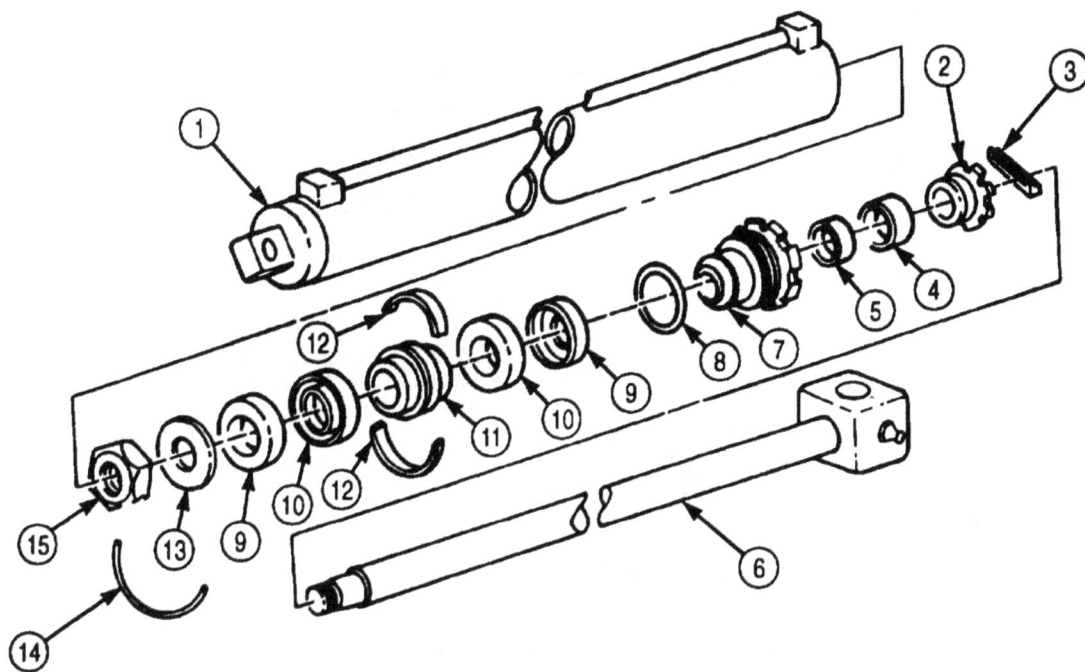

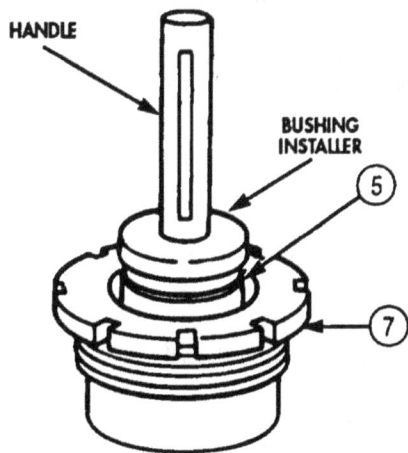

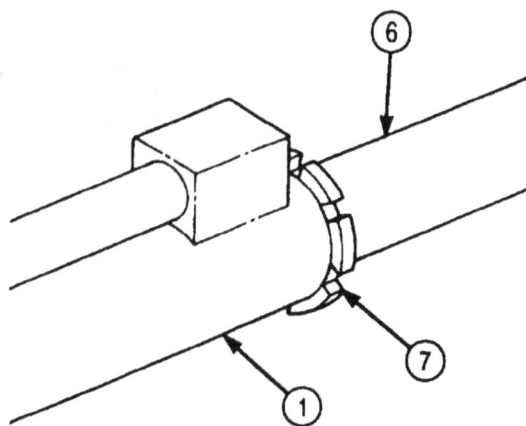

FOLLOW-ON TASKS:• Lubricate extension cylinder (LO 9-2320-272-12).
• Install extension cylinder (para. 4-196).

5-107. CRANE SWINGER GEARCASE REPAIR

THIS TASK COVERS:

a. Disassembly   c. Assembly
b. Cleaning, Inspection, and Repair

INITIAL SETUP:

APPUCABLE MODELS
M936/A1/A2

TOOLS
General mechanic's tool kit (Appendix E, Item 1)
Torque wrench (Appendix E, Item 144)
Arbor press
Mandrel

MATERIALS/PARTS
Two gearcase gaskets (Appendix D, Item 233)
Gearcase seal (Appendix D, Item 6241
Fifteen lockwashers (Appendix D, Item 350)
Two woodruff keys (Appendix D, Item 734)
Gasket (Appendix D, Item 234)
Two woodruff keys (Appendix D, Item 736)
Woodruff key (Appendix D, Item 735)
Drycleaning solvent (Appendix C, Item 71)

REFERENCES (TM)
LO 9-2320-272-12
TM 9-2320-272-24P

EQUIPMENT CONDITION
Crane swinger gearcase removed (para. 4-200).

GENERAL SAFETY INSTRUCTIONS
• Keep fire extinguisher nearby when using drycleaning solvent.
• Drycleaning solvent is flammable and toxic. Do not use near an open flame.

---

a. Disassembly

WARNING

Drycleaning solvent is flammable and toxic. Do not use near open flame and always have a fire extinguisher nearby when solvents are used. Use only in well-ventilated places, wear protective clothing, and dispose of cleaning rags in approved container. Failure to do this may result in injury or death to personnel and/or damage to equipment.

1. Thoroughly clean exterior with drycleaning solvent (para. 2-14).

NOTE
Have drainage container ready to catch oil.

2. Remove two drainplugs (8) from crane swinger gearcase (13) and drain lubricant from crane swinger gearcase (13).

3. Reinstall two drainplugs (8) in crane swinger gearcase (13).

4. Remove four screws (11), lockwashers (12), shaft bearing cap (10), and gasket (9) from crane swinger gearcase (13). Discard lockwashers (12) and gasket (9).

5. Remove four screws (5), lockwashers (4), gearcase motor mounting cap (6), and gasket (7) from crane swinger gearcase (13). Discard lockwashers (4) and gasket (7).

6. Remove six screws (l), lockwashers (14), cover (2), and gasket (3) from crane swinger gearcase (13). Discard lockwashers (14) and gasket (3).

7. Using arbor press and mandrel, remove gear shaft assembly (20) from crane swinger gearcase (13).

8. Remove screw (15), lockwasher (25), and retaining washer (24) from gear shaft assembly (20). Discard lockwasher (25).

5-107. CRANE SWINGER GEARCASE REPAIR (Contd)

NOTE

Use puller to remove bearing and gear.

9. Remove upper bearing (23), thrust washer (22), gear (21), gear washer (16), lower bearing (17), seal (18), and two woodruff keys (19) from gear shaft assembly (20). Discard woodruff keys (19) and seal (18).

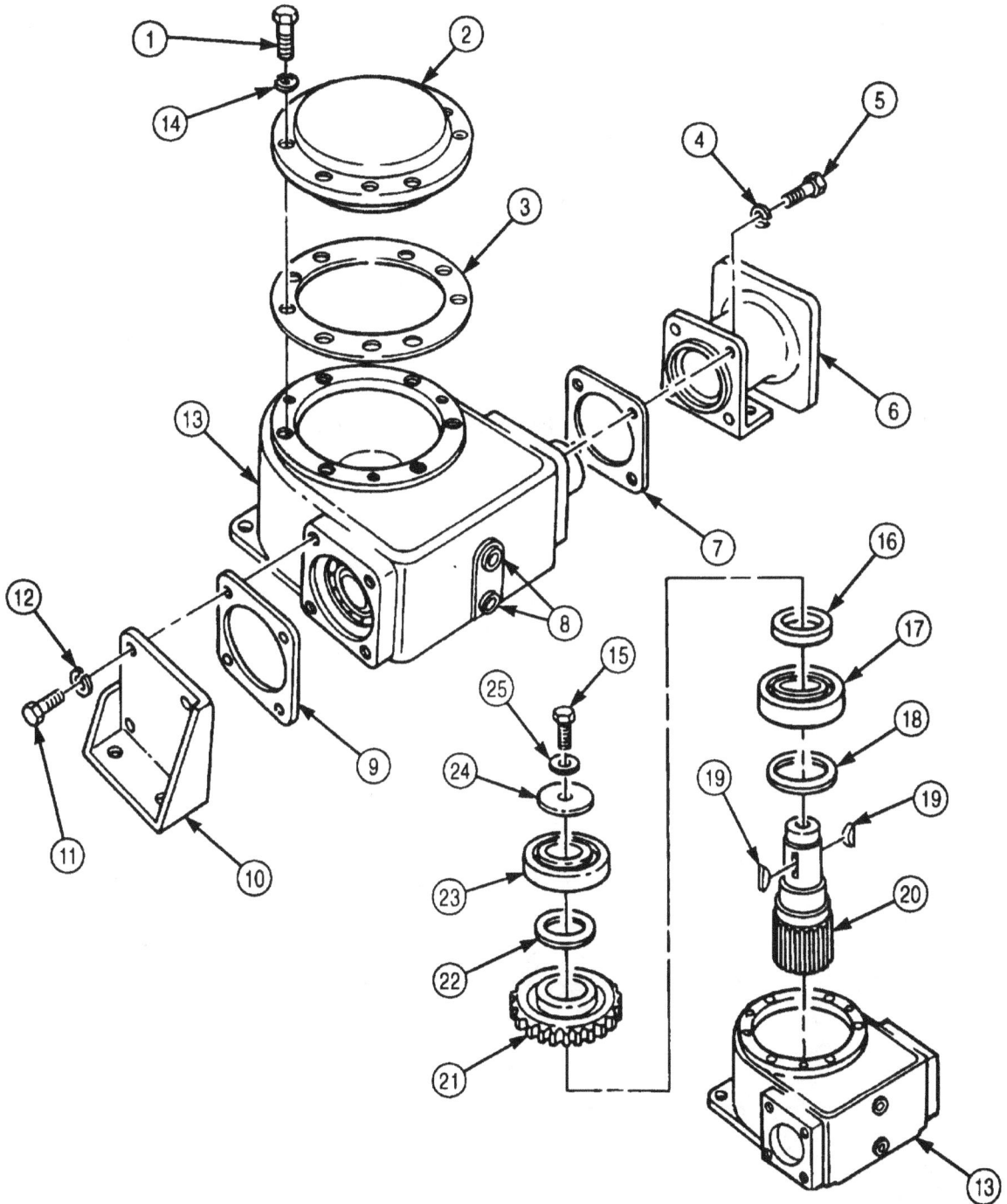

5-107. CRANE SWINGER GEARCASE REPAIR (Contd)

10. Using arbor press and mandrel, remove coupling (11) and woodruff key (8) from shoulder shaft (10). Discard woodruff key (8).

11. Using arbor press and mandrel, press crane swinger gearcase (5) shoulder shaft (10) down until bearing (1) is removed.

12. Turn crane swinger gearcase (5) over and press bearing (1) and shoulder shaft (10) assembly from crane swinger gearcase (5).

13. Remove two bushings (2), springs (3), spacers (4), washers (6), and woodruff keys (9) from shoulder shaft (10). Discard woodruff keys (9).

14. Remove worm gear (7) from shoulder shaft (10).

b. Cleaning, Inspection, and Repair

WARNING

Drycleaning solvent is flammable and toxic. Do not use near open flame and always have a fire extinguisher nearby when solvents are used. Use only in well-ventilated places, wear protective clothing, and dispose of cleaning rags in approved container. Failure to do this may result in injury or death to personnel and/or damage to equipment.

CAUTION

Do not spin-dry bearings with compressed air.

1. Clean all crane swinger gearcase (5) components with drycleaning solvent.

2. Inspect four bearings (1). Replace if scored, pitted, or broken.

3. Inspect crane swinger gearcase (5), gearcase motor mounting cap (14), shaft bearing cap (15), and gearcase cover (12) for cracks and breaks. Replace if cracked or broken.

4. Inspect shoulder shaft (10), worm gear (7), and gearshaft (13). Replace if broken, cracked, chipped, or out-of-round.

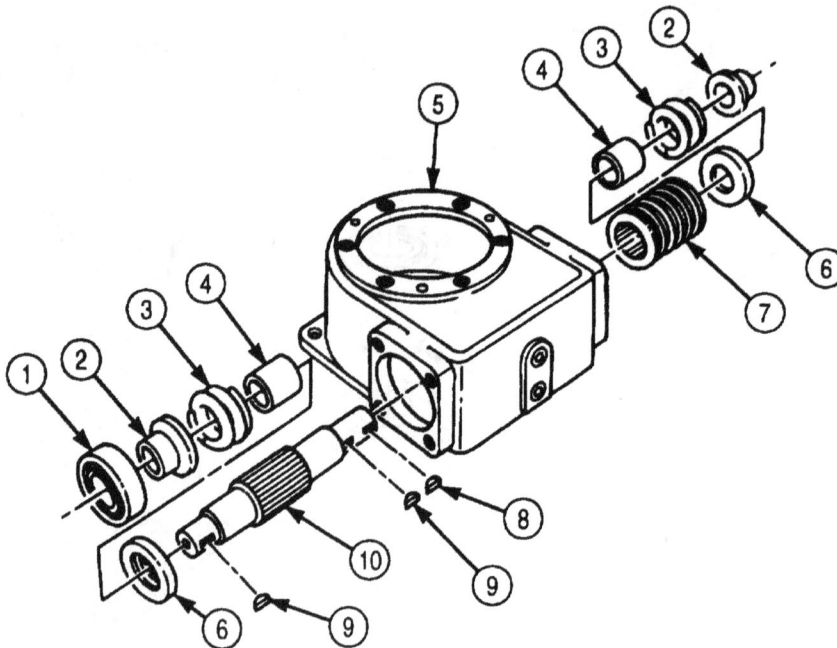

5-107. CRANE SWINGER GEARCASE REPAIR (Contd)

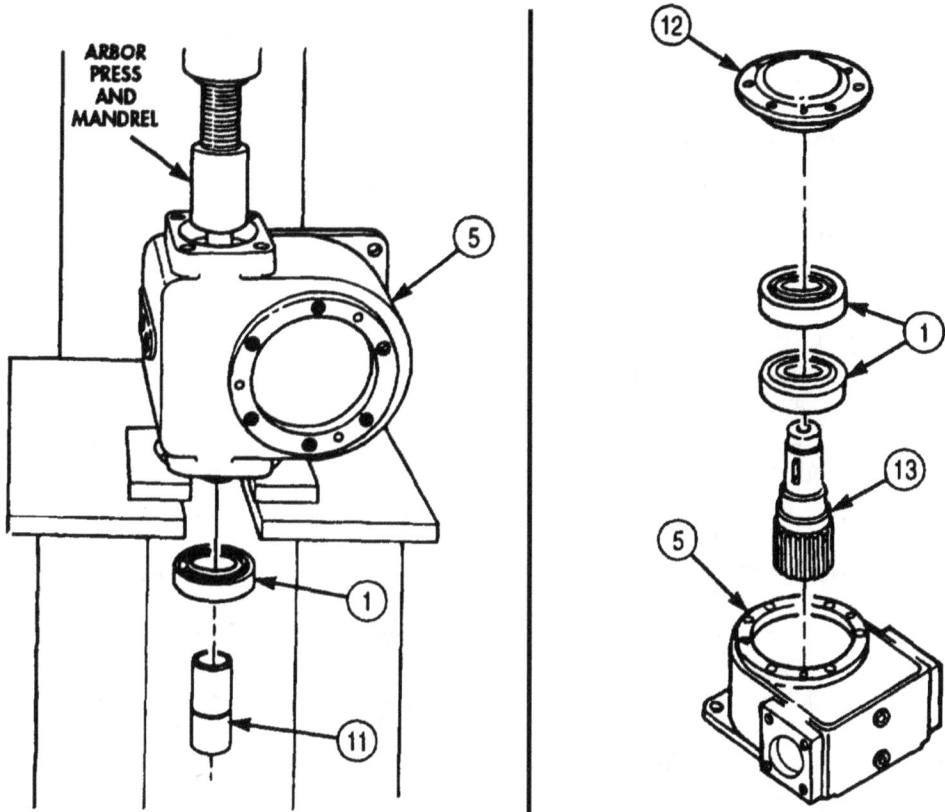

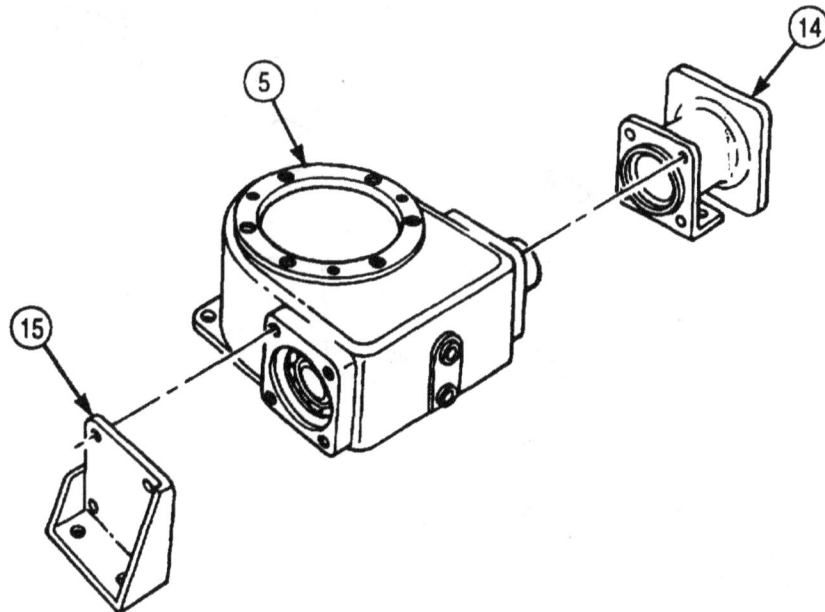

ARBOR
PRESS
AND
MANDREL

5-107. CRANE SWINGER GEARCASE REPAIR (Contd)

c. Assembly

1. Install two new woodruff keys (9), washers (6), spacers (4), springs (3), and bushings (2) on shoulder shaft (10).
2. Install worm gear (7) on shoulder shaft (10).
3. Using arbor press and mandrel, install shoulder shaft bearings (1) on shoulder shaft (10) and press shoulder shaft (10) into crane swinger gearcase (5) until bearing (1) is flush with end of crane swinger gearcase (5).
4. Turn crane swinger gearcase (5) over and, using arbor press and mandrel, press second shoulder shaft bearing (1) into crane swinger gearcase (5). Ensure bearing (1) is flush with crane swinger gearcase (5).
5. Install new woodruff key (8) on shoulder shaft (10).
6. Install new seal (14) on gear shaft (16).
7. Using arbor press and mandrel, press bearing (13) onto gear shaft (16).
8. Position lower thrust washer (12) and two new woodruff keys (15) on gear shaft (16) and, using arbor press and mandrel, press gear (17) onto gear shaft (16) over woodruff keys (15).
9. Position upper thrust washer (18) on gear shaft (16) and, using arbor press and mandrel, press bearing (19) onto gear shaft (16).
10. Install retaining washer (20), new lockwasher (21), and screw (22) on gear shaft (16). Tighten screw (22) 44-61 lb-ft (60-83 N•m).
11. Install new gasket (23) and gearcase cover (24) on crane swinger gearcase (5) with six new lockwashers (25) and screws (11). Tighten screws (11) 44-61 lb-ft (60-83 N•m).
12. Install new gasket (28) and shaft bearing cap (27) on crane swinger gearcase (5) with four new lockwashers (26) and screws (34). Tighten screws (34) 44-61 lb-ft (60-83 N•m).
13. Install coupling (29) in crane swinger gearcase (5).
14. Install new gasket (30) and mounting cap (31) on crane swinger gearcase (5) with four new lockwashers (32) and screws (33). Tighten screws (33) 44-61 lb-ft (60-83 N•m).

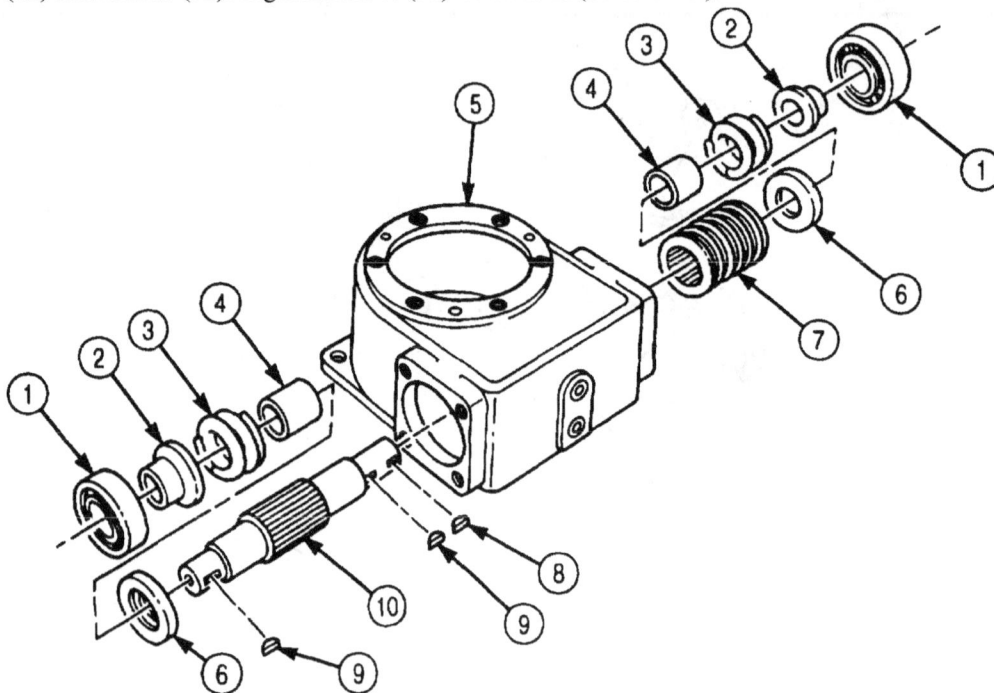

5-107. CRANE SWINGER GEARCASE REPAIR (Contd)

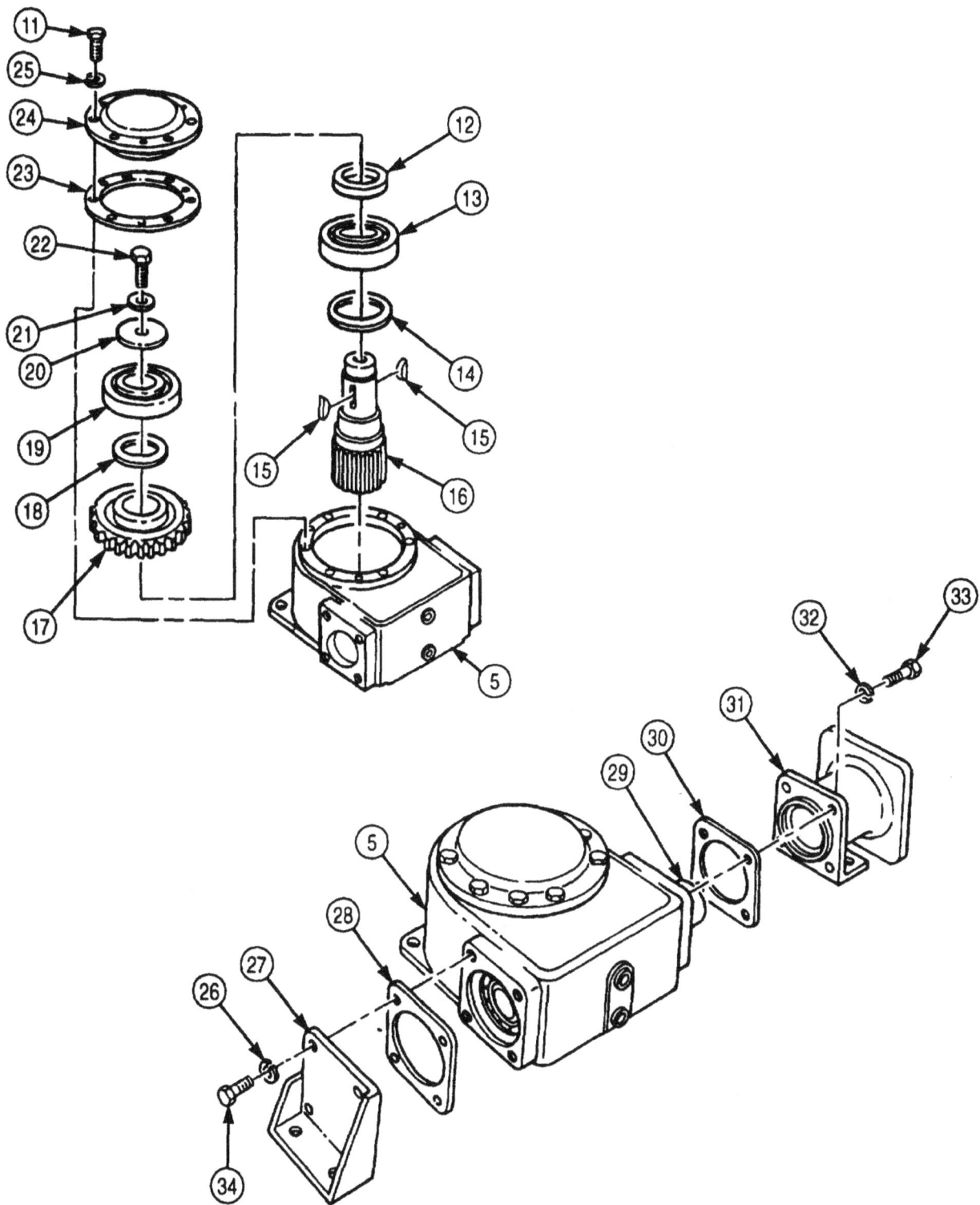

FOLLOW-ON TASKS: • Fill crane swinger gearcase with lubricant (LO 9-2320-272-12).
• Install crane swinger gearcase (para. 4-200).

5-108. CRANE CONTROL VALVE REPAIR

THIS TASK COVERS:

a. Disassembly
b. Cleaning, Inspection, and Repair

c. Assembly

INITIAL SETUP:

APPLICABLE MODELS
M936/A1/A2

SPECIAL TOOLS
Seal retainer remover and replacer
  (Appendix E, Item 117)

TOOLS
General mechanic's tool kit (Appendix E, Item 1)
Soft-jawed vise

MATERIALS/PARTS
Relief valve kit (Appendix D, Item 533)
Load check valve kit (Appendix D, Item 270)
Eight snaprings (Appendix D, Item 665)
O-ring (Appendix D, Item 469)
O-ring (Appendix D, Item 468)
Four O-rings (Appendix D, Item 489)
Four lockwashers (Appendix D, Item 350)
Drycleaning solvent (Appendix C, Item 71)
Lint-free cloth (Appendix C, Item 21)

REFERENCES (TM)
TM 9-2320-272-24P

EQUIPMENT CONDITION
Crane control valve removed (para. 4-201).

GENERAL SAFETY INSTRUCT IONS
• Keep fire extinguisher nearby when using drycleaning solvent.
• Drycleaning solvent is flammable and toxic. Do not use near an open flame.

NOTE
Do not perform this procedure unless relief valve kit and load check valve kit are available.

a. Disassembly

WARNING

Drycleaning solvent is flammable and toxic. Do not use near open flame and always have a fire extinguisher nearby when solvents are used. Use only in well-ventilated places, wear protective clothing, and dispose of cleaning rags in approved container. Failure to do this may result in injury or death to personnel and/or damage to equipment.

1. Thoroughly clean exterior of crane control valve (3) with drycleaning solvent (para. 2-14).

2. Remove eight snaprings (21), pins (20), and four handles (19) from base (2) and crane control valve (3). Discard snaprings (21).

3. Remove four nuts (13), lockwashers (14), screws (1), control brackets (15) and (7), and base (2) from crane control valve (3). Discard lockwashers (14).

4. Remove nipple adapter (17) and O-ring (16) from crane control valve (3). Discard O-ring (16).

5. Remove nipple (18) from nipple adapter (17).

NOTE
Mark position and angle of tube and all elbows for assembly.

6. Remove tube (6), connector (5), and O-ring (4) from crane control valve (3). Discard O-ring (4)

7. Remove two elbows (10), three adapters (11), and two O-rings (12) from crane control valve (3). Discard O-rings (12).

8. Remove four elbows (9) and O-rings (8) from crane control valve (3). Discard O-rings (8).

5-108. CRANE CONTROL VALVE REPAIR (Contd)

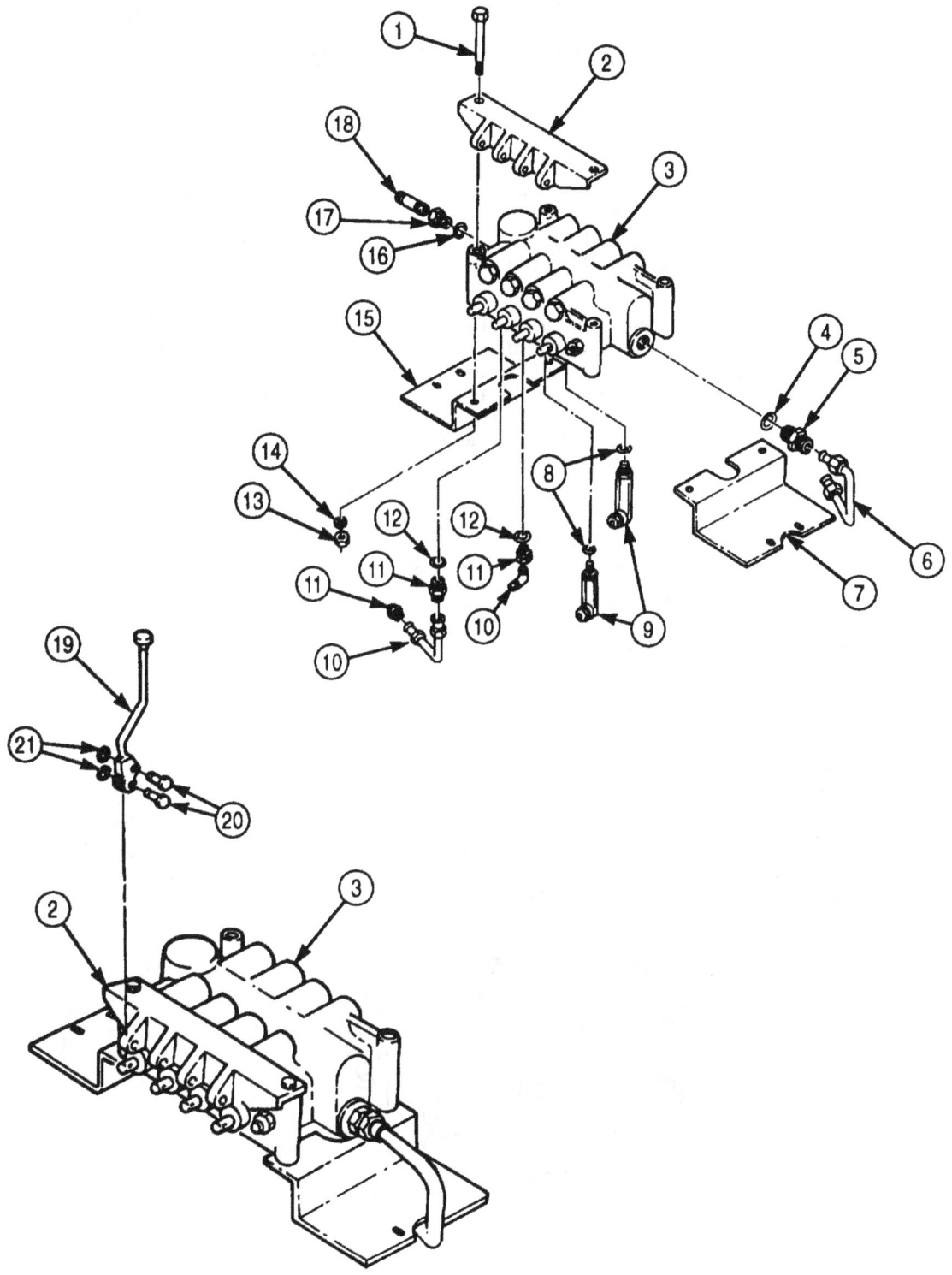

5-108. CRANE CONTROL VALVE REPAIR (Contd)

NOTE

All spring-centered spool valves in the control valve are removed and disassembled the same way. Three spool valves are stamped with a D (double action) on the shank of the spool valve. The fourth is stamped with an S (single action) and is located next to the oil outlet port in the control valve. Steps 9 through 12 will cover a double-action valve only.

9. Remove spool valve cap (3) and spool valve (2) from crane control valve (1).

10. Clamp spool valve (2) in soft-jawed vise.

11. Using improvised compression tool, compress spring (6) and remove snapring (4), outer spacer (5), spring (6), travel limit washer (7), and inner spacer (8) from spool valve (2).

12. Using seal retainer remover and replacer, remove retainer (24), seal (25), and O-ring (26) from crane control valve (1). Discard seal (25) and O-ring (26).

NOTE

- All check valves in the control valve are removed and assembled the same way. Check valves located adjacent to valve spools marked with a D (double action) are identical. The check valve located by the valve spool marked S (single action) contains only one spring and poppet. All other components are the same as the D-type spools. Steps 13 and 14 cover a check valve adjacent to a D-type spool.

- Do not remove check seat sleeve from control valve. Check valve parts should be identified so they may be returned to their original position during assembly.

13. Remove two check valve caps (9), springs (17), and poppets (14) from crane control valve (1).

14. Remove backup ring (16), O-ring (15), backup ring (13), backup ring (12), O-ring (11), and O-ring (10) from each check valve cap (9). Discard backup ring (16), O-ring (15), backup ring (13), backup ring (12), O-ring (11), and O-ring (10).

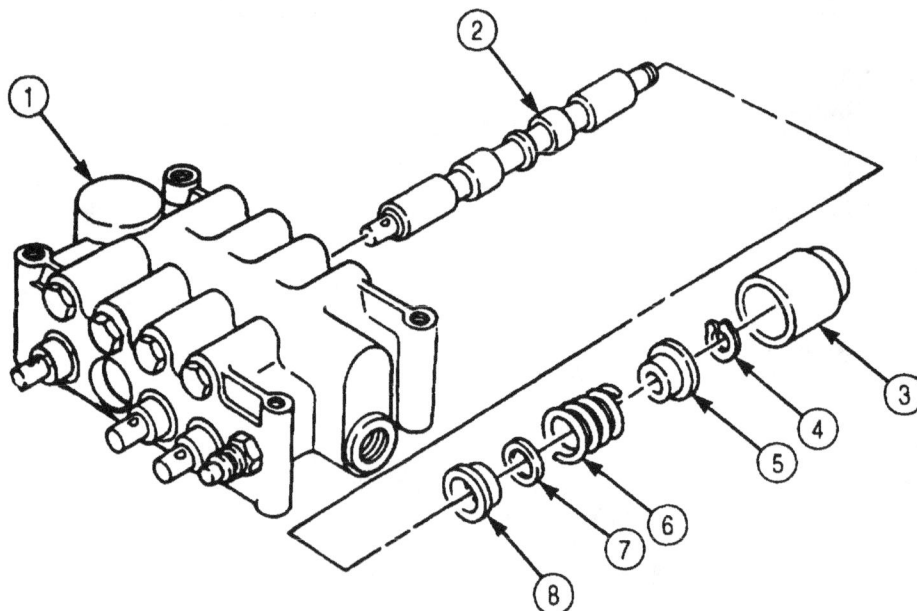

5-108. CRANE CONTROL VALVE REPAIR (Contd)

15. Remove relief valve (23) from crane control valve (1). Discard relief valve (23).

16. Remove seat retainer plug (18), drain sleeve (20), and seat retainer (19) from crane control valve (1). Discard seat retainer (19).

17. Remove poppet spring (22) and poppet assembly (21) from crane control valve (1). Discard poppet assembly (21) and poppet spring (22).

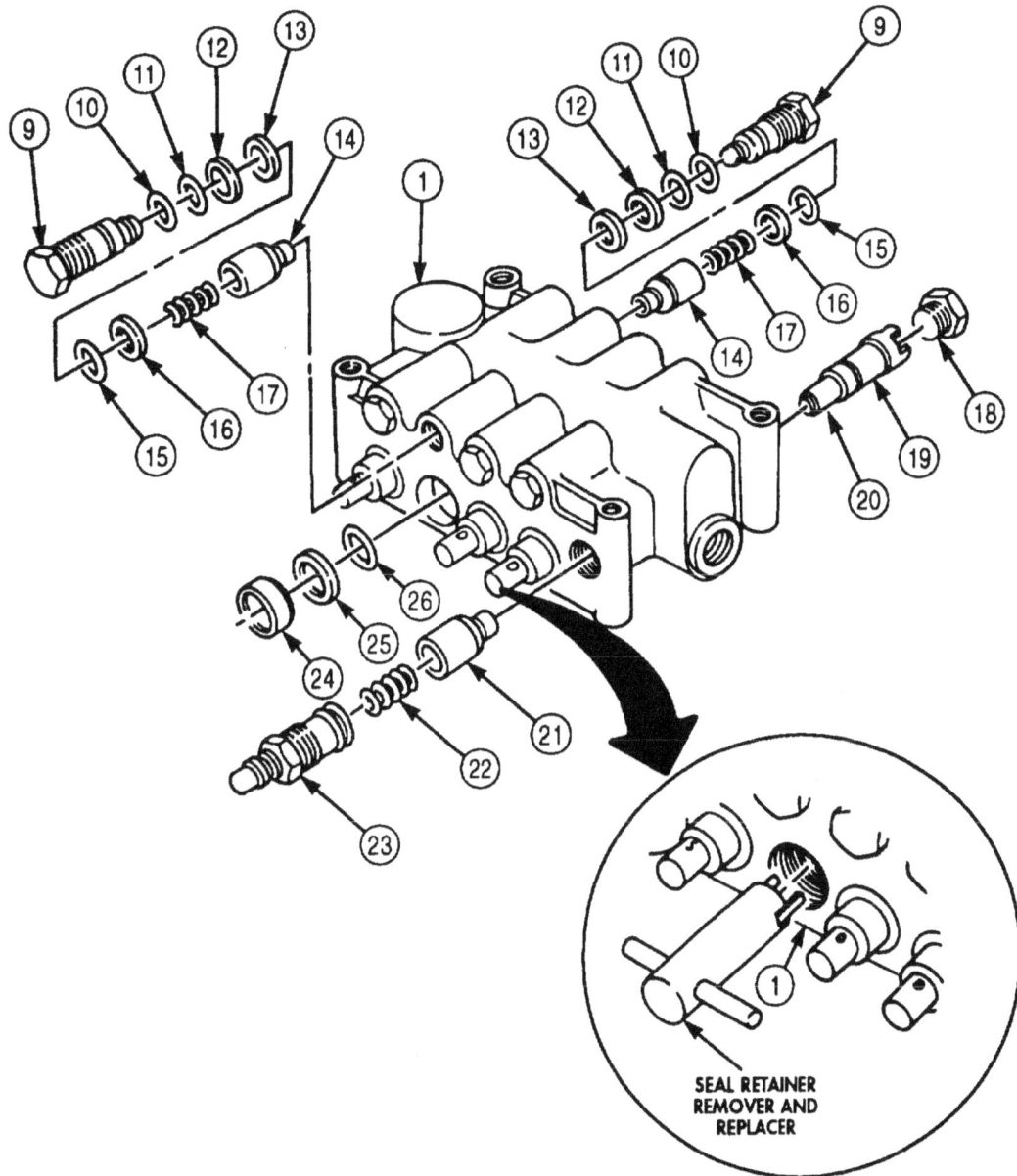

SEAL RETAINER
REMOVER AND
REPLACER

5-108. CRANE CONTROL VALVE REPAIR (Contd)

b. Cleaning, Inspection, and Repair

WARNING

Drycleaning solvent is flammable and toxic. Do not use near open flame and always have a tire extinguisher nearby when solvents are used. Use only in well-ventilated places, wear protective clothing, and dispose of cleaning rags in approved container. Failure to do this may result in injury or death to personnel and/or damage to equipment.

1. Clean all control valve components with drycleaning solvent and dry with lint-free cloth (para. 2-14).
2. Inspect spool valve (1) and crane control valve (2). Replace crane control valve (2) if grooved, scratched, cracked, or broken.

c. Assembly

NOTE

Drain sleeve should go into control valve far enough to leave two or three full threads exposed. Installing seat retainer plug at this point sets the seat retainer to the proper depth.

1. Install new seat retainer (13) and drain sleeve (14) in crane control valve (2).
2. Install seat retainer plug (12) in crane control valve (2).
3. Install new poppet assembly (15) and poppet spring (16) in crane control valve (2) so poppet assembly (15) points toward seat retainer (13).
4. Install new relief valve (17) on crane control valve (2).
5. Install new O-rings (4) and (5), new backup rings (6) and (7), new O-ring (9), and new backup ring (10) on each check valve cap (3).
6. Install two poppets (8), springs (ll), and check valve caps (3) into crane control valve (2).

NOTE

The remaining check valve caps are assembled the same way.

7. Clamp spool valve (1) in soft-jawed vise.
8. Using improvised compression tool, compress spring (21) and install inner spacer (23), travel limit washer (22), spring (21), and outer spacer (20) on spool valve (1) with new snapring (19).
9. Install spool valve (1) and spool valve cap (18) in crane control valve (2).
10. Using seal retainer remover and replacer, install new O-ring (26), new seal (25), and retainer (24) on crane control valve (2).

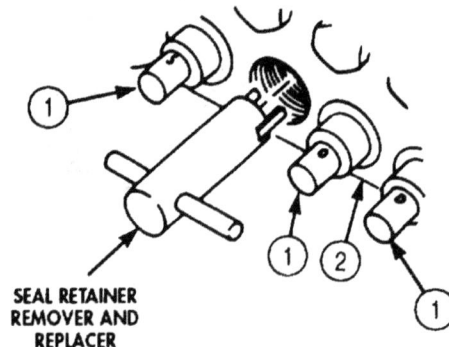

SEAL RETAINER
REMOVER AND
REPLACER

5-108. CRANE CONTROL VALVE REPAIR (Contd)

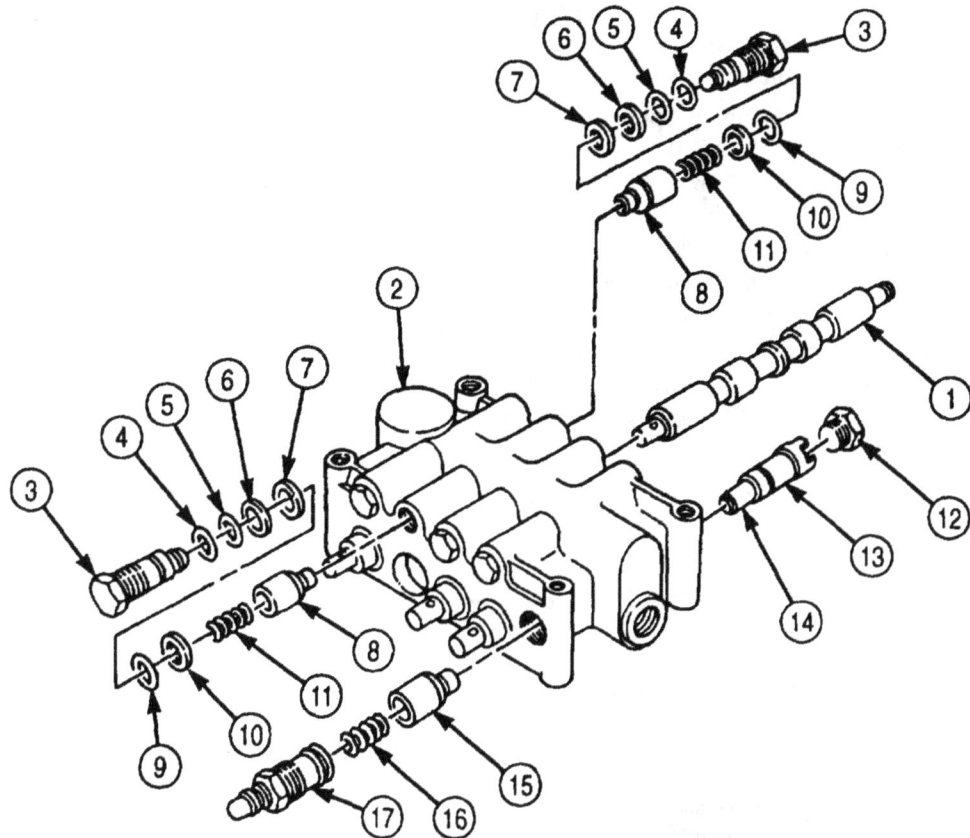

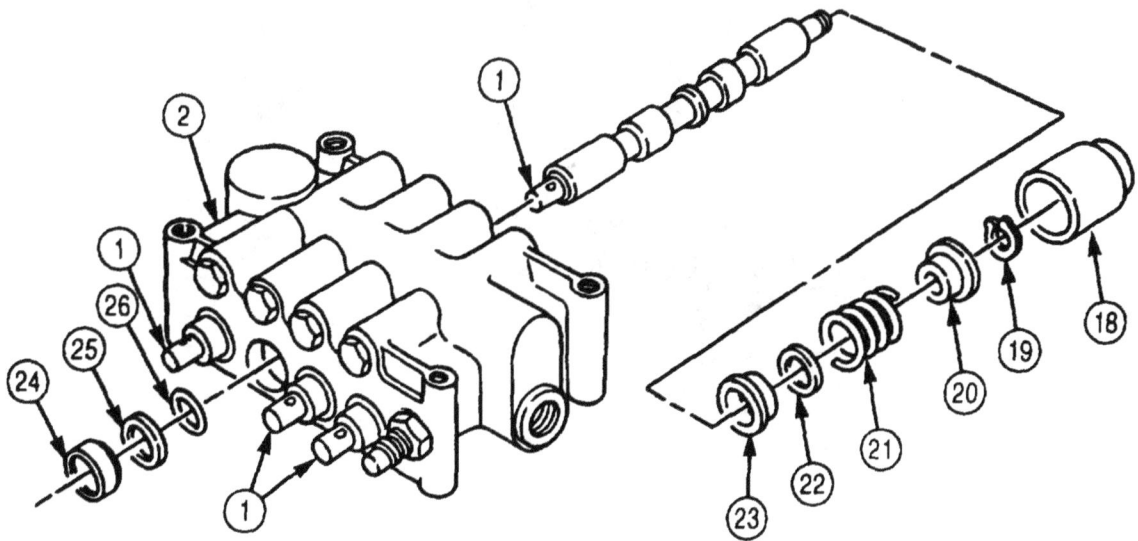

5-108. CRANE CONTROL VALVE REPAIR (Contd)

NOTE
- Install all parts in steps 11 through 13 in their marked location and angle on control valve.
- The three remaining spool valves are assembled into the control valve in the same way.

11. Install four new O-rings (8), elbows (9), two new O-rings (12), three adapters (ll), two elbows (10), new O-ring (4), connector (5), and tube (6) on crane control valve (3).

12. Install nipple adapter (17) on nipple (18).

13. Install new O-ring (16) and nipple adapter (17) on crane control valve (3).

14. Install right control bracket (7), left control bracket (15) and base (2) on crane control valve (3) with four screws (1), new lockwashers (14), and nuts (13).

15. Align four operating handles (19) with holes in base (2) and control spool valves (21), and install four operating handles (19) with eight pins (20) and new snaprings (22).

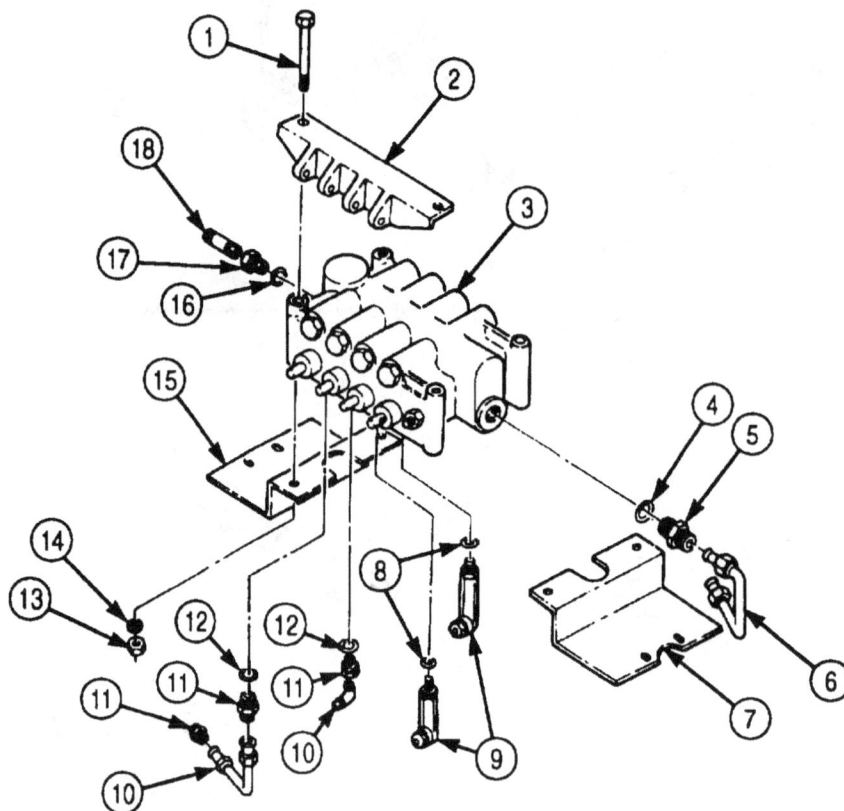

5-108. CRANE CONTROL VALVE REPAIR (Contd)

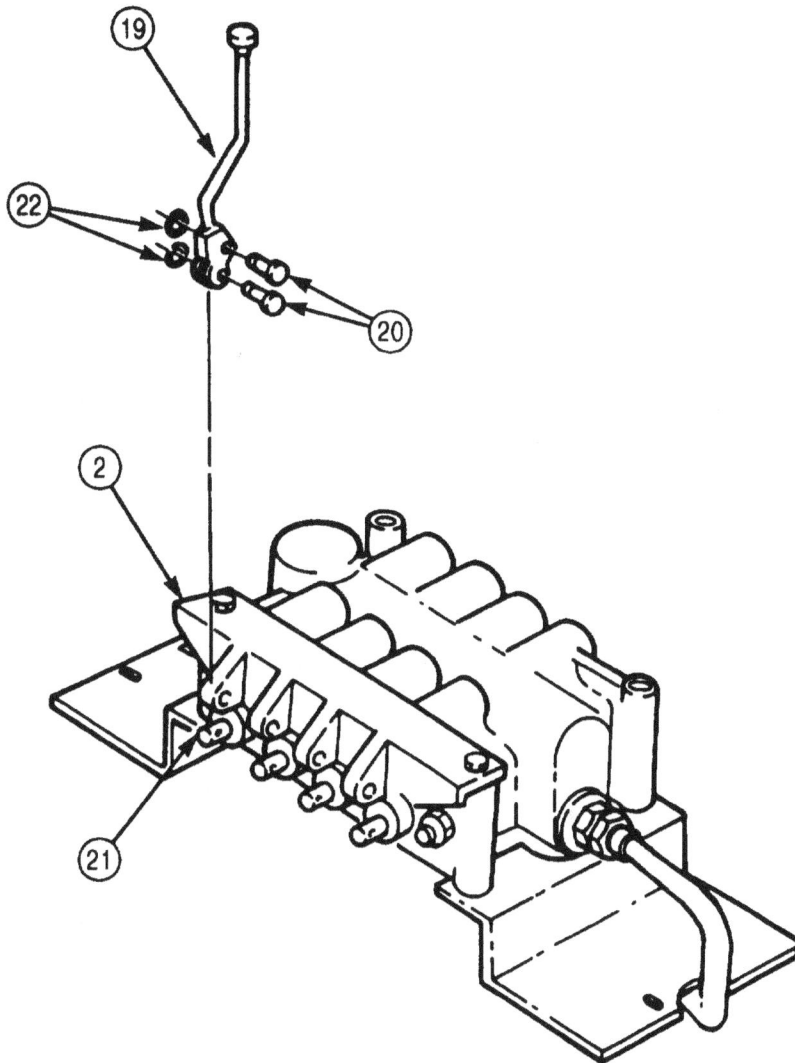

FOLLOW-ON TASK: Install crane control valve (para. 4-201).

CHAPTER 6
SHIPMENT AND LIMITED STORAGE

Section I. GENERAL PREPARATION OF VEHICLE FOR SHIPMENT

6-1. SCOPE

a. This section provides instructions on preserving and protecting vehicles for shipment.

b. Protection for vehicles and accompanying equipment must be sufficient to protect the material against deterioration and physical damage.

6-2. CLEANING

WARNING

Drycleaning solvent is flammable and will not be used near open flame. Use only in well-ventilated places. Failure to do so may result in injury to personnel.

CAUTION

Cleaning materials or paints containing chlorinated hydrocarbon class solvents are not to be used on composite taillights and parking lights. Damage to taillight and parking light lenses may result.

Prior to application of preservatives, surfaces must be cleaned to ensure removal of corrosion, soil, grease, or vehicle acid and alkali residues.

a. **Interior of Vehicle.** Remove all dirt and other foreign matter from all painted metal surfaces of vehicle by scrubbing with cloths soaked in drycleaning solvent (appendix C, item 71). DO NOT apply solvent to electrical equipment or rubber parts of any nature. Use trichloroethylene (appendix C, item 76) to clean electrical parts and electrical contact points. Use warm water for cleaning rubber parts. Apply preservative compounds to rubber parts as required (TM 9-247).

b. **Exterior of Vehicle.** Clean exterior surfaces of vehicle to ensure removal of all dirt and foreign matter. After cleaning, immediately dry parts to remove excess cleaning solutions or residual moisture. Allow parts to air-dry or wipe with clean, dry, lint-free cloth (appendix C, item 21).

6-3. LUBRICATION

WARNING

Drycleaning solvent is flammable and will not be used near open flame. Use only in well-ventilated places. Failure to do so may result in injury to personnel.

After cleaning has been accomplished, wipe all grease fittings clean with drycleaning solvent (appendix C, item 71) and lubricate vehicle in accordance with LO 9-2320-272-12. Remove excess grease after lubrication and before processing.

6-4. PRESERVATION

All critical unpainted metal surfaces must be protected during shipment. Use procedures and materials listed in steps a. and b. below. If the preservatives listed below are not available, oil or grease listed in LO 9-2320-272-12 may be used for this purpose, but is effective for only a few days; therefore, equipment protected must be closely watched for signs of corrosion. When selecting preservatives, use only those that will not damage the surface to which they are applied.

a. Battery Leads. Disconnect both batteries (para. 3-126). Each battery lead terminal, including the jumper lead ends, must be wrapped with tape (appendix C, item 73).

b. Miscellaneous Preservation. Coat all unpainted, exposed, or machined metal surfaces on the exterior of the vehicle with corrosion-preventive compound (appendix C, item 25).

6-5. PACKAGING

Electrical Openings. Cover all electrical receptacles with tape (appendix C, item 73) or with plastic caps which will afford the same degree of protection.

6-6. PACKING

Pack all Basic Issue Items (BII) and Additional Authorization List (AAL) items to prevent mechanical damage.

6-7. SHIPMENT OF ARMY DOCUMENTS

Prepare all Army shipping documents accompanying vehicle in accordance with DA Pam 738-750.

Section II. LOADING AND MOVEMENT

6-8. LOADING AND MOVEMENT

For transportability guidance handling and movement of vehicles, refer to TM 55-2320-272-15-1.

Section III. LIMITED STORAGE

6-9. SCOPE

Commanders are responsible for ensuring that all vehicles issued or assigned to their command are maintained in a serviceable condition and properly cared for and that personnel under their command comply with technical instructions. Lack of time, trained personnel, or proper tools may result in a unit being incapable of performing maintenance for which it is responsible. In such cases, unit commanders may, with the approval of major commanders, place a vehicle that is beyond the maintenance capability of the unit in administrative storage. For detailed maintenance information, refer to AB 750-1.

6-10. LIMITED STORAGE INSTRUCTIONS

a. Time Limitations. Administrative storage is restricted to a period of 90 days and must not be extended unless the vehicle is reprocessed in accordance with step b.

6-10. LIMITED STORAGE INSTRUCTIONS (Contd)

b. Storage Procedure. Perform disassembly only as required to clean and preserve exposed surfaces. Except as otherwise noted, and to the maximum extent consistent with safe storage, place the vehicle in administrative storage in as nearly a completely assembled condition as possible. Install and adjust equipment so that the vehicle may be placed in service and operated with minimum delay.

(1) The vehicle should be stored on level ground in the most favorable location available, preferably one which affords protection from exposure to the elements and from pilferage.

(2) Perform semiannual Preventive Maintenance Checks and Services (PMCS) (para. 2-12) on vehicle intended for administrative storage. This maintenance consists of inspecting, cleaning, servicing, preserving, lubricating, adjusting, and replacing minor repair parts as required.

(3) Remove both batteries (para. 3-125), place in covered storage, and maintain a charged condition.

(4) Provide access to the vehicle to permit inspection, servicing, and subsequent removal from storage.

6-11. INSPECTION IN LIMITED STORAGE

a. Conduct visual inspection of vehicles in limited storage at least once a month and immediately following hard rains, heavy snowstorms, windstorms, or other severe weather conditions. Perform disassembly as required to fully ascertain the extent of any discovered deterioration or damage. Maintain a record of these inspections for each vehicle. Attach record to vehicle so it is protected from the weather,

b. Perform necessary reprocessing for limited storage when rust or deterioration is found on any unpainted area. Immediately repair damage caused to vehicle by severe weather conditions. Repair damage to On-Equipment Materiel (OEM) as necessary. Thoroughly clean, dry, and repaint painted surfaces showing evidence of wear.

6-12. REMOVAL FROM LIMITED STORAGE

Materiel removed from administrative storage will be:

(1) Restored to normal operating conditions.

(2) Repaired as required.

(3) Returned to normal PMCS schedule using last type service completed as a starting point.

(4) Calibrate equipment as required (TM 43-180).

APPENDIX A
REFERENCES

A-1. SCOPE

This appendix lists all forms, field manuals, and technical manuals for use with this manual.

A-2. PUBLICATIONS INDEX

The following indexes should be consulted frequently for latest changes or revisions and for new publications relating to material covered in this manual.

Consolidated Index of Army Publications and Blank Forms . . . . DA Pam 25-30

Functional User's Manual for The Army Maintenance Management System (TAMMS) . . . . DA Pam 738-750

A-3. FORMS

The following forms pertain to this manual. See DA Pam 25-30, Consolidated Index of Army Publications and Blank Forms, for index of blank forms. See DA Pam 738-750, The Army Maintenance Management System (TAMMS), for instructions on the use of maintenance forms pertaining to this manual.

Unit Status Reporting . . . . AR 220-1

U.S. Army Accident Report . . . . DA Form 285

Equipment Operator's Qualification Record (Except Aircraft) . . . . DA Form 348

Recommended Changes to Publications and Blank Forms . . . . DA Form 2028

Recommended Changes to Equipment Technical Publications . . . . DA Form 2028-2

Organizational Control Record for Equipment . . . . DA Form 2401

Exchange Tag . . . . DA Form 2402

Equipment Inspection and Maintenance Worksheet . . . . DA Form 2404

Maintenance Request Register . . . . DA Form 2405

Materiel Condition Status Report . . . . DA Form 2406

Maintenance Request . . . . DA Form 2407

Maintenance Request - Continuation Sheet . . . . DA Form 2407-1

Equipment Log Assembly (Records) . . . . DA Form 2408

Equipment Control Record . . . . DA Form 2408-9

Equipment Maintenance Log (Consolidated) . . . . DA Form 2409

Packaging and Improvement Report . . . . DD Form 6

Preventive Maintenance Schedule and Record . . . . DD Form 314

Accident Identification Card . . . . DD Form 518

Processing and Deprocessing Record for Shipment, Storage, and
Issue of Vehicles and Spare Engines . . . . DD Form 1397

U.S. Government Motor Vehicle Operator's Identity Card . . . . OF Form 346

Operator Report on Motor Vehicle Accidents . . . . SF Form 91

Report of Discrepancy (ROD) . . . . SF Form 364

Product Quality Deficiency Report . . . . SF Form 368

A-4. FIELD MANUALS

A-5. TECHNICAL MANUALS

A-5. TECHNICAL MANUALS (Contd)

Operator's Manual for Lathe, Brake Drum, Floor Mounted, 60-Inch Rated Swing;
25 Inch Maximum Drum Diameter, 115 Volt, 60 Cycle, Single Phase . . . . TM 9-4910-482-10

Operator's Manual: Grinding Machine, Valve Face, Bench Mounting,
9/32 Inch to 11/16 Inch Chuck, 0 to 45 Degree Valve Face Angle Adjustment, 115-Volt,
AC/DC, 60-Cycle, Single Phase. . . . . TM 9-4910-484-10

Operator's and Organizational Maintenance Manual Including Repair Parts and
Special Tools List for Simplified Test Equipment for Internal Combustion Engines TM 9-4910-571-12&P

Operator's, Unit, Intermediate Direct Support, and Intermediate General Support
Maintenance Manual for Lead-Acid Storage Batteries; 4HN, 24-Volt; 2HN, 12-Volt TM 9-6140-200-14

Principles of Automotive Vehicles. . . . . TM 9-8000

Packaging of Materiel: Preservation Vol. I . . . . TM 38-230-1

Packaging of Materiel: Packing Vol. II . . . . TM 38-230-2

Painting Instructions for Army Materiel . . . . TM 43-0139

Transportability Guidance (Trucks, 5-Ton, 6x6, M809 series)
Truck, Cargo, WWN, M813; Truck Cargo, Dropside, WWN, M813A1 . . . . TM 55-2320-260-15-1

Administrative Storage of Equipment . . . . TM 740-90-1

Storage and Materials Handling . . . . TM 743-200-1

General Packing Instructions for Field Units . . . . TM 746-10

Procedures for Destruction of Tank-Automotive Equipment to Prevent Enemy Use
(U.S. Army Tank-Automotive Command). . . . . TM 750-244-6

Cooling Systems: Tactical Vehicles (Reprinted W/Basic Including C1 and C2) . . . . TM 750-254

A-6. TECHNICAL BULLETINS

Load-Testing Vehicles Used to Handle Missiles and Rockets: Medium Wrecker: M62;
Medium Wrecker, Truck: M543 Series and M816 and Wrecker, Truck Tractor,
M246 Series and M819 . . . . TB 9-352

Tactical Wheeled Vehicles: Repair of Frames . . . . TB 9-2300-247-40

Standards for Oversea Shipment or Domestic Issue of Special Purpose Vehicles,
Combat, Tactical, Construction, and Selected Industrial and Troop Support,
US Army Tank-Automotive Materiel Readiness Command . . . . TB 9-2300-281-35

Truck, 5-Ton, 6x6 M939 Series Truck, Warranty Procedures for Cummins Engine,
Model NHC 250 (NSN 2815-01-111-2262) and Allison Transmission
Model MT654CR (2520-01-117-30101 . . . . TB 9-2300-295-15/21

Warranty Program for Truck, 5-Ton, 6x6 M939A2 Series Truck, Cargo:
5-Ton, 6x6, Dropside, M923A2, M925A2 Truck, Cargo: 5-Ton, 6x6, XLWB, M927A2 . TB 9-2300-358-24

Mandatory Brake Hose Inspection and Replacement-Tactical Vehicles. . . . . TB 9-2300-405-14

Security of Tactical Wheeled Vehicles . . . . TB 9-2300-422-20

Calibration and Repair Requirements for the Maintenance of Army Materiel. . . . . TB 43-180

Equipment Improvement Report and Maintenance Digest
(US Army Tank-Automotive Command) Tank and Automotive Equipment . . . . TB 43-0001-39-1

Safety, Inspection, and Testing of Lifting Devices. . . . . TB 43-0142

Color, Marking, and Camouflage Painting of Military Vehicles, Construction Equipment,
and Materials Handling Equipment. . . . . TB 43-0209

Non-Aeronautical Equipment, Army Oil Analysis Program (AOAP) . . . . TB 43-0210

A-6. TECHNICAL BULLETINS (Contd)

A-7. OTHER PUBLICATIONS

APPENDIX B
MAINTENANCE ALLOCATION CHART

Section I. INTRODUCTION

B-1. GENERAL

a. This section provides a general explanation of all maintenance and repair functions authorized at various maintenance categories.

b. The Maintenance Allocation Chart (MAC) in section II designates overall authority and responsibility for the performance of maintenance functions on the identified end item or component. The application of the maintenance functions to the end item or component will be consistent with the capacities and capabilities of the designated maintenance categories.

c. Section III lists the tools and test equipment (both special tools and common tool sets) required for each maintenance function as referenced from section II.

d. Section IV contains supplemental instructions and explanatory notes for a particular maintenance function.

B-2. MAINTENANCE FUNCTIONS

Maintenance functions will be limited to and defined as follows:

a. **Inspect.** To determine the serviceability of an item by comparing its physical, mechanical, and/or electrical characteristics with established standards through examination (e.g., by sight, sound, or feel).

b. **Test.** To verify serviceability by measuring the mechanical, pneumatic, hydraulic, or electrical characteristics of an item and comparing those characteristics with prescribed standards.

c. **Service.** Operations required periodically to keep an item in proper operating condition; i.e., to clean (includes decontaminate, when required), to preserve, to drain, to paint, or to replenish fuel, lubricants, chemical fluids, or gasses.

d. **Adjust.** To maintain or regulate, within prescribed limits, by bringing into proper or exact position, or by setting the operating characteristics to specified parameters.

e. **Align.** To adjust specified variable elements of an item to bring about optimum or desired performance.

f. **Calibrate.** To determine and cause corrections to be made or to be adjusted on instruments or test, measuring, and diagnostic equipment used in precision measurement. Consists of comparisons of two instruments, one of which is a certified standard of known accuracy, to detect and adjust any discrepancy in the accuracy of the instrument being compared.

g. **Remove/Install.** To remove and install the same item when required to perform service or other maintenance functions. Install may be the act of emplacing, seating, or fixing into position a spare, repair part, or module (component or assembly) in a manner to allow the proper functioning of an equipment or system.

h. **Replace.** To remove an unserviceable item and install a serviceable counterpart in its place. "Replace" is authorized by the MAC and is shown as the 3d position code of the SMR code.

i. **Repair.** The application of maintenance services, including fault location/troubleshooting, removal/installation, and disassembly/assembly procedures, and maintenance actions to identify troubles and restore serviceability to an item by correcting specific damage, fault, malfunction, or failure in a part, subassembly, modules (component or assembly), end item, or system.

B-2. MAINTENANCE FUNCTIONS (Contd)

j. **Overhaul.** That maintenance effort (service/action) prescribed to restore an item to a completely serviceable/operational condition as required by maintenance standards in appropriate technical publications. Overhaul is normally the highest degree of maintenance performed by the Army. Overhaul does not normally return an item to like-new condition.

k. **Rebuild.** Consists of those services/actions necessary for the restoration of unserviceable equipment to a like-new condition in accordance with original manufacturing standards. Rebuild is the highest degree of materiel maintenance applied to Army equipment. The rebuild operation includes the act of returning to zero those age measurement (hours/miles, etc.) considered in classifying Army equipment/components.

B-3. EXPLANATION OF COLUMNS IN THE MAC, SECTION II

a. **Column (1) - Group Number.** Column 1 lists functional group code numbers, the purpose of which is to identity maintenance significant components, assemblies, subassemblies, and modules with the next higher assembly. End item group number shall be "00."

b. **Column (2) - Component/Assembly.** Column 2 contains the names of components, assemblies, subassemblies, and modules for which maintenance is authorized.

NOTE

Those components that are unique to M939A2 series vehicles are identified in column 2 by an asterisk (*).

c. **Column (3) - Maintenance Function.** Column 3 lists the functions to be performed on the item listed in column 2. (For detailed explanation of these functions, see para. B-2.)

d. **Column (4) - Maintenance Category.** Column 4 specifies, by the listing of a work-time figure in the appropriate subcolumn(s), the category of maintenance authorized to perform the function listed in column 3. This figure represents the active time required to perform that maintenance function at the indicated category of maintenance. If the number or complexity of the tasks within the listed maintenance function varies at different maintenance categories, appropriate work-time figures will be shown for each category. The work-time figure represents the average time required to restore an item (assembly, subassembly, component, module, end item, or system) to a serviceable condition under typical field operating conditions. This time includes preparation time (including any necessary disassembly/assembly time), troubleshooting/fault location time, and quality assurance/quality control time in addition to the time required to perform the specific tasks identified for the maintenance functions authorized in the MAC. The symbol designations for the various maintenance categories are as follows:

C . . . . Operator or Crew

0 . . . . Unit Maintenance

F . . . . Direct Support Maintenance

H . . . . General Support Maintenance

e. **Column (5) - Tools and Equipment.** Column 5 specifies, by code, those common tool sets (not individual tools) and special tools, TMDE, and support equipment required to perform the designated function.

f. **Column (6) - Remarks.** This column shall, when applicable, contain a letter code, in alphabetic order, which shall be keyed to the remarks contained in section IV

B-4. EXPLANATION OF COLUMNS IN TOOL AND TEST EQUIPMENT REQUIREMENTS, SECTION III

a. Column (1) - Reference Code. The tool and test equipment reference code correlates with a code used in the MAC, section II, column 5.

b. Column (2) - Maintenance Category. The lowest category of maintenance authorized to use the tool or test equipment.

c. Column (3) - Nomenclature. Name or identification of the tool or test equipment.

d. Column (4) - National/NATO Stock Number. The National Stock Number (NSN) of the tool or test equipment.

e. Column (5) - Tool Number. The manufacturer's part number.

B-5. EXPLANATION OF COLUMNS IN REMARKS, SECTION IV

a. Column (1) - Reference Code. The code recorded in column 6, section II.

b. Column (2) - Remarks. This column lists information pertinent to the maintenance function being performed as indicated in the MAC, section II.

Section II. MAINTENANCE ALLOCATION CHART

(1) Group Number	(2) component/Assembly	(3) Maintenance Function	(4) Maintenance Category Unit C	O	Direct Support F	General Support H	(5) Tools and Equipment	(6) Remarks
01	**ENGINE**							
0100	Engine Assembly	Inspect	0.1					A
		Test		1.5			2 thru 5	
		Service		2.0			2 thru 5	A
		Replace			8.0		7 thru 9	
		Repair				16.0	7, 11 thru 13	
		Overhaul				40.0		B
	Engine Assembly*	Inspect	0.1					A
		Test		1.5			2 thru 5	
		Service		2.0			2 thru 5	
		Replace			8.0		7 thru 9	
		Repair				14.0	12, 13,	
		Overhaul				40.0	34 thru 37	V
	Mount, Engine Lifting	Inspect		0.2				
		Replace			0.2		6 thru 9	
	Bracket, Engine Mounting	Inspect		0.2				
		Replace			2.0		6 thru 9	
0101	Head, Cylinder Assembly	Inspect		0.2				
		Replace		4.0			6 thru 9	
		Repair				5.0	6 thru 9	
	Sleeve, Cylinder	Inspect				0.3		
		Replace				2.0	7, 11 thru 13	
	Liner, Cylinder*	Inspect				0.3		
		Replace				2.0	7, 11, 13 23,34 thru 36	
0102	Crankshaft	Inspect				1.5		
		Replace				5.0	7, 11 thru 13	
		Repair				6.0	7, 11 thru 13	
	Damper, Vibration	Inspect		0.2				
		Replace			1.0		6 thru 9	
	Flange, Crankshaft	Inspect		0.2				
		Adjust		0.5			6 thru 9	
		Replace		0.4			6 thru 9	
0103	Flywheel, Ring Gear	Inspect		1.0				
		Replace		1.0			6 thru 9	
	Housing, Flywheel Gear	Inspect		0.5				
		Replace		4.0			6 thru 9	
	Ring Gear, Flexplate*	Inspect		1.0				
		Replace thru 9, 23		1.0				6

*(M939A2 only)

Section II. MAINTENANCE ALLOCATION CHART (Contd)

(1) Group Number	(2) Component/Assembly	(3) Maintenance Function	(4) Maintenance Category				(5) Tools and Equipment	(6) Remarks
			Unit		Direct Support	General Support		
			C	O	F	H		
0103 (Contd)	ENGINE (Contd) Housing, Flywheel*	Inspect Replace		0.5	4.0		6 thru 9, 23	
	Cover, Rear*	Inspect Replace		0.1	0.2 1.0		6 thru 9, 23	
	Seal, Rear Oil*	Inspect Replace		0.1	6.0		6 thru 9, 23	A
0104	Piston Assembly	Inspect Replace Repair				0.3 5.0 2.0	9 thru 11 9 thru 11	
	Piston Assembly*	Inspect Replace Repair				0.3 5.0 2.0	9 thru 11, 23 9 thru 11	
	Rod, Connecting	Inspect Replace				0.3 2.0	7, 11 thru 13	
0105	Valve Cover	Inspect Replace		0.1	1.0		9 thru 11	A
	Valve Cover*	Inspect Replace		0.1 1.0			2 thru 5	A
	Assembly, Rocker Lever*	Inspect Adjust Replace			0.5 0.5 1.5		6 thru 9 6 thru 9	
	Rod, Push, Intake/Exhaust*	Inspect Replace			0.1 0.5		6 thru 9	
	Shaft, Rocker Lever*	Inspect Replace			0.3 2.0		6 thru 9	
	Tube, Oil Manifold*	Inspect Replace			0.1 2.0		6 thru 9	
	Spring, Valve	Inspect Replace			0.3 0.5		6 thru 9	
	Valves, Intake and Exhaust	Inspect Replace Repair			0.2	1.5 0.5	7, 11 thru 13 9 thru 13	
	Cover, Front Gear	Inspect Replace			0.2 2.0		6 thru 9	
	Seal, Front Oil	Inspect Replace			2.0		6 thru 9	A

* (M939A2 only)

Section II. MAINTENANCE ALLOCATION CHART (Contd)

(1) Group Number	(2) Component/Assembly	(3) Maintenance Function	(4) Maintenance Category				(5) Tools and Equipment	(6) Remarks
			Unit		Direct Support	General Support		
			C	O	F	H		
0105 (Contd)	ENGINE (Contd) Housing, Front Gear*	Inspect Replace		0.1	0.2 4.0		6 thru 9	A
	Shaft Assembly, Rocker Arm	Inspect Replace			0.3 2.0		6 thru 9	
	Bearing, Camshaft	Inspect Replace				0.5 2.0	7, 11 thru 13	
	Camshaft	Inspect Replace				0.2 1.0	7, 11 thru 13	
	Camshaft, Gear	Inspect Replace				0.2 2.0	7, 11 thru 13	
	Camshaft*	Inspect Replace				0.5 1.5	7, 11 thru 13, 20	
	Camshaft, Gear*	Inspect Replace				0.2 2.0	7, 11 thru 13, 20	
	Bushings, Camshaft*	Inspect Replace				0.2 6.5	7, 11 thru 13, 20	
	Tappets*	Inspect Replace				0.1 4.0	7, 11 thru 13, 20	
0106	Crankcase Breather	Inspect Replace		0.5 1.0			2 thru 5	
	Pump, Oil*	Inspect Replace			0.5 4.0		2 thru 5	
	Oil (Dipstick) Tube	Inspect Replace	0.2	0.3			2 thru 5	A
	Cooler, Engine Oil*	Replace Repair		1.0 1.5			2 thru 5	
	Head, Filter*	Inspect Replace		0.1 0.5			2 thru 5	
	Oil Pan*	Inspect Replace		0.1	4.2		6 thru 9	A
	Tube, Oil Suction*	Inspect Replace			0.1 0.5		6 thru 9	
	Pump, Oil	Inspect Replace Repair			0.5 2.0	4.0	6 thru 9 7, 10 thru 13	

* (M939A2 only)

Section II. MAINTENANCE ALLOCATION CHART (Contd)

(1) Group Number	(2) Component/Assembly	(3) Maintenance Function	(4) Maintenance Category				(5) Tools and Equipment	(6) Remarks
			Unit		Direct Support	General Support		
			C	O	F	H		
0106 (Contd)	**ENGINE (Contd)** Oil Pan	Inspect		0.2				
		Replace			1.0		6 thru 9	
	Oil Filter	Inspect		0.1				
		Replace		0.5			2 thru 5	
	Oil Cooler, Engine	Inspect			0.5			
		Replace			0.7		6 thru 9	
		Repair			1.0		6 thru 9	
0108	Manifold, Intake	Inspect			0.5			
		Replace			3.0		6 thru 9	
	Manifold, Exhaust	Inspect			0.5			
		Replace			2.0		6 thru 9	
03	**FUEL SYSTEM**							
0301	Injectors, Fuel	Inspect			0.5			
		Test				0.5	7, 11 thru 13	
		Adjust				0.5	7, 11 thru 13	
		Replace			1.5		6 thru 9	
		Repair				1.0	7, 11 thru 13	
		Calibrate				2.0	7, 11 thru 13	
		Overhaul				1.0	7, 11 thru 13	
	Injectors, Fuel*	Inspect			0.5			
		Test				1.5	7, 11 thru 13, 22, 38	
		Replace			2.0		6 thru 9	
		Repair				1.0	7, 11 thru 13, 22, 25	
0302	Pump, Fuel Supply	Inspect			0.5			
		Replace			2.0		6 thru 9	
		Adjust				2.0	7, 11 thru 13	
		Repair				4.0	7, 11 thru 13	
		Calibrate				2.0	7, 11 thru 13	
		Overhaul				4.0	7, 11 thru 13	
	Lines and Fittings, Injection Pump*	Inspect	0.1					A
		Replace		0.5			2 thru 5	D
	Tubes, Fuel Injector*	Inspect		0.1				
		Replace		1.0			2 thru 5	

* (M939A2 only)

Section II. MAINTENANCE ALLOCATION CHART (Contd)

(1) Group Number	(2) Component/Assembly	(3) Maintenance Function	(4) Maintenance Category				(5) Tools and Equipment	(6) Remarks
			Unit		Direct Support	General Support		
			C	O	F	H		
	FUEL SYSTEM (Contd)							
0302 (Contd)	Fuel Injection Pump*	Inspect			0.5			
		Test				5.0	7, 11 thru 13, 33	
		Adjust				1.0	7, 11 thru 13, 33	
		Calibrate				6.0	7, 11 thru 13, 33	U
		Replace			2.0		2 thru 5, 23, 39	
		Repair				4.0	7, 11 thru 13, 27 thru 32	
	Fuel Transfer Pump*	Inspect		0.2				
		Replace		0.3			2 thru 5	
0304	Cleaner, Air	Inspect	0.2					A
		Replace		0.5			2 thru 5	
	Element	Service	0.5					A
		Replace		0.5				
	Lines and Connections, Vent	Inspect		0.5				D
		Replace		1.0			2 thru 5	
	Indicator, Air Cleaner	Inspect	0.2					A
		Test		0.2				
		Replace		0.5			2 thru 5	
	Hoses and Clamps	Inspect	0.1					A
		Replace		0.5			2 thru 5	
0305	Hoses and Clamps, Turbocharger*	Inspect	0.2					A
		Replace		1.0			2 thru 5	
	Tube, Oil Drain*	Inspect	0.2					A
		Replace		1.0			2 thru 5	
	Turbocharger Assembly*	Inspect		0.6				
		Replace		1.0			2 thru 5	
		Repair			1.5		6 thru 9	
0306	Tank, Fuel	Inspect	0.1					A
		Replace		1.0			2 thru 5	
		Repair			1.0		6 thru 9	C
	Bracket, Fuel Tank Mounting	Inspect	0.1					A
		Replace		1.0			2 thru 5	
	Lines and Fittings, Fuel	Inspect	0.1					A
		Replace		1.5			2 thru 5	D
	Valve, Fuel Selector	Inspect	0.1					A
		Replace		1.0			2 thru 5	
	Lines and Fittings, Fuel Pump to Engine	Inspect	0.1					A
		Replace		1.0			2 thru 5	D

*(M939A2 only)

Section II. MAINTENANCE ALLOCATION CHART (Contd)

(1) Group Number	(2) Component/Assembly	(3) Maintenance Function	(4) Maintenance Category Unit C	O	Direct Support F	General Support H	(5) Tools and Equipment	(6) Remarks
	FUEL SYSTEM (Cod)							
0308	Governor, Fuel Pump (AFC)	Inspect				0.5		
		Test				1.2	7, 11 thru 13	
		Replace				2.5	7, 11 thru 13	
	Spring Pack, Fuel Pump	Inspect				0.5		
		Test				1.0	7, 11 thru 13	
		Replace				2.0	7, 11 thru 13	
	Governor, Fuel Pump (VS)	Inspect				0.5		
		Test				1.0	7, 11 thru 13	
		Replace				2.0	7, 11 thru 13	
	Lower Spring Pack, Fuel Pump	Inspect				0.5		
		Test				1.0	7, 11 thru 13	
		Replace				2.0	7, 11 thru 13	
	Upper Spring Pack, Fuel Pump	Inspect				0.5		
		Test				1.0	7, 11 thru 13	
		Replace				2.0	7, 11 thru 13	
	Governor, Fuel Injection Pump*	Inspect				0.2		
		Replace				0.1	7, 11 thru 13	
		Repair				0.5	7, 11 thru 13	
0309	Fuel Filter	Service	0.1					A
		Replace		0.5			2 thru 5	
	Filter, Fuel Water Separator	Service	0.1					A
		Replace		0.5			2 thru 5	
0311	Aids, Engine Starting Lines and Fittings	Inspect		0.5				
		Replace		1.0			2 thru 5	D
	Pump, Hand Primer	Inspect		0.1				
		Replace		1.0			2 thru 5	
	Harness and Switch	Inspect		0.1				
		Test		0.2			2 thru 5	
		Replace		1.0			2 thru 15	
0312	Control, Accelerator and Throttle Linkage	Inspect		0.2				
		Adjust		0.5			2 thru 5	
		Replace		1.0			2 thru 5	

* (M939A2 only)

Section II. MAINTENANCE ALLOCATION CHART (Contd)

(1) Group Number	(2) Component/Assembly	(3) Maintenance Function	(4) Maintenance Category Unit C	O	Direct Support F	General Support H	(5) Tools and Equipment	(6) Remarks
04	**EXHAUST SYSTEM**							
0401	Pipe, Exhaust and Tail	Inspect	0.2					
		Replace		2.0			2 thru 5	
	Muffler	Inspect		0.2				
		Replace		2.0			2 thru 5	
	Shield, Vertical Exhaust	Inspect		0.1				
		Replace		1.0			2 thru 5	
05	**COOLING SYSTEM**							
0501	Radiator	Inspect	0.2	0.2				A
		Service	0.3					
		Replace		2.0			2 thru 5	
		Repair			3.0		6 thru 9	E
	Tank, Surge	Inspect	0.2					
		Service	0.1	0 2				A
		Replace		1.0			2 thru 5	
	Radiator Hoses	Inspect	0.2					A
		Replace		0.5			2 thru 5	
	Canister, Thermostat*	Inspect	0.1					A
		Replace		0.5				
0502	Shroud, Radiator Fan	Inspect	0.1					A
		Replace		1.0			2 thru 5	
	Water Manifold and Headers	Inspect		0.5				
		Replace			1.0		6 thru 9	
	Coolant Lines, Air Compressor	Inspect		0.5				
		Replace			1.0		6 thru 9	
	Actuator, Fan Clutch	Inspect		0.2				
		Replace		0.4			2 thru 9	
0503	Thermostats and Housing	Inspect	0.1	0.2				A
		Replace		1.0			2 thru 5	
0504	Pump, Water*	Inspect	0.1					A
		Replace		1.5			2 thru 5	
	Pump, Water	Inspect		0.5				
		Replace		1.0			6 thru 9	
		Repair			1.5		6 thru 9	
	Belt, Water Pump	Inspect	0.1	0.1				
		Adjust		0.3			2 thru 5	
		Replace		0.5			2 thru 5	
0505	Fan Assembly	Inspect	0.1	0.1				A
		Replace		1.0			2 thru 5	
	Clutch, Fan Drive	Inspect		0.2				
		Replace		0.5			2 thru 5	
		Repair			1.0		6 thru 9	
	Pulley, Belt Tensioner*	Inspect		0.1				
		Replace		0.5			2 thru 5	

*(M939A2 only)

Section II. MAINTENANCE ALLOCATION CHART (Contd)

(1) Group Number	(2) Component/Assembly	(3) Maintenance Function	(4) Maintenance Category				(5) Tools and Equipment	(6) Remarks
			Unit		Direct Support	General Support		
			C	O	F	H		
0505 (Contd)	COOLING SYSTEM (Contd)							
	Drivebelt, Fan	Inspect	0.1					A
		Replace		0.3			2 thru 5	
	Fan Actuator*	Inspect	0.1					A
		Replace		0.8			2 thru 5	
0507	Aftercooler, Engine*	Inspect		0.3				
		Replace		1.0			2 thru 5	
06	**ELECTRICAL SYSTEM**							
0601	Alternator	Inspect		0.1				
		Adjust		0.6			6 thru 9	
		Test		0.5			2 thru 5	
		Replace		1.0			2 thru 5	
		Repair			1.0		6 thru 9	
	Belt, Alternator	Inspect	0.1					A
		Adjust		0.2			2 thru 5	
		Replace		0.3			2 thru 5	
0603	Starter	Inspect		0.1				
		Test		0.5			2 thru 5	
		Replace		1.5			2 thru 5	
		Repair		1.0			6 thru 9	G
0606	Valve, Fuel Shutoff Electrical	Inspect		1.0				
		Replace			0.7		6 thru 9	
0607	Switches and Circuit Breakers	Inspect	0.2					A
		Test		0.7			2 thru 5	
		Replace		1.0			2 thru 5	
	Instruments and Gauges	Inspect	0.1					A
		Test		0.2			2 thru 5	
		Replace		1.0			2 thru 5	
0608	Control, Directional Turn Indicator	Inspect	0.1					A
		Test		0.2			2 thru 5	
		Replace		0.5			2 thru 5	
		Repair		0.2			2 thru 5	
	Flasher	Inspect	0.1					A
		Replace		0.5			2 thru 5	
	Box, Protective Control	Test		0.3			2 thru 5	
		Replace		0.7			2 thru 5	
0609	Lights	Inspect	0.1				2 thru 5	A
		Adjust		0.2			2 thru 5	
		Replace		0.5			2 thru 5	
0610	Unit, Sending	Inspect		0.1				
		Replace		0.3			2 thru 5	
	Switch, Stoplight	Inspect		0.1				
		Replace		0.5			2 thru 5	
	Buzzer, Warning Control	Inspect	0.1					A
		Replace		0.3			2 thru 5	

* (M939A2 only)

Section II. MAINTENANCE ALLOCATION CHART (Contd)

(1) Group Number	(2) Component/Assembly	(3) Maintenance Function	(4) Maintenance Category				(5) Tools and Equipment	(6) Remarks
			Unit		Direct Support	General Support		
			C	O	F	H		
	ELECTRICAL SYSTEM (Contd)							
0610 (Contd)	Transponder, Fuel Pump	Inspect		0.2				
		Replace		0.3			2 thru 5	
0611	Horn	Inspect	0.1					A
		Replace		0.5			2 thru 5	
	Switch, Horn	Inspect		0.1				
		Replace		0.5			2 thru 5	
0612	Battery	Inspect	0.1					A
		Test		0.5			2 thru 5	
		Service	0.5					A
		Replace		0.5			2 thru 5	
		Repair			1.0		2 thru 5	H
	Cables, Battery	Inspect	0.1					A
		Replace		0.8			2 thru 5	
		Repair		0.5			2 thru 5	
	Box, Battery	Inspect	0.1					A
		Replace		1.8			2 thru 5	
		Repair		1.5			2 thru 5	
0613	Harness, Chassis Wiring	Inspect	0.1					A
		Test		0.5			2 thru 5	
		Replace			4.5		6 thru 9	
		Repair		1.0			2 thru 5	
07	**TRANSMISSION**							
0705	Modulator Control Cable	Adjust		0.5			2 thru 5	
		Replace		0.7			2 thru 5	
0708	Torque Converter/Flywheel Assembly	Replace			8.0		6 thru 9	
		Repair				5.0	7, 11 thru 14	
0710	Transmission Assembly	Inspect	0.2					A
		Test			1.0		6 thru 9	
		Service	0.5	1.0			2 thru 5	
		Adjust			2.5		6 thru 9	
		Replace			6.0		6 thru 9	
		Repair			8.0	10.0	6 thru 9, 11 thru 14	
		Overhaul				20.0	I	
	Linkage, Shift Cable	Inspect		0.1				
		Adjust			1.0		6 thru 9	
		Replace			1.0		6 thru 9	
0721	Transmission Oil Cooler	Inspect	0.1					A
		Replace		2.0			6 thru 9	
	Lines and Fittings	Inspect	0.2					A
		Replace		1.0			6 thru 9	D
	Transmission Oil Cooler*	Inspect	0.2					
		Replace		1.5			2 thru 5	

* (M939A2 only)

Section II. MAINTENANCE ALLOCATION CHART (Contd)

(1) Group Number	(2) Component/Assembly	(3) Maintenance Function	(4) Maintenance Category Unit C	O	Direct Support F	General Support H	(5) Tools and Equipment	(6) Remarks
0721 (Contd)	TRANSMISSION (Contd) Lines and Fittings*	Inspect	0.1					A
		Replace		1.0			2 thru 5	
08	TRANSFER CASE ASSEMBLY							
0801	Transfer Case Assembly	Inspect		0.3				
		Service		0.5			2 thru 5	A
		Replace			5.0		6 thru 9	
		Repair			3.0	4.0	6 thru 14	
		Overhaul				9.5		J
	Coupling, Yoke Input and output	Inspect		0.3				
		Replace			1.0		6 thru 9	
	Flange, Output	Inspect		0.3				
		Replace			1.0		6 thru 9	
	Cushion, Mounting	Inspect		0.2				
		Replace			1.0		6 thru 9	
	Gear, Speedometer	Inspect		0.2				
		Replace			0.5		6 thru 9	
	Seals, Input and Output Shaft	Inspect		0.5				
		Replace			2.0		6 thru 9	
	Bearings, Gears, and Shafts	Inspect				0.5		
		Replace				2.0	7, 11 thru 14	
		Repair				3.0	7, 11 thru 14	
0803	Controls and Linkage	Inspect		0.5				
		Adjust			0.3		6 thru 9	
		Replace			1.5		6 thru 9	
		Repair			1.0		6 thru 9	
	Cylinder, Air Shift	Inspect		0.5				
		Replace			2.0		6 thru 9	
		Repair			0.7		6 thru 9	
09	PROPELLER SHAFTS							
0900	Shaft Assembly, Propeller	Inspect		0.3				
		Service		0.5			2 thru 5	A
		Replace		1.5			2 thru 5	
		Repair		1.0				
	Joint, Universal	Inspect		0.2				A
		Service		0.3			2 thru 5	
		Replace		1.5			2 thru 5	
	Dampener, Vibration	Service		0.1				
		Replace		0.7			2 thru 5	
	Bearing, Center	Inspect		0.2				
		Replace		1.5			2 thru 5	

* (M939A2 only)

Section II. MAINTENANCE ALLOCATION CHART (Contd)

(1) Group Number	(2) Component/ Assembly	(3) Maintenance Function	(4) Maintenance Category Unit C	(4) Unit O	(4) Direct Support F	(4) General Support H	(5) Tools and Equipment	(6) Remark
10	**FRONT AXLE**							
1000	Front Axle Assembly	Inspect		0.5				
		Service		1.0			2 thru 5	A
		Replace			5.0		6 thru 9	
		Overhaul				10.0		K
1002	Carrier Assembly, Differential	Inspect		0.5				
		Service		0.5			2 thru 5	A
		Replace			7.0		6 thru 9	
		Repair				4.0	7, 11 thru 13	
	Seal, Pinion	Inspect		0.2				
		Replace		2.0			6 thru 9	
	Flange, Companion	Inspect		0.3				
		Replace			2.0		6 thru 9	
		Repair			1.6		6 thru 9	
1004	Arms, Steering	Inspect		0.2				
		Replace			2.5		6 thru 9	
	Boot, Dust (CV)	Inspect		0.3				
		Replace		0.7			2 thru 5	
	Knuckle, Steering	Inspect		0.1				
		Service		0.2			2 thru 5	A
		Replace			2.5		6 thru 9	
	Axle Shaft and Universal Joint	Inspect		0.2			2 thru 5	
		Service		0.3			2 thru 5	
		Replace		1.5				
11	**REAR AXLE**							
1100	Rear Axle	Inspect	0.1	0.5				A
		Service		1.0			2	
		Replace			4.0		2 thru 5	
		Repair				4.0	7, 11 thru 13	
		Overhaul				14.0		I
	Rear Axle Assembly	Inspect		0.3				
		Service		0.5			2 thru 5	
		Replace			4.0		6 thru 9	
		Repair			4.0		6 thru 9	
		Overhaul				14.0		K
1102	Carrier Assembly, Differential	Inspect		0.5				
		Service		0.5			2 thru 5	
		Replace			7.0		6 thru 9	
		Repair				4.0	6, 11 thru 13	
	Seal, Pinion	Inspect		0.2				
		Replace		1.0			6 thru 9	
	Flange, Companion	Inspect		0.2				
		Replace			1.0		6 thru 9	
		Repair			1.6		6 thru 9	

Section II. MAINTENANCE ALLOCATION CHART (Contd)

(1) Group Number	(2) Component/Assembly	(3) Maintenance Function	(4) Maintenance Category Unit C	O	Direct Support F	General Support H	(5) Tools and Equipment	(6) Remarks
12	**BRAKES**							
1201	Drum, Handbrake	Inspect		0.3				
		Replace		1.5			2 thru 5	
	Shoes, Handbrake	Inspect		0.3				
		Adjust		0.5			2 thru 5	
		Replace		2.0			2 thru 5	
		Repair			1.0		6 thru 9	
	Linkage, Handbrake	Inspect		0.3				
		Adjust		0.5			2 thru 5	
		Replace		1.0			2 thru 5	
	Brakeshoes, Parking	Inspect		0.5				
		Adjust		1.0			2 thru 5 19	
		Replace		3.0			2 thru 5, 22	
		Repair			0.5		6 thru 9	
1202	Brakeshoes, Service	Inspect		0.5				
		Adjust		1.0			2 thru 5, 18	
		Replace		1.5			2 thru 5, 21	
		Repair			1.0		6 thru 8	
	Air Manifold*	Inspect		0.1				
		Replace		0.5			2 thru 5	
1206	Valve, Treadle	Inspect	0.1					A
		Test		0.5			2 thru 5	
		Replace		1.0			2 thru 5	
		Repair			1.5		6 thru 9	
1208	Chamber, Service Brake	Inspect	0.1					A
		Replace		0.7			2 thru 5	
		Repair			1.0		6 thru 9	
	Chamber, Spring Brake	Inspect	0.1					A
		Replace		1.0			2 thru 5	
		Repair			1.0		6 thru 9	
	Valve, Safety Air Pressure	Inspect		0.1				
		Replace		0.5			2 thru 5	
	Valve, Front Airbrake Chamber (Limiting Valve)	Inspect		0.1				
		Replace		0.5			2 thru 5	
		Repair			0.3		6 thru 9	
	Reservoir, Air Primary/ Secondary	Inspect		0.1				
		Replace		1.0			2 thru 5	
	Reservoir, Air Wet Tank Pressure and Spring Brake	Inspect		0.2				
		Replace		1.5			2 thru 5	
	Valve, Hand Control	Test		0.5			2 thru 5	
		Replace		1.0			2 thru 5	
	Valves, Brake Air Control Miscellaneous	Inspect		0.5			2 thru 5	
		Replace		1.0			2 thru 5	

* (M939A2 only)

Section II. MAINTENANCE ALLOCATION CHART (Contd)

(1) Group Number	(2) Component/Assembly	(3) Maintenance Function	(4) Maintenance Category				(5) Tools and Equipment	(6) Remarks
			Unit		Direct Support	General Support		
			C	O	F	H		
1208 (Contd)	**BRAKES (Contd)** Valve, Brake Lock Control	Test		0.5			2 thru 5	
		Replace		0.5			2 thru 5	
	Lines and Fittings, Airbrake System	Inspect	0.1	0.5				A, Q
		Replace		1.0			2 thru 5	
1209	Compressor, Air	Inspect	0.1					A
		Test		0.5			2 thru 5	
		Replace			1.5		6 thru 9	
		Repair			2.5		6 thru 9	
	Governor, Air	Inspect		0.1				
		Test		0.5			2 thru 5	
		Adjust		0.3			2 thru 5	
		Replace		0.5			2 thru 5	
	Evaporator, Alcohol	Inspect	0.1					A
		Service	0.2					
		Replace		0.5			2 thru 5	
	Mounting and Lines, Air Compressor*	Inspect	0.1	0.1				A
		Replace		1.0			2 thru 5	G
	Compressor, Air*	Inspect	0.1					A
		Replace		3.0			2 thru 5	
		Repair			3.0		6 thru 9	
1211	Trailer Airbrake Connections and Controls	Inspect	0.1					A
		Replace		0.3			2 thru 5	
	Hose, Trailer Airbrake	Inspect	0.1					A
		Replace		0.5			2 thru 5	
13	**WHEELS AND HUBS**							
1311	Hub. Wheel*	Inspect	0.1					A
		Replace		1.5			2 thru 5	
		Repair		2.0			2 thru 5	
	Bearings, Wheel Hub	Inspect		0.5				
		Service		1.0			2	
		Adjust		0.5			2 thru 5	
		Replace		2.5			2 thru 5, 15	
	Wheel (M939A1/A2)	Inspect	0.1					A
		Replace		1.0			2 thru 5, 19, 20	
	Drum, Brake	Inspect		0.2				
		Replace		1.5			2 thru 5	
		Repair			2.0		6 thru 9	
	Seals, Inner and Outer Hub*	Inspect		0.5			2 thru 5	
		Replace		2.5			2 thru 5	
	Valve Assembly, Wheel* W/Filter	Inspect		0.1				
		Service		0.5			2 thru 5	
		Replace		1.0			2 thru 5	
		Repair		1.5			2 thru 5	

*(M939A2 only)

Section II. MAINTENANCE ALLOCATION CHART (Contd)

(1) Group Number	(2) Component/Assembly	(3) Maintenance Function	(4) Maintenance Category Unit C	O	Direct Support F	GENERAL Support H	(5) Tools and Equipment	(6) Remarks
	WHEELS AND HUBS (Contd)							
1311 (Contd)	Wheel (M939)	Inspect	0.1					A
		Replace		1.0			2 thru 5	
	Hub, Wheel	Inspect		0.2				
		Replace		1.5			2 thru 5	
1313	Tires	Inspect	0.2	0.2				A
		Service	0.2					A,L
		Replace		1.0			2 thru 5 20, 21	L
		Repair		1.0			2 thru 5	L
		Rebuild					2.0	L
	Tubes	Replace		0.5			2 thru 5	
		Repair		1.0			2 thru 5	L
14	**STEERING**							
1401	Link, Front Drag	Inspect		0.2				
		Service		0.2			2 thru 5	
		Replace		1.0			2 thru 5	
	Tie Rod Assembly	Inspect		0.2				
		Service		0.2			2 thru 5	
		Replace		1.5			2 thru 5	
		Repair		1.5			2 thru 5	
	Arm, Pitman Steering	Inspect		0.1				
		Replace		1.0			1 thru 5	
	Column, Steering (Lower)	Inspect			0.2			
		Replace			2.5		6 thru 9	
		Repair			2.0		6 thru 9	
	Column, Steering (Upper)	Replace			2.5		6 thru 9	
		Repair			2.0		6 thru 9	
	Wheel, Steering	Inspect		0.2				
		Replace		1.0			2 thru 5	
1407	Gear, Steering Assembly	Inspect		0.2				
		Service		0.5			1 thru 5	
		Adjust			0.5		6 thru 9	
		Replace			3.0		6 thru 9	
		Repair			2.5		6 thru 9	
1410	Pump, Hydraulic and Reservoir	Inspect		0.2				
		Service	0.1					A
		Replace		1.5			6 thru 9	
		Repair			1.0		6 thru 9	
	Power Steering Pump Assembly	Inspect	0.1					A
		Replace		1.0			2 thru 5	
		Repair			1.5		6 thru 9	
1411	Lines and Fittings, Power Steering	Inspect	0.1					A
		Replace		1.0			2 thru 5	D

* (M939A2 only)

Section II. MAINTENANCE ALLOCATION CHART (Contd)

(1) Group Number	(2) Component/Assembly	(3) Maintenance Function	(4) Maintenance Category Unit C	(4) Maintenance Category Unit O	(4) Maintenance Category Direct Support F	(4) Maintenance Category General Support H	(5) Tools and Equipment	(6) Remarks
	STEERING (Contd)							
1412	Cylinder, Steering Assist	Inspect		0.2				
		Adjust		0.5			2 thru 5	
		Replace		1.5			2 thru 5	
		Repair			2.0		6 thru 9	
15	**FRAME AND TOWING ATTACHMENTS**							
1501	Frame	Inspect		0.5				
		Repair			1.0	2.5	6 thru 14	M
	Bumper, Front	Inspect		0.2				
		Replace		2.0			2 thru 5	
	Brackets, Frame	Inspect	0.1					A
		Replace		1.0			2 thru 5	
1503	Pintle	Inspect	0.1					A
		Service		0.1			2 thru 5	
		Replace		0.5			2 thru 5	
1504	Carrier, Spare Wheel	Inspect	0.1					A
		Replace		1.0			2 thru 5	
1506	Wheel, 5th	Inspect	0.2					A
		Service	0.5					A
		Replace		2.5			2 thru 5	
		Repair			2.0		6 thru 9	
16	**SPRINGS AND SHOCK ABSORBERS**							
1601	Front Spring	Inspect	0.2					A
		Replace		3.0			1 thru 5	
		Repair		2.0			2 thru 5	
	Shackles	Inspect	0.2					A
		Replace		1.0			1 thru 5	
		Repair		0.4			2 thru 5	
	Springs, Rear and Seat	Inspect	0.1					A
		Replace		4.0			1 thru 5	
		Repair		2.0			2 thru 5	
1604	Absorber, Front Shock	Inspect	0.1					A
		Replace		0.5			2 thru 5	
1605	Rod, Rear Torque	Inspect		0.1				
		Replace		1.5			6 thru 9	
		Repair			2.0		6 thru 9	
18	**BODY, HOOD, AND CAB**							
1801	Doors	Inspect	0.1					A
		Service		0.1			2 thru 5	
		Adjust		0.5			2 thru 5	
		Replace		1.0			2 thru 5	
		Repair			1.0		6 thru 10, 26	
	Cab	Inspect	0.1					A
		Replace			6.0		6 thru 9	P

Section II. MAINTENANCE ALLOCATION CHART (Contd)

(1) Group Number	(2) Component/Assembly	(3) Maintenance Function	(4) Maintenance Category Unit C	0	F	General Support H	(5) Tools and Equipment	(6) Remarks
	BODY, HOOD, AND CAB (Contd)							
1801 (Contd)	Hood	Inspect	0.1					A
		Adjust		0.5			2 thru 5	
		Replace			2.0		6 thru 9	
		Repair			1.0		6 thru 10	N
	Cab Engine Cover	Inspect	0.1					A
		Replace			0.5		6 thru 9	
1802	Fenders	Inspect	0.1					A
		Replace			2.0		6 thru 9	
		Repair			1.0		6 thru 10	N
	Boards, Running	Inspect	0.5					A
		Replace		2.0			2 thru 5	
1806	Cab Insulation	Inspect	0.1					A
		Replace		2.0			2 thru 5	
	Seats	Inspect	0.1					A
		Replace		1.0			2 thru 5	
		Repair			1.0		6 thru 9	T
1810	Body, Cargo	Inspect	0.2					A
		Replace			3.0		6 thru 9	
		Repair			10.0		6 thru 10	C, N
	Troop Seat and Racks	Inspect	0.2					A
		Replace		1.0			2 thru 5	
		Repair		1.0			2 thru 5	
	Tailgate	Inspect	0.2					
		Replace		1.0			2 thru 5	
		Repair			2.0		6 thru 9	C, N
1812	Doors	Replace		2.5			2 thru 5	
		Repair			5.0		6 thru 10, 26	N
	Roof, Ceiling, Sides, Floors and Underframe	Replace				8.0	7, 11 thru 13	
		Repair				16.0	7, 10 thru 13	N
	Mount, Front Cab	Inspect	0.1					
		Replace		2.0			2 thru 5	
	Mount, Rear Cab	Inspect	0.1					
		Replace		2.0			2 thru 5	
	Counterbalance	Service		1.0			2 thru 5	
		Replace		8.0			2 thru 5	
	Electrical Wiring	Replace			4.0		6 thru 9	
		Repair			4.0		6 thru 9	
	Heater	Replace		4.0			2 thru 5	
		Repair			1.0		6 thru 9	S
	Air Conditioner	Inspect	0.2				6 thru 9	
		Replace			4.0		6 thru 9	

Section II. MAINTENANCE ALLOCATION CHART (Contd)

(1) Group Number	(2) Component/Assembly	(3) Maintenance Function	(4) Maintenance Category Unit C	(4) Unit O	(4) Direct Support F	(4) General Support H	(5) Tools and Equipment	(6) Remarks
	BODY, HOOD, AND CAB (Contd)							
1812 (Contd)	Boarding Ladders and Hardware	Inspect Replace	0.1 0.3					A
20	**HOIST, WINCH, AND POWER TAKEOFF (PTO)**							
2001	Winch, Front and Rear	Inspect Service Adjust Replace Repair Overhaul	0.1	0.5 0.5 3.0	4.0	11.0	2 thru 5 2 thru 5 2 thru 5 6 thru 9	A, R R O
	Band, Automatic	Adjust Replace		0.5	2.0		2 thru 5 6 thru 9	
	Cable, Winch	Inspect Service Replace Repair	0.5 0.5	1.0 0.5			2 thru 5 2 thru 5	A R R R
	Drum, Brake (Front Only)	Adjust Replace		1.0	2.0		2 thru 5 6 thru 9	
	Shaft, Hydraulic Pump Drive Assembly	Inspect Replace Repair		0.3 1.0 2.0			2 thru 5 2 thru 5	
	Lines and Fittings, Hydraulic	Inspect Replace		0.2	0.5		6 thru 9	D
	Boom, Assembly	Inspect Test Service Replace Repair		0.2 0.3	0.5 15.0 8.0		6 thru 9 2 thru 5 6 thru 9 6 thru 9	
	Hoist Cylinders and Power Controls Hoist Cylinder	Replace Repair			8.0	6.0	6 thru 9 7, 11 thru 13	
	Valve, Control	Replace Repair			2.0	4.0	6 thru 9 7, 11 thru 13	
	Cylinder, Boom Elevating	Inspect Replace Repair	0.2		2.0	3.0	6 thru 9 7, 11 thru 13	A
	Motor, Winch, Front	Inspect Repair		0.2 2.0			2 thru 5	
	Cylinder, Extension	Replace Repair			3.5	4.0	6 thru 9 7, 11 thru 13	
	Motor and Gearbox, Hydraulic Swing	Replace Repair			2.0	5.0	6 thru 9 7, 11 thru 13	

Section II. MAINTENANCE ALLOCATION CHART (Contd)

(1) Group Number	(2) Component/Assembly	(3) Maintenance Function	(4) Maintenance Category Unit C	O	Direct Support F	General Support H	(5) Tools and Equipment	(6) Remarks
2001 (Contd)	**HOIST, WINCH, AND PTO (Contd)** Winch Hoist Assembly	Inspect		0.2				
		Service		0.5			2 thru 5	A, R
		Adjust		0.5			2 thru 5	
		Replace			4.5		6 thru 9	
		Repair			4.0		6 thru 9	
	Cable, Winch Hoist	Inspect	0.5					A, R
		Service	0.5					A, R
		Replace		2.0				
		Repair		0.5				
	Valve, Control Assembly	Replace			2.5		6 thru 9	
		Repair			4.0		6 thru 9	
	Motor, Hydraulic	Inspect		0.2				
		Replace			2.0		6 thru 9	
	Reservoir, Hydraulic Oil	Inspect	0.1					A
		Service	0.1					A
		Replace			2.0		6 thru 9	
	PTO and Winch Control Cables*	Inspect	0.1					A
		Replace			1.0		6 thru 8	
	Hydraulic Tubing, Winch Pump*	Inspect	0.1	0.2				
		Replace			0.5		6 thru 8	
	Propeller Shaft, Front Winch Dump Controls*	Inspect	0.1					A
		Replace		0.8			2 thru 5	
		Repair		1.4			6 thru 8	
2004	PTO, Transfer Case	Inspect	0.1					A
		Replace			5.0		2 thru 5	
		Repair			2.5		6 thru 8	
	PTO, Transmission	Inspect		0.2				
		Replace			1.0		6 thru 9	
		Repair			3.0		6 thru 9	
	PTO, Shift Linkage	Adjust		0.3			2 thru 5	
		Replace			1.0		6 thru 9	
22	**BODY CHASSIS AND ACCESSORY ITEMS**							
2201	Bows	Inspect	0.1					A
		Replace		1.0			2 thru 5	
	Cover, Cab Top	Inspect	0.1					A
		Replace		0.5			2 thru 5	
		Repair			1.5		6 thru 9	T
	Curtains, Body Cover	Inspect	0.1					A
		Replace		1.0			2 thru 5	
		Repair			1.0		6 thru 9	T
2202	Motor, Windshield Wiper	Inspect	0.1					A
		Replace		0.7			2 thru 5	

* (M939A2 only)

Section II. MAINTENANCE ALLOCATION CHART (Contd)

(1) Group Number	(2) Component/Assembly	(3) Maintenance Function	(4) Maintenance Category				(5) Tools and Equipment	(6) Remarks
			Unit		Direct Support	General Support		
			C	O	F	H		
2202 (Contd)	**BODY CHASSIS AND ACCESSORY ITEMS (Contd)** Arm and Blade, Windshield Wiper	Inspect Adjust Replace	0.1	0.1 0.2			 2 thru 5 2 thru 5	A
	Washer Bottle and Controls	Service Replace	0.1	1.0			 2 thru 5	A
	Mirror, Rearview	Inspect Replace	0.1	0.5			 2 thru 5	A
	Spotlight	Replace Repair		0.5 0.5			2 thru 5 2 thru 5	
	Windshield	Inspect Replace	0.1	0.5			 2 thru 5	A
	Heater Valves	Inspect Replace	0.1	1.0			 2 thru 5	A
2207	Heater, Personnel	Inspect Replace	0.2	3.0			 2 thru 5	A
33	**SPECIAL PURPOSE KITS**							
3303	Winterization Kits							
	Kit, Engine Coolant Heater	Inspect Install	0.2		10.0		 6 thru 9	A P
	Kit, Radiator Cover	Inspect Install	0.1		1.2		 6 thru 9	A P
	Kit, Hardtop Cab	Inspect Install	0.1	3.5			 2 thru 5	A P
	Kit, Fuel Burning Personnel Heater	Inspect Install	0.2		6.0		 6 thru 9	A P
	Kit, Pioneer Tool Bracket	Inspect		1.0			2 thru 5	P
	Kit, Swingfire Heater	Inspect Install	0.2		3.6		 6 thru 9	A P
3305	Kit, Deepwater Fording	Inspect Install	0.2		3.0		 6 thru 9	A P
3307	Special Purpose Kits							
	Kit, Troop Seat and Rack	Inspect Install	0.1		1.0		 6 thru 9	A P
	Kit, Air Dryer	Inspect Install	0.2		8.0		 6 thru 9	A P
	Kit, A-Frame	Inspect Install	0.2	1.0			 2 thru 5	A P
	Kit, Fire Extinguisher Mounting	Install		1.0			2 thru 5	P
	Kit, Chemical Agent Alarm Mounting	Install		4.0			2 thru 5	P

Section II. MAINTENANCE ALLOCATION CHART (Contd)

(1) Group Number	(2) Component/Assembly	(3) Maintenance Function	(4) Maintenance Category				(5)	(6)
			Unit		Direct Support	General Support		
			C	O	F	H		
3307 (Contd)	SPECIAL PURPOSE KITS (Contd)							
	Kit, Machine Gun Mounting	Install		3.5			2 thru 15	P
	Kit, Decontamination (M13) Apparatus Mounting	Install		3.0			2 thru 5	P
	Kit, Mud Guard	Install		0.5			2 thru 5	P
	Kit, Rifle Mounting	Inspect Install	0.1		2.0		6 thru 9	A P
	Kit, Hand Airbrake	Inspect Install	0.1	4.0			2 thru 5	A P
	Kit, 100-Amp Alternator	Install		2.0			2 thru 5	P
	Kit, Convoy Warning Light	Install		4.0			2 thru 5	P
	Kit, European Mini-Lighting	Install		1.4			2 thru 5	P
	Kit, Automatic Throttle	Install		2.5			2 thru 5	P
	Kit, Atmospheric Fuel Tank Vent System	Install		5.0			2 thru 5	P
	Kit, Vehicle Tiedown	Install		2.0			2 thru 5	P
	Kit, Hydraulic Hose Chafe Guard	Install		2.0			2 thru 5	P
43	**CTIS SYSTEM**							
4316	Lines and Fittings, Air*	Inspect Replace	0.1	0.5			2 thru 5	A D
	Dryer, Air W/Filter*	Inspect Service Replace Repair	0.1	0.2 0.5 0.5 1.5			2 thru 5 2 thru 5 2 thru 5	
	Control Unit (ECU),* Electronic	Inspect Replace	0.1	0.5			2 thru 5	A
4317	Valve, Relief Safety*	Inspect Replace	0.1	0.5			2 thru 5	A
	Controller, Pneumatic*	Inspect Replace Repair	0.1	0.1 1.0 1.5			2 thru 5 2 thru 5	
	Wiring Harness, Electrical*	Inspect Replace Repair		0.1 1.5 1.5			2 thru 5 2 thru 5	D
	Switch, Pressure*	Inspect Replace	0.1	0.1			2 thru 5	A
	Warning Light, Amber*	Inspect Replace	0.1	0.1			2 thru 5	A
	Seals, Air (Front and Rear)*	Test Replace		0.5 1.0			2 thru 5 2 thru 5 15 thru 17	

* (M939A2 only)

Section II. MAINTENANCE ALLOCATION CHART (Contd)

(1) Group Number	(2) Component/Assembly	(3) Maintenance Function	(4) Maintenance Category Unit — C	O	Direct Support F	General Support H	(5) Tools and Equipment	(6) Remarks
4317 (Contd)	**CTIS SYSTEM (Contd)** Filter, Compressed Air*	Inspect Replace Repair	0.8	1.0 1.0			2 thru 5 2 thru 5	A
47	**GAUGES (NON-ELECTRICAL)**							
4701	Drive, Tachometer*	Inspect Replace	0.1	0.6			2 thru 5	A
	Speedometer	Inspect Replace	0.1	1.0			2 thru 5	A
	Tachometer	Inspect Replace	0.1	1.0			2 thru 5	A
4702	Gauge, Air Pressure	Inspect Replace	0.1	0.5			2 thru 5	A

*(M939A2 only)

Section III. TOOL AND TEST EQUIPMENT REQUIREMENTS

(1) REFERENCE CODE	(2) MAINTENANCE CATEGORY	(3) NOMENCLATURE	(4) NATIONAL/NATO STOCK NUMBER	(5) TOOL NUMBER
1	O	Organizational Maintenance Tool Kit	5180-00-762-1737	5704499
2	O	No. 1 Common Organizational Maintenance Tool Kit	4910-00-754-0654	SC4910-95-CL-A74
3	O	No. 1 Supplemental Organizational Maintenance Tool Kit	4910-00-754-0643	SC4910-95-CL-A73
4	O	No. 2 Common Organizational Maintenance Tool Kit	4910-00-754-0650	SC4910-95-CL-A72
5	O	General Mechanic's Tool Kit	5180-00-177-7033	SC5180-90-CL-N26
6	F	Tool Kit, Direct Support	5180-00-762-1740	5704500
7	F	Automotive Maintenance and Repair Tool Kit	4910-00-754-0705	SC4910-95-CL-A31
8	F	Automotive Maintenance Shop Equipment, Wheeled Vehicles, Set A	4910-00-348-7696	SC4910-95-CL-A02
9	F	Tool Kit, General Mechanic's	5180-00-699-5273	SC5180-90-CL-N05
10	F	Shop Equipment, Field Maintenance Welding	3470-00-357-7268	SC3470-95-CL-A08
11	H	General Support Tool Kit	5180-00-762-1741	5704501
12	H	No. 1 Supplemental Automotive Maintenance and Repair Tool Set	4910-00-754-0706	SC4910-95-CL-A62
13	H	No. 2 Supplemental Automotive Maintenance and Repair Tool Set	4910-00-754-0707	SC4910-95-CL-A63
14	H	Tool Kit, General Support	5180-01-147-5824	5704171
15	O	Bearing Punch	5120-01-285-5192	20511262
16	O	Air Seal Installer	5120-01-285-7620	20511263
17	O	Air Gauge Assembly	5220-01-298-5730	20511320
18	O	Adjusting Tool, Brakeshoe	5120-01-154-3029	J34061
19	O	Tool, Wheel Assembly	4910-01-218-4490	J35193
20	O	Bolt, Inserter Tool	4910-01-220-1512	J35198
21	O	Pliers, Brake Repair	5120-01-152-2318	J33111
22	F	Tube Reducer	4730-01-284-9086	23622
23	F	Engine Barring Tool	5120-01-285-5193	3377371
24	F	Wrench Box	5120-01-178-5351	CXM 1519
25	F	Nozzle Cleaning Kit	2915-01-285-2527	3376947
26	F	Tool Kit, Glass Cutting	5180-00-357-7737	SC4940-95-CL-A18
27	H	Tube, Separation	4910-01-336-8204	KDEP-1052
28	H	Puller, Side Plug	5120-01-343-2585	KDEP-1056

Section III. TOOL AND TEST EQUIPMENT REQUIREMENTS (Contd)

(1) REFERENCE CODE	(2) MAINTENANCE CATEGORY	(3) NOMENCLATURE	(4) NATIONAL/NATO STOCK NUMBER	(5) TOOL NUMBER
29	H	Tappet Holder	5120-01-345-2586	KDEP-1068
30	H	Spring Compressor	5120-01-341-6000	KDEP-1505
31	H	Plunger Lift Device	4910-01-338-6241	1-688-130-135
32	H	Rack Extension	5340-01-341-6572	9-681-233-100
33	H	Fuel Pump Test Stand	4910-01-194-7667	DFP 156
34	H	Cylinder Liner Clamps	5120-01-262-7309	3822503
35	H	Mechanical Puller	5120-01-291-5769	3822786
36	H	Cylinder Liner Puller	5120-01-143-2032	3376015
37	H	Torque Angle Gauge	5120-01-386-5992	3823878
38	H	Tester, Injector Nozzle	4910-00-255-8641	7551255
39	H	Puller Kit	5180-00-999-4053	J24420C

Section IV. REMARKS

REFERENCE CODE	REMARKS
A	Perform PMCS as shown in TM 9-2320-272-10.
B	Engine overhaul will be in accordance with DMWR 9-2815-500.
C	Welding will be in accordance with TM 9-237.
D	Repair of lines and fittings will be in accordance with TM 9-243.
E	Test and repair of radiator will be in accordance with TM 750-254.
F	Repair of alternator will be in accordance with TM 9-2920-225-34.
G	Repair of starter will be in accordance with TM 9-2920-243-34.
H	Repair of batteries will be in accordance with TM 9-6140-200-14.
I	Transmission overhaul will be in accordance with DMWR 9-2520-522.
J	Transfer overhaul will be in accordance with DMWR 9-2520-530.
K	Overhaul of front and rear axle will be in accordance with DMWR 9-2520-508.
L	Tires/Tubes: Repair and Inspection TM 9-2610-200-14 Storage TM 743-200-1
M	Repair of frames will be in accordance with TB 2300-247-40.
N	Metal body repair will be in accordance with FM 43-2.
O	Overhaul of front and rear winches will be in accordance with DMWR 9-3830-501.
P	Refer to kit installation instructions for kit installation.
Q	Inspection of brake lines will be in accordance with TB 9-2300-405-14.
R	Service/inspection of winch/hoist wire rope/cables will be in accordance with TB 43-0142 and TB 9-0352.
S	Repair of heaters will be in accordance with TM 9-2540-205-24&P.
T	Repair of canvas will be in accordance with FM 10-16.
U	Operation of fuel pump test stand, P/N DFP 156, will be in accordance with TM 9-4910-778-14&P.
V	Engine overhaul will be in accordance with DMWR 9-2815-358.

APPENDIX C
EXPENDABLE/DURABLE SUPPLIES AND MATERIALS LIST

Section I. Introduction (page C-l).
Section II. Expendable/Durable Supplies and Materials List (page C-2).

Section I. INTRODUCTION

C-1. SCOPE

This appendix lists expendable/durable supplies and materials you will need to maintain M939 series vehicles. This listing is for informational purpose only and is not authority to requisition listed items.

C-2. EXPLANATION OF COLUMNS

Column (1) - Item Number. This number is assigned to the entry in the listing and is referenced in the INITIAL SETUP of applicable tasks under the heading of MATERIALS/PARTS.

b. Column (2) - Level. This column identifies the lowest level of maintenance that requires the listed item.

C - Operator/Crew
O - Organizational Maintenance
F - Direct Support Maintenance

c. **Column (3) - National Stock Number.** This is the National Stock Number assigned to the item; use it to request or requisition the item.

d. Column (4) - Description, CAGEC, and Part Number. Indicates the federal item name and, if required, a description to identify the item. The last line for each item indicates the Commercial and Government Entity Code (CAGEC) (in parentheses) followed by a part number. These codes are identified as:

CAGEC	MANUFACTURER
05972	Loctite Corporation
19200	U.S. Army Research and Development Command
19207	U.S. Army Tank-automotive and Armaments Command, AMSTA-IM-MM
58536	Federal Commercial Item Promulgated by General Services Administration
62377	Permatex Industrial
71984	Dow Corning Corp.
72932	Gulf Oil Corp.
75037	Minnesota Mining and Manufacturing Company
77220	Dana Corp.
78500	Rockwell Heavy Vehicle Systems Inc.
79819	BT Office Products Intl. Inc.
80064	Naval Ship Systems Command
80244	General Services Administration
81348	Federal Specifications

C-2. EXPLANATION OF COLUMNS (Contd)

81349	Military Specification Promulgated by Military
81755	Lockheed Martin Company
96599	Fischbein-Dave Company
96906	Military Standards
96980	American Grease Stick Company
97403	U.S. Army Communications - Electronic Command Center
98308	Castrol Industrial North America

e. **Column (5) - Unit of Issue.** This column indicates the size of container issued for each NSN. This measure is expressed by an abbreviation [such as ea (each), oz (ounce), gal. (gallon)]. If the unit of measure differs from the unit of issue, requisition the lowest unit of issue that will satisfy your requirements.

EXPENDABLE/DURABLES SUPPLIES AND MATERIALS LIST

(1) REFERENCE CODE NO.	(2) LEVEL	(3) NATIONAL STOCK NUMBER	(4) DESCRIPITON	(5) UNIT OF ISSUE
	O	6830-00-264-6751	ACETYLENE, TECHNICAL: gas filled acetylene, (81348) BB-A-106 225 cubic feet	CU-FT
2	F	0040-00-142-9193	ADHESIVE, LIQUID: ethyl cyanacrylic resin, type II (81349) MILA-46050 10 1 ounce bottles per box	BX
3	O	8040-00-262-9062	ADHESIVE, LIQUID: silicone, rubber, Silastic 732RTV (clear) non-hardening type I (81348) MMM-A-139 1 pint	PT
4	F	8030-01-014-5869	ADHESIVE SEALANT: anaerobic threadlock, medium strength, MIL-S-46163, type II, grade N, Loctite 242, (05972) 24231 50 milliliter	ML
5	O	8040-00-833-9563	ADHESIVE: silicone rubber, MIL-A-46106, type I, (97403) 1 kit	KT
6	O	8040-00-543-7170	ADHESIVE: synthetic rubber, class II, (80244) MMM-A-189 CL2 1 pint can	PT
7	O	8040-00-262-9005	ADHESIVE: synthetic rubber, MMM-A-1617, type II, (79819) 1357 1 gallon can	GAL.
8	C	6850-00-174-1806	ANTIFREEZE: arctic grade, permanent type, fluorescent yellow, (81349) MILA-11755 55 gallon drum	GAL.
9	C	6850-00-181-7929 6850-00-181-7933 6850-00-181-7940	ANTIFREEZE: ethylene glycol, permanent type, inhibited {-60°F (-51°C)} blue-green in color, (81349) MILA-46153 1 gallon container 5 gallon container 55 gallon drum	 GAL. GAL. GAL.
10	O	8030-00-251-3980	ANTISEIZE COMPOUND: temperature resistant lubricant, MIL-A-907, Loctite, antiseize, 1050°F (566°C), (05972) 76764 1 pound can with brush top	LB

EXPENDABLE/DURABLES SUPPLIES AND MATERIALS LIST (Contd)

(1) REFERENCE CODE NO.	(2) LEVEL	(3) NATIONAL STOCK NUMBER	(4) DESCRIPITON	(5) UNIT OF ISSUE
11	O	8105-00-837-7754	BAG: plastic, polyethylene, (58536) A-A-1799 500 bags per box	BX
12	0	8125-01-082-9697	BOTTLES: oil sample, (81996) PD 8125-1 120 bottles per box	BX
13	F	6850-00-974-3738	CALIBRATING OIL: gulf 45A, (72932) 45A 55 gallon drum	GAL.
14	O	5340-00-450-5718	CAP AND PLUG SET: (19207) 10935405 1 set	SET
15	F	8030-00-682-6745	CAULKING COMPOUND: rubber, synthetic (81349) MIL-C-18255 1 kit	KT
16	0	5120-00-273-9793	CHALK LINE AND REEL: self-chalking, hand crank rewind, (81349) GGG-C-291 50 feet	EA
17	0	7510-00-164-8893	CHALK: marking, white, (81348) SS-C-266 1 gross	CR
18	0	6850-00-926-2275	CLEANING COMPOUND: windshield, concentrated (81348) 16 ounce bottle	OZ
19	0	6850-00-598-7328	CLEANING COMPOUND KIT engine cooling system (81349) MIL-C-10597 1 kit	KT
20	0	5350-00-221-0872	CLOTH, ABRASIVE (crocus): 9 inch x 11 inch sheets, (58536) A-A-1206 50 sheets/package	PG
21	0	7920-00-044-9281	CLOTH, CLEANING: lint-free, general purpose, white, (81349) MIL-C-85043 10 pound box	LB
22	0	8010-01-160-6741	COATING: polyurethane, chemical agent resistant, green 383, MIL-C-46168C, type II, (19207) 5584154 1 gallon can	GAL.
23	O	8010-00-959-4661	COATING: battery box, non-drying epoxy, (81349) MIL-C-22750 1 kit	KT

EXPENDABLE/DURABLES SUPPLIES AND MATERIALS LIST (Contd)

(1) REFERENCE CODE NO.	(2) LEVEL	(3) NATIONAL STOCK NUMBER	(4) DESCRIPITON	(5) UNIT OF ISSUE
24	F	5330-00-069-3321	CORK Type IB, Class 2 (81348) Spec HH-C-576 36 inch x 36 inch	EA
25	O	8030-00-244-1297	CORROSION PREVENTIVE COMPOUND: grade II, soft film (81349) MIL-C-16173 1 gallon can	GAL.
26	F	9150-00-265-9406	CUTTING FLUID: (81348) C-O-376 1 gallon can	GAL.
27	O	7930-00-282-9699	DETERGENT: nonsudsing, general purpose, liquid, (80244) MIL-D-16791 type I 1 gallon	GAL.
28	O	9150-00-935-1017	GAA GREASE: automotive and artillery, MIL-G-10924, (98308) BRAYCOTE610 14 ounce can	OZ
29	O	8040-01-378-0235 8040-00-664-4134	GASKET COMPOUND: shellac, heavy-bodied, hard setting, (62377) 2 ounce bottle 1 pint bottle	OZ PT
30	O	8040-00-728-3088	GASKET SEALANT: silicone rubber, MIL-A-46106, type 1, (78500) 1199-T-3842 1 kit	KT
31	C	9620-00-233-6712	GRAPHITE, POWDERED: MIL-SS-G-659 (81348) SSG659 1 pound kit	KT
32	C	9150-01-197-7698 9150-01-197-7690 9150-01-197-7689 9150-01-197-7692 9150-00-530-7369	GREASE, AUTOMOTIVE AND ARTILLERY (MIL-G-10924) (81349) 2-1/4 ounce tube 1-3/4 pound can 6-1/2 pound can 35 pound can 120 pound drum	OZ LB LB LB LB
33	C	9150-00-027-2954	GREASE, SILICONE INSULATED: electric motor, (71984) Molykote DC44 5.3 ounce tube	OZ
34	O	4720-00-845-0630	HOSE: nonmetallic, rubber synthetic, black, fuel/oil resistant (23040) C2AZ9324C 3/8 inch ID, 5/8 inch OD	FT

EXPENDABLE/DURABLES SUPPLIES AND MATERIALS LIST (Contd)

(1) REFERENCE CODE NO.	(2) LEVEL	(3) NATIONAL STOCK NUMBER	(4) DESCRIPITON	(5) UNIT OF ISSUE
35	C		HYDRAULIC FLUID: SAE 15W40 (O-1236) (81349) MIL-L-46167	
		9150-00-402-2372	5 gallon can	GAL.
36	0		INHIBITOR, CORROSION: liquid cooling system; powder (81348) MIL-A-54009	
		6850-01-160-3868	1 quart bottle	Q T
		6850-01-287-8067	1 gallon bottle	GAL.
37	F		LAPPING AND GRINDING COMPOUND: valve-grinding compound, grease-mixed, grit 220 fine, (58536) A-A-1203 type I	
		5350-00-193-1341	1 pound can	LB
38	F		LAPPING AND GRINDING COMPOUND: valve-grinding compound, grease-mixed, grit 120 coarse, (58536) A-A-1203 type I	
		5350-00-271-5966	1 pound can	LB
39	O		LUBRICANT, TIRE AND RIM: liquid, (96980) AA18	
		2640-00-256-5527	1 gallon	GAL.
40	0		METHYL ALCOHOL, METHANOL: (81348) O-M-232	
		6810-00-597-3608	1 gallon can	GAL.
		6810-00-275-6010	5 gallon can	GAL.
41	C		OIL, FUEL, DIESEL DF-1: Winter (81349) (W-F-800)	
			gauge	GAL.
		9140-00-286-5289	55 gallon drum, 18gauge	GAL.
		9140-00-286-5286	bulk	GAL:
42	C		OIL, FUEL, DIESEL DF-2: Regular (81348) (W-F-800)	
			gauge	GAL.
		9140-00-286-5297	55 gallon drum, 18gauge	GAL.
		9140-00-286-5294	bulk	GAL:
43	C		OIL, FUEL, DIESEL DF-A: Arctic (81348) (VV-F-800)	
		9140-00-286-5284	55 gallon drum, 16gauge	GAL.
		9140-00-286-5285	55 gallon drum, 18gauge	GAL.
		9140-00-286-5283	bulk	GAL.

EXPENDABLE/DURABLES SUPPLIES AND MATERIALS LIST (Contd)

(1) REFERENCE CODE NO.	(2) LEVEL	(3) NATIONAL STOCK NUMBER	(4) DESCRIPITON	(5) UNIT OF ISSUE
44	C		OIL, LUBRICATING, ENGINE: arctic (ice, sub-zero) OEA (SEA OW-20) (19200) (9377757)	
		9150-00-402-4478	1 quart can	QT GAL.
		9150-00-49 1-7197	55 gallon drum, 16gauge	GAL.
45	C		OIL, LUBRICATING, GEAR, EXPOSED: CW (81348) (W-L-751)	
		9150-00-234-5197	5 pound can	LB
46	C		OIL, LUBRICATING, GEAR, MULTI-PURPOSE: GO 75 (81349) (MIL-L-2105)	
		9150-01-035-5390	1 quart can	QT GAL.
		9150-01-422-9342	55 gallon drum, 16gauge	GAL.
		9150-00-183-7807	bulk	GAL:
47	C		OIL, LUBRICATING, GEAR, MULTI-PURPOSE: GO 80/90 (81349) (MIL-L-2105)	QT
		9150-01-035-5393	5 gallon can	GAL.
		9150-01-035-5394	55 gallon drum, 16gauge	GAL:
48	C		OIL, LUBRICATING, OE/HDO: 1OW (81349) (MIL-L-2104)	QT
		9150-00-188-9858	5 gallon drum	GAL.
		9150-00-191-2772	55 gallon drum, 16gauge	GAL:
49	C		OIL, LUBRICATING: internal combustion engine, tactical service, OE/HDO 10, (81349) MIL-L-2104	
		9150-00-186-6668	5 gallon drum	GAL.
50	C		OIL, LUBRICATING, OE/HDO: 30W (81349) (MIL-L-2104)	
		9150-00-189-6729	55 gallon drum, 16gauge	GAL.
		9150-00-183-7808	bulk	GAL.
51	C		OIL, TURBINE FUEL, AVIATION: grade JP-8 (81349)	
		9130-01-031-5816	bulk	GAL.
52	O		OXYGEN, TECHNICAIgas filled oxygen, (81348) BB-A-925 240 cubic feet (to be filled/refilled	
		6830-00-292-0129	locally)	CU-FT

EXPENDABLE/DURABLES SUPPLIES AND MATERIALS LIST (Contd)

(1) REFERENCE CODE NO.	(2) LEVEL	(3) NATIONAL STOCK NUMBER	(4) DESCRIPITON	(5) UNIT OF ISSUE
53	F		PETROLATUM: technical, oil-soluble grease (Vaseline) (81348) W-P-236	
		9150-00-250-0933	7-1/2 pound can	LB
54	F		PIGMENT: iron, blue, oil base (substitute for Prussian blue), (81348) TT-P-381	
		8010-00-247-4334	1/2 pint can	PT
55	F		PLASTIGAGE: clearance range green (77220) PG-1	
		5210-00-640-6177	12 each box	BX
56	O		POWDEREDGRAPHITE: MIL-SS-G-659 (81348) SSG659	
		9620-00-233-6712	1 pound can	LB
57	F		PRIMER: zinc chromate (81348) Spec TT-P-1757	
		0010-00-145-0312	1 pint can	P-r
58	O		RAG: wiping, unbleached cotton and cotton-synthetic, mixed colors (58536) A-A-531	
		7920-00-205-1711	50 pound bale	LB
59	O		ROPE, FIBROUS: 3 strand, 3/8 inch diameter, 1-1/8 inch circumference, 1,350 pound capacity (81348) TR605	
		4020-00-23 1-2581	406.667 yards, minimum	YD
60	O		SACK: shipping, water resistant, ten pound load capacity, (58536) A-A-160	
		8105-00-290-0340	250 each box	BX
61	O		SEALING COMPOUND: plastic epoxy, resin, MIL-R-46082, type II, Loctite RC/640, (05972) 64031	
		8030-00-111-6404	50 cubic centimeters	CC
62	O		SEALING COMPOUND: plastic, tetrafluoroethylene, Loctite 277, (05972) 59241	
		8030-00-204-9149	250 cubic centimeters	CC
63	C		SEALING COMPOUND; polyester and plastic, tetrafluoroethylene, Loctite 592-31, (05972)	
		8030-01-054-0740	50 cubic centimeters	CC

EXPENDABLE/DURABLES SUPPLIES AND MATERIALS LIST (Contd)

(1) REFERENCE CODE NO.	(2) LEVEL	(3) NATIONAL STOCK NUMBER	(4) DESCRIPITON	(5) UNIT OF ISSUE
		8030-00-247-2525	11 ounce tube	OZ
65	O		SEALING COMPOUND non-hardening, MIL-S-45180, type II, Permatex no. 2, (80064) 1756371	
		8030-00-252-3391	11 ounce tube	OZ
66	O		SEALING COMPOUND resin, synthetic, MIL-R-46082, type I, Loctite 75, (05972) 669-31	
		8030-00-180-6222	50 cubic centimeters	c c
67	F		SEALING COMPOUND: non-hardening, Permatex no. 51H (pipe joint compound), (77247) 51H	
		8030-00-503-0316	4 ounce can	OZ
68	F		SEALING COMPOUND: liquid resin, nonhardening, type III, (77247) MIL-S-45180	
		8030-00-656-1426	1 pint can	PT
69	O		SODIUM BICARBONATE: technical (81348) 0-5-576	
		6810-00-264-6618	1 pound box	LB
70	O		SOLDER: rosin core, 60/40, 0.094 inch diagnostic, (81348) QQ-S-571	
		3439-00-224-3567	5 pound spool	LB
71	C		SOLVENT, DRYCLEANING: type III, biodegradable (81348) 134 Hi-Solv	
		6850-01-277-0595	5 gallon	GAL.
		6850-01-244-3207	55 gallon drum	GAL.
72	O		TAPE, ANTISEIZING: white, MIL-T-27730, (81755) P5025-2R	
		8030-00-889-3535	1/2 inch wide x 260 inches long x 0.0035 inch thick with snap-on shell	EA
73	O		TAPE, INSULATION: electrical MIL-I-24391, 3/4 inch wide, black (75037)	
		5970-00-4 19-4291	108 feet roll	FT

EXPENDABLE/DURABLES SUPPLIES AND MATERIALS LIST (Contd)

(1) REFERENCE CODE NO.	(2) LEVEL	(3) NATIONAL STOCK NUMBER	(4) DESCRIPITON	(5) UNIT OF ISSUE
74	O	5970-01-189-6927	TAPE, INSULATION: electrical (81349) MIL-I-24391 3/4 inch wide x 10 yards long x 0.0085 inch thick	EA
75	F	7510-00-290-2023	TAPE, PRESSURE SENSITIVE: masking, 1/2 inch wide, tan, A-A-883 type II (81348) PPP-T-42 60 yards per roll	EA
76	O	6810-00-678-4418	TRICHLOROETHYLENE: liquid (81349) O-T-634 1 gallon can	GAL.
77	O	4020-00-291-5901	TWINE: cotton (string), 16 ply (81348) A-A-1451 375 yard ball	YD
78	F	8010-00-239-5736	WHITE LEAD: basic carbonate, paste in oil, (96906) MS35599-1 type B 1 pound can	LB
79	O	9525-00-990-7799	WIRE, NON-ELECTRICAL: safety wire (96906) QQ-N-281, Class A 1 pound roll	LB
80	F	9505-00-198-9125	WIRE: non-electric, iron, (80244) 22-W-1642-125-36 1 pound roll	LB
81	F	8030-00-222-0503	WOOD PRESERVATIVE: (81349) MIL-S-13518 1 gallon can	GAL.
82	F	5510-00-270-6031	WOOD, LAMINATED DECKING: red or white oak, ungraded, MIL-W-3912, treated with preservative TT-W-572, (97403) 12319E0079 228 inches X 11.5 inches X 1.12 inches	EA

APPENDIX D
MANDATORY REPLACEMENT PARTS

Section I. INTRODUCTION

D-1. SCOPE

This appendix list mandatory replacement parts you will need to maintain M939, M939A1, and M939A2 series vehicles.

D-2. EXPLANATION OF COLUMNS

a. **Column (1) - Item Number.** This number is assigned to each entry in the listing and is referenced in the Initial Setup of applicable task under the heading of MATERIALS/PARTS.

b. **Column (2) - Nomenclature.** Name or identification of part.

c. **Column (3) - Part Number.** The manufacturer's part number.

d. **Column (4) - National/NATO Stock Number.** The national stock number of the part.

Section II. MANDATORY REPLACEMENT PARTS

(1) ITEM NO	(2) NOMENCLATURE	(3) PART NUMBER	(6) NATIONAL/NATO STOCK NUMBER
1	Adapter	3903845	5307-01-196-4246
2	Adapter	ER22663	4730-01-127-6697
3	Adjuster Parts Kit	1164	2530-01-278-7364
4	Air Compressor Repair Kit	3801808	4910-01-272-5374
5	Anchor Bolt	231013398	5360-01-145-6923
6	Anchor Plunger Kit (Left)	1173	
7	Anchor Plunger Kit (Right)	1174	
8	Backing Ring	032590	5330-01-135-4069
9	Banjo Seal	3903380	5330-01-195-5268
10	Basic Overhaul Kit	6884259	2520-01-140-2376
11	Bearing	6071-2RS	3110-01-126-1287
12	Bearing	69867	3110-00-100-2368
13	Bearing Retainer	2411521	5365-01-235-2580
14	Bearing Shell	214950 214951 214952 214953	3120-01-087-3004 3120-01-155-4442 3120-01-157-3316 3120-01-155-8707
15	Brake Band	7409663	2530-00-740-9663
16	Brakeshoe Lining	2740-D-1122	2530-01-135-0187
17	Brakeshoe Lining and Assembly Kit	2000-P-1446	2530-01-326-6127
18	Breakoff Screw	2 423 450 005	5305-01-336-0006
19	Breakoff Screw	1 423 450 056	5305-01-301-5112
20	Breakoff Screw	2 910 172 197	5305-01-335-9965
21	Bushing	S1003A	5326-00-598-5255
22	Bushing	55602H	3120-00-740-9344
23	Bushing	S1077	5365-00-362-1880
24	Bushing	7411156	3120-00-741-1156
25	Bushing	7409666	3120-00-740-9666
26	Bushing	7954534	3120-01-120-8450
27	Bushing	23011924	4730-01-078-2732
28	Bypass Disc	200819	4820-00-400-5189
29	Bypass Seat	153526	2815-00-131-1700
30	Bypass Spring	251152	5360-00-932-7452
31	Bypass Valve	3902338	2815-01-211-5270

Section II. MANDATORY REPLACEMENT PARTS (Contd)

(1) ITEM NO	(2) NOMENCLATURE	(3) PART NUMBER	(6) NATIONAL/NATO STOCK NUMBER
32	Camshaft Bore Cork Gasket	9333-1	5330-00-729-4427
33	Camshaft Bushing	100670	3120-00-573-0391
34	Camshaft Bushing	157870	3120-00-906-6657
35	Chamfered Washer	23014094	5310-01-145-6923
36	Channel Seal	7373291-2	5330-01-026-7512
37	Checkball	8622757	2520-00-008-7306
38	Compression Ring	187350	4310-01-197-1882
39	Compression Ring	650330	4310-01-079-5245
40	Connecting Rod Bearing	3901434	3120-01-275-7665
41	Connector	MS27144-1	5935-00-167-7775
42	Copper Washer	007603014106	5330-12-156-4523
43	Cooper Washer	3912889	5310-01-271-5706
44	Copper Washer	2916710613	5325-01-276-8488
45	Copper Washer	2916710619	5330-01-301-1763
46	Cotter Pin	MS24665-283	5310-00-842-3044
47	Cotter Pin	MS24665-500	5310-00-187-9567
48	Cotter Pin	210492	5310-01-246-4339
49	Cotter Pin	MS24665-362	5310-00-298-1498
50	Cotter Pin	MS24665-134	5315-00-839-5820
51	Cotter Pin	IF316	5315-00-816-1794
52	Cotter Pin	MS2466-172	5315-00-187-9370
53	Cotter Pin	AN415-4	5315-01-057-8371
54	Cotter Pin	MS24655-490	5315-00-059-0205
55	Cotter Pin	MS24665-214	5315-00-080-3503
56	Cotter Pin	MS24665-353	5315-00-839-5822
57	Cotter Pin	MS24665-351	5315-00-893-5821
58	Cotter Pin	MS24665-369	5315-00-059-0187
59	Cotter Pin	MS24665-498	5315-00-849-9854
60	Cotter Pin	MS24665-335	5315-00-012-0123
61	Cotter Pin	MS24665-361	5315-00-059-0184
62	Cotter Pin	MS24665-300	5315-00-234-1863
63	Cotter Pin	MS24665-655	5315-00-187-9414
64	Cotter Pin	MS24665-493	5315-00-018-7988
65	Cotter Pin	AN415-6	5315-01-018-9991

Section II. MANDATORY REPLACEMENT PARTS (Contd)

(1) ITEM NO	(2) NOMENCLATURE	(3) PART NUMBER	(6) NATIONAL/NATO STOCK NUMBER
66	Cotter Pin	MS24665-132, A82-1	5315-00-839-2325
67	Cotter Pin	MS34665-627	5315-00-013-7308
68	Cotter Pin	MS24665-502	5315-00-849-5582
69	Cotter Pin	MS24665-423	5315-00-013-7228
70	Cotter Pin	MS24665-381	5315-00-839-2326
71	Cotter Pin	MS24665-427	5315-00-879-2910
72	Cotter Pin	MS24665-491	5315-00-059-0206
73	Cotter Pin	L6451-101	5315-00-187-9591
74	Cotter Pin	MS24665-631	5315-00-597-7399
75	Cotter Pin	MS24665-49	5315-01-136-4542
76	Cotter Pin	MS24665-359	5315-00-013-7214
77	Cotter Pin	MS24665-628	5315-00-846-0126
78	Cotter Pin	MS24665-238	5315-00-239-8027
79	Cotter Pin	137159	5315-00-013-7159
80	Cotter Pin	CO2-40-4	5315-01-135-9506
81	Cotter Pin	MS24665-5	5315-00-236-8345
82	Cotter Pin	MS24665-285	5315-01-359-1451
83	Cotter Pin	MS24665-153	5315-00-185-0037
84	Cotter Pin	MS24665-151	5315-00-815-1405
85	Cotter Pin	MS24665-357	5315-00-298-1481
86	Cotter Pin	20511322Z	2530-01-272-2912
87	Cotter Pin	3801260	3120-01-132-9339
88	Cotter Pin	215090	5330-00-064-4399
89	Cotter Pin	175831	5340-00-485-0945
90	Cotter Pin	7084738	2590-00-471-5343
91	Cotter Pin	8380498	2590-00-471-5344
92	Cotter Pin	A-3261-S-253	3040-01-149-1111
93	Cotter Pin	390415	
94	Cotter Pin	3076189	5330-01-080-5021
95	Cotter Pin	3917737	5330-01-272-1282
96	Cotter Pin	69519	5315-00-475-2514
97	Cotter Pin	70550	2815-01-124-0232
98	Cotter Pin	3917883	5315-01-270-8286

Section II. MANDATORY REPLACEMENT PARTS (Contd)

(1) ITEM NO.	(2) NOMENCLATURE	(3) PART NUMBER	(4) NATIONAL/NATO STOCK NUMBER
99	Dowel Pin	3904483	5315-01-270-8285
100	Dowel Ring	60575	5365-00-428-6201
101	Drive Pin	S-2286	5305-00-804-6318
102	Drivebelt	3912004	3030-01-217-3754
103	Dustcap Assy	DCL6N-3	
104	Dustcap Assy	SERUR14-16	2530-01-286-0108
105	Expansion Plug	3032693	5340-01-271-2420
106	Expansion Plug	156075	5340-01-271-2419
107	Expansion Plug	3900965	5340-01-194-8936
108	Fan Drive Repair Kit	F212028	
109	Fastener	7529309	2540-00-562-0422
110	Felt Seal	7411160	5330-00-741-1160
111	Felt Washer	5X625	5330-00-740-9312
112	Felt Washer	7409928	5310-00-740-9928
113	Felt Washer	7417093	5330-00-741-7093
114	Felt Washer	7409929	5330-00-740-9929
115	Felt Washer	7411154	5330-00-741-1154
116	Felt Washer	7411159	5330-00-741-1159
117	Felt Washer	7417094	5330-00-741-7094
118	Ferrule	2297-N-5630	4730-01-272-0582
119	Fiber Washer	14079550	5330-00-107-3925
120	Filter	3313281	2940-01-157-6309
121	Filter	146483	2910-00-790-8736
122	Filter	599791	4460-01-284-2344
123	Filter Element Kit	ERS-28001	4330-01-272-2937
124	Filter Element Kit	12503	4330-01-284-6203
125	Filter Kit	AR 51480	2940-00-404-3057
126	Filter Screen	23010654	2940-01-140-8227
127	Filter Strip	12277066	9320-01-109-5696
128	Fluid Pressure Kit	3652-11	4330-01-243-0055
129	Fluid Pressure Kit	256476	2910-00-152-2033
130	Flywheel Seal Ring	6770492	5330-00-999-3760
131	Frame Seal	7373300	9390-00-737-3300
132	Freeze Plug	213394	5310-01-087-0682

Section II. MANDATORY REPLACEMENT PARTS (Contd)

(1) ITEM NO.	(2) NOMENCLATURE	(3) PART NUMBER	(4) NATIONAL/NATO STOCK NUMBER
133	Freeze Plug	213395	5340-01-087-0681
134	Freeze Plug	216524	5340-012-086-6193
135	Fuel Filter	33472	2910-01-201-7719
136	Fuel Injection Pump Maintenance Kit	57k0144	2910-01-339-0423
137	Fuel Injection Pump Repair Kit	1417010008	2910-01-339-7912
138	Gasket	3048341	5330-01-262-5118
139	Gasket	3923054	5330-01-190-9555
140	Gasket	3911941	5330-01-272-1146
141	Gasket	3914388	5330-01-190-1905
142	Gasket	1235675	5330-01-299-6616
143	Gasket	7539072	5330-00-753-9072
144	Gasket	A5711	5310-01-133-5847
145	Gasket	153518	5330-01-044-2096
146	Gasket	154088	5330-00-961-9470
147	Gasket	70441	5330-00-508-0411
148	Gasket	3060912	5330-01-272-1142
149	Gasket	3914310	5330-01-287-8656
150	GASKET	3913032	5330-01-901-1828
151	GASKET	3913025	5330-01-302-0780
152	GASKET	35913027	5330-01-301-1829
153	GASKET	130226	5330-00-106-6370
154	GASKET	3914391	5330-01-272-1138
155	GASKET	1164431	5330-00-143-7797
156	GASKET	11664480	5330-00-252-3274
157	Gasket	33069823	5330-01-137-4487
158	Gasket	3031007	5330-01-165-2314
159	Gasket	29501160	5330-01-120-8090
160	Gasket	3905449	5330-01-271-8307
161	Gasket	7979274	5330-00-740-9600
162	Gasket	3910642	5330-01-271-8306
163	Gasket	154916	5330-01-071-5727
164	Gasket	7979275	5330-00-734-6993
165	Gasket	1227128	5330-01-115-0604

Section II. MANDATORY REPLACEMENT PARTS (Contd)

(1) ITEM NO.	(2) NOMENCLATURE	(3) PART NUMBER	(4) NATIONAL/NATO STOCK NUMBER
166	Gasket	1089995	5330-00-182-3489
167	Gasket	65274	5330-00-246-0309
168	Gasket	142234	5330-00-659-3178
169	Gasket	134276	5330-00-193-7652
170	Gasket	12255817	5330-01-379-4345
171	Gasket	3011273	5330-01-267-0399
172	Gasket	3017750	5330-00-861-8592
173	Gasket	307613	5330-01-147-4071
174	Gasket	151911	5330-00-961-6314
175	Gasket	12256082-1	5330-01-374-0400
176	Gasket	134285	5330-00-465-5818
177	Gasket	110453	5330-00-143-8371
178	Gasket	157551	5330-00-143-8376
179	Gasket	68210	5330-00-328-8656
180	Gasket	3008017	5330-01-079-6514
181	Gasket	3008591	5330-01-086-3523
182	Gasket	3012972	5330-01-131-2967
183	Gasket	3069101	5330-00-026-2931
184	Gasket	70089-1	5330-00-537-2382
185	Gasket	320-1850	5330-01-181-0631
186	Gasket	154018	5330-00-852-7347
187	Gasket	3054841	5330-01-285-4827
188	Gasket	3921852	5330-01-272-1145
189	Gasket	3908096	5330-01-266-3294
190	Gasket	3914301	5330-01-271-4308
191	Gasket	3929253	5330-01-317-3213
192	Gasket	3915772	5330-01-263-6179
193	Gasket	3914017	5330-01-289-3135
194	Gasket	3914302	5330-01-272-1143
195	Gasket	3917780	5330-01-321-2053
196	Gasket	3911942	5330-01-281-9013
197	Gasket	3915800	5330-01-270-8144
198	Gasket	3201386	5330-01-181-0630
199	Gasket	176027	5330-00-129-9389

Section II. MANDATORY REPLACEMENT PARTS (Contd)

(1) ITEM NO.	(2) NOMENCLATURE	(3) PART NUMBER	(4) NATIONAL/NATO STOCK NUMBER
200	Gasket	3047159	5330-00-131-7072
201	Gasket	12302621	5330-01-232-1487
202	Gasket	7376584	5330-00-737-6584
203	Gasket	3021735	5330-01-082-6985
204	Gasket	50161-2	5330-01-285-1601
205	Gasket	6884872	5330-01-111-9291
206	Gasket	10900396	5330-00-419-5872
207	Gasket	7346886	5330-00-734-6886
208	Gasket	7346896	5330-00-641-2466
209	Gasket	291882	5330-01-123-6409
210	Gasket	20510093-4Z	5330-01-271-9407
211	Gasket	11663365	5330-01-054-4011
212	Gasket	7973339	5330-00-895-3424
213	Gasket	7409821	5330-00-740-9821
214	Gasket	7409822	5330-00-057-3823
215	Gasket	3914308	5330-01-271-6404
216	Gasket	7409933	5330-00-740-9933
217	Gasket	7409931	5330-00-166-4333
218	Gasket	7409932	5330-00-740-9932
219	Gasket	20511420	5330-01-361-5600
220	Gasket	2208-U-697	5330-00-549-7694
221	Gasket	35-P-8	5330-00-485-0895
222	Gasket	22-P-24-2	5330-00-485-0865
223	Gasket	70705	5330-00-562-1176
224	Gasket	12288013	5330-00-781-7774
225	Gasket	6771366	5330-00-911-9411
226	Gasket	23016347	2840-01-068-1713
227	Gasket	23014221	5330-00-557-6518
228	Gasket	23045099	5330-01-219-2555
229	Gasket	11592566	5330-00-414-6695
230	Gasket	3918174	5330-01-271-5791
231	Gasket	7535583	5330-00-415-1484
232	Gasket	830431	5330-00-415-1488
233	Gasket	10876133	5330-00-826-5202

Section II. MANDATORY REPLACEMENT PARTS (Contd)

(1) ITEM NO.	(2) NOMENCLATURE	(3) PART NUMBER	(4) NATIONAL/NATO STOCK NUMBER
234	Gasket	10876132	5330-00-826-5203
235	Gasket	7529300	5365-01-129-0399
236	Gasket	MS51071-7	6680-00-882-0965
237	Gasket	981072	5220-00-982-4259
238	Gasket	35-P-41	5330-01-133-0205
239	Gasket	100764	5330-00-506-4866
240	Gasket	173086	5330-00-132-0247
241	Gasket	23011670	5330-00-001-1984
242	Gasket	118394	
243	Gasket	3008947	5330-01-129-6541
244	Gasket	HFB642010-A1	
245	Gasket	3780-Q-381	5330-01-292-9575
246	Gasket	2208-S-1033	5330-01-272-1148
247	Gasket and Seal Set	5518441	5330-01-341-6583
248	Gasket and Shim Set	7346807	5330-00-513-1443
249	Governor Filter	6882687	4330-01-074-9642
250	Governor Filter O-ring	6882689	5330-01-080-3254
251	Grommet	7035447, 747R	5340-00-264-7182
252	Grommet	MS35489-135	5325-00-263-6648
253	Half-Keeper	127554	5340-01-143-6048
254	Hi-Lo Shaft Seal	A-1205-P-1758	5330-01-132-8346
255	Injector Overhaul Kit	AR-51522	2910-00-117-3689
256	Injector Sleeve	3909886	2910-01-271-9826
257	Injector Sleeve	3011934	2910-01-146-0048
258	Inner Seal Ring	23011471	4310-01-006-4952
259	Inner Seal Ring	6883031	5330-01-083-3065
260	Insulation	12302774	
261	Insulator	11669319-1	5970-01-114-3753
262	Insulator	12256707	5340-01-104-7843
263	Isolator	11669109	5340-01-101-0005
264	Jamnut	MS51967-18 51967-18	5310-00-763-8919
265	Key	8327444	5315-00-281-7652

Section II. MANDATORY REPLACEMENT PARTS (Contd)

(1) ITEM NO.	(2) NOMENCLATURE	(3) PART NUMBER	(4) NATIONAL/NATO STOCK NUMBER
266	Key	7535631	5315-01-217-2269
267	Key Washer	114638	5310-00-887-8325
268	Keyway Insert	5-X-663	5330-01-133-7262
269	Leather Washer	77121-1	5310-00-760-7493
270	Load Check ValveKit	5704273	4820-01-093-5785
271	Lock Pin	10166	5315-01-284-9812
272	Locknut	MS21045-4	5310-00-061-7325
273	Locknut	MS51943-46	5310-00-935-3569
274	Locknut	MS21045-3	5310-00-061-7326
275	Locknut	MS51922-5	5310-00-959-7600
276	Locknut	MS21044N4	5310-00-877-5796
277	Locknut	MS51922-13	5310-00-984-3807
278	Locknut	5590560	5310-01-126-9404
279	Locknut	942279	5310-01-193-6884
280	Locknut	MS51922-9	5310-00-984-3806
281	Locknut	MS21045-7	5310-00-274-9364
282	Locknut	MS21245-L10	5310-00-449-2381
283	Locknut	MS21044-N3	5310-00-877-5797
284	Locknut	MS51922-17	5310-00-087-4652
285	Locknut	MS21045-8	5310-00-062-4952
286	Locknut	MS51922-61	5310-00-832-9719
287	Locknut	1779-Z-260	5310-00-949-6280
288	Locknut	MS51922-21	5310-00-959-1488
289	Locknut	3913371	5310-01-287-5742
290	Locknut	12301125	5310-01-210-0199
291	Locknut	MS51943-36	5310-00-814-0672
292	Locknut	MS21045-C3	5310-00-263-2862
293	Locknut	MS17830-06C	5310-00-176-6341
294	Locknut	MS51943-40	5310-00-488-3888
295	Locknut	MS51943-39	5310-00-488-3889
296	Locknut	MS51922-53	5310-00-225-6408
297	Locknut	MS51922-1	5310-00-088-1251
298	Locknut	MS51968-8	5310-00-732-0559
299	Locknut	MS51943-34	5310-00-241-6658

Section II. MANDATORY REPLACEMENT PARTS (Contd)

(1) ITEM NO.	(2) NOMENCLATURE	(3) PART NUMBER	(4) NATIONAL/NATO STOCK NUMBER
300	Locknut	MS51922-6	5310-00-143-6102
301	Locknut	9422305	5310-01-130-4274
302	Locknut	8712289-5	5310-00-044-3342
303	Locknut	3906216	5310-01-270-8251
304	Locknut	MS51943-33	5310-00-814-0673
305	Locknut	FCO30005	5310-01-270-8342
306	Locknut	MS51943-35	5310-00-935-9021
307	Locknut	AN365-1024A	5310-00-208-1918
308	Locknut	7373244	5310-00-269-7044
309	Locknut	MS21044-N8	5310-00-877-5795
310	Locknut	MS21045-10	5310-00-982-5009
311	Locknut	MS21045-12	5310-00-982-5012
312	Locknut	MS51922-37	5310-00-067-9507
313	Locknut	MS51943-32	5310-00-935-9022
314	Locknut	MS21083C12	5310-00-923-4219
315	Locknut	MS51943-38	5310-00-994-1006
316	Locknut	MS21045-14	5310-00-982-5014
317	Locknut	11609727-2	5310-00-176-6690
318	Locknut	MS21045-5	5310-00-982-4912
319	Locknut	9422301	5310-01-149-4407
320	Locknut	1227-K-1051	5310-01-099-0397
321	Locknut	MS51943-44	5310-00-241-6664
322	Locknut	MS51943-2	5310-01-374-0508
323	Locknut	NAS1021-N17	5310-00-325-1900
324	Locknut	3311501033	5310-01-143-0512
325	Locknut	2304890	5310-00-748-0548
326	Locknut	S-212	5310-00-011-7051
327	Locknut	MS21045-6, 456702	5310-00-982-4908
328	Locknut	9419479	5310-01-409-1642
329	Locknut	MS51943-5	5310-01-249-0904
330	Locknut	9422298	5310-01-150-5914
331	Locknut	MS21045-18	5310-00-057-7153
332	Locknut	G-9415992	5310-00-421-3991

Section II. MANDATORY REPLACEMENT PARTS (Contd)

(1) ITEM NO.	(2) NOMENCLATURE	(3) PART NUMBER	(4) NATIONAL/NATO STOCK NUMBER
333	Locknut	MS519221-1	5315-00-088-1251
334	Locknut	131245	5310-00-013-1245
335	Locknut	MS21042-8	5310-00-807-1468
336	Locknut	MS51943-4	5310-00-935-3750
337	Locknut	MS51943-6	5310-01-344-8250
338	Locknut	MS51943-10	5310-00-455-9967
339	Locknut	109319	2518-00-406-8936
340	Lockplate	3914708	5310-01-330-8313
341	Lockplate	3039305	
342	Lockstrap	6880899	5340-01-056-0037
343	Locktab	028426	5310-01-135-6758
344	Locktab Washer	8758258	5310-00-147-3274
345	Lockwasher	MS35335-33	5310-00-209-0786
346	Lockwasher	MS35338-62	5310-00-274-8710
347	Lockwasher	MS35338-63	5310-00-274-8715
348	Lockwasher	MS12203-2	5310-00-159-6209
349	Lockwasher	MS35338-8, S-604	5310-00-261-7340
350	Lockwasher	210104-8S	5310-00-003-4094
351	Lockwasher	MS35338-42	5310-00-045-3299
352	Lockwasher	MS35338-43	5310-00-045-3296
353	Lockwasher	MS45904-73	
354	Lockwasher	MS35338-46	5310-00-130-9065
355	Lockwasher	MS35333-44	5310-00-194-1483
356	Lockwasher	MS35335-42	5310-00-595-7237
357	Lockwasher	MS35335-32	5310-00-596-7691
358	Lockwasher	MS27183-14	5310-00-080-6004
359	Lockwasher	138485	
360	Lockwasher	MS353335-35, 138489	5310-00-627-6128
361	Lockwasher	WA-LH8-3	5310-00-031-2673
362	Lockwasher	MS35338-52	5310-00-754-2005
363	Lockwasher	MS35043-53	5310-00-926-5885
364	Lockwasher	7410218, MS35338-45	5310-00-407-9566

Section II. MANDATORY REPLACEMENT PARTS (Contd)

(1) ITEM NO.	(2) NOMENCLATURE	(3) PART NUMBER	(4) NATIONAL/NATO STOCK NUMBER
365	Lockwasher	1229-S-513-C	5310-01-062-3384
366	Lockwasher	MS35333-49	5310-00-582-6714
367	Lockwasher	MS35333-41	5310-00-167-0721
368	Lockwasher	MS35340-48	5310-00-834-7606
369	Lockwasher	MS35335-31	5310-00-596-7693
370	Lockwasher	2379-10ZF	5310-01-135-4828
371	Lockwasher	MS35333-40	5310-00-550-1130
372	Lockwasher	MS45904-84	5310-00-935-8984
373	Lockwasher	MS35335-30	5310-00-209-0788
374	Lockwasher	MS35338-41	5310-00-045-4007
375	Lockwasher	MS35333-47	5310-00-550-3714
376	Lockwasher	MS35340-43	5310-00-721-7809
377	Lockwasher	MS35338-50	5310-00-820-6653
378	Lockwasher	MS35333-78	5310-00-261-7156
379	Lockwasher	MS45904-76	5310-00-061-1258
380	Lockwasher	3910266	5310-01-270-8423
381	Lockwasher	MS35338-49	5310-00-167-0680
382	Lockwasher	MS35338-47	5310-00-209-0965
383	Lockwasher	3912897	5310-01-270-8405
384	Lockwasher	MS35336-9	5310-00-550-0248
385	Lockwasher	MS35335-34	5310-00-510-6674
386	Lockwasher	96906	5310-00-959-4679
387	Lockwasher	06853	5310-00-119-4864
388	Lockwasher	MS35335-40	5310-00-275-3683
389	Lockwasher	MS35333-39	5310-00-576-5752
390	Lockwasher	AN3066-12	5975-00-793-5550
391	Lockwasher	MS35233-38	5310-00-559-0070
392	Lockwasher	MS35338-50	5310-00-004-5034
393	Lockwasher	MS35338-51	5310-00-584-7888
394	Lockwaasher	378003	5310-00-838-1490
395	Lockwasher	9411417	5310-00-799-4910
396	Lockwasher	2 916 699 092	5310-01-300-7037
397	Lockwasher	2 916 699 085	5310-01-301-1875
398	Lockwasher	2 916 699 083	5310-01-301-7811

Section II. MANDATORY REPLACEMENT PARTS (Contd)

(1) ITEM NO.	(2) NOMENCLATURE	(3) PART NUMBER	(4) NATIONAL/NATO STOCK NUMBER
399	Lockwasher	MS35335-39	5310-00-800-0695
400	Lockwasher	MS35338-44	5310-00-582-5965
401	Lockwasher	26X-3074	
402	Lockwasher	MS45904-68	5310-00-889-2528
403	Lockwasher	20X-0196	
404	Lockwasher	MS45904-72	5310-00-889-2527
405	Lockwasher	MS35335-36	5310-00-550-3503
406	Lockwasher	MS35335-60	5310-00-209-1239
407	Lockwasher	181466	5310-00-484-1718
408	Lockwasher	S-622	5310-00-562-6557
409	Lockwasher	WA-LM7-2	5310-00-486-5355
410	Lockwasher	MS35336-39	5310-00-194-9213
411	Lockwasher	S-610	5310-01-300-8400
412	Lockwasher	S-606	5310-00-410-6756
413	Lockwasher	178556	5310-00-017-8556
414	Lockwasher	120380	5310-00-209-2946
415	Lockwasher	11500207	5310-01-206-7306
416	Lockwasher	1613, MS35338-48	5310-00-584-5272
417	Lockwasher	MS51848-14	5310-00-171-1735
418	Lockwasher	MS45904-74	
419	Lockwire	22-W-1642-100	9505-00-554-0098
420	Lockwire	MS20995F9H2	
421	Lockwire	MS20995F91	9505-00-846-0941
422	Lubrication Valve	6834624	2520-00-557-5900
423	Manifold Pressure Compensator Maintenance Kit	57K0143	2910-01-339-0422
424	O-ring	145504	5330-01-051-4243
425	O-ring	130240	5330-00-106-6969
426	O-ring	15434	5330-01-272-1123
427	O-ring	3910824	5330-01-281-8997
428	O-ring	MS28775-032	5330-01-049-7374
429	O-ring	6762127	5330-01-010-9693
430	O-ring	MS28778-10	5330-00-285-9842

Section II. MANDATORY REPLACEMENT PARTS (Contd)

(1) ITEM NO	(2) NOMENCLATURE	(3) PART NUMBER	(6) NATIONAL/NATO STOCK NUMBER
431	O-ring	3909397	5330-01-272-1120
432	O-ring	12301126	5330-01-210-2155
433	O-ring	MS35489-6	5325-00-263-6632
434	O-ring	N72259	5330-00-152-1759
435	O-ring	N72260	5330-01-166-3662
436	O-ring	MS28775-238	5330-00-579-75745
437	O-ring	AN6290-6	5330-00-804-5695
438	O-ring	131026	5330-00-143-8485
439	O-ring	137075	5330-00-420-9624
440	O-ring	67270	5330-00-171-3879
441	O-ring	3019116	5330-01-160-7458
442	O-ring	70624	5330-00-506-4874
443	O-ring	3024789	5330-01-145-5381
444	O-ring	3910260	5330-01-272-1124
445	O-ring	3913994 TS33-016 70	5330-01-291-6537
446	O-ring	3910503	5330-01-272-1121
447	O-ring	3916284	5330-01-272-1122
448	O-ring	FCO30020	5330-01-271-9372
449	O-ring	FD76A	5330-01-123-2832
450	O-ring	FD145A	5330-01-271-9374
451	O-ring	FD0077	5330-01-079-6513
452	O-ring	233955	5330-00-074-2692
453	O-ring	ER-82141	5330-01-129-0384
454	O-ring	MS29513-115	5330-00-248-3847
455	O-ring	MS28775-219	5330-00-579-7925
456	O-ring	MS28775-225	5330-00-579-7927
457	O-ring	MS28778-12	5330-00-251-8839
458	O-ring	154129	5330-00-948-6482
459	O-ring	MS28775-113	5330-00-582-2855
460	O-ring	3032874	5330-01-220-2389
461	O-ring	68061-A	5330-00-970-3461
462	O-ring	100478	5330-00-081-9289
463	O-ring	213768	5330-01-072-8983

Section II. MANDATORY REPLACEMENT PARTS (Contd)

(1) ITEM NO	(2) NOMENCLATURE	(3) PART NUMBER	(6) NATIONAL/NATO STOCK NUMBER
464	O-ring	582826	5330-00-522-8544
465	O-ring	MS28775-249	5330-01-019-2448
466	O-ring	M83461/1-427	5330-01-183-0985
467	O-ring	MS28775-243	5330-00-579-7544
468	O-ring	MS28778-20	5330-00-816-3546
469	O-ring	MA28775-128	5330-00-702-5643
470	O-ring	AS3551-12	5330-00-776-2830
471	O-ring	MS28775-232	5330-00-585-8247
472	O-ring	MS27183	5310-01-317-1812
473	O-ring	7374401	5330-00-984-3756
474	O-ring	20510736	4730-01-279-1519
475	O-ring	154087	5330-00-772-7657
476	O-ring	3037236	5330-01-331-9293
477	O-ring	MS28778-8	5330-00-808-0794
478	O-ring	3007759	5330-01-072-4436
479	O-ring	282818	5330-01-217-0734
480	O-ring	18048	5330-01-280-6503
481	O-ring	028466	5310-01-135-6754
482	O-ring	032586	5330-01-143-2780
483	O-ring	032616	5330-01-149-7229
484	O-ring	032571	5330-01-135-4809
485	O-ring	032615	5330-01-135-4068
486	O-ring	M83461/1-012	5330-01-046-3300
487	O-ring	MS28775-206	5330-01-133-5858
488	O-ring	M832484/1-324	5330-01-005-3704
489	O-ring	MS28778-16	5330-00-804-5694
490	O-ring	M83461/1-219	5330-01-128-3954
491	Oil Filter	PH3519	2940-01-110-2489
492	Oil Filter Element	PF297	2940-00-950-8410
493	Oil Pan Gasket	3032861	5330-01-147-0748
494	Oil Pump Kit	3802278	2815-01-268-8753
495	Oil Seal	A-1205-E2137	5330-01-271-9362
496	Oil Seal	10938292	5330-00-145-8355

Section II. MANDATORY REPLACEMENT PARTS (Contd)

(1) ITEM NO	(2) NOMENCLATURE	(3) PART NUMBER	(6) NATIONAL/NATO STOCK NUMBER
497	Oil Seal	4591SCR, A-1205-Z-2132	5330-01-271-9410
498	Oil Seal	211255	5330-00-135-6382
499	Oil Seal	208069	5300-00-006-2529
500	Oil Seal	A-1205-U-1737	5330-01-023-0269
501	Oil Seal	28-P-52	5330-00-237-7828
502	Oil Seal	23016643	5330-01-219-2375
503	Oil Seal	6773311	5330-00-999-3752
504	Oil Seal	A-1205-U-1633	5330-01-126-0565
505	Oil Seal Kit	3802387	
506	Oil Seal Kit	3802389	5330-01-344-0567
507	Oil Sleeve	1844-J-634	2520-01-132-6841
508	Outer Bearing Oil Seal	7431447	5330-00-961-3596
509	Outer Seal Ring	6883035	5330-01-104-8934
510	Packing	10900300	5330-00-523-4235
511	Packing	11609215	5330-00-269-4953
512	Packing	032634-A1	5330-01-135-0682
513	Packing Ring	032361	5330-01-195-5757
514	Pin	7412376	2510-00-741-2376
515	Piston Inner Seal Ring	23015880	5330-01-146-6053
516	Piston Outer Seal Ring	6833981	5365-01-010-9689
517	Piston Ring	180810	2815-01-079-3290
518	Piston Rings	3802110	2815-01-271-9792
519	Piston Seal	10900304	5340-00-523-4305
520	Piston Seal Ring	6758740	5330-00-582-0456
521	Plastic Backup Washer	028435	5310-01-161-6131
522	Power Steering Filter Kit	ERS27788	2530-01-137-5921
523	Power Steering Parts Kit	20510093-26Z	2530-01-272-2910
524	Pressure Regulator Plunger	391852	5340-01-331-9625
525	Pumping Element Kit	20510093-25Z	5330-01-271-9544
526	Rear Cover Plate Gasket	151623	5330-01-082-1906
527	Rear Cover Plate Seal	M39807	5330-00-005-0858
528	Regulator Channel Filler	10906350	2510-00-179-5708
529	Regulator Channel Seal	7373301	9390-00-737-3301

Section II. MANDATORY REPLACEMENT PARTS (Contd)

(1) ITEM NO	(2) NOMENCLATURE	(3) PART NUMBER	(6) NATIONAL/NATO STOCK NUMBER
530	Relay Valve Kit	599913	2530-01-284-4287
531	Relay Valve Kit	599911	2530-01-284-4288
532	Relief Fitting	70295	4730-00-011-3175
533	Relief Valve Kit	5704274	2590-00-606-2383
534	Repair Kit	3011472	5330-00-480-6133
535	Repair Kit	AR73350	2815-00-913-2074
536	Resilient Mount	7521436	5340-00-040-2073
537	Retaining Ring	110827	5330-00-785-7894
538	Retaining Ring	MS16625-1100	5365-00-807-2636
539	Retaining Ring	3901996	5325-01-280-5592
540	Retaining Ring	401309	5325-00-613-7796
541	Ring Seal	900877	2590-01-119-4103
542	Rivet	MS20600AD6W4	5320-00-528-3276
543	Rivet	7B5049	5320-00-262-6492
544	Rivet	RV876	5320-01-146-9582
545	Rivet	MS20600-MP8W/4	5320-01-068-2340
546	Rivet	MS20600AD5W2	5320-00-582-3302
547	Rivet	RV200-6-3	5320-00-582-3268
548	Rivet	MS20600AD8W7	5320-00-721-5384
549	Rivet	MS24662-234	5320-00-930-7865
550	Rivet	RV200-6-4	5320-00-582-3276
551	Rivet	RV200-6-2	5320-00-584-1285
552	Rivet	MS20470A6-6	5320-00-242-1580
553	Rivet	ADJUST644	5320-00-956-7355
554	Rivet	MS20470A5-8	5320-00-234-8557
555	Rivet	MS20473A6-9	5320-00-264-3266
556	Rivet	CR9163-6-6	5320-00-582-3499
557	Rivet	RV200-6-1	5320-00-616-4350
558	Rivet	MS24661-226	5320-00-231-3663
559	Rivet	RV200-6-5	5320-00-582-3301
560	Rod Seal	MS28775-222	5330-00-338-4460
561	RQV Governor Cover Housing Maintenance Kit	57K0141	2910-01-338-4460
562	RQV Governor Cover Maintenance Kit	57K0142	2910-01-338-6473

Section II. MANDATORY REPLACEMENT PARTS (Contd)

(1) ITEM NO	(2) NOMENCLATURE	(3) PART NUMBER	(6) NATIONAL/NATO STOCK NUMBER
563	Rubber Bumper	7535643	5340-00-766-3330
564	Rubber Bushing	7409618	5365-00-740-9618
565	Rubber Grommet	MS35489-19	5325-00-276-6091
566	Rubber Seal	12256106	5330-01-120-8454
567	Safety Wire	ASTM AS41	9505-00-248-9842
568	Safety Wire	22W1642125	9505-00-198-9125
569	Screw	MS90726-97	5305-00-225-9092
570	Screw	MS51106-421	5305-00-940-9517
571	Screw	522875	
572	Screw-Assembled Lockwasher	425841	5306-00-042-5841
573	Screw-Assembled Lockwasher	187995, 7373271	5305-00-696-5285
574	Screw-Assembled Lockwasher	455176	5305-01-225-2106
575	Screw-Assembled Lockwasher	425648	5305-00-042-5648
576	Screw-Assembled Lockwasher	9414109	5305-01-229-9587
577	Screw-Assembled Lockwasher	423569	5306-01-226-0798
578	Screw Assembled Lockwasher	70772	5305-00-477-6769
579	Screw-Assembled Lockwasher	3021470	5305-01-144-6233
580	Screw-Assembled Lockwasher	3012472	5305-01-112-4312
581	Screw-Assembled Lockwasher	7372083-1	5305-01-090-7626
582	Screw Assembled Lockwasher	MS90726-59	5305-00-912-5113
583	Screw Assembled Lockwasher	423568	5305-00-042-3568
584	Screw Assembled Lockwasher	11663070	5305-01-104-9018
585	Screw Assembled Lockwasher	11664479-1	5306-01-106-3850
586	Screw Assembled Lockwasher	3010593	5305-01-197-3449
587	Screw Assembled Lockwasher	423518	5305-00-638-0714
588	Screw Assembled Lockwasher	30105494	5305-01-130-6100
589	Screw Assembled Lockwasher	3010590	5305-01-119-8621
590	Screw Assembled Lockwasher	3010596	5305-01-088-6019
591	Screw Assembled Lockwasher	3010592	5305-01-176-8018
592	Screw Assembled Lockwasher	3010597	5305-01-086-7036
593	Screw Assembled Washer	425567	5305-00-042-5597
594	Seal	67946	5365-00-197-9327
595	Seal	3900216	5310-01-188-0997

Section II. MANDATORY REPLACEMENT PARTS (Contd)

(1) ITEM NO	(2) NOMENCLATURE	(3) PART NUMBER	(6) NATIONAL/NATO STOCK NUMBER
596	Seal	8265	5310-00-246-0221
597	Seal	186780	5330-00-864-5422
598	Seal	188318	5330-01-301-1761
599	Seal	145530	5330-01-201-3623
600	Seal	3902466	5330-01-272-1246
601	Seal	7059240	5330-00-414-3754
602	Seal	3909410	5330-01-192-2037
603	Seal	3903475	5330-01-791-8047
604	Seal	032579	5330-01-143-4186
605	Seal	11607302-7	5330-00-340-3637
606	Seal	10937693-3	2510-01-197-4200
607	Seal	17657/55-542465	5330-01-150-9691
608	Seal	10875107-7	5330-01-119-5801
609	Seal	7409940	5330-00-292-1600
610	Seal	500207	5330-00-585-3210
611	Seal	E-450121VG	5330-01-131-5416
612	Seal	3003156	5330-01-072-8830
613	Seal	3915707	5340-01-281-7792
614	Seal	8735035	5330-00-470-2115
615	Seal	8735034	5330-00-419-9468
616	Seal	87350536	5330-00-419-9469
617	Seal	10608E44S	5330-00-020-5375
618	Seal	10937640	5330-00-338-0774
619	Seal	10915159	9390-00-405-0215
620	Seal	7373291-3	5330-00-152-3217
621	Seal	10937691	5330-01-098-6555
622	Seal	10937727	5330-01-016-1245
623	Seal	12302744	5330-01-209-7354
624	Seal	500163	5330-00-178-2191
625	Seal	S-1003-A	5365-00-598-5255
626	Seal	23010610	2840-01-141-9503
627	Seal	1205-Y-1633	5330-01-137-4799
628	Seal	C-366005	2590-01-134-9834
629	Seal	A-1205-D-2162	5330-01-308-0175

Section II. MANDATORY REPLACEMENT PARTS (Contd)

(1) ITEM NO	(2) NOMENCLATURE	(3) PART NUMBER	(6) NATIONAL/NATO STOCK NUMBER
630	Seal	A-1205-N-2120	5330-01-272-1147
631	Seal	11607267-2	9320-00-421-7230
632	Seal	10937683-2	9390-00-158-2408
633	Seal	8380420	9320-00-451-8080
634	Seal Assembly	12375801	5330-00-740-9550
635	Seal Assembly	3033677, 3071085	5330-00-005-0407
636	Seal Ring	6880389	5330-01-141-9579
637	Seal Ring	6830187	2520-00-405-1842
638	Seal Ring	23014632	2520-00-557-6211
639	Seal Ring	6839163	5330-00-374-4873
640	Seal Ring	23017696	2520-01-130-5770
641	Seal Ring	MS28775-111	5330-00-579-8108
642	Seal Ring	032570	5330-01-135-4067
643	Seal Ring	035552	5330-01-135-4808
644	Seal Washer	3918191	5310-01-340-8469
645	Seal Washer	3903037	5310-01-195-1441
646	Seal Washer	3914896	5310-01-331-9411
647	Seal-Locking Retainer	3909063	5310-01-143-0542
648	Self-Locking Screw	940937	5305-00-292-4594
649	Self-Locking Screw	MS35764-1297	5306-01-052-2402
650	Self-Locking Screw	9409225	5305-00-638-2362
651	Shim	68192	5365-00-378-2885
652	Shim	12256738-1	5365-01-110-8183
653	Side Cover Gasket	7346899	5330-00-734-6899
654	Sleeve	C5165X4	4730-00-969-6941
655	Sleeve	8120115B, 120115B	4730-00-054-2571
656	Sleeve	8 12011B	
657	Sleeve Bearing	7346983	3120-00-537-0614
658	Slotted Nut	7979183	5310-00-740-9621
659	Snapring	N5002-500MD	5325-00-914-5837
660	Snapring	S-16255	2815-00-815-0355
661	Snapring	1229-H-2816	5325-01-162-7624

Section II. MANDATORY REPLACEMENT PARTS (Contd)

(1) ITEM NO	(2) NOMENCLATURE	(3) PART NUMBER	(6) NATIONAL/NATO STOCK NUMBER
662	Snapring	112302	5325-00-420-9696
663	Snapring	1229-U-2829	5325-01-129-6849
664	Snapring	378391	5325-00-477-0304
665	Snapring	MS90707-1050	5325-00-419-3322
666	Soft Ball Bearing	213769	3110-01-079-8190
667	Speed Nut	17-10015-13	5310-00-885-7734
668	Spring	68274	5360-00-664-5343
669	Spring Center Nut	MS51968-14	5310-00-732-0560
670	Spring Center Nut	10883157	5305-00-139-7074
671	Spring Guide	23045085	4710-00-557-5885
672	Spring Nut	7951891	5310-01-122-2060
673	Spring Nut	5305-18	5310-01-122-2060
674	Spring Nut	MS13532-35	5315-00-814-3530
675	Springtite Assembly	1164243	5306-00-238-5661
676	Steering Knuckle Boot Replacement Kit	5704510	2530-01-25-9272
677	Steering Parts Kit	1790522K	2530-01-339-7913
678	Steering Parts Kits	5223281	2530-01-340-0365
679	Stop Screw	5225875	5306-00-281-1651
680	Suction Flange Gasket	67963	5330-00-171-7267
681	Terminal	MS20659-127	5940-00-113-3147
682	Thrust Ring Set	157280	
683	Thrust Washer	31P-27	5310-00-469-4039
684	Tiedown Strap	MS3367-1-0	5975-00-984-6582
685	Tiedown Strap	MS3367-3-9	5975-00-451-5001
686	Tiedown Strap	MS3367-2-0	5975-00-899-4606
687	Tiedown Strap	MS3367-1-9, SST2SC	5975-00-074-2072
688	Tiedown Strap	MS3367-6-9	5975-00-133-8696
689	Tiedown Strap	12256372	5340-01-104-9012
690	Tiedown Strap	MS3367-7-9	5975-00-570-9598
691	Tiedown Strap	MS3367-1-9	5975-01-013-2742
692	Tiedown Strap	MS3367-4	5975-00-727-5153
693	Tiedown Strap	11669079-5	4730-01-331-6630

Section II. MANDATORY REPLACEMENT PARTS (Contd)

(1) ITEM NO	(2) NOMENCLATURE	(3) PART NUMBER	(6) NATIONAL/NATO STOCK NUMBER
694	Tiedown Strap	MS3367-5-9	5975-00-111-3208
695	Tiedown Strap	MS2367-202	
696	Tiedown Strap	MS3367-2-9	5975-00-156-3253
697	Tiedown Strap	MS3367-3-0	5975-00-985-6630
698	Top Cover Gasket	7535079	5330-00-138-8388
699	Transmission Oil Filter	6883044	2520-01-124-6469
700	Turbocharger Repair Kit	3802149	2950-01-271-2345
701	Two-Piece Seal	032791A1	2530-01-131-7445
702	Universal Joint Kit	CP85WB62	2520-01-280-4129
703	Universal Joint Parts Kit	5704528	2520-00-460-6477
704	Universal Plate	7335053	5340-01-119-5682
705	Valve Guide Pin	6834410	2520-01-011-1068
706	Valve Repair Kit	289352	2530-01-134-1834
707	Valve Repair Kit	17BV361	4820-01-034-0971
708	Valve Seat Inseat	170296	5340-00-933-3009
709	Valve Spring	6836928	5360-01-123-5483
710	Vent Door Weatherseal	7373317	9390-00-737-3317
711	Washer	3902425	5310-01-209-0508
712	Washer	3906659	5310-01-270-8417
713	Washer	6834908	5310-00-557-5942
714	Washer	11511514	5310-01-286-5452
715	Washer	65003-S	5310-01-099-2550
716	Washer	3901798	5310-01-270-8388
717	Wear Sleeve Kit	99293	4910-01-313-4621
718	Weatherseal	12368265	9390-01-285-9623
719	Wiper	7409553	2590-00-740-9553
720	Wiper Strip	2012993-4	5330-00-972-2635
721	Wood	11682324-1	
722	Wood	11682324-2	
723	Wood	11682323-1	
724	Wood	11682323-2	
725	Wood	11665738	2510-01-180-6164
726	Woodruff Key	MS35756-8	5315-00-616-5526
727	Woodruff Key	MS35756-6	5315-00-616-5514-

Section II. MANDATORY REPLACEMENT PARTS (Contd)

(1) ITEM NO	(2) NOMENCLATURE	(3) PART NUMBER	(6) NATIONAL/NATO STOCK NUMBER
728	Woodruff Key	MS35756-14	5315-00-616-5520
729	Woodruff Key	MS35756-17	5315-00-012-4553
730	Woodruff Key	MS20067-270	5315-00-042-3293
731	Woodruff Key	MS35756-13	5315-00-616-5521
732	Woodruff Key	MS35756-1	5315-00-616-5519
733	Woodruff Key	MS35756-38	5315-00-043-1789
734	Woodruff Key	8328341	5315-00-281-7651
735	Woodruff Key	MS35756-21	5315-00-616-5500
736	Woodruff Key	MS35156-20	5315-00-616-5501
737	Woodruff Key	M20067-493	5315-01-119-5239
738	Woodruff Key	9300-1	5315-00-276-4438

APPENDIX E
COMMON AND SPECIAL TOOL IDENTIFICATION LIST

E-1. SCOPE

This appendix list common and special tools outside of the general mechanic's toolkit which you will need when servicing M939 series vehicles.

E-2. EXPLANATION OF COLUMNS

a. **Column (1) - Item Number.** This number is assigned to each entry in the listing and is referenced in the Initial Setup of applicable task under the heading of special tools, test equipment, or tools.

b. **Column (2) - Item Name.** Name or identification of special or common tool.

c. **Column (3) - National Stock Number.** This identifies the manufacturer's part number or catalog number assigned to each tool or kit.

d. **Column (4) - Part Number.** This is the National Stock Number assigned to each tool or kit; use it to request or requisition the tool or kit.

e. **Column (5) - Reference.** This is the supply catalog number in which the common tool can be found. Special tools are not found in supply catalogs and are referenced individually in the reference column.

COMMON AND SPECIAL TOOL IDENTIFICATION LIST (Contd)

(1) REFERENCE CODE NO.	(2) NOMENCLATURE	(3) PART NUMBER	(4) NATIONAL/NATO STOCK NUMBER	(5) REFERENCE
1	General's Mechanic Tool Kit	SC5180-90CL-N26	5180-00-177-7033	
2	Adapters	J-24459	5120-01-054-4045	
3	Adjusting Wrench	3375165	4910-01-097-6927	
4	Air Gauge Assembly	20511320	5220-01-298-5730	
5	Air Seal Installer	20511263	5120-01-285-7620	
6	Angle Cutter, 60	GGG-C-613	5133-00-228-2323	4940-95-CL-BO2
7	Ball Joint Vise	1160002854	4910-00-999-1506	Special Tool
8	Barring Tool	ST-747	4910-00-150-5798	Special Tool
9	Bead Cutting Tool	ST-788	5110-00-932-2089	Special Tool
10	Bearing Installer	J-25393A	5120-01-132-5448	Special Tool
11	Bearing Installer, Front Support	J-24457	5120-01-141-9459	Special Tool
12	Bearing Punch	20511262	5120-01-285-5192	
13	Bearing Replacer	7950082	5120-00-795-0082	Special Tool
14	Bearing Replacer	7950159	5120-00-795-0159	Special Tool
15	Bearing Replacer	J-28435	5120-01-134-8449	
16	Belt Tension Gauge	ST-1136	5220-01-141-5776	
17	Bolt Inserter Tool	J-35198	4910-01-220-1512	
18	Boring Tool, Oil Pump	3375206	4910-01-085-7824	
19	Brake Reliner	MILR 13495TYCR1	4910-00-173-5310	4910-95-CL-A31
20	Brake Repair Pliers	J-33111	5120-01-152-2318	
21	Brakeshoe Adjusting Tool	J-34061	5120-01-154-3029	
22	Burnishing Tool	ST-708	4910-00-999-1503	Special Tool
23	Bushing Installer	3005319	5120-00-792-1612	
24	Bushing Installer, Center Support	J-24794	5120-01-132-5470	Special Tool
25	Bushing Replacer	J-36376	5120-01-132-5469	
26	Cam Busing Replacement Tool	ST-782	5120-00-953-9664	
27	Center Support Lifter	J-24455	5120-01-054-4001	Special Tool
28	Cleaning Brush	ST-876	7920-00-168-3244	Special Tool
29	Compressor Bar and Screw	J-24475	5120-01-132-5449	Special Tool
30	Compressor Base	J-24475-A	5120-01-132-5449	Special Tool
31	Compressor, Low and First Spring	J-24452	5120-01-054-7221	Special Tool
32	Compressor, Piston Ring	ST-755	5120-00-116-7676	Special Tool
33	Crosshead Guide Spacer	ST-633	4910-00-150-5797	

COMMON AND SPECIAL TOOL IDENTIFICATION LIST(contd)

(1) REFERENCE CODE NO.	(2) NOMENCLATURE	(3) PART NUMBER	(4) NATIONAL/NATO STOCK NUMBER	(5) REFERENCE
34	Cutter Seat	ST-662	4910-00-999-1208	
35	Cylinder Liner Clamps	3822503	5120-01-262-7309	
36	Dial Indicator	196A	5120-00-277-8840	4910-95-CL-A63
37	Driver Handle	J-24202-4	5120-01-054-4042	Special Tool
38	Driver Handle	J-8092	5120-00-677-2259	Special Tool
39	Dust Shield Installer	J-24198	5120-01-054-4052	Special Tool
40	Electrical Tool Kit	7550526	5180-00-876-9336	4910-95-CL-A31
41	Engine and Transmission Sling	J-36130-812	4910-01-353-2519	Special Tool
42	Engine Barring Tool	3377321	5120-01-262-7307	
43	Engine Barring Tool	3377371	5120-02-285-5193	
44	Engine Repair Stand	3375193	4910-00-977-7506	
45	First Clutch Clearance Gauge	J-26914	5210-01-065-9030	Special Tool
46	Forward Clutch Clearance Gauge	J-29146	5210-01-134-8224	
47	Fourth Clutch Clearance Gauge	J-29156	5120-01-134-8225	Special Tool
48	Front Support Lifter	J-24473	5120-01-054-4056	Special Tool
49	Fuel Injecion Tester (Test Stand)	11020200	4910-00-817-7431	
50	Fuel Pump Test Stand	DFP 156	4910-01-194-7667	
51	Gauge, Hydraulic Pressure	3005456	4910-00-792-8304	Special Tool
52	Gauge Block	ST-547	5210-01-157-3091	
53	Gauge, Injector Protrusion	3376220	5210-00-690-7949	Special Tool
54	Gauge, Keystone	J-24599	5220-01-028-1109	
55	Gauge, Mechanical Force		5210-01-018-2832	4910-95-CL-A31
56	Gauge, Wheel Alignment Toe-In/Out	WA361	5210-00-529-1205	4910-95-CL-A74
57	Gear Pump Block Plate	ST-844	5365-00-904-2159	
58	Gear Unit Lifter	J-24454	5120-01-054-4057	Special Tool
59	Guide Pin	J-1927-1	5120-01-144-4483	Special Tool
60	Guide Pin	J-24315-1	5315-01-141-9458	
61	Handle	7010321	5120-00-601-2234	Special Tool
62	Heat Gun	500A	4940-00-561-1002	
63	Holding Fixture	J-23642	5120-01-132-5468	
64	Holding Fixture Adapter Set	J-24462	5120-01-054-4043	Special Tool
65	Holding Fixture Base	J-3289-20	5120-01-144-4484	

COMMON AND SPECIAL TOOL IDENTIFICATION LIST (contd)

(1) REFERENCE CODE NO.	(2) NOMENCLATURE	(3) PART NUMBER	(4) NATIONAL/NATO STOCK NUMBER	(5) REFERENCE
66	Indicator, Level and Angle	3375855	4910-00-074-0020	
67	Injection Sleeve Extractor	ST-1140	4910-00-150-5858	Special Tool
68	Injector Leakage Detector	3375375	5120-01-029-6861	Special Tool
69	Injector and Valve Adjustment Kit	3375842	4910-00-548-7984	Special Tool
70	Injector Body Wrench	ST-1298	5120-00-033-2738	
71	Injector Nozzle Tester	7551255	4910-00-255-8641	
72	Injector Sleeve Cutter	ST-884	4910-00-981-3105	Special Tool
73	Injector Sleeve Expander Tool/Cutter	ST-880	3441-00-922-6699	Special Tool
74	Injector Sleeve Holding Tool	ST-1179	5120-00-104-1795	
75	Injector Sleeve Installation Mandrel/Driver	ST-1227	5120-00-981-3108	Special Tool
76	Injector Timing Fixture	3375522	4910-01-394-0391	
77	Inner Wheel Socket	5120-00-378-4411	5120-00-378-4411	Special Tool
78	Liner Clamp Set	3376669	5120-00-104-1816	
79	Liner Driver, Engine Cylinder Sleeve	ST-1229	5120-00-999-1206	Special Tool
80	Micrometer Set, Outside, 0-6 in.	GGG-C-105 TYICLISTA	5210-00-554-7134	4910-95-CL-A63
81	Micrometer, Depth	445BS-6RL	5120-00-619-4045	4910-95-CL-A63
82	Micrometer, Inside, 0-1 In.	GGG-C-105	5210-00-221-1918	
83	Micrometer, Inside, 2-12 In.	124B	5120-00-221-1921	4910-95-CL-A63
84	Mounting Plate	11600040	4910-00-977-7505	
85	Mounting Plate	3375133	4910-01-128-2685	Special Tool
86	Multimeter	AN/PSM-45	6625-01-139-2512	4910-95-CL-A72
87	Nozzle Cleaning Kit	3376947	2915-01-285-2527	
88	Oil Filter Wrench	J-29927	5120-01-037-1595	
89	Oil Primer Pump	1P0540	5120-01-217-9642	Special Tool
90	Oil Pump Body Centering Band	J-24461	5120-01-054-4044	Special Tool
91	Output Shaft Oil Seal Installer	J-24620	5120-01-132-5466	Special Tool
92	Output Shaft Seal and Dust Shield Remover	J-24171	5120-01-048-2153	Special Tool
93	Output Shaft Seal and Dust Shield Remover	J-29355	5120-01-135-1978	Special Tool

COMMON AND SPECIAL TOOL IDENTIFICATION LIST (Contd)

(1) REFERENCE CODE NO.	(2) NOMENCLATURE	(3) PART NUMBER	(4) NATIONAL/NATO STOCK NUMBER	(5) REFERENCE
94	Pin Remover	J-28708	5120-01-176-3893	Special Tool
95	Piston Return Spring Compressor	J-6438-01	5120-01-132-5465	
96	Piston Ring Expander	ST-763	5120-00-150-7486	Special Tool
97	Piston Ring Groove Gauge	ST-560	5210-00-999-1209	
98	Plunger Lift Device	1-688-130-135	4910-01-338-6241	
99	Portable Magnetic Tester	ST-1166	6635-01-128-2676	
100	Power Steering Pressure Gauge Kit	7010267	4910-00-627-7043	Special Tool
101	Puller Kit	1178	5120-00-313-9496	4910-95-CL-A72
102	Puller Kit, Mechanical	GGG-P 781	5180-00-423-1596	4910-95-CL-A31
103	Puller Kit, Universal	J-24420	5180-00-999-4053	Special Tool
104	Puller Screw	8366689	5120-00-836-6689	Special Tool
105	Puller Set, Bearing Converter Turbine	J-26956	5120-01-134-2338	
106	Puller, Crosshead Guide	ST-1134	4910-00-150-5848	
107	Puller, Cylinder Liner	3376015	5120-01-143-2032	
108	Puller, Main Bearing Cap	ST-1178	5120-01-141-5777	
109	Puller, Mechanical	3822786	5120-01-291-5769	
110	Puller, Mechanical	CG60DB	5120-00-620-0020	
111	Puller, Side Plug	KDEP-1056	5120-01-343-2585	
112	Rack Extension	9-681-233-100	5340-01-341-6572	
113	Rear Axle Oil Seal Wiper Replacer	7950136	5120-00-795-0136	Special Tool
114	Retainer Ring Depth Tool	J-24453	5120-01-054-4050	
115	Ring Bearing Installer	J-24447	5120-01-054-4054	Special Tool
116	Rocker Lever Bushing Block And Mandrel	ST-691	3460-00-999-1210	Special Tool
117	Seal Retainer Remover and Replacer	Y-56205		
118	Second Clutch Clearance Gauge	J-26918	5210-01-132-5467	Special Tool
119	Separation Tube	KDEP-1052	4910-10-336-8204	
120	Shaft Installation Tool	3375204	4910-01-118-3747	Special Tool
121	Slide Hammer	3376617	5120-01-187-3626	
122	Slide Hammer	J-6125-1	5120-01-112-2165	Special Tool
123	Snap Gauge	A210605	5220-00-449-7013	4910-95-CL-A63

COMMON AND SPECIAL TOOL IDENTIFICATION LIST (contd)

(1) REFERENCE CODE NO.	(2) NOMENCLATURE	(3) PART NUMBER	(4) NATIONAL/NATO STOCK NUMBER	(5) REFERENCE
124	Snapring Gauge	J-24208-4	5120-01-116-5016	Special Tool
125	Snapring Gauge	J-6843-01	5120-00-293-0186	4910-95-CL-A62
126	Soldering Torch Kit	LP-999	3439-00-542-0531	4910-95-CL-A74
127	Spindle Bearing Sleeve Remover	7950127	5120-00-378-4301	Special Tool
128	Spindle Bearing Sleeve Replacer (Front Axle Oil Seal)	7950129	5120-00-795-0129	Special Tool
129	Spring Compressor Main Regulator and Lockup	J-24459	5120-01-054-4045	Special Tool
130	Spring Pack Adjusting Tool	ST-984	4910-00-150-5805	
131	Spring Tester	J-22738-02	4940-01-138-8259	
132	Spring Tester	SPT	6635-00-641-7346	4910-95-CL-A63
133	Spring Tester	3375182	4910-01-142-4929	
134	Tappet Holder	KDEP-1068	5120-01-345-2586	
135	Tappet Spring Compressor	KDEP-1505	5120-01-341-6000	
136	Telescoping Gauge Set	GGG-G-17	5210-00-473-9350	4910-95-CL-A63
137	Test Kit	MILT 13011B	4910-00-250-2423	Special Tool
138	Tester, Injector Spray Angle	11600056	4910-00-999-1501	Special Tool
139	Third Cluth Clearance Gauge	J-26916	5210-01-141-9457	Special Tool
140	Tool Driver	ST-1122	4910-00-150-5843	Special Tool
141	Torque Angle Gauge	3823878	5120-01-386-5992	
142	Torque Wrench	ST-753-1	2815-00-972-9661	Special Tool
143	Torque Wrench Adapter	ST-669	5120-00-103-4987	Special Tool
144	Torque Wrench, 1/2-In. Drive	A-A-2411	5120-00-640-6364	4910-95-CL-A31
145	Torque Wrench, 3/4 In. Drive	TES18000A	5120-01-118-3679	4910-95-CL-A31
146	Torque Wrench, 3/8-In Drive	TE-12A	5120-00-230-6380	4910-95-CL-A31
147	Transmission Jack	9037-20BM	4910-00-585-3622	4910-95-CL-A62
148	Transmission Shift Lever Seal Installer	J-26282	5120-01-115-1161	Special Tool
149	Transmission Shift Lever Seal Remover	J-26401	5120-01-118-6264	Special Tool
150	Travel Template	3375355	4910-01-074-0020	
151	Tube Reducer	23622	4730-01-284-9086	
152	Vacuum Tester	ST-1257	4910-01-144-3837	
153	Valve Guide Arbor	ST-954	4910-00-150-5804	
154	Valve Guide Arbor (Mandrel Set)	ST-663	3460-00-999-1173	

COMMON AND SPECIAL TOOL IDENTIFICATION LIST (contd)

(1) REFERENCE CODE NO.	(2) NOMENCLATURE	(3) PART NUMBER	(4) NATIONAL/NATO STOCK NUMBER	(5) REFERENCE
155	Valve Pin Remover	J-24412-2	5120-01-048-3128	Special Tool
156	Valve Seat Insert Extractor	ST-1279	5120-01-128-2679	
157	Valve Seat Insert Staking Tool	ST-1124	4910-00-150-5844	Special Tool
158	Valve Seat Insert Tool	ST-257	4910-00-345-3708	
159	Vernier Caliper, 0-6 In.	GGG-C-111	5210-00-113-1548	4910-95-CL-A31
160	Wheel Assembly Tool	J-35193	4910-01-218-4490	Special Tool
161	Wrench Box	CXM 1519	5120-01-178-5351	
162	Wrench, Cup Retainer	ST-995	5120-00-155-7492	Special Tool
163	Wrench, Crowfoot	GGG-C-1507	5120-00-184-8390	Special Tool
164	Wrench, Crowfoot Injector	ST-1072	4910-00-185-8511	Special Tool
165	Wrench, Locknut	3375166	4910-01-097-6928	
166	Wrench, Spanner	4C4734	5120-01-440-1710	1U6680
167	Wrench, Spanner	8747917		Special Tool
168	Gauge, Pressure	PDSS-1P-210A-004		
169	Flowmeter, 2-30 GPM	701-030	6680-01-188-0289	
170	Flowmeter, 10-100 GPM	801-100		

APPENDIX F
ILLUSTRATED LIST OF MANUFACTURED ITEMS

Section I. INTRODUCTION

This appendix includes complete instructions for making items authorized to be manufactured or fabricated at Unit, Direct Support and General Support Maintenance levels.

All bulk materials needed to manufacture the item are listed by part number or specification number in a tabular list on the illustration.

Section II. MANUFACTURED ITEMS PART NUMBER INDEX

MATERIALS	
DESCRIPTION	NSN
Mild Tool Steel Bar	9510-00-042-1493

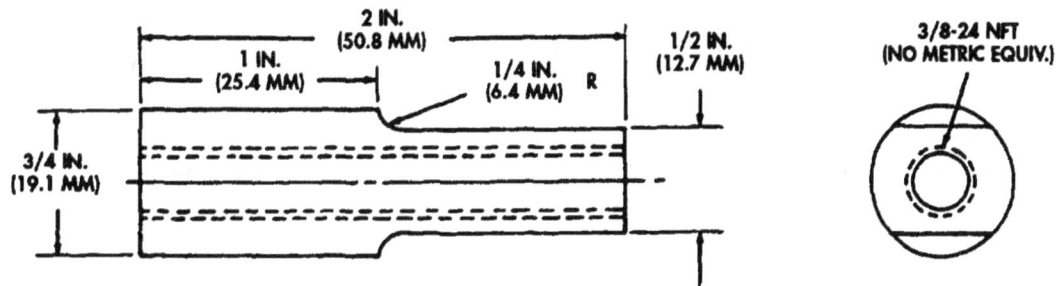

Figure F-l. Flange Puller Standoff.

NOTE:
1. Dimensions shown are in inches with metric equivalent.
2. All dimensions are +/- 0.30 in. (0.76 mm).

INSTRUCTIONS:
Fabricate and finish per diagram above.

Section II. MANUFACTURED ITEMS PART NUMBER INDEX (Contd)

MATERIALS	
DESCRIPTION	NSN
Screw	5305-01-072-4270

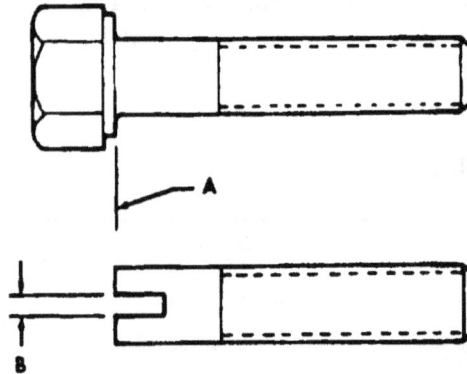

Figure F-2. Guide Screw.

NOTES:

1. Two guide screws are required.
2. Fabricate from screw part number 177734 (NSN 5305-01-072-4270) or equivalent.

PROCEDURE:

1. Remove head of screw at A.
2. Cut slot in screw at B. Slot shall be large enough to accommodate screwdriver tip.

APPENDIX G
TORQUE LIMITS

G-1. GENERAL

This appendix provides general torque limits for screws used on M939, M939A1, and M939A2 vehicles. Special torque limits are indicated in the maintenance procedures for applicable components. The general torque limits given in this appendix shall be used when specific torque limits are not indicated in the maintenance procedure. These general torque limits cannot be applied to screws that retain rubber components. The rubber components will be damaged before the correct torque limit is reached. If a special torque limit is not given in the maintenance instructions, tighten the screw or nut until it touches the metal bracket, then tighten it one more turn.

G-2. TORQUE LIMITS

Table G-l lists dry torque limits. Dry torque limits are used on screws that do not have lubricants applied to the threads. Table G-2 lists wet torque limits. Wet torque limits are used on screws that have high-pressure lubricants applied to the threads. For metric fasteners, refer to table G-3 for torque limit requirements.

G-3. HOW TO USE TORQUE TABLE

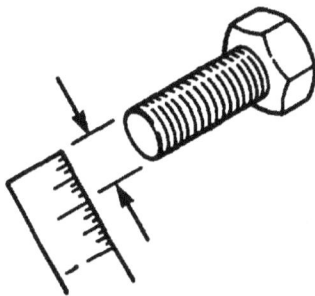

a. Measure the diameter of the screw you are installing.

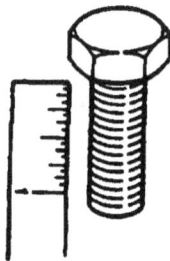

b. Count the number of threads per inch.

c. Under the heading SIZE, look down the left-hand column until you find the diameter of the screw you are installing (there will usually be two lines beginning with the same size).

d. In the second column under SIZE, find the number of threads per inch that matches the number of threads you counted in step b.

CAPSCREW HEAD MARKINGS

Manufacturer's marks may vary. These are all SAE Grade 5 (3-line).

Metric screws are of three grades: 8.8, 10.9, and 12.9. Grades & Manufacturer's marks appear on the screw head.

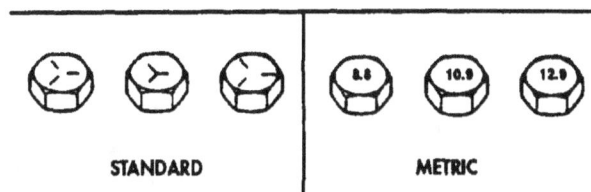

STANDARD   METRIC

e. To find the grade screw you are installing, match the markings on the head to the correct picture of CAPSCREW HEAD MARKINGS on the torque table.

f. Look down the column under the picture you found in step e. until you find the torque limit (in lb-ft or N°m) for the diameter and threads per inch of the screw.

G-1

Table G-1. Torque Limits for Dry Fasteners.

CAPSCREW HEAD MARKINGS

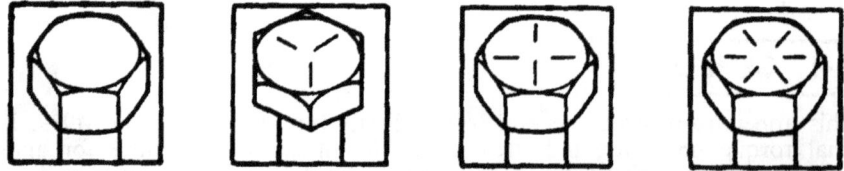

DIA. INCHES	THREADS PER INCH	DIA. MILLIMETERS	SAE GRADE NO.1or2 POUND FEET	NEWTON METERS	SAE GRADE NO. 5 POUND FEET	NEWTON METERS	SAE GRADE NO.6 or 7 POUND FEET	NEWTON METERS	SAE GRADE NO. 8 POUND FEET	NEWTON METERS
1/4	20	6.35	5	6.78	8	10.85	10	13.56	12	16.27
1/4	28	6.35	6	8.14	10	13.56	----	----	14	18.98
5/16	18	7.94	11	14.92	17	23.05	19	25.76	24	32.54
5/16	24	7.94	13	17.63	19	25.76	----	- - -	27	36.61
3/8	16	9.53	18	24.41	31	42.04	34	46.10	44	59.66
3/8	24	9.53	20	27.12	35	47.46	----	- - - -	49	66.44
7/16	14	11.11	28	37.97	49	66.44	55	74.58	70	94.92
7/16	20	11.11	30	40.68	55	74.58	----	----	78	105.77
1/2	13	12.70	39	52.88	75	101.70	85	115.26	105	142.38
1/2	20	12.70	41	55.60	85	115.26	- - -	- - - -	120	162.72
9/16	12	14.29	51	69.16	110	149.16	120	162.72	155	210.18
9/16	18	14.29	55	74.58	120	162.72	- - -	----	170	230.52
5/8	11	15.88	63	85.43	150	203.40	167	226.45	210	284.76
5/8	18	15.88	95	128.82	170	230.52	- - -	- - - -	240	325.44
3/4	10	19.05	105	142.38	270	366.12	280	379.68	375	508.50
3/4	16	19.05	115	155.94	295	400.02	----	- - - -	420	569.52
7/8	9	22.23	160	216.96	395	535.62	440	596.64	605	820.38
7/8	14	22.23	175	237.30	435	589.86	----	- - -	675	915.30
1	8	25.40	235	318.66	590	800.04	660	894.96	910	1,233.96
1	14	25.40	250	339.00	660	894.96	- - - -	----	990	1,342.44
1-1/8	----	28.58	—	—	800.-880	1,084.8-1,193.3	----	- - - -	1,280-1,440	1,735.68 1,952.64
1-1/4	----	31.75	—	—	—	-----	----	- - -	1,820 2,000	2467.92 2,712.00
1-3/8	----	34.93	—	—	1,460-1,680	1,979.8. 2,278.1	----	- - - -	2,380-2,720	3,227.28 3,688.32
1-1/2	----	38.10	----	----	1,940-2,200	2,630.6-2,983.2	----	- - - -	3,160-3,560	4,284.96 4,827.36

Table G-2. Torque Limits for Wet Fasteners.

CAPSCREW HEAD MARKINGS

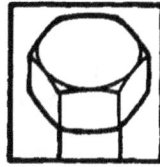

SIZE			TORQUE							
			SAE GRADE No.1 or 2		SAE GRADE No .5		SAE GRADE N0.6 or 7		SAE GRADE No.8	
DIA. INCHES	THREADS PER INCH	DIA. MILLIMETERS	POUND FEET	NEWTON METERS	POUND FEET	NEWTON METERS	POUND FEET	NEWTON METERS	POUND FEET	NEWTON METERS
1/4	20	6.35	4.5	6.1	7.2	9.76	9.0	12.20	10.8	14.64
1/4	28	6.35	5.4	7.32	9.0	12.20	—	—	12.6	17.09
5/16	18	7.94	9.9	13.42	15.3	20.75	17.1	23.19	21.6	29.29
5/16	24	7.94	11.7	15.87	17.1	23.19	----	----	24.3	32.95
3/8	16	9.53	16.2	21.97	27.9	37.83	30.6	41.49	39.6	53.70
3/8	24	9.53	18.0	24.41	31.5	42.71	- - -	- - - -	44.1	59.80
7/16	14	11.11	25.2	34.17	44.1	59.80	49.5	67.12	63.0	85.43
7/16	20	11.11	27.0	36.61	49.5	67.12	- - -	- - -	70.2	95.19
1/2	13	12.70	35.1	47.60	67.5	91.53	76.5	103.73	94.5	128.14
1/2	20	12.70	36.9	50.04	76.5	103.73	-----	-----	108.0	146.45
9/16	12	14.29	45.9	62.24	99.0	134.24	108.0	146.45	139.5	189.16
9/16	18	14.29	49.5	67.12	108.0	146.45	----	-----	153.0	207.47
5/8	11	15.88	56.7	76.89	135.0	183.06	150.3	203.81	189.0	256.28
5/8	18	15.88	85.5	115.94	153.0	207.47	----	-----	216.0	292.90
3/4	10	19.05	94.5	128.14	243.0	329.51	252.0	341.71	337.5	457.65
3/4	16	19.05	103.5	140.35	265.5	360.02	-----	------	378.0	512.57
7/8	9	22.23	144.0	195.26	355.5	482.06	396.0	536.98	544.5	738.34
7/8	14	22.23	157.5	213.57	391.5	530.87	—	-----	607.5	823.77
1	8	25.40	211.5	286.79	531.0	720.04	594.0	805.46	819.0	1,110.56
1	14	25.40	225.0	305.10	594.0	805.46	----	-----	891.0	1,208.20
1-1/8	----	28.58	-----	------	720.0-792.0	976.32-1,073.95	-----	-----	1,152.0-1,296.0	1,562.11-1,757.38
1-1/4	----	31.75	----	—	—	----	—	----	1,637.99-1,800.00	2,221.11-2,440.80
1-3/8	----	34.93	—	-----	1,314.0-1,512.0	1,781.78-2,050.27	----	----	2,142.0.2,448.0	2,904.55-3,319.49
1-1/2	----	38.10	-----	-----	1,746.0.1,980.0	2,367.58-2,684.88	----	- - -	2,844.0.3,204.0	3,856.46-4,344.62

Table G-3.. Torque Limits for Metric Fasteners.

CAPSCREW HEAD MARKINGS

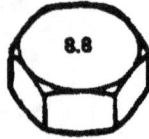

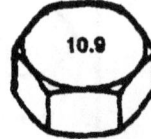

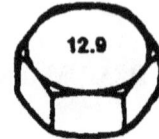

SIZE	TORQUE	
	POUND FEET	NEWTON METERS
M6	7-11	9.49 - 14.92
M8	21 - 35	28.48-47.46
M10	45 - 65	61.02 - 88.14
M12	80-120	108.48 - 162.72
M14	130-190	176.28 - 257.64
M16	200-280	271.20 - 379.68
M20	400-520	542.40 - 705.12
M24	700-900	949.20 - 1,220.40
M30	1,400 - 1,800	1,898.40 - 2,440.80
M36	2,400 - 3,000	3.254.40 - 4,068.OO

APPENDIX G (Contd)

Tubing Application Tightening Assembly Instructions.

Slide tubing over barbed insert until it bottoms on fitting.

MINI-BARB

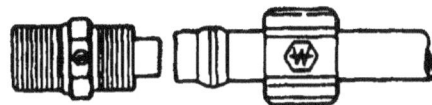

1. Slide nut and sleeve on tubing.
2. Slide I.D. of tubing onto fitting insert until it bottoms.
3. Assemble nut to fitting body.
4. Tighten assembly finger-tight to cover body threads.

KNURL-ON

1. Slide nut and sleeve on tubing.
2. Slide I.D. of tubing onto fitting insert until it bottoms.
3. Assemble nut to fitting body.
4. Finger-tighten nut. From that point, tighten with a wrench two complete turns.

SELF-ALIGN-PTF

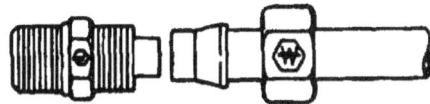

1. Cut tubing to desired length. Ensure ends are cut reasonably square.
2. Slide tubing into preassembled fitting and push until tube bottoms.
3. Tighten nut as indicated in chart. Another check on proper assembly is dimension A, when nut is fully tightened.

NYLON TUBING FOR AIRBRAKE

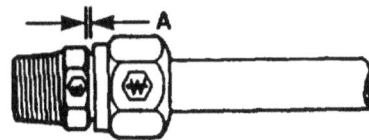

DISASSEMBLY - Remove nut and pull tubing out of fitting body. Insert will remain on tubing.

REASSEMBLY - Push tubing and insert into fitting body until it bottoms. Thread nut onto fitting body and tighten as in step 3.

TUBE O.D.	TIGHTEN NUT TO:	A
1/4	85 - 115 lb-in. (9.6 - 13.0 N°m)	.085/.105
3/8	12 - 17 lb-ft (16.3 - 23.1 N°m)	.125/.145
1/2	25 - 33 lb-ft (33.9 - 44.7 N°m)	.100/.120
5/8	26 - 35 lb-ft (35.3 - 47.5 N°m)	.115/.135
3/4	38 - 50 lb-ft (51.5 - 67.8 N°m)	.180/.200

APPENDIX G (Contd)

Tubing Application Tightening Assembly Instructions (Contd).

1. Slide nut and sleeve on tubing. Threaded end of nut (1) must face out.
2. Insert tubing into fitting. Ensure tubing is bottomed on fitting shoulder.
3. Thread nut onto fitting body until it is hand tight.
4. From that point, tighten with a wrench the number of turns indicated at right.

COPPER TUBING FOR HAND AIRBRAKE

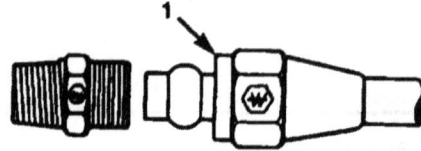

TUBE SIZE	ADDITIONAL NUMBER OF TURNS FROM HAND-TIGHT
1/4, 3/8	1-3/4
1/2, 5/8, 3/4	3-1/4

G-4. TORQUE WRENCH ADAPTERS

Some tasks require the use of a torque wrench adapter when the nut or screw cannot be reached with a regular socket on the end of the torque wrench. These adapters add to the overall length of the torque wrench and make the dial or scale reading less than the actual torque applied to the nut or screw. To prevent overtorquing and damage to equipment, calculate correct dial or scale reading using Conversion Formula (para. G-5).

APPENDIX G (Contd)

G-5. CONVERSION FORMULA

Corrected dial or scale readings are determined by the use of the following formula:

$$\text{Corrected reading} = \text{Required torque value} + \frac{\text{Length of torque wrench} + \text{Length of adapter}}{\text{Length of torque wrench}}$$

NOTE

The length of the torque wrench is measured from the center of the handle to the center of the drive. The length of the adapter is measured from the center of the drive to the center of the wrench.

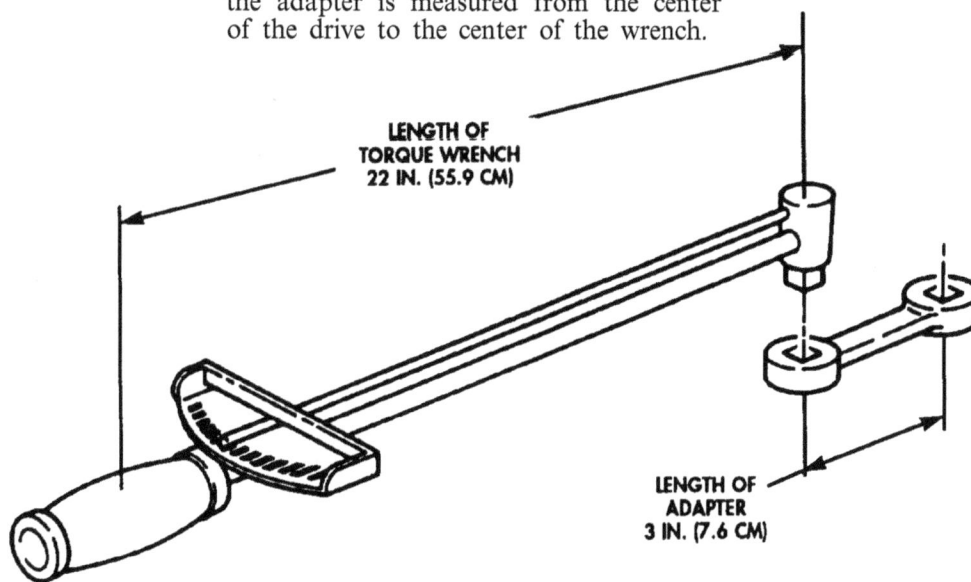

LENGTH OF
TORQUE WRENCH
22 IN. (55.9 CM)

LENGTH OF
ADAPTER
3 IN. (7.6 CM)

In this example, the torque wrench measures 22 in. (55.9 cm) and the adapter is 3 in. (7.6 cm). The required torque is 19 lb-ft (25.8 N°m).

$$\text{Corrected reading} = 19 \text{ lb-ft (25.8 N•m)} + \frac{22 \text{ in. (55.9 cm)} + 3 \text{ in. (7.6 cm)}}{22 \text{ in. (55.9 cm)}}$$

$$\text{Corrected reading} = 19 \text{ lb-ft (25.8 N•m)} + \frac{25 \text{ in. (63.5 cm)}}{22 \text{ in. (55.9 cm)}}$$

$$\text{Corrected reading} = 19 \text{ lb-ft (25.8 N•m)} + 1.14$$

$$\text{Corrected reading} = 17 \text{ lb-ft (23.1 N•m)}$$

APPENDIX H
SCHEMATIC AND WIRING DIAGRAMS

INDEX

INDEX (Contd)

INDEX (Contd)

INDEX (Contd)

INDEX (Contd)

INDEX (Contd)

INDEX (Contd)

INDEX (Contd)

INDEX (Contd)

INDEX (Contd)

INDEX (Contd)

INDEX (Contd)

INDEX (Contd)

INDEX (Contd)

INDEX (Contd)

INDEX (Contd)

INDEX (Contd)

INDEX (Contd)

INDEX (Contd)

INDEX (Contd)

INDEX (Contd)

INDEX (Contd)

INDEX (Contd)

INDEX (Contd)

INDEX (Contd)

INDEX (Contd)

INDEX (Contd)

INDEX (Contd)

INDEX (Contd)

INDEX (Contd)

INDEX (Contd)

INDEX (Contd)

INDEX (Contd)

INDEX (Contd)

INDEX (Contd)

INDEX (Contd)

INDEX (Contd)

INDEX (Contd)

INDEX (Contd)

INDEX (Contd)

INDEX (Contd)

INDEX (Contd)

INDEX (Contd)

INDEX (Contd)

INDEX (Contd)

INDEX (Contd)

INDEX (Contd)

INDEX (Contd)

INDEX (Contd)

INDEX (Contd)

INDEX (Contd)

INDEX (Contd)

INDEX (Contd)

INDEX (Contd)

INDEX (Contd)

INDEX (Contd)

INDEX (Contd)

INDEX (Contd)

INDEX (Contd)

INDEX (Contd)

INDEX (Contd)

INDEX (Contd)

INDEX (Contd)

INDEX (Contd)

INDEX (Contd)

INDEX (Contd)

INDEX (Contd)

INDEX (Contd)

INDEX (Contd)

INDEX (Contd)

INDEX (Contd)

INDEX (Contd)

INDEX (Contd)

INDEX (Contd)

www.ingramcontent.com/pod-product-compliance
Lightning Source LLC
Chambersburg PA
CBHW080413030426
42335CB00020B/2435